Recent Advances in Fourier Analysis and Its Applications

NATO ASI Series

Advanced Science Institutes Series

A Series presenting the results of activities sponsored by the NATO Science Committee, which aims at the dissemination of advanced scientific and technological knowledge, with a view to strengthening links between scientific communities.

The Series is published by an international board of publishers in conjunction with the NATO Scientific Affairs Division

A Life Sciences **B Physics**	Plenum Publishing Corporation London and New York
C Mathematical and Physical Sciences **D Behavioural and Social Sciences** **E Applied Sciences**	Kluwer Academic Publishers Dordrecht, Boston and London
F Computer and Systems Sciences **G Ecological Sciences** **H Cell Biology**	Springer-Verlag Berlin, Heidelberg, New York, London, Paris and Tokyo

Series C: Mathematical and Physical Sciences - Vol. 315

Recent Advances in Fourier Analysis and Its Applications

edited by

J. S. Byrnes
Prometheus Inc., Newport, RI, and
University of Massachusetts at Boston,
Boston, MA, U.S.A.

and

Jennifer L. Byrnes
Harvard University,
Cambridge, MA, U.S.A.

Kluwer Academic Publishers

Dordrecht / Boston / London

Published in cooperation with NATO Scientific Affairs Division

Proceedings of the NATO Advanced Study Institute on
Fourier Analysis and Its Applications
Il Ciocco, Italy
July 16–29, 1989

Library of Congress Cataloging in Publication Data

NATO Advanced Study Institute on Fourier Analysis and Its Applications (1989 : Ilocco, Italy)
Recent advances in Fourier analysis and its applications : proceedings of the NATO Advanced Study Institute on Fourier Analysis and Its Applications, held in Il Ciocco, Italy, July 16-19, 1989 / edited by J.S. Byrnes and Jennifer L. Byrnes.
p. cm. -- (NATO ASI series. Series C, Mathematical and physical sciences ; vol. 315)
"Published in cooperation with NATO Scientific Affairs Division."
Includes bibliographical references and index.
ISBN-13:978-94-010-6784-3 e-ISBN-13:978-94-009-0665-5
DOI: 10.1007/978-94-009-0665-5

1. Fourier analysis--Congresses. I. Byrnes, J. S. II. Byrnes, Jennifer L. III. North Atlantic Treaty Organization. Scientific Affairs Division. IV. Title. V. Series: NATO ASI series. Series C, Mathematical and physical sciences ; no. 315.
QA403.5.N37 1989
515'.2433--dc20 90-4964

ISBN-13:978-94-010-6784-3

Published by Kluwer Academic Publishers,
P.O. Box 17, 3300 AA Dordrecht, The Netherlands.

Kluwer Academic Publishers incorporates the publishing programmes of
D. Reidel, Martinus Nijhoff, Dr W. Junk and MTP Press.

Sold and distributed in the U.S.A. and Canada
by Kluwer Academic Publishers,
101 Philip Drive, Norwell, MA 02061, U.S.A.

In all other countries, sold and distributed
by Kluwer Academic Publishers Group,
P.O. Box 322, 3300 AH Dordrecht, The Netherlands.

Printed on acid-free paper

Contents

Part IV: Complex Analysis and Fourier Analysis

Part V: Problems

Preface

This volume contains papers presented at the July, 1989 NATO Advanced Study Institute on Fourier Analysis and its Applications. The conference, held at the beautiful Il Ciocco resort near Lucca, in the glorious Tuscany region of northern Italy, created a dynamic interaction between world-renowned scientists working in the usually disparate communities of pure and applied Fourier analysts.

The papers to be found herein include important new results in x-ray crystallography by Nobel Laureate Herbert Hauptman, the application of the new concept of bispectrum to system identification by renowned probabilist Athanasios Papoulis, fascinating applications of number theory in Fourier analysis by eminent electrical engineer Manfred R. Schroeder, and exciting concepts regarding polynomials with restricted coefficients by foremost mathematical problem solver Donald J. Newman. The remaining papers further illustrate the inherent power and beauty of classical Fourier analysis, whether the results presented were sought as an end in themselves, or whether these classical methods were employed as a tool in illustrating and solving a particular applied problem. From antenna design to concert hall acoustics to image and speech processing to unimodular polynomials, each conference participant benefited significantly from his or her exposure, in many cases for the first time, to those scientists on the other end of the spectrum from themselves. The purpose of this volume is to pass those benefits on to the reader.

The conference, and this volume, are divided into five sections, which naturally contain a great deal of overlap. Chemical, physical, and biomedical applications are the subjects of the initial part. Important new results in x-ray crystallography, including the possibility of applying a recently developed crystallographic technique to the solution of very large systems of equations arising in other fields, are presented here, as well as the application of Fourier techniques to study anatomic structures and thermal storage problems.

The second section deals with significant recent progress in the study of polynomials with restricted coefficients. In addition to fascinating theoretical results, applications presented here include concert hall acoustics and other uses of reflection phase gratings, and the design of new algorithms for notch filters, null steering, and low peak factor signals. The mathematicians in attendance at the conference were surprised and delighted to learn of the important applications of their "esoteric" ideas to problems in radar, sonar, filtering, speech processing, etc. The engineers and physicists, on the other hand, were equally surprised and delighted to learn of the existence of a large body of deep and applicable mathematics.

Fourier analysis in engineering is the subject of part three. Deserving special mention here is the paper on holography by Walter Schempp, which generated a great deal of interest and enthusiasm at the meeting. Also included are new results in spectrum estimation, antenna theory and design, pattern recognition, and sampling theory, particularly irregular sampling. Several papers by participants who are in the forefront of the exciting, newly rediscovered field of wavelets are found here as well.

The fourth section addresses the interaction between Fourier analysis and complex analysis. Among the topics covered here are new Wiener-Plancherel formulas, weighted estimates for the Laplace operator and the Fourier transform, spectra in various function spaces, and new complex function theory results. Several unsolved problems of current interest comprise the fifth and final section.

The cooperation of many individuals and organizations was required in order to make the conference the success that it was. First and foremost we wish to thank NATO, and especially Dr. Louis V. da Cunha and his staff, for the initial grant and subsequent help. Financial support was also received from the US National Science Foundation, the European Office of the US Air Force, and Prometheus Inc. This additional support is gratefully acknowledged.

We also wish to express our sincere appreciation to the director's assistants, Marcia Byrnes, Alice Lang-Smith, and Kevin J. Spratt, as well as Harold S. Shapiro of the organizing committee and John J. Benedetto, for their invaluable aid. We are also grateful to Kathryn Hargreaves and Karl Berry, our TeXnicians, for their superlative work on the conference program, table of contents, preface, and problem section. Finally, our heartfelt thanks to the Il Ciocco staff, especially Bruno Giannasi and Alberto Suffredini, for offering an ideal setting, not to mention the magnificent meals, that promoted the productive interaction between the participants of the conference. All of the above, the other speakers, and the remaining conferees, made it possible for our Advanced Study Institute, and this volume, to fulfill the stated NATO objectives of disseminating advanced knowledge and fostering international scientific contacts.

March 24, 1990

J.S. Byrnes, Newport, Rhode Island
Jennifer L. Byrnes, Cambridge, Massachusetts

Part I: Chemical, Physical and Biomedical Applications

A MINIMAL PRINCIPLE IN THE DIRECT METHODS OF X-RAY CRYSTALLOGRAPHY

by

Dr. Herbert Hauptman

Medical Foundation of Buffalo

73 High Street

Buffalo, NY 14203

Abstract

The electron density function, $\rho(r)$, in a crystal determines its diffraction pattern, that is, both the magnitudes and phases of its X-ray diffraction maxima, and conversely. If, however, as is always the case, only magnitudes are available from the diffraction experiment, then the density function $\rho(r)$ cannot be recovered. If one invokes prior structural knowledge, usually that the crystal is composed of discrete atoms of known atomic numbers, then the observed magnitudes are, in general, sufficient to determine the positions of the atoms, that is, the crystal structure.

The intensities of a sufficient number of X-ray diffraction maxima determine the structure of a crystal. The available intensities usually exceed the number of parameters needed to describe the structure. From these intensities a set of members $|E_H|$ can be derived, one corresponding to each intensity. However, the elucidation of the crystal structure also requires a knowledge of the complex numbers $E_H = |E_H| \exp(i\phi_H)$, the normalized structure factors, of which only the magnitudes $|E_H|$ can be determined from

J. S. Byrnes and J. F. Byrnes (eds.), Recent Advances in Fourier Analysis and Its Applications, 3–15.

experiment. Thus, a "phase" ϕ_H, unobtainable from the diffraction experiment, must be assigned to each $|E_H|$, and the problem of determining the phases when only the magnitudes $|E_H|$ are known is called "the phase problem". Owing to the known atomicity of crystal structures and the redundancy of observed magnitudes $|E_H|$, the phase problem is solvable in principle.

PART I

Traditional Techniques of Direct Methods

The electron density function $\rho(\mathbf{r})$ in a crystal, a function of the position vector $\mathbf{r}$, is a triply periodic function of $\mathbf{r}$ and may therefore be represented by means of the triple Fourier Series

$$\rho(\mathbf{r}) = \frac{1}{V} \sum_{\mathbf{H}} F_{\mathbf{H}} \exp\left(-2\pi i \mathbf{H}\cdot\mathbf{r}\right) \tag{I.1}$$

where V is the volume of the fundamental parallelepiped and the three components of the vector $\mathbf{H}$ range over all the integers. The Fourier coefficient $F_{\mathbf{H}}$, said to be the structure factor corresponding to the reciprocal lattice vector $\mathbf{H}$, may then be calculated in the usual way:

$$F_{\mathbf{H}} = \int_V \rho(\mathbf{r}) \exp\left(2\pi i \mathbf{H}\cdot\mathbf{r}\right) dV \tag{I.2}$$

where the integration is carried out over the unit cell V. Clearly each $F_{\mathbf{H}}$ is a complex number which may be written in polar form:

$$F_{\mathbf{H}} = |F_{\mathbf{H}}| \exp\left(i\phi_{\mathbf{H}}\right) \tag{I.3}$$

where $|F_{\mathbf{H}}|$ is the magnitude and $\phi_{\mathbf{H}}$ is said to be the phase of the structure factor F_H. Substituting from (3) into (1), one obtains

$$\rho(r) = \frac{1}{V} \sum_{H} |F_H| \exp\left(i\phi_H - 2\pi i H\cdot r\right) \qquad (I.4)$$

from which it is clear that knowledge of the magnitudes and phases of the structure factors leads directly to the electron density function $\rho(r)$ the positions of the maxima of which coincide with the atomic position vectors and therefore yield the crystal structure. The value of the density function $\rho(r)$ at any of its maxima serves to identify the kind of atom located at the corresponding position.

When a beam of x-rays is incident on a crystal it is scattered in a large number of well-defined directions which are determined by the lattice parameters. In this way one obtains the diffraction pattern: the directions, intensities, and phases of the scattered rays.

The electron density function $\rho(r)$ determines completely the nature of the diffraction pattern via Eq. (I.2) because the intensity of the ray scattered in the direction labeled by the reciprocal lattice vector H is obtainable directly (after taking into account certain experimental factors) from the magnitude $|F_H|$ of the corresponding structure factor F_H and the phase of the scattered ray is simply the phase ϕ_H of F_H (Eq. (I.3)). Conversely, diffraction intensities (or magnitudes $|F_H|$) and phases ϕ_H lead directly, via Eq. (I.4), to the electron density function $\rho(r)$ and thence to the crystal structure. Thus it is clear that the diffraction experiment holds the key to the determination of crystal and molecular structures.

While the intensities of the diffraction maxima are readily obtained from experiment, it turns out that the required phases cannot be experimentally determined; they are lost in the diffraction experiment. Presumably then, one may use arbitrary values for the phases ϕ_H together with the measured values for the magnitudes $|F_H|$ in Eq. (I.4) and in this way obtain a myriad of density functions $\rho(\mathbf{r})$, all consistent with the measured magnitudes $|F_H|$. The pessimistic conclusion then was that, despite its initial promise, the diffraction experiment could not, after all, lead unambiguously to unique crystal structures, even in principle.

The belief that diffraction intensities alone were insufficient to determine crystal structures unambiguously was finally refuted in the early 1950's by the recognition that *a priori* structural knowledge, when combined with the measurable diffraction intensities, did in fact provide sufficient information to lead, in general, to unique crystal and molecular structures. The methods for doing this (so-called direct methods) show that, because real crystal structures must satisfy certain restrictive conditions, the phases may not be arbitrarily specified, and relationships exist among the intensities and phases of the scattered X-rays which permit the phases to be recovered once the intensities are known. Thus the phase information, which is lost in the diffraction experiment, is in fact to be found among the measurable intensities. In short, the phase problem, which is to determine the values of the missing phases from the known diffraction intensities, is a solvable one, at least in principle.

The restrictive condition which all crystal structures must satisfy is atomicity, and it is this property which constitutes the foundation on which the structure of direct methods is based. Molecules consist of atoms. Hence the electron density function in a crystal is non-negative everywhere and reaches maximum positive values at the positions of the atoms and drops down to small values between the atomic positions. This property of crystal structures, together with the large number of diffraction intensities available from experiment, is, in general, sufficiently restrictive to determine unique values for the phases of the scattered X-rays. It turns out in fact that the measured intensities are more than sufficient for this purpose so that the phase problem, far from being unsolvable, even in principle, is actually a greatly overdetermined one.

By exploiting the atomicity property of real crystal structures and deriving the joint conditional probability distributions of certain well-defined collections of structure factors, assuming as known appropriate sets of measured diffraction intensities (or magnitudes $|F|$), one obtains relationships among the known magnitudes $|F|$ and the desired phases ϕ having probabilistic validity. If a sufficient number of such relationships are available, which is usually the case for "small" molecular structures, say those having fewer than some 100 non-hydrogen atoms in the molecule, then one can calculate the values of the unknown phases ϕ in terms of the known magnitudes $|F|$. The Fourier synthesis, Eq. (I.4), then yields the crystal structure, as already explained.

PART II

A New Minimal Principle in X-Ray Crystallography

Summary

The problem of determining crystal and molecular structures by the techniques of X-ray crystallography is the problem of determining the positions of the atoms in the crystal when only the intensities of the diffraction maxima are available from experiment. However the associated phases, which are also needed if one is to deduce the structure unambiguously from the experimental observations, are lost in the diffraction experiment. By exploiting prior structural information, usually the atomicity property of real crystal structures and the non-negativity of the corresponding electron density function, it can be shown that the lost phase information is contained in the measured diffraction intensities and can be recovered provided that the molecular structure is not too large. To this end one introduces special linear combinations of the phases, the so-called structure invariants, whose values are uniquely determined by the structure alone, independently of the choice of origin. Estimates of the structure invariants in terms of measured intensities then lead unambiguously to the values of the individual phases. The method is to derive the conditional probability distributions of the structure invariants assuming as known certain suitably chosen sets of intensities and to use the mode as an estimate of the corresponding structure invariant. It is now proposed to employ, not merely the modes of these distributions, but their conditional expectation values and variances. Specifically, one determines the values of a set of phases as those which generate structure invariants whose conditional expectation values and variances are in agreement with their theoretical values as determined from their known distributions. One is thus led to a function $R(\phi)$ of the phases, dependent on measured intensities, which assumes a minimum when the phases are equal to their true values (the minimal principle). In this way it is expected that by using all measured intensities to estimate the values of a large block of phases simultaneously one will strengthen existing techniques of phase determination so as to be routinely applicable to macromolecular structures. It should be stressed however that there still remains outstanding the problem of finding the minimum of the function $R(\phi)$.

1. Introduction

All existing algorithms for using direct methods to extract phase information from the measured intensities have certain common features and inherent deficiencies. The values of special linear combinations of phases, the structure invariants, are uniquely determined by the structure alone, independently of the choice of origin. Conditional probability distributions of these structure invariants are derived assuming as known certain suitably chosen sets of intensities, and the mode is used as an estimate of the corresponding structure invariant. Values of the individual phases are then found one at a time in step-by-step procedures which have the inherent weakness that a mistake at an early stage resulting from an improperly estimated invariant may cause the entire process to go awry.

The work described here is an attempt to improve upon what has already been done by developing *ab initio* phasing techniques which (a) avoid this sequential determination of phases and (b) are applicable to all structures regardless of size. In order to do this, the full details of the known three- and four-phase structure invariant distributions are exploited more effectively. The conditional expectation values and variances of these distributions are used. The values of a set of phases are determined as those which generate structure invariants whose conditional expectation values and variances are in agreement with their theoretical values. In this way, all measured intensities are used to estimate the values of a large block of phases simultaneously. Thus, it is expected that the weakness inherent in the usual sequential, step-by-step procedures will be eliminated.

The goal of this work is to modify and strengthen existing techniques for determining crystal and molecular structures of complex biologically important compounds directly from the observed intensities in the X-ray diffraction experiment by minimizing dependence on chemical modification, such as the introduction of heavy atoms, and without requiring any detailed knowledge concerning molecular connectivity, configuration or conformation. Existing methods are able to determine routinely structures having as many as 100 non-hydrogen atoms in the molecule, and occasionally structures having as many as 200 or even 300 non-hydrogen atoms in the molecule have been solved.

The present goal is to devise a new *ab initio* phasing strategy which, it is anticipated, will be capable of routinely solving structures of much greater complexity, perhaps even in the macromolecular range. In order to do this, the properties of the conditional probability distributions of the structure invariants have been exploited to derive formulas which it is hoped will permit the simultaneous evaluation of many phases, thereby eliminating the sequential phase determination which is an inherent weakness of the direct methods techniques currently in use and which has restricted their application to relatively small molecules.

A direct phasing procedure based on a novel minimal function of the phases is described in this account. This function is based on an analysis of the conditional probability distributions of the triplet and quartet structure invariants and takes into consideration their conditional expectation values and variances. Use of this function may result in a global process for *ab initio* phase determination rather than the sequential phase determination methods characteristic of the commonly employed direct methods, and it is anticipated that it will be applicable to much larger structures. The theoretical background of the minimal function is described below.

2. Theoretical Background

Introduction to Direct Methods

When X-rays are scattered by a crystalline solid, each of the diffracted waves (reflections) has an associated amplitude and phase angle which together constitute the structure factor. Although the diffraction amplitudes are easily found by measuring the intensities of the diffracted waves, no experimental technique exists whereby the structure factor phases can be directly measured. The problem of calculating the phases, when only the amplitudes are known, is called the phase problem. Two general classes of methods have evolved for solving this problem. These are structure deconvolution methods and direct phasing methods. The deconvolution methods circumvent the phase problem by the substitution of a related problem concerning the deconvolution of a function, the so-called Patterson Function, related to the electron density. These methods are optimally suited to crystal structures containing some elements having high atomic numbers, "heavy atoms". However, these methods are less effective in solving structures having molecules composed of first row elements alone, which often have the greatest biochemical interest. Direct methods are those procedures which attempt to solve the phase problem by reconstructing the lost phase information directly from the observed structure factor amplitudes. These methods rely on the existence of relationships among the structure factors which express the values of certain linear combinations of phases, called structure invariants, in terms of normalized structure factor magnitudes.

The Normalized Structure Factors E. The normalized structure factors are defined by

$$E_H = |E_H| \exp(i\phi p) = \frac{1}{\sigma_2^{1/2}} \sum_{j=1}^{N} Z_j \exp\left(2\pi i H \cdot r_j\right) \qquad \text{(II.1)}$$

where H is an arbitrary reciprocal lattice vector which labels the diffraction intensity, N is the number of atoms in the unit cell, Z_j is the atomic number of the atom labeled j, r_j is its position vector, and

$$\sigma_n = \sum_{j=1}^{N} Z_j^n, \; n = 2, 3, 4, \ldots . \qquad \text{(II.2)}$$

The magnitudes $|E|$ are directly obtainable from the diffraction intensities, but the phases ϕ are lost in the diffraction experiment.

The Structure Invariants. Although the values of the individual phases are known to depend on the structure and the choice of origin, there exist certain linear combinations of the phases whose values are determined by the structure alone and are independent of the choice of origin. These linear combinations of the phases are called the structure invariants. The most important classes of structure invariants are the three-phase structure invariants (triplets).

$$\phi_{HK} = \phi_H + \phi_K + \phi_{-H-K} , \tag{II.3}$$

and the four-phase structure invariants (quartets),

$$\phi_{LMN} = \phi_L + \phi_M + \phi_N + \phi_{-L-M-N} . \tag{II.4}$$

The Probabilistic Background. It is assumed that the position vectors r_j are random variables which are uniformly and independently distributed. Then the structure invariants, as functions of random variables via Eq. (II.1), (II.3), and (II.4), are themselves random variables, and their conditional probability distributions, assuming as known certain magnitudes $|E|$, may then be found.

The Conditional Probability Distributions of the Triplet and the Quartet. For fixed reciprocal lattice vectors H and K, the conditional probability distribution of the triplet ϕ_{HK} (Eq. (II.3)), assuming as known the three magnitudes

$$|E_H|, \ |E_K|, \ |E_{H+K}|, \tag{II.5}$$

is known. From this distribution the conditional expected value and variance of cos ϕ_{HK}, given the three magnitudes (II.5) can be deduced.

In a similar way the conditional expected value and variance of $\cos \phi_{LMN}$, where ϕ_{LMN} is the quartet (II.4), assuming as known the seven magnitudes

$$|E_L|, \ |E_M|, \ |E_N|, \ |E_{L+M+N}|, \ |E_{L+M}|, \ |E_{M+N}|, \ |E_{N+L}|, \qquad \text{(II.6)}$$

can be calculated.

3. The Minimal Principle

The Heuristic Background. The mode of the triplet distribution is zero and the variance of the cosine is small if the three magnitudes $|E_H|$, $|E_K|$, $|E_{H+K}|$, (II.5), are large. In this way one obtains the estimate for the triplet ϕ_{HK} (Eq. II.3):

$$\phi_{HK} = \phi_H + \phi_K + \phi_{-H-K} \simeq 0 \qquad \text{(II.7)}$$

which is particularly good in the favorable case that $|E_H|$, $|E_K|$, and $|E_{H+K}|$ are all large. The estimate given by Eq. (II.7) is one of the cornerstones of current techniques of direct methods. It is surprising how useful (II.7) has proven to be in the applications especially since it yields only the zero estimate of the triplet, and only those estimates are reliable for which $|E_H|$, $|E_K|$, $|E_{H+K}|$ are all large. It turns out that the relationship (II.7) becomes increasingly unreliable for larger structures, and the current step-by-step sequential direct methods procedures dependent on (II.7) eventually fail.

By their almost exclusive reliance on the triplet relationship (II.7), the traditional direct methods techniques do not fully exploit our detailed knowledge of the triplet distribution and ignore almost completely the quartet distribution. We propose now to determine the values of the phases ϕ in such a way that they generate triplets and quartets which have distributions in accord with their theoretical distributions. More specifically, one determines the values of a set of phases as those which generate triplets ϕ_{HK} and quartets ϕ_{LMN} whose cosines have conditional expectation values and variances in agreement with their theoretical values. In this connection it should be noted that, for a sufficiently large basis set of phases, say more than some 300 phases in the base, the number of structure invariants which they generate exceeds by far (two or three orders of magnitude at least) the number of unknown phases ϕ. Owing to this great redundancy, a large number of identities among the structure invariants, equal to the difference between the number of structure invariants and the number of phases, must be satisfied. An important aspect of our present formulation is that all identities among the structure invariants, which must of necessity hold, will in fact be satisfied.

The Minimal Principle. Carrying out the program described in the previous paragraph one is led to a function $R(\phi)$ of the phases, dependent on known magnitudes $|E|$ but too complicated to be reproduced here, and having the property that it is a minimum when the phases take on their true values. There remains the problem, by no means trivial, of devising techniques for minimizing the known function $R(\phi)$ of phases.

Acknowledgement. This work has been supported by Grant No. CHE8822296 from the National Science Foundation.

FOURIER SHAPE ANALYSIS OF ANATOMIC STRUCTURES

D. N. Kennedy[†*], P. A. Filipek[†] and V. S. Caviness[†]
[†]*Pediatric Neurology Service and* [*]*Department of Radiology*
Massachusetts General Hospital
Harvard Medical School
Boston, MA
U.S.A.

ABSTRACT. In this paper we explore quantitative shape analysis by Fourier methods. Applications and examples of the techniques discussed are demonstrated on anatomic data as presented in three-dimensional Nuclear Magnetic Resonance (NMR) imaging. There is a wide variety of common problems that can benefit from a quantitative method of shape analysis. Indeed, shape constitutes one of four major independent parameters of structural morphometry (volume, shape, geometric localization and composition), each of which require quantitative measures in order to provide a sensitive means of analysis. The Fourier method of shape analysis can be applied to two-dimensional structural surfaces. In short, the method involves representation of the surface in polar coordinates and Fourier transformation to obtain a measure of the spatial frequency constituents that comprise the structure. The limitations and technical considerations imposed both by the method and the primary data itself are discussed. Finally, two-dimensional applications of the method are presented, for example, in the evaluation of cerebral asymmetry of normal adults.

1. Introduction

Fourier analysis can be applied to the quantitative analysis of shape. The goal of the work discussed here is to explore the methods by which we can take the mathematical *tool* of Fourier shape analysis and pose some of the questions of morphometry (the study of structure and form) in such a way that the application of this tool becomes appropriate.

The study of shape forms one of the bases in the study of morphometry. In our conceptual model of morphometry, we hypothesize that there are at least four potentially independent properties: volume, shape, geometric localization and composition. Geometric localization refers to the absolute position of a structure in a given coordinate system. Composition refers to properties of the material within a given structure, and, in the case of magnetic resonance images, is reflected by the signal intensity characteristics of the structure. Each of these properties requires an objective means of quantification.

J. S. Byrnes and J. F. Byrnes (eds.), Recent Advances in Fourier Analysis and Its Applications, 17–28.

We have applied the methods of structural morphometry to the analysis of anatomic structures in general and to the human brain in particular. The human brain is a highly complex organ, composed of numerous highly integrated substructures. Overall measures of brain size have been shown to be useful in some instances, but are known to lack sensitivity and specificity [1]. As far as shape is concerned, since the brain is encased in the solid super-structure of the skull, alterations in shape are almost invariably due to intrinsic factors (development, pathology, etc.) as opposed to external factors (such as posture, positioning, etc.). In addition, relatively little is known about the importance of the geometric integration of structures in the functioning of the brain. What is known, however, is that insults incurred during brain development are much more likely to be compensated for than are insults incurred later in life [1]. This plasticity of the brain permits, therefore, the development of widely varying structural conformations that prove to be equally successful at carrying out "normal" function.

In general, the practice of morphometry requires either direct access to the structure of interest itself, or access to a faithful representation of that structure. Direct access to the human brain is possible, and indeed has been shown to be important, in postmortem examination [1]. The limitations of such examinations include the variable amount of time between the clinical and pathologic observations, and the variable nature of the changes occurring during death and tissue preparation. Recent advances in *in-vivo* diagnostic imaging have provided a rapid, accurate and safe method of indirect observation. In particular, the use of Nuclear Magnetic Resonance Imaging provides a method of high spatial and anatomic resolution in the human brain that is sufficient for morphometric analysis.

There are three general domains in which we have found neuroanatomic shape analysis to be applicable:

1) detection of specific structural shape abnormalities in disease states relative to the normal population,
2) temporal changes in structures being transformed by development or disease and
3) evaluation of symmetry (or asymmetry) between paired homologous structures.

The first domain requires the characterization of normal shape and its associated natural biological variation in order to differentiate normal and abnormal. The second is also related to shape changes that occur in moving structures (such as the heart wall in systole and diastole) and have been previously discussed by other investigators [2]. The third domain is applicable when a priori information leads to expectations regarding structural homology as in the cerebral hemispheres an its component substructures.

2. Methods

There are two main categories of methodologies that must be addressed to properly approach the problem of quantitative shape analysis. First, the methods of the Fourier shape analysis itself and second, the methods of data preparation that produce data suitable for shape analysis.

2.1 TWO-DIMENSIONAL FOURIER SHAPE ANALYSIS

The Fourier shape analysis is designed to operate on structural surface outlines which are represented as a directed list of **N** *x,y* Cartesian coordinate pairs. In practice, **N** can range from 50 to 1000 points depending on the size of the structure. The *x,y* representation of the outline points is converted into polar coordinates (r,θ) representing the radial distance of the outline point relative to some suitably chosen reference point (the center of mass, for example) as a function of its angle relative to x axis. The two-dimensional analysis expands the radial function into its circular harmonics.

$$R(\theta) = \sum_{n=0}^{\infty} [A_n \operatorname{Sin}(n\theta) + B_n \operatorname{Cos}(n\theta)] .$$

The A_n and B_n coefficients can be obtained by solving:

$$A_n = (1/\pi) \int_0^{2\pi} R(\theta) \operatorname{Cos}(n\theta)\, d\theta$$

$$B_n = (1/\pi) \int_0^{2\pi} R(\theta) \operatorname{Sin}(n\theta)\, d\theta$$

where $R(\theta)$ is represented as a linear function of θ between the sampled outline points. Solution of these equations yield [3]:

$$A_n = 1/\pi \sum_i^N \left[\frac{[r_i - r_{i+1}]\,[\cos(n\theta_i) - \cos(n\theta_{i+1})]}{\theta_i - \theta_{i+1}} + \frac{[r_{i+1}\sin(n\theta_{i+1}) - r_i\sin(n\theta_i)]}{n} \right]$$

$$B_n = 1/\pi \sum_i^N \left[\frac{[r_i - r_{i+1}]\,[\sin(n\theta_i) - \sin(n\theta_{i+1})]}{\theta_i - \theta_{i+1}} + \frac{[r_{i+1}\cos(n\theta_{i+1}) - r_i\cos(n\theta_i)]}{n} \right]$$

An alternative representation of the solution is to express each of the circular harmonics as a function of a magnitude, P, and phase angle, ϕ:

$$R(\theta) = R_0 + \sum_{n=1}^{\infty} [P_n \operatorname{Cos}(n\theta + \phi_n)]$$

so that

$$P_n = (A_n^2 + B_n^2)^{1/2}$$

and

$$\phi_n = (|B| / B) \, Cos^{-1}(A_n / P_n) .$$

Thus the results of the shape analyses are represented as *spectra* which show the value of nth coefficient (A_n and B_n, or P_n and ϕ_n) as a function of the coefficient number, n.

2.2 DATA PREPARATION

The application of two-dimensional shape analysis to three-dimensional structures requires careful selection of representative standardizable and reproducible structural outlines. By the use of three-dimensional magnetic resonance imaging [4] which encompasses the entire head, we routinely obtain approximately 60 cross-sectional coronal images with in-plane resolution of 1.1 mm and a slice thickness of 3.1 mm. Knowledge and visualization of the brain anatomy allows the establishment of a standardized coordinate system within the imaged brain itself. This coordinate system is determined by the localization of three points of reference including the midpoints of the decussations of the anterior (**AC**) and posterior (**PC**) commissures, and the midpoint of the genu of the corpus callosum. The origin of this coordinate system is located at the midpoint of the **AC-PC** line segment. A rectilinear coordinate system for the standard orientation is defined as: ***x*** axis increasing laterally from the subjects right to left, ***y*** axis increasing from posterior to anterior, and ***z*** axis increasing from superior to inferior (***xy*** = transaxial, ***xz*** = coronal, and ***yz*** = sagittal orientations). We apply a homogeneous transformation between the coordinates in the original image data set and the standardized presentation of the brain. This transformation consists of dimensional scaling, three-dimensional translation, and a three-dimensional rotation. This transition is imposed such that the **AC-PC** line is oriented along the y axis, and the mid-sagittal plane is in the ***yz*** plane (the ***x*** axis is, by definition, normal to the mid-sagittal plane). This new data set is then interpolated into a new set of image planes parallel to ***xz*** plane. We can then produce two-dimensional anatomic regions by image segmentation [1,4,5] of comparable standardized slices across patients (or within same patient across time).

3. Results

One of the advantages of the Fourier shape analysis is that the results have an easily interpretable form. In general, the nth harmonic terms represents contribution by a *n* 'lobed' structure. Thus, the second harmonic represents an elongation (ellipse-like), the third harmonic is triangular (trefoil), etc. In the A_n, B_n representation, the *A* and *B* terms represent the sine and cosine contributions respectively. The P_n, ϕ_n represents the magnitude and phase angle of the nth harmonic component. Figure 1 shows an illustrative example of a *pure* shape, created with $R_0 = 100$, $A_3 = B_3 = 7$ and the remaining coefficients set to zero. This corresponds to a $P_3 = 19.80$ with $\phi_3 = 45°$. This shape outline is shown in Figure 1a, and the calculated *A* and *P* spectra are shown and compared to the known values in Figure 1b-c.

The accuracy of a shape reconstruction can be controlled by the number of coefficients used to perform the reconstruction. An original outline

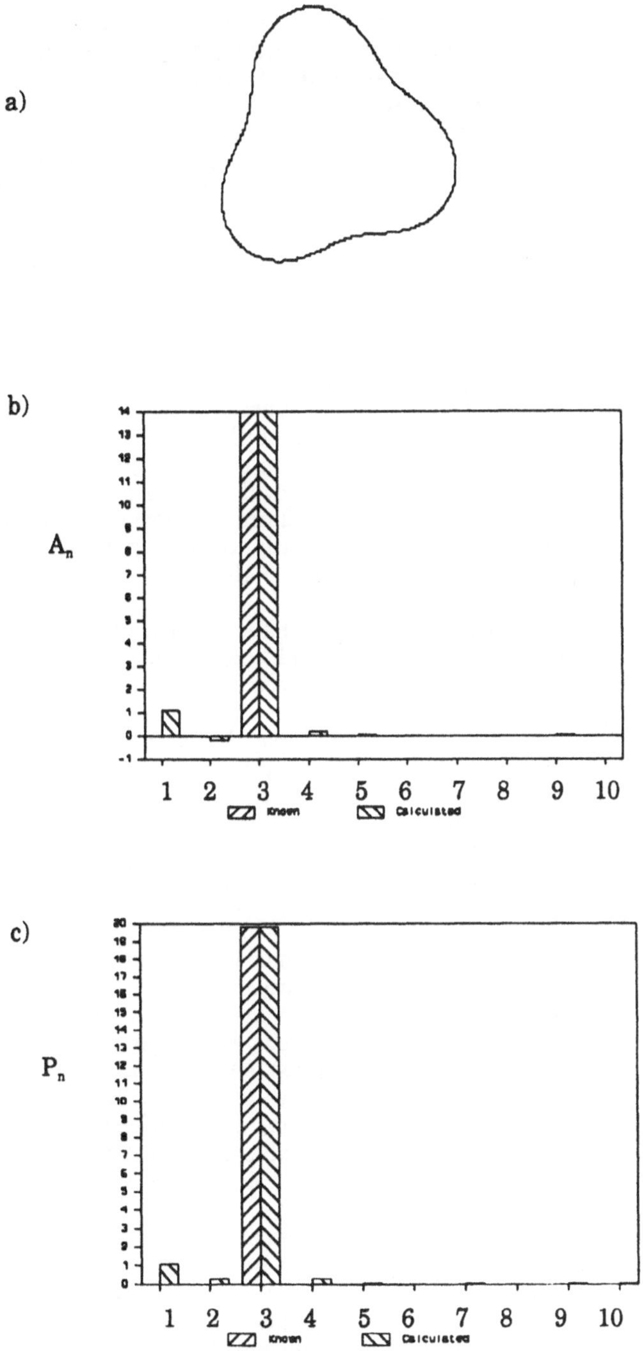

Figure 1: a) Test outline, R_0 and third harmonic only. b) Known and calculated A coefficients. c) Known and calculated P coefficients.

of the exterior of a cerebral hemisphere is shown in Figure 2a, and the reconstruction from two, ten and forty coefficients, respectively, is shown in Figures 2b-d. As the number of coefficients is increased, the higher order spatial characteristics of the structure are visualized.

Figure 3a shows a representative set of cross-sectional outlines of brain structures including the right and left cerebral exterior, the right and left cerebellar exterior and the right and left lateral ventricle. Representative *P* shape spectra for the right cerebral and left cerebellar exterior are shown in Figures 3b-c. The general characteristics of these two different types of shapes are reflected by differences in the form of the shape spectra.

In addressing the question of symmetry one may ask, in Figure 3a, "How similar are the right and left cerebral structures in this plane?". If the left hemisphere were to be purely a reflection of the right hemisphere about the *y* axis, the question becomes "How similar are these shapes, independent of the reflection?". This question can be addressed in two different ways. First, examination of the magnitude spectra of the two shapes will indicate the similarity in the amplitude of the harmonics of the original structures (see Figure 4c). The reflection through the *y* axis will only, in general, result in a phase shift of inversion for each of the harmonic terms. Alternatively, representing the left cerebral outline, explicitly correcting for the reflection about the *y* axis yields the outline shown in 4b. The resulting *A* and *B* coefficients can now be compared directly.

Finally, in Figure 5 we compare the shape spectra of a pair of sibling subject with suspected reversal of the typical cerebral asymmetry (based on volumetric analysis) to the average shape spectrum compiled from a population of 5 *normal* individuals [6]. The difference in the second Fourier coefficient between the siblings and the normal population was statistically significant. This finding is consistent with a perturbation in cerebral symmetry since the second coefficient contributes significantly to the general cerebral shape in this coronal presentation.

4. Discussion

There are a number of potential methods for the quantitative analysis of shape characteristics. These include Fourier Shape Analysis [3,7,8] and Fractal Shape Dimension [9] to name two. As a first approach, we have chosen to concentrate on the Fourier methods of shape description. The evaluation of any method of shape description should consider the following criteria [9]: uniqueness, independence (low cross-correlation between parameters), invariance (to rotation, scale and reflection) and efficiency. From a data reduction point of view, it is important to be able to reconstruct an acceptable representation of the shape while retaining less data than required for a point by point description of the surface.

As can be seen from the material presented in the **Results** section, Fourier shape analysis is useful in providing a quantitative measure of shape. An advantage of the Fourier method includes the intuitive and understandable nature of the results. Figure 3 highlights the reasonable nature of the Fourier result, showing a strong second coefficient for the somewhat elliptical cerebral exterior and a strong third coefficient for the more triangular cerebellar exterior. A major disadvantage of the Fourier method results from the

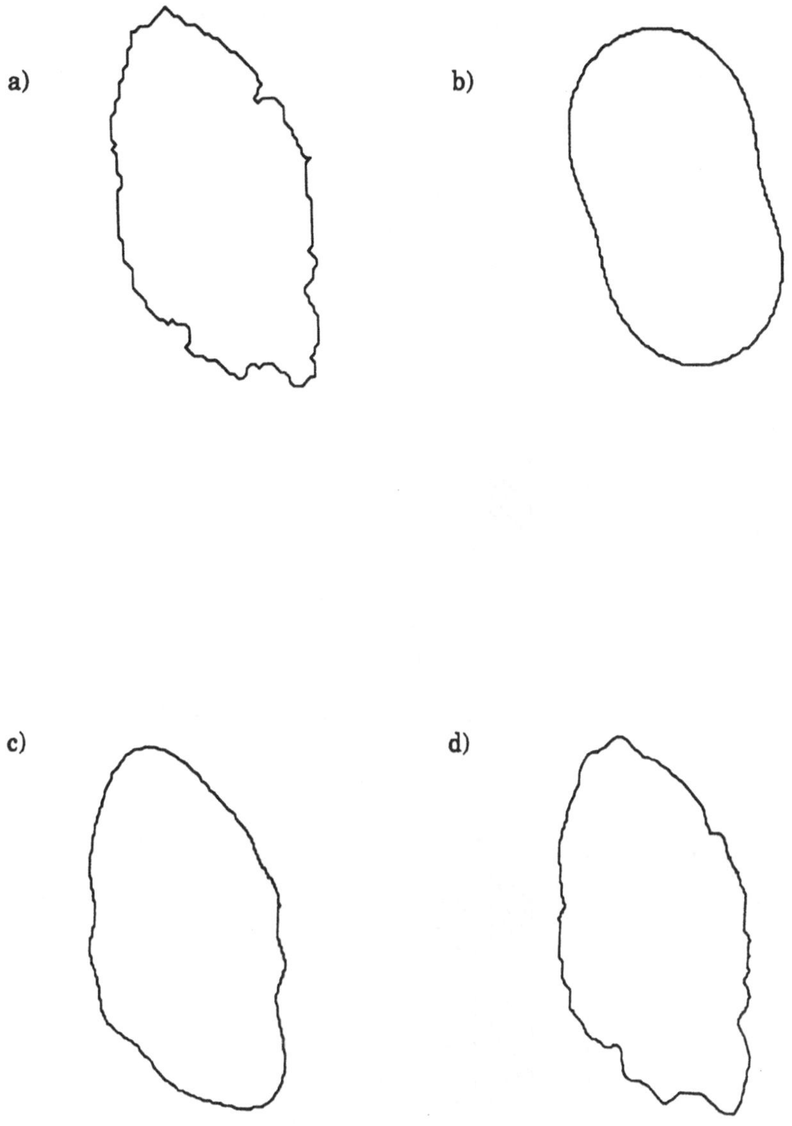

Figure 2: Outline reconstruction. a) original outline. b) - d) reconstruction with 2, 10 and 40 Fourier coefficients respectively.

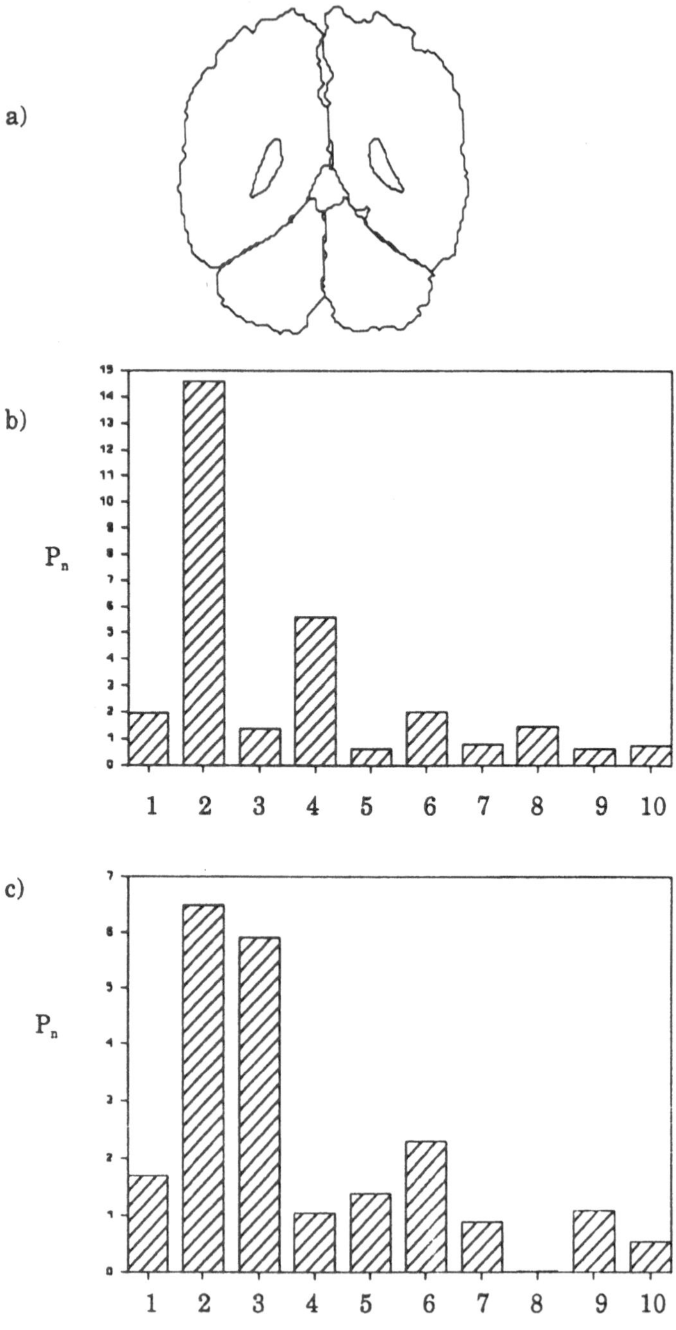

Figure 3: a) Representitive brain structural outlines: right and left cerebral exterior (RCE, LCE), cerebellar exterior (RCbE, LCbE) and lateral ventricle (RLV, LLV). P spectrum for b) right cerebral exterior and c) left cerebellar exterior.

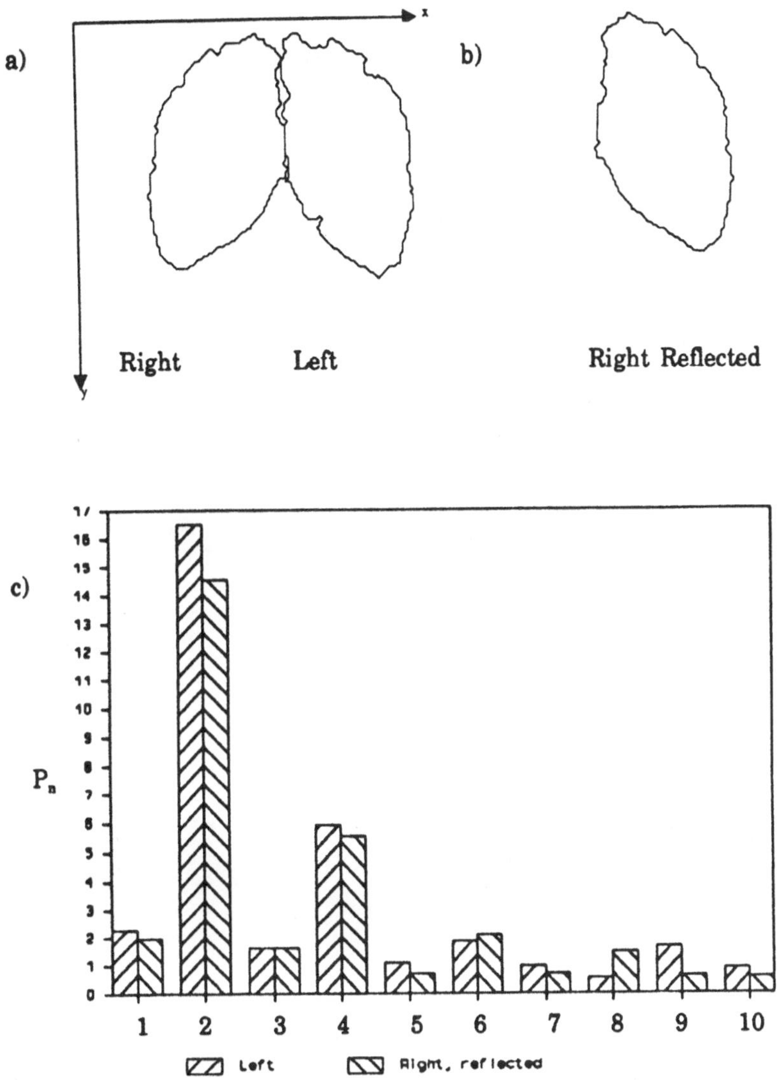

Figure 4: Comparison of right-left cerebral symmetry. a) Right and left cerebral exterior. b) right cerebral exterior reflected about y axis. c) Comparison of P spectra from left and right-reflected outlines.

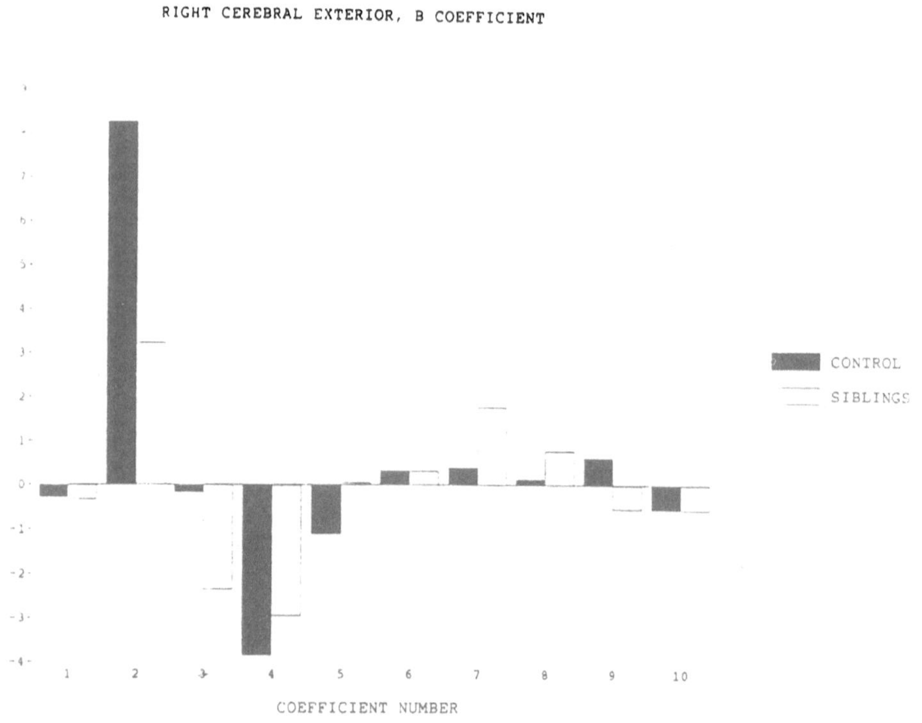

Figure 5: B shape spectrum for right cerebral exterior of two siblings in order to evaluate symmetry reversal when compared to normal controls.

limitation that the analysis is not properly defined when the radial function being evaluated is not single valued, thus prohibiting proper analysis of structures with inclusions by this formulation.

The Fourier shape analysis is **not** invariant with respect to rotation or reflection. However, the presence of both of these effects can be determined from the phase data. Both reflections and rotations of structures will yield magnitude spectra that are identical to the original unperturbed structure. Rotation by an angle α causes a phase change of n times α for each of the ϕ_n terms. Reflection about the y axis causes a phase inversion for all harmonics. Reflection about the x axis causes a phase inversion of the even harmonic terms and modulates the odd harmonic terms such that $\phi_{new} = 2\pi - \phi_{old}$. In the evaluation of spatial similarity (Figure 4) the results, as expected, indicate that the structures being compared are similar but not identical. This is consistent with the fact that the right and left hemispheres are known to be both functionally and structurally asymmetric. Similar differences are obtained when evaluating temporal similarity and can be attributed to growth which can occur during the time interval between the scans.

Finally, the most compelling result from the brain science point of view is the ability to accurately quantify changes in cerebral symmetry. The concordance of the shape (and volume) measures **between** the siblings shown in Figure 5 was remarkably high (underscoring the magnitude of the genetic factor in brain development) and yet they possessed a different spatial symmetry than the normal population. The detection of such phenomena is exceedingly important since symmetry of the cerebral hemispheres has been correlated, for example, with handedness, cerebral dominance and the occurrence of developmental dyslexia [1].

5. Conclusions

In concluding it is possible to state that Fourier shape analysis can be used as a tool in the quantitative evaluation of shape and its change over time. In particular, applications of Fourier shape analysis to the evaluation of neuroanatomic structures can provide new insights into the functional geometry of the human brain. Successful application of this technology requires the use of a standardized image data acquisition that allows reproducible identification of structures. Work is proceeding on the extension of the two-dimensional methodology to three dimensions as well as to the application of this type of analysis in the development of 'normal' population shape statistics for the identifiable anatomic structures.

6. Acknowledgements

The authors wish to thank the following people for various types of assistance and discussions: Stephen Kennedy, Jonathan Sacks, Robert McKinstry and A. Greg Sorensen. This work was supported in part by the National Institute of Neurological, Communicative Disorders and Stroke, under grants NS24279 and NS07170.

7. References

1. Caviness, V.S., Filipek, P.A. and Kennedy, D.N. (1989) 'Magnetic resonance technology in human brain science: Blueprint for a program based upon morphometry', Brain and Development 11,1-13.
2. Jouan, A. (1987) 'Analysis of sequences of cardiac contours by Fourier descriptors for plane closed curves', IEEE Transactions on Medical Imaging MI-6(2), 176-180.
3. Ehrlich, R. and Weinberg, B. (1970) 'An exact method for characterization of grain shape', Journal of Sedimentary Petrology 40(1), 205-212.
4. Filipek P.A., Kennedy, D.N., Caviness, V.S. et al. (1989), 'Magnetic resonance imaging-based brain morphometry: Development and application to normal subjects', Annals of Neurology 25(1), 61-67.
5. Kennedy,D.N., Filipek, P.A. and Caviness, V.S. (1989) 'Anatomic segmentation and volumetric calculations in nuclear magnetic resonance imaging', IEEE Transactions on Medical Imaging MI-8(1), 1-7.
6. Filipek P.A, Kennedy, D.N., Kennedy, S.K. and Caviness, V.S. (1988) 'Shape analysis of the brain of two siblings based upon magnetic resonance imaging', Annals of Neurology 24, 355-356.
7. Kennedy, D.N., Kennedy, S.K., Filipek, P.A. and Caviness, V.S. (1988) 'Morphometric characterization of brain shape by Fourier analysis', Proc Soc Mag Res Med 7, 993.
8. Porter, G.A. and Ehrlich, R. (1979) 'Sources and nonsourses of beach sand along southern Monterey Bay, CA - Fourier shape analysis', Journal of Sedimentary Petrology 49, 727-732.
9. Kennedy, S.K. and Lin, W.H. (1986) 'FRACT - a FORTRAN sub-routine to calculate the variables necessary to determining the fractal dimension of closed forms', Computers in Geosciences 12, 705-712.

HARMONIC ANALYSIS OF THE SEMI-CLASSICAL LIMIT

KETILL INGØLFSSON
Department of Mathematics
Saint Joseph's University
5600 City Line Avenue
Philadelphia, Pennsylvania 19131
U.S.A.

ABSTRACT. In the harmonic analysis of quantum-dynamical notions a method is presented that already in a simple perturbation-setting can successfully be used in order to obtain pointmechanical laws under the so-called classical nonrelativistic limit. In quantum mechanics a state sometimes is developed in terms of two-parameter families of unitary operators applied on the Hilbert space of absolutely square integrable functions in R^3. Through a similar procedure one can introduce such two-parameter families of the unitary operators $\mathbf{U}(r,s)$ and $\mathbf{Q}(r,s)$ for all positive real values of r and s that $\mathbf{U}\,\mathbf{Q}$ will represent a semigroup generated by a selfadjoint extension of the Laplace operator. In certain dynamical integrals one can moreover evaluate consistent limits when $s \to \infty$. By correspondence relations one may establish an isomorphy between the structure presented here and quantum dynamics. The existence of classical limits implies that $s \to \infty$. Reversely, $s \to \infty$ does not uniquely determine physical limits, and it is therefore necessary to specify the exact meanings of the classical and the semi-classical limits.

1. INTRODUCTION

The classical limit of quantum theories is mostly understood as an ensemble of dynamical expressions, which analytically may be obtained as genuine limits when $\hbar$, Planck's constant, tends to zero. The existence of such limits has been considered by many noted physicists, for example Landau [1] and Messiah [2], as necessary characteristics of a consistent form in a new theory. The argumentation is as old as the quantum mechanical approach itself. Already when the physical operator-algebra was first formulated, Heisenberg [3] pointed out that commutation relations in any of the components of the n-tuples of canonical position and momentum conveniently converted the noncommutative algebra into the commutative form of analytical point-mechanics, when the $\hbar$ became zero.

Despite its fruitfulness, this general idea has always been met with much scepticism. Some of the critical argumentations will be

J. S. Byrnes and J. F. Byrnes (eds.), Recent Advances in Fourier Analysis and Its Applications, 29–40.

referred to in the following as formal contradictions. The main purpose of this contribution, however, is to establish a mathematically consistent form for the desired transition to classical theories. The suggestions made here might likewise be useful, when other challenges are met.

It was recognized by many of the early authors, for example in the "Sources" edited by B.L.van der Waerden [4], that a natural constant of the fundamental importance of Planck's constant should not be subject to a continuous change. Therefore alternative formulations like the transition to large quantum numbers and exclusion rules became a possible choice (as one for example may see in Cohen-Tannoudji [5]). Moreover, once an algebra had been well defined, calculated results could not easily be tied to the continuous transition through $\hbar$ in a new algebra. Some early authors like Heitler [6] emphasized the insensitivity of Planck's constant to any manipulations of the mathematical form. Most notably, however, there were still considerable difficulties in achieving the correct analytical limits. This explains the large number of research papers in this area until now.

The present contribution does not follow up with recognizing the recent excellent research concerning the semi-classical limit. This would require more physically oriented discussion. For the more formal consideration it is sufficient to refer to some older yet successful calculations of the limit.

In a contribution by Rosen [7] one may find the following situation: Suppose that subscripts for the n-tuples are disregarded and that one considers a dynamical system described by a Hamiltonian of the form $\mathbf{H} = \mathbf{U}(p) + \mathbf{V}(q)$. For the quantum statistical description of the system there is an ensemble of wave functions, $\mathbf{g}_j(q,t)$ when j is the enumerator index, that satisfy the Schroedinger equation

$$i\hbar\frac{\partial}{\partial t}\,\mathbf{g}_j = \left(\mathbf{U}(-i\hbar\frac{\partial}{\partial q}) + \mathbf{V}(q)\right) \mathbf{g}_j \quad .$$

Expectation-values of the observables $\mathbf{O}(\mathbf{q},\mathbf{p})$, which as mappings of $\mathbf{q}$ and $\mathbf{p}$ in general are differential operators in q, are then properly defined in the form

$$[\mathbf{O}]_\hbar = \sum_j w_j \int \mathbf{g}_j^*(q,t)\,\mathbf{O}(q,-i\hbar\frac{\partial}{\partial q})\,\mathbf{g}_j(q,t)\,d^nq \;.$$

They can be reformulated by the use of so-called hypercharacteristic functions expressed (in a notion by Wigner) as

$$\mathbf{f}_\hbar(u,v,t) \;\sum_j w_j \int \exp(iuq)\mathbf{g}_j^*(q-\hbar v/2,t)\mathbf{g}_j(q+\hbar v/2,t)\;d^nq,$$

and one obtains, for example, as the probability density function of the observable $\mathbf{q}$ the expressions

$$P_\hbar(q,t) = \sum_j w_j\,|\mathbf{g}_j(q,t)|^2 \;=\; (2\pi)^{-n}\int \exp(-iuq)\mathbf{f}_\hbar(u,0,t)d^nu.$$

As v is 0 under the mapping $\mathbf{f}_h$ in the above integral the transition by $\hbar \to 0$ is trivial. The classical limits can also be carried out in several other related characteristic statistical expressions.

A careful analysis of these "successful" cases seems to indicate that a well defined limit just as well can be obtained when some characteristic parameter tends to zero, while $\hbar$ at least is bounded above. This is a much weaker demand than made in the beginning. It may therefore eliminate much of the objections raised by physical reasons. From the formal point of view it gives not alone one of many consistent analytical approaches. Indeed, it will even provide the opportunity to give the so-called semi-classical limit an appropriate meaning. From the physical literature one may not always obtain the concept distinguished from the classical limit.

The following integral relations, which may be found in the Fourier analysis of one-particle dynamics, will be considered under the semi-classical limit. The first of these expressions is the Dollard [8] relation for the so-called free particle,

$$\mathbf{g}_t(x) = (\exp(-i\mathbf{H}_o t\hbar^{-1})\mathbf{g}_o)(x) = (\mathbf{U}_{t,\hbar}\mathbf{Q}_{t,\hbar}\,\mathbf{g}_o(x) =$$
$$(2\pi\hbar)^{-3/2} \int_{R^3} \exp(ipx\hbar^{-1})\exp(-ip^2 t/2m\hbar)\, \hat{\mathbf{g}}_o(p)\, d^3p \quad .$$

The operator $\mathbf{H}_o$ goes directly with the selfadjoint extension of the Laplace operator while the proportionality factor is $-\hbar^2/2m$. The operators $\mathbf{Q}$ and $\mathbf{U}$ are for any absolutely square integrable function, $\mathbf{g}$, well defined by the identities

$$(\mathbf{Q}_{t,\hbar}\,\mathbf{g})(x) = \exp(imx^2/2t\hbar)\,\mathbf{g}(x)$$
$$(\mathbf{U}_{t,\hbar}\,\mathbf{g})(x) = (m/it)^{3/2}\,\exp(imx^2/2t\hbar)\,\hat{\mathbf{g}}(mx/2),$$

and the Fourier-transform of $\mathbf{g}$ is denoted by $\hat{\mathbf{g}}$. The second expression considered under the semi-classical limit will be the Ehrenfest [9] relation for a single wave-package,

$$\frac{d}{dt} \int_W \mathbf{g}_t^*(x)\,(-i\hbar\frac{\partial}{\partial x_j})\,\mathbf{g}_t(x)\,d^3x = -\int_W \mathbf{g}_t^*(x)\,\frac{\partial V}{\partial x_j}\,\mathbf{g}_t(x)\,d^3x\ .$$

If the integral-domain W is bounded, one must now assume that the wave-function $\mathbf{g}$ has a compact support. From the semi-classical limit of the integral-expressions one will finally be able to obtain the equivalent of Newton's classical laws of motion.

2. DYNAMICAL OPERATORS IN A SINGULAR PERTURBATION SETTING

In the analytical structure developed in this section two-parameter families of operators will be used in a similar integral-form as the operators $\mathbf{Q}_{t,\hbar}$ and $\mathbf{U}_{t,\hbar}$ introduced above. The new dynamical operators will, however, be considered without any reference made to physical measurements. Instead, the necessary elements in this structure may be introduced by the following definitions:

By the letter r will be meant a nonnegative and by the letter s a positive element of the R^1. By the letters u and v are denoted elements of the R^3, and by uv is therefore meant the inner product of these vectors in R^3. Suppose that two mutually isometrically isomorphic complex Hilbert spaces, $\hat{\mathbf{F}}_s$ and $\mathbf{F}_s$, of square integrable functions, say

$\hat{\mathbf{f}}$ and $\mathbf{f}$ in the variables v and u respectively, are assigned to any positive value of s. The domains of the functions $\hat{\mathbf{f}}(v)$ and $\mathbf{f}(u)$ are "almost everywhere" within the Euclidean spaces V^3 and U^3. If $\hat{\mathbf{F}}_s$ is the image under the Fourier-transform of $\mathbf{F}_s$, then is U^3 called a "pre-local space" and V^3 is called a "pre-momenta space". Moreover, as both the Fourier-transform as well as its inverse represent wave-packages in the pre-momenta and in the pre-local spaces respectively, then are the Hilbert spaces $\hat{\mathbf{F}}_s$ and $\mathbf{F}_s$ accordingly named "wave-function" spaces. The inner-product norms of $\mathbf{f}$ or $\hat{\mathbf{f}}$ in $\mathbf{F}_s$ or $\hat{\mathbf{F}}_s$ will be denoted in this section by $\| \mathbf{f} \|_s$ or by $\| \hat{\mathbf{f}} \|_s$ respectively. Suppose that the Hilbert space, which other elements of the structure generally are embedded in, may for any positive value of s be determined up to an isometrical isomorphism as one of the wave-function spaces $\hat{\mathbf{F}}_s$ or $\mathbf{F}_s$. The elements of the one-parameter family $\{ \mathbf{F}_s ;\ s>0 \}$ are now named "unitarily equivalent" because there exists for any s and s' such that $0 < s \leq s' < \infty$ an isometry $\mathbf{I}^s_{s'}$, taking $\mathbf{f}$ to $\mathbf{f}'$ with $\| \mathbf{f} \|_s = \| \mathbf{f}' \|_s$.

The concepts introduced above may in particular be be considered as elements of a formal dynamical structure appropriately called "properly posed pre-quantum dynamics", when the following properties are fulfilled: a) When s has been determined, an one-to-one mapping is well defined for any positive value of r by the relation

$$v = (rs)^{-1}\ u$$

from the pre-local space U^3 onto the pre-momenta space V^3. b) There exists an one parameter family, $\{ \mathbf{H}_s ;\ s>0 \}$, of linear operators in the general Hilbert space, such that each of the operators can be densely defined for any positive s in the wave function space $\mathbf{F}_s$. c) There exists an one-parameter strongly continuous semigroup, $\{ \mathbf{U}(r);\ r \geq 0 \}$, of unitary operators with the skew-selfadjoint generator $-i\mathbf{H}$ defined in the general Hilbert space such that

$$\mathbf{E}_s(k)\ \mathbf{H} \subset \mathbf{H}\ \mathbf{E}_s(k) = \int_{-\infty}^{k} k'\ d\mathbf{E}_s(k'),$$

when one assumes that every $\mathbf{H}_s$ is selfadjoint in $\mathbf{F}_s$ with the spectral resolution $\{ \mathbf{E}_s(k) \}$ for any given positive value of s. d) Suppose that $\mathbf{f}$ is a square integrable function in the variable u, which "almost everywhere" in U^3 takes complex values and belongs to a $\mathbf{F}_s$ for any positive s. If $U^s_{s'}$ is a unitary transformation taking some $\mathbf{f}_{(s)}$ of the general Hilbert space to $\mathbf{f}_{(s')}$, when the $\mathbf{f}_{(s)}$ is represented by $\mathbf{f}$ in $\mathbf{F}_s$ and $\mathbf{f}_{(s')}$ is represented by the same $\mathbf{f}$ in $\mathbf{F}_{s'}$, then is the path given by the set $\{ \mathbf{U}^s_{s'}\ \mathbf{f}_{(s)};\ s' \geq s \}$ continuous in the general Hilbert on any closed interval in s'.

The evolution of the properly posed pre-quantum dynamical system in the above sense will now be given by one of the two one-parameter families $\{ \mathbf{f}_r(u);\ r \geq 0 \}$ and $\{ \hat{\mathbf{f}}_r(v);\ r \geq 0 \}$ of wave functions in $\mathbf{F}_s$ and $\hat{\mathbf{F}}_s$, when these elements in particular are defined as follows:
1) Independent of s, the Fourier transform is given by the relation

$$\hat{\mathbf{f}}_r(v) = \text{l.i.m.}\ (2\pi)^{-3/2} \int_{U^3} \exp(-iuv)\ \mathbf{f}_r(u)\ d^3u$$

for all positive r, but l.i.m. stands for the "limit in the mean".
2) For at least one positive value of s there exists a $\mathbf{f}$ with $\|\mathbf{f}\|_s=1$ such that $\|\mathbf{H}_{s'}\mathbf{f}\|_{s'}<\infty$ for all $s'\geq s$ and such that the path given by $[\mathbf{U}^s_{s'}\mathbf{f}_{(s)};\ s'\geq s]$ in accordance with the property d) has become a single point (element) in the general Hilbert space.
3) The element $\mathbf{f}_0$ is given as $\mathbf{f}$ and the function $\mathbf{f}_r(u)$ is then well defined by the expression $(\exp(-i\mathbf{H}_s r)\ \mathbf{f}_0)(u)$ for all nonnegative r, when the exponential expression is given by the Stieltjes-integral

$$\int_{-\infty}^{\infty} \exp(-ikr)\ d\mathbf{E}_s(k) \quad .$$

In accordance with Stone's theorem on generators of unitary groups it now follows that $\mathbf{f}_r$ and $\hat{\mathbf{f}}_r$ are strongly continuously differentiable in the variable r and are therefore solutions of the wave equation for the appropriate choice of the operators $\mathbf{H}_s$ and the given initial value.

In the sequel it will be shown, how the evolution of such system can be developed for positive values of r by using two strongly continuous two-parameter families of partially isometric operators, $[\mathbf{Q}(r,s);\ r>0\wedge s>0]$ and $[\mathbf{U}(r,s);\ r>0\wedge s>0]$, in the square integrable $\mathbf{L}^2$-space. For a particular s the operator $\mathbf{Q}(r,s)$ is defined as an unitary operator on $\mathbf{F}_s$, which takes a $\mathbf{f}(u)$ to $(\mathbf{Q}(r,s)\mathbf{f})(u)$ in accordance with the formula

$$(\mathbf{Q}(r,s)\ \mathbf{f})(u) = \exp(iu^2/2rs)\ \mathbf{f}(u) \qquad (1)$$

for all u in U^3. Similarly $\mathbf{U}(r,s)$ is for a particular s defined as an unitary operator on $\mathbf{F}_s$, which takes a $\mathbf{f}(u)$ to $(\mathbf{U}(r,s)\mathbf{f})(u)$ in accordance with the Fourier-transform formula

$$(\mathbf{U}(r,s)\ \mathbf{f})(u) = (irs)^{-3/2}\ \exp(iu^2/2rs)\ \hat{\mathbf{f}}\,((rs)^{-1}u) \qquad (2)\ .$$

The partially isometric character of the operators $\mathbf{Q}(r,s)$ and $\mathbf{U}(r,s)$ is found in the definition of the orthogonal complement to $\mathbf{F}_s$ with respect to $\mathbf{L}^2$ as the nullspace of each operator for any of the positive values of s and of r.

In the examples taken for the $\mathbf{H}_s$ in the following the orthogonal complements to the $\mathbf{F}_s$ with respect to $\mathbf{L}^2$ are assumed to be empty. Then are also $\mathbf{Q}$ and $\mathbf{U}$, which were introduced above, unitary on $\mathbf{L}^2$, The first such example is the so-called "free operator"

$$\mathbf{H}^0_s \;=\; -(s/2)\ \text{"}\,\Delta\,\text{"} \qquad (3)\ .$$

By the operator " Δ " is then meant the selfadjoint extension of the Laplace operator, which as follows is well defined in any unitary space of absolutely square integrable functions: If $(-v^2)\mathbf{f}(v)$ belongs to $\mathbf{L}^2$ on V^3, then the inverse Fourier-transform of this function converges to $(\Delta\ \mathbf{f})(u)$, uniformly in u when $\mathbf{f}$ is twice continuously differentiable in u. In a more general case of $\mathbf{f}$ one may use the fact that the relation

$$\mathbf{f}^{\mathbf{D}}(u) \equiv (\text{"}\,\Delta\,\text{"}\mathbf{f})(u) = (2\pi)^{-3/2}\int_{V^3} \exp(iuv)\ \hat{\mathbf{f}}^{\mathbf{D}}(v)\ d^3v$$

implies an identity on U^3, when the right side converges as l.i.m.

If this is true when $\hat{\mathbf{f}}^{D}(v)$ is $(-v^2)\hat{\mathbf{f}}^0$, then is the inverse transform $\mathbf{f}$ of $\hat{\mathbf{f}}$ an element in the domain of $\mathbf{H}^0_s$.

By the same methods as several authors have proved the relation, which in the introduction was attributed to Dollard [8], one can prove that the simple relation

$$\exp(-i\mathbf{H}^0_s r) = \mathbf{U}(r,s)\ \mathbf{Q}(r,s) \tag{4}$$

is identically true on an everywhere dense subspace of $\mathbf{L}^2$, when the r is positive, irrespective of the positive value taken for s. The fact that the operators are all unitary implies that the relation (4) is then identically true on $\mathbf{L}^2$. The relation (4) in particular implies

$$\mathbf{f}^0_r(u) \equiv (\exp(-i\mathbf{H}^0_s r)\ \mathbf{f})(u) = (\mathbf{U}(r,s)\ \mathbf{Q}(r,s)\ \mathbf{f})(u) =$$

$$(2\pi)^{-3/2} \int_{V^3} \exp(i\ uv)\ \exp(-iv^2 r/2s)\ \hat{\mathbf{f}}(v)\ d^3v \tag{5}$$,

which is the expression for a "free" evolution of a properly posed prequantum dynamical system, when $\mathbf{f} = \mathbf{f}^0$ belongs to the domain of $\mathbf{H}^0_s$.

At this point it is important to take note of the fact that the existence of the semigroup $[\ \exp(-i\mathbf{H}^0_s r);\ r \geq 0\ ]$ is sufficient in order to show that $\mathbf{f}$ is the s-lim of $\mathbf{f}^0_r$ when $r \to 0$, but the right side of the equation (5) still can not be considered under the limit when $r \to 0$. This will determine the singular character of the perturbation presented at the conclusion of this section. It is of course possible to write the standard singular perturbation form , but it will be omitted as it substantially increases the length of these notes. It may also be mentioned that the expressions (1) and (2) make a good sense, if the sign of s is reversed. It is then even true that

$$\begin{aligned} \mathbf{Q}^{\#}(r,s) &= \mathbf{Q}(r,-s) \\ \mathbf{U}^{\#}(r,s) &= \mathbf{U}(r,-s) \end{aligned} \tag{6}$$

for any s different from zero. The relations (6) would replace the rules of hypercharacteristic extensions in Rosen's calculations [7],G5.

After this discussion of the "free" Hamiltonian $\mathbf{H}^0_s$ one may now consider a second example of a Hamiltonian, which does not reduce $\mathbf{L}^2$. It is the selfadjoint operator

$$\mathbf{H}_s = \mathbf{H}^0_s + \mathbf{V}_s \tag{7}$$

in which the operator $\mathbf{V}_s$ is a real valued multiplicative function. Suppose that V(w) is a continuously differentiable function in the real variable w. Then is $\mathbf{V}_s$ a mapping taking any $\mathbf{f}(u)$ of $D(\mathbf{H}^0_s)$, the domain of the free Hamiltonian, to $s^{-1}V(|u|/s)\mathbf{f}(u)$ such that the new function is in $\mathbf{L}^2$. If the domain of $\mathbf{V}_s$ is denoted by $D(\mathbf{V}_s)$, then is this constraint simply expressed by the inclusion

$$D(\mathbf{H}^0_s) \subset D(\mathbf{V}_s) \tag{8}$$.

By using the selfadjoint operators $\mathbf{H}_s^0$ and $\mathbf{H}_s$ under the consideration of the relation (8), one may finally formulate the following two identically true relations in $D(\mathbf{H}_s)$:

$$\mathbf{U}(r,s) = \exp(-i\mathbf{H}_s^0 r) - \mathbf{U}(r,s)(\mathbf{Q}(r,s) - 1) \tag{9}$$

$$\mathbf{U}(r,s)(1 - i\mathbf{I}(r,s)) = \exp(-i\mathbf{H}_s r) - \mathbf{U}(r,s)(\mathbf{Q}(r,s)-1)(1-i\mathbf{I}(r,s)) \tag{10}$$

The operator $\mathbf{I}(r,s)$ is only then different from zero when the "interaction" $\mathbf{V}_s$ of the pre-quantum dynamical system is different from zero. The norm $\| \mathbf{I}(r,s)\mathbf{f} \|_s$ will uniformly in s tend to zero when r tends to zero. If written out in further detail one will obtain for the operator $\mathbf{I}(r,s)$ the integral expression

$$\mathbf{I}(r,s) = \int_0^r \mathbf{Q}(r',-s)\mathbf{U}(r',-s)\, \mathbf{V}_s \exp(-i\mathbf{H}_s r')\, dr' \quad ,$$

which as improper integral (for small r') converges on any $\mathbf{f}$ in $D(\mathbf{H}_s^0)$. The relation (10) is derived from the formulation

$$\exp(-i\mathbf{H}_s^0 r) = \exp(-i\mathbf{H}_s r) + i\exp(-i\mathbf{H}_s^0 r) \int_0^r \exp(i\mathbf{H}_s^0 r')\mathbf{V}_s \exp(-i\mathbf{H}_s r')dr' ,$$

which frequently is the basis for holomorphic studies (Kato [10]) or for perturbation series (Ingólfsson [11]) through iteration when $r \geq 0$.

The relations (9) and (10) only should be considered when $r > 0$. They then represent singular perturbations of the elements of the semigroups $\{ \exp(-i\mathbf{H}_s^0 r); r \geq 0 \}$ and $\{ \exp(-i\mathbf{H}_s r); r \geq 0 \}$ by operatorterms, not well defined when r=0 and when $r > 0$ not analytic in any open interval containing s=0. The operators $\mathbf{U}(r,s)$ and $\mathbf{U}(r,s)(1-i\mathbf{I}(r,s))$, however, asymptotically converge in the strong sense when s is very large to these semigroup-elements respectively. This is true in the sense that the operator- mamimumnorm $\| \mathbf{Q}(r,s)-1 \|_s$ then tends to zero. Indeed, it is possible to show that the least upper bound of the expression

$$\| \mathbf{f} \|_s^{-2} \cdot \| (\mathbf{Q}(r,s)-1)\mathbf{f} \|_s^2 = \int_{U^3} |\exp(iu^2/2rs)-1|^2 \frac{|\mathbf{f}(u)|^2}{\| \mathbf{f} \|_s^2} d^3u$$

for all $\mathbf{f}$ of $\mathbf{L}^2$ is well defined if the rs is a finite, nonzero number. The subscript s is not significant for the vectornorms in the given example and will therefore from now on be omitted. The integrand on the right in the above expression is always bounded by $4 |\mathbf{f}(u)|^2 \| \mathbf{f} \|^{-2}$ and tends to zero, when s becomes very large by constant, positive r. This means in accordance with the dominated convergence theorem of Lebesgue that the whole integral now is tending to zero. The operatornorm itself therefore tends to zero.

The relations (9) and (10) can, finally, be used in order to formulate two alternating series, which asymptotically converge for positive r and large s to elements of the semigroups $\{ \exp(-i\mathbf{H}_s^0 r); r \geq 0 \}$ and $\{ \exp(-i\mathbf{H}_s r); r \geq 0 \}$ respectively. One will obtain the two series

$$\exp(-i\mathbf{H}_s^0 r) \sum_{n=0}^{N} (-1)^n (\mathbf{Q}(r,s)-1)^n \quad ,$$

by the iteration of $\mathbf{U}(r,s)$ in the relation (9), and

$$\sum_{n=0}^{N} (-\mathbf{U}(r,s)(\mathbf{Q}(r,s)-1)\mathbf{U}(r,-s))^n \exp(-i\mathbf{H}_s r) \quad ,$$

by the iteration of $\mathbf{U}(r,s)(1-i\mathbf{I}(r,s))$ in the relation (10). By using these expressions one may show the singular relationship between the unitary operators, which are used in the perturbation relations (9) and (10). The formulation of the two series will otherwise not be an essential part of the discussion in the following section.

3. THE SEMI-CLASSICAL LIMIT

The analytical structure of the last section is largely based on identities, which alone by themselves do not define any physical concepts. They are therefore not applicable to a physical theory until there has been given some appropriate selection of correspondence rules to supplement them with.

In this section the same Hamiltonians will be considered as in the previous section. They do not reduce $\mathbf{L}^2$ and vectornorms therefore again may be written without the subscript s. In the general Hilbert space the unitary transformation $\mathbf{U}^s_{s'}$ is an automorphism which takes any element of the unit sphere to a continuous path in the same sphere when $s' \geq s$. Such progression is not a part of a physical evolution, but it might reveal, how a quantum setting evolves into a classical one, when the number s becomes very large.

In accordance with the property 3) in the previous section there exists a $\mathbf{f}$ in $\mathbf{L}^2$, which belongs to any $\mathbf{F}_s$ and as $\mathbf{f}_{(s')}$, under the inverse isomorphism in a general Hilbert space, is mapped onto itself by $\mathbf{U}^s_{s'}$ for all $s' \geq s$. Such $\mathbf{f}$ represents an (initial) observation of the dynamical system. The set of all $\mathbf{f}$ with this property may be named $\mathbf{F}_o$. It depends on the particular physical situation, but it always is a linear space and contained in $D(\mathbf{H}_s)$.

Suppose that $\mathbf{f}^s_o$ is selected in the following sense as a so-called reference scale in the $\mathbf{F}_o$: A ball of radius b and with the center 0 in R^3 is denoted by B(b). The number $b_{\frac{1}{2}}$ is defined as

$$\sup \left[b;\ \int_{U'} |\mathbf{f}_o(u)|^2 d^3u = \tfrac{1}{2}\ \forall \text{ L-measurable } U' \subset B(b) \text{ in } U^3 \right]$$

and in regard to the Fourier transform $\hat{\mathbf{f}}_o$ the number $\hat{b}_{\frac{1}{2}}$ is defined as

$$\sup \left[b;\ \int_{V'} |\hat{\mathbf{f}}_o(v)|^2 d^3v = \tfrac{1}{2}\ \forall \text{ L-measurable } V' \subset B(b) \text{ in } V^3 \right] ,$$

when $\|\mathbf{f}_o\|$ is 1. The function $\mathbf{f}_o$ is a reference scale if the numbers $b_{\frac{1}{2}}$ and $\hat{b}_{\frac{1}{2}}$ in their physically dimensionsless units can be interpreted as the equivalent of unit magnitudes of distance and momentum respectively. Suppose that the numbers e and ê are selected such that $b_{\frac{1}{2}}e = \hat{b}_{\frac{1}{2}}\hat{e} = 1$ and represent distance and momentum in the appropriate physical units. Then is in general $x = eu$ and $p = \hat{e}v$. The procedure is analogous to formulating a time-unit as the half-life of a spontaneous decay process.

If x is the position vector in a physical center-of-mass reference system, U^3, and p is the corresponding physical momentum vector in V^3, then are the real numbers t, $\hbar$, and m the other physically relevant quantities of the system, i.e. t as physical time, $\hbar$ as the value of Planck's constant, and m as the mass of the system.

The following relations may be understood as correspondence rules giving a physical meaning to the variables of the formal structure, which was presented in the previous section. The positive real numbers e, $\hat{e}$, y, and $\hat{y}$ now all represent magnitudes of physical observables, which are measured in the same units as x, p, t, and $\hbar$. While the first two relations have been already justified, the other relations will tie the previous dynamical structure to the formal solution of the Schroedinger equation. One will then obtain

$$1)\quad u = x/e, \quad v = p/\hat{e}, \quad r = t/y, \quad s = \hbar/\hat{y}$$

$$2)\quad \hat{y} = m\, e^2/y, \quad \hbar = e\,\hat{e} \quad .$$

By given u, v, r, and s and under consideration of the relations 2), there is still a considerable freedom in selecting the values of the physical constants. This is reflected in the following three different possibilities to calculate the value of s:

$$s = (\hbar y)/(me^2) = (\hat{e}y)/(me) = (\hat{e}^2 y)/(m\hbar) \quad . \qquad (11)$$

One can, finally, select the physical wavefunctions $\mathbf{g}(x)$ and $\hat{\mathbf{g}}(p)$, still complying with the above mentioned freedom, by stating as the third correspondence rule that

$$3)\quad \mathbf{g}(x) = e^{-3/2}\, \mathbf{f}(x/e), \quad \hat{\mathbf{g}}(p) = \hat{e}^{-3/2}\, \mathbf{f}(p/\hat{e}) \quad .$$

One obtains therewith the following relations, which always are the starting point of any quantum wave-theory, once through the interpretation of the quantization and then again of statistical reasons:

$$\begin{aligned} &\| \hat{\mathbf{g}} \| = \| \mathbf{g} \| = 1 \iff \| \mathbf{f} \| = \| \mathbf{f} \| = 1 \\ &\hat{\mathbf{g}}(p) = (2\pi\hbar)^{-3/2} \int_{R^3} \exp(-ipx\hbar^{-1})\, \mathbf{g}(x)\, d^3x \quad . \end{aligned} \qquad (12)$$

The Hilbert space is again determined up to isometrical isomorphisms as one of the two physical function spaces, in which $\hat{\mathbf{g}}$ and $\mathbf{g}$ respectively are embedded and mutually related by the Fourier-transform.

It is possible consistently to achieve useful properties of the dynamical form under the limit $s \to \infty$ in two different ways. A general procedure is determined in accordance with the

Definition 1: If limits exist in some concepts of the dynamical structure, when s tends to infinity, where a) $\hbar$ is bounded above, b) $\hat{y}$ tends to zero, then such limits will be named a "semi-classical" structure.

In a semi-classical structure is included a genuinly restricted class of dynamical concepts, which may be obtained in accordance with the

Definition 2: If limits exist in some of the concepts contributing to a semi-classical structure, when s tends to infinity, where a) $\hbar$ tends to zero, b) $\hat{y}\hat{e}^2/m$ is bounded above while $\hat{y}$ tends to zero, then such limits will be named a "classical" structure.

When $\hbar$ is a constant and e tends to zero while me/y is bounded above, then is given an example of a semi-classical limit under $s \to \infty$, which contradicts the concept of a classical limit. It seems, however, that this case is consistent with many situations in a physical theory.

The approximation formulas (9) and (10) will now be formulated in terms of concepts used for the physical space alone. In order to develop the necessary operator-expressions one may use the definitions

$$(\mathbf{H}(\hbar)\, \hbar^{-1} t\mathbf{g})(x) = e^{-3/2}(\mathbf{H}_s r\mathbf{f})(x/e),$$

$$(\mathbf{H}_o(\hbar)\hbar^{-1} t\mathbf{g})(x) = e^{-3/2}(\mathbf{H}^o_s r\mathbf{f})(x/e),$$

$$(\mathbf{V}(\hbar)\, \hbar^{-1} t\mathbf{g})(x) = e^{-3/2}(\mathbf{V}_s r\mathbf{f})(x/e),$$

$$(\mathbf{Q}_{t,\hbar}\mathbf{g})(x) = e^{-3/2}(\mathbf{Q}(r,s)\mathbf{f})(x/e),$$

$$(\mathbf{U}_{t,\hbar}\mathbf{g})(x) = e^{-3/2}(\mathbf{U}(r,s)\mathbf{f})(x/e).$$

It has already been established that $\mathbf{H}^0_s + \mathbf{V}_s$, with $(\mathbf{H}^0_s\, \mathbf{f})(u) = -(s/2)\cdot$ ("Δ"$\mathbf{f}$)(u) and $(\mathbf{V}_s\, \mathbf{f})(u) = s^{-1}V(|u|/s)\mathbf{f}(\hat{u})$, is the Hamiltonian $\mathbf{H}_s$ of the pre-quantum dynamical setting. Then is accordingly in the new physical setting $\mathbf{H}(\hbar) = \mathbf{H}_o(\hbar) + \mathbf{V}(\hbar)$ with $(\mathbf{H}_o(\hbar)\mathbf{g})(x) = (-\hbar^2/2m)$("$\Delta$"$\mathbf{g}$)(x) and $(\mathbf{V}(\hbar)\, \mathbf{g})(x) = \hat{y}/\hat{y}\; V(|x|\hat{y}/e\hbar)\mathbf{g}(x)$. Suppose that $\mathbf{g}_t(x) = e^{-3/2}\mathbf{f}_{t/y}(x/e)$, when $\mathbf{f}_o \equiv \mathbf{f} \in \mathbf{F}_o$. Then one will obtain the relations

$$(\mathbf{Q}_{t,\hbar}\, \mathbf{g}_o)(x) = \exp(imx^2/2t\hbar)\; \mathbf{g}_o(x)$$

$$(\mathbf{U}_{t,\hbar}\, \mathbf{g}_o)(x) = (m/it)^{3/2} \exp(imx^2/2t\hbar)\; \hat{\mathbf{g}}_o(mx/t) ,$$

which are the formulas for the **Q** and **U** operators given in the introduction. Relations equivalent to (9) and (10) will now have the form

$$\mathbf{U}_{t,\hbar} = \exp(-i\mathbf{H}_o(\hbar)\, t) - \mathbf{U}_{t,\hbar}(\mathbf{Q}_{t,\hbar} - 1)$$

$$\mathbf{U}_{t,\hbar}(1 - i\mathbf{I}_{t,\hbar}) = \exp(-i\mathbf{H}(\hbar)t) - \mathbf{U}_{t,\hbar}(\mathbf{Q}_{t,\hbar} - 1)(1 - i\mathbf{I}_{t,\hbar}),$$

$$\mathbf{I}_{t,\hbar} = \int_0^t \mathbf{Q}_{t',-\hbar}\, \mathbf{U}_{t',-\hbar}\, \mathbf{V}(\hbar) \exp(-i\mathbf{H}(\hbar)t')\, dt' \quad .$$

The asymptotic pattern for the approximation already shown for the corresponding terms in the second section, when s tends to infinity, is now given in precisely the same way mathematically, when $\hat{y}$ tends to zero in accordance with Definition 1. These results can also be explained physically. If $\mathbf{g}_o$ is normed to one, the inequality

$$\int_{S^3} \left| |(\exp(-i\mathbf{H}_o(\hbar)t\; \mathbf{g}_o)(x)|^2 - (\mathbf{U}_{t,\hbar}\mathbf{g}_o)(x)|^2 \right| d^3x \leq$$

$$2\; \| (\mathbf{U}_{t,-\hbar} \exp(-i\mathbf{H}(\hbar)t) - 1)\, \mathbf{g}_o \|$$

is correct, if S^3 is any Lebesgue-measurable set in R^3. One may, of course integrate on the left in the above inequality over any unbounded integraldomain as well. It could then be asked: What is $P(\mathbf{g}_o, K)$, the probability that a particle at the time t is somewhere on the cone K, if it was in its center at 0 o'clock ? The answer is

$$P(\mathbf{g}_o, K) = \lim_{\hat{y} \to o} \int_K |(\exp(-i\mathbf{H}_o(\hbar)t)\, \mathbf{g}_o)(x)|^2\, d^3x =$$

$$\lim_{\hat{y} \to \pm o} |m/t|^3 \int_K |\mathbf{g}_o(mx/\tilde{t})|^2\, d^3x = \int_{\pm K} |\mathbf{g}_o(p)|^2\, d^3p. \quad (13)$$

The number $ts|s|^{-1}$ has been denoted as $\tilde{t}$ in one of the expressions (13). The index $\hat{y}$ takes negative values when s is negative. The cone K is mapped onto itself by $p = mx/\tilde{t}$, when $\tilde{t}$ is positive, and it is inverted when $\tilde{t}$ is negative. Therefore the signs $\pm$ are used with K. The equations (13) express the fact that $P(\mathbf{g}_o, K)$ is the same as the probability that the momentum is in K, if $\hat{y}$ becomes very small. This reminds us of the classical free particle-mechanics, but it tells us, however, somewhat less than Newton's second law of motion, if S^3 is the integral-domain in local space and if $S^3 \subset K$ is a bounded region on the cone. In this latter case the relations comply with the uncertainty principle.

Finally, one may consider an interacting system under the semi-classical limit, if the Hamiltonian $\mathbf{H}(\hbar)$ implies the use of a $\mathbf{V}(\hbar)$, which satisfies necessary analytical requirements. The Ehrenfest-relation, which was given in the introduction, represents Newton's third law of motion. The compact integral domain W must have a positive L-measure, but otherwise it is in the quantum setting not important, how large it is. If one considers by a constant $\hbar$ the semi-classical limit, then it is sufficient for the existence of the limits that the function $V(|w|)$ for any $i \in [1,2,3]$ satisfies the conditions

1) The $\lim_{L_i \to \infty} L^{-1} (\frac{\partial}{\partial w_i} V)(|x|/L)$ exists $\forall x$ and is named $f^i_\infty(x)$,
2) The f^i_∞ are bounded and piecewise continuous functions, which moreover are strictly continuous in a neighborhood of the zero.

J.Neggers [12] has recently classified many such functions, as so much as they are not the trivial Coulomb case, but still one-to-one as a value field under a given order. His class is so wide that many physically relevant potentials seem to fit in it without further restrictions, while $D(\mathbf{H}_o(\hbar)) \subset D(\mathbf{V}(\hbar))$.

REFERENCES

1. Landau, L.D. and Lifschitz, E.M.(1947) Quantum Mechanics, first edition Moscow. In Quantenmechanik, Akademie-Verlag Berlin (1965) see the pp.21-23 and 56.
2. Messiah, A. (1962) Mécanique Quantique, Dunod, Paris. See in vol. 1, chap.VI, §1: La limite classique de la mécanique ondulatoire, pp.180-194.

3. Heisenberg, W. (1958) Physikalische Prinzipien der Quantentheorie, Hirzel, Stuttgart.
4. van der Waerden, B.L. (1967) Sources of Quantum Mechanics, North-Holland Publ.Co., Amsterdam.
5. Cohen-Tannoudji, C., Diu, B., Laloë, F. (1973) Mecanique Quantique, Hermann, Paris. Complément pp.372-382.
6. Heitler, W. (1954) The Quantum Theory of Radiation, 3rd ed. Oxford at the Clarendon Press. See p. 115.
7. Rosen, G. (1969) Formulations of Classical and Quantum Dynamical Theory, Academic Press, New York and London. Appendix G.
8. Dollard, J. (1964) An introduction to Potential Scattering Theory, Conference Notes, Flagstaff.
9. Ehrenfest, P. (1927) Zeitschrift für Physik,45, 455.
10. Kato, T. (1966) Perturbation Theory for Linear Operators, Springer, New York. See for example Chapter X,§5.
11. Ingólfsson, K. (1980) Notes on Asymptotic Series by H^o-strictly Singular Perturbations, Int.Journal of Quantum Chem., 17, 99-106.
12. Neggers, J. (1989) Value Fields and Symmetry, a personal communication and report from the University of Alabama.

FOURIER TRANSFORM IN NON-EUCLIDEAN SPACE AND ITS APPLICATIONS TO PHYSICS

M. Kovalyov and M. Legare
Department of Mathematics
University of Alberta
Edmonton, Canada T6G 2G1

ABSTRACT. In this paper we shortly discuss the theory of Fourier transform on the non-Euclidean space and show how to generalize it to enable us to solve the Dirac equations on the Friedmann - Robertson - Walker space-times. We also indicate possible application of the theory to the description of strong interactions.

1. Hyperbolic space

In this article we would like to discuss Fourier transform on the hyperbolic space and present some applications to physics. We start by recalling the definition of hyperbolic space. There are three standard models of the hyperbolic space $\mathbb{H}^3$ which we are going to describe now.

I. The first model, known as the upper half-space model, consists of the points (x, y), $x \in \mathbb{R}^2$, $y > 0$, endowed with the metric

$$ds^2 = \frac{dx^{12} + dx^{22} + dy^2}{k^2 y^2} \tag{1.1a}$$

where the parameter k is a constant and $-|k|$ equals to the sectional curvature of $\mathbb{H}^3$.

II. The second model is known as the geodesic model and consists of the points (r, θ, φ) $r > 0$, $0 \leq \theta \leq \pi$, $0 \leq \varphi \leq 2\pi$. The metric form is given by the formula

$$ds^2 = dr^2 + \left(\frac{\sinh kr}{k}\right)^2 (d\theta^2 + \sin^2\theta d\varphi^2). \tag{1.1b}$$

This model is the equivalent of the spherical coordinates on $\mathbb{H}^3$. As $k \to 0$ we obtain ordinary spherical coordinates on $\mathbb{R}^3$.

J. S. Byrnes and J. F. Byrnes (eds.), Recent Advances in Fourier Analysis and Its Applications, 41–57.

III. The third model consists of the points (ξ, τ), $\xi \in \mathbb{R}^2$, $\tau > 0$ connected by the relationship $|\xi|^2 - \tau^2 = k^{-2}$ and the metric form is given by the formula:

$$ds^2 = d\tau^2 - d\xi^{12} - d\xi^{22} - d\xi^{32}. \tag{1.1c}$$

The three models are related to each other by the following coordinate transformations:

$$\begin{aligned}
x^1 &= \frac{\sin\theta\cos\varphi}{\coth(kr) - \cos\theta} = \frac{\xi^1}{\tau - \xi^3}\\
x^2 &= \frac{\sin\theta\sin\varphi}{\coth(kr) - \cos\theta} = \frac{\xi^2}{\tau - \xi^3}\\
y &= \frac{1}{\cosh(kr) - \sinh(kr)\cos\theta} = \frac{1}{\tau - \xi^3}\\
\xi^1 &= \frac{\sinh kr}{k}\sin\theta\cos\varphi = \frac{x^1}{ky}\\
\xi^2 &= \frac{\sinh kr}{k}\sin\theta\sin\varphi = \frac{x^2}{ky}\\
\xi^3 &= \frac{\sinh kr}{k}\cos\theta = \frac{1}{2k}\Big(\frac{1}{y} - y - \frac{x^{12} + x^{22}}{y}\Big)\\
\tau &= \frac{\cosh kr}{k} = \frac{2}{k}\Big(\frac{1}{y} + y + \frac{x^{12} + x^{22}}{y}\Big).
\end{aligned} \tag{1.2}$$

The group of isometries of $\mathbb{H}^3$ is $PSL(2, \mathbb{C})$. It can be realized using quaternionic notation for the upper half-space model coordinates:

$$q = x^1 \cdot 1 + x^2 \cdot i + y \cdot j.$$

The action of $PSL(2, \mathbb{C})$ on q corresponds to a fractional linear transformation [1]:

$$q' = (Aq + B)(Cq + \mathcal{D})^{-1}$$

where the matrix

$$\begin{bmatrix} A & B \\ C & \mathcal{D} \end{bmatrix}$$

belongs to $PSL(2, \mathbb{C})$.

Later on we will need a special subgroup of $PSL(2, C)$ which is $SU(2)$ quotiented by its center and whose elements are given by the matrices

$$\begin{bmatrix} ae^{i\chi} & (e_1 + ie_2)b \\ (-e_1 + ie_2)b & ae^{i\chi} \end{bmatrix}$$

where $a^2 + b^2 = e_1^2 + e_2^2 = 1, \quad 0 \le \chi \le 2\pi$.
The latter subgroup keeps the point $(x^1 = x^2 = 0,\ y = 1)$ fixed and thus corresponds to the rotations of the ordinary Euclidean space. Its action on a point $(x^1, x^2, y) \in \mathbb{H}^3$ can be written explicitly in terms of the variables x^1, x^2, y as:

$$x'^1 = \tfrac{1}{\Delta}\left[x^1 - 2(e_1x^1 + e_2x^2)e_1 + \tfrac{a}{b}e_1\right] - \tfrac{a}{b}e_1$$

$$x'^2 = \tfrac{1}{\Delta}\left[x^2 - 2(e_1x^1 + e_2x^2)e_2 + \tfrac{a}{b}e_2\right] - \tfrac{a}{b}e_2$$

$$y' = \tfrac{y}{\Delta} \tag{1.3}$$

with

$$\Delta = (bx^1 - ae_1)^2 + (bx^2 - ae_2)^2 + b^2y^2.$$

The $\mathbb{H}^3$-analogues of the Euclidean planes are the so-called horospheres, which are given by the equation

$$\left(\tilde{b}y - \tfrac{c}{\tilde{b}}\right)^2 + \left(\tilde{b}x^1 - \tilde{a}\tilde{e_1}\right)^2 + \left(\widetilde{bx}^2 - \tilde{a}\tilde{e_2}\right)^2 = \left(\tfrac{c}{\tilde{b}}\right)^2$$

where

$$c > 0, \quad \tilde{a}^2 + \tilde{b}^2 = \tilde{e}_1^2 + \tilde{e}_2^2 = 1.$$

If we fix $\tilde{a}, \tilde{b}, \tilde{e}_1, \tilde{e}_2$ we obtain a one-parameter family of surfaces which we call "parallel" horospheres with direction $\beta = \left(\frac{\tilde{a}}{\tilde{b}}\tilde{e}_1,\ \frac{\tilde{a}}{\tilde{b}}\tilde{e}_2\right)$. When $\tilde{b} \neq 0$, this family consists of the spheres of radius $\frac{c}{\tilde{b}}$ tangent to the plane $y = 0$ at the point $\beta = \left(\frac{\tilde{a}}{\tilde{b}}\tilde{e}_1,\ \frac{\tilde{a}}{\tilde{b}}\tilde{e}_2\right)$. If $\tilde{b} = 0$ we have a family of parallel planes $y = \frac{1}{2c}$, which can be thought of as spheres with infinite radii tangent to $\partial\mathbb{H}^3$ at the point $\beta_\infty = (\infty, \infty)$.
Moreover for each set of parallel spheres with direction β we can define a family of geodesics with direction β by the equations:

$$\frac{1}{\Delta}\left[x^1 - 2(\tilde{e}_1x^1 + \tilde{e}_2x^2)\tilde{e}_1 + \frac{\tilde{a}}{\tilde{b}}\tilde{e}_1\right] - \frac{\tilde{a}}{\tilde{b}}\tilde{e}_1 = c$$
$$\frac{1}{\Delta}\left[x^2 - 2(\tilde{e}_1x^1 + \tilde{e}_2x^2)\tilde{e}_2 + \frac{\tilde{a}}{\tilde{b}}\tilde{e}_2\right] - \frac{\tilde{a}}{\tilde{b}}\tilde{e}_2 = d$$

with c and d being constants.
The transformation (1.3) maps the set of parallel spheres with direction $\beta = \left(\frac{a}{b}e_1,\ \frac{a}{b}e_2\right)$ and the set of geodesics with direction $\beta = \left(\frac{a}{b}e_1,\ \frac{a}{b}e_2\right)$ onto correspondingly the set of planes $y =$ const and the set of lines $x_1 =$ const, $x_2 =$ const.

2. The conformally invariant wave-~ and the Dirac operators on the open Friedmann - Robertson - Walker space-times.

The Friedmann - Robertson - Walker space-time is a space-time endowed with the metric

$$d\sigma^2 = R^2(t)\{dt^2 - dr^2 - (\frac{\sinh kr}{k})^2(d\theta^2 + \sin^2\theta d\varphi^2)\} \tag{2.1}$$

where $R(t)$ is the so-called "expansion factor".
The Friedmann - Robertson - Walker space-times describe homogeneous and isotropic expanding universes. When k is real, imaginary or zero we correspondingly obtain the open, closed and flat cases. There are two important conformally invariant partial differential operators on the space-time; the wave operator

$$\Box_{\mathbb{H}^3,R(t)} = L + \frac{\rho}{6}$$

where L is the Laplace - Beltrami operator corresponding to (2.1) and ρ is the scalar curvature, and the Dirac operator:

$$\mathcal{D}_{\mathbb{H}^3,R(t)} = \gamma^0\left(\frac{\partial}{\partial t} + \frac{3R'}{2R}\right) + \gamma^1\left(\frac{\partial}{\partial r} + k\text{cotanh } kr\right) +$$
$$+ \gamma^2\frac{k}{\sinh kr}\left(\frac{\partial}{\partial\theta} + \frac{\cot\theta}{2}\right) + \gamma^3\,\frac{k}{\sinh kr\sin\theta}\,\frac{\partial}{\partial\varphi}$$

whose solutions correspondingly are scalar and spinor fields.
Because of conformal invariance, finding solutions of

$$\Box_{\mathbb{H}^3,R(t)}\tilde{u} = 0$$

and

$$\mathcal{D}_{\mathbb{H}^3,R(t)}\tilde{\psi} = 0$$

is reduced to finding solutions of

$$\Box_{\mathbb{H}^3}u = 0$$

and

$$\mathcal{D}_{\mathbb{H}^3}\psi = 0$$

where

$$\Box_{\mathbb{H}^3} = \Box_{\mathbb{H}^3, R(t)=1}$$
$$\mathcal{D}_{\mathbb{H}^3} = \mathcal{D}_{\mathbb{H}^3, R(t)=1}.$$

Then $\tilde{u}$ and $\tilde{\psi}$ can be expressed in terms of u and ψ as [3]:

$$\tilde{u} = R(t)u$$
$$\tilde{\psi} = R^{3/2}\psi.$$

3. Solving $\Box_{\mathbb{H}^3} u = 0$ by Fourier transform.

There are at least two ways to solve the Cauchy problem for

$$\Box_{\mathbb{H}^3} u = 0.$$

The first method is based on developing the theory of Fourier transform on $\mathbb{H}^3$, we discuss it below. The second method is generalization of the method of spherical means of F. John as described in [4] and is mentioned here only for the sake of completeness. To introduce the Fourier transform we use model I. The idea of the Euclidean Fourier transform is to express given function $f(x),\ x \in \mathbb{R}^3$ as a linear combination of the eigenfunctions of the Laplacian which are $e^{i(x\cdot\beta)p},\ p > 0,\ |\beta| = 1.$

$$f(x) = \int_0^\infty \int_{S_1} \hat{f}(p) e^{i(x\cdot\beta)p} p^2 dp\, dS_\beta$$

where

$$\hat{f}(p) = \int_{\mathbb{R}^3} f(x) e^{-i(x\cdot\beta)p} dx.$$

with S_1 and dS_β being correspondingly the unit sphere and its Haar measure. Generalization of the Laplacian for $\mathbb{H}^3$ is the Laplace - Beltrami operator on $\mathbb{H}^3$:

$$y^2\left(\frac{\partial^2}{\partial x^{12}} + \frac{\partial^2}{\partial x^{22}} + \frac{\partial^2}{\partial y^2}\right) - y\,\frac{\partial}{\partial y}.$$

Though for our purposes it is more convenient to use the conformally invariant operator

$$L = y^2\left(\frac{\partial^2}{\partial x^{12}} + \frac{\partial^2}{\partial x^{22}} + \frac{\partial^2}{\partial y^2}\right) - y\,\frac{\partial}{\partial y} + 1.$$

One can easily find at least one eigenfunction of L which is y^α, indeed

$$L\, y^\alpha = (\alpha - 1)^2 y^\alpha.$$

Since isometries (1.3) preserve the Laplace - Beltrami operator and therefore L, we immediately obtain whole bunch of eigenfunctions of L in the form:

$$\left\{\frac{y(|\beta|^2+1)}{(x^1-\beta^1)^2+(x^2-\beta^2)^2+y^2}\right\}^\alpha$$

where $\beta = (\beta^1 = \frac{ae_1}{b},\ \beta^2 = \frac{ae_2}{b}) \in \partial \mathbb{H}^3 = \mathbb{R}^2$.
Similarly to the Euclidean case, one can write out any "decent" function on $\mathbb{H}^3$ as:

$$f(x^1,x^2,y) = \int\limits_{\substack{0\le\lambda<+\infty \\ \beta\in\partial\mathbb{H}^3}} \tilde{f}(\lambda,\beta)\left\{\frac{y(|\beta|^2+1)}{(x^1-\beta^1)^2+(x^2-\beta^2)^2+y^2}\right\}^{1+i\lambda} \times$$

$$\lambda^2 d\lambda\ \frac{d\beta_1 d\beta_2}{(1+|\beta|^2)^2}$$

where $\tilde{f}(\lambda,\beta)$ is the non-Euclidean Fourier transform of f, and the measure

$$\frac{d\beta_1 d\beta_2}{(1+|\beta|^2)^2}$$

is the Haar measure on $\partial\mathbb{H}^3$. The function $\tilde{f}(\lambda,\beta)$ is given by the formula

$$\tilde{f}(\lambda,\beta) = W \int\limits_{\mathbb{H}^3} f(x^1,x^2,y)\left\{\frac{y(|\beta|^2+1)}{(x^1-\beta^1)^2+(x^2-\beta^2)^2+y^2}\right\}^{1-i\lambda} \frac{dx^1 dx^2 dy}{y^3}$$

where W is a constant.
The result has become quite classical by now and is described in quite a few textbooks.
Using non-Euclidean Fourier transform we can solve the Cauchy problem for $\Box_{\mathbb{H}^3} u = 0$ in pretty much the same way as we do it in the Euclidean case.
As the final remark in this section we mention that $\left[\frac{y(|\beta|^2+1)}{(x^1-\beta^1)^2+(x^2-\beta^2)^2+y^2}\right]$ has one more similarity with the Euclidean function $e^{ix\cdot\beta}$, $|\beta| = 1$:
$\frac{y(|\beta|^2+1)}{(x^1-\beta^1)^2+(x^2-\beta^2)^2+y^2} =$ const on the set of parallel horospheres of $\mathbb{H}^3$ just like $e^{ix\cdot\beta} =$ const on the set of parallel planes of $\mathbb{R}^3$.

4. Solving $\mathcal{D}_{\mathbb{H}^3}\psi = 0$ by Fourier transform.

In the Euclidean case the Fourier transform can also be used to solve the initial value problem for the Dirac equations. We want to develop a similar method for the Dirac equations on $\mathbb{H}^3$. First of all we write down the Dirac equations on $\mathbb{H}^3$ in terms of model I [5]:

$$\left(\gamma^0 \frac{\partial}{\partial t} + ky\gamma^1 \frac{\partial}{\partial x^1} + ky\gamma^2 \frac{\partial}{\partial x^2} + dy\gamma^3 \frac{\partial}{\partial y} - k\gamma^3\right)\psi = 0$$

where we choose the following set of Dirac matrices:

$$\gamma^0 = \begin{bmatrix} 0_2 & \sigma_2 \\ \sigma_2 & 0_2 \end{bmatrix}, \quad \gamma^1 = \begin{bmatrix} 0_2 & -i\sigma_3 \\ -i\sigma_3 & 0_2 \end{bmatrix},$$

$$\gamma^2 = \begin{bmatrix} 0_2 & -\mathbb{I}_2 \\ \mathbb{I}_2 & 0_2 \end{bmatrix}, \quad \gamma^3 = \begin{bmatrix} 0_2 & i\sigma_1 \\ i\sigma_1 & 0 \end{bmatrix}$$

We can look for a solution in the form $e^{iwt}\varphi(y)$. Substituting it into the Dirac equations we obtain a solution in the form

$$e^{iwt}\begin{pmatrix} -iwy^{1-\frac{iw}{k}} & 0 & 0 & 0 \\ 0 & iwy^{1+\frac{iw}{k}} & 0 & 0 \\ 0 & 0 & -iwy^{1-\frac{iw}{k}} & 0 \\ 0 & 0 & 0 & iwy^{1+\frac{iw}{k}} \end{pmatrix}\begin{pmatrix} C_1 \\ C_2 \\ C_3 \\ C_4 \end{pmatrix}$$

where the C_i's are arbitrary constants. We can think of a solution of this form as propagating in the direction $\beta = (\infty, \infty)$.
In order to obtain spinor solutions moving in an arbitrary direction $(\frac{a}{b}\, e_1,\ \frac{a}{b}\, e_2)$, we apply the transformation (1.3) which maps the family of horospheres with direction $(\frac{a}{b}\, e_1,\ \frac{a}{b}\, e_2)$ onto the family of horospheres with direction (∞, ∞). Since these transformations leave the metric invariant, the solutions to the Dirac equations in the direction $(\frac{a}{b}\, e_1,\ \frac{a}{b}\, e_2)$ are:

$$\Psi(t, x^1, x^2, y) = S^{\dagger}(a, b, e_1, e_2)\psi(t, y'),$$

where S represents the $D^{(\frac{1}{2},0)} \oplus D^{(0,\frac{1}{2})} SL(2, \mathbb{C})$ representation of the rotation $\mathcal{R}$ of the orthonormal frames induced by (1.3), that is:

$$S\gamma^i S^{\dagger} = (\mathcal{R}^T)^i_j \gamma^j \quad (i, j = 1, 2, 3),$$

with:

$$\mathcal{R} = \frac{y}{y'}\mathcal{J},$$

$$[\mathcal{R}]^i_j = \begin{bmatrix} 1 - \frac{2}{|z|^2+y^2}[(\mathrm{Im}\ e^*z)^2 + e_1^2 y^2], \\ \frac{-2}{|z|^2+y^2}[e_1e_2((z^2)^2 - (z^1)^2 + y^2) + (e_1^2 - e_2^2)z^1z^2], \\ \frac{-2yz^1}{|z|^2+y^2}, \end{bmatrix}$$

$$\begin{matrix} \frac{-2}{|z|^2+y^2}[e_1e_2((z^1)^2 - (z^2)^2 + y^2) + z^1z^2(e_2^2 - e_1^2)], \\ 1 - \frac{2}{|z|^2+y^2}[(\mathrm{Im}\ e^*z)^2 + e_2^2 y^2], \\ \frac{-2yz^2}{|z|^2+y^2}, \end{matrix}$$

$$\begin{matrix} \frac{-2y}{|z|^2+y^2}[(e_2^2 - e_1^2)z^1 - 2e_1e_2z^2] \\ \frac{-2y}{|z|^2+y^2}[(e_1^2 - e_2^2)z^2 - 2e_1e_2z^1] \\ 1 - \frac{2y^2}{|z|^2+y^2} \end{matrix}$$

where $\mathcal{R}^T$ is the transpose of the rotation matrix $\mathcal{R}$, $\mathcal{J}$ is the jacobian of the transformation (1.3), $z^1 \equiv x^1 - \frac{a}{b}\ e_1$, $z^2 \equiv x^2 - \frac{a}{b}e_2$, $z \equiv z^1 + iz^2$, and $e \equiv e_1 + ie_2$.

Explicitly:*

$$S(a, b, e_1, e_2) = \left[\begin{array}{c|c} u & O_2 \\ \hline O_2 & u^* \end{array}\right],$$

with:

$$u = \frac{1}{\sqrt{|z|^2 + y^2}}\begin{bmatrix} z^*e, & -ye^* \\ ye, & ze^* \end{bmatrix} \in SU(2).$$

Hence, the spinor solutions with direction $\beta = \left(\frac{a}{b}e_1,\ \frac{a}{b}e_2\right)$ to the Dirac equations have the following form:

$$\Psi(t, x^1, x^2, y) = e^{iwt}\left(\begin{array}{c|c} u^\dagger & O_2 \\ \hline O_2 & u^T \end{array}\right).$$

*The spinor transformation S has been determined from the normal eigenvector ($\boldsymbol{n}$) corresponding to the eigenvalue 1 and the two other eigenvalues ($e^{\pm i\alpha}$) of R. The canonical homomorphism between $SU(2)$ and $SO(3)$ is then used to arrive to $u(= \cos\frac{\alpha}{2} - i\boldsymbol{n}\cdot\boldsymbol{\sigma}\sin\frac{\alpha}{2})$.

$$\begin{pmatrix} -(\frac{y}{\Delta})^{1-\frac{iw}{k}} & O & O & O \\ O & (\frac{y}{\Delta})^{1+\frac{iw}{k}} & O & O \\ O & O & -(\frac{y}{\Delta})^{1-\frac{iw}{k}} & O \\ O & O & O & (\frac{y}{\Delta})^{1+\frac{iw}{k}} \end{pmatrix} \begin{pmatrix} C_1 \\ C_2 \\ C_3 \\ C_4 \end{pmatrix}$$

where the C_i's are constants and Δ is as before.

It turns out that the solutions of this form form a complete set, i.e. any solution of the Dirac equations can be expressed through them.

We define Fourier transform of a spinor ψ to be

$$\hat{\psi}(t,\beta,w) = \frac{1}{16\pi^3} \int\limits_{\mathbb{H}^3} \times$$

$$\begin{pmatrix} \left[\frac{y(|\beta|^2+1)}{|x-\beta|^2+y^2}\right]^{1+iw} & O & O & O \\ O & \left[\frac{y(|\beta|^2+1)}{|x-\beta|^2+y^2}\right]^{1-iw} & O & O \\ O & O & \left[\frac{y(|\beta|^2+1)}{|x-\beta|^2+y^2}\right]^{1+iw} & O \\ O & O & O & \left[\frac{y(|\beta|^2+1)}{|x-\beta|^2+y^2}\right]^{1-iw} \end{pmatrix}$$

$$\times \begin{pmatrix} u(x_1,x_2,y,\beta) & | & O_2 \\ --- & -|- & --- \\ O_2 & | & u^*(x_1,x_2,y,\beta) \end{pmatrix} \psi(t,x_1,x_2,y)\, \frac{dx_1dx_2dy}{y^3} \tag{4.1}$$

where $\beta = \frac{a}{b}\, e \in \partial\mathbb{H}^3$, $x = x_1 + ix_2$, and u, u^* are as given before. Then the following inversion formula holds:

$$\psi(t,x_1',x_2',y) = \int\limits_{\mathbb{R}} \int\limits_{\partial\mathbb{H}^3} \begin{pmatrix} u^\dagger(x_1',x_2',y',\beta) & | & O_2 \\ --- & -|- & --- \\ O_2 & | & u^T(x_1',x_2',y',\beta) \end{pmatrix} \times$$

$$\begin{pmatrix} \left[\frac{y'(|\beta|^2+1)}{|x'-\beta|^2+y'^2}\right]^{1-iw} & O & O & O \\ O & \left[\frac{y'(|\beta|^2+1)}{|x'-\beta|^2+y'^2}\right]^{1+iw} & O & O \\ O & O & \left[\frac{y'(|\beta|^2+1)}{|x'-\beta|^2+y'^2}\right]^{1-iw} & O \\ O & O & O & \left[\frac{y'(|\beta|^2+1)}{|x'-\beta|^2+y'^2}\right]^{1+iw} \end{pmatrix} \times$$

$$\hat{\psi}(t,\beta,w)\,\frac{d\beta_1 d\beta_2 w^2 dw}{(1+|\beta|^2)^2}. \tag{4.2}$$

The formulas (4.1) and (4.2) constitute the spinor-version of the non-Euclidean Fourier transform. Representation (4.2) implies the statement that any solution of the Dirac equations can be represented as

$$\psi(t,x_1',x_2',y') = \int_{\mathbb{R}}\int_{\partial\mathbb{H}^3} e^{iwt} \begin{pmatrix} u^\dagger(x_1',x_2',y',\beta) & | & O_2 \\ --- & -|- & --- \\ O_2 & | & u^T(x_1',x_2',y',\beta) \end{pmatrix} \times$$

$$\begin{pmatrix} \left[\frac{y'(|\beta|^2+1)}{|x'-\beta|^2+y'^2}\right]^{1-iw} & O & O & O \\ O & \left[\frac{y'(|\beta|^2+1)}{|x'-\beta|^2+y'^2}\right]^{1+iw} & O & O \\ O & O & \left[\frac{y'(|\beta|^2+1)}{|x'-\beta|^2+y'^2}\right]^{1-iw} & O \\ O & O & O & \left[\frac{y'(|\beta|^2+1)}{|x'-\beta|^2+y'^2}\right]^{1+iw} \end{pmatrix} \times$$

$$\hat{\psi}(0,\beta,w)\,\frac{w^2 dw d\beta_1 d\beta_2}{(1+|\beta|^2)^2}$$

where $\hat{\psi}(0,\beta,w)$ is the transform given by (4.1) of the initial data.

In order to prove representation (4.2) it suffices to show that

$$\int_R \int_{\partial \mathbb{H}^3} u^\dagger(x_1', x_2', y', \beta) \times$$

$$\begin{pmatrix} \left[\frac{y'(|\beta|^2+1)}{|x'-\beta|^2+y'^2}\right]^{1-iw}\left[\frac{y(|\beta|^2+1)}{|x-\beta|^2+y^2}\right]^{1+iw} & \Big| & \bigcirc \\ --- & -|- & --- \\ \bigcirc & \Big| & \left[\frac{y'(|\beta|^2+1)}{|x'-\beta|^2+y'^2}\right]^{1+iw}\left[\frac{y(|\beta|^2+1)}{|x-\beta|^2+y^2}\right]^{1-iw} \end{pmatrix} \times$$

$$\times\, u(x_1, x_2, y, \beta)\, \frac{w^2 dw d\beta_1 d\beta_2}{(1+|\beta|^2)^2} = 16\pi^3 y'^3 \delta(x-x', y-y')\mathbb{I}_2. \tag{4.3}$$

The diagonal matrix on the left-hand side can be written as a sum

$$\frac{1}{2}\Big\{\Big[\frac{y'(|\beta|^2+1)}{|x'-\beta|^2+y'^2}\Big]^{1-iw}\Big[\frac{y(|\beta|^2+1)}{|x-\beta|^2+y^2}\Big]^{1+iw} + \Big[\frac{y(|\beta|^2+1)}{|x'-\beta|^2+y'^2}\Big]^{1+iw} \times$$
$$\Big[\frac{y(|\beta|^2+1)}{|x-\beta|^2+y^2}\Big]^{1-iw}\Big\}$$

$$\times \begin{pmatrix} 1 & 0 \\ 0 & 1 \end{pmatrix} + \frac{1}{2}\Big\{\Big[\frac{y'(|\beta|^2+1)}{|x'-\beta|^2+y'^2}\Big]^{1-iw}\Big[\frac{y(|\beta|^2+1)}{|x-\beta|^2+y^2}\Big]^{1+iw} -$$

$$\Big[\frac{y'(|\beta|^2+1)}{|x'-\beta|^2+y'^2}\Big]^{1+iw}\Big[\frac{y(|\beta|^2+1)}{|x-\beta|^2+y^2}\Big]^{1-iw}\Big\} \begin{pmatrix} 1 & 0 \\ 0 & -1 \end{pmatrix}. \tag{4.4}$$

The second term is an odd function of w, thus when we substitute (4.4) into (4.3) and integrate in w over $\mathbb{R}$ the contribution from the second term will be zero. Therefore the integral in (4.3) will not change if we replace the middle term in (4.3) by its even part. Having done this we can rewrite (4.3) as

$$\frac{1}{2}\int_{-\infty}^{+\infty} w^2 dw \int_{\partial H^3} \Big\{\Big[\frac{y'(|\beta|^2+1)}{|x'-\beta|^2+y'^2}\Big]^{1-iw}\Big[\frac{y(|\beta|^2+1)}{|x-\beta|^2+y'^2}\Big]^{1+iw}$$

$$+\Big[\frac{y'(|\beta|^2+1)}{|x'-\beta|^2+y'^2}\Big]^{1+iw}\Big[\frac{y(|\beta|^2+1)}{|x-\beta|^2+y^2}\Big]^{1-iw}\Big\}$$

$$\times\, u^\dagger(x_1', x_2', y', \beta) u(x_1, x_2, y, \beta)\, \frac{d\beta_1 d\beta_2}{(1+|\beta|^2)^2} = 16\pi^3 y'^3 \delta(x-x', y-y'). \tag{4.5}$$

Since the contribution from both terms in the braces is essentially the same it is sufficient to show:

$$\int_{-\infty}^{+\infty} w^2 dw \int_{\partial\mathbb{H}^3} \left[\frac{y'(|\beta|^2+1)}{|x'-\beta|^2+y'^2}\right]^{1-iw} \left[\frac{y(|\beta|^2+1)}{|x-\beta|^2+y^2}\right]^{1+iw} \times$$

$$u^\dagger(x_1', x_2', y', \beta) u(x_1, x_2, y, \beta) \frac{d\beta_1 d\beta_2}{(1+|\beta^2)^2}$$

$$= 16\pi^3 y'^3 \delta(x-x',\ y-y'). \tag{4.6}$$

Though (4.6) is not invariant with respect to isometries of $\mathbb{H}^3$ it is equivalent to

$$\int_{-\infty}^{+\infty} w^2 dw \int_{\partial\mathbb{H}^3} \left[\frac{y'(|\beta|^2+1)}{|x'-\beta|^2+y'^2}\right]^{1-iw} \left[\frac{y(|\beta|^2+1)}{|x-\beta|^2+y^2}\right]^{1+iw} \times$$

$$\times\, u(x_1', x_2', y', \beta_0) u^\dagger(x_1', x_2', y', \beta)$$

$$\times\, u(x_1, x_2, y, \beta) u^\dagger(x_1, x_2, y, \beta_0) \frac{d\beta_1 d\beta_2}{(1+|\beta|^2)^2} = 16\pi^3 y^3 \delta(x-x',\ y-y') \tag{4.7}$$

which is invariant with respect to isometries of $\mathbb{H}^3$, because both expressions in the brackets and products $u(x_1, x_2, y, \beta)u^\dagger(x_1, x_2, y, \beta_0)$ are invariant for any two directions β, β_0. β_0 in (4.7) can be any direction. In particular (4.6) is obtained from (4.7) when $\beta_0 = \infty$.

The invariance with respect to isometries allows us to assume without loss of generality that $x_1 = x_2 = x_1' = x_2' = 0,\ y' = 1$ (otherwise we can always find an isometry which can map (x_1', x_2', y') into $(0,0,1)$ and (x_1, x_2, y) into a point on the y-axis). The assumption allows us to simplify (4.6) to the form

$$\int_{-\infty}^{+\infty} w^2 dw \int_{\partial\mathbb{H}^3} \left[\frac{y(|\beta|^2+1)}{|\beta|^2+y^2}\right]^{1+iw} u^\dagger(0,0,1,\beta) u(0,0,y,\beta) \frac{d\beta_1 d\beta_2}{(1+|\beta|^2)^2}$$

$$= 16\pi^3 y^3 \delta(x, y-1). \tag{4.8}$$

According to the definition

$$u(0,0,y,\beta)=\frac{1}{\sqrt{|\beta|^2+y^2}}\begin{pmatrix}-\beta^* e & -ye^*\\ ye & -\beta e^*\end{pmatrix}=\frac{1}{\sqrt{|\beta|^2+y^2}}\begin{pmatrix}-|\beta| & -ye^*\\ ye & -|\beta|\end{pmatrix}$$

$$u^\dagger(0,0,1,\beta)=\frac{1}{\sqrt{|\beta|^2+1}}\begin{pmatrix}-\beta e^* & e^*\\ -e & -\beta^* e\end{pmatrix}=\frac{1}{\sqrt{|\beta|^2+1}}\begin{pmatrix}-|\beta| & -e^*\\ -e & -|\beta|\end{pmatrix}$$

$$u^\dagger(0,0,1,\beta)u(0,0,y,\beta)=\frac{1}{\sqrt{(|\beta|^2+1)(|\beta|^2+y^2)}}\begin{pmatrix}|\beta|^2+y & -\beta^*(1-y)\\ \beta(1-y) & |\beta|^2+y\end{pmatrix}$$

where we used that $|\beta|e=\beta$.
Using change of variables $p=|\beta|$, $\varphi=\arctan\frac{\beta_2}{\beta_1}$ we rewrite (4.8) as

$$\int_{-\infty}^{+\infty} w^2 dw \int_{\partial H^3}\left[\frac{y(p^2+1)}{p^2+y^2}\right]^{1+iw}\frac{1}{\sqrt{(p^2+1)(p^2+y^2)}}$$
$$\times\begin{pmatrix}p^2+y & -pe^{-i\varphi}(1-y)\\ pe^{i\varphi}(1-y) & p^2+y\end{pmatrix}\times\frac{pdp\, d\varphi}{(1+p^2)^2}=$$
$$=16\pi^3y^3\delta(x,y-1). \tag{4.9}$$

Integration in φ yields that the off-diagonal elements are equal to zero. To estimate the diagonal elements we first need to evaluate:

$$I=\int_0^\infty\left[\frac{y(p^2+1)}{p^2+y^2}\right]^{1+iw}\frac{1}{\sqrt{(p^2+1)(p^2+y^2)}}(p^2+y)\frac{pdp}{(1+p^2)^2}.$$

Using change of variable $p=\sqrt{\frac{1+t}{1-t}}$, $p^2+1=\frac{2}{1-t}$ we obtain

$$I=\int_{-1}^{1}\left[\frac{2y}{1+y^2+(1-y^2)t}\right]^{1+iw}\cdot\frac{1+y+(1-y)t\ dt}{\sqrt{2(1+y^2+t(1-y^2))}}=$$

$$=\frac{1}{(1+y^2)(1-y)}\,Re\left\{\frac{y^{1+iw}}{0.5-iw}-\frac{y^{2-iw}}{0.5-iw}\right\}.$$

Having changed variables as $y=e^{-r}$ we can rewrite the diagonal elements on the left-hand side of (4.9) as

$$\frac{2\pi y^{1.5}}{(1+y)^2(1-y)}\,Re\,\mathcal{J}, \tag{4.10}$$

where

$$\mathcal{J} = \int\limits_{-\infty}^{+\infty} \frac{e^{ir(w+\frac{1}{2}i)} - e^{-ir(w+\frac{1}{2}i)}}{(-i)(w+\frac{1}{2}i)} w^2 dw$$

Differentiating, we obtain

$$\frac{\partial \mathcal{J}}{\partial r} = \int\limits_{-\infty}^{+\infty} \left[e^{ir(w+\frac{1}{2}i)} + e^{-ir(w+\frac{1}{2}i)}\right](-w^2)dw$$

$$= e^{-\frac{r}{2}} \int\limits_{-\infty}^{+\infty} e^{irw}(-w^2)dw + e^{\frac{r}{2}} \int\limits_{-\infty}^{+\infty} e^{-irw}(-w^2)dw =$$

$$= 4\pi \cosh \frac{r}{2} \delta''(r).$$

The functional $2\cosh\frac{r}{2}\delta''(r)$ acts just like $\delta''(r)$. The absolute value of the variable r can be interpreted as the spherical coordinate r introduced in (1.18). If we consider test functions $f(r)$, the following relations are satisfied respectively:

$$\int_{-\infty}^{+\infty} f(r) \cosh \frac{r}{2} \delta''(r) \sinh^2 r dr = \int_{-\infty}^{+\infty} f(r)\delta''(r) \sinh^2 r dr,$$

and

$$\int_{-\infty}^{+\infty} f(r)\left(\frac{\delta'(r)}{\sinh r \cosh \frac{r}{2}}\right) \sinh^2 r dr = \int_{-\infty}^{+\infty} f(r)\left(\frac{\delta'(r)}{\sinh r}\right) \sinh^2 r dr$$

and thus we conclude that $\mathcal{J} - \delta'(r) =$ const. The functional on the left-hand side is odd in r, and therefore the constant must be odd, i.e. $= 0$ and thus the diagonal elements (4.10) are equal to

$$\frac{8\pi}{1-y}\delta'(r) = \frac{8\pi}{1-e^r}\delta'(r) = \frac{8\pi}{\sinh r}\delta'(r)$$

which completes the rough sketch of the proof, because the functional $-\frac{1}{8\pi \sinh r}\delta'(r)$ acts in model II in exactly the same manner as the functional $\delta(x, y-1)$ acts in model I.

5. Possible applications to physics

A first application concerns nuclear physics. Analogously to the description of the electromagnetic interaction by the exchange of massless particles called photons, H. Yukawa proposed in 1935 [6] to describe the (short-ranged) strong interactions by the exchange of massive particles of zero spin; for instance, the later discovered π-mesons (or pions). In this setting, which is nowadays considered as a fruitful phenomenological model, the particles carrying the nuclear interaction obey to the Klein-Gordon equation in Minkowski space:

$$\frac{\partial^2 u}{\partial t^2} - \sum_{i=1}^{3} \frac{\partial^2 u}{\partial x^{i2}} + m^2 u = 0, \tag{5.1}$$

where $x \in \mathbb{R}^3$, and m is the mass of the particle responsible for the interaction. As for the electromagnetic case, the time-independent spherically symmetric solution:

$$u(r) = \frac{e^{-mr}}{r},$$

can be interpreted as the potential function. It provides a theoretical basis for the experimentally found exponential decay of strong interactions.
In [7], one of the authors extended this idea by replacing (5.1) with the wave equation in hyperbolic space. Instead of (5.1), we have the following equation:

$$\frac{\partial^2 u}{\partial t^2} - \frac{\partial^2 u}{\partial r^2} - 2k\coth(kr)\frac{\partial u}{\partial r} - \frac{1}{\sinh^2 r}\Delta_s u - k^2 u = 0 \tag{5.2}$$

where the hyperbolic space is described by model II, Δ_s is the Laplace-Beltrami operator on the unit sphere and k is the mass of the spinless massive particle. The spherically symmetric static solution of (5.2) reads:

$$u(r) = \frac{(A - Br)k}{\sinh(kr)},$$

where A, B are real constants.
It also decays exponentially. Moreover, it agrees with the repulsive behaviour of the nuclear potential near $r = 0$.
Using Fourier transform method we can expand any solution of (5.2) in terms of plane-waves whose typical representative is

$$u = e^{-iwt} y^{1+i\frac{w}{ck}}. \tag{5.3}$$

It was shown [7] that the plane-wave solutions satisfy the Einstein's momentum - energy - mass conservation law

$$\mathcal{E}^2 - c^2\mathcal{P}^2 = k^2 c^4,$$

where $\mathcal{E}^2$ and $\mathcal{P}^2$ are the averages of energy and momentum operators squared.

Though (5.2) does provide an opportunity to model some properties of the strong interactions, it is far from perfect. One can hope that using a system of first order equations we can obtain much better model. The system of Dirac equations on $\mathbb{H}^3$ might be a good candidate.

The second application is related to cosmology. Let us mention that the Fourier modes or components of the general solutions to the Kleim - Gordon and Dirac equation can play an important role in quantam field considerations of scalar and Dirac spinor fields in curved space-times [8]. For instance, they can be useful in the evaluation of the creation of particles in expanding universes and in the calculation of the renormalized expectation value of the energy-momentum tensor, which is desired to estimate back reaction effects of quantum fields on gravitational fields [8 – 17]. However, let us note that since the Friedman - Robertson - Walker space-times are conformally equivalent to Minkowski space-time, it follows that massless scalar and Dirac spinor fields, which obey the above-mentioned conformally invariant equations, will not give rise in these backgrounds to creation of particles, but still they might influence the gravitational field through their energy-momentum tensor [8].

References

1. A.F. Beardon, The Geometry of Discrete Groups, *Graduate Texts in Mathematics*, vol. 91, Springer-Verlag, N.Y., 1983.

2. S.W. Hawking and G.F.R. Ellis, The Large Scale Structure of Space-Time, Cambridge Univrsity Press, Cambridge, 1973.

3. Y. Choquet - Bruhat and D. Christodoulou, Ann. Scient. Ec. Norm. Sup., 4ème Série, **t. 14**, 481 (1981).

4. F. John, Partial Differential Equations, Springer-Verlag, N.Y. 1978.

5. A. Lichnerowicz, Bull. Soc. Math. France **92**, 11 (1964).

6. H. Yukawa, Proc. Phys. -Math. Soc. (Japan) **17**, 48 (1935).

7. M. Kovalyov, Hadronic J., **12** (1989).

8. N.D. Birrell and P.C.W. Davies, Quantum Fields in Curved Space, Cambridge University Press, Cambridge, 1982 and references therein.

9. L. Parker, *Phys. Rev.* **D3**, 346 (1971).

10. L.H. Ford, *Phys. Rev.* **D14**, 3304 (1976).

11. G. Schäfer and H. Dehnen, *Astron. Astrophys.* **54**, 823 (1977).

12. J. Audretsch and G. Schäfer, *J. Phys. A: Math. Gen.* **11**, 1583 (1978).

13. A.A. Grib, S.G. Mamayev, V.M. Mostepanenko, *Fortsch. Phys.* **28**, 173 (1980).

14. M. Castagnino, L. Chimento, D.D. Harari and C.A. Núñez, *J. Math. Phys.* **25**, 360 (1984).

15. A.H. Najmi and A.C. Ottewill, *Phys. Rev.* **D30**, 2573 (1984).

16. L.P. Chimento and M.S. Mollerach, *Phys. Rev.* **D34**, 3689 (1986).

17. C.J. Isham and J.E. Nelson, *Phys. Rev.* **D10**, 3226 (1974).

COMPLEX FINITE FOURIER TRANSFORM TECHNIQUES APPLIED TO THERMAL ENERGY STORAGE PROBLEMS

MAZHAR ÜNSAL
Department of Mechanical Engineering
University of Gaziantep
27310 Gaziantep, Turkey

ABSTRACT. Fourier series techniques have been applied in the past to many heat conduction problems governed by elliptic and parabolic partial differential equations. Fourier transform techniques may be used when a problem is not suitable for application of the separation of variables technique. Underground thermal energy storage problems typically lead to problems governed by parabolic type partial differential equations having solutions of periodic behaviour for large values of the time coordinate. Asymptotic solutions of thermal energy storage problems valid for large times can be properly obtained in closed form by application of the complex finite Fourier transform technique. Exact periodic solutions for such transient heat conduction problems are reported in this study. Solutions presented can be utilized in seasonal thermal energy storage system simulation studies in cylinderical and spherical coordinates.

1. INTRODUCTION

Complex finite Fourier transform techniques are applied in this study to transient heat conduction problems of seasonal thermal energy storage. Recent literature on the subject indicates need for much theoretical research in this subject where very little previous analytical work is available.
The problem considered is the prediction of the transient temperature field during annually periodic operation of seasonal thermal energy storage systems. Analyses presented leads to closed form solutions to the annual periodic temperature outside cylinderically and spherically shaped underground seasonal thermal energy stores.

1.1. Practical Importance of Seasonal Thermal Energy Storage

World oil crises of the 1970's has increased research on renewable energy systems. Research on solar energy gained momentum and solar water heating found commerciality in many countries. Solar aided space heating systems, on the other hand, are not economically feasible in comparison to solar domestic water heating systems and therefore solar aided space heating systems are not in wide commercial use today. Availability of excess solar energy in summer months and energy need for heating of residences during

J. S. Byrnes and J. F. Byrnes (eds.), Recent Advances in Fourier Analysis and Its Applications, 59–72.

winter months is a situation of mismatch between building energy demand and solar energy supply and this is an important factor rendering solar aided space heating systems economically unfeasible. Seasonal thermal energy storage is therefore a subject of technical importance which may lead to the realization of economically feasible solar aided heating systems.
The concept of using undisturbed ground as an energy source for heat pump systems has found practical application in the past. Storing solar energy in earth by circulating a solar heated fluid through pipes buried in earth and using this energy during the winter season as a heat source for heat pump systems was first suggested in 1956 and is considered to be a relatively new concept. A pioneering pilot plant based on this principle was designed only recently in 1969[1]. There has been considerable interest in seasonal energy storage systems within the last decade and economics of seasonal thermal energy storage systems has been noted attractive in a recent survey article published in the Mechanical Engineering magazine[2]. Boreholes in rock, rock caverns, underground steel tanks, abandoned mines, aquifers, open pits on earth, abandoned hydropower tunnels, pipes embedded in earth, underground concrete tanks and excavated bedrock are examples of physical space for seasonal thermal energy storage.
Currently, there is only a very small number of system simulation models available for predicting the long term unsteady behaviour of seasonal thermal energy storage systems. Lund and Östman[3] have developed a numerical procedure to predict the performance of seasonal thermal energy stores charged by solar energy. Their model was used to predict the annual characteristics of a heat pump operated district solar aided heating system. Their model was found to predict an annual solar fraction of 70 percent for a 500 house community with 35 m^2 solar collector area per house and a rockbed store size of 550 m^3 per house. Gabrielsson[4] has reviewed research and development work on seasonal thermal energy storage conducted in Sweden and indicated need for research directed to improve systems for the small project sector in order to bring down costs and improve economics of small size seasonal thermal energy storage projects. Bankston[5] gave a comprehensive review of literature on the subject emphasizing practical applications in different countries.
It appears that much research is required for development of advanced mathematical models and computer codes for seasonal thermal energy storage systems. It is expected that the results presented in this study be used at small computational costs in computer simulation studies of seasonal solar energy storage systems.

2. ANNUAL PERIODIC TEMPERATURE OUTSIDE A SPHERICAL STORE

Transient heat transfer problem in earth outside a spherical seasonal solar energy store is analysed in this section. The problem is formulated considering a spatially lumped time dependent store temperature $T_W(t)$. A closed form analytical solution to the periodic transient heat transfer problem is obtained by application of the complex finite Fourier transform technique.

2.1. Formulation of the Spherical Store Problem

Let us consider a spherical thermal energy store with radius, $r = a$, which is located at a sufficient depth from ground level. The farfield temperature outside the store will be constant and equal to the deep ground temperature T_∞. Temperature of earth surrounding the store is a function of the radial coordinate measured from the center of the spherical store and time, $T=T(r,t)$. The problem formulation for annually periodic transient temperature in earth around the store is given by the following partial differential equation, initial and boundary conditions written in the spherical coordinate system.

$$\frac{\partial^2 T}{\partial r^2} + \frac{2}{r}\frac{\partial T}{\partial r} = \frac{1}{\alpha}\frac{\partial T}{\partial t} \tag{1}$$

$$T(a,t) = T_W(t) \tag{2}$$

$$T(\infty,t) = T_\infty \tag{3}$$

$$T(r,0) = T(r,\text{one year}) \tag{4}$$

α, in equation (1), is the thermal diffusivity of earth surrounding the thermal energy store. An energy balance relating energy charged to the store, energy accumulated in the store and energy conducted to the surrounding earth gives:

$$Q = \rho_w V_w c_w \frac{dT_w}{dt} - kA\frac{\partial T}{\partial r}(a,t) \tag{5}$$

Q is net energy charge rate to the store. ρ_W, V_W, c_W are density of water in the store, volume of water in the store and specific heat of water. k is thermal conductivity of soil and A is surface area of the store. The problem consisting of equations (1)-(5) will be put in dimensionless form by introducing the following dimensionless variables

$$x = r/a,\ t = \alpha t/a^2,\ \phi = (T-T_\infty)/T_\infty,\ \phi_W = (T_W-T_\infty)/T_\infty$$

$$\psi(x,t) = x\phi(x,\tau),\ q = Q/(4\pi a k T_\infty),\ p = \rho_W c_W/(3\rho c)$$

$$Y = \alpha(\text{one year})/a^2 \tag{6}$$

where ρc is density times specific heat of earth surrounding the store. Dimensionless problem formulation now becomes:

$$\frac{\partial^2 \psi}{\partial x^2} = \frac{\partial \psi}{\partial \tau} \tag{7}$$

$$\psi(1,\tau) = \phi_W(\tau) \tag{8}$$

$$\psi(\infty,\tau) = 0 \tag{9}$$

$$\psi(x,0) = \psi(x,Y) \tag{10}$$

$$q = p\frac{d\phi_w}{d\tau} - \frac{\partial\psi}{\partial x}(1,\tau) + \psi(1,\tau) \tag{11}$$

2.2. Application of the Complex Finite Fourier Transform Technique

The problem given by equations (7)-(11) will be analysed via the complex finite Fourier transform defined by :

$$\psi(x,\tau) = \sum_{n=-\infty}^{+\infty} \psi_n(x)\exp\left\{\frac{i2\pi n\tau}{Y}\right\} \tag{12}$$

$$\psi_n(x) = \frac{1}{Y}\int_{-Y/2}^{+Y/2} \psi(x,\tau)\exp\left\{-\frac{i2\pi n\tau}{Y}\right\} \tag{13}$$

Application of this transform to (7)-(11) yields :

$$\frac{d^2\psi_n}{dx^2} - \frac{i2\pi n}{Y}\psi_n = 0 \tag{14}$$

$$\psi_n(1) = \phi_{wn} \tag{15}$$

$$\psi_n(\infty) = 0 \tag{16}$$

$$q_n = p\frac{i2\pi n}{Y}\phi_{wn} - \frac{d\psi_n}{dx}(1) + \psi_n(1) \tag{17}$$

Solution of (14)-(16) is given by:

$$\psi_n(x) = \phi_{wn}\exp\left\{-(1+i)\frac{\sqrt{n\pi}\,(x-1)}{\sqrt{Y}}\right\} \tag{18}$$

Substitution of (18) into (17) yields :

$$\phi_{wn} = \frac{q_n(\eta_1 - i\eta_2)}{\eta_1^2 + \eta_2^2} \tag{19}$$

where

$$\eta_1 = 1 + \frac{\sqrt{n\pi}}{\sqrt{Y}} \quad \text{and} \quad \eta_2 = \frac{\sqrt{n\pi}}{\sqrt{Y}} + p\frac{2\pi n}{Y} \tag{20}$$

Equations (12),(18),(19) and (20) give the annual periodic temperature in earth outside a spherical seasonal thermal energy store as a function of the complex finite Fourier coefficients, q_n, of the dimensionless heat input rate to the store. The solution presented in this section will be utilized in section 5 to obtain a closed form solution for the case when q is a step function.

3. ANNUAL PERIODIC TEMPERATURE OUTSIDE A CYLINDERICAL STORE

An analytical solution for the transient temperature field outside an underground cylinderical thermal store with an arbitrarily varying surface temperature is presented in this section. The solution to the transient heat transfer problem is obtained by application of an integral Hankel transform technique. Formulation of the problem is based on hypothesis similar to those listed in the previous section for the spherical store problem.

3.1. Formulation of the Cylinderical Store Problem

A cylinderical store which is located in considerable depth from ground level is considered. The farfield temperature around the store is equal to the deep ground temperature T_∞. Water in the store is assumed fully mixed and at a spatially lumped time varying temperature $T_W(t)$. This problem is first analysed to obtain a general (nonperiodic) solution valid for an initial temperature distribution in earth equal to T_∞. The periodic solution is then obtained by an application of the complex finite Fourier Transform technique. The problem formulation for the transient temperature outside a cylinderical thermal energy store is given by:

$$\frac{1}{r}\frac{\partial}{\partial r}\left[r\frac{\partial T}{\partial r}\right]=\frac{1}{\alpha}\frac{\partial T}{\partial t} \tag{21}$$

$$T(a,t) = T_W(t) \tag{22}$$

$$T(\infty,t) = T_\infty \tag{23}$$

$$T(r,0) = T_\infty \tag{24}$$

This problem will be put in dimensionless form by introducing

$$x = r/a,\ \tau = \alpha t/a^2,\ \phi = (T - T_\infty)/T_\infty,\ \phi_W = (T_W - T_\infty)/T_\infty$$
$$q = Q/(2\pi L k T_\infty),\ p = \rho_W c_W/(2\rho c),\ Y = \alpha(\text{one year})/a^2 \tag{25}$$

where L is the lenght of the cylinderical store. It is noted that definitions for q and p are different from those used in the first section for the spherical store problem. Dimensionless problem formulation for the cylinderical thermal energy store problem is given by :

$$\frac{1}{x}\frac{\partial}{\partial x}\left[x\frac{\partial\phi}{\partial x}\right]=\frac{\partial\phi}{\partial\tau} \tag{26}$$

$$\phi(1,\tau)=\phi_w(\tau) \tag{27}$$

$$\phi(\infty,\tau)=0 \tag{28}$$

$$\phi(x,0)=0 \quad \text{for } x>1 \tag{29}$$

3.2. Application of the Integral Hankel Transform

Periodic solutions of the problem posed by equation (26) and boundary conditions (27)-(28) can, in principle, be obtained by a direct application of the complex finite Fourier transform technique to (26)-(28). Such a procedure will yield a solution for ϕ in terms of Bessel functions with complex argument. It is then very difficult, if not impossible, to evaluate ϕ numerically using a digital computer. It is therefore advantageous to follow an alternate procedure which is described in this section. The problem posed by equations (26)-(29) will be first analysed for a constant value of ϕ_w. Solution for arbitrary time varying ϕ_w will then be obtained via an application of Duhamel's superposition principle. Replacing ϕ_w with the constant ϕ_o, and introducing the transformation

$$\phi(x,\tau)=\frac{\phi_o}{x}+\psi(x,\tau) \tag{30}$$

into (26)-(29), one obtains:

$$\frac{\phi_o}{x^3}+\frac{1}{x}\frac{\partial}{\partial x}\left[x\frac{\partial\psi}{\partial x}\right]=\frac{\partial\psi}{\partial\tau} \tag{31}$$

$$\psi(1,\tau)=0 \tag{32}$$

$$\psi(\infty,\tau)=0 \tag{33}$$

$$\psi(x,0)=-\frac{\phi_o}{x} \tag{34}$$

The problem given by (31)-(34) can be solved by an application of the following one-sided integral Hankel transform.

$$\psi(x,\tau)=\int_1^\infty K(x,\lambda)\ \psi_n(\lambda,\tau)\lambda\, d\lambda \tag{35}$$

$$\psi_n(\lambda,\tau)=\int_1^\infty K(x,\lambda)\psi(x,\tau)x\, dx \tag{36}$$

A proper kernel for this integral transform is obtained from the solution of the following problem:

$$\frac{1}{x}\frac{d}{dx}\left[x\frac{dK}{dx}\right]+\lambda^2 K=0 \tag{37}$$

$$K(1,\tau) = 0 \tag{38}$$

$$K(\infty,\tau) = 0 \tag{39}$$

Solution of (37)-(39) yields the following kernel function for the integral transform:

$$K(x,\lambda) = J_0(\lambda x)Y_0(\lambda) - Y_0(\lambda x)J_0(\lambda) \tag{40}$$

Application of this transform to (31)-(33) yields:

$$\psi_n(\lambda,\tau)=\alpha(\lambda)+\beta(\lambda)\, e^{-\lambda^2\tau} \tag{41}$$

where

$$\alpha(\lambda)=\phi_0\int_1^\infty x^{-2}K(x,\lambda)dx \tag{42}$$

Application of the initial condition, equation (34), gives:

$$-\frac{\phi_0}{x}=\int_1^\infty \{\alpha(\lambda)+\beta(\lambda)\}K(x,\lambda)\lambda d\lambda \tag{43}$$

or

$$\alpha(\lambda)+\beta(\lambda)=-\phi_0\int_1^\infty K(x,\lambda)dx \tag{44}$$

which yields the following expression for $\beta(\lambda)$:

$$\beta(\lambda)=-\phi_0\left[\int_1^\infty x^{-2}K(x,\lambda)dx+\int_1^\infty K(x,\lambda)dx\right] \tag{45}$$

Integration of equation (37) and substitution from (42) and (44) yields:

$$\beta(\lambda)=\left[\frac{1}{\lambda^2}-1\right]\alpha(\lambda)-\frac{\phi_0}{\lambda^2}\frac{dK}{dx}(1,\lambda) \tag{46}$$

Multiplying equation (37) by x^{-2n} and integrating by parts twice, one can show that:

$$\int_1^\infty x^{-2n}K(x,\lambda)dx=\frac{1}{\lambda^2}\frac{dK}{dx}(1,\lambda)-\frac{1}{\lambda^2}(2n+2)^2\int_1^\infty x^{-2n-2}K(x,\lambda)dx \tag{47}$$

Integrating equation (42) by parts and substitution from equation (47) yields the following closed form formula for $\alpha(\lambda)$:

$$\alpha(\lambda)=\frac{\phi_0}{\lambda^2}\frac{dK}{dx}(1,\lambda)\sum_{n=1}^{\infty}(-1)^n\frac{n^2}{\lambda^{2n}} \tag{48}$$

with $\alpha(\lambda)$ and $\beta(\lambda)$ given by equations (48) and (46), the solution for $\psi(x,\tau)$ is now given by:

$$\psi(x,\tau)=\int_1^\infty\left\{\alpha(\lambda)+\beta(\lambda)e^{-\lambda^2\tau}\right\}K(x,\lambda)\lambda d\lambda \tag{49}$$

or

$$\psi(x,\tau)=\phi_0\int_1^\infty\frac{K(x,\lambda)}{\lambda}\frac{dK}{dx}(1,\lambda)\left\{g(\lambda)+h(\lambda)e^{-\lambda^2\tau}\right\}d\lambda \tag{50}$$

where

$$g(\lambda)=\sum_{n=1}^{\infty}\frac{(-1)^n n^2}{\lambda^{2n}} \tag{51}$$

and

$$h(\lambda)=\left\{\frac{1}{\lambda^2}-1\right\}g(\lambda)-1 \tag{52}$$

Application of Duhamel's superposition principle[6], now yields the following solution for the dimensionless temperature outside the store.

$$\phi(x,\tau)=\frac{\phi_w(\tau)}{x}+\phi_w(0)\int_1^{\infty}\frac{K(x,\lambda)}{\lambda}\frac{dK}{dx}(1,\lambda)\left\{g(\lambda)+h(\lambda)e^{-\lambda^2\tau}\right\}d\lambda$$

$$+\int_0^{\tau}\frac{d\phi_w(\xi)}{d\xi}\int_1^{\infty}\frac{K(x,\lambda)}{\lambda}\frac{dK}{dx}(1,\lambda)\left\{g(\lambda)+h(\lambda)e^{-\lambda^2(\tau-\xi)}\right\}d\lambda d\xi \quad (53)$$

Analytical solution of the transient heat conduction problem given by (53) will now be utilized to obtain an integro-differential equation for the annual periodic temperature of the thermal energy store. The finite complex Fourier transform technique will then be applied yielding a closed form solution for the dimensionless periodic store temperature. The second term on the right hand side of equation (53) vanishes by (29).

Heat may be added to the store at an arbitrary time varying rate Q(t). This will be partially accumulated in the thermal energy store and partially transmitted into earth surrounding the store. In terms of dimensionless variables, the energy balance equation for the cylinderical thermal energy store is given by:

$$q=p\frac{d\phi_w}{d\tau}-\frac{\partial\phi}{\partial x}(1,\tau) \quad (54)$$

Substitution of the solution for ϕ from equation (53) into (54) yields the following integro-differential equation for the dimensionless water temperature in the thermal energy store.

$$q=p\frac{d\phi_w}{d\tau}+(1-d_0)\phi_w-\int_1^{\infty}\frac{h(\lambda)}{\lambda}\left[\frac{dK}{dx}(1,\lambda)\right]^2 e^{-\lambda^2\tau}\int_0^{\tau}\frac{d\phi_w}{d\xi}e^{\lambda^2\xi}d\xi d\lambda \quad (55)$$

where d_0 is given by

$$d_0=\int_1^{\infty}\frac{g(\lambda)}{\lambda}\left[\frac{dK}{dx}(1,\lambda)\right]^2 d\lambda \quad (56)$$

3.3. Application of the Complex Finite Fourier Transform Technique

The last term on the right hand side of equation (55) may be integrated by parts. Equation (55) then takes the following form:

$$q=p\frac{d\phi_w}{d\tau}+(1-d_0)\phi_w+\sum_{n=1}^{\infty}\left\{d_n\frac{d^n\phi_w}{d\tau^n}-e_n\frac{d^n\phi_w}{d\tau^n}(0)\right\} \quad (57)$$

where

$$d_n = (-1)^n \int_1^\infty \left[\frac{dK}{dx}(1,\lambda)\right]^2 \frac{h(\lambda)}{\lambda^{2n+1}} d\lambda \tag{58}$$

and

$$e_n = (-1)^n \int_1^\infty \left[\frac{dK}{dx}(1,\lambda)\right]^2 \frac{h(\lambda)e^{-\lambda^2\tau}}{\lambda^{2n+1}} d\lambda \tag{59}$$

We can seek periodic solutions of equation (57) valid for large times by first letting $\tau \Rightarrow \infty$ in equation (57). By this means, terms representing history of the water temperature in the store disappears from the equation. Letting

$$e_n = 0 \text{ for } n \geq 1 \tag{60}$$

equation (57) takes the following form.

$$q = p\frac{d\phi_w}{d\tau} + (1 - d_o)\phi_w + \sum_{n=1}^{\infty} d_n \frac{d^n\phi_w}{d\tau^n} \tag{61}$$

Application of the complex finite Fourier transform defined by equations (12) and (13) to equation (61) yields:

$$q_n = \phi_{wn}\left\{p\frac{i2\pi n}{Y} + 1 + d_o + \sum_{j=1}^{\infty} d_j\left[\frac{i2\pi n}{Y}\right]^j\right\} \tag{62}$$

which may be solved for ϕ_{wn} to obtain:

$$\phi_{wn} = \frac{q_n(\gamma_1 - i\gamma_2)}{\gamma_1^2 + \gamma_2^2} \tag{63}$$

This equation is similar to equation (19). γ_1 and γ_2 are given by the following expressions:

$$\gamma_1 = 1 + d_o + \sum_{j=1}^{\infty} d_{2j}(-1)^j\left[\frac{2\pi n}{Y}\right]^{2j} \tag{64}$$

$$\gamma_2 = \frac{2\pi np}{Y} + \sum_{j=1}^{\infty} d_{2j-1}(-1)^{j+1}\left[\frac{2\pi n}{Y}\right]^{2j-1} \tag{65}$$

First twenty coefficients, d_i, appearing in these expressions are universal constants independent of thermal properties of earth. These constants, defined by formulas (56) and (58), were computed numerically using a digital computer and are listed in Table 1.

TABLE 1. List of computed coefficients, d_i

i	0	1	2	3	4
d_i	0.04936	0.19278	-.09896	0.06452	-.05015
i	5	6	7	8	9
d_i	0.03825	-.03357	0.02683	-.02523	0.02043
i	10	11	12	13	14
d_i	-.02018	0.01644	-.01682	0.01360	-.01441
i	15	16	17	18	19
d_i	0.01158	-.01259	0.01001	-.01114	0.00882

Equations (63)-(65) give the complex Fourier coefficients for water temperature in the cylinderical store. Substitution of the complex Fourier series for $\phi_w(\tau)$ into equation (53) will give the periodic solution for the transient temperature in earth outside the cylinderical store.

4. A FOURIER SERIES REPRESENTATION FOR HEAT INPUT RATE

Solutions obtained for the dimensionless temperature of the spherical and the cylinderical thermal energy stores are functions of the dimensionless heat input rate. Hence, the temperature distribution in earth depends on the dimensionless energy input function, q. This function may be expressed by a Fourier series with a period of one year. Defining g_i to be monthly components of the dimensionless annual energy gain rate to the store, q will be expressed by the following vector.

$$q = (g_1, g_2, g_3, g_4, g_5, g_6, g_7, g_8, g_9, g_{10}, g_{11}, g_{12}) \tag{66}$$

If monthly components of the q function are expressed as depicted in equation (66), then the dimensionless heat input rate can be expressed by:

$$q = \sum_{n=-\infty}^{+\infty} q_n \exp\left\{\frac{i2\pi n\tau}{Y}\right\} \tag{67}$$

where

$$q_o = \frac{1}{12}\sum_{j=1}^{12} q_j \tag{68}$$

$$q_n = \frac{1}{2\pi n}\sum_{j=1}^{12} q_j\{\eta_{3,j} + i\eta_{4,j}\} \quad \text{for} \quad n \geq 1 \tag{69}$$

and

$$\eta_{3,j} = \sin(2\pi n r_j) - \sin(2\pi n r_{j-1}) \tag{70}$$

$$\eta_{4,j} = \cos(2\pi n r_j) - \cos(2\pi n r_{j-1}) \tag{71}$$

with $r_j = (j-6)/12$. When the dimensionless heat input is represented by a step function, as given in equation (66), the complex Fourier coefficients of the dimensionless water temperature in the store is determined by substitution of q_n given in equation (68)-(69) into equation (19) for the case of a spherical store and into equation (63) for the case of a cylinderical store.

5. A CLOSED FORM SOLUTION FOR THE SPHERICAL STORE PROBLEM WITH A STEP FUNCTION TYPE HEAT INPUT RATE

Solar energy addition to the store over the whole year and energy extraction from the store for the purpose of space heating during winter months are incorporated into the thermal system model via the dimensionless heat source term, q. Substitution of the complex Fourier coefficients q_n from (68) and (69) into (19) gives complex Fourier coefficients for $\phi_w(\tau)$ in the spherical store problem. Complex Fourier coefficients of ψ follow by substitution of ϕ_{wn} into (18). ψ can then be calculated from (12).

Letting a_n and b_n be the real Fourier coefficients of ϕ_w, noting that $\phi_{wo} = a_o$ and $\phi_{wn} = (a_n - ib_n)/2$ when $n > 0$, one finds from equations (19),(68) and (69) that

$$a_o = \frac{1}{12}\sum_{i=1}^{12} g_i \tag{72}$$

$$a_n = \frac{1}{\pi n[\eta_1^2 + \eta_2^2]}\sum_{i=1}^{12} g_i\{\eta_1\eta_{3,i} + \eta_2\eta_{4,i}\} \tag{73}$$

$$b_n = \frac{1}{\pi n[\eta_1^2 + \eta_2^2]} \sum_{i=1}^{12} g_i\{\eta_2\eta_{3,i} - \eta_1\eta_{4,i}\} \tag{74}$$

ψ can now be expressed by

$$\psi(x,\tau) = A_0 + \sum_{n=1}^{\infty} [A_n \cos(2\pi n r_n) + B_n \sin(2\pi n r_n)] \tag{75}$$

Coefficients in (75) are given by the following relationships:

$$A_0 = a_0 \tag{76}$$
$$A_n = (a_n \cos\omega_n - b_n \sin\omega_n)\exp(-\omega_n) \tag{77}$$
$$B_n = (a_n \sin\omega_n + b_n \cos\omega_n)\exp(-\omega_n) \tag{78}$$

where

$$\omega_n = \frac{\sqrt{n\pi}(x-1)}{\sqrt{Y}} \tag{79}$$

Dimensionless temperature of earth adjacent to the spherical store can be calculated noting that $\phi = \psi/x$.

6. A CLOSED FORM SOLUTION FOR THE CYLINDERICAL STORE PROBLEM WITH A STEP FUNCTION TYPE HEAT INPUT RATE

Substitution of the complex Fourier coefficients q_n from (68) and (69) into (63) gives complex Fourier coefficients for $\phi_w(\tau)$ for the cylinderical store problem. Letting a'_n and b'_n be the Fourier coefficients of ϕ_w , one finds from equations (63)-(65),(68) and (69) that

$$a'_0 = \frac{1}{12(1+d_0)} \sum_{i=1}^{12} g_i \tag{80}$$

$$a'_n = \frac{1}{\pi n[\gamma_1^2 + \gamma_2^2]} \sum_{i=1}^{12} g_i\{\gamma_1\eta_{3,i} + \gamma_2\eta_{4,i}\} \tag{81}$$

$$b'_n = \frac{1}{\pi n[\gamma_1^2 + \gamma_2^2]} \sum_{i=1}^{12} g_i\{\gamma_2\eta_{3,i} - \gamma_1\eta_{4,i}\} \tag{82}$$

Periodic solution for the dimensionless transient temperature in earth follows from substitution of the Fourier series for ϕ_w into (53).

7. CONCLUSIONS

Utilization of the complex form of the Fourier transform technique drastically decreases the amount of algebra leading to the closed form solutions presented. The formulas presented in this study can be effectively used at low computational costs to investigate long term performance of seasonal thermal energy storage systems. Solutions reported can be utilized in system simulation studies where heat may be added to the store over one year from a source such as solar energy and heat may be extracted from the store for winter heating of a residential building.
A number of physical observations can be deduced from the results of the present study. The annual average value of the dimensionless store temperature for the spherical store problem is equal to the annual average value of the dimensionless heat input rate as seen by setting $n = 0$ in equation (19). If we let $n = 0$ in (63), we see that annual average value of the dimensionless store temperature for the cylinderical store is given by $\phi_{wo} = q_o/(1+d_o)$. It is noted that the dimensionless annual average temperature in both problems depend on the dimensionless heat input rate and are not affected by dimensionless system parameters such as p and Y.

8. REFERENCES

[1] Svec, O.J. (1987) ' Potential of Ground Heat Source Systems', International Journal of Energy Research 11,573-581.
[2] ME Staff Report (1983) 'Seasonal Thermal Energy Storage', Mechanical Engineering 105, 28-34.
[3] Lund, P.D. and Östman, M.B. (1985) ' A Numerical Model for Seasonal Storage of Solar Heat in the Ground by Vertical Pipes', Solar Energy 34,351-366.
[4] Gabrielsson, Eric (1988) ' Seasonal Storage of Thermal Energy-Swedish Experience , Journal of Solar Energy Engineering 110, 202-207.
[5] Bankston, Charles A. (1988) ' The Status and Potential of Central Solar Heating Plants with Seasonal Storage: An International Report', in K.W. Böer(Ed.), Advances in Solar Energy, Plenum Press, pp.352-444.
[6] Arpacı, Vedat (1966) Conduction Heat Transfer, Addison Wesley, Palo Alto, p.307.

SPLINE INTERPOLATION OF DATA OF POWER GROWTH

A. JAKIMOVSKI
School of Mathematical Sciences
Tel-Aviv University
Israel

ABSTRACT. *

1. The case of simple knots

Let $\mathbf{x} := (x_i)_{i\in Z}$ denote a strictely increasing real sequence satisfying $x_i < x_{i+1}$ $(\forall i)$, $\lim_{i\to-\infty} x_i = -\infty$, *and* $\lim_{i\to+\infty} x_i = +\infty$. For m=1,2,... the space $\mathcal{S}_{\mathbf{m},\mathbf{x}}$ of *spline functions* of degree *m-1* (or order m) with *simple knots* $\mathbf{x}$, defined by

$$\mathcal{S}_{\mathbf{m},\mathbf{x}} := \{S(.) \mid S \in \mathbf{C}^{\mathbf{m}-2}(I\!R) \text{ and } S|_{(x_i,x_{i+1}]} \in \pi_{\mathbf{m}-1} \quad (\forall i \in Z\!\!Z)\} ,$$

the continuity condition $S \in \mathbf{C}^{\mathbf{m}-2}(I\!R)$ being omitted if $\mathbf{m} = 1$.

For $\rho \geq 0$ define the class $\mathbf{F}_\rho$ of *power-dominated functions* $f : I\!R \to \mathbf{C}$ by $\mathbf{F}_\rho := \{f(.) = \mathcal{O}(|t|^\rho) \text{ as } |t| \to \infty\}$. For $\rho \geq 0$ the class $\mathbf{Y}_{\rho,x}$ of *power dominated sequences* $\mathbf{y}:=(y_i)_{i\in Z} \in \omega$ (ω denotes the space of all bi-infinite complex-valued sequences) is defined by $\mathbf{Y}_{\rho,x} := \{\mathbf{y} \mid y_i = \mathcal{O}(|x_i|^\rho) \text{ as } |i| \to \infty \}$.

Schoenberg (1972) posed the following problem: Given a power-dominated *data-sequence* $y \in \mathbf{Y}_{\rho,x}$, investigate the existence, uniqueness, and representation of a power-dominated spline function which will *interpolate y at the knots* x namely a spline $S(.) \in \mathcal{S}_{m,x} \bigcap \mathbf{F}_\rho$ such that $S(x_i)=y_i$ $(\forall i \in Z\!\!Z)$.

Schoenberg's methods depends in several places on the translation invariance of the sequence of integers, a property which failes for more general knot-sequences, and accordingly a different treatment must be sought.

We were, however, able to utilize several significant ideas from a paper of de Boor (1976) to get the following result.

THEOREM 1. *Let* $\rho \geq 0$, $y \in \omega$ *and m=1,2,.... Assume that the sequence* $\mathbf{x}$ *satisfies the following two conditions:* $x_i < x_{i+1}$ $(\forall i)$, $\lim_{i\to-\infty} x_i = -\infty$ *and* $\lim_{i\to+\infty} x_i = +\infty$; and *for every* $\epsilon > 0, \exists \kappa_\epsilon > 0$ *such that* $\forall i,j \in Z\!\!Z$, $\frac{x_{i+1}-x_i}{x_{j+1}-x_j} \leq \kappa_\epsilon\, e^{\epsilon|i-j|}$. *In order that a function S(.) should exist, with the properties*

$$S(.) \in \mathcal{S}_{2\mathbf{m},\mathbf{x}} , \quad S(t) = \mathcal{O}(|t|^\rho) \text{ as } |t| \to +\infty \quad \text{and} \quad S(x_i) = y_i \quad (\forall i \in Z\!\!Z)$$

* For the complete text see A. Jakimovski, D.C. Russell and M. Stieglitz (1983) and (1984).

J. S. Byrnes and J. F. Byrnes (eds.), Recent Advances in Fourier Analysis and Its Applications, 73–75.

it is necessary and sufficient that $y_i = \mathcal{O}(|x_i|^\rho)$ *as* $|i| \to +\infty$; *and then the spline* $S(.)$ *is unique and has the representation* $S(t) = \sum_{k\in\mathbb{Z}} y_k L_{k,2m}(t;x) \quad (\forall t \in I\!R)$, *where the series convergece uniformly on any compact subset of* $\boldsymbol{R}$.

The functions $L_{k,2m}$, called *fundamental splines* are defined by the next theorem.

THEOREM 2. *Let* $\mathbf{x}$ *satisfy* $x_i < x_{i+1}$ $(\forall i)$, $\lim_{i\to-\infty} x_i = -\infty$ *and* $\lim_{i\to+\infty} x_i = +\infty$; *and let m=1,2,.... Then there is a unique spline function* $\mathbf{L}_{k,2m}$ *satisfying* $\mathbf{L}_{k,2m} \in S_{2m,x}$, $\mathbf{L}^{(m)}_{k,2m} \in \mathbf{L}_2(I\!R)$, $\mathbf{L}_{k,2m}(x_i) = \delta_{k,i}$ $(\forall i \in \mathbb{Z})$. *Moreover, if for some* $\epsilon > 0$ $\exists\kappa_\epsilon > 0$ *such that* $\forall i,j \in \mathbb{Z}$, $\frac{x_{i+1}-x_i}{x_{j+1}-x_j} \le \kappa_\epsilon\, e^{\epsilon|i-j|}$, *then* $\exists c_\epsilon > 0$ *and* $d_\epsilon > 0$ *such that* $|L_{k,2m}(t)| \le c_\epsilon e^{-d_\epsilon|k-j|}$ $(\forall t \in [x_j, x_{j+1}],\ \forall k,j \in \mathbb{Z})$. *i.e.* $L_{k,2m}(t)$ *decays exponentialy as* $|t| \to \infty$.

I. J.Schoenberg (1972) raised and solved this problem for *splines of odd degree* in the *cardinal case* $x_i = i$, for any $\rho \in I\!R_+$. Carl De Boor (1976) obtained the above result for splines of odd degree, when $\rho = 0$ (namely for a *bounded* data sequence) and for knots with a *finite global mesh ratio*, namely satisfying $\mathbf{M}_\mathbf{x} := \sup_{i,j} \frac{x_{i+1}-x_i}{x_{j+1}-x_j} < +\infty$; which is equivalent to $0 < \underline{c} \le x_{i+1} - x_i \le \bar{c} < \infty$ $(\underline{c}, \bar{c}$ *are fixed numbers*$)$, $\forall i \in \mathbb{Z}$.

Rong-qing Jia (1983) has examined in more detail the interpolation of *bounded* data by cubic splines, and in this case he is also able to relax de Boor's condition on the sequence of knots. We shall look into his condition later.

In the hypothesis of Theorem 1 it was assumed that the sequence $\mathbf{x}$ satisfies the assumption *for every* $\epsilon > 0, \exists\kappa_\epsilon > 0$ *such that* $\forall i,j \in \mathbb{Z}$, $\frac{x_{i+1}-x_i}{x_{j+1}-x_j} \le \kappa_\epsilon\, e^{\epsilon|i-j|}$. From the proof of Theorem 1 it follows that there exists a number $\epsilon_0 > 0$ with the property that if for *some* ϵ, $\epsilon_0 > \epsilon > 0$ the following condition holds: *for the given* $\epsilon > 0, \exists\kappa_\epsilon > 0$ *such that* $\forall i,j \in \mathbb{Z}$, $\frac{x_{i+1}-x_i}{x_{j+1}-x_j} \le \kappa_\epsilon\, e^{\epsilon|i-j|}$; then the conclusion of the theorem still holds. Since it was somewhat difficult to determine this ϵ_0 we assumed in the theorem that the condition holds $\forall\epsilon \ge 0$. Rong-qing Jia (1983) considered the case ρ=0, m=2 and showed that it is possible to take $0 < \epsilon_0 < \log\left[\frac{3+\sqrt{5}}{2}\right]$.

2. The case of multiple knots

A sequence of pairs $(z_i, \mu_i)_{i\in\mathbb{Z}}$ satisfying $z_i < z_{i+1}$, $\lim_{i\to-\infty} z_i = -\infty$, $\lim_{i\to+\infty} z_i = +\infty$ *and* $\mu_i \in \{1,2,\ldots,r\}$ $(\forall i \in \mathbb{Z})$ *where* $r = \max_{i\in\mathbb{Z}} < +\infty$. The number r is called *the maximal multiplicity of the pair* $(\mathbf{z}, \mu)$.

For a given m=1,2,... the space $S_{(\mathbf{z},\mu)}$ of *spline functions of degree* m-1 *with multiple knots* $(\mathbf{z}, \mu)$. when r≤m is defined by

$$S_{(\mathbf{z},\mu)} := \{S(.) \in C^{m-1-\mu_i}(z_{i-1}, z_{i+1}) \quad and \quad S|_{(z_i,z_{i+1})} \in \pi_{m-1} \quad (\forall i \in \mathbb{Z})\}.$$

The continuity condition involving $C^{m-1-\mu_i}$ is to be omitted if $\mu_i = r = m$.

We shall stretch out the pair $(\mathbf{z}, \mu)$ into a single sequence of the form $\ldots, z_{i-1}, \underbrace{z_i, \ldots, z_i}_{\mu_i \ times}, z_{i+1}, \ldots$; this sequence will be denoted by $\mathbf{x} \equiv (x_i)_{i\in\mathbb{Z}}$ which is assumed to satisfy $x_i \le x_{i+1}$, $x_i < x_{i+r}$ $(\forall i \in \mathbb{Z})$, *and* $x_i \to \pm\infty$ *as* $i \to \pm\infty$.

For each $i \in \mathbb{Z}$ write $i^- := min\{j | x_j = x_i\}$, $i^+ := max\{j | x_j = x_i\}$. We shall assume that the sequence $\mathbf{x}$ satisfies the *Condition* $\mathbf{K}$, namely

$$\forall \epsilon > 0, \quad \exists K_\epsilon > 0 \quad such \quad that \quad \forall i,j \in \mathbb{Z}, \quad \frac{x_{i+ +1} - x_{i+}}{x_{j+ +1} - x_{j+}} \leq K_\epsilon e^{\epsilon|i-j|}.$$

THEOREM 3. *Let m=1,2...* and $\rho \geq 0$. *Let* $(\mathbf{z}, \mu)$ *be a sequence of multiple knots with maximal multiplicity* r *and satisfying Condition* **K**. *If a sequence* $\mathbf{y} \equiv (y_{ij})$ $(i \in \mathbb{Z}, \quad j = 0,1,\ldots,\mu_i - 1)$ *satisfies* $y_{ij} = \mathcal{O}(|z_i|^\rho \quad as \quad |i| \to +\infty \quad for \quad j = 0,1,\ldots,\mu_i - 1$, *then there is one and only one function S(.) satisfying* $S(.) \in \mathcal{S}_{2m,(\mathbf{z},\mu)}$, $S(.) = \mathcal{O}(|t|^\sigma) \quad as \quad |t| \to +\infty \quad for \quad some \quad \sigma \geq 0, \quad and \quad S^{(j)}(z_i) = y_{ij} \quad (\forall i \in \mathbb{Z}, \quad j = 0,1,\ldots,\mu_i - 1)$. *This unique spline S(.) and its derivatives have the representation* $S^{(q)}(t) = \sum_{i \in \mathbf{Z}} \sum_{j=0}^{\mu_i - 1} y_{ij} L^{(q)}_{i,j,(\mathbf{z},\mu),2m}(t) \quad \forall t \in \mathbb{R}, \quad q = 0,1,\ldots,m-1$. *The above series are uniformly convergent on compact sets in* **R**. *The function* $L^{(q)}_{i,j,(\mathbf{z},\mu),2m}(t)$ $(i \in \mathbb{Z}, j = 0,1,\ldots,\mu_i - 1)$ *is the unique spline function satisfying* $L_{i,j,(\mathbf{z},\mu),2m}(.) \in \mathcal{S}_{2m,(\mathbf{z},\mu)}$, $L_{i,j,(\mathbf{z},\mu),2m}(.) \in L^2(\mathbb{R}), \quad and \quad L^{(k)}_{i,j,(\mathbf{z},\mu),2m}(t) = \delta_{kj}$. Also $\exists \alpha \in [0, r-1]$ such *that* $(x_{j+m} - x_j)^q S^{(q)}(t) = \mathcal{O}(|t|^{\rho+\alpha}), \quad z_i \leq t \leq z_{i+1}, \quad |j| \to +\infty$.

Lipow and Schoenberg (1973) posed and solved this Hermite interpolation problem in the cardinal case $z_i = i \quad (i \in \mathbb{Z})$, for each $\rho \geq 0$ when each knot has the same multiplicity r (r<m).

References

de Boor, C. (1976) *Odd-degree spline interpolation at a bi-infinite knot sequence*, Approximation Theory, Bonn. Springer Lecture Notes No. 556. 30-53.

Jakimovski, A., Russell, D.C. and Stieglitz, M. (1983) Spline interpolation of power-dominated data, Functional Analysis and Approximation, ed. Butzer, P.L. and B.Sz.-Nagy, Birkhäuser Verlag 1983.

Jakimovski, A., Russell, D.C. and Stieglitz, M. (1984) Hermite spline interpolation of data of power growth, Constructive Theory of Functions, Sofia 1984, 430-439.

Jia, Rong-qing, (1983) On a conjecture of C.A. Michelli concerning cubic spline interpolation at a bi-infinite knot-sequence, J. Approximation theory 38, 284-292.

Lipow, P.R. and Schoenberg, I.J. (1973) *Cardinal interpolation and spline functions III. Cardinal Hermite interpolation*, Linear Algebra & Appl. 6, 273-304.

Schoenberg, I.J. (1972) *Cardinal interpolation and spline functions:II. Interpolation of data of power growth*, J. Approximation Theory 6, 404-420.

Part II: Polynomials with Restricted Coefficients

Properties on the Unit Circle of Polynomials with Unimodular Coefficients†

By Donald J. Newman and André Giroux‡

Prometheus Inc.
21 Arnold Ave.
Newport, RI 02840

Abstract

A concrete explicit construction of a unimodular polynomial with prescribed zeros on the unit circle is given. More precisely a polynomial $P(z) = a_0 + a_1 z + \ldots a_N z^N$ is produced for which $|a_i| = 1$ for all $i = 0, 1, \ldots, N$ and for which $P(\alpha_j) = 0$ for a given set of $\alpha_j, j = 1, 2, \ldots, n, |\alpha_j| = 1$, and $P(z) \neq 0$ elsewhere on $|z| = 1$. It is further shown how to extend this construction so as to maintain these properties and force the maximum of $|P(z)|$ to occur at any given number $\beta \neq \alpha_j, j = 1, 2, \ldots, n$ and $|\beta| = 1$. The dependence of N on n is exponential, but there is reason to believe that this is actually necessary and not just a weakness of the method.

AMS (MOS) classifications 42A16, 42A28

The following problem is posed in [1]: given the magnitude of the coefficients of a polynomial P, a finite subset S of the unit circle C, and a point p on C distinct from those in S, choose the phases of the coefficients so that $P(z) = 0$ for all z in S, the maximum on C of $|P(z)|$ occurs at $z = p$, and the maximum of $|P(z)|$ on a subset of C excluding an appropriate interval around p (the "beamwidth") is as small as possible. As explained in [1], this problem arises naturally in linear antenna theory, filter theory, and other classical electrical engineering applications.

Our main result is encompassed in the following theorem:

Theorem. Given $\alpha_1, \alpha_2, \ldots, \alpha_n, \beta$ on C, $\beta \neq \alpha_i$, there exists a polynomial P with coefficients of modulus one such that the zeros of P on C are precisely the points $\alpha_1, \alpha_2, \ldots, \alpha_n$ and that

$$||P|| \equiv \max\{|P(z)| : z \in C\} = |P(\beta)|.$$

(The points α_j are not assumed to be distinct: multiple zeros are allowed.)

We shall prove the theorem in two steps, first constructing a polynomial with coefficients of modulus one with the prescribed zeros (lemma 1) and then modifying it so that it also has the prescribed maximum (lemma 4).

Let U denote the (non-linear) class of polynomials with coefficients of modulus one. We introduce in U the operation of "encapsulation" which we denote by $\otimes$.

Definition. Let M be the degree of P. Then

$$(P \otimes Q)(z) \equiv P(z)Q(z^{M+1}).$$

Explicitly, then, if $P(z) = \sum_{k=0}^{M} a_k z^k$ and $Q(z) = \sum_{j=0}^{N} b_j z^j$, then $P \otimes Q(z) = \sum_{i=0}^{N(M+1)+M} c_i z^i$, where $c_{j(M+1)+k} = a_k b_j$ for $0 \leq k \leq M$ and $0 \leq j \leq N$.

† Also to appear in the Proceedings of the American Math Society.

‡ The authors are also with Temple University and the University of Montreal respectively.

J. S. Byrnes and J. F. Byrnes (eds.), Recent Advances in Fourier Analysis and Its Applications, 79–81.

From this expression, it is clear that $\otimes$ is associative and that U is indeed closed under it. Also, if Q does not vanish on C, then the zeros of $P \otimes Q$ are precisely those of P.

Lemma 1 . Given $\alpha_1, \alpha_2, \ldots, \alpha_n$ on C, there exists P_n in U which vanishes at the α_j and nowhere else on C.

Proof. Let $\omega = \exp 2\pi i/3$. We shall prove a little more, namely that such a polynomial can be found with constant term one and leading term $-\omega$. We use induction on n. Let

$$Q_1(z) = 1 - z/\alpha_1$$

and let

$$P_1(z) = Q_1(z) \quad \text{if} \quad \alpha_1\omega = 1$$

and

$$\begin{aligned} P_1(z) &= Q_1(z) \otimes (1 + z + \alpha_1\omega z^2) \\ &= 1 - (1/\alpha_1)z + z^2 - (1/\alpha_1)z^3 + \alpha_1\omega z^4 - \omega z^5 \end{aligned}$$

otherwise (the condition $\alpha_1\omega \neq 1$ also ensures that the polynomial $1 + z + \alpha_1\omega z^2 \neq 0$ on C). Assume now that $R(z) = 1 + \cdots - \omega z^N$ is a polynomial in U of which the zeros on C are precisely the points $\beta_j = \alpha_j/\alpha_n$ for $1 \leq j \leq n-1$. Set

$$\begin{aligned} S(z) &= R(z)(1 - z^N + z^{2N} - z^{3N+1}) \\ &= 1 + \cdots + \omega z^{4N+1}. \end{aligned}$$

To verify that S is in U, we need only check the modulus of the coefficients of z^N and z^{2N}. But, the coefficient of z^N is $-\omega - 1 = \omega^2$ and that of z^{2N} is $\omega + 1 = -\omega^2$. We now claim that

$$1 - z^N + z^{2N} - z^{3N+1}$$

does not vanish on C except for a simple zero at $z = 1$. Indeed, for the sum of four points on C to vanish, it is necessary and sufficient that they cancel in pairs. There are thus three possibilities:

a) $1 - z^N = 0$ and $z^{2N} - z^{3N+1} = 0$; this gives $z^N = 1$ and $z^{N+1} = 1$, hence $z = 1$;
b) $1 + z^{2N} = 0, -z^N - z^{3N+1} = 0$; this implies $z^{2N} = -1, z^{2N+1} = -1$, which is impossible;
c) $1 - z^{3N+1} = 0, -z^N + z^{2N} = 0$; in this case, $z^{3N+1} = 1$ and $z^N = 1$, therefore $z^{2N+1} = z^{N+1} = z^N = 1$, hence $z = 1$ again.

Setting

$$Q_n(z) = S(z/\alpha_n) = 1 + \cdots + (\omega/\alpha_n^{4N+1})z^{4N+1}$$

we obtain a polynomial in U vanishing on C precisely at the points $\alpha_1, \alpha_2, \cdots, \alpha_n$. Then

$$P_n(z) = Q_n(z) \quad \text{if} \quad \alpha_n^{4N+1} = -1$$

and

$$P_n(z) = Q_n(z) \otimes (1 + z - \alpha_n^{4N+1}z^2) \quad \text{otherwise.}$$

This completes the induction and the proof of the lemma.

Lemma 2. If $n > 3$, the polynomial $p_n(z) = 1 + z + z^2 + \cdots + z^{n-1} - z^n + z^{n+1} + \cdots + z^{2n-1}$ does not vanish on C and $\max\{|p_n(z)| : z \in C\}$ is attained if and only if $z = 1$.

Proof. Since

$$p_n(z) = \frac{z^{2n} - 1}{z - 1} - 2z^n$$

we see that

$$p_n(e^{i2\theta})e^{-i(2n-1)\theta} = \frac{\sin 2n\theta}{\sin\theta} - 2e^{i\theta}$$

so that $p_n(e^{i2\theta}) = 0$ implies that $e^{i\theta}$ is real which implies in turn that $p_n(e^{i2\theta}) = p_n(1) = 2n - 2$, a contradiction.

Our statement about the maximum modulus property of $p_n(z)$ follows from the fact that $p_n{}^2(z)$ has positive coefficients.

Lemma 3. If $n > 0$, the polynomial $q_n(z) = 1 + z + z^2 + \cdots + z^n - z^{n+1}$ does not vanish on C.

Proof. One has $(1-z)q_n(z) = 1 - 2z^{n+1} + z^{n+2}$, which, by the triangle inequality, can vanish only if $1, z^{n+1}$ and z^{n+2} are in the same direction; that is, here, $z = 1$. But, obviously, $q_n(1) \neq 0$.

Lemma 4. If P is any polynomial such that $P(\beta) \neq 0$, then there exists a polynomial T in U which does not vanish on C and is such that $P \otimes T$ attains its maximum modulus on C at $z = \beta$.

Proof. We can assume that $\beta = 1$. let $T = q_m \otimes r \otimes p_n$ where q_m and p_n are as in lemmas 3 and 2 respectively, and $r(z) = 1 + e^{i\theta}z - z^2$ with $\theta \neq \pm\pi/2$: $r(z)$ does not vanish on C and $\operatorname{Im} r'(1)/r(1) = 2\sin\theta$. Consider first $P^* = P \otimes q_m \otimes r$. If M is the degree of P, then

$$\frac{P^{*\prime}(1)}{P^*(1)} = \frac{P'(1)}{P(1)} + (M+1)\frac{q_m'(1)}{q_m(1)} + (M+1)(m+2)\frac{r'(1)}{r(1)}$$

and therefore

$$\operatorname{Im}\frac{P^{*\prime}(1)}{P^*(1)} = \operatorname{Im}\frac{P'(1)}{P(1)} + 0 + (M+1)(m+2)\operatorname{Im}\frac{r'(1)}{r(1)} = \operatorname{Im}\frac{P'(1)}{P(1)} + (M+1)(m+2)2\sin\theta.$$

A proper choice of m and θ will cancel this expression, making zero a stationary point of the function $|P^*(e^{i\theta})|^2$. Since this function does not vanish at $\theta = 0$, this stationary point can be transformed into the maximum point over all θ by multiplying P^* by a delta-like function, in our case by encapsulating it with p_n for suitably large n. Indeed, one has

$$|p_n(e^{i\theta})|^2 = 4 - 4\frac{\sin n\theta}{\sin\theta/2}\cos\theta/2 + \left(\frac{\sin n\theta}{\sin\theta/2}\right)^2, \quad .$$

By computing the second derivative, it is easily seen that $p_n(\theta)$ falls off like $e^{-cn\theta^2}$, and this shows that it is an adequate delta function for our purposes.

This completes the proof.

Reference.

[1] James S. Byrnes, Donald J.Newman, "Null Steering Employing Polynomials with Restricted Coefficents", *IEEE Transactions on Antennas and Propagation*, **36**(1988), 301–303.

ON THE MAXIMUM MODULUS FOR A CERTAIN CLASS OF UNIMODULAR TRIGONOMETRIC POLYNOMIALS

GEORGE BENKE *
The MITRE Corporation
7525 Colshire Drive
McLean, VA 22102
USA

ABSTRACT. A certain class of trigonometric polynomials of degree N squared is defined. The polynomials in this class have coefficients which are piecewise complex exponentials and therefore have unit modulus. The supremum norm for the polynomials in this class are studied. The cases where the frequencies in the coefficient sections are a permutation of the integers from 1 to N, and where the frequencies are independent uniformly distributed random variables are treated.

1. Introduction

In this paper we investigate the supremum norm of a certain class of unimodular trigonometric polynomials. Consideration of this class is motivated by the so-called "unimodular polynomial problem". The problem is this: given a trigonometric polynomial $P_N(x) = \sum_1^N a_n e^{inx}$, how close can we get to the situation where $|P_N(x)|$ is constant and the $|a_n|$ are constant? If $|a_n| = 1$ for $n = 1, \ldots, N$, then $\int_0^{2\pi} |P_N(x)|^2 dx = 2\pi N$ so that a constant $|P_N(x)|$ would equal $N^{1/2}$. If $a_n = \exp(i\pi n^2/N)$ then the polynomials are denoted by G_N and are called the Gauss polynomials. Littlewood [7] showed that $\|G_N\|_\infty \leq 1.37N^{1/2}$ and that $|G_N(x)|/N^{1/2}$ converges to 1 uniformly except on an interval of length less than $2N^{-\frac{1}{2}+\delta}$.

In [3] Byrnes constructed polynomials B_{N^2} which are similar to the Gauss polynomials in that $|B_{N^2}| \leq 2.6N$ except on an interval of length $2N^{-1}$. Figure 1 gives plots of the moduli of the Gauss and Byrnes polynomials for degree 100.

Rudin [8] and Shapiro [9] constructed a pair of polynomials P_N and Q_N of degree 2^N with coefficients which are ± 1 and which satisfy $|P_N(x)|^2 + |Q_N(x)|^2 = 2^{N+1}$ for all x. From this it follows that $\|P_N\|_\infty \leq \sqrt{2}(\deg P_N)^{1/2}$. These poly-

* The author is also on the faculty of the Mathematics Department, Georgetown University, Washington DC 20057

J. S. Byrnes and J. F. Byrnes (eds.), Recent Advances in Fourier Analysis and Its Applications, 83–100.

nomials do not satisfy a lower bound property similar to that of the Gauss and Byrnes polynomials. Figure 2 gives plots of $|P_N|$ and $|Q_N|$ for $N = 7$.

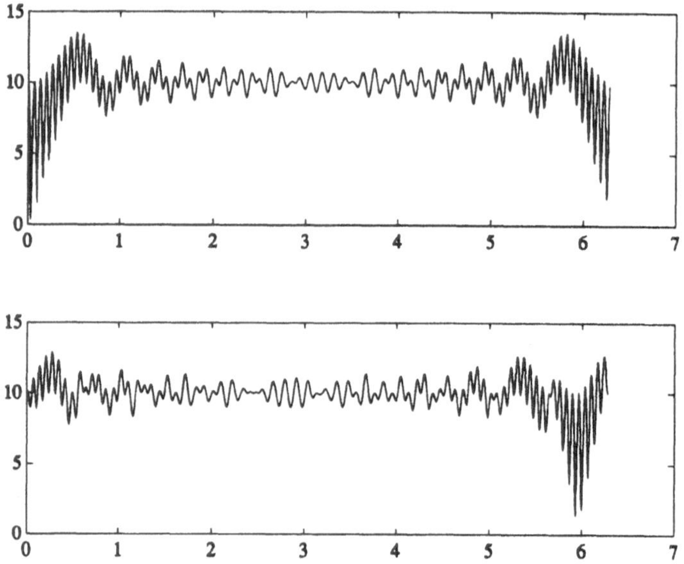

Figure 1. The moduli of the Gauss polynomial $G_{100}(x)$ (top) and the Byrnes polynomial $B_{100}(x)$ (bottom) for 1024 equally spaced values of x between 0 and 2π.

Elaborating on the work of Byrnes, Körner [6] used probabilistic techniques to show the existence of polynomials P_N which satisfy

$$AN^{1/2} \leq |P_N(x)| \leq BN^{1/2}$$

where $0 < A < B < \infty$. This line of investigation culminated in the dramatic results of Kahane [4], where he used probabilistic techniques to show the following. Suppose ϵ_N is a positive sequence converging to 0. Then for N sufficiently large, there exists a polynomial P_N of degree N with coefficients of modulus 1 so that

$$(1 - \epsilon_N)N^{1/2} \leq |P_N(x)| \leq (1 + \epsilon_N)N^{1/2}.$$

However, explicit non-probabilistic constructions which give the results of Kahane or even those of Körner are not known. If the strict unimodular condition $|a_n| = 1$ is replaced by $|a_n| \leq 1$, then Beller and Newman [1] give an explicit polynomial P_N of degree N such that

$$(0.02)N^{1/2} \leq |P_N(x)| \leq N^{1/2}.$$

A different construction is given by Benke [2], where a polynomial P_N of degree $N(N+1)$ with coefficients $|a_n| \leq 1$ is constructed satisfying

$$(0.47)N \leq |P_N(x)| \leq N.$$

In this paper we note a similarity in structure between the Gauss polynomials and the Byrnes polynomials. This similarity motivates our definition of a class of unimodular polynomials of degree N^2 which contain the Byrnes polynomial B_N and the Gauss polynomial G_{N^2}. We will then investigate the $\| \|_\infty$-norm behavior of polynomials in this class in both a deterministic and probabilistic setting.

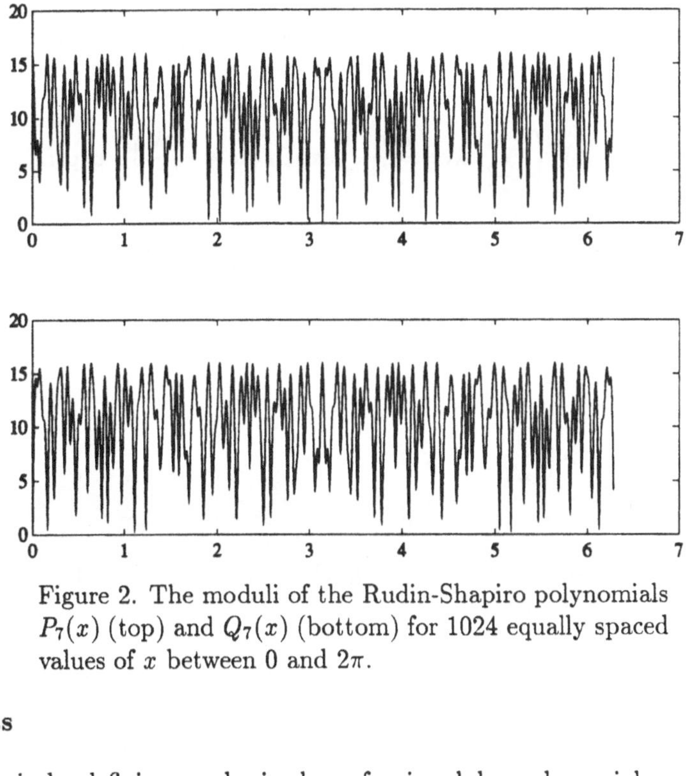

Figure 2. The moduli of the Rudin-Shapiro polynomials $P_7(x)$ (top) and $Q_7(x)$ (bottom) for 1024 equally spaced values of x between 0 and 2π.

2. Results

We begin by defining our basic class of unimodular polynomials.

Definition. *Let W_N denote the set of all polynomials of the form*

$$P_N(x) = \sum_{k=0}^{N-1} a_k Q_N\left(x + \frac{2\pi c_k}{N}\right) e^{iNkx}$$

where $|a_k| = 1$ for $k = 0, \ldots, N-1$, where $Q_N(x) = \sum_{n=0}^{N-1} q_n e^{inx}$ with $|q_n| = 1$, and where $c_n \in [0, N]$ for $n = 0, \ldots, N-1$.

Remark. *The Gauss polynomial G_{N^2} and the Byrnes polynomial B_{N^2} belong to the unimodular class W_N.*

This can be seen as follows. The Fourier transform of a polynomial P_N in the class W_N is

$$\hat{P}_N(n) = \sum_{k=0}^{N-1} a_k \hat{Q}_N(n - Nk) \exp\left(2\pi i c_k \left(\frac{n}{N} - k\right)\right).$$

Since $\hat{Q}_N(n) = 0$ except for $n = 0, \ldots, N-1$, $|\hat{P}_N(n)| = 1$, for $n = 0, \ldots, N^2 - 1$, and $|\hat{P}_N(n)| = 0$ otherwise. If we let $a_k = 1$, $q_k = 1$, and $c_n = n$, we get the Byrnes polynomials. If $a_n = (-1)^n$, $q_n = \exp i\pi(n/N)^2$, and $c_n = n$ we get the Gauss polynomials. This can be seen as follows. Note that $(-1)^k = \exp\left(i\pi k^2\right)$ and consider

$$\begin{aligned}
&\sum_{k=0}^{N-1}(-1)^k \left(\sum_{n=0}^{N-1} \exp\left(\frac{i\pi n^2}{N^2}\right) \exp in\left(x + \frac{2\pi k}{N}\right)\right) \exp(iNkx)\\
&= \sum_{k=0}^{N-1}\sum_{n=0}^{N-1} \exp i\left(\pi k^2 + \frac{\pi n^2}{N^2} + nx + \frac{2\pi kn}{N} + Nkx\right)\\
&= \sum_{k=0}^{N-1}\sum_{n=0}^{N-1} \exp i\pi\left(\frac{Nk+n}{N}\right)^2 \exp ix(Nk+n)\\
&= \sum_{j=0}^{N^2-1} \exp\left(\frac{i\pi n^2}{N}\right) \exp(inx)\ .
\end{aligned}$$

Proposition 1. *Suppose P_N belongs to the unimodular class W_N. If the defining constants c_k are a permutation of $\{0, \ldots, N-1\}$ then*

$$\|P_N\|_\infty \leq N^{\frac{3}{2}}.$$

Proof: If $Q_N(x) = \sum_{n=0}^{N-1} q_n \exp(inx)$ then it is easily verified that

$$\frac{1}{N}\sum_{k=0}^{N-1} Q_N\left(x + \frac{2\pi k}{N}\right) = q_0. \tag{1}$$

Replacing Q_N by $|Q_N|^2$ gives

$$\frac{1}{N}\sum_{k=0}^{N-1} \left|Q_N\left(x + \frac{2\pi k}{N}\right)\right|^2 = \sum_{n=0}^{N-1} |q_n|^2. \tag{2}$$

For $P_N \in W_N$ we have, using the Schwarz inequality,

$$\left| \sum_{k=0}^{N-1} a_k Q_N \left(x + \frac{2\pi c_k}{N} \right) e^{iNkx} \right|$$

$$\leq \left(\sum_{k=0}^{N-1} \left| a_k e^{iNkx} \right|^2 \right)^{\frac{1}{2}} \left(\sum_{k=0}^{N-1} \left| Q_N \left(x + \frac{2\pi c_k}{N} \right) \right|^2 \right)^{\frac{1}{2}}.$$

Since $|a_k| = 1$, the first sum equals N. From the second equation and the fact that the c_k are a permutation of $\{0, \ldots, N-1\}$, the second sum equals N^2. Therefore $|P_N(x)| \leq N^{\frac{3}{2}}$. ∎

Proposition 2. *Suppose P_N belongs to the unimodular class W_N. Let c_k be a permutation of $\{0, \ldots, N-1\}$, and let $a_k = 1$ for all k. Then*

$$\left| P_N \left(\frac{2\pi j}{N} \right) \right| = N \quad \textit{for } j = 0, \ldots, N-1.$$

Proof:

$$\begin{aligned} P_N \left(\frac{2\pi j}{N} \right) &= \sum_{k=0}^{N-1} Q_N \left(\frac{2\pi (j + c_k)}{N} \right) \\ &= \sum_{k=0}^{N-1} \sum_{n=0}^{N-1} q_n \exp \left(\frac{2\pi i (j + c_k) n}{N} \right) \\ &= \sum_{n=0}^{N-1} q_n \left(\sum_{k=0}^{N-1} \exp \left(\frac{2\pi i k n}{N} \right) \right) \exp \left(\frac{2\pi i j n}{N} \right) \end{aligned}$$

where the inside sum has been reindexed. Since the inner sum vanishes unless $n = 0$, we get

$$P_N \left(\frac{2\pi j}{N} \right) = N q_0. \; ∎$$

Proposition 3. *Suppose P_N is in the unimodular class W_N. Let c_k be a permutation of $\{0, \ldots, N-1\}$ and let $q_n = \exp(2\pi i n m / N)$ for some fixed integer $m \in \{0, \ldots, N-1\}$. Then*

$$\left| P_N \left(\frac{2\pi j}{N} \right) \right| = N \quad \textit{for } j = 0, \ldots, N-1.$$

Proof:

$$P_N \left(\frac{2\pi j}{N} \right) = \sum_{k=0}^{N-1} a_k Q_N \left(\frac{2\pi (j + c_k)}{N} \right)$$

and

$$Q_N\left(\frac{2\pi(j+c_k)}{N}\right)=\sum_{n=0}^{N-1}\exp\frac{2\pi in}{N}(j+m+c_k).$$

This sum vanishes unless $c_k \equiv -(j+m) \bmod N$. Since there is exactly one $k \in \{0,\ldots,N-1\}$ for which this congruence holds, say k_0, we have

$$P_N\left(\frac{2\pi j}{N}\right)=a_{k_0}N. \blacksquare$$

Let us now focus on the polynomials in W_N for which $a_k = 1$ and $q_n = 1$, and the c_k are a permutation of $\{0,\ldots,N-1\}$. For $\sigma \in S_N$ (the group of permutations of $\{0,\ldots,N-1\}$), let

$$f_\sigma(x)=\sum_{k=0}^{N-1}D_N\left(x+\frac{2\pi\sigma(k)}{N}\right)e^{iNkx}$$

where

$$D_N(x)=\sum_{n=0}^{N-1}e^{inx}.$$

Proposition 4. *For $\sigma \in S_N$, and f_σ as defined above*

$$\|f_\sigma\|_\infty \le 3N + N\log N.$$

Proof: We have

$$|f_\sigma(x)| \le \sum_{k=0}^{N-1}\left|D_N\left(x+\frac{2\pi k}{N}\right)\right|.$$

Let I_k denote the half open interval $[2\pi k/N, 2\pi(k+1)/N)$. Since the right hand member of the previous inequality is periodic of period $2\pi/N$,

$$\begin{aligned}\|f_\sigma\|_\infty &\le \sup_{x\in I_0}\sum_{k=0}^{N-1}\left|D_n\left(x+\frac{2\pi k}{N}\right)\right| \\ &\le \sum_{k=0}^{N-1}\sup_{x\in I_k}|D_N(x)|.\end{aligned}$$

Using the simple estimate that $\sin(x/2) > (x/\pi)$ for $0 \le x \le \pi$, we have

$$|D_N(x)| = \frac{\sin\frac{1}{2}Nx}{\sin\frac{1}{2}x} \le \begin{cases} N & \text{if } |x| \le \dfrac{\pi}{N} \\ \dfrac{\pi}{x} & \text{if } \dfrac{\pi}{N} \le |x| \le \pi. \end{cases}$$

This gives

$$\|f_\sigma\|_\infty \le 2\left(N + \pi \sum_{k=1}^{[N/2]+1} \frac{N}{2\pi k}\right)$$

where [ ] denotes the greatest whole integer function. Thus

$$\|f_\sigma\|_\infty \le 2\left(N + \left(\frac{N}{2}\right)(1+\log N)\right)$$

which gives the result. ▮

Proposition 5. *There exist $\sigma \in S_N$ such that*

$$\|f_\sigma\|_\infty \ge CN\log N$$

where C is a constant.

Proof: Consider the permutation

$$\sigma^{-1} = 0\ 2\ 4\ \ldots\ N-2\ N-1\ \ldots\ 3\ 1 \qquad \text{for even } N$$

and

$$\sigma^{-1} = 0\ 2\ 4\ \ldots\ N-1\ N-2\ \ldots\ 3\ 1 \qquad \text{for odd } N.$$

We treat the case where N is even. The odd case is done similarly. Let us write $f_\sigma = f_1 + f_2$ where

$$f_1(x) = \sum_{k=0}^{(N/2)-1} D_N\left(x + \frac{2\pi k}{N}\right) e^{i2kNx}$$

$$f_2(x) = \sum_{k=N/2}^{N-1} D_N\left(x + \frac{2\pi k}{N}\right) e^{i(2(N-k)-1)Nx}$$

$$= e^{i(2N-1)Nx} \sum_{k=N/2}^{N-1} D_N\left(x + \frac{2\pi k}{N}\right) e^{-i2Nkx}$$

Suppose $x = \pi/N$. We have

$$f_1\left(\frac{\pi}{N}\right) = \sum_{k=0}^{(N/2)-1} D_N\left(\frac{\pi}{N} + \frac{2\pi k}{N}\right)$$

$$f_2\left(\frac{\pi}{N}\right) = -\sum_{k=N/2}^{N-1} D_N\left(\frac{\pi}{N} + \frac{2\pi k}{N}\right).$$

Since

$$D_N(x) = \frac{\sin \frac{1}{2}Nx}{\sin \frac{1}{2}x} e^{i\frac{1}{2}(N-1)x}$$

we have

$$\begin{aligned} f_1\left(\frac{\pi}{N}\right) &= e^{i\frac{1}{2}\pi(1-\frac{1}{N})} \sum_{k=0}^{(N/2)-1} \frac{e^{-i\pi k/N}}{\sin \frac{\pi}{N}\left(k+\frac{1}{2}\right)} \\ f_2\left(\frac{\pi}{N}\right) &= -e^{i\frac{1}{2}\pi(1-\frac{1}{N})} \sum_{k=N/2}^{N-1} \frac{e^{-i\pi k/N}}{\sin \frac{\pi}{N}\left(k+\frac{1}{2}\right)}. \end{aligned}$$

Changing the index of summation in this last sum by putting $l = N - 1 - k$, we find

$$f_2\left(\frac{\pi}{N}\right) = -e^{-i\frac{1}{2}\pi(1-\frac{1}{N})} \sum_{l=0}^{(N/2)-1} \frac{e^{-i\pi l/N}}{\sin \frac{\pi}{N}\left(l+\frac{1}{2}\right)} = -\overline{f_1\left(\frac{\pi}{N}\right)}.$$

Thus $f_\sigma(\pi/N) = 2i\Im f_1(\pi/N)$. By a simple calculation

$$\Im f_1\left(\frac{\pi}{N}\right) = \sum_{k=0}^{(N/2)-1} \cot \frac{\pi}{N}\left(k+\frac{1}{2}\right).$$

Since cot is a decreasing function on $(0, \pi)$, we have for $k = 0, \ldots, (N/2) - 1$,

$$\cot \frac{\pi}{N}\left(k+\frac{1}{2}\right) > \frac{N}{\pi} \int_{\pi\left(k+\frac{1}{2}\right)/N}^{\pi\left(k+\frac{3}{2}\right)/N} \cot x \; dx \; .$$

Hence

$$\begin{aligned} \Im f_1\left(\frac{\pi}{N}\right) &> \frac{N}{\pi} \int_{\pi/2N}^{\pi(N+1)/2N} \cot x \; dx \\ &= \frac{N}{\pi}\left(\log\left(\cos\frac{\pi}{2N}\right) - \log\left(\sin\frac{\pi}{2N}\right)\right) \\ &> CN \log N \end{aligned}$$

for some constant C. This proves the proposition. ∎

When σ is the identity permutation, f_σ is the Byrnes polynomial and has $\|f_\sigma\|_\infty < CN$ for some constant C. The question arises: what is the distribution of $\|f_\sigma\|_\infty$ as σ runs through S_N? This is a difficult question which we have not been able to answer. The modulus of a typical f_σ is plotted in figure 3 for $N = 10$.

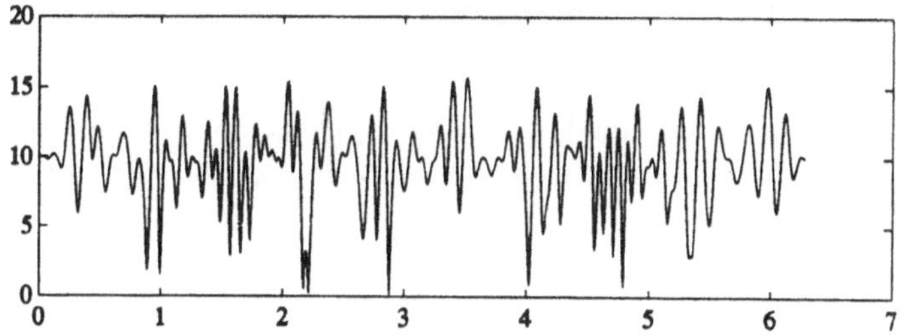

Figure 3. The modulus of $|f_\sigma(x)|$ for $N = 10$ and $\sigma =$ 4591672083 plotted for 1024 evenly spaced values of x between 0 and 2π.

For fixed x, we can get some information concerning the distribution of $f_\sigma(x)$. Let us consider S_N with the uniform discrete probability measure. Letting $\mathcal{E}$ denote expectation, we have the following result.

Proposition 6. *For each fixed x,*

$$\mathcal{E}(f_\sigma(x)) = D_N(Nx)$$

and

$$\mathcal{E}(|f_\sigma(x)|^2) = N^2$$

Proof: For a fixed k, as σ runs through S_N, $\sigma(k)$ assumes every value in the set $\{0, 1, \ldots, N-1\}$ exactly $(N-1)!$ times. Therefore

$$\sum_{\sigma\in S_N} D_N\left(x + \frac{2\pi\sigma(k)}{N}\right) = (N-1)! \sum_{j=0}^{N-1} D_N\left(x + \frac{2\pi j}{N}\right).$$

Using (1) in the proof of Proposition 1, we have

$$\sum_{j=0}^{N-1} D_N\left(x + \frac{2\pi j}{N}\right) = N$$

and therefore

$$\mathcal{E}\left(D_N\left(x + \frac{2\pi\sigma(k)}{N}\right)\right) = 1.$$

It follows that

$$\mathcal{E}(f_\sigma(x)) = \sum_{k=0}^{N-1} \mathcal{E}\left(D_N\left(x + \frac{2\pi\sigma(k)}{N}\right)\right) e^{iNkx} = D_N(Nx).$$

Next consider

$$\mathcal{E}\left(\sum_{k\neq j} D_N\left(x+\frac{2\pi\sigma(k)}{N}\right)\overline{D_N\left(x+\frac{2\pi\sigma(j)}{N}\right)}\right).$$

Since for fixed k and j not equal, $\sigma(k)$ and $\sigma(j)$ take on any fixed (unequal) pair of values exactly $(N-2)!$ times as σ runs through S_N, this expectation equals

$$\frac{(N-2)!}{N!}\sum_{k\neq j} D_N\left(x+\frac{2\pi k}{N}\right)\overline{D_N\left(x+\frac{2\pi j}{N}\right)}.$$

Consider the equation

$$\left|\sum_{k=0}^{N-1} D_N\left(x+\frac{2\pi k}{N}\right)\right|^2 = \sum_{k=0}^{N-1}\left|D_N\left(x+\frac{2\pi k}{N}\right)\right|^2 + \sum_{k\neq j} D_N\left(x+\frac{2\pi k}{N}\right)\overline{D_N\left(x+\frac{2\pi j}{N}\right)}.$$

By (1) and (2) in the proof of Proposition 1 this equation becomes

$$N^2 = N^2 + \sum_{k\neq j} D_N\left(x+\frac{2\pi k}{N}\right)\overline{D_N\left(x+\frac{2\pi j}{N}\right)}.$$

Therefore

$$\mathcal{E}\left(\sum_{k\neq j} D_N\left(x+\frac{2\pi\sigma(k)}{N}\right)\overline{D_N\left(x+\frac{2\pi\sigma(j)}{N}\right)}\right) = 0. \tag{1}$$

Next consider

$$\mathcal{E}\left(\sum_{k=0}^{N-1}\left|D_N\left(x+\frac{2\pi\sigma(k)}{N}\right)\right|^2\right) = \sum_{k=0}^{N-1}\frac{1}{N!}\sum_{\sigma\in S_N}\left|D_N\left(x+\frac{2\pi\sigma(k)}{N}\right)\right|^2$$

which by the same argument used in the beginning of this proof equals

$$\sum_{k=0}^{N-1}\frac{(N-1)!}{N!}\sum_{j=0}^{N-1}\left|D_N\left(x+\frac{2\pi j}{N}\right)\right|^2.$$

Using (2) from the proof of Proposition 1 gives

$$\mathcal{E}\left(\sum_{k=0}^{N-1}\left|D_N\left(x+\frac{2\pi\sigma(k)}{N}\right)\right|^2\right) = \sum_{k=0}^{N-1}\frac{1}{N}N^2 = N^2. \tag{2}$$

Finally consider

$$\mathcal{E}\left(|f_\sigma(x)|^2\right) = \sum_{k=0}^{N-1} \mathcal{E}\left(\left|D_N\left(x+\frac{2\pi\sigma(k)}{N}\right)\right|^2\right)$$
$$+\sum_{k\neq j} \mathcal{E}\left(D_N\left(x+\frac{2\pi\sigma(k)}{N}\right)\overline{D_N\left(x+\frac{2\pi\sigma(j)}{N}\right)}\right) e^{iN(k-j)x}.$$

Using (1) and (2) then gives $\mathcal{E}\left(|f_\sigma(x)|^2\right) = N^2$ which proves the proposition. ∎

If $\sigma(k)$ is regarded as a random variable on the symmetric group S_N, then $\{\sigma(0), \ldots, \sigma(N-1)\}$ is not a set of independent random variables. This contributes to the difficulty in analysing the distribution of $\|f_\sigma\|_\infty$. We will therefore consider the random polynomials

$$P_N(x) = \sum_{k=0}^{N-1} D_N\left(x+\frac{2\pi R_k}{N}\right) e^{iNkx}$$

where $R_0, \ldots, R_{N-1}$ are independent uniformly distributed random variables on $[0, N]$. In order to prove the main theorem we need several lemmas.

Lemma 1. *For $k \geq 2$ and $N \geq 1$*

$$\frac{1}{2\pi}\int_0^{2\pi} |D_N(x) - 1|^k \, dx \leq 2N^{k-1}.$$

Proof: We have

$$|D_N(x) - 1| = \frac{\sin\frac{1}{2}(N-1)x}{\sin\frac{1}{2}x} \leq \begin{cases} (N-1) & \text{if } |x| \leq \dfrac{\pi}{N-1} \\ \dfrac{\pi}{x} & \text{if } \dfrac{\pi}{N-1} \leq |x| \leq \pi. \end{cases}$$

Therefore

$$\frac{1}{2\pi}\int_0^{2\pi} |D_N(x)-1|^k dx \leq \frac{1}{\pi}\int_0^{\pi/(N-1)} (N-1)^k dx$$
$$+\frac{1}{\pi}\int_{\pi/(N-1)}^{\pi} \left(\frac{\pi}{x}\right)^k dx < 2N^{k-1}. \blacksquare$$

Lemma 2. *For $N \geq 2$ and $x \leq 3/2$*

$$1 + \frac{1}{N}(e^x - x - 1) \leq e^{(x^2 \log N)/N}.$$

Proof: Let us write

$$e^{(x^2 \log N)/N} - \left(1 + \frac{1}{N}(e^x - x - 1)\right) = \frac{x^2 \log N}{N}(g_N(x) - f_N(x))$$

where

$$g_N(x) = \frac{e^{(x^2 \log N)/N} - 1}{x^2 \frac{\log N}{N}}$$

and

$$f_N(x) = \frac{e^x - x - 1}{x^2 \log N}.$$

We can write

$$g_N(x) = 1 + \frac{1}{2!}\left(x^2 \frac{\log N}{N}\right) + \frac{1}{3!}\left(x^2 \frac{\log N}{N}\right)^2 + \ldots$$

from which it follows that $g_N(x) \geq 1$ for all x. Next we will show that $f_N(x)$ is monotone increasing for all x. Let $h(x) = f_N(x) \log N$, then

$$h'(x) = \frac{1}{3!} + \frac{2x}{4!} + \frac{3x^2}{5!} + \ldots \tag{1}$$

which shows that$f'_N(x) > 0$ for $x \geq 0$. Since the terms in (1) are decreasing in magnitude for $|x| \leq 2$ we have for $-2 \leq x \leq 0$

$$h'(x) \geq \frac{1}{3!} + \frac{2x}{4!} \geq 0.$$

We also have for $x \leq 0$

$$h'(x) = \frac{e^x + 1}{x^2} + \frac{2(1 - e^x)}{x^3} \geq \frac{1}{x^2}\left(1 + \frac{2}{x}\right).$$

It follows that $f'_N(x) \geq 0$ for $x \leq -2$. Hence $f_N(x)$ is montone increasing for all x.

For $N \geq 3$ we have

$$f_N(\frac{3}{2}) \leq \frac{e^{\frac{3}{2}} - \frac{3}{2} - 1}{\frac{9}{4}\log 3} < 1$$

from which it follows that for $x \leq \frac{3}{2}$

$$g_N(x) - f_N(x) \geq 1 - f_N\left(\frac{3}{2}\right) > 0$$

and the result follows for $N \geq 3$.

For $N = 2$, the result can also be verified by elementary methods augmented by some detailed computations. ▮

Lemma 3. *If $f(x) = \sum_{n=0}^{N-1} a_n e^{inx}$, then there is an interval I of length $\geq \frac{3}{2N}$ such that $|f(x)| \geq \frac{1}{2}\|f\|_\infty$ for $x \in I$.*

Proof: Let x_0 be such that $|f(x_0)| = \|f\|_\infty$, let $g(x) = |f(x)|^2$ and let $I = (x_0 - \frac{3}{4N}, x_0 + \frac{3}{4N})$. By Bernstein's inequality [10; p11], $\|g'\|_\infty \leq N\|g\|_\infty$. Then for $x \in I$

$$\|g\|_\infty - g(x) = g(x_0) - g(x) \leq \|g'\|_\infty |x - x_0| \leq N\|g\|_\infty \frac{3}{4N}.$$

Hence $\frac{1}{4}\|g\|_\infty \leq g(x)$ which gives $|f(x) \geq \frac{1}{2}\|f\|_\infty$ for $x \in I$. ▮

The next lemma is a variant of Theorem 1 in [5; p68].

Lemma 4. *Let (E, μ) be a positive finite measure space. Let $(\Omega, \mathcal{P})$ be a probability space. Let $\{f_n\}_{n=1}^N$ be a set of complex measurable functions on E. Let $\{\xi_n\}_{n=1}^N$ be a set of complex measurable functions on $\Omega \times E$, such that for each $x \in E$, $\{\xi_n(\cdot, x)\}_{n=1}^N$ is a set of independent random variables satisfying $\mathcal{E}(\xi_n(\cdot, x)) = 0$. Define*

$$P(\omega, x) = \sum_{n=1}^N \xi_n(\omega, x) f_n(x).$$

Suppose the following conditions hold:
(i) There exists a $\rho > 0$ such that for all $\omega \in \Omega$,

$$\mu\{x \in E \,:\, |P(\omega, x)| \geq \frac{1}{2}\, \|P(\omega, \cdot)\|_\infty\} \geq \frac{1}{\rho}\mu(E).$$

(ii) There exist positive numbers K and ν such that for all $x \in E$ and for $|\zeta| \leq C\sqrt{(\log 2\rho\nu)/Kr}$

$$\mathcal{E}\left(e^{\zeta \Re \xi_n(\cdot, x)}\right) \leq e^{\zeta^2 K}$$

$$\mathcal{E}\left(e^{\zeta \Im \xi_n(\cdot, x)}\right) \leq e^{\zeta^2 K}$$

where $C = \max \|f_n\|_\infty$ for $n = 1, \ldots, N$ and $r = \sum_{n=1}^N \|f_n\|_\infty^2$.
Then

$$\mathcal{P}\left(\|P(\omega, \cdot)\|_\infty \geq 8\left(K \sum_{n=1}^N \|f_n\|_\infty^2 \log(2\rho\nu)\right)^{\frac{1}{2}}\right) \leq \frac{4}{\nu}.$$

Proof: Suppose first that f_n and ξ_n are real. Let $\lambda = \sqrt{(\log 2\rho\nu)/Kr}$ and consider

$$\mathcal{E}\left(e^{\lambda P(\cdot,x)}\right) = \mathcal{E}\left(\prod_{n=1}^{N} e^{\lambda\xi_n(\cdot,x)f_n(x)}\right) = \prod_{n=1}^{N} \mathcal{E}\left(e^{\lambda\xi_n(\cdot,x)f_n(x)}\right).$$

Using condition (ii) of the hypothesis with $\zeta = \lambda f_n(x)$ we have

$$\mathcal{E}\left(e^{\lambda P(\cdot,x)}\right) \le \prod_{n=1}^{N} e^{\lambda^2 K f_n^2(x)} = e^{\lambda^2 K \sum_{n=1}^{N} f_n^2(x)} \le e^{\lambda^2 K r}.$$

Let $M(\omega) = \|P(\omega,\cdot)\|_\infty$ and define the random sets $I_1(\omega)$ and $I_2(\omega)$ by

$$\begin{aligned} I_1(\omega) &= \{x \in E \ : \ P(\omega,x) \ge M(\omega)/2\} \\ I_2(\omega) &= \{x \in E \ : \ P(\omega,x) \le -M(\omega)/2\}. \end{aligned}$$

Then for each $\omega \in \Omega$

$$e^{\lambda M(\omega)/2}(\mu(I_1(\omega)) + \mu(I_2(\omega))) \le \int_{I_1} e^{\lambda P(\omega,x)}d\mu(x) + \int_{I_2} e^{-\lambda P(\omega,x)}d\mu(x).$$

Hence

$$e^{\lambda M(\omega)/2} \le \frac{1}{\mu(I_1(\omega)) + \mu(I_2(\omega))}\left(\int_{I_1} e^{\lambda P(\omega,x)}d\mu(x) + \int_{I_2} e^{-\lambda P(\omega,x)}d\mu(x)\right).$$

Using condition (i) of the hypothesis gives

$$e^{\lambda M(\omega)/2} \le \frac{\rho}{\mu(E)}\int_E e^{\lambda P(\omega,x)} + e^{-\lambda P(\omega,x)}d\mu(x).$$

Now taking expectations and using the above estimate on $\mathcal{E}\left(e^{\lambda P(\cdot,x)}\right)$ gives

$$\mathcal{E}\left(e^{\lambda M/2}\right) \le 2\rho e^{\lambda^2 K r} = \frac{1}{\nu}\exp\left(\log 2\rho\nu + \lambda^2 K r\right).$$

This inequality can be expressed as

$$\mathcal{E}\left(\exp\frac{\lambda}{2}\left[M - \frac{2}{\lambda}\log 2\rho\nu - 2\lambda K r\right]\right) \le \frac{1}{\nu}.$$

Denoting the term in the square brackets by B, we have $\mathcal{P}(\exp\frac{\lambda}{2}B \ge 1) \le \frac{1}{\nu}$, or $\mathcal{P}(B \ge 0) \le \frac{1}{\nu}$. Substituting $\sqrt{(\log 2\rho\nu)/Kr}$ for λ gives

$$\mathcal{P}\left(M \ge 4(Kr\log 2\rho\nu)^{\frac{1}{2}}\right) \le \frac{1}{\nu}.$$

If f_n is complex and ξ_n is real, we have

$$\mathcal{P}\left(\|\Re P\|_\infty \geq 4\left(K\sum_{n=1}^{N}\|\Re f_n\|_\infty^2 \log 2\rho\nu\right)^{\frac{1}{2}}\right) \leq \frac{1}{\nu}$$

and

$$\mathcal{P}\left(\|\Im P\|_\infty \geq 4\left(K\sum_{n=1}^{N}\|\Im f_n\|_\infty^2 \log 2\rho\nu\right)^{\frac{1}{2}}\right) \leq \frac{1}{\nu}.$$

Hence

$$\mathcal{P}\left(\|P\|_\infty \geq 4\left(K\sum_{n=1}^{N}\|f_n\|_\infty^2 \log 2\rho\nu\right)^{\frac{1}{2}}\right) \leq \frac{2}{\nu}.$$

This follows since if $\|P\|_\infty^2 \geq |z|^2$ for some complex number z, then $\|\Re P\|_\infty^2 \geq |\Re z|^2$ or $\|\Im P\|_\infty^2 \geq |\Im z|^2$.

Next suppose that f_n and ξ_n are both complex. Let

$$R(\omega, x) = \sum_{n=1}^{N} \Re\xi_n(\omega, x) f_n(x)$$

and

$$S(\omega, x) = \sum_{n=1}^{N} \Im\xi_n(\omega, x) f_n(x).$$

Note that if $\|R+iS\|_\infty \geq 2\gamma$ for some positive γ then either $\|R\|_\infty \geq \gamma$ or $\|S\|_\infty \geq \gamma$. Hence we have

$$\mathcal{P}\left(\|R\|_\infty \geq 4\left(K\sum_{n=1}^{N}\|f_n\|_\infty^2 \log 2\rho\nu\right)^{\frac{1}{2}}\right) \leq \frac{2}{\nu}$$

and

$$\mathcal{P}\left(\|S\|_\infty \geq 4\left(K\sum_{n=1}^{N}\|f_n\|_\infty^2 \log 2\rho\nu\right)^{\frac{1}{2}}\right) \leq \frac{2}{\nu}$$

which implies

$$\mathcal{P}\left(\|P\|_\infty \geq 8\left(K\sum_{n=1}^{N}\|f_n\|_\infty^2 \log 2\rho\nu\right)^{\frac{1}{2}}\right) \leq \frac{4}{\nu}$$

This completes the proof of the lemma. ∎

Theorem. *Let*

$$Q_N(\omega, x) = \sum_{k=0}^{N-1} D_N\left(x + \frac{2\pi R_k(\omega)}{N}\right) e^{iNkx} - D_N(Nx)$$

where $R_0, \ldots, R_{N-1}$ are independent random variables with uniform distributions on $[0, N]$. For $N \geq 2$, and any $\beta \geq 24/\sqrt{2}$

$$\mathcal{P}(\|Q_N\|_\infty \geq \beta N \log N) \leq \frac{32\pi}{3} N^{2-(3\sqrt{2}\beta/32)}.$$

Proof: We apply Lemma 4 as follows. Let $E = [0, 2\pi)$ with μ being Lebesgue measure. Let

$$\xi_{k+1}(\omega, x) = D_N\left(x + \frac{2\pi R_k(\omega)}{N}\right) - 1$$

and $f_{k+1}(x) = e^{iNkx}$. Then for each ω, $P(\omega, x)$ has the form $\sum_{n=0}^{N^2-1} a_n e^{inx}$, $C = 1$, $r = N$, and $\mathcal{E}(\xi_k(\cdot, x)) = 0$ for all x and k. Using Lemma 3 we have condition (i) of Lemma 4 satisfied with $\rho = 4\pi N^2/3$. Since $R_0, \ldots, R_{N-1}$ are independent uniformly distributed on $[0, N]$

$$\mathcal{E}\left(|\xi_k(\cdot, x)|^j\right) = \frac{1}{N}\int_0^N \left|D_N\left(x + \frac{2\pi y}{N}\right) - 1\right|^j dy.$$

Therefore by Lemma 1

$$\mathcal{E}\,|\xi_k(\cdot, x)|^j \leq 2N^{j-1}.$$

Using this estimate, Lemma 2, and the fact that $\mathcal{E}\left(\xi_k(\cdot, x)\right) = 0$ we have

$$\begin{aligned}
\mathcal{E}\left(e^{\lambda \Re \xi_k(\cdot, x)}\right) &= 1 + \frac{\lambda^2}{2!}\mathcal{E}\left((\Re\xi_k(\cdot, x))^2\right) + \frac{\lambda^3}{3!}\mathcal{E}\left((\Re\xi_k(\cdot, x))^3\right) + \cdots \\
&\leq 1 + |\lambda|^2 \frac{2N}{2!} + |\lambda|^3 \frac{2N^2}{3!} + \cdots \\
&\leq 1 + \frac{1}{N}\left(e^{\sqrt{2}|\lambda| N} - \sqrt{2}|\lambda| N - 1\right) \\
&\leq e^{2\lambda^2 N \log N}
\end{aligned}$$

for $\sqrt{2}|\lambda|N \leq 3/2$. Let $K = 2\alpha N \log N$ for any $\alpha \geq 1$. Then

$$\mathcal{E}\left(e^{\lambda \Re \xi_k(\cdot, x)}\right) \leq e^{\lambda^2 K} \tag{1}$$

and by the same argument

$$\mathcal{E}\left(e^{\lambda\Im\xi_k(\cdot,x)}\right) \le e^{\lambda^2 K} \tag{2}$$

for $N \ge 2$ and $\sqrt{2}|\lambda|N \le 3/2$. Let $\nu = 3N^{(9\alpha/4)-2}/8\pi$ then a straightforward calculation gives $\sqrt{\log 2\rho\nu/Kr} = 3/(2\sqrt{2}N)$. Since (1) and (2) hold for $|\lambda| \le 3/(2\sqrt{2}N)$, they hold for

$$|\lambda| \le C\left(\frac{\log 2\rho\nu}{Kr}\right)^{\frac{1}{2}}$$

so that hypothesis (ii) of Lemma 4 is satisfied. Consequently

$$\mathcal{P}\left(\|Q_N\|_\infty \ge 8(Kr\log 2\rho\nu)^{\frac{1}{2}}\right) \le \frac{4}{\nu}.$$

Making the appropriate substitutions for K, r, ρ and ν gives

$$\mathcal{P}(\|Q_N\|_\infty \ge \frac{24\alpha}{\sqrt{2}}N\log N) \le \frac{32\pi}{3}N^{2-(9\alpha/4)}.$$

Finally letting $\beta = 24\alpha/\sqrt{2}$ gives the result. ∎

REFERENCES

[1] E. Beller and D.J. Newman, "The minimum modulus of polynomials," *Proc. Amer. Math. Soc.* **45** number 3 (1974) 463–465.

[2] G. Benke, "On the minimum modulus of trigonometric polynomials," to be published

[3] J.S. Byrnes, "On polynomials with coefficients of modulus one," *Bull. London Math. Soc.* **36** (1961) 171–176.

[4] J.-P. Kahane, "Sur les polynomes a coefficients unimodulaires," *Bull. London Math. Soc.* **12** (1960) 321–342.

[5] J.-P. Kahane, "Some random series of functions (second edition)," *Cambridge University Press* (1985)

[6] T. Körner, "On a polynomial of J.S. Byrnes," *Bull. London Math. Soc.* **12** (1980) 219–224.

[7] J.E. Littlewood, "On the mean values of certain trigonometrical polynomials," *J. London Math. Soc.* **36** (1961) 307–334.

[8] W.G. Rudin, "Some theorems on Fourier Coefficients," *Proc. Amer. Math. Soc.* **10** (1959) 855–859.

[9] H.S. Shapiro, "Extremal problems for polynomials and power series," *Thesis M.I.T.* (1951)

[10] A. Zygmund, "Trigonometric Series II (second edition)," *Cambridge University Press* (1968)

NUMBER THEORY AND FOURIER ANALYSIS APPLICATIONS IN PHYSICS, ACOUSTICS AND COMPUTER SCIENCE*

M.R. SCHROEDER
Drittes Physikalisches Institut
Universität Göttingen
Bürgerstraße 42–44
D–3400 Göttingen
F.R. Germany

ABSTRACT. Number theory is traditionally considered a rather abstract field, far removed from practical applications. In the recent past, however, the "higher arithmetic" has provided highly useful answers to numerous real–world problems. Many of these uses depend on the special correlation and Fourier Transform properties of certain real and complex sequences derived from different branches of number theory, particularly finite fields and quadratic residues. The applications include the design of new musical scales, powerful cryptographic systems, and diffraction gratings for acoustic and electromagnetic waves with unusually broad scatter, with applications in radar camouflage, laser speckle removal, noise abatement, and concert hall acoustics. Another prime domain of number theory is the construction of very effective error–correction codes, such as those used for picture transmission from space vehicles and in compact discs (CDs). Other new uses include schemes for spread–spectrum communication, "error–free" computing, fast computational algorithms, and precision measurements (of interplanetary distances, for example) at extremely low signal–to–noise ratios. In this manner the "fourth prediction" of General Relativity (the slowing of electromagnetic radiation in gravitation fields, predicted by Einstein as early as 1907) has been fully confirmed. In contemporary physics the quasiperiodic route to chaos of nonlinear dynamical systems (the double–pendulum and the three–body problem, to mention two simple examples) are being analyzed in terms of such number-

---

*Based on the authors book *Number Theory in Science and Communication*, (Springer-Verlag, 1986)

J. S. Byrnes and J. F. Byrnes (eds.), Recent Advances in Fourier Analysis and Its Applications, 101–129.

theoretic concepts as continued fractions, Fibonacci numbers, the golden mean and Farey trees. Even the recently discovered new state of matter, christened quasicrystals, is most effectively described in terms of arithmetic principles. And last not least, prime numbers, whose distribution combines predictable regularity and surprising randomness, are a rich source of pleasing artistic design - either directly or through the Fourier Transform.

1. Introduction

Number theory has been considered since time immemorial to be the very paradigm of pure (some would say useless) mathematics. According to Carl Friedrich Gauss, the "Princeps Mathematicorum", "mathematics is the queen of sciences – and number theory is the queen of mathematics". What could be more beautiful than a deep, satisfying relation between whole numbers? (One is almost tempted to call them whole*some* numbers.) Indeed, it is hard to come up with a more appropriate designation than their learned name: the integers, meaning the "untouched ones". How high they rank, in the realms of pure thought and aesthetics, above their lesser brethren: the real and complex numbers, whose first names virtually exude unsavory involvement with the complex realities of everyday life!

Yet the theory of integers can provide totally unexpected answers to real-world problems. In fact, discrete mathematics is taking on an ever more important role. If nothing else, the advent of the digital computer and digital communication has seen to that. But even earlier, in physics, the emergence of quantum mechanics and discrete elementary particles put a premium on the methods and, indeed, the spirit of discrete mathematics.

In mathematics proper, Hermann Minkowski, in the preface to his introductory book on number theory, *Diophantische Approximationen*, published in 1907 (the year he gave special relativity its proper four–dimensional clothing in preparation for its journey into general covariance and cosmology) expressed his conviction that the "deepest interrelationships in analysis are of an arithmetical nature". Or, on another occasion: "The primary source (Urquell) of all of mathematics are the integers".

Yet much of our schooling concentrates on analysis and other branches of continuum mathematics to the virtual exclusion of number theory, group theory, combinatorics and graph theory. As an illustration, at a recent symposium on information theory, the author met several young mathematicians, working in the field of *primality testing*, who, in all their studies up to the Ph.D., had not heard a single lecture on number theory!

Or, to give an earlier example, when Werner Heisenberg discovered "matrix" mechanics in 1925, he didn't know what a matrix was (Max Born had to tell him), and neither Heisenberg nor Born knew what to make of the appearance of matrices in the context of the atom. (David Hilbert is reported to have told them to go look for a differential equation with the same eigenvalues, if that would make them happier. They did not follow Hilbert's well–meant advice and thereby may have missed discovering the Schrödinger wave equation.)

Integers have repeatedly played a crucial role in the evolution of the natural sciences. Thus, in the 18th century, Lavoisier discovered that chemical compounds are composed of fixed proportions of their constituents which, when expressed in proper weights, correspond to the ratios of *small integers*. This was one of the strongest hints to the existence of atoms; but chemists, for a long time, ignored the evidence and continued to treat atoms merely as a conceptual convenience devoid of physical meaning. (Ironically, it was from the statistical laws of *large* numbers, in Einstein's analysis of Brownian motion at the beginning of our century, that the irrefutable reality of atoms and molecules finally emerged.)

In the analysis of optical spectra, certain integer relationships between the wavelengths of spectral lines emitted by excited atoms gave early clues to the *structure of atoms*, culminating in the creation of matrix mechanics in 1925, an important year in the growth of *integer physics*.

Later, the near–integer ratios of atomic weights suggested to physicists that the *atomic nucleus* must be made up of an *integer* number of similar nucleons. The *deviations* from integer ratios led to the discovery of elemental *isotopes*.

And finally, small divergencies in the atomic weight of pure isotopes from exact integers constituted an early confirmation of Einstein's famous equation $E = mc^2$, long before the "mass defects" implied by these integer discrepancies blew up into the widely noticed and infamous mushroom clouds.

On a more harmonious theme, the role of integer ratios in musical scales has been appreciated ever since Pythagoras first pointed out their importance. The occurrence of integers in biology – from plant morphology to the genetic code – is pervasive. It has even been hypothesized that the North American 17–year cicada selected its life cycle because 17 is a prime number, prime cycles offering better protection from predators than nonprime cycles. (The suggestion that the 17–year cicada "knows" that 17 is a *Fermat* prime has yet to be touted though.)

Another reason for the resurrection of the integers is the penetration of our lives by that 20th–century descendant of the abacus, the *digital computer*. (Where did all the slide rules go? Ruled out of most significant places by the ubiquitous pocket calculator, they are sliding fast into restful oblivion.)

An equally important reason for the recent revival of the integer is the congruence of *congruential arithmetic* with numerous modern developments in the natural sciences and digital communications – especially "secure" communication by cryptographic systems. Last not least, the proper protection and security of computer systems and data files rests largely on "keys" based on congruence relationships.

In congruential arithmetic, what counts is not a numerical value per se, but rather its remainder or *residue* after division by a *modulus*. Similarly, in wave interference (be it of ripples on a lake or of electromagnetic fields on a hologram plate) it is not path differences *per se* that determine the resulting interference pattern, but rather the residues after dividing by the wavelength. For perfectly periodic events, there is no difference between a path difference of half a wavelength or one–and–a–half wavelengths: in either case the interference will be destructive.

One of the most dramatic consequences of congruential arithmetic is the existence of the chemical elements as we know them. In 1913, Niels Bohr postulated that certain integrals associated with electrons in "orbit" around the atomic nucleus should have integer values, a requirement that ten years later became comprehensible as a wave interference phenomenon of the newly discovered de Broglie *matter* waves: In essence, integer–valued integrals meant that path differences are divisible by the electron's wavelength without leaving a remainder.

2. Music and Numbers

2.1 MUSICAL SCALES

Ever since Pythagoras, small integers and their ratios have played a fundamental role in the construction of musical scales. There are good reasons for this preponderance of small integers both in the production and perception of music. String instruments, as abundant in antiquity as today, produce simple frequency ratios when their strings are subdivided into equal lengths: Shortening a string by one half produces the frequency ratio 1:2, the *Octave*; and shortening it by one third gives the frequency ratio 3:2, the perfect *Fifth*.

In perception, ratios of small integers avoid unpleasant beats between harmonics. Apart from the frequency ratio 1:1 ("Unison"), the Octave is the most easily perceived interval. Next in importance comes the perfect Fifth. Unfortunately, as a consequence of the fundamental theorem of arithmetic, musical scales exactly congruent modulo the Octave cannot be constructed from the Fifth alone because there are no positive integers k and m such that

$$\left[\frac{3}{2}\right]^m = \left[\frac{2}{1}\right]^k .$$

In fact, number theory tells us that $3^m \neq 2^n \pm 1$, except for $m = 1$, $n = 1$ and $m = 2$, $n = 3$. However, there are good *approximations*: if we write $3^m \approx 2^n$, or $\log_2 3 \approx \frac{n}{m}$. Now, as every information theorist remembers, $\log_2 3 = 1.58$; but that decimal knowledge is not very helpful. Rather, we need the *continued fraction* expansion $\log_2 3 = [1, 1, 1, 2, 2, \ldots]$, which yields the close approximation $m = 12$(!), $n = 19$. In other words, if we want to make a good Fifth with an equal tempered (equal frequency ratio) scale, the basic interval $1:2^{1/12}$, the *Semitone*, recommends itself. In fact, the Semitone interval has come to dominate much of Western music. The equal tempered Fifth comes out as $2^{7/12} = 1.498\ldots$, which approaches 3:2 with an error of only one part in one thousand!

Another fortunate number–theoretic coincidence is the fact that 7 is coprime with 12. As a consequence, we can reach all 12 notes of the octave by repeating the Fifth (modulo the Octave). This is the famous Circle of Fifths.

2.2 CONCERT HALL AND QUADRATIC RESIDUES

There is another connection between music and numbers: concert hall acoustics. Extensive physical tests and psychophysical evaluation of the acoustic qualities of concert halls around the world have established the importance of *laterally* traveling sound waves. Such waves produce dissimilar signals at a listener's two ears, a kind of stereophonic condition that is widely preferred for music listening.

In order to convert sound waves travelling longitudinally (from the stage via the ceiling to the back wall) into lateral waves, the author has recommended ceiling structures that scatter sound waves, without absorption, into broad lateral patterns. In the physicist's language, concert hall ceilings should be "reflection phase gratings" with nearly equal energies going into the different diffraction orders (lateral directions).

How should one go about designing such an ideal scatterer for sound (or light or radar) waves? Curiously, the answers come from number theory and its connection with Fourier Analysis.

3. Number–Theoretic Phase Arrays and Diffraction Gratings with Broad Radiation (Scattering) Characteristics

3.1 WHY NUMBER THEORY?

Transducer arrays are among the most useful inventions for focusing waves of all kinds: TV antennas on roof tops, strings of hydrophones in the ocean, rows of "dishes" for listening to the far end of the universe, loudspeaker columns in lecture halls, and microphone arrays on the conference table. In all of these applications the aim is to focus the waves into (or from) a single direction for better angular discrimination and improved signal–to–noise ratio. However, arrays with low or even *minimum directivity* and *gratings* that scatter an incoming wave as broadly as possible are also frequently needed. Here I shall attempt to elucidate the design principles for certain phase gratings and phase arrays that scatter or radiate waves of one or many wavelengths into broad directional patterns.

There are numerous uses for such arrays and gratings in physics and electrical engineering in which coherent waves (laser, radar, sonar, audible sound, and even electrons or neutrons) are desired to be scattered as broadly as possible. Such broad scatterers would be helpful in radar and sonar camouflage, for minimizing bothersome speckles in laser work, or for diffusing sound waves to achieve better acoustics in concert halls, recording studios, lecture halls, and even private parlors. In other applications, instead of phase gratings, phase *arrays* with broad radiation characteristics are often needed.

Curiously (but rationally) the best methods for accomplishing these goals are rooted in number theory – albeit different branches of the "higher arithmetic". In all cases, the grating and array designs described here are based on the Fourier properties of real or complex periodic sequences fashioned from quadratic residues, the Legendre symbol, primitive roots, primitive polynomials in finite number fields, the index function (or discrete logarithm), and another number-theoretic logarithm: the Zech logarithm.

The reasons for the "monopoly" that number theory seems to hold in the tasks at hand are twofold:

(1) Diffraction from gratings and radiation from arrays are governed by wave interference phenomena. In wave interference, pathlength differences that differ only by *integer multiples* of the wavelength λ are equivalent. In fact, a path difference of $n\lambda$ is equivalent to a *no* path difference at all. What counts is not the path difference *per se*, but the *residue* after subtracting an integer number of wavelengths. Thus, the appropriate mathematics to describe wave interference must be *residue arithmetic*, a branch of number theory. In residue arithmetic all numbers that leave the same remainder after division by some other number are put into the same "equivalence class." (For example, 2 and 9 are in the same equivalence class "modulo 7," i.e. they leave the same remainder after dividing by 7.)

(2) The second reason for the importance of number theory for broad scattering and radiation is the possibility of constructing periodic constant–magnitude sequences, using different principles of the theory, that have an autocorrelation function for nonzero shifts that is small compared to the value for zero shift. Sequences with this correlation property have "broad" power spectra; in fact, in most cases discussed here, all a.c. components have the *same* power. If these sequences are realized as *spatial* distributions (as antenna arrays or diffraction

gratings), then the Fourier–transform variable is *spatial frequency*, which is related to the direction of travel of the waves. Specifically, a flat power spectrum means a flat distribution of the radiated (or diffracted) power over the different directions.

It is interesting to note that low–correlation sequences are also useful for the design of low peak–factor waveforms for speech synthesizers and many other applications such as pulse compression in radar and sonar.

Some of these sequences come in "families", in which case they can also be used in the design of "signature codes" for spread–spectrum communication.

3.2 THE CONCEPT OF SPATIAL FREQUENCY

A *periodic* array radiates energy into several discrete directions α_k, given by the formula

$$L \sin \alpha_k = k\lambda, \tag{1}$$

where L is the period length of the array and $L \sin \alpha_k$ is the phase difference of wavelets from two array elements spaced one period L apart and travelling in the azimuthal direction α_k; the angle $\alpha_k = 0$ corresponds to the "broadside" direction of the array. For constructive interference to occur, this phase difference must be an integer multiple, $k\lambda$, of the wavelength λ. It is this simple "in phase" condition for wavelets travelling in the preferred directions that is expressed in Eq. (1).

In modern usage, the angle α_k is replaced by the *spatial frequency*,

$$f_k := \frac{1}{\lambda} \sin \alpha_k. \tag{2}$$

Just like temporal frequency, which is the reciprocal of the smallest nonzero time difference between instants of equal phases in a wave at a fixed point in space, so spatial frequency is the reciprocal of the smallest nonzero distance between points of equal phases in a spatial wave field at a fixed time. If the space considered has more than one dimension, the spatial frequency is not a scalar but has more than one component, which can often be combined into a spatial frequency *vector*.

The possible values for f_k are given by (1):

$$f_k = \frac{k}{L}. \tag{3}$$

This equation brings out the analogy with a periodic waveform of period T and its harmonic frequencies $f_k = k/T$. In the spatial case, the period is a length L and the spatial frequencies f_k are "harmonics" (integer multiples) of the *fundamental* spatial frequency $1/L$.

The enormously fruitful concept of spatial frequency is akin to the physicist's "wavenumber", usually measured in cm^{-1}.

There is also a simple geometric interpretation of spatial frequency: It is the number of wave crests per unit length (say meter) measured in a given direction α from the wave front. In the direction in which the wave travels ($\alpha = \pi/2$), the distance between successive crests is, by definition, λ. Thus the spatial frequency in that direction is $1/\lambda$. For other directions ($\alpha \neq \pi/2$), the crest spacing is increased to $\lambda/\sin\alpha$, an effect well known from the propagation of microwaves in waveguides or light in optical fibers. Thus the spatial frequency becomes $(1/\lambda)\sin\alpha$. in accordance with the definition Eq. (2). Since $|\sin\alpha| \leq 1$, the spatial frequency of a propagating wave is always less than or equal to the reciprocal wavelength.

3.3 QUADRATIC–RESIDUE ARRAYS

Suppose our periodic array has N equidistant radiating elements (dipoles, loudspeakers, or other kind of transducers) per period, strung out along a straight line, each element radiating wavelets with complex amplitudes a_n. Periodicity of the a_n means

$$a_n = a_{n+mN}, \tag{4}$$

where m, like n and N, is an integer. Since we want to consider *phase* arrays, we also impose the condition that $|a_n|$ be constant or, without loss of generality,

$$|a_n| = 1 \tag{5}$$

for all nonvanishing a_n. (We shall occasionally admit arrays with "missing" elements, i.e., $a_n = 0$, for one or possibly several values of n per period.)

For such arrays we can write, for all nonvanishing a_n,

$$a_n = e^{i\phi_n}, \tag{6}$$

where the (real) phase angles ϕ_n bring into evidence that we are considering *phase* arrays.

How should we choose the ϕ_n such that the radiation pattern has the same magnitude in all possible directions a_k given by Eq. (1) or, equivalently, at all spatial frequencies f_k given by Eq. (3)? Linear superposition yields, for the complex amplitude A_k in the direction α_k,

$$A_k = \sum_{n=0}^{N-1} a_n e^{-2\pi i n w \sin \alpha_k / \lambda}, \tag{7}$$

where $w = L/N$ is the spacing between adjacent elements. Replacing the angle α_k by the spatial frequency f_k, from Eqs. (2) and (3), we obtain

$$A_k = \sum_{n=0}^{N-1} a_n e^{-2\pi i n k/N}, \quad k = 0, 1,..., N-1, \tag{8}$$

i.e. the A_k are the *discrete Fourier transform* (DTF) of the a_n.

How do we choose the a_n to yield constant $|A_k|$? A well–known result from number theory relating to "Gauss sums" (see Schroeder (1986) for a simple introduction) states that properly scaled *quadratic* phases ϕ_n will produce the desired constant $|A_k|$, provided N is a *prime* number $p > 2$. With

$$\phi_n = \frac{2\pi n^2}{p}, \tag{9}$$

we obtain

$$a_n = e^{2\pi i n^2/p}, \tag{10}$$

where n^2 can be replaced by its *remainder* modulo the prime p. For example, for $p = 7$, $3^2 = 9$ can be replaced by 2 and $5^2 = 25$ by 4, etc.; it makes no difference in Eq. (10).

To show that, for this choice of ϕ_n, the Fourier coefficients A_k have constant magnitude

$$|A_k| = p, \qquad k = 0, 1, \ldots, p-1 \tag{11}$$

is straightforward although slightly tedious (unless one uses Gauss' results pertaining to "his" sums). A more generally useful method is to calculate the (periodic) correlation sequences c_n of the a_n, defined by

$$c_m := \sum_{n=0}^{p-1} a_n a^*_{n+m}\ , \tag{12}$$

where a^*_n is the conjugate complex of a_n. The power spectrum $|A_k|^2$ is then given by the Discrete Fourier Transform (DFT) of the c_m

$$|A_k|^2 = \sum_{m=0}^{p-1} c_m\, e^{-2\pi imk/p}\ . \tag{13}$$

With Eqs. (9) and (10) we obtain

$$c_m = e^{-2\pi im^2/p} \sum_{n=0}^{p-1} e^{-4\pi inm/p}\ . \tag{14}$$

For $m = 0$ or, more generally, $m \equiv 0 \bmod p$, the result is

$$c_m = p, \qquad \text{for } m \equiv 0 \bmod p \tag{15}$$

For $m \not\equiv 0 \bmod p$, the sum in Eq. (14) runs through a complete set of pth roots of 1 and therefore vanishes

$$c_m = 0 \qquad \text{for } m \not\equiv 0 \bmod p. \tag{16}$$

Now, as is well known, a periodic correlation function that vanishes everywhere except for $m = 0$ (and $m \equiv 0 \bmod p$) has a "flat", i.e. constant, power spectrum. In fact, this follows directly from Eq. (13), which yields, with c_m from Eqs. (15) and (16), $|A_k| = p$ for all k, as stated in Eq. (11).

It is interesting to note that the phases in Eq. (9) can be multiplied by an integer $r \not\equiv 0 \bmod p$ without affecting the flat–spectrum property. This is important for the design of phase *gratings* (see below) that scatter well at many different frequencies.

For *finite* arrays, the radiation characteristic becomes a *continuous* function of the spatial frequency variable f or the continuous direction angle α. Such patterns are obtained by convolving the complex sequence A_k with the properly scaled sinc–function (the Fourier transform of the rectangular window):

$$\mathrm{sinc}(fD) := \frac{\sin(\pi fD)}{\pi fD}, \tag{17}$$

where D is the *total* length of the array. Figure 1 shows the radiated energy as a function of the direction angle α, in the range $-\pi/2 \leq \alpha \leq \pi/2$, for an array with $p = 7$ having only 2 periods (thus comprising a total of 14 elements). The elements spacing is half a wavelength. Although a 2–period array is far from an infinite array, the uniformity of the 7 radiation maxima is striking. The *angles* α_k for the maxima, too, are very near the infinite–array values. From Eqs. (2) and (3), we obtain

$$\alpha_k = \arcsin\left[\frac{\lambda}{L}k\right],$$

which, for $L = 7\lambda/2$, yields the 7 angles

$$\alpha_k = 0^0, \pm 17^0, \pm 35^0, \text{ and } \pm 59^0,$$

in excellent agreement with Fig. 1.

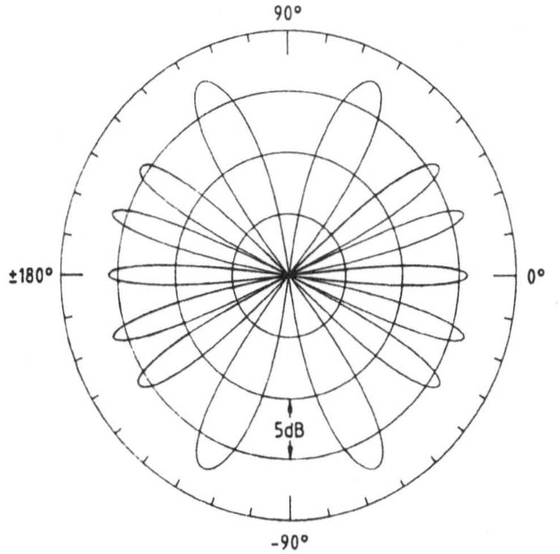

Figure 1: Radiation pattern for a quadratic–residue phase array based on the prime number $p = 7$ and comprising 14 elements (two periods). Element spacing equals half a wavelength. Note uniformity of energy flow into 14 different directions.

3.4 TWO–DIMENSIONAL QUADRATIC RESIDUE ARRAYS

Consider a two–dimensional array with equidistant element spacings, w_1 and w_2, in two orthogonal directions and complex amplitudes a_{nm}. For constant magnitudes, $|a_{nm}| = 1$, the array amplitudes are characterized by the phase angles ϕ_{nm}. For a quadratic residue array these are given by

$$\phi_{nm} = 2\pi \left[\frac{n^2}{p_1} + \frac{m^2}{p_2}\right], \tag{20}$$

where p_1 and p_2 are the same or different primes greater than 2. As before, n^2 and m^2 can be replaced by their residues modulo p_1 and p_2, respectively.

There are discrete spatial frequencies, f_{mn}, for the infinite double–periodic array are the 2–component vectors

$$f_{nm} = \left[\frac{n}{w_1 p_1}, \frac{m}{w_2 p_2}\right] . \tag{21}$$

The corresponding polar angles Θ_{nm} and azimuthal angles α_{nm} are given by

$$\sin \Theta_{nm} = \lambda |f| \tag{22}$$

and

$$\sin \alpha_{nm} = \frac{n}{|f| w_1 p_1} , \tag{23}$$

where $|f|$ is the length of spatial frequency vector f_{nm} and n and m are restricted to those integers for which $|f| \leq 1/\lambda$.

3.5 QUADRATIC–RESIDUE PHASE GRATINGS

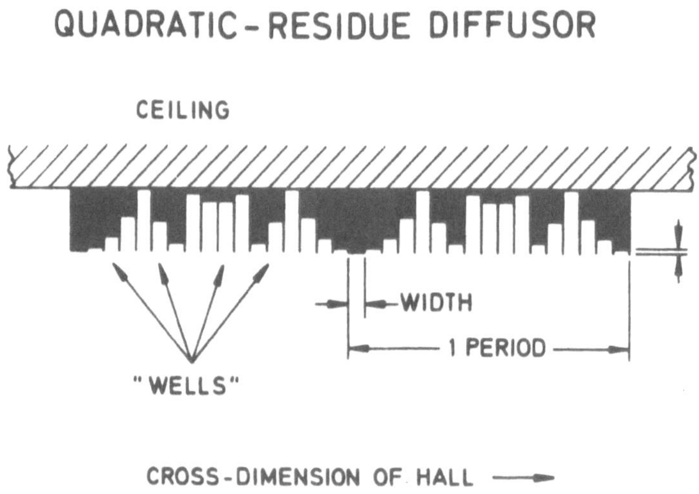

Figure 2: Quadratic–residue reflection phase-grating, based on the prime number $p = 17$, for effective scattering of waves. An incident wave of wavelength λ is reflected in the "wells", which have different depths, $d_n = n^2\lambda/2p$, causing phase shifts equal to $2\pi n^2/p$, where n^2 is taken modulo p. Note the separators between adjacent wells or "troughs". Such wave-diffusing structures are useful in radar and sonar camouflage, diffusion of coherent light, noise abatement – and concert hall acoustics.

The principles for designing quadratic–residue phase arrays are also applicable to the design of quadratic–residue phase gratings (Schroeder, 1986). The different phase angles ϕ_n are realized by the reflection of an incident wave from surfaces at different depths, see Fig. 3. In fact, a plane wave of wavelength λ, reflected from a surface recessed by d_n, suffers a phase shift equal to $\phi_n = 2d_n \cdot 2\pi/\lambda$. To achieve the phase shifts according to Eq. (9), the depths must therefore equal

$$d_n = \frac{\lambda}{2} \cdot \frac{n^2}{p}, \tag{24}$$

where, to keep the d_n as small as possible, n^2 is replaced by its residue mod p. Figure 3 shows the angular distribution of the energy of a phase incident wave diffracted from the grating, with $p = 17$, shown in Fig. 2. Note the broad scatter, which is highly desirable in many applications.

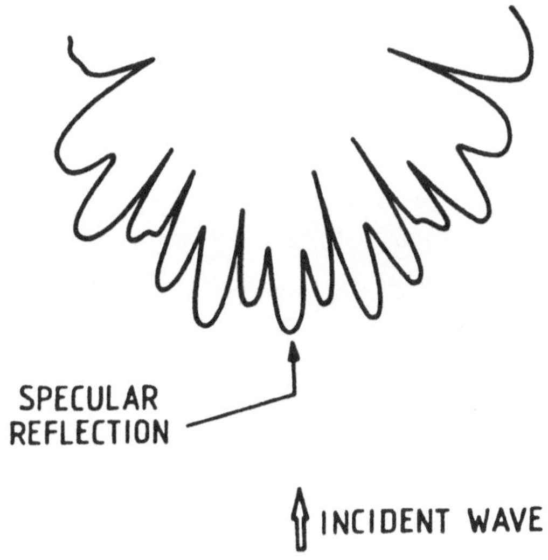

Figure 3: Diffraction pattern for the phase grating shown in Fig. 2. Note near–uniformity, within ± 3 dB, of the intensities of the different diffraction orders. The measurement was made with 3-cm microwaves using a metallic grating.

For a wavelength λ_m that is an integer fraction m of λ, $\lambda_m = \lambda/m$, the phase shift effected by the troughs of the grating are multiplied by a factor m. As long as $m \not\equiv 0 \bmod p$, the new phases will also lead to the flat–spectrum property. Thus, a reflection phase grating based on the prime number p = 17, for example, will scatter well not only at the "fundamental frequency" f_1 (corresponding to the wavelength λ) but at all integer multiples of that frequency up to $(p-1)f_1$, or $16f_1$ in the example. At pf_1, though, the grating will be substantially behave as a plane mirror! An accurate diffraction theory (Strube, 1980) shows that the energy going into the different directions is not quite uniform.

The *grating constant*, *w*, must be chosen smaller than $\lambda/2$ if $\pm 90^0$ scatter is desired and smaller than $\lambda/4$ if, for a grazing–incidence, good back–scatter is the goal.

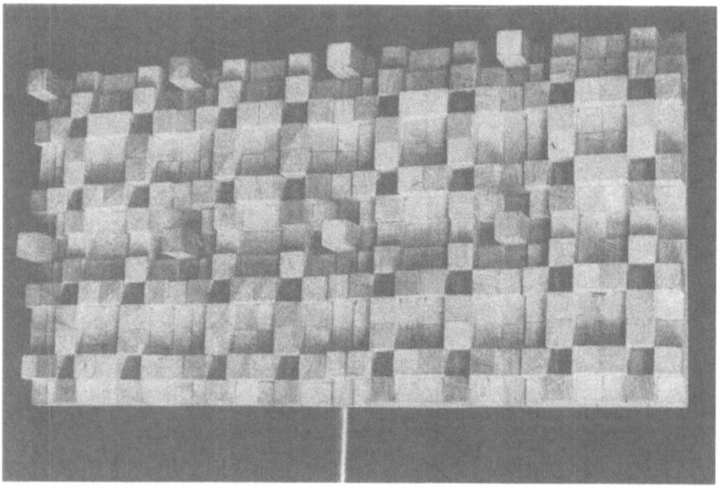

Fig. 4. Wood–block model of two–dimensional phase grating (without separators) based on the prime number $p = 7$, both horizontally and vertically. Such a grating scatters an incident wave broadly over the solid angle. The absence of separators makes the theoretical treatment more difficult but does not abrogate the excellent scattering characteristics of the grating.

These gratings were originally proposed to improve the acoustics of concert halls by providing more laterally travelling sound waves and to eliminate harsh echoes (Schroeder, 1979). They are now widely used in recording studios, and even homes, to improve sound reproduction (D'Antonio, 1984). Other uses are in noise abatement (a noise dispersed into many directions is less loud or even inaudible because of auditory masking), sonar and radar camouflage, and coherent optics to "randomize phases". Quadratic–residue phase gratings are, in a sense, the ultimate in frosted glass.

Figure 4 shows a wood–block model of two–dimensional phase grating in which both prime numbers in Eq. (20) are equal to 7. It will scatter an incident wave into 49 different directions given by Eqs. (22) and (23).

3.6 GALOIS ARRAYS

In some applications, arrays or gratings that *suppress* one direction are needed. A preferred solution to this problem is given by surface designs based on *primitive elements* from finite number (Galois) fields, $GF(p^m)$.

It is well known (Golomb, 1982) that linear recursion obtained from primitive polynomials over finite fields generate periodic sequences whose Fourier transform has constant magnitude for all nonzero frequencies. For example, for $GF(2^3)$, the primitive polynomial $x^3 + x + 1$ gives the recursion

$$b_{n+3} = b_{n+1} + b_n \mod 2, \tag{25}$$

which, together with the initial condition $b_0 = b_1 = b_2 = 1$, leads to a sequence of period length $P = 7$:

$$\{b_n\} = 1\,1\,1\,0\,0\,1\,0;\ \ 1\,1\,1\,\ldots\ . \tag{26}$$

The mapping $0 \rightarrow 1$ and $1 \rightarrow -1$, or $a_n = e^{i\pi b_n}$, leads to the following constant magnitude sequence:

$$\{a_n\} = -1\ -1\ -1\ 1\ 1\ 1\ -1\ 1;\ \ -1\ -1\ -1\ \ldots\ . \tag{27}$$

One of the outstanding properties of the sequence $\{a_n\}$ is that any product of the sequence with a shifted version of itself will generate the sequence $\{a_n\}$ with some shift. This property is intimately related to the fact that the sequence was generated from a primitive element. For example,

$$\{a_n\}\cdot\{a_{n-1}\} = 1\ 1\ {-1}\ 1\ {-1}\ {-1}\ {-1}\ \ldots = \{a_{n-4}\}.$$

Thus, the autocorrelation sequence

$$c_m := \sum_{n=0}^{P-1} a_n a_{n+m} , \tag{28}$$

where the sum is extended over one period, $P = 2^m - 1$, equals

$$c_m = \sum_{n=0}^{P-1} a_n = -1 , \qquad \text{for } m \not\equiv 0 \bmod P , \tag{29}$$

there being an excess of one -1 in one period of $\{a_n\}$.

Of course, for $m = 0$,

$$c_0 = P . \tag{30}$$

Thus, c_m has precisely 2 different values, and it is easy to show (Golomb, 1982) that the Discrete Fourier Transform of such a two–valued periodic sequence has a.c. components of equal magnitude, $|A_k|$, while the d.c. component is partially suppressed. In fact,

$$|A_k|^2 = P + 1 \qquad \text{for } k \not\equiv 0 \bmod P , \tag{31}$$

and

$$|A_k|^2 = 1 \qquad \text{r } k \equiv 0 \bmod P . \tag{32}$$

These statements are true for any m and $P = 2^m - 1$, as long as the recursion is based on a primitive polynomial in $GF(2^m)$, where *primitive* means that the polynomial of degree m cannot be factored over $GF(2)$ and is itself not a factor of a polynomial $x^r + 1$ for any $r < 2^m - 1$. Primitive polynomials exist for all p and m and many are listed in the literature (see MacWilliams and Sloane, 1978).

3.7 PHASE GRATINGS BASED ON $p = 2$

Figure 5 shows a phase grating based on the prime number $p = 2$ and the primitive polynomial $x^3 + x + 1$ in $GF(2^3)$. The grating has 2 different "depths" that correspond to be 0s and 1s in the Galois sequence Eq. (26) generated by the linear recursion Eq. (25), which corresponds to $x^3 + x + 1$. A 1 in the sequence corresponds to a reflection factor of -1 and a 0 corresponds to reflection factor of 1.

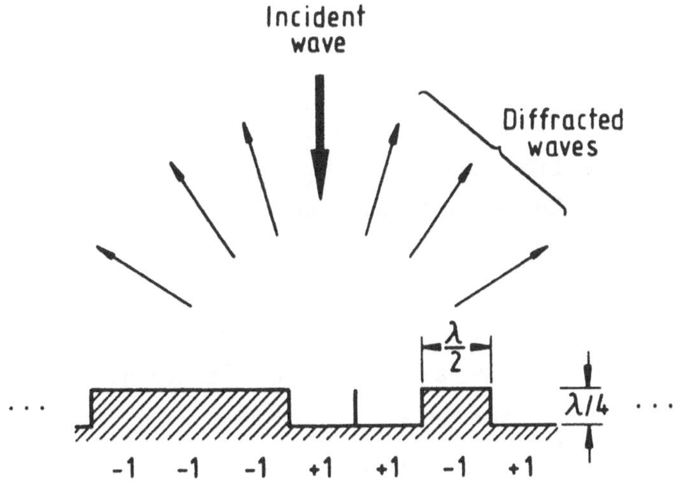

Fig. 5. A corrugated surface acting as a reflection phase grating. If the design is based on a Galois sequence, then the reflected energy is broadly scattered as shown in Fig. 6.

Thus, Eq. (27) is the sequence of reflection factors. They are realized in a reflection phase grating by "grooves" or troughs of depth $\lambda/4$, where λ is the wavelength of the incident wave. Thus, the wavelet reflected from the bottom of a trough is retarded by $2\lambda/4 = \lambda/2$ when it reemerges from the grating (compared to the wavelets reflected from the top level). The corresponding phase shift is π, and the reflection factor is multiplied by -1.

Since the spectrum of this sequence has constant magnitude, the corresponding far–field diffraction pattern is expected to have lobes of (roughly) equal energies. The conversion of spatial frequency to azimuthal angle is governed by Eqs. (2) and (3). In the grating shown in Fig. 6, the spatial sampling interval or grating constant w equals $\lambda/2$, so that $L = (2^3-1)\lambda/2$. The theoretical diffraction angles are therefore (as in the case of the quadratic–residue grating of the same period length)

$$\{a_k\} = 0^0, \quad \pm 17^0, \quad \pm 35^0, \quad \pm 59^0 . \tag{33}$$

The diffraction pattern of the grating shown in Fig. 5 for a (nearly) plane, normally incident wave is shown in Fig. 6. The grating consisted of sheet metal shaped like Fig. 5. The radiation was a 3 cm electromagnetic wave. The angles of the 7 major lobes are in good agreement with Eq. (33). The center lobe is, however, larger than expected: According to Eqs. (31) and (32) its energy should be only $1/(P+1) = 1/8$ of the side lobes. The discrepancy that we see here does not reflect (no pun) a failure of the Kirchhoff approximation, but the fact that the distances between the grating and source and receiver are *finite*. Fourier–transform diffraction patterns, called *Fraunhofer diffraction*, occur at an infinite distance from the grating. To observe Fraunhofer patterns at finite distances, the grating must be "curved" (i.e., mounted on a circular cylinder) with a radius of curvature R that is the *harmonic mean* of the distances of source and receiver from the grating, R_s and R_r, respectively

$$R = \frac{2R_sR_r}{R_s+R_r} . \tag{34}$$

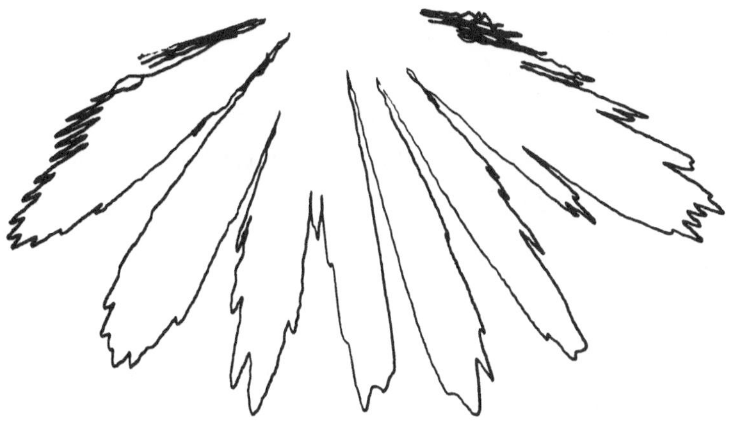

Fig. 6. Diffraction pattern from the Galois reflection phase grating shown in Fig. 5. Note nearly equal energies being scatterd into the seven diffraction orders.

3.8 PHASE GRATINGS FOR $p > 2$

Phase gratings based on $p = 2$ reflect almost mirrorlike ("specularly") when the frequency exceeds the "design frequency" by a factor 2. The reason is that for twice the design frequency, the depths of the troughs equals $\lambda/2$ (instead of $\lambda/4$). Thus, the "round trip" delay of wavelets reflected from the bottom of the troughs is $2\cdot\lambda/2 = \lambda$, which is congruent 0 modulo the wavelength λ. Hence, the irregular surface, Fig. 5, acts almost like a plane mirror.

For $p > 2$, gratings can be designed that scatter effectively at more than one frequency, namely $p-1$ frequencies. For $p = 3$, for example, a primitive polynomial with the 3 elements 0, 1, 2, over $GF(3^2)$ is $x^2 + x + 2$. It leads to the linear recursion

$$b_{n+2} = 2b_{n+1} + b_n \mod 3, \tag{35}$$

which generates the 3–valued Galois sequence of period length $3^2-1 = 8$,

$$\{b_k\} = 1\,1\,0\,1\,2\,2\,0\,2;\;\; 1 \ldots, \tag{36}$$

where each doublet, except 0 0, occurs exactly once per period.

The corresponding sequence of the reflection coefficients is obtained by the mapping $r_k = \exp(i2\pi b_k/3)$. With $\omega = \exp(i2\pi/3)$, one has

$$\{r_k\} = \omega, \omega, 1, \omega, \omega^2, \omega^2, 1, \omega^2;\;\; \omega \ldots\;\; . \tag{37}$$

Such an array, which has 3 different depths, scatters radiation efficiently at the design frequency and at twice the design frequency. However, at 3 times the design frequency it acts as much like a plane mirror.

3.9 PRIMITIVE–ROOT GRATINGS

Arrays or gratings that correspond to $GF(p^m)$ with $m = 1$ are especially simple to design. In these cases the primitive element is simply a primitive root g (Schroeder, 1986, Chapter 13). For example, for $p = 11$, $g = 2$ is a primitive root. This means that the power of g generate all 10 nonzero elements of $GF(11)$. In fact, for $p = 11$ and $g = 2$,

$$\{g^k\} \equiv 2, 4, 8, 5, 10, 9, 7, 6, 1; 2 \ldots \bmod 11. \tag{38}$$

The corresponding phase angles are

$$\phi_k = 2\pi g^k/p \bmod 2\pi. \tag{39}$$

Figure 7 shows the diffraction pattern of a primitive–root grating for $p = 7$ and $g = 3$ with 14 troughs (2 periods). Note the large nonzero–order diffraction lobes and the small specular reflection, whose energy is reduced by a factor $p = 7$ compared to the stronger lobes – as predicted and desirable in some applications.

A prime number p has precisely $\phi(p-1)$ primitive roots, where $\phi(n)$ is Euler's "totient function", defined as the number of positive integers smaller than n and coprime to it. For $p = 11$, $p-1 = 10$ and $\phi(10) = 4$, because there are precisely 4 integers below 10 that are coprime to it (1, 3, 7, 9). Knowing $g = 2$, the other primitive roots of 11 are given by $2^3 = 8$, $2^7 \equiv 7$ and $2^9 \equiv 6 \bmod 11$. It seems that only permutations like Eq. (38), generated from primitive roots, lead to the desired spectral property of constant–magnitude sequences with phase angles given by Eq. (39).

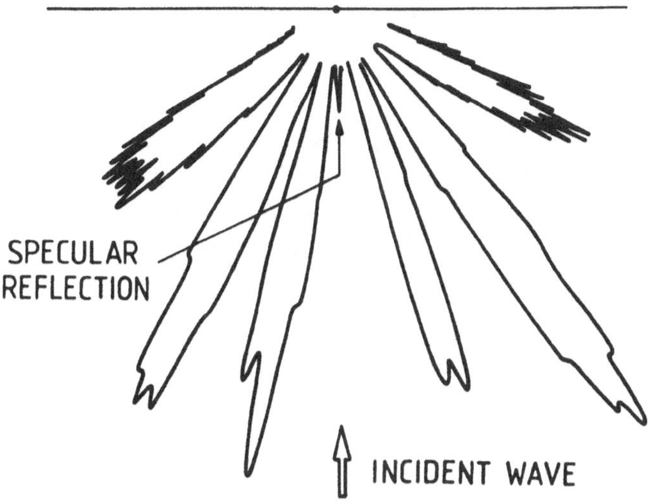

Fig. 7. Diffraction pattern from a primitive–root grating design based on the prime number $p = 7$. Note the low energy specular scatter.

Primitive–root phases are also useful for phase arrays with low–amplitude broadside lobes and equal amplitudes for the non–broadside directions. The radiation pattern of a phase array based on $p = 11$ and $g = 2$ and comprising 2 periods (20 elements) is shown in Fig. 8. The element spacing is $\lambda/2$ and the phase angles are given by Eq. (39). Note the closeness of the pattern to the theoretical prediction.

3.10 COMPLEX LEGENDRE GRATING

Gratings that scatter broadly but completely suppress the specular reflection are based on "complexified" Legendre sequences[1] (Schroeder, 1986)

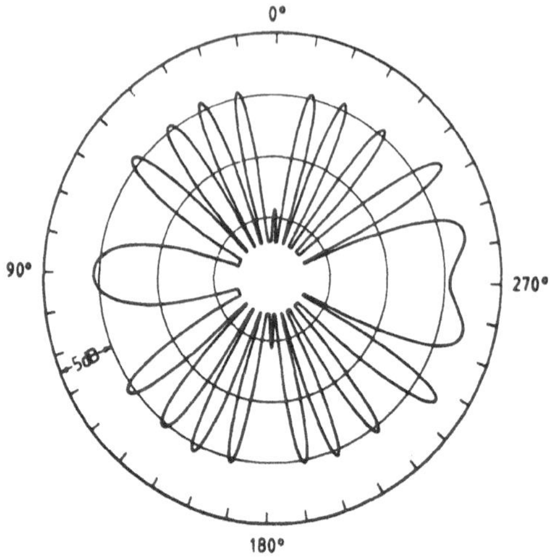

Fig. 8. Radiation pattern of 20–element primitive–root array with half–wave-length spacing between individual elements. Signals supplied to individual elements have constant amplitudes across the array, but phase angles vary according to a primitive–root sequence based on the prime number $p = 11$. Note uniformity of radiation into different major lobes, except the broadside directions ($\phi = 0^0$ and 180^0) where the radiation is low.

$$r_0 = 0, \quad r_n = e^{2\pi i(\text{ind}_g\, n)/(p-1)}, \quad n \not\equiv 0 \bmod p \tag{40}$$

where ind_g is the number–theoretic logarithm or index function defined by

$$g^{\text{ind}_g n} \equiv n \bmod p. \tag{41}$$

[1]For the original (real) Legendre sequences, r_n equals 1 (or −1) depending on whether n is (or is not) a quadratic–residue of p. They can be obtained from (40) by multiplying the exponent by $(p-1)/2$.

Note that, while all nonzero amplitudes r_n have equal magnitudes, r_0 is *zero*, meaning total energy *absorption* rather than 100 % reflection.

It is not difficult to show that the Discrete Fourier Transform R_m of the sequence Eq. (40) is given by

$$R_0 = 0 \quad \text{and} \quad |R_m|^2 = p-1 \quad \text{for} \quad m \not\equiv \bmod p, \tag{42}$$

i.e. it has equal magnitudes for all a.c. components and is *zero* for the d.c. component, i.e. the specular reflection is completely suppressed.

Interestingly, the Discrete Fourier Transform of a Legendre sequence equals, within a constant factor, the conjugate complex of the sequence itself

$$R_m = R_1 r_m^* , \tag{43}$$

a remarkable property, reminiscent of the fact that the Fourier transform of a Gauss pulse is a Gauss pulse!

The flat–spectrum property is maintained if the exponent in Eq. (40) is multiplied by an integer $k = 1, 2, \dots, p-1$. Thus, the grating will scatter broadly at $p-1$ different frequencies (the design frequency multiplied by $k = 1, 2, \dots, p-1$).

However, if the depths of the grating troughs are increased by a factor k, the r_n will then assume only $(p-1)/\mathrm{GCD}(p-1,k)$ different values, where $\mathrm{GCD}(x,y)$ is the greatest common divisor of c and y. As a consequence, the grating will scatter broadly only at $(p-1)/\mathrm{GCD}(p-1,k)$ consecutive integer–multiples of the design frequency.

3.11 ZECH GRATINGS

Another number-theoretic logarithm, the Zech logarithm (Schroeder, 1986, Chapter 25) allows the suppression of *two* directions while scattering nearly equal energies into all remaining directions. The Zech logarithm, $z(n)$, may be defined by

$$\alpha^{z(n)} := 1 - \alpha^n , \qquad n = 1, 1, \dots, p^m - 2 \tag{44}$$

where α is a primitive element of the Galois field $GF(p^m)$. Such a finite number field can be represented by p^m different polynomials of maximal degree m-1 with coefficients drawn from 0, 1,..., p-1. For $p = 2$, the coefficients are limited to 0 and 1 and obey the rules for modulo 2 addition, e.g. 1+1 = 0. As a shorthand notation, such polynomials are represented by the m-tuples of the coefficients. Thus, for example, the polynomial $1 + x + x^2$ is written as 111, while the polynomial x^2 becomes 001, etc.

To illustrate the use of the Zech logarithm, let us calculate $z(n)$ for the Galois field $GF(2^3)$ with the primitive element $\alpha = x = 010$ and the modulus condition $x^3 = 1 + x$.

$$
\begin{array}{ll}
\alpha^{-\infty} = 000 & 1-\alpha^{-\infty} = 100 = \alpha^0 \\
\alpha^0 = 100 & 1-\alpha^0 = 000 = \alpha^{-\infty} \\
\alpha^1 = 010 & 1-\alpha^1 = 110 = \alpha^3 \\
\alpha^2 = 001 & 1-\alpha^2 = 101 = \alpha^6 \\
\alpha^3 = 110 & 1-\alpha^3 = 010 = \alpha^1 \\
\alpha^4 = 011 & 1-\alpha^4 = 111 = \alpha^5 \\
\alpha^5 = 111 & 1-\alpha^5 = 011 = \alpha^4 \\
\alpha^6 = 101 & 1-\alpha^6 = 001 = \alpha^2
\end{array}
\qquad (45)
$$

Thus, with Eq. (44) we have, for n = 1, 2, 3, 4, 5, 6, the Zech logarithms $z(n)$ = 3, 6, 1, 5, 4, 2.

It is interesting to note that the Zech logarithm generates a unique permutation of the numbers 1, 2,...,p^m-2 with interesting applications in "frequency hopping" schemes (Schroeder, 1986, Chapter 25) for spread-spectrum communications and Doppler radar and sonar with minimum ambiguity between range (target distance) and range rate (target velocity).

The trough depths of the grating are chosen to give reflection factors

$$
r_n = \exp[2\pi i z(n)/(p^m-1)] \text{ , for } n = 1,2,...p^m-2 \qquad (46)
$$

Again, as in the case of the complex Legendre sequence, all nonzero reflection factors have magnitude 1, except r_0 which is zero, meaning energy absorption.

The DFT R_m of the r_n is given by

$$|R_0|^2 = |R_0|^2 = 1,$$
$$|R_k|^2 = p^m, \quad k = 2, 3, \ldots, p^m - 2. \tag{47}$$

Note that *two* Fourier components, R_0 *and* R_1, are small.

A diffraction pattern for a Zech *grating*, with reflection factors according to Eq. (46), is shown in Fig. 9. Note that both the specular reflection *and* one of the first-order diffractions is attenuated, resulting in a pronounced lateral scatter that may be useful in a number of applications. (To observe these attenuated diffraction orders at finite distances, the grating has to be "bent" or mounted on a curved surface with radius of curvature given by Eq. (34); see also the discussion of Fraunhofer diffraction in the section on Galois gratings.)

Fig. 9. Diffraction pattern from a grating based on the Zech logarithm for the finite number field $GF(2^3)$. Note the suppression of *two* diffraction orders, the specular one and one first-order diffraction, creating an even more lateral energy flow pattern than quadratic-residue, Galois, and primitive-root gratings.

If phase angle appearing in Eq. (46) are multiplied by an integer $k = 2, 3, \ldots p^m - 2$, then, instead of $|R_1|$, $|R_k|$ will be small. This is illustrated for a Zech *array* with $k = 2$ in Fig. 10.

For the same reason, a Zech *grating*, operated at k times the design frequency, will have a low-energy kth order diffraction.

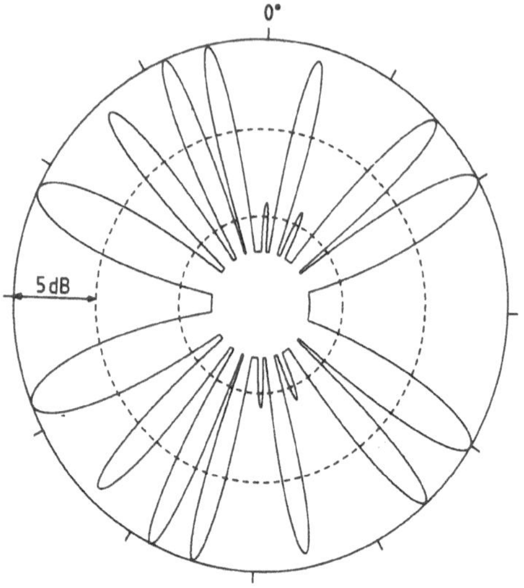

Fig. 10. Radiation patter of constant-amplitude phase array with phase angles based on Zech logarithm for $GF(2^3)$. The array has two period length 8 (total number of nonzero array elements: 14). The spacing between individual elements equals 9/16 wavelengths. Note small lobes in the broadside (0^0) and second-order directions (26^0, corresponding to $k = 2$). The attenuation of the small lobes corresponds closely to the theoretical value for the infinite array $1/p^m$, i.e. -9.5 dB. By choosing other values for the parameter k, any of the other nonbroadside lobes can be made small.

3.12 CONCLUSION

Number theory has played an important (and initially surprising) role in the design of phase arrays and grating with broad directional characteristics. The question is sometimes asked "what about random phases?" After all, such random media as white paper and frosted glass are excellent scatters for visible light.

To answer the question, A. Steingrube and the author have simulated phase gratings with *random* trough depths (see Schroeder, 1979). The result was that the best of several random gratings was never as good a scatterer as the grating designed by number theory.

Another approach would be to design gratings and arrays by inverse Fourier methods and iterative optimization. This can indeed be done for one or two frequencies but to achieve equal scatter into many directions at *many* frequencies would require massive computation, using perhaps such advanced methods as simulated annealing, and even then it is not clear whether the results would be superior to the instant answers provided by number theory.

References

D'Antonio, P. and Konnert, J.H. (1984) "The Reflection Phase Grating Diffusor: Design Theory and Application", J. Audio Eng. Soc. **32**, 228-238.

Golomb, S.W. (1982) Shift Register Sequences, Aegean Park Press, Laguna Hills, California.

MacWilliams, F.J. and Sloane, N.J.A. (1978) The Theory of Error Correcting Codes, North Holland, Amsterdam.

Scholtz, R.A. (1982) "The Origins of Spread-Spectrum Communications", IEEE Trans. Communications **30**, 822-852.

Schroeder, M.R. (1979) "Binaural Dissimilarity and Optimum Ceilings for Concert Halls: More Lateral Sound Diffusion", J. Acoust. Soc. Am. **65**, 958-963.

Schroeder, M.R. (1986) Number Theory in Science and Communication, 2nd Enlarged Ed., Springer, Berlin, New York.

Strube, H.W. (1980) "Scattering of a Plane Wave by a Schroeder Diffusor: A Mode-Matching Approach", J. Acoust. Soc. Am. **67**, 453-459 and "Diffraction by a Planar, Locally Reacting, Scattering Surface", 460-469.

FUNCTIONS OF MODULUS 1 ON $\mathbf{Z}_n$ WHOSE FOURIER TRANSFORMS HAVE CONSTANT MODULUS, AND "CYCLIC n-ROOTS"

GÖRAN BJÖRCK
Department of Mathematics
University of Stockholm
Box 6701
S-113 85 Stockholm, Sweden

ABSTRACT. We consider on one hand elements x in $\mathbf{C}^n$ as functions on the cyclic group $\mathbf{Z}_n$. A function which has a constant modulus and whose Fourier transform also has a constant modulus, is called *bi-equimodular*. On the other hand, we study the set $CR(n)$ of "cyclic n-roots", i. e. solutions of a certain system (2.1) of algebraic equations. We prove that there is essentially a one-one correspondence between the bi-equimodular functions and the cyclic n-roots of modulus one. For prime p, we give preliminary results about the structure of the bi-equimodular p-functions, starting from our negative answer in [2] to the question asked by Enflo whether the "classical" ones known already to Gauss are the only ones existing. The structure is related to the subgroups of the multiplicative group on $\mathbf{Z}_p \setminus \{0\}$, e. g. the group of quadratic residues. For $n \leq 7$, $CR(n)$ is completely known. The proofs (in [3]) depend on computer work, using symbolic algebra. Related material can be found in the book [7] by Schroeder and in the paper [6] by McGehee.

1. Notation and definitions related to Fourier transforms on $\mathbf{Z}_n$.

In this paper n will denote an integer > 1. Whenever it is required that n is a *prime*, we will instead use p. Let $\mathbf{Z}_n = \{0, 1, \dots, n-1\}$ be the cyclic group. Considering any $x = (x_0, \dots, x_{n-1}) \in \mathbf{C}^n$ as a complex-valued function on $\mathbf{Z}_n$, we will define its Fourier transform $\hat{x} = (\hat{x}_0, \dots, \hat{x}_{n-1})$ by $\hat{x}_\nu = \sum_{j=0}^{n-1} x_j \omega^{j\nu}$, where $\omega = \exp(\frac{2\pi i}{n})$. When dealing with an element x of $\mathbf{C}^n$ from this point of view, we call it a *n-function* or just *function*. By an *equimodular n-function* we mean an $x \neq 0$ such that $|x_j|$ is independent of j. By a *bi-equimodular n-function* we mean an x such that x and $\hat{x}$ both are equimodular. From the Fourier inversion formula follows that $\hat{x}$ is bi-equimodular if and only if x is. One of the aims of this paper is to give some ideas about the structure of the bi-equimodular n-functions.

We will next (in the prime case) give definitions related to the idea of "p-functions taking few values". Let G be a subgroup of the multiplicative group $\mathbf{Z}_p^*$ on $\mathbf{Z}_p \setminus \{0\}$, and suppose that G has order s and index t (so that $p - 1 = s \cdot t$). Let g be a generator for $\mathbf{Z}_p^*$, and let $G_0, \dots, G_{t-1}$ be the cosets of G, numbered in such a way that $G_k = \{g^{k+mt}; m = 0, 1, \dots, s-1\}$. In particular we will have $G_0 = G$. We will call $x \in \mathbf{C}^p$ *a simple function with pre-index* t if it has the following form:

$$\begin{cases} x_0 = c, \\ x_j = c_k \text{ if } j \in G_k, \end{cases}$$

where $c, c_0, \dots, c_{t-1}$ are fixed numbers. Trivially, every x is a simple function with pre-index $(p-1)$. We will now show that if x is a simple function, then $\hat{x}$ is also a simple function with the same

J. S. Byrnes and J. F. Byrnes (eds.), Recent Advances in Fourier Analysis and Its Applications, 131–140.

pre-index. In fact, for $k = 0, \ldots, t-1$, let η_k be the *Gaussian sum*

$$\eta_k = \sum_{j \in G_k} \omega^j.$$

Suppose now that $\nu \in G_l$, i.e. that $\nu \equiv g^{l+tn} \pmod p$ for some n. Then

$$\hat{x}_\nu = \sum_k c_k \sum_{j \in G_k} \omega^{j g^{l+tn}} = \sum_k c_k \sum_{m=0}^{t-1} \omega^{g^{k+l+tn+tm}} = \sum_k c_k \eta_{k+l},$$

and thus

$$\begin{cases} \hat{x}_0 = d, \\ \hat{x}_\nu = d_l \text{ if } \nu \in G_l, \end{cases}$$

where we have defined

$$(1.1) \qquad \begin{cases} d = c + sc_0 + sc_1 + \cdots + sc_{t-2} + sc_{t-1}, \\ d_0 = c + \eta_0 c_0 + \eta_1 c_1 + \cdots + \eta_{t-2} c_{t-2} + \eta_{t-1} c_{t-1}, \\ d_1 = c + \eta_1 c_0 + \eta_2 c_1 + \cdots + \eta_{t-1} c_{t-2} + \eta_0 c_{t-1}, \\ \quad \vdots \\ d_{t-1} = c + \eta_{t-1} c_0 + \eta_0 c_1 + \cdots + \eta_{t-3} c_{t-2} + \eta_{t-2} c_{t-1}. \end{cases}$$

More generally, x is called *a function with pre-index t* if for some fixed integers k and l and some *simple* function y with pre-index t we have

$$x_j = \omega^{kj} y_{j-l}.$$

This definition is inspired by the well-known properties of the Fourier transform with respect to translation and to multiplication by an exponential. It is clear that $\hat{x}$ has pre-index t if and only if x has pre-index t, the roles played by k and l being interchanged.

A simple calculation now shows that if x has pre-index t_1 and pre-index t_2, then x also has pre-index $t := \gcd(t_1, t_2)$. Thus the following definition makes sense. The *index of a function x* is the greatest common divisor of the pre-indices of x. Thus the index of x is also the smallest pre-index of x. Clearly a simple function with a small index takes few different values.

2. Notation, definitions, and simple examples related to "cyclic n-roots".

Let n be an integer > 1, and let $z = (z_0, \ldots, z_{n-1}) \in \mathbf{C}^n$. We will call z a *"cyclic n-root"*, if z satisfies the following system of n algebraic equations:

$$(2.1) \qquad \begin{cases} z_0 + z_1 + \cdots + z_{n-1} = 0, \\ z_0 z_1 + z_1 z_2 + \cdots + z_{n-1} z_0 = 0, \\ \quad \vdots \\ z_0 z_1 \cdots z_{n-2} + z_1 z_2 \cdots z_{n-1} + \cdots + z_{n-1} z_0 \cdots z_{n-3} = 0, \\ z_0 z_1 \cdots z_{n-1} = 1. \end{cases}$$

Note that the sums are cyclic and contain just n terms and are *not* the elementary symmetric functions when $n > 3$. A second aim of this paper is to present some ideas and results about the structure of the cyclic n-roots. The connection between this aim and that stated in Section 1 will appear in Section 3.

We will denote the set of cyclic n-roots by $CR(n)$. It seems probable that $CR(p)$ is finite for every *prime* p, but the author does not know if this is true. We prove in Section 5.3 that the number of cyclic p-roots of index three (defined in Section 3) is finite. Considering $CR(n)$ for non-prime n, we remark that Fröberg and the author have proved in [3] that $CR(6)$ has 156 elements, whereas Backelin has proved in [1] that if $CR(n)$ is finite, then n is square-free.

Finally, we will call $z \in \mathbf{C}^n$ *unimodular* if $|z_j| = 1$ for all j. Unimodular cyclic n-roots are closely related to bi-equimodular n-functions, as explained in Section 3. We will call $z \in \mathbf{C}^n$ *essentially real*, if $z = Cx$ for some real vector x and some (complex) constant C.

The following examples of cyclic n-roots, referred to by name in the sequel, can be checked by simple calculations left to the reader:

For any n, the two *ϵ-roots* are defined by

$$z_{n-1} = \epsilon,\ z_0 = \epsilon^{-1},\ z_1 = z_2 = \cdots = z_{n-2} = 1 \text{ with } \epsilon + \epsilon^{-1} + n - 2 = 0.$$

Let $p > 2$ be prime, let $a \neq 0$ and b be elements in $\mathbf{Z}_p$ and let $\omega = \exp(\frac{2\pi i}{p})$. Then a *classical* cyclic p-root z is defined by $z_j = \omega^{aj+b}$. It is unimodular.

3. Relations between bi-equimodular functions and unimodular cyclic n-roots. The classical example.

Let x be an element of $(\mathbf{C} \setminus \{0\})^n$ and z an element of $\mathbf{C}^n$, related by

$$z_j = \frac{x_{j+1}}{x_j},$$

(where, as always, the subscript n is to be read as 0). Then we have the fundamental

Proposition 1. *x is bi-equimodular if and only if z is a unimodular cyclic n-root.*

Proof (from [2]). It is clear that that x is equimodular if and only if $|z_j| = 1$ for all j and $\prod z_j = 1$. Thus, to complete the proof we may assume that x is equimodular and prove that $\hat{x}$ is equimodular if and only if z is a cyclic n-root. Without loss of generality, we may assume that $x_0 = 1$ and thus all $|x_j| = 1$. We will now rearrange the double sum $|\hat{x}_\nu|^2 = \sum_j x_j \omega^{\nu j} \cdot \sum_k \bar{x}_k \omega^{-\nu k}$, writing $j - k = s$, and using the relation $x_{k+s} = x_k z_k z_{k+1} \cdots z_{k+s-1}$. Since all $x_k \bar{x}_k = 1$, we get

$$|\hat{x}_\nu|^2 = \sum_{s=0}^{n-1} \omega^{\nu s} \sum_{k=0}^{n-1} x_k \bar{x}_k z_k z_{k+1} \cdots z_{k+s-1} = n + \sum_{s=1}^{n-1} a_s \omega^{\nu s},$$

where we have written $a_s = \sum_{s=0}^{n-1} z_k z_{k+1} \cdots z_{k+s-1}$ for $s = 1, 2, \ldots, n-1$. Consider now the polynomial $P(\zeta) = \sum_{s=1}^{n-1} a_s \zeta^s$. It has degree at most $(n-1)$, and $P(\omega^\nu) = |\hat{x}_\nu|^2 - n$ for $\nu = 0, 1, \ldots, n-1$. Hence $\hat{x}$ is equimodular if and only if P takes the same value in the n points ω^ν, which occurs if and only if P is constant. This completes the proof, since P is constant if and only if all $a_s = 0$, that is, if and only if z is a cyclic n-root.

This result gives a good framework for studying bi-equimodular functions. It is fruitful to work sometimes on the "x-level" and sometimes on the "z-level". In particular, in the prime case, if z is a cyclic p-root we define its *pre-indices* and its *index* as the pre-indices and the index of a corresponding x, respectively. Also, if z has pre-index t and the corresponding x is a *simple* function with pre-index t, we will say that z is a *simple* cyclic p-root with pre-index t.

By *the classical bi-equimodular function* we will (in the prime case) denote the following x, known by Gauss to be bi-equimodular:

$$x_j = \omega^{\frac{1}{2}aj(j-1)+bj},$$

where $a \neq 0$ and b are elements in $\mathbf{Z}_p$. That this x is bi-equimodular follows by Proposition 1 from the last example in Section 2. This x, and thus the corresponding z, has index $\frac{1}{2}(p-1)$.

4. Equivalence relations.

Already in our definition of functions with pre-index t, the idea of equivalence (with respect to two properties of the Fourier transform) is inherent. But we will go one step further, and we prefer to define the concept on the z-level. Let $\mathbf{C}_1^p$ denote the set $\{z \in \mathbf{C}^p; z_0 z_1 \cdots z_{p-1} = 1\}$.

We will consider a finite group S of mappings $T\colon \mathbf{C}_1^p \to \mathbf{C}_1^p$ such that $Tz \in CR(p)$ whenever $z \in CR(p)$. S is defined by the generators K (conjugation), P (permutation), B ("backwards"), M (multiplication by ω), and A ("association") and certain obvious relations, as follows (where indices are taken modulo p, and where g is the generator for $\mathbf{Z}_p^*$ chosen in Section 1):

$$\begin{aligned}
(Kz)_j &= \bar{z}_j, \\
(Pz)_j &= z_{j-1}, \\
(Bz)_j &= z_{p-1-j}, \\
(Mz)_j &= \omega z_j, \\
(Az)_j &= z_{jg} z_{jg+1} \cdots z_{jg+g-1}.
\end{aligned}$$

It is clear that K and B have order 2, P and M have order p, and A has order $(p-1)$, that K commutes with P, B and A, that B commutes with A and M, and that the following additional relations hold: $KM = M^{p-1}K$, $PA = AP^g$, and $M^g A = AM$. We will now define a *root class* as an orbit in $CR(p)$ under the action of the group S, that is, it consists of a cyclic p-root and all its images under the elements T of the group. For a fixed p, the *size* of a root class (its number of elements) can have different values. The sum of the sizes of the classical root classes is $p(p-1)$. Since clearly all elements in a root class are unimodular if one of them is, and similarly for essentially real elements, we can define *unimodular* and *essentially real root classes* in an obvious way. It is also clear that all elements in a root class have the same index, which we will call the *index* of the root class.

Since it sometimes is less natural to consider two cyclic p—roots as equivalent when A is involved than when K, P, M, and B suffice, we will consider that subgroup S_1 of S which is generated by K, P, M, and B, and define the *complexity* of a root class C as the number of those disjoint orbits of S_1 which make up C. Clearly, the complexity of the ϵ-root class is one, whereas the situation for the classical root classes is more complicated: For $p = 5$, there are two classical root classes, each of size 10 and complexity one, but for $p = 7$, there is one classical root class of index 42 and complexity three.

To be able to express our results concisely on the x-level, we define a *bi-class* as a set of bi-equimodular functions, normalized by prescribing $x_0 = 1$, such that the set of corresponding z is a root class.

Finally, we remark that it seems natural to look for a an alternative equivalence relation where all classical roots are in the same root class. It also remains to define a suitable equivalence relation in the *non-prime* case.

5. Cyclic p-roots of small index.

Since our z_k compares the values of x_k and x_{k+1}, it will be natural to consider the following concept for a fixed prime p and a fixed pre-index t: For $d = 1, \ldots, p-1$, and for i and $k = 0, 1, \ldots, t-1$, we define the *transition number* $n_{ik}(d)$ as the number of elements b in $\{1, 2, \ldots, (p-1)\}$ for which $b \in G_i$ and $b + d \in G_k$. (Subscripts are taken modulo t. We do not count $b = p - d$).

Suppose now that $d \in G_a$, i.e. that $d \equiv g^{a+lt}$ for some l (all congruences are modulo p). Since for each b which contributes to $n_{ik}(1)$, we have $b \equiv g^{i+ut}$ and $b + 1 \equiv g^{k+vt}$ for some u and v. Thus

$$g^{k+a+(l+v)t} \equiv g^{i+a+(l+u)t} + d. \tag{5.1}$$

Writing n_{ik} instead of $n_{ik}(1)$, we thus get

$$n_{i+a,k+a}(d) = n_{ik}. \tag{5.2}$$

Let us now define m by $(p-1) \in G_m$. Consider a cyclic p-root z with pre-index t. Fix a d such that $d \in G_a$, and consider the individual products in the degree d equation of (2.1). These products will take the values c_k/c_i with the frequency $n_{i+a,k+a}(d)$, the value c_a/c once, and the value c/c_{m+a} once (since $p - d \in G_{m+a}$). Thus (5.2) implies that all equations whose degrees d belong to the same coset G_a, are identical. We have proved:

Proposition 2. *For a simple cyclic p-root z with pre-index t, the system (2.1) consists of the following t equations (where $n_{ik} = n_{ik}(1)$ are the transition numbers, and m is defined by $p-1 \in G_m$, and the subscripts of the c:s are counted modulo t)*

$$\frac{c_a}{c} + \frac{c}{c_{a+m}} + \sum_{k=0}^{t-1}\sum_{i=0}^{t-1} n_{ik}\frac{c_{k+a}}{c_{i+a}} = 0, \quad (a = 0, 1, 2, \ldots, t-1). \tag{5.3}$$

Using the same ideas, we will now prove a property of the association operation A, to be used later to determine the complexity of certain root classes. If $b \in G_i$ and $b + 1 \in G_k$, we get (as in (5.1)), that $gb \in G_{i+1}$ and $g(b+1) \in G_{k+1}$. Thus by the definition of A, we have $(Az)_b = c_{k+1}/c_{i+1}$, whereas $z_b = c_k/c_i$. Considering separately the values of $(Az)_0$ and $(Az)_{p-1}$, we have proved:

Proposition 3. *If z is a simple cyclic p-root with pre-index t then we have (with m defined by $p - 1 \in G_m$, and with the subscripts of the c:s counted modulo t)*

$$(Az)_0 = \frac{c_1}{c}, \; (Az)_{p-1} = \frac{c}{c_{m+1}}, \quad \text{and } (Az)_b = \frac{c_{k+1}}{c_{i+1}} \text{ if } b \in G_i \text{ and } b + 1 \in G_k. \tag{5.4}$$

We will now establish some simple relations between the n_{ik}. Considering n_{ik} as the entries of a matrix N, we see that the k-th column sum of N equals the number of times $(b+1) \in G_k$, and the i:th row sum equals the number of times $b \in G_i$ (in the definition of the transition numbers). Since the cosets have s elements but we "loose" one element because $1 \in G_0$, and one because $p - 1 \in G_m$, we get

$$\sum_{k=0}^{t-1} n_{ik} = \begin{cases} s, & \text{if } i \neq m; \\ s-1, & \text{if } i = m, \end{cases} \tag{5.5}$$

$$\sum_{i=0}^{t-1} n_{ik} = \begin{cases} s, & \text{if } k \neq 0; \\ s-1, & \text{if } k = 0. \end{cases} \tag{5.6}$$

For the t "diagonal sums", we use (5.3) to get

$$\sum_{d=1}^{p-1} n_{ik}(d) = \sum_{d=1}^{p-1} n_{i-a,k-a} = s \cdot \sum_{a=0}^{t-1} n_{i-a,k-a}. \tag{5.7}$$

Since in the main diagonal we "loose" s out of a total of s^2 transitions, (5.7) leads to

$$\sum_{a=0}^{t-1} n_{i-a,k-a} = \begin{cases} s, & \text{if } i \neq k; \\ s-1, & \text{if } i = k. \end{cases} \tag{5.8}$$

5.1. INDEX ONE.

The *simple* cyclic p-roots of (pre-)index one are given by (5.3), which now takes the form

$$\frac{c_0}{c} + \frac{c}{c_0} + p - 2 = 0.$$

Taking $c_0 = 1$ and $c = \epsilon$, we recognize the ϵ-roots (Section 3). Thus, for any p there is precisely one root class of index one, the *ϵ-class*. It has complexity one. For $p = 2$, it is unimodular and has size two. For $p = 3$, it is unimodular, has size six, and coincides with the classical class. For $p > 3$, it is essentially real and has size $2p^2$.

5.2. INDEX TWO.

Here of course G_0 is the set of quadratic residues modulo p. We will have to consider separately the cases $p \equiv 1$ and $p \equiv 3 \pmod 4$.

Let us first take $p = 4k + 1$. As is well known, $(p - 1)$ is then a quadratic residue, and thus $m = 0$. Then from (5.5), (5.6), and (5.8) follows that $n_{00} = k - 1$ and $n_{01} = n_{10} = n_{11} = k$. Hence (5.3), the condition for (c, c_0, c_1) to give a simple cyclic p-root with pre-index two, is

$$\begin{cases} \dfrac{c_0}{c} + \dfrac{c}{c_0} + (2k-1) + k(\dfrac{c_1}{c_0} + \dfrac{c_0}{c_1}) = 0, \\ \dfrac{c_1}{c} + \dfrac{c}{c_1} + (2k-1) + k(\dfrac{c_1}{c_0} + \dfrac{c_0}{c_1}) = 0. \end{cases}$$

Taking $c = 1$ and subtracting the two equations, we get $c_0 - c_1 + (1/c_0) - (1/c_1) = 0$. Discarding the case $c_0 = c_1$, which would give index one, we get $c_0 c_1 = 1$. Writing $c_0 + c_1 = u$, we have that c_0 and c_1 are the solutions γ of the equation $\gamma^2 - u\gamma + 1 = 0$, where $ku^2 + u - 1 = 0$. After some calculation, this gives (with $c = 1$, $c_1 = 1/c_0$, and $\delta = \pm 1$)

$$c_0 = \frac{\delta\sqrt{p} - 1 + i\sqrt{p^2 - 3p + 2\delta\sqrt{p}}}{p-1}, \tag{5.9}$$

or the same relation where c_0 and c_1 have replaced each other. For each $p \equiv 1 \pmod 4$, we thus get two root classes C_1 and C_2 of index two (one for each value of u, i.e. for each value of δ, and both C_j are unimodular. For $p = 5$, we see that each C_j coincides with a classical root class (note that (5.9) gives the well-known exact values of $\exp(\frac{2\pi i}{5})$ and $\exp(\frac{4\pi i}{5})$, respectively) and thus has size 10 and complexity one. For $p > 5$, the size of each C_j is $2p^2$. Finally, we will prove that C_j always has complexity one. In fact, if $z \in C_j$, then since $c_0 c_1 = 1$, Proposition 3 gives that $(Az)_j = 1/z_j = \bar{z}_j$, and hence $Az = Kz$. Thus S_1 acts on C_j as S does (see notation in Section 4), and the result follows.

Let us now take $p = 4k+3$. Then $p-1$ is not a quadratic residue, that is $m = 1$. This time we get $n_{01} = k+1$ and $n_{00} = n_{10} = n_{11} = k$, and (5.3) takes the form

$$\begin{cases} \dfrac{c_0}{c} + \dfrac{c}{c_1} + 2k + (k+1)\dfrac{c_1}{c_0} + k\dfrac{c_0}{c_1} = 0, \\ \dfrac{c_1}{c} + \dfrac{c}{c_0} + 2k + (k+1)\dfrac{c_0}{c_1} + k\dfrac{c_1}{c_0} = 0. \end{cases}$$

Again discarding the case $c_0 = c_1$, we now have $c = c_0$ or $c = c_1$. This time we get, for $p > 3$, exactly one root class C of index two. It is unimodular and has size $4p^2$. (For $p = 3$, we get again $c_0 = c_1$, and we are back in the classical root class which has index one.) One member of C has $c = c_1 = 1$ and

$$(5.10) \qquad c_0 = \frac{1-p}{1+p} + i\frac{2\sqrt{p}}{1+p}.$$

We now prove that the complexity is again one. Let $z \in C$. Since $z_0 = c_0/c$ and $z_{p-1} = c/c_1$, we get by Proposition 3 that $(Az)_0 = c_1/c = 1/z_{p-1} = \bar{z}_{p-1}$ and $(Az)_{p-1} = c/c_0 = 1/z_0 = \bar{z}_0$. Let $1 \le b \le p-2$, and suppose that $b \in G_i$ and $b+1 \in G_k$. Then by Proposition 3, $(Az)_b = c_{k+1}/c_{i+1}$. But, since $-1 \in G_1$, we also have that $-b$ and thus $p-b \in G_{i+1}$ and $p-1-b \in G_{k+1}$, which gives that $z_{p-b-1} = c_{i+1}/c_{k+1}$. Thus $(Az)_b = 1/z_{p-b-1} = \bar{z}_{p-b-1}$. Altogether, we have found that $Az = KBz$, and the result follows as above.

Index two is treated on the "x-level" in Section 5.4.

Finally, it may be mentioned that among the first examples of non-classical bi-equimodular p-functions found numerically by the author and presented in exact form in [2] were those given in (5.10) for $p = 7$ and 11, with no idea that quadratic residues, let alone more general subgroups of $\mathbf{Z}_p^*$, are involved. When finding this structure in the fall of 1985, the author was still unaware of related work [6] by McGehee and related material available in the book [7] by Schroeder.

5.3. INDEX THREE.

Now G_0 is the set of cubic residues modulo p. The relations (5.5), (5.6), and (5.8) do not suffice to determine the transition numbers uniquely but leave two parameters, e.g. n_{01} and n_{02}. It seems to follow from counting the points on elliptic curves over a finite field, that there is essentially only *one*, seemingly stochastic, parameter. We will, however, not discuss this further in this paper but be content with the formulas $n_{10} = n_{22} = n_{01}$, $n_{20} = n_{11} = n_{02}$, $n_{21} = n_{12} = s - n_{01} - n_{02}$, and $n_{00} = s - 1 - n_{01} - n_{02}$. Since -1 is always a cubic residue, we have $m = 0$. Hence (5.3) takes the form

$$(5.11) \qquad \begin{cases} \dfrac{c_0}{c} + \dfrac{c}{c_0} + n_{01}(\dfrac{c_1}{c_0} + \dfrac{c_0}{c_1}) + n_{02}(\dfrac{c_0}{c_2} + \dfrac{c_2}{c_0}) + (s - n_{01} - n_{02})(\dfrac{c_1}{c_2} + \dfrac{c_2}{c_1}) = 1 - s, \\ \dfrac{c_1}{c} + \dfrac{c}{c_1} + n_{02}(\dfrac{c_1}{c_0} + \dfrac{c_0}{c_1}) + (s - n_{01} - n_{02})(\dfrac{c_0}{c_2} + \dfrac{c_2}{c_0}) + n_{01}(\dfrac{c_1}{c_2} + \dfrac{c_2}{c_1}) = 1 - s, \\ \dfrac{c_2}{c} + \dfrac{c}{c_2} + (s - n_{01} - n_{02})(\dfrac{c_1}{c_0} + \dfrac{c_0}{c_1}) + n_{01}(\dfrac{c_0}{c_2} + \dfrac{c_2}{c_0}) + n_{02}(\dfrac{c_1}{c_2} + \dfrac{c_2}{c_1}) = 1 - s. \end{cases}$$

In [3], equations (5.11) are considered for $p = 7$ (denoting c_0/c by a, c_2/c_0 by b, and (unfortunately) c_1/c_2 by c). We have $s = 2$, $n_{01} = 0$, and $n_{02} = 1$. The computer calculations (several hours) show that there are three root classes of index three. One is the classical class, which has size 42 and complexity three. The two others both have size 294 and complexity three. One of them is unimodular and one is essentially real. The *simple* functions in these two classes come from the equation $d^6 + 9d^5 - 3d^4 - 118d^3 - 180d^2 - 3d + 43 = 0$, where we have put $d = c_1/c_2 + c_2/c_1$.

For any fixed p, one could calculate n_{01} and n_{02}, feed the values into (5.11) and then find the number of cyclic p-roots of index three by a similar computer calculation. We now want to get from *one* calculation a similar result for *all* p. Instead of an exact count we have to be satisfied with the following result which is only of interest as long as we do not know that $CR(p)$ is finite for all prime p:

Proposition 4. *For any prime p, the number of cyclic p-roots of pre-index three is finite.*

Proof. We apply the method of [3] to the field $\mathbf{Z}_2$ instead of $\mathbf{Q}$. Since "pre-index three" makes sense only if $p \equiv 1 \pmod 6$, we will have an even s, that is, $s = 0$ in $\mathbf{Z}_2$. We only get 4 versions of (5.11) to check, depending on the parities of n_{01} and n_{02}, rather than infinitely many. If (5.11) has finitely many solutions over $\mathbf{Z}_2$, it also has finitely many solutions over $\mathbf{Q}$. This is so, since the number of solutions is finite if and only if certain resultants are non-zero. If they are non-zero over $\mathbf{Z}_2$, they are certainly non-zero over $\mathbf{Q}$. Introducing, as mentioned above, a, b, and c in (5.11), clearing denominators, and homogenizing with the help of a new variable z, we have to consider the ideal I in $P = \mathbf{Z}_2[a, b, c, z]$, generated by the three polynomials

$$(a^2 + z^2)bcz + n_{01}(b^2c^2 + z^4)a + n_{02}(b^2 + z^2)acz + n_+(c^2 + z^2)abz + abcz^2$$
$$(a^2b^2c^2 + z^6) + n_{02}(b^2c^2 + z^4)az + n_+(b^2 + z^2)acz^2 + n_{01}(c^2 + z^2)abz^2 + abcz^3, \text{ and}$$
$$(a^2b^2 + z^4)c + n_+(b^2c^2 + z^4)a + n_{01}(b^2 + z^2)acz + n_{02}(c^2 + z^2)abz + abcz^2,$$

where (in $\mathbf{Z}_2$) $n_+ = n_{01} + n_{02}$, and n_{01} and n_{02} are 0 or 1. If $n_{01} = n_{02} = 0$, we do not need the methods of [3], since the first equation of (5.11) is $a + a^{-1} = 1$, which gives two values for a, and then we get two valus for b from the third equation, and finally two values of c from the second equation, so there are 8 solutions.

As in [3], we define $H_m(t) = (\text{Hilb}(P/I;t) - \text{Hilb}(P/I : (z^m);t))/t^m$, where "Hilb" is a Hilbert series. Using the Macaulay program, we get in a few seconds (note that the subscripts of H are "eleven" etc, not "one-one"):

$$\text{for } n_{01} = 0 \text{ and } n_{02} = 1,\; H_{11}(t) = H_{12}(t) = (1 + 3t + 6t^2 + 9t^3 + t^4)/(1 - t),$$
$$\text{for } n_{01} = 1 \text{ and } n_{02} = 0,\; H_9(t) = H_{10}(t) = (1 + 3t + 6t^2 + 10t^3)/(1 - t),$$
$$\text{for } n_{01} = 1 \text{ and } n_{02} = 1,\; H_{11}(t) = H_{12}(t) = (1 + 3t + 5t^2 + t^3)/(1 - t).$$

Hence in each case there are finitely many solutions, by the proposition in [3]. The proof is complete. We remark that unless we learn more about the n_{ik} than what (5.5), (5.6), and (5.8) tell us, a similar proof for pre-index 5 would involve 8192 cases.

5.4 INDEX TWO AND EQUIMODULAR EIGENVECTORS OF THE FOURIER TRANSFORM.

A special case of *bi-equimodular* p-vector is of course any *equimodular* eigenfunction of the Fourier transform F. By the Fourier inversion formula, $F^4 = p^2I$. Hence, the only possible eigenvalues are $\pm\sqrt{p}$ and $\pm i\sqrt{p}$. For a fixed pre-index t, our formula (1.1) defines for each p of the form $p = st + 1$ a linear operation on $\mathbf{C}^{t+1}$. Let us for the moment call this operation *the p-transform*. In this section we will for any p discuss the bi-equimodular p-functions with pre-index two from this point of view. We will first consider the case $p = 4k + 1$. In this case, the Gaussian sums are $\eta_0 = -\frac{1}{2} + \frac{1}{2}\sqrt{p}$ and $\eta_1 = -\frac{1}{2} - \frac{1}{2}\sqrt{p}$. The p-transform (sending (c, c_0, c_1) to (d, d_0, d_1)), has matrix

$$A = \begin{pmatrix} 1 & 2k & 2k \\ 1 & \eta_0 & \eta_1 \\ 1 & \eta_1 & \eta_0 \end{pmatrix}.$$

Since the trace of A is $\sqrt{p}$, it follows that $\sqrt{p}$ is a double and $-\sqrt{p}$ a single eigenvalue. We find that (c, c_0, c_1) is an eigenvector corresponding to the double eigenvalue iff $-2c + (1 + \sqrt{p})(c_0 + c_1) = 0$. Taking $c = 1$, and requiring $|c_0| = |c_1| = 1$, we will get $c_1 = \bar{c}_0 = 1/c_0$. This gives us (5.9) with $\delta = +1$.

Taking $p = 4k + 3$, we have $\eta_0 = -\frac{1}{2} + \frac{i}{2}\sqrt{p}$ and $\eta_1 = -\frac{1}{2} - \frac{i}{2}\sqrt{p}$. None of the index two solutions is an eigenvector. But we can "twist" the idea of eigenvector and ask for a vector $(1, c_0, 1)$, such that its p-transform equals $-i\sqrt{p}(c_0, 1, c_0)$. If $|c_0| = 1$, such a vector would of course also serve our purpose. An easy calculation shows that (5.10) gives such a c_0.

6. The structure of $CR(n)$ for some small n.

Unless otherwise noted, the results in the following list are taken from [3], where they are proved and given in more detail. We will notice that when n grows, new phenomena tend to appear, whereas different types of general cyclic n-roots tend to coincide for small n. We state the results mostly on the "z-level", remembering that for p prime, unimodular root classes correspond to bi-classes on the "x-level". In particular, for $p > 2$, the classical root classes are unimodular and have index $\frac{1}{2}(p-1)$, and the sum of there sizes is $p(p-1)$, whereas the ϵ-class is essentially real and not unimodular for $p > 3$ and has size $2p^2$ and complexity and index one. At least for $n = 2, 3, 5, 6$, and 7, all the elements of $CR(n)$ have multiplicity 1 as solutions of (2.1). When n is not prime, let us for the moment use (with quotation marks) the notions of "size" and "root class" as defined in Section 4, replacing A by the identity operator. Then the "size" of the ϵ-"class" (which of course exists also in this case) is $2n^2$.

In the trivial case $n = 2$, there is only one root class, the ϵ-class. It is unimodular.

In the equally trivial case $n = 3$, there is only one root class, the ϵ-class = the classical class. It is unimodular.

There are infinitely many cyclic 4-roots, namely $(t, \delta/t, -t, -\delta/t)$ with $\delta = \pm 1$.

For $n = 5$, Lovász [5] answered Enflo's question by proving that there is no other *bi-class* than the two classical ones. These have size 10 and complexity one. There is just one more *root class*, namely the ϵ-class. Thus $CR(5)$ has 70 elements.

$CR(6)$ has 156 elements. There are two essentially real "root classes", namely one of "size" 72, (the ϵ-"class") and one of "size" 36, one unimodular "root class" of "size" 12 (see Section 7), and finally one "root class" of "size" 36, which is neither unimodular nor essentially real. The two "root classes" of "size" 36 were found by D. Lazard [4], correcting a mistake in [3]. They contain the solution $(c, ce, -ce, -c, -c/e, c/e)$, where $c^6 = -1$ and $e^2 - fe + 1 = 0$ with $f^2 - 2f - 2 = 0$. (Each of the two values of f gives one "root class").

The largest prime p for which $CR(p)$ is known, is 7. There are 924 cyclic 7-roots. These belong to five root classes, one of index one and complexity one (Section 5.1), one (unimodular) of index two and complexity one (Section 5.2), and three (two of which are unimodular) of index three and complexity three (Section 5.3).

For $n = 11$, the author has numerically found a unimodular root class of index 10. This rules out the hypothesis (true for prime p with $3 \leq p \leq 7$) that the index of every bi-equimodular p-function is less than $(p-1)$.

We know 150906 elements in $CR(42)$. In fact, applying Proposition 5 below to $CR(6)$ and $CR(7)$, and to the ϵ-"classes" for certain other factors of 42 than 6 and 7, we find $924 \cdot 156 + 2 \cdot 42^2 + 2 \cdot 21^2 + 6 \cdot 2 \cdot 14^2$ different elements in $CR(42)$.

7. A subset of $CR(n)$ for every n which is not a prime power.

Comparing $CR(2)$ and $CR(3)$ with the "root class" of "size" $12 = 2 \cdot 6$ in $CR(6)$ mentioned in Section 6, we find a pattern which can be generalized to give the following result. The proof is an easy calculation, left to the reader.

Proposition 5. *Let $n = k \cdot l$, where k and l are relatively prime. Then, for each $u \in CR(k)$ and each $v \in CR(l)$, we get a cyclic n-root z from the formula $z_j = u_{j \bmod k} \cdot v_{j \bmod l}$.*

Acknowledgement.

The author thanks Professor D. Lazard for correcting mistakes in [3] and in a preliminar version of the present paper.

References.

[1] J. Backelin, *Square multiples n give infinitely many cyclic n–roots*, Reports, Matematiska Institutionen, Stockholms Universitet, 1989–No 8.

[2] G. Björck, *Functions of modulus one on $\mathbf{Z}_p$ whose Fourier transforms have constant modulus*, Proceedings of A. Haar Memorial Conference, Budapest, 1985, Colloq. Math. Soc. János Bolyai **49**, 193-197 (1985).

[3] G. Björck, R. Fröberg, *A faster way to count the solutions of inhomogeneous systems of algebraic equations, with applications to cyclic n-roots* (submitted to J. Symbolic Comput.), Reports, Matematiska Institutionen, Stockholms Universitet, 1989–No 7.

[4] D. Lazard, private communication.

[5] L. Lovász, private communication.

[6] O.C. McGehee, *Gaussian Sums and Harmonic Analysis on Finite Fields*, Contemp. Math. (AMS) **91**, 177-191 (1989).

[7] M.R.Schroeder, *Number Theory in Science and Communication*, Springer, Berlin - New York, 1984 and 1986.

Part III: Fourier Analysis in Engineering

UNCERTAINTY PRINCIPLE INEQUALITIES AND SPECTRUM ESTIMATION

JOHN J. BENEDETTO
Department of Mathematics
and
Systems Research Center
University of Maryland
College Park, MD 20742

ABSTRACT. Uncertainty principle inequalities are useful devices for estimating signal duration and power spectra in signal analysis. The classical uncertainty principle of quantum mechanics, formulated by Heisenberg, Pauli, Weyl, and Wiener, is an example of such an inequality; and it was applied to signal analysis by Gabor.

Weighted and local versions of the classical uncertainty principle inequality are proved in order to provide fine estimation of signal duration. Further, upper bounds in the case of the classical inequality are analyzed for the wavelet bases or frames in which the signals are decomposed. Finally, in the context of the Bell Labs uncertainty principle inequality, a power spectrum estimation method is introduced; and this leads to a multidimensional signal reconstruction technique. The one-dimensional version of this technique is due to Wiener and Wintner and was developed by Bass and Bertrandias.

The setting throughout is $\mathbb{R}^d$; and the methods include weighted Fourier transform norm inequalities, wavelet theory, and Wiener's generalized harmonic analysis.

INTRODUCTION

We shall present a group of weighted Fourier transform norm inequalities. These inequalities are unified by their common heritage from the classical uncertainty principle of quantum mechanics and by their form which provides estimates of signal energy in terms of both time and frequency information. Weighted Fourier transform norm inequalities are motivated by some of the central issues of signal analysis. For example, in linear system theory weights correspond to various filters in energy concentration problems, and in prediction theory weighted L^p-spaces arise for weights corresponding to power spectra of stationary stochastic processes [B1].

Our presentation is organized as follows:

1. The classical uncertainty principle inequality;
2. Weighted uncertainty principle inequalities;

J. S. Byrnes and J. F. Byrnes (eds.), Recent Advances in Fourier Analysis and Its Applications, 143–182.

3. Local uncertainty principle inequalities;
4. The classical uncertainty principle and wavelet theory;
5. Spectrum estimation and the Wiener-Wintner theorem;
A. Closure theorems;
B. Notation.

We shall forego a summary of results in this Introduction, but we do make the following remarks.

a. Section 1 not only provides a statement and background of the classical uncertainty principle inequality, but motivates the topics and points of view of the subsequent sections, e.g., the goals of modern wavelet theory were formulated in terms of the classical uncertainty principle during the 1940s, cf., Sections 1.2.4-1.2.6 and Section 4.

b. The inequalities proved in Sections 2 and 3 allow for further analysis to determine maximal spaces for which they are valid. Appendix A indicates the procedure.

c. The important Bell Labs uncertainty principle [LPS] is not one of our topics, but it does play a role in the manner we view power spectrum estimation, e.g., Section 5.

d. Our treatment of uncertainty principle inequalities does not depend on the beautiful work of Cowling and Price [CP], but we would be remiss not to reference their contributions.

e. Finally, Section 2 is part of an ongoing project with Hans Heinig.

1. The Classical Uncertainty Principle Inequality

1.1. STATEMENT AND PROOF OF THE INEQUALITY

The *classical uncertainty principle inequality* is

1.1.1. *Theorem.* Given $(t_0, \gamma_0) \in \mathbb{R}\times\hat{\mathbb{R}}$. Then

$$\forall f \in \mathcal{S}(\mathbb{R}), \quad \|f\|_2^2 \le 4\pi \|(t-t_0)f(t)\|_2 \|(\gamma-\gamma_0)\hat{f}(\gamma)\|_2 \tag{1.1.1}$$

and there is equality in (1.1.1) if

$$f(t) = C\, e^{-s(t-t_0)^2}\, e^{2\pi i t\gamma_0} \tag{1.1.2}$$

for $C \in \mathbb{C}$ and $s > 0$.

Proof: The mapping, $f(t) \longmapsto f(t+t_0)\, e^{-2\pi i t\gamma_0}$, shows that it is sufficient to verify (1.1.1) for $(t_0, \gamma_0) = (0,0)$.

The following calculation gives (1.1.1) for $(t_0,\gamma_0) = (0,0)$:

$$\|f\|_2^4 = \left[\int t|f(t)^2|' dt\right]^2 \le \left[\int |t||f(t)^2|' dt\right]^2 \le 4\left[\int |t\bar{f}(t)f'(t)|dt\right]^2$$

(1.1.3) $$\le 4\|tf(t)\|_2^2\|f'(t)\|_2^2 = 16\pi^2\|tf(t)\|_2^2\|\gamma\hat{f}(\gamma)\|_2^2.$$

If f is defined by (1.1.2) then equality is obtain in (1.1.1) by direct calculation.

q.e.d.

1.1.2. *An Elementary Closure Question.* Is (1.1.1) valid for each $f \in L^2(\mathbb{R})$? Obviously, the answer is "yes" if either of the factors on the right side is infinite. If $\|tf(t)\|_2 + \|\gamma\hat{f}(\gamma)\|_2 < \infty$ there is something to check, but the answer is still "yes". The proof follows from Theorem A.1.2 in Appendix A.1 for the case $L^2_{1,1}(\mathbb{R})$. As we shall see, the problem of extending an inequality, proved for functions in a convenient space, to all appropriate functions is not always routine.

1.1.3. *The Classical Inequality for Odd Functions.* DeBruijn [De, Theorem 3.4] has proved that

$$3\|f(t)\|_2^2 \le 4\pi\|tf(t)\|_2\ \|\gamma\hat{f}(\gamma)\|_2$$

for all odd functions $f \in \mathcal{S}(\mathbb{R})$.

1.1.4. *Underlying Inequalities.* The main ingredients in (1.1.3) are integration by parts, Hölder's inequality, and the Plancherel theorem. We shall extend and refine Theorem 1.1.1 in several ways in Sections 2 and 3. The main ingredients of our proofs will be the same: integration by parts or conceptually similar ideas such as generalizations of Hardy's inequality; Hölder's inequality; and weighted norm inequalities for the Fourier transform, of which the Plancherel theorem is a special case.

1.2. HISTORY AND MOTIVATION

The classical uncertainty principle inequality was developed in the context of quantum mechanics. Our results in Sections 2 and 3 will be interpreted in the spirit of Gabor's view of communication theory [G]. We now address this transition.

1.2.1. *Concert Pitch and the Classical Uncertainty Principle Inequality.* In I am a mathematician (pp. 105-107), Wiener comments on his 1925 lecture at Göttingen. It was apparently at this lecture that Theorem 1.1.1 was first proved [Ba]; and, of course, Heisenberg's (as well as Pauli's, Schrödinger's, Weyl's, etc.) profound contributions on the classical uncertainty principle were being made during the same period - a human counterexample to the very same principle!

In any case, Wiener explicitly conceived of the analogy between the laws of physics and musical notation in the sense of normal behavior becoming unpredictable when normal time intervals are sufficiently compressed. For example, let us consider the following idealized piano experiment. The standard for concert pitch is that the A above middle C should have 440 vibrations per second. Thus, the A four octaves down (and the last key on the piano) should have 27.5 vibrations per second. Suppose we could strike this last key for a time interval of 1/30 seconds, i.e., the hammer strikes the string and 1/30 seconds later the damper returns to the string, thereby stopping the sound. We have very precise time information but correspondingly imprecise frequency information since the emitted sound is anything but the desired pure periodic pitch of this low A.

This piano experiment has the flavor of the classical uncertainty principle inequality, and we can quantify it in terms of Theorem 1.1.1. In fact, if a sound (signal) f is emitted at time t_0 and lasts a very short time, then (1.1.1) asserts that the frequency range for f is quite broad. In particular, f is not close to a pure tone of frequency γ_0, for, if it were, then $\|(\gamma-\gamma_0)\hat{f}(\gamma)\|_2$, as well as $\|(t-t_0)f(t)\|_2$, would be small in contrast to the "loudness" $\|f\|_2$.

By comparison, the relevance of Theorem 1.1.1 for quantum mechanics can be illustrated by considering a freely moving mass point with varying location $x \in \mathbb{R}$. The term $\|tf(t)\|_2^2$ represents the average distance of x from its expected value $t_0 = 0$. In fact, the position x is interpreted as a random variable depending on the state function f; more precisely, the probability that x is in a given region $A \subseteq \mathbb{R}$ is defined as $\int_A |f(t)|^2 dt$, and $\|tf(t)\|_2^2$ is the variance of x.

1.2.2. *Wiener on Linear Operators and Quadratic Means.* Quantum mechanics has been the spawning ground for the two topics in our title: uncertainty principle inequalities and spectrum estimation.

We commented on the former well-known relationship in Section 1.2.1. For the latter we require Wiener's brilliant insights from the late 1920's. He and Max Born were among the first (some say the first) to associate linear operators on function spaces to physical quantities. They did this in terms of arithmetic mean integral operators in their study of the Heisenberg theory [W, Volume 3]. Wiener was then able to apply their idea for quantum mechanics to many other topics including filtering and prediction. The mathematical theory for this point of view is Wiener's generalized harmonic analysis, i.e., the Fourier analysis of non-periodic undamped signals [W, Volume 2]. A critical component for effecting this analysis is the space of functions having bounded quadratic means. Windowing methods in the spectrum estimation problem can be viewed as applications of Wiener's theory. This is the topic of Section 5, where we also illustrate the role of the uncertainty principle in spectrum estimation.

1.2.3. *Gabor and the Fundamental Principle of Communication.* In 1946, Dennis Gabor formulated and analyzed the "fundamental principle of communication" [G]. The initial idea is much like Wiener's described in Section 1.2.1: the more a signal f is concentrated, the longer the bandwidth. Thus, local information about f at any given time is inextricably contained in all of the bandwidth. More precisely, if a good description of f in terms of $\hat{f}$ is required then we must have good information about $\hat{f}$ on all of its domain. This point of view can be quantified to a certain extent by effective sampling; but the principle with which Gabor initiates his study is the intrinsic irreconcilability of achieving high definition reconstruction or transmission (of f) and of obtaining sufficient bandwidth sampling (of $\hat{f}$).

As an example, suppose the mode of transmission is by means of Fourier series and that we wish to transmit f on $[-T,T]$ and have available the frequency band $[-\Omega,\Omega]$. The Fourier series of f on $[-T,T]$ considered as a $2T$-periodic function on $\mathbb{R}$ has the form $\Sigma c_n e^{\pi int/T}$, so that the transmission of f is equivalent to the transmission of $\{c_n\}$. Because of the prescribed bandwidth there are effectively $2T\Omega$ "spectral lines" $\gamma_n = n/T$ available to transmit f on $[-T,T]$. Since each datum c_n is associated with a specific spectral line (in the Fourier series), we can only expect to transmit $2T\Omega$ data. The problem (and the principle) is that high resolution of f by N points in $[-T,T]$, say, may require many more than these $2T\Omega$ data and their independent combinations.

Gabor's landmark paper completed a line of research begun by Carson, Nyquist, Küpfmüller, and Hartley, concurrent with the quantum mechanics formulation of the classical uncertainty principle. One of Gabor's basic arguments is that Theorem 1.1.1 is at the root of the "fundamental principle of communication."

1.2.4. *Gabor on the Classical Uncertainty - Principle and Wavelets.* Because of its minimization property in the classical uncertainty principle inequality, Gabor reasoned that the modulated probability pulse in (1.1.2) is the "natural basis on which to build up an analysis of signals in which both time and frequency are recognized as references" [G, p.435]. His analysis of signals, the "Gabor representation," is the origin of the important Weyl-Heisenberg frame - decompositions [DGM] which play such an important role in wavelet theory, e.g., [FG; FJ; HW; M1; M2] for diverse ideas and technology as well as extensive bibliographies on wavelets. Wavelet theory provides discrete signal reconstruction by means of time and frequency localization to the extent allowable by the uncertainty principle, cf., Section 4. The joint localization stands in contrast to the point of view effected by the "fundamental principle of communication." Because of the uncertainty principle there are no contradictions in this contrast.

1.2.5. *Definition/Representation.* a. Given $g \in L^1_{loc}(\mathbb{R})$. The Gabor wavelet $\psi = \psi_g$ is defined as

$$\psi_g(u;t,\gamma) = g(u-t)e^{2\pi i(u\gamma - ct\gamma)}$$

for fixed $c \in \mathbb{R}$. The <u>Gabor wavelet transform</u> of $f \in L^1_{loc}(\mathbb{R})$ is the function

$$F_\psi(f)(t,\gamma) = F_\psi(t,\gamma) = \int f(u)\,\overline{\psi(u;t,\gamma)}du$$

defined on $\mathbb{R}\times\hat{\mathbb{R}}$.

b. As indicated in <u>Section 1.2.4</u>, Gabor considered the case $g(u) = e^{-su^2}$, $s > 0$. The resulting wavelet can be interpreted in terms of the parameters: s is the sharpness of the pulse, t the epoch of its peak, and γ and $\varphi = ct\gamma$ are the frequency and phase constants of modulating oscillation.

c. Given $g \in L^2(\mathbb{R})$ and $a > 0$. Let $\alpha = 1/a$ and $c = 1$. Then $\psi_{m,n}(u) = \psi_g(u;ma,n\alpha)$ is an orthonormal basis of $L^2(\mathbb{R})$ under certain conditions on g, e.g., if the Zak transform of g has modulus 1. In this case we have the <u>Gabor representation</u>,

$$\forall f \in L^2(\mathbb{R}),\ f = \sum_{m,n} c_{m,n}(f)\,\psi_{m,n} \text{ in } L^2(\mathbb{R}). \tag{1.2.1}$$

d. The orthonormal case in part <u>c</u> gives a false impression of versatility since orthonormality is never obtained for compactly supported continuous g or for the Gaussian. On the other hand, the von Neumann lattices $\{(ma,n\alpha)\}$, $\alpha = 1/a$, can be perturbed and/or the orthonormality can be weakened so that the Gabor representation is valid for many different Gabor wavelets.

In light of our discussion in <u>Section 5</u>, we have proved the following continuous Gabor representation for bounded Gabor wavelets ("Gabor representations and wavelets" <u>AMS Contemporary Math. Series</u>, 1989).

1.2.6. *Theorem.* Given $g \in L^\infty(\mathbb{R})$ with continuous and not identically zero autocorrelation,

$$G(u) = \lim_{T\to\infty} \frac{1}{2T}\int_{-T}^{T} g(u+v)\overline{g(v)}dv.$$

Then, for each $f \in L^1(\mathbb{R})$, $\lim_{n\to\infty} \|f_n - f\|_1 = 0$ where

$$f_n(u) = \frac{1}{G(0)} \lim_{T\to\infty} \frac{1}{2T} \iint_{-T}^{T} F_\psi(f)(t,\gamma)\psi(u;t,\gamma)\rho_n(\gamma)dtd\gamma$$

and $\{\check{\rho}_n\} \subseteq L^1(\mathbb{R})$ is an L^1-approximate identity.

1.2.7. *Remark.* Implicit in considering non-Gaussian Gabor wavelets and in veering towards signal analysis with its host of data windows and filters and prediction theoretic weights, we are viewing Theorem 1.1.1 as one of many possible uncertainty principle inequalities. One expects analogous inequalities and interpretations for other norms and weights besides $\|\cdots\|_2$ and t^2 or γ^2; and this is the subject of Sections 2 and 3.

1.3. EXAMPLES

1.3.1. *Unboundedness in the Classical Uncertainty Principle Inequality.* Suppose $(t_0, \gamma_0) = (0,0)$. If $f(t) = \chi_{(T)}(t)$ then $\hat{f}(\gamma) = [\sin(2\pi T\gamma)]/[\pi\gamma]$ and the right side of (1.1.1) is infinite. Similarly, if $f \in L^2(\mathbb{R})$ behaves like $|t|^a$ as $|t| \longrightarrow \infty$, where $a \in [-3/2, -1/2)$ then the right side of (1.1.1) is infinite. ($a < -1/2$ ensures $f \in L^2(\mathbb{R})$ and $a \geq -3/2$ ensures $\|tf(t)\|_2 = \infty$.)

Theorem 1.1.1 does not provide useful information for these signals. This fact, and the importance of comparing time and frequency energy concentrations in signal analysis, lead to the Bell Labs uncertainty principle [LPS], cf., Proposition 5.1.5.

1.3.2. *The Classical Uncertainty Principle Inequality and Energy Concentration.* Given $T, \Omega > 0$. For each $f \in L^2(\mathbb{R})$,

$$E_T(f) = \left(\int_{-T}^{T} |f(t)|^2 dt\right) \Big/ \|f\|_2^2 \quad \left(\text{resp.,} \quad E_\Omega(\hat{f}) = \left(\int_{-\Omega}^{\Omega} |\hat{f}(\gamma)|^2 d\gamma\right) \Big/ \|f\|_2^2\right)$$

represents the proportion of the total energy of f in $[-T,T]$ (resp., of $\hat{f}$ in $[-\Omega,\Omega]$). It is clear that $E_T(f)\, E_\Omega(\hat{f}) \leq 1$ and that

$$\forall \varepsilon > 0, \ \exists f_\varepsilon \in L^2(\mathbb{R}) \text{ such that } E_T(f_\varepsilon)\, E_\Omega(\hat{f}_\varepsilon) < \varepsilon. \tag{1.3.1}$$

Since $\operatorname{supp} f_\varepsilon$ can be taken as a subset of $[-T,T]$ in the verification of (1.3.1), we consider the following ratios vis a vis Theorem 1.1.1.

For each T-time limited $f \in L^2(\mathbb{R})$, let $V(\hat{f}) = \|\gamma\hat{f}(\gamma)\|_2^2 / \|f\|_2^2$. Then, for such f, we have

$$\frac{1}{16\pi^2 T^2} < \inf\{V(\hat{f})\} \leq \sup\{V(\hat{f})\} = \infty$$

by Theorem 1.1.1.

1.3.3. *The Classical Uncertainty Principle Inequality for Hilbert Spaces.* We close Section 1, as we began it, with a statement of the classical uncertainty principle inequality. We give the conventional Hilbert space formulation, cf., the remarks in Section 1.2.2.

Theorem. Let A, B be self-adjoint operators on a Hilbert space H (A and B need not be continuous). Define the commutator $[A,B] = AB - BA$, the expectation $E_f(A) = (Af,f)$ of A at $f \in D(A)$ ($D(A)$ is the domain of A), and the variance $\sigma_f^2(A) = E_f(A^2) - \{E_f(A)\}^2$ of A at $f \in D(A^2)$. If $f \in D(A^2) \cap D(B^2) \cap D(i[A,B])$ and $\|f\| = 1$ then

$$\{E_f(i[A,B])\}^2 \le 4\, \sigma_f^2(A)\, \sigma_f^2(B).$$

The verification is routine, and (1.1.1) is a corollary for $H = L^2(\mathbb{R})$ and for $A(f)(t) = tf(t)$ and $B(f)(t) = i(2\pi i\gamma\hat{f}(\gamma))^{\vee}(t)$, $f \in \mathcal{S}(\mathbb{R})$.

2. Weighted Uncertainty Principle Inequalities

2.1. FIRST METHOD: HARDY AND FOURIER TRANSFORM NORM INEQUALITIES

2.1.1. *Two Fundamental Inequalities.* a. The Hardy operator is the positive linear operator P_d defined as

$$P_d(f)(x) = \int_0^{x_d} \cdots \int_0^{x_1} f(t_1, \cdots, t_d)\,dt_1 \cdots dt_d = \int_{\langle 0,x\rangle} f(t)\,dt$$

for Borel measurable functions f on $\mathbb{R}^{+d}$. The dual Hardy operator P'_d is defined as

$$P'(f)(x) = \int_{x_d}^{\infty} \cdots \int_{x_1}^{\infty} f(t_1, \cdots, t_d)\,dt_1 \cdots dt_d = \int_{\langle x,\infty\rangle} f(t)\,dt,$$

where $x > 0$, i.e., each $x_j > 0$ for $x = (x_1, \cdots, x_d)$.

Hardy's inequality (1920) is

$$\int_0^{\infty} P_1(f)(t)^p t^{-p} dt < \left(\frac{p}{p-1}\right)^p \int_0^{\infty} f(t)^p dt, \tag{2.1.1}$$

where $p > 1$ and $f \ge 0$ ($f \not\equiv 0$) is Borel measurable.

b. The Hausdorff-Young inequality is

$$\forall f \in \mathcal{S}(\mathbb{R}^d), \quad \|\hat{f}\|_{p'} \le B_d(p)\|f\|_p, \tag{2.1.2}$$

where $B_d(p) = (p^{1/p}(p')^{-1/p'})^{d/2}$ and $1 < p \le 2$.

The extension of Hausdorff-Young's inequality for Fourier series to the case of Fourier transforms is due to Titchmarsh (1924). $B_d(p)$ is the best constant and (2.1.2) is an equality for $f(t) = e^{-\pi|t|^2}$

(Babenko (1961) and Beckner (1975)). Finally, (2.1.2) allows an extension to $L^p(\mathbb{R}^d)$ since $\mathscr{S}(\mathbb{R}^d)$ is dense in $L^p(\mathbb{R}^d)$; in particular, the Fourier transform is well-defined for each $f \in L^p(\mathbb{R}^d)$, $1 < p \le 2$.

2.1.2. *Proposition.* Given $1 < p \le 2$.

$$\forall f \in \mathscr{S}(\mathbb{R}), \quad \|f\|_2^2 \le 4\pi B_1(p)\|tf(t)\|_p \|\gamma\hat{f}(\gamma)\|_p. \tag{2.1.3}$$

The proof is similar to the proof of Theorem 1.1.1: the L^p-version of Hölder's inequality is used instead of the L^2-version, and the Hausdorff-Young inequality replaces the Plancherel theorem.

2.1.3. *Proposition* [HS, Theorem 1.1]. Given $1 < p \le 2$.

$$\forall f \in \mathscr{S}_0(\mathbb{R}), \quad \|f\|_2^2 \le 2\pi p B_1(p)\|tf(t)\|_p \|\gamma\hat{f}(\gamma)\|_p. \tag{2.1.4}$$

The constant in (2.1.4) is sharper than that in (2.1.3) for $1 < p < 2$. The closure properties of $\mathscr{S}_0(\mathbb{R})$ are discussed in Appendix A.3. The proof of (2.1.4) is similar to the proof of Theorem 1.1.1 but depends on Hardy's inequality in the following way:

$$\int_0^\infty |\hat{f}(\gamma)|^2 d\gamma \le \left[\int_0^\infty |\gamma\hat{f}(\gamma)|^p d\gamma\right]^{1/p}\left[\int_0^\infty |\tfrac{1}{\gamma}\hat{f}(\gamma)|^{p'} d\gamma\right]^{1/p'}$$

$$= \left[\int_0^\infty |\gamma\hat{f}(\gamma)|^p d\gamma\right]^{1/p}\left[\int_0^\infty |P_1((\hat{f})')(\gamma)|^{p'} \gamma^{-p'} d\gamma\right]^{1/p'}$$

$$\le p\left[\int_0^\infty |\gamma\hat{f}(\gamma)|^p d\gamma\right]^{1/p}\left[\int_0^\infty |(\hat{f})'(\gamma)|^{p'} d\gamma\right]^{1/p'},$$

cf., Sections 2.1.7-2.1.9 for a fuller treatment in a more general setting.

2.1.4. *Definition/Theorem.* Definition. Given even non-negative Borel measurable functions u and v on $\hat{\mathbb{R}}$ and $\mathbb{R}$, respectively. Suppose there is $K > 0$ such that

$$\sup_{s>0} \left[\int_0^{1/s} u(x)dx\right]^{1/q}\left[\int_0^s v(x)^{-p'/p}dx\right]^{1/p'} = K, \tag{2.1.5}$$

where $1 < p \le q < \infty$. In this case we write $(u,v) \in F(p,q)$.

We have the following generalization of the Hausdorff-Young inequality.

Theorem [BH1]. Given $1 < p \le q < \infty$ and even weights u and v for which $(u,v) \in F(p,q)$. Assume $1/u$ and v are increasing on $(0,\infty)$. Then there is $C = C(K)$ such that

$$\forall f \in \mathcal{S}(\mathbb{R}^d) \cap L^p_v(\mathbb{R}^d), \quad \|\hat{f}\|_{q,u} \le C\|f\|_{p,v} < \infty.$$

After completing [BH1] in 1982, we realized we could use Theorem 2.1.4 to prove the following weighted uncertainty principle inequality, e.g., [B2, p.408; HS].

2.1.5. *Theorem.* Given $1 < p \le q < \infty$ and even non-negative Borel measurable functions v and w which are increasing on $(0,\infty)$. Assume $(1/w, v) \in F(p,q)$ with constant K (as in (2.1.5)). Then there is $C = C(K)$ (the same as in Theorem 2.1.4) such that

$$\text{(2.1.6)} \qquad \forall f \in \mathcal{S}(\mathbb{R}), \quad \|f\|_2^2 \le 4\pi C(K)\|tf(t)\|_{p,v}\,\|\gamma\hat{f}(\gamma)\|_{q',w^{q'/q}}.$$

Proof: By means of the first part of (1.1.3) (for $\hat{f}$ instead of f), Hölder's inequality, and Theorem 2.1.4 we have the estimate,

$$\|f\|_2^2 = \|\hat{f}\|_2^2 \le 2\int|\gamma(\hat{f})^{-}(\gamma)(\hat{f})'(\gamma)|\,d\gamma$$

$$= 2\int|\gamma(\hat{f})^{-}(\gamma)w(\gamma)^{1/q}|\,|(\hat{f})'(\gamma)w(\gamma)^{-1/q}|\,d\gamma$$

$$\le 2\left[\int|\gamma(\hat{f})^{-}(\gamma)|^{q'}\,w(\gamma)^{q'/q}d\gamma\right]^{1/q'}\left[\int|(\hat{f})'(\gamma)|^{q}\,w(\gamma)^{-1}d\gamma\right]^{1/q}$$

$$\le 2C\|((\hat{f})')^{\vee}(t)\|_{p,v}\,\|\gamma\hat{f}(\gamma)\|_{q',w^{q'/q}}$$

and the result is obtained since $((\hat{f})')^{\vee}(t) = 2\pi i t f(t)$. q.e.d.

2.1.6. *Remark.* a. Naturally, the right side of (2.1.6) is not necessarily finite, although it is easy to specify a convenient subspace of $\mathcal{S}(\mathbb{R})$ where it is finite. We shall not discuss the closure question for (2.1.6), that is, the question of characterizing the appropriate space of functions for which (2.1.6) is valid and the right side is finite, cf., Section 1.1.2.

b. The monotonicity and symmetry hypotheses in Theorem 2.1.4 and 2.1.5 can be weakened at the expense of complicating condition (2.1.5) with criteria formulated in terms of rearrangements; [BH2] provides remarks and bibliography for contributions in this direction, as well as for higher dimensions.

If $p = 1$ and $q > 1$ then Theorem 2.1.4 is true for any positive Borel measurable weight u. In this case, the proof is routine and the constant C is explicit [BH1, pp.272-273]. If $p > 1$ the constant C is less explicit, but it can be estimated by examining the proof of Calderón's rearrangement inequality (Studia Math. 26(1966), 273-299) which we use in proving Theorem 2.1.4.

2.1.7. *Weighted Hardy Inequalities in* $\mathbb{R}^d$.

Lemma [He, Theorem 3.1]. Given $1 < p \le q < \infty$ and non-negative Borel measurable functions u and v on $X \subseteq \mathbb{R}^d$. Suppose $P : L^p_v(X) \longrightarrow L^q_u(X)$ is a positive linear operator with canonical dual operator P' : $L^{q'}_{u^{-q'/q}}(X) \longrightarrow L^{q'}_{v^{-p'/p}}(X)$ defined by the duality $\int_X P(f)(x)g(x)dx = \int_X f(x)P'(g)(x)dx$. Assume there exist $K_1, K_2 > 0$ such that

$$\forall g \in L^{(q/p)'}(X), \quad \text{for which} \quad g \ge 0 \quad \text{and} \quad \|g\|_{(g/p)'} \le 1,$$

there are non-negative functions,

$$f_1 \in L^p_v(X), \quad h_1 \in L^p_{u^{p/q}g}(X), \quad f_2 \in L^{p'}_{u^{-p'/q}g}(X), \quad h_2 \in L^{p'}_{v^{-p'/p}}(X),$$

with the properties

$$P(f_1) \le K_1 h_1 \quad \text{and} \quad P'(f_2 g) \le K_2 h_2 \tag{2.1.7}$$

and

$$v = f_1^{-p/p'} h_2 \quad \text{and} \quad u = h_1^{-q/p'} f_2^{q/p}.$$

Then $P \in \mathcal{L}(L^p_v(X),\ L^q_u(X))$, $P' \in \mathcal{L}(L^{q'}_{u^{-q'/q}}(X),\ L^{p'}_{v^{-p'/p}}(X))$, and $\|P\|$, $\|P'\| \le K_1^{1/p'}\ K_2^{1/p}$.

Setting

$$f_1 = v^{-p'/p}\ P_d(v^{-p'/p})^{-1/p},$$

$$h_1 = P_d(v^{-p'/p})^{-1/p'},$$

$$f_2 = u^{p/q}\ P'_d(u)^{-p/(qp')},$$

$$h_2 = P_d(v^{-p'/p})^{-1/p'},$$

it is easy to verify (2.1.7) for any non-negative $g \in L^{(q/p)'}(\mathbb{R}^{+d})$, for which $\|g\|_{(q/p)'} \le 1$, as long as (2.1.8), (2.1.9), and (2.1.10) are assumed. As a result, Hernandez obtained the following version of Hardy's inequality on $\mathbb{R}^{+d}$.

Theorem [He, Section 4.2]. Given $1 < p \le q < \infty$ and non-negative Borel measurable functions u and v on $\mathbb{R}^{+d}$. Assume there exist K, $C_1(p)$, $C_2(p) > 0$ such that

$$\sup_{s>0} \left[\int_{\langle s,\infty\rangle} u(x)dx\right]^{1/q} \left[\int_{\langle 0,s\rangle} v(x)^{-p'/p}\,dx\right]^{1/p'} = K, \tag{2.1.8}$$

(2.1.9) $\quad \forall x \in \mathbb{R}^{+d}, \quad P_d(v^{-p'/p}(P_d v^{-p'/p})^{-1/p})(x)$

$$\leq C_1(p)\ P_d(v^{-p'/p})(x)^{1/p'},$$

and

(2.1.10) $\quad \forall x \in \mathbb{R}^{+d}, \quad P_d'(u(P_d' u)^{-1/p'})(x)$

$$\leq C_2(p)^{q/p}\ P_d'(u)(x)^{1/p}.$$

Then $P_d \in \mathcal{L}(L^p_v(\mathbb{R}^{+d}),\ L^q_u(\mathbb{R}^{+d}))$, $P_d' \in \mathcal{L}(L^{q'}_{u^{-q'/q}}(\mathbb{R}^{+d}),\ L^{p'}_{v^{-p'/p}}(\mathbb{R}^{+d}))$, and $\|P_d\|,\ \|P_d'\| \leq KC_1(p)^{1/p'}C_2(p)^{1/p}$.

Remark. Condition (2.1.8) is necessary and sufficient for weighted Hardy inequalities on $\mathbb{R}$ and necessary on $\mathbb{R}^d$, $d > 1$. Conditions (2.1.9) and (2.1.10) are automatically satisfied on $\mathbb{R}$. Conditions (2.1.8), (2.1.9), and (2.1.10) are sufficient but not necessary on $\mathbb{R}^d$, $d > 1$. Sawyer has given a characterization for $d = 2$.

2.1.8. *Regrouping Lemma*. Let Ω be the subgroup of the orthogonal group whose corresponding matrices with respect to the standard basis are diagonal with ± 1 entries. Each element $\omega \in \Omega$ can be identified with an element $(\omega_1, \cdots, \omega_d) \in \{-1, 1\}^d$, and $\omega\gamma = (\omega_1\gamma_1, \cdots, \omega_d\gamma_d)$. Thus,

$$\int F(\gamma)d\gamma = \sum_{\omega \in \Omega} \int_{\hat{\mathbb{R}}^{+d}} F(\omega\gamma)d\gamma,$$

and since

$$\sum_{\omega \in \Omega} a_\omega^{1/r}\, b_\omega^{1/r'} \leq \left(\sum_{\omega \in \Omega} a_\omega\right)^{1/r} \left(\sum_{\omega \in \Omega} b_\omega\right)^{1/r'},$$

for $1 < r < \infty$ and $a_\omega, b_\omega \geq 0$, we have -

Lemma. Given $1 < r < \infty$ and suppose $F \in L^r(\hat{\mathbb{R}}^d)$, $G \in L^{r'}(\hat{\mathbb{R}}^d)$. Then

$$\sum_{\omega \in \Omega} \left(\int_{\hat{\mathbb{R}}^{+d}} |F(\omega\gamma)|^r d\gamma\right)^{1/r} \left(\int_{\hat{\mathbb{R}}^{+d}} |G(\omega\gamma)|^{r'} d\gamma\right)^{1/r'} \leq \|F\|_r\ \|G\|_{r'}.$$

2.1.9. *Definition/Uncertainty Principle Inequality*.

Definition. $\mathcal{S}_{oa}(\mathbb{R}^d) = \{f \in \mathcal{S}(\mathbb{R}^d):\ \hat{f}(\gamma) = 0$ if some $\gamma_j = 0\} \subseteq \mathcal{S}_o(\mathbb{R}^d)$. Thus, $f \in \mathcal{S}(\mathbb{R}^d)$ is an element of $\mathcal{S}_{oa}(\mathbb{R}^d)$ if $\hat{f} = 0$ on the coordinate axes.

Combining Hardy's inequality (Theorem 2.1.7) and the regrouping lemma (Lemma 2.1.8) we obtain the following uncertainty principle inequality.

Theorem. Given $1 < r < \infty$ and non-negative Borel measurable weights v and w. Suppose $u = w^{-r'/r}$, and assume that, for all $\omega \in \Omega$, the weights $u(\omega\gamma)$ and $v(\omega\gamma)$ satisfy conditions (2.1.8), (2.1.9) and (2.1.10) on $\hat{\mathbb{R}}^{+d}$ for $p = q = r'$ and constants $K(\omega)$, $C_1(p,\omega)$, and $C_2(p,\omega)$. If $C = \sup_{\omega\in\Omega} K(\omega)C_1(p,\omega)^{1/p'}C_2(p,\omega)^{1/p}$ then

$$\text{(2.1.11)} \qquad \forall f \in \mathcal{S}_{oa}(\mathbb{R}^d), \quad \|f\|_2^2 \le C\|\hat{f}\|_{r,w}\|\partial_1,\cdots,\partial_d\hat{f}\|_{r',v}.$$

2.1.10. *Method.* At this point, generalizations of Proposition 2.1.3 and Theorem 2.1.5 can be stated by applying d-dimensional versions of Theorem 2.1.4 to the factor $\|\partial_1,\cdots,\partial_d\hat{f}\|_{r',v}$, on the right side of (2.1.11), e.g., Remark 2.1.6b. We shall avoid a baroque extravaganza with all forms of rearrangements, and confine ourselves to the following section.

2.1.11. *Corollaries.*

If $v = 1$ and $p = q = r'$ then (2.1.8) has the form,

$$\text{(2.1.12)} \qquad \sup_{s>0} (s_1\cdots s_d)^{1/r}\left[\int_{\langle s,\infty\rangle} u(y)dy\right]^{1/r'} = K,$$

and (2.1.9) is satisfied for $C_1(r') = r^d$.

Corollary. Given $1 < r \le 2$ and let the non-negative Borel measurable weight w be invariant under the action of Ω. Assume $K < \infty$ (in (2.1.12)) for $u = w^{-r'/r}$ and that

$$\text{(2.1.13)} \qquad P'_d(w^{-r'/r}(P'_d w^{-r'/r})^{-1/r}) \le C_2(r')(P'_d w^{-r'/r})^{1/r'}.$$

Then

$$\forall f \in \mathcal{S}_{oa}(\mathbb{R}^d), \quad \|f\|_2^2 \le (2\pi)^d r^{d/r} K C_2(r')^{1/r'} B_d(r)\|t_1\cdots t_d f(t)\|_r\|\hat{f}\|_{r,w}$$

$$\le (2\pi)^d r^{d/r} d^{-d/2} K C_2(r')^{1/r'} B_d(r)\|\,|t|^d f(t)\|_r\|\hat{f}\|_{r,w}.$$

The weight $w(\gamma) = |\gamma_1\cdots\gamma_d|^r$, $1 < r \le 2$, is Ω-invariant, $K = (r' - 1)^{-d/r'}$ in (2.1.12), and (2.1.13) is satisfied for $C_2(r') = (r(r' - 1))^d$. Consequently, we obtain the following d-dimensional generalization of Proposition 2.1.3.

Corollary. Given $1 < r \le 2$. Then

$$\forall f \in \mathcal{S}_{oa}(\mathbb{R}^d), \quad \|f\|_2^2 \le (2\pi r)^d B_d(r)\|t_1\cdots t_d f(t)\|_r\|\gamma_1\cdots\gamma_d\hat{f}(\gamma)\|_r.$$

2.2. SECOND METHOD: A_p-WEIGHTS, AND WEIGHTED GRADIENT AND RIESZ TRANSFORM INEQUALITIES

2.2.1. *Weighted Gradient Inequalities.*

Theorem [Si, Theorem 4.1]. Given $1 < q < \infty$ and non-negative Borel measurable functions u and v on $\mathbb{R}^d$.

a. There is a constant $C > 0$ such that

$$\forall g \in C_c^\infty(\mathbb{R}^d), \quad \|g\|_{q,u} \le C\|t\cdot\nabla g(t)\|_{q,v} \tag{2.2.1}$$

if and only if

$$\sup_{s\in\mathbb{R}^d}\left[\int_0^1 u(xs)x^{d-1}dx\right]^{1/q}\left[\int_1^\infty (v(xs)x^d)^{-q'/q}x^{-1}dx\right]^{1/q'} = K < \infty. \tag{2.2.2}$$

The constants C and K satisfy the inequalities,

$$K \le C \le K\, q^{1/q}(q')^{1/q'}.$$

b. There is a constant $C > 0$ such that

$$\forall g \in C_c^\infty(\mathbb{R}^d) \text{ for which } g(0) = 0, \quad \|g\|_{q,u} \le C\|t\cdot\nabla g(t)\|_{q,v} \tag{2.2.3}$$

if and only if

$$\sup_{s\in\mathbb{R}^d}\left[\int_1^\infty u(xs)x^{d-1}dx\right]^{1/q}\left[\int_0^1 (v(xs)x^d)^{-q'/q}x^{-1}dx\right]^{1/q'} = K < \infty. \tag{2.2.4}$$

2.2.2. *A_p-weights and Fourier Transform Norm Inequalities.*

Definition. Given $1 < p < \infty$ and a non-negative Borel measurable function w on $\mathbb{R}^d$. w is an A_p-weight, written $w \in A_p$, if

$$\sup_Q \left[\frac{1}{|Q|}\int_Q w(x)dx\right]\left[\frac{1}{|Q|}\int_Q w(x)^{-p'/p}dx\right]^{p/p'} = K < \infty.$$

Theorem. Given $1 < p \le q \le p' < \infty$ and let w be a non-negative Borel measurable radial function on $\mathbb{R}^d$. Assume $w(|t|)$ is increasing on $(0,\infty)$. There is a constant $C > 0$ such that

$$\forall f \in C_c^\infty(\mathbb{R}^d), \quad \left[\int |\hat{f}(\gamma)|^q\, |\gamma|^{d(\frac{q}{p'}-1)}\, w\left(\frac{1}{|\gamma|}\right)^{q/p}d\gamma\right]^{1/q} \le C\|f\|_{p,w} \tag{2.2.5}$$

if and only if $w \in A_p$.

(The reference for the case d = 1 is J. Benedetto, H. Heinig, R. Johnson, "Fourier inequalities with A_p-weights" ISNM 80 (1987), 217-232. The d > 1 version stated above is in [HSi, Theorem 2.10].)

Remark. We view Theorem 2.2.2 as the culmination of some interesting classical analysis. Take d = 1. In case p = q and w = 1, (2.2.5) is the Hardy, Littlewood, Paley theorem (1931),

$$\int |\hat{f}(\gamma)|^p |\gamma|^{p-2} d\gamma \le C\|f\|_p.$$

If $q = p'$ and $w = 1$, (2.2.5) is the Hausdorff-Young theorem. If $w(t) = |t|^\alpha$, $0 \le \alpha < p-1$, then (2.2.6) reduces to Pitt's theorem (1937),

$$\left(\int |\hat{f}(\gamma)|^q |\gamma|^{-\beta} d\gamma\right)^{1/q} \le C\left(\int |f(t)|^p |t|^\alpha dt\right)^{1/p},$$

where $\beta = \frac{q}{p}(\alpha+1) + 1 - q$. The fact that Fourier transform inequalities are characterized in terms of A_p-weights was initially surprising since the A_p-condition was associated with maximal function and singular integral norm inequalities.

2.2.3. *Definition/Theorem on Riesz Transforms.*

Definition. The d-dimensional Riesz transforms are the d singular integral operators $R_1, \cdots, R_d$ defined by the odd kernels $k_j(x) = \Omega_j(x)/|x|^d$, $j = 1, \cdots, d$, where $\Omega_j(x) = c_d x_j/|x_j|$ and $c_d = \Gamma(\frac{d+1}{2})/\pi^{(d+1)/2}$. In fact,

$$(R_j f)(x) = \lim_{T^{-1}, \varepsilon \to 0} \int_{\varepsilon \le |t| \le T} f(x-t) k_j(t) dt$$

exists a.e. for each $f \in L^p(\mathbb{R}^d)$, $1 < p < \infty$, and there is $C = C(p)$ such that

$$\forall f \in L^p(\mathbb{R}^d), \quad \|R_j f\|_p \le C\|f\|_p,$$

$j = 1, \cdots, d$. $C = C(p)$ does not depend on d [GR, p.223]. Also, we compute

$$\hat{k}_j(\gamma) = -i \frac{\gamma_j}{|\gamma|}, \quad j = 1, \cdots, d$$

Theorem (Hunt, Muckenhoupt, and Wheeden, 1973). Given $1 < p < \infty$ and suppose $w \in A_p$. Then $R_j \in \mathcal{L}(L^p_w(\mathbb{R}^d), L^p_w(\mathbb{R}^d))$, $j = 1, \cdots, d$, e.g., [GR, pp.196, 204, 411-413].

2.2.4. *Uncertainty Principle Inequality.*

Theorem. Given $1 < r \le 2$ and a non-negative radial weight $w \in A_r$ on $\mathbb{R}^d$ for which $w(|t|)$ is increasing on $(0,\infty)$. Assume

(2.2.6)
$$\sup_{s\in\mathbb{R}^d}\left(\int_0^1 \frac{w(xs)^{-r'/r}}{|xs|^{r'}}\, x^{d-1}dx\right)^{1/r'} \times$$
$$\times \left(\int_1^\infty w\left(\frac{1}{|xs|}\right)^{-1}|xs|^r\, x^{-\frac{dr}{r'}-1}\, dx\right)^{1/r} = K < \infty.$$

Then there is a constant $C = C(K) > 0$ such that

(2.2.7)
$$\forall f \in C_c^\infty(\mathbb{R}^d),\quad \|f\|_2^2 \le C\| \,|t|f(t)\|_{r,w}\| \,|\gamma|\hat{f}(\gamma)\|_{r,w}.$$

Proof: a. For $1 < r < \infty$ we have

(2.2.8)
$$\|f\|_2^2 \le \| \,|t|f(t)\|_{r,w}\|f\|_{r',u}.$$

where
$$u(t) = |t|^{-r'}\, w(t)^{-r'/r}.$$

b. The second factor on the right side of (2.2.8) is estimated by means of Theorem 2.2.1a where q and v in (2.2.1) are $q = r'$ and
$$v(t) = |t|^{-r'}\, w\left(\frac{1}{|t|}\right)^{r'/r},$$
respectively. Thus,

(2.2.9)
$$\|f\|_{r',u} \le C_1\|t\cdot\nabla f(t)\|_{r',v}$$

if and only if (2.2.6) holds.

c. By Minkowski's inequality the right side of (2.2.9) is bounded by

(2.2.10)
$$C_1\sum_{j=1}^{d}\left(\int |(R_jG_j)^{\vee}(t)|^{r'}\, w\left(\frac{1}{|t|}\right)^{r'/r} dt\right)^{1/r'},$$

where $G_j^{\vee}(t) = \partial_j f(t)$. Combining (2.2.9) and (2.2.10), and applying Theorem 2.2.2 for the case $p = r$ and $q = p'$ (so that $1 < r \le 2$), we obtain

(2.2.11)
$$\|f\|_{r',u} \le C_1C_2\sum_{j=1}^{d}\|R_jG_j\|_{r,w}.$$

d. Finally, combining (2.2.8) and (2.2.11) and applying Theorem 2.2.3 to the right side of (2.2.11) we have the estimate

$$\|f\|_2^2 \le C_1C_2C_3\| |t|f(t)\|_{r,w} \sum_{j=1}^{d} \|(\partial_j f)^\wedge(\gamma)\|_{r,w} \tag{2.2.12}$$

$$\le 2\pi d^{1/r'} C_1C_2C_3\| |t|f(t)\|_{r,w} \left[\int |\hat{f}(\gamma)|^r \left[\sum_{j=1}^{d} |\gamma_j|^r\right] w(\gamma)d\gamma\right]^{1/r}$$

$$\le 2\pi d^{1/2} C_1C_2C_3\| |t|f(t)\|_{r,w} \| |\gamma|\hat{f}(\gamma)\|_{r,w}.$$

q.e.d.

2.2.5. *Corollary*. Given $1 < r \le 2$ and $d > r'$. Then there is $C > 0$ such that

$$\forall f \in \mathcal{S}(\mathbb{R}^d), \quad \|f\|_2^2 \le C\| |t|f(t)\|_r \| |\gamma|\hat{f}(\gamma)\|_r. \tag{2.1.13}$$

2.2.6. *Remark*. a. In the notation of (2.2.12) the constant C in Corollary 2.2.5 is of the form

$$C = 2\pi d^{1/2} C_1(r,d)B_d(r)C_3(r).$$

Since it is of interest to measure the growth of C as d increases, we note that $C_1(r,d)$ can be estimated in terms of K in (2.2.6) for any w, cf., the second corollary of Section 2.1.11.

b. Theorem 2.2.1b gives rise to an analogue of Theorem 2.2.4 which, for $w = 1$, yields (2.2.13) for $d < r'$.

3. Local Uncertainty Principle Inequalities

3.1. THE RESULTS OF FARIS AND PRICE

Faris' local uncertainty principle inequality is -

3.1.1. *Theorem* [F, (3.2)]. Given a Borel measurable set $E \subseteq \hat{\mathbb{R}}$. Then

$$\forall f \in \mathcal{S}(\mathbb{R}), \quad \int_E |\hat{f}(\gamma)|^2 d\gamma \le 2\pi|E|\|tf(t)\|_2\|f\|_2, \tag{3.1.1}$$

cf., Remark 3.1.5b.

Price's generalization of Faris' result is -

3.1.2. *Theorem* [P, Theorem 1.1]. Given a Borel measurable set $E \subseteq \hat{\mathbb{R}}^d$ and $\alpha > d/2$. Then

$$\forall f \in \mathcal{S}(\mathbb{R}^d), \quad \int_E |\hat{f}(\gamma)|^2 d\gamma \le C|E|\| |t|^\alpha f(t)\|_2^{\frac{d}{\alpha}}\|f\|_2^{2-\frac{d}{\alpha}} \tag{3.1.2}$$

where

$$C = \frac{\omega_{d-1}}{2\alpha}\,\Gamma\left(\frac{d}{2\alpha}\right)\Gamma\left(1-\frac{d}{2\alpha}\right)\left(\frac{2\alpha}{d}-1\right)^{\frac{d}{2\alpha}}\left(1-\frac{d}{2\alpha}\right)^{-1}$$

and $C|E|$ is the smallest possible constant.

3.1.3. *A Local Uncertainty Principle.* Theorem 1.1.1 asserts that if f is concentrated then $\hat{f}$ is spread out. Faris' and Price's theorems quantify the folklore that this spread is smooth, and is not the result of $\hat{f}$ having isolated peaks far from the origin. For example, if $f_n = \sqrt{n}\,\chi_{(1/(2n))}$ then the spectral energy in any fixed band E tends to 0 as $n \to \infty$ since $\|f_n\|_2 = 1$ and $\|tf_n(t)\|_2 = 1/(2n\sqrt{3})$. Similarly, if $\hat{f}$ had a peak about the point γ_0, then $\|\hat{f}\|_{2,E_j}$ would be more or less constant as $|E_j| \to 0$, where $\gamma_0 \in E_j$, thereby contradicting (3.1.1).

Theorem 3.1.2 has been generalized to Heisenberg groups [PS].

3.1.4. *Proposition* (Carlson, 1934).

$$\forall f \in \mathcal{S}(\mathbb{R}),\quad \|f\|_1^2 \le 2\pi\|tf(t)\|_2\|f\|_2.$$

Proof: By Hölder's inequality,

$$\frac{1}{\pi}\|f\|_1^2 \le \frac{1}{\sqrt{t}}\int (t+u^2)|f(u)|^2du = t^{1/2}\|f\|_2^2 + t^{-1/2}\|uf(u)\|_2^2;$$

and the result is obtained by minimizing this function of t.

q.e.d.

3.1.5. *Remark* a. Carlson proved his inequality for power series and obtained the smallest possible constant. This inequality and related ones are used in [B2] in conjunction with the Bell Labs uncertainty principle, e.g., Proposition 5.1.5. Beurling (1938) found a new proof of Carlson's inequality allowing an extension to $\mathbb{R}^d$ [K]. For example, in $\mathbb{R}^2$ the inequality has the form,

$$\|f\|_1^2 \le C(\|\partial_1^2\hat{f}\|_2 + \|\partial_2^2\hat{f}\|_2)\|f\|_2.$$

b. Our reason for discussing Carlson's inequality is to suggest alternative proofs of Theorem 3.1.1 and 3.1.2. (3.1.1) is an immediate consequence of Proposition 3.1.4:

$$\|\hat{f}\|_{2,E}^2 \le |E|\,\|\hat{f}\|_\infty^2 \le |E|\,\|f\|_1^2 \le 2\pi|E|\,\|tf(t)\|_2\|f\|_2.$$

Theorem 3.1.2 would require a weighted version of Carlson's inequality on $\mathbb{R}^d$.

3.2. LOCAL UNCERTAINTY PRINCIPLE INEQUALITIES ON $\mathbb{R}$

3.2.1. *Hardy's Inequality.* We shall use Theorem 2.1.7 for $d = 1$. In this case (2.1.9) and (2.1.10) are automatically satisfied and we can take $C_1(p) = p'$ and $C_2(p) = p^{p/q}$. Also, if $p = 1$ or $q = \infty$ then Theorem 2.1.7 is valid and $\|P_1\|$ can be taken as K (defined in (2.1.8)).

The weight u in Theorem 2.1.7 can be replaced by $\mu \in M_+(\mathbb{R})$ [BH2].

Finally, we should mention that Hardy's inequality can be formulated on rather general spaces, not just $\mathbb{R}^{+d}$. In particular, there are formulations on $\mathbb{R}^d$. The reason we have chosen $\mathbb{R}^{+d}$ is illustrated in the following result. In fact, it is natural to apply Theorem 3.2.2 to the weight $v = 1$; and in this case the constant defined by (3.2.1) would be infinite if we employed Hardy's inequality on $\mathbb{R}$. Consequently, we deal with $\mathcal{S}_0(\mathbb{R})$ and $\mathbb{R}^+$ instead of $\mathcal{S}(\mathbb{R})$ and $\mathbb{R}$.

3.2.2. *Theorem.* Given $1 < r < \infty$ and non-negative Borel measurable weights u, v, and w. Define $u_{\pm}(\gamma) = u(\pm\gamma)$ with similar notation for v and w. Assume

$$\text{(3.2.1)} \qquad \sup_{s>0} \left(\int_s^\infty \frac{u_{\pm}(x)}{w_{\pm}(x)^{r'}}dx\right)^{1/r'} \left(\int_0^s v_{\pm}(x)^{-r/r'}dx\right)^{1/r} = K_{\pm} < \infty.$$

There exist constants $C_{\pm} > 0$ for which $K_{\pm} \le C_{\pm} \le K_{\pm}(r)^{1/r}(r')^{1/r'}$ and such that

$$\text{(3.2.2)} \qquad \forall f \in \mathcal{S}_0(\mathbb{R}), \quad \|\hat{f}\|_{2,u}^2 \le C\|(\hat{f})'\|_{r',v}\|\hat{f}w\|_{r,u},$$

where $C = \max\{C_+, C_-\}$.

"Proof": The result follows from the estimate,

$$\int_0^\infty |\hat{f}(\gamma)|^2 u(\gamma)d\gamma = \int_0^\infty |\hat{f}(\gamma)w(\gamma)\hat{f}(\gamma)\frac{1}{w(\gamma)}|u(\gamma)d\gamma$$

$$\le \left(\int_0^\infty |\hat{f}(\gamma)w(\gamma)|^r u(\gamma)d\gamma\right)^{1/r} \left(\int_0^\infty P_1|(\hat{f})'|(\gamma)^{r'}\frac{u(\gamma)}{w(\gamma)^{r'}}d\gamma\right)^{1/r'},$$

Hardy's inequality, and the regrouping lemma.

"q.e.d."

The following result reduces to Proposition 2.1.3 in case $u(\gamma) = 1$ and $w(\gamma) = |\gamma|$.

3.2.3. *Corollary*. Given $1 < r \le 2$ and non-negative Borel measurable weights u and w. Define $u_{\pm}(\gamma) = u(\pm\gamma)$ and $w_{\pm}(\gamma) = w(\pm\gamma)$. Assume

$$\sup_{s>0} s^{1/r}\left(\int_s^{\infty} \frac{u_{\pm}(x)}{w_{\pm}(x)^{r'}}\,dx\right)^{1/r'} = K_{\pm} < \infty. \tag{3.2.3}$$

Then

$$\forall f \in \mathcal{S}_0(\mathbb{R}),\ \|\hat{f}\|_{2,u}^2 \le 2\pi B_1(r)K(r)^{1/r}(r')^{1/r'}\|tf(t)\|_r\|\hat{f}w\|_{r,u}, \tag{3.2.4}$$

where $K = \max\{K_+, K_-\}$.

3.2.4. *Example*. a. If $u = \chi_{(\Omega)}$ and $w = 1$, then Corollary 3.2.3 yields the inequality,

$$\int_{-\Omega}^{\Omega} |\hat{f}(\gamma)|^2 d\gamma \le 2\pi B_1(r)\Omega\left(\int |tf(t)|^r dt\right)^{1/r}\left(\int_{-\Omega}^{\Omega} |\hat{f}(\gamma)|^r d\gamma\right)^{1/r}, \tag{3.2.5}$$

for $1 < r \le 2$ and $f \in \mathcal{S}_0(\mathbb{R})$. (3.2.5) improves on Theorem 3.1.1 for $f \in \mathcal{S}_0(\mathbb{R})$.

b. If $u = \chi_E$ and $w(\gamma) = |\gamma|$ then Corollary 3.2.3 yields the inequality,

$$\int_E |\hat{f}(\gamma)|^2 d\gamma \le 2\pi r B_1(r)\left(\int |tf(t)|^r dt\right)^{1/r}\left(\int_E |\gamma\hat{f}(\gamma)|^r d\gamma\right)^{1/r}, \tag{3.2.6}$$

for $1 < r \le 2$ and $f \in \mathcal{S}_0(\mathbb{R})$. (3.2.6) is a local version of Proposition 2.1.3.

c. In searching for smallest constants in (3.2.6), a direct calculation gives the following result for $E = [-\Omega, \Omega]$ and for the Gaussian $g(t) = (1/\sqrt{\pi})e^{-t^2}$ with dilation $g_\lambda(t) = \lambda g(\lambda t)$:

$$\forall \varepsilon > 0,\ \exists\, \lambda(\varepsilon,\Omega) > 0 \text{ such that } \forall \lambda > \lambda(\varepsilon,\Omega),$$

$$16\pi^2 \int |tg_\lambda(t)|^2 dt \int_{-\Omega}^{\Omega} |\gamma\hat{g}_\lambda(\gamma)|^2 d\gamma \le \left(\int_{-\Omega}^{\Omega} |\hat{g}_\lambda(\gamma)|^2 d\gamma\right)^2 + \varepsilon.$$

3.2.5. *Example*. a. Given $1 < r \le 2$, $1 < p < \infty$, a symmetric set $E \subseteq \hat{\mathbb{R}}$ for which $|E| < \infty$, and $f \in \mathcal{S}_0(\mathbb{R})$. Then

(3.2.7)

$$\int_E |\hat{f}(\gamma)|^2 d\gamma \le C(p,r)|E|^{\frac{1}{p'r'}}\left(\int |tf(t)|^r dt\right)^{1/r}\left(\int_E |\gamma|^{1+\frac{r}{pr'}} |\hat{f}(\gamma)|^r d\gamma\right)^{1/r},$$

where

$$C(p,r) = 2^{1-\frac{1}{p'r'}} \pi \left[\frac{r}{pr'}\right]^{\frac{1}{pr'}} (r)^{1/r}(r')^{1/r'} B_1(r).$$

(3.2.7) is a consequence of Corollary 3.2.3 by using Hölder's inequality (for p and p') on (3.2.3) and then setting $u = \chi_E$ and $w(\gamma) = |\gamma|^{\alpha}$ for $\alpha = \frac{1}{r} + \frac{1}{pr'}$.

Naturally, (3.2.7) is just another of many special cases of (3.2.4). We mention it since it involves the measure of E as does Example 3.2.4a and a power weight factor of $\hat{f}$ as does Example 3.2.4b.

b. Letting $p \to \infty$, $p \to 1$, and $r \to 1$, (3.2.7) gives rise to the following inequalities:

$$\int_E |\hat{f}(\gamma)|^2 d\gamma \le 2^{1/r}\pi(r)^{1/r}(r')^{1/r'}B_1(r)|E|^{1/r'} \times$$

$$\times \left[\int |tf(t)|^r dt\right]^{1/r} \left[\int_E |\gamma|\,|\hat{f}(\gamma)|^r d\gamma\right]^{1/r};$$

$$\int_E |\hat{f}(\gamma)|^2 d\gamma \le 2\pi r B_1(r) \left[\int |tf(t)|^r dt\right]^{1/r} \left[\int_E |\gamma|^{1+\frac{r}{r'}} |\hat{f}(\gamma)|^r d\gamma\right]^{1/r};$$

$$\int_E |\hat{f}(\gamma)|^2 d\gamma \le 2\pi \int |tf(t)|dt \int_E |\gamma\hat{f}(\gamma)|d\gamma.$$

3.3. LOCAL UNCERTAINTY PRINCIPLE INEQUALITIES ON $\mathbb{R}^d$

Instead of pursuing the method of Section 3.2 where Hölder, Hardy and Fourier transform inequalities were involved in that order, we shall state a weighted Fourier transform norm inequality (Theorem 3.3.1) particularly suited to making local estimates of spectral energy. The resulting method to obtain local uncertainty principle inequalities is first to use this weighted norm inequality Theorem 3.3.1 to estimate $\|\hat{f}\|^2_{2,u}$, and then to implement Hölder's inequality in the usual ways. We omit statements of the uncertainty principle inequalities which follow from this procedure.

3.3.1. *Theorem* [BH2, Theorem 4.3]. Given $1 < r \le 2$ and non-negative radial weights $u, v \in L^1_{loc}(\mathbb{R}^d)$. Suppose $v^{1-r'} \in L^1_{loc}(\mathbb{R}^d\backslash B(0,T))\backslash L^1(\mathbb{R}^d)$ for each $T > 0$. Assume

$$\sup_{s>0} \left[\int_0^s x^{d+1} u\left(\frac{x}{\pi}\right)dx\right]^{1/2} \left[\int_0^{1/s} x^{d-1+r'} v(x)^{1-r'} dx\right]^{1/r'} = K_1 < \infty$$

and

$$\sup_{s>0} \left[\int_s^{\infty} x^{d-1}\, u\left(\frac{x}{\pi}\right)dx\right]^{1/2} \left[\int_{1/s}^{\infty} x^{d-1} v(x)^{1-r'}\,dx\right]^{1/r'} = K_2 < \infty.$$

There is a constant $C > 0$ such that

$$\text{(3.3.1)} \qquad \forall f \in L^r_v(\mathbb{R}^d), \quad \|\hat{f}\|_{2,u} \le C\|f\|_{r,v},$$

and C can be chosen as

$$C = 2\pi^{-(d-1)/2}\, \omega_{d-1}^{\frac{1}{2}+\frac{1}{r'}} (r)^{1/2}(r')^{1/r'}(K_1 + K_2).$$

The integrability condition on $v^{1-r'}$ allows us to obtain the inequality (3.3.1) for each $f \in L^r_v(\mathbb{R}^d)$, e.g., Appendix A.2. If supp u is compact and $v^{1-r'}$ is integrable off of a neighborhood of the origin then $K_2 < \infty$, and it is for this reason that Theorem 3.3.1 can be used to obtain local inequalities.

4. The Classical Uncertainty Principle and Wavelet Theory

4.1. WAVELET BASES AND WEYL-HEISENBERG FRAMES

The first problem of wavelet theory is to construct best possible orthonormal bases $\{\psi_n\}$ of $L^2(\mathbb{R})$. "Best possible" means that each ψ_n should be as smooth as possible and should have controllable, preferably compact, support. It also means that each basis should be sparse and "localizable", a notion which indicates that local changes or fine tuning in a signal can be made by adjustments to a small number of basis elements. The ψ_n are wavelets, cf., Remark 4.1.2c.

More general notions to effect decompositions are implemented when orthonormal bases are too complicated for applicability or too restrictive, e.g., Section 4.1.3, cf., Section 1.2.5d.

4.1.1. *Theorem* (Daubechies [D2]). For each $r \ge 1$ there is a compactly supported function $\psi \in C^{(r)}(\mathbb{R})$ (the space of r-times continuously differentiable functions) with the property that $\{\psi_{m,n}\} \subseteq L^2(\mathbb{R})$ is an orthonormal basis of $L^2(\mathbb{R})$ where

$$\text{(4.1.1)} \qquad \forall\, m,n \in \mathbb{Z}, \quad \psi_{m,n}(t) = 2^{m/2}\psi(2^m t - n).$$

4.1.2. *Remark*. a. Theorem 4.1.1 is difficult to prove. The first result of this type is due to Meyer [M1]. The Meyer "analyzing wavelet" ψ is an element of $\mathcal{S}(\mathbb{R})$ for which supp $\hat{\psi}$ is compact; and the resulting orthonormal basis $\{\psi_{m,n}\}$ is an unconditional basis of Sobolev spaces, Besov spaces, etc. Related results are due to Lemarié and Battle.

cf., Section 1.2.5a,c for $c = 0$. If $\{\psi_{m,n}\}$ is a frame it is designated a Weyl-Heisenberg frame for $L^2(\mathbb{R})$.

4.1.4. *Theorem* [D1], cf., [HW]. Given $g \in L^2(\mathbb{R})$ and $a > 0$. Assume

$$(4.1.4) \qquad 0 < A = \operatorname*{ess\,inf}_{t\in\mathbb{R}} \sum |g(t-ma)|^2 \le \operatorname*{ess\,sup}_{t\in\mathbb{R}} \sum |g(t-ma)|^2 = B < \infty$$

and

$$(4.1.5) \qquad \lim_{\alpha\to 0} \sum_{k\neq 0} \beta\left(\frac{k}{\alpha}\right) = 0,$$

where

$$\beta(s) = \operatorname*{ess\,sup}_{t\in\mathbb{R}} \sum |g(t-ma)||g(t-s-ma)|.$$

Then there is $\alpha_0 > 0$ such that for each $\alpha \in (0,\alpha_0)$, $\{\psi_g(t;ma,n\alpha)\}$ is a Weyl-Heisenberg frame for $L^2(\mathbb{R})$ with frame bounds

$$\alpha^{-1}A - \alpha^{-1}\sum_{k\neq 0}\beta\left(\frac{k}{\alpha}\right) \quad \text{and} \quad \alpha^{-1}B + \alpha^{-1}\sum_{k\neq 0}\beta\left(\frac{k}{\alpha}\right).$$

4.1.5. *Remark.* a. Condition (4.1.4) is a necessary condition in order that $\{\psi_g(t;ma,n\alpha)\}$ be a Weyl-Heisenberg frame. The sufficient condition (4.1.5) has been the subject of an important analysis in terms of Wiener-type spaces by D. Walnut, cf., Section 4.2.2d.

b. The relative merits of the translations/modulations of (4.1.3) and of the translations/dilations of (4.1.1) have been intensely editorialized and analyzed.

4.2. BALIAN'S *THEOREM*

4.2.1. *Theorem* (Balian). Given $g \in L^2(\mathbb{R})$ and $a,\alpha > 0$ for which $a\alpha = 1$. If $\{\psi_g(t;ma,n\alpha)\}$ is a Weyl-Heisenberg frame then

$$(4.2.1) \qquad tg(t) \notin L^2(\mathbb{R}) \quad \text{or} \quad \gamma\hat{g}(\gamma) \notin L^2(\hat{\mathbb{R}}),$$

cf., Section 4.2.5.

4.2.2. *The Zak Transform: Definition/Remarks.*

a. Given $g \in L^2(\mathbb{R})$ and $a,\alpha > 0$. The Zak transform of g is the function,

$$Z(g)(t,\gamma) = a^{1/2}\sum g(t-ka)e^{2\pi ik\gamma/\alpha},$$

defined on the rectangle $R_{a,\alpha} = [-a/2,a/2)\times[-\alpha/2,\alpha/2)$. $Z(g)$ extends to a function on $\mathbb{R}\times\hat{\mathbb{R}}$ satisfying the quasi-double periodicity condition,

$$\forall\, m,n \in \mathbb{Z},\ Z(g)(t+ma,\gamma+n\alpha) = Z(g)(t,\gamma)e^{2\pi i m\gamma/\alpha}.$$

Properties of the Zak transform are exposited in [D1; DGM; HW], and A.J.E.M. Janssen has made some of the most important recent contributions.

b. The Zak transform is a unitary (linear bijective isometry) map, $Z : L^2(\mathbb{R}) \rightarrow L^2(R_{a,\alpha})$.

c. Given $g \in L^2(\mathbb{R})$ and $a,\alpha > 0$. If Zg is continuous on $\mathbb{R}\times\hat{\mathbb{R}}$ then Zg has a zero in $R_{a,\alpha}$. Examples of continuous Zak transforms Zg on $\mathbb{R}\times\hat{\mathbb{R}}$ are provided by $g \in C_c(\mathbb{R})$. C. Heil has proved that a large class of functions g with continuous Zak transform Zg on $\mathbb{R}\times\hat{\mathbb{R}}$ is the original Segal algebra (due to Wiener),

$$W(\mathbb{R}) = \{f \in L^1(\mathbb{R}) \cap C_b(\mathbb{R}) : \sum \|f\|_{\infty,[n,n+1]} < \infty\},$$

where $\|\ldots\|_{\infty,[n,n+1]}$ is the usual L^∞-norm on the interval $[n,n+1]$. Clearly, the elements of $W(\mathbb{R})$ vanish at $\pm\infty$ and belong to $L^2(\mathbb{R})$. $W(\mathbb{R})$ is also the first of the Wiener-type Banach spaces mentioned in Remark 4.1.5a defined locally and then globally; in the case of $W(\mathbb{R})$ the local norms are $\|\ldots\|_{\infty,[n,n+1]}$ and the global norm is $\|\ldots\|_1$.

d. If $g \in L^2(\mathbb{R})$ and $a\alpha = 1$ then $\{\psi_g(t;ma,n\alpha)\}$ is a Weyl-Heisenberg frame with frame bounds A and B if and only if $0 < A \le |Z(g)(t,\gamma)|^2 \le B < \infty$ a.e. on $R_{a,\alpha}$, cf., part c and Section 1.2.5d.

4.2.3. *Example.* Given $g = \chi_{[0,1)}$ and $a = \alpha = 1$. Define $\psi_{m,n}(t) = \psi_g(t;m,n)$ as in (4.1.3). Then Zg is defined a.e. on $\mathbb{R}\times\hat{\mathbb{R}}$ by the property,

$$\forall n,\quad \forall t \in (n,n+1),\quad \text{and}\quad \forall\gamma,\quad Z(g)(t,\gamma) = e^{2\pi i n\gamma}$$

Since $|Z(g)(t,\gamma)| = 1$ a.e. on $R_{1,1}$ it is easy to check that $\{\psi_{m,n}\}$ is an orthonormal basis of $L^2(\mathbb{R})$, and hence it is a Weyl-Heisenberg frame. Further, $tg(t) \in L^2(\mathbb{R})$ and g', which exists a.e., is an element of $L^2(\mathbb{R})$, cf., (4.2.1). In this example, $\gamma\hat{g}(\gamma) \notin L^2(\mathbb{R})$ which corroborates Balian's theorem. However, this function g provides a counterexample to the proposed "proof" of Balian's theorem for g' that we've reproduced in Section 4.2.4.

With regard to the previous observation about g' and $\gamma\hat{g}(\gamma)$, recall that if $f \in L^2(\mathbb{R})$ and $\gamma\hat{f}(\gamma) \in L^2(\hat{\mathbb{R}})$ then f' exists a.e., is an element of $L^2(\mathbb{R})$, and $f'(t) = (2\pi i\gamma\hat{f}(\gamma))^\vee(t)$ a.e.

4.2.4. *Analysis.* The following is a plan to verify Balian's theorem for f, $tf(t)$, $f' \in L^2(\mathbb{R})$, $a\alpha = 1$, and $\{\psi_f(t;ma,n\alpha)\}$.

Step 1. Prove that $\partial_1 Zf, \partial_2 Zf \in L^2_{loc}(\mathbb{R}^2)$.

The notion of "multiresolution analysis", due to S. Mallat, is the most important idea associated with the difficult constructions of "analyzing wavelets" ψ which give rise to bases $\{\psi_{m,n}\}$ defined by (4.1.1), e.g., [M2] or the forthcoming book by Meyer. The original constructions did not benefit from this unifying and underlying concept. Intuitively a system of dilations and translations defined by (4.1.1) can be thought of as a deblurring mechanism. For example, if $f \in L^2(\mathbb{R})$ then

$$f_M = \sum_{m\leq M} \sum_n (f, \psi_{m,n})\psi_{m,n}$$

is a blurred vision of f in the following sense. If $\operatorname{supp} \psi \subseteq [-\frac{1}{2}, \frac{1}{2}]$ then $\operatorname{supp} \psi_{m,n} \subseteq [n2^{-m}-2^{-(m+1)}, n2^{-m}+2^{-(m+1)}] = I_{m,n}$ and $|I_{m,n}| = 2^{-m}$; and so if we add the $M+1$ term to f_M we have the effect of bringing to light the behavior of f at intervals of length $2^{-(M+1)}$, and thereby deblurring f_M.

b. Theorem 4.1.1 and the other results mentioned in part a have analogues in $\mathbb{R}^d$.

c. For a given $\psi \in L^2(\mathbb{R})$, the set $\{\psi_{m,n}\}$ defined by (4.1.1) is called an affine system because of the role of the underlying $ax+b$ group, cf., Section 4.1.3d where the Heisenberg group plays an analogous role for the Weyl-Heisenberg system. "Analyzing wavelets" ψ for affine systems satisfy the "wave condition" $\int\psi(t)dt = 0$. Consequently, the corresponding wavelets $\psi_{m,n}$ (defined by (4.1.1)) are more appropriately and usually called "wavelets" than are the elements of the Weyl-Heisenberg system.

4.1.3. *Definitions/Representation.* a. Let H be a separable Hilbert space with inner product $(\ldots,\ldots)$. A sequence $\{\psi_n\} \subseteq H$ is a frame for H with frame bounds A and B if

$$\exists\, A, B > 0 \text{ such that } \forall\, f \in H,$$

$$A\|f\|^2 \leq \sum |(f,\psi_n)|^2 \leq B\|f\|^2.$$

b. Orthonormal bases in H are bounded unconditional bases which, in turn, are frames.

c. For a given frame $\{\psi_n\}$, the S-operator is the map $S : H \longrightarrow H$, $f \longmapsto \sum(f,\psi_n)\psi_n$. It is a basic fact that $S : H \longrightarrow H$ is an isomorphism (linear bijective topological isomorphism), from which we obtain the frame representation,

$$(4.1.2) \qquad \forall\, f \in H,\ f = \sum(f,\psi_n)S^{-1}\psi_n = \sum(S^{-1}f,\psi_n)\psi_n.$$

d. Given $g \in L^2(\mathbb{R})$ and $a, \alpha > 0$; we define the Weyl-Heisenberg system,

$$(4.1.3) \qquad \forall\, m,n \in \mathbb{Z},\ \psi_{m,n}(t) = \psi_g(t; ma, n\alpha) = g(t-ma)e^{2\pi i t n\alpha},$$

Step 2. Define $Z_r(f)(t,\gamma)$ to be the continuous mean of Zf over a square of side $2r$ centered at (t,γ).

Step 3. Assume $\{\psi_f(t;ma,n\alpha)\}$ is a Weyl-Heisenberg frame with frame bounds A and B. Use Step 1 to prove that $A^{1/2}/2 \leq |Z_r(f)(t,\gamma)| \leq 2B^{1/2}$ on a specified square R containing $R_{1,1}$.

Step 4. Obtain a contradiction by considering $\log Z_r(f)$.

Most details for these steps are correct and can be found in the literature. The above outline would give the result since the hypotheses $tf(t)$, $f' \in L^2(\mathbb{R})$ allow a verification of Step 1. (Recall that g of Example 4.2.3 satisfies $tg(t)$, $g' \in L^2(\mathbb{R})$.) The problem with this outline is part of the proof of Step 3 where the following calculation is made for $Z = Z(f)$:

$$Z(t',\gamma') - \frac{1}{4r^2}\int_I\!\!\int Z(t'',\gamma'')dt''d\gamma''$$

$$= -\frac{1}{4r^2}\int_I\!\!\int (Z(t'',\gamma'')-Z(t',\gamma'')+Z(t',\gamma'')-Z(t',\gamma')dt''d\gamma''$$

$$= -\frac{1}{4r^2}\int_I\!\!\int\left[\int_{t'}^{t''}\partial_1 Z(t''',\gamma'')dt''' + \int_{\gamma'}^{\gamma''}\partial_2 Z(t',\gamma''')d\gamma'''\right]dt''d\gamma'',$$

where $I = \{(t'',\gamma'') : |t''-t'| \leq r$ and $|\gamma''-\gamma'| \leq r\}$. Generally, this calculation fails since the fundamental theorem of calculus is not applicable. For example, in the case of Example 4.2.3,

$$\int_{t'}^{t''} \partial_1 Z(g)(t''',\gamma'')dt''' = 0$$

and

$$Z(g)(t'',\gamma'') - Z(g)(t',\gamma'') = e^{2\pi i n''\gamma''} - e^{2\pi i n'\gamma''};$$

and this causes problems on R when $|n'' - n'| > 0$.

4.2.5. *Discussion of Balian's Theorem.* Balian's original treatment was in terms of orthonormal bases [Bal], cf., [HJJ]. An ingenious simple proof is due to Battle [Bat]. It has been possible, by using Battle's idea or by a more delicate touch in dealing with Step 3 (of Section 4.2.4), for several groups to verify Theorem 4.2.1, either by formal calculation or correct proof.

Using the uncertainty principle, Theorem 1.3.3, and assuming $tg(t)$, $\gamma\hat{g}(\gamma) \in L^2$ and an orthonormal basis for the Weyl-Heisenberg system, Battle obtains a contradiction. The uncertainty principle is invoked in the sense that he calculates $E_g(i[A,B]) = 0$, noting that $E_g(i[A,B]) = \|g\|_2^2$.

C. Heil has observed, in light of his result quoted in Section 4.2.2, that Balian's theorem is immediate with the further hypothesis, $g \in W(\mathbb{R})$.

Finally, we note that Balian's theorem leads one to quantify further the notion of functions g being far away from the Gaussian such as reflected by condition (4.2.1). In this regard, we mention DeBruijn's theorem to the effect that if (1.1.1) is almost an equality for a function g then g is almost equal to a function of the form (1.1.2) [De, Theorem 3.3].

4.2.6. *Strong Uncertainty for Weyl-Heisenberg Systems.* Given $g \in L^2(\mathbb{R})\setminus\{0\}$ and $a, \alpha > 0$. Define the Weyl-Heisenberg system $\psi_{m,n}(t) = \psi_g(t; ma, n\alpha)$, $m, n \in \mathbb{Z}$. For any fixed $(t_0, \gamma_0) \in \mathbb{R}\times\hat{\mathbb{R}}$, we have the strong uncertainty property

$$\sup_{m,n} \|(t-t_0)\psi_{m,n}(t)\|_2 \|(\gamma-\gamma_0)\hat{\psi}_{m,n}(\gamma)\|_2 = \infty, \tag{4.2.2}$$

cf., (4.3.2).

To verify (4.2.2) we first observe by basic function theory associated with Jensen's or Carleman's theorem and the Paley-Wiener theorem that if g (resp., $\hat{g}$) vanishes on a half-line then $\hat{g}$ (resp., g) cannot vanish on intervals without being identically zero. Next, consider the estimate,

$$\|(t-t_0)\psi_{m,n}(t)\|_2^2 \|(\gamma-\gamma_0)\hat{\psi}_{m,n}(\gamma)\|_2^2$$

$$\geq \int((t-t_0)+ma)^2|g(t)|^2dt \int_{\gamma_0}^{\infty} ((\gamma-\gamma_0)+n\alpha)^2|\hat{g}(\gamma)|^2d\gamma \tag{4.2.3}$$

$$\geq (n\alpha)^2\int((t-t_0)+ma)^2|g(t)|^2dt \int_{\gamma_0}^{\infty} |\hat{g}(\gamma)|^2d\gamma,$$

where the second inequality is valid for $n \geq 0$. Thus, if supp g is contained in a half-line then (4.2.2) follows from (4.2.3). If supp g is not contained in a half-line then the obvious adjustment of (4.2.3) yields (4.2.2).

4.3. BOURGAIN'S THEOREM

4.3.1. *Notation for Expected Values.* Given $\psi \in L^2(\mathbb{R})$ and consider the affine system $\{\psi_{m,n}\}$ defined in (4.1.1). For each m, n the expected values $(t_{m,n}, \gamma_{m,n}) \in \mathbb{R}\times\hat{\mathbb{R}}$ are

$$t_{m,n} = \int t|\psi_{m,n}(t)|^2dt \quad \text{and} \quad \gamma_{m,n} = \int \gamma|\hat{\psi}_{m,n}(\gamma)|^2d\gamma,$$

cf., the end of Section 1.2.1 for the origin of this terminology from quantum mechanics.

4.3.2. *Weak Uncertainty for Affine Systems.* Given $\psi \in L^2(\mathbb{R})$ and consider the notation from Section 4.3.1. We have the weak uncertainty property,

$$\sup_{m,n} \|(t-t_{m,n})\psi_{m,n}(t)\|_2 \|(\gamma-\gamma_{m,n})\hat{\psi}_{m,n}(\gamma)\|_2 < \infty, \tag{4.3.2}$$

cf., (4.2.2). The inequality (4.3.2) is a consequence of the equalities

$$\begin{aligned} \|(t-t_{m,n})\psi_{m,n}(t)\|_2 &= 2^{-m}\|(t-t_{0,0})\psi(t)\|_2 \\ \|(\gamma-\gamma_{m,n})\hat{\psi}_{m,n}(\gamma)\|_2 &= 2^{m}\|(\gamma-\gamma_{0,0})\hat{\psi}(\gamma)\|_2, \end{aligned} \tag{4.3.3}$$

which, in turn, follow from easy calculations.

4.3.3. *Problem.* "Weak uncertainty" for wavelet bases such as the affine system described in Theorem 4.1.1 provides simultaneous control of time and frequency information among all the basis elements. It is natural to ask how precise this simultaneity can be. To quantify this question we ask specifically if there is an orthonormal basis $\{\psi_n\}$ of $L^2(\mathbb{R})$ having expected values $(t_n,\gamma_n) \in \mathbb{R}\times\hat{\mathbb{R}}$ so that

$$\sup_n \|(t-t_n)\psi_n(t)\|_2 < \infty \quad \text{and} \quad \sup_n \|(\gamma-\gamma_n)\hat{\psi}_n(\gamma)\|_2 < \infty.$$

Because of (4.3.3) any such refinement of (4.3.2) would have to go outside the realm of affine systems. In any case the following result gives a strong solution to this problem.

4.3.4. *Theorem* (Bourgain [Bo]). For every $\varepsilon > 0$ there is an orthonormal basis $\{\psi_n\}$ of $L^2(\mathbb{R})$ having expected values $(t_n,\gamma_n) \in \mathbb{R}\times\hat{\mathbb{R}}$ and satisfying the inequalities,

$$\sup_n \|(t-t_n)\psi_n(t)\|_2 < \frac{1}{2\sqrt{\pi}}+\varepsilon \quad \text{and} \quad \sup_n \|(\gamma-\gamma_n)\hat{\psi}_n(\gamma)\|_2 < \frac{1}{2\sqrt{\pi}}+\varepsilon.$$

Thus,

$$\forall\ n,\ 1 = \|\psi_n\|_2^2 \le 4\pi\|(t-t_n)\psi_n(t)\|_2\|(\gamma-\gamma_n)\hat{\psi}_n(\gamma)\|_2 < 4\pi\left[\frac{1}{2\sqrt{\pi}}+\varepsilon\right]^2.$$

5. Spectrum Estimation and the Wiener-Wintner Theorem

5.1. SPECTRUM ESTIMATORS AND THE UNCERTAINTY PRINCIPLE

5.1.1. *Problem.* The spectrum estimation problem is to clarify and quantify the statement: find periodicities in a signal $f(t)$ recorded over a fixed time interval $[-T,T]$. In more picturesque language, we want to filter the noise from the incoming signal f in order to determine the intelligent message (periodicities) therein.

5.1.2. *Definition.* In order to quantify this problem we introduce the following mathematical setting.

Let $f(t,\alpha)$ be a stationary stochastic process (SSP), where the sample functions $f(\cdot,\alpha)$ on $\mathbb{R}$ are indexed by α in the underlying probability space X. (For our purposes, an SSP f is characterized by the conditions that the expected value $E\{f(t)\}$ is constant, $E\{f(t+h)\overline{f(u+h)}\} = E\{f(t)\overline{f(u)}\}$ for all $h, t, u \in \mathbb{R}$, and $\lim_{h\to 0} E\{|f(h) - f(0)|^2\} = 0$.)

The autocorrelation of the SSP f is the continuous positive definite function

$$R_f(t) = E\{f(t+u)\overline{f(u)}\};$$

and the power spectrum of f is the positive measure $S = S_f$ for which $\hat{R}_f = S_f$ (a distributional Fourier transform).

5.1.3. *Definition/Remark.* Given $f \in L^2_{loc}(\mathbb{R}^d)$ and define

$$\forall T > 0, \quad P_{f,T} = \frac{1}{|B(T)|} (f\chi_{B(T)})*(f\chi_{B(T)})^{\sim}$$

so that $P_{f,T} \in L^1_{loc}(\mathbb{R}^d) \subseteq M(\mathbb{R}^d)$. Suppose that there is a continuous positive definite function P_f for which $\lim_{T\to\infty} P_{f,T} = P_f$ in the vague topology $\sigma(M(\mathbb{R}^d), C_c(\mathbb{R}^d))$. Then $P_f \in L^\infty(\mathbb{R}^d)$ is the autocorrelation of f and $\hat{P}_f = \mu_f \in M_{b+}(\hat{\mathbb{R}}^d)$ is the power spectrum of f.

In order to reconcile the two apparently different definitions on $\mathbb{R}$ of both autocorrelation and power spectrum we shall assume that

$$\text{(5.1.1)} \qquad \forall t \in \mathbb{R}, \quad \lim_{T\to\infty} \frac{1}{2T} \int_{-T}^{T} f(t+u,\alpha)\overline{f(u,\alpha)}du = R_f(t)$$

converges in measure. Processes satisfying condition (5.1.1) are correlation ergodic processes; and verification of correlation ergodicity of $f(t,\alpha)$ requires knowledge of its fourth order moments.

5.1.4. *Definition/Fact.* Given $v \in L^1(\mathbb{R})$ and suppose f is an SSP for which each sample function $f(\cdot,\alpha)$ is an element of $L^\infty(\mathbb{R})$. Then

$$S_v(\gamma,\alpha) = \left|\int f(t,\alpha)v(t)\, e^{-2\pi i t\gamma}dt\right|^2$$

is the periodogram associated with the process f and data window v.

If f is a real SSP and $v = v^{\sim}$ then it is not difficult to verify that

$$\text{(5.1.2)} \qquad E\{S_v(\gamma)\} = S*V^2,$$

where $\hat{v} = V$. Further if $\{V_T^2 : T > 0\}$ is an L^1-approximate identity and $\hat{v}_T = V_T$ then

(5.1.3) $$\lim_{T\to\infty} E\{S_{v_T}(\gamma)\} = S$$

in the canonical weak topology.

Because of (5.1.3) we can assert that $E\{S_{v_T}\}$ is an asymptotically unbiased estimator of S, thereby providing a relatively naive solution to our vaguely posed spectrum estimator problem.

A special case of the Bell Labs uncertainty principle is -

5.1.5. *Proposition* [LSP]. Given T, $\Omega > 0$. There is $c = c(T\Omega) \in (0,1)$ such that for each Ω-band limited function f, i.e., $\text{supp}\,\hat{f} \subseteq [-\Omega,\Omega]$,

(5.1.4) $$\|f\chi_{(T)}\|_2 \le c(T\Omega)\|\hat{f}\|_2;$$

$c(T\Omega)^2$ is the largest eigenvalue of the operator BA, where $Af = f\chi_{(T)}$ and $Bg = (\hat{g}\chi_{(\Omega)})^{\vee}$. ($\chi_{(T)}$ is the characteristic function of $[-T,T]$.)

Once the values $c(T\Omega)$ are known, Proposition 5.1.5 answers the question: what is an upper bound of the energies $\|f\chi_{(T)}\|_2^2$ as f ranges through the Ω-band limited signals having a fixed finite energy? The classical uncertainty principle says that the variances of $|f|^2$ and $|\hat{f}|^2$ cannot both be small; Proposition 5.1.5 says that the energies of Ω-band limited signals and of their restrictions to $[-T,T]$ cannot be arbitrarily close, in spite of the Plancherel theorem.

5.1.6. *Calculation.* The uncertainty principle embodied in Proposition 5.1.5 allows us to quantify energy loss in spectrum estimators as the following calculation shows for the estimator $E\{S_v\}$. Besides the above hypotheses we add the realistic conditions that $\text{supp}\, v \subseteq [-T,T]$, $\|v\|_2 = 1$, $v > 0$ on $(-T,T)$, and $\text{supp}\, S \subseteq [-\Omega,\Omega]$ for fixed T, $\Omega > 0$. We compute

$$\begin{aligned}\|E\{S_v\}\|_2 &= \|S * V^2\|_2 = \|S^{\vee}(v*v)\|_2\\ &\le \|S^{\vee}(v*v)\,\frac{1}{v*v}\|_{L^2[-2T,2T]} = \|S^{\vee}\chi_{(2T)}\|_2\\ &\le c(T\Omega)\ \|S\|_2.\end{aligned}$$

One can make similar calculations involving weighted L^2-spaces (and therefore reminiscent of the classical uncertainty principle), where the uncertainty components are high resolution and precision of estimator, respectively [Gr].

5.2. DISCUSSION OF SPECTRUM ESTIMATION

5.2.1 *Deterministic Assumptions.* The path we have chosen in Section

5.1 to quantify the loosely posed Problem 5.1.1 is based on the following deterministic assumptions.

i. The signal f is defined on the product space $[-T,T]\times X$, where X is a probability space and f is the restriction to $[-T,T]\times X$ of some SSP g.

ii. The expectation of the periodogram $S_v(\gamma,\alpha)$ is known, where $\mathrm{supp}\, v \subseteq [-T,T]$ and $\|v\|_2 = 1$.

iii. The power spectrum S of f is uniquely determined.

Assumption iii is a theorem in many cases involving the experimentally reasonable hypothesis that $\mathrm{supp}\, S_g$ is compact, where S_g is the power spectrum of the SSP g which, in turn, is an extension of f defined on $[-T,T]\times X$. Mathematically, this assumption allows us to specify S in (5.1.2) and, ultimately, leads to the Beurling - Malliavin theory [B1].

On the other hand, Assumption iii is not universally accepted. In fact, the maximum entropy method (MEM) of spectrum estimation is based on a point of view opposite that of a uniquely determined power spectrum. MEM does not assert the existence of a unique power spectrum and then estimate it; instead, given f on $[-T,T]\times X$ or the autocorrelation R on $[-T,T]$, MEM models autocorrelation data outside $[-T,T]$ by maximizing a certain entropy integral.

5.2.2. *Statistical Assumptions.* Besides the questions surrounding Assumption iii(of Section 5.2.1), we must also note that Assumption i preempts all genuine statistical problems.

A different point of view from that of Section 5.2.1 is the following. We are given some sample paths of finite duration $[-T,T]$ corresponding to a specific experiment. The only stochastic process available may be an idealized process representing the potential output of some underlying mechanism. For example, in speech analysis the generation of a specific sound varies with time and person but is subject to statistical regularities, and this mechanism is the ultimate source of realistic spectrum estimation of frequencies corresponding to the sound.

5.2.3. *Remark.* Spectrum estimation is a multi-faceted, basic, and deep problem with points of view bordering on the philosophical and viable techniques ranging from sophisticated periodogram analysis to various "high resolution" methods such as MEM. We refer to [PI] and [L] for a scholarly presentation of diverse methods and a brilliant new insight, respectively.

The purpose of our discussion has been to present the remaining parts of Section 5 which provide theorems susceptible to transition and interpretation as spectrum estimation algorithms.

5.3. SIGNAL DETERMINISTIC ESTIMATORS ON $\mathbb{R}^d$

Suppose the data characterizing a given signal f is known. In the following result, V can be thought of as a properly shaped window function so that the left side of (5.3.1) represents the power of f in the region supp V. Formula (5.3.1) provides a method for computing this power in terms of the known functions f and v. In practice, then, numerical estimates of the right side of (5.3.1), in case f is only available on a compact set, lead to a spectrum estimation algorithm.

5.3.1. *Theorem* [B3]. Given $f \in L^2_{loc}(\mathbb{R}^d)$ with autocorrelation P_f and power spectrum μ_f. Assume there is an increasing function $i(R)$ on $(0,\infty)$ for which $\sup_{|t|\leq T} |f(t)| \leq i(T)$ and $\lim_{T\to\infty} i(T)^2/T = 0$. Then

$$(5.3.1) \qquad \forall v \in C_c(\mathbb{R}^d), \quad \int |V(\gamma)|^2 d\mu_f(\gamma) = \lim_{T\to\infty} \frac{1}{|B(T)|} \int_{B(T)} |f*v(t)|^2 \, dt,$$

where $\hat{v} = V$.

5.3.2. *Corollary*. Given the hypotheses of Theorem 5.3.1. If $v \in L^p(\mathbb{R}^d)$ and $f \in L^{p'}(\mathbb{R}^d)$, $1 \leq p < \infty$, then the tempered distribution $\hat{v} = V$ is a well-defined element of $L^2_{\mu_f}(\mathbb{R}^d)$ and

$$\|V\|_{2,\mu_f} \leq \|f\|_{p'} \|v\|_p.$$

5.4. AUTOCORRELATION DETERMINISTIC ESTIMATORS ON $\mathbb{R}^d$

Suppose the autocorrelation data for a signal is known and the signal itself is not explicitly known. This is a typical situation and one of the reasons autocorrelations are so important, e.g., [B1]. In the following result, V can be thought of as a properly shaped window function so that the left side of (5.4.1) represents the power of f in the region supp V. In practice, then, numerical estimates of the right side of (5.4.1), in case P is only available on a compact set, lead to a spectrum estimation algorithm.

5.4.1. *Proposition* [B3]. Given $\mu \in M_{b+}(\hat{\mathbb{R}}^d)$ for which $\check{\mu} = P \in L^{p'}(\mathbb{R}^d)$, $p \in [1,\infty]$. Then

$$(5.4.1) \qquad \begin{gathered} \forall v \in L^1(\mathbb{R}^d) \cap L^p(\mathbb{R}^d), \\ \|V\|_{2,\mu} \leq \|P\|_{p'}^{1/2} (\|v\|_1 \|v\|_p)^{1/2} \leq \tfrac{1}{2} \|P\|_{p'}^{1/2} (\|v\|_1 + \|v\|_p), \end{gathered}$$

where $\hat{v} = V$ and $(L^1(\mathbb{R}^d) \cap L^p(\mathbb{R}^d), \|\cdots\|_1 + \|\cdots\|_p)$ is a Banach space.

5.4.2. *Corollary* [B3]. Given $d \geq 2$ and $1 \leq p < 2d/(d+1)$. Then $\|\mu_{d-1}\|_{p'} < \infty$ and, or each $v \in L^1(\mathbb{R}^d) \cap L^p(\mathbb{R}^d)$,

$$(5.4.2) \qquad \left(\int_{\Sigma_{d-1}} |V(\theta)|^2 \, d\sigma_{d-1}(\theta)\right)^{1/2} \le \tfrac{1}{2}\|\check{\mu}_{d-1}\|_{p'}^{1/2}(\|v\|_1 + \|v\|_p),$$

where $\hat{v} = V$.

5.4.3. *Remark.* The corollary is immediate from elementary properties of Bessel functions and Proposition 5.4.1, which itself follows from an elementary calculation. We stress the simplicity of (5.4.2) to compare it with the much deeper Tomas-Stein restriction theorem:

$$(5.4.3) \qquad \forall v \in L^1(\mathbb{R}^d) \cap L^p(\mathbb{R}^d),$$
$$\left(\int_{\Sigma_{d-1}} |V(\theta)|^2 \, d\sigma_{d-1}(\theta)\right)^{1/2} \le c(p) \, \|v\|_p,$$

where $\hat{v} = V$ and $1 \le p \le 2(d+1)/(d+3)$. In (5.4.3) the constant $c(p)$ is not explicit, $p = 2(d+1)/(d+3)$ is largest possible, and the right hand norm is $\|v\|_p$. In (5.4.2) the constant is explicit and the values of p extend beyond $2(d+1)/(d+3)$, but the right hand norm is $\|v\|_1 + \|v\|_p$.

5.5. THE WIENER-WINTNER THEOREM IN $\mathbb{R}^d$

Underlying the results in Section 5.3 and 5.4 is the question: given $\mu \in M_+(\hat{\mathbb{R}}^d)$, does there exist $f \in L^2_{loc}(\mathbb{R}^d)$ such that $\hat{P}_f = \mu$? The Wiener-Wintner theorem provides an answer to this question and a means of reconstructing signals f on $\mathbb{R}^d$ corresponding to a given power spectrum.

5.5.1. *Preliminaries.* Given $\mu \in M_{b+}(\hat{\mathbb{R}}^d)$ and let δ_ω be the Dirac measure supported by $\{\omega\}$. It is well-known that there is a sequence $\{\mu_n\} \subseteq M_{b+}(\text{supp}\,\mu)$ of positive discrete measures,

$$\mu_n = \sum_{j=1}^{N_n} a_{j,n} \, \delta_{\omega_{j,n}}, \quad a_{j,n} > 0,$$

such that $\{\omega_{j,n} : \ j = 1, \cdots, N_n\} \subseteq \text{supp}\,\mu$ for each n,

$$(5.5.1) \qquad \lim_{n\to\infty} \langle \mu_n, 1\rangle = \langle \mu, 1\rangle,$$

and $\lim_{n\to\infty} \mu_n = \mu$ in the (vague) topology $\sigma(M_b(\hat{\mathbb{R}}^d), C_c(\hat{\mathbb{R}}^d))$. Actually, (5.5.1) and the $\sigma(M_b(\hat{\mathbb{R}}^d), C_c(\hat{\mathbb{R}}^d))$ convergence allow us to conclude that $\lim_{n\to\infty} \mu_n = \mu$ in the "Levy" topology $\sigma(M_b(\hat{\mathbb{R}}^d), C_b(\hat{\mathbb{R}}^d))$.

For a given $\mu \in M_{b+}(\hat{\mathbb{R}}^d)$ and sequence $\{\mu_n\} \subseteq M_{b+}(\text{supp}\,\mu)$ as above, we define

$$f_n(t) = \sum_{j=1}^{N_n} a_{j,n}^{1/2} e^{2\pi i t\cdot\omega_{j,n}}$$

so that

$$\|f_n\|_\infty \le \sum_{j=1}^{N_n} a_{j,n}^{1/2} \le N_n \sup_{1\le j\le N_n} a_{j,n}^{1/2}.$$

5.5.2. *Lemma/Notation.* The proof of the following result depends on elementary properties of Bessel functions.

Lemma. For each n,

$$\lim_{T\to\infty} \frac{1}{|B(T)|} \int_{B(T)} f_n(t+x)\overline{f_n(x)}dx = \check{\mu}_n(t)$$

uniformly on $\mathbb{R}^d$.

Notation. We use the *Lemma* to define a specific sequence $\{T_n\}$ and a specific function f in the following way. From the uniform convergence we know that

$$\forall n \ge 1,\ \exists A_n \ge A_{n-1} \text{ such that } \forall t \in \mathbb{R}^d \text{ and } \forall T \ge A_n,$$

$$\left| \frac{1}{|B(T)|} \int_{B(T)} f_n(t+x)\overline{f_n(x)}dx - \check{\mu}_n(t) \right| < \frac{1}{2^{n+1}}.$$

We set $T_n = (A_1+1)(A_2+2)\cdots(A_n+n)$ so that $T_n \ge n!$ and the sequences $\{T_n\}$, $\{T_{n+1}/T_n\} = \{A_{n+1}+n+1\}$, and $\{T_{n+1}-T_n\} = \{(A_1+1)\cdots(A_n+n)(A_{n+1}+n)\}$ increase to infinity. For this sequence $\{T_n\}$ and for $\{f_n\}$ defined above we define f on $\mathbb{R}^d$ by setting $f(t) = f_n(t)$ for $T_n<|t|<T_{n+1}$, $n \ge 1$, and letting $f(t) = 0$ for $|t| < T_1$. Clearly, $f \in L^\infty_{loc}(\mathbb{R}^d)$ and the values of $f(t)$ for $|t| = T_n$ are not important.

The proof of the following result is long and intricate.

5.5.3. *Theorem* [B3]. Given $\mu \in M_{b+}(\hat{\mathbb{R}}^d)$ with corresponding functions $\{f_n\}$ and $f \in L^\infty_{loc}(\mathbb{R}^d)$. Assume there is $C > 0$ such that for all n

(5.5.2) $$\|f_n\|_\infty^2 \le CT_n.$$

Then, for each $t \in \mathbb{R}^d$,

$$\lim_{T\to\infty} \frac{1}{|B(T)|} \int_{B(T)} f(t+x)\overline{f(x)}dx = \check{\mu}(t).$$

5.5.4. *Remark.* The original proof of Theorem 5.5.3 by Wiener and Wintner on $\mathbb{R}$ is cryptic and there have been important contributions on $\mathbb{R}$ by J. Bass and Bertrandias.

There is left unanswered the problem of characterizing those μ for which $f \in L^\infty(\mathbb{R}^d)$. This problem is not yet solved in case $d = 1$.

A. Closure Theorems

A.1. BI-SOBOLEV SPACES

A.1.1. *Definition.* Given integers $m,n \ge 0$ and $1 \le p \le \infty$. The Sobolev space $L^p_m = L^p_m(\mathbb{R}^d)$ is the Banach space of functions $f \in L^p(\mathbb{R}^d)$ for which

$$\|f\|_{m,p} = \sum_{|\alpha|\le m} \|\partial^\alpha f\|_p < \infty.$$

The weighted space $L^p_{o,n} = L^p_{o,n}(\mathbb{R}^d)$ is the Banach space of functions $f \in L^p(\mathbb{R}^d)$ for which

$$\|f\|_{p,n} = \sum_{|\beta|\le n} \|t^\beta f(t)\|_p < \infty.$$

The Bi-Sobolev space $L^p_{m,n} = L^p_{m,n}(\mathbb{R}^d)$ is the Banach space of functions $f \in L^p_m \cap L^p_{o,n}$ for which

$$\|f\|_{m,p,n} = \|f\|_{m,p} + \|f\|_{p,n} < \infty. \tag{A.1.1}$$

A.1.2. *Theorem.* Given integers $m,n \ge 0$. $C_c^\infty(\mathbb{R}^d)$ is dense in the Hilbert space $(L^2_{m,n}, \|\cdots\|_{m,2,n})$ with inner product

$$[f,g] = \sum_{|\alpha|\le m} (\partial^\alpha f, \partial^\alpha g) + \sum_{|\beta|\le n} (t^\beta f, t^\beta g),$$

where $(\cdots,\cdots)$ is the usual inner product on $L^2(\mathbb{R}^d)$.

Proof: a. It is well-known that $C_c^\infty(\mathbb{R}^d)$ is dense in L^p_m, $1 \le p < \infty$. To see this, choose $f \in L^p_m$, let $\{h_j\} \subseteq C_c^\infty(\mathbb{R}^d)$ be an L^1-approximate identity with each $\operatorname{supp} h_j \subseteq B(0,1)$, and take $u_j \in C_c^\infty(\mathbb{R}^d)$ defined by $u_j(t) = u(t/j)$ where $0 \le u \le 1$ and $u = 1$ on $B(0,1)$.

Fix $|\alpha| \le m$. Not only does $\partial^\alpha(f*h_j) = f*\partial^\alpha h_j$ but, by integration by parts, $\partial^\alpha(f*h_j) = (\partial^\alpha f)*h_j$. Consequently, we can apply Young's theorem to obtain $\|\partial^\alpha(f*h_j)\|_p \le \|\partial^\alpha f\|_p \|h_j\|_1 \le K\|\partial^\alpha f\|_p$. Thus, each $f*h_j$ is an element of L^p_m, as is each $u_j(f*h_j)$.

The desired density will follow from the triangle inequality once we prove

$$\text{(A.1.2)} \qquad \lim_{j\to\infty} \|\partial^{\alpha}[(f*h_j)(u_j-1)]\|_p = 0$$

for each $|\alpha| \le m$. To this end we first use Leibniz's formula for the estimate,

$$\text{(A.1.3)} \qquad \begin{aligned} \|\partial^{\alpha}[(f*h_j)(u_j-1)]\|_p &\le \|(u_j-1)\partial^{\alpha}(f*h_j)\|_p + \\ &\sum_{\substack{\beta\le\alpha\\ |\beta|\ge 1}} |C_{\alpha\beta}|j^{-|\beta|}\|\partial^{\alpha-\beta}(f*h_j)(t)\partial^{\beta}u|_{(t/j)}\|_p. \end{aligned}$$

The dominated convergence theorem and Young's theorem allow us to prove that the first term on the right side of (A.1.3) tends to 0 as $j\to\infty$. Young's theorem and the fact that $\lim_{j\to\infty} j^{-|\beta|}\|\partial^{\beta}u\|_{\infty} = 0$ for $|\beta| \ge 1$ show that the remaining terms on the right side of (A.1.3) tend to 0 as $j\to\infty$. (A.1.2) is proved.

This density in L^p_m can also be verified by an equicontinuity argument much like the one we give for $L^2_{o,n}$.

b. It is sufficient to prove that

$$\text{(A.1.4)} \qquad \forall f \in L^2_{o,n}, \quad \lim_{j\to\infty} \|F_{\beta j}(f)\|_2 = 0$$

for each $|\beta| \le n$, where $F_{\beta j}(f) = F_j(f) = t^{\beta}(u_j(f*h_j) - f)$. To this end we first show that

$$\text{(A.1.5)} \qquad \sup_j \|F_j(f)\|_2 = C(f) < \infty.$$

This is accomplished by the estimate,

$$\begin{aligned} \|F_j(f)\|_2 - \|t^{\beta}f(t)\|_2 &\le C(\beta)\ \|\partial^{\beta}(\hat{f}\hat{h}_j)\|_2 \\ \le \sum_{\gamma\le\beta} |C_{\beta\gamma}|\|\partial^{\beta-\gamma}\hat{f}\,\partial^{\gamma}\hat{h}_j\|_2 &\le \sum_{\gamma\le\beta} |C_{\beta\gamma}|\|\partial^{\beta-\gamma}\hat{f}\|_2 \sup_{\gamma\le\beta} \|\partial^{\gamma}\hat{h}_j\|_{\infty}, \end{aligned}$$

and the fact, in the case h_j is the dilation $j^d h(jt)$, that

$$|\partial^{\gamma}\hat{h}_j(\lambda)| = |C(\gamma)\int h(u)u^{\gamma}\, j^{-|\gamma|}\, e^{-2\pi i(u/j)\cdot\lambda}du| \le K(\gamma)j^{-|\gamma|}$$

since $\operatorname{supp} h$ is compact. The estimate used the Plancherel theorem and so we note the fact that the distribution $\partial^{\beta}(\hat{f}\hat{h}_j)$ is an element of $L^2(\mathbb{R}^d)$.

It is easy to check that the elements of $L^2_{o,n}$ having compact support are dense in $L^2_{o,n}$ and that $\{F_{\beta j}\}$ is contained in $\mathcal{L}(L^2_{o,n}, L^2(\mathbb{R}^d))$ for each $|\beta| \le n$. Because of (A.1.5) we can invoke the uniform boundedness principle and obtain $\sup \|F_j\| = C < \infty$. Thus, $\{F_j\}$ is equicontinuous. On the other hand it is routine to check that $\lim_{j\to\infty} \|F_j(f)\|_2 = 0$ for compactly supported functions $f \in L^2_{o,n}$. This convergence on a dense subset of $L^2_{o,n}$ combined with the equicontinuity yield convergence on $L^2_{o,n}$, and the resulting limit $F(f)$ for $f \in L^2_{o,n}$ determines an element $F \in \mathcal{L}(L^2_{o,n}, L^2(\mathbb{R}^d))$. Therefore, (A.1.4) is obtained.

q.e.d.

A.1.3. *Remark*. Instead of defining $L^p_{m,n}$ to deal with closure questions for the uncertainty principle, we could define the <u>weighted Sobolev space</u> $L^p_{m,w}(\mathbb{R}^d)$ consisting of functions $f \in L^1_{loc}(\mathbb{R}^d)$ for which

$$\|f\|_{m,p,w} = \sum_{|\alpha| \le m} \|\partial^\alpha f\|_{p,w(\alpha)} < \infty.$$

In case $d = 1$, $p = 2$, $m = 1$ (and so $\alpha = 0, 1$), and $w = (w(0), w(1))$ with $w(0)(t) = (1+t^2)$ and $w(1)(t) = 1$, we see that $L^2_{1,w}(\mathbb{R})$ is the Bi-Sobolev space $L^2_{1,1}(\mathbb{R})$.

A.2. L^p_v AND DENSE MOMENT SPACES

A.2.1. *Theorem* [BH2]. Given $v \in L^1_{loc}(\mathbb{R}^d)$ where $v > 0$ a.e. and choose $p \in (1, \infty)$.

a. If $h \in L^p_v(\mathbb{R}^d)$ annihilates $\mathcal{S}_o(\mathbb{R}^d) \cap L^p_v(\mathbb{R}^d)$ then h is a <u>constant function</u>.

b. $\overline{\mathcal{S}_o(\mathbb{R}^d) \cap L^p_v(\mathbb{R}^d)} = L^p_v(\mathbb{R}^d)$ or $L^p_v(\mathbb{R}^d) \subseteq L^1(\mathbb{R}^d)$.

c. If $v^{1-p'} \notin L^1(\mathbb{R}^d)$ then $\overline{\mathcal{S}_o(\mathbb{R}^d) \cap L^p_v(\mathbb{R}^d)} = L^p_v(\mathbb{R}^d)$.

A.2.2. *Remark*. a. <u>Theorem A.2.1</u> requires some effort to prove, but is considerably easier if $L^{p'}_{v^{1-p'}}(\mathbb{R}^d) \subseteq \mathcal{S}'(\mathbb{R}^d)$.

b. The condition, $p > 1$, is necessary in <u>Theorem A.2.1</u>. In fact, if $p = 1$ and $v = 1$ then by a standard spectral synthesis result, the L^1-closure of $\mathcal{S}_o(\mathbb{R}^d)$ is the (closed) maximal ideal $\{f \in L^1(\mathbb{R}^d)\colon \hat{f}(0) = 0\}$.

A.3. $L^2_{1,1}$ AND PROPER MOMENT SPACES

Consider $\mathcal{S}_o(\mathbb{R})$ as a subspace of $L^2_{1,1}(\mathbb{R})$.

A.3.1. *Proposition.* $\mathcal{S}_o(\mathbb{R})^{\perp}$ (as a subspace of $L^2_{1,1}(\mathbb{R})'$) is the set of constant functions on $\mathbb{R}$.

Using this fact and the inclusion, $L^2_{1,1}(\mathbb{R}) \subseteq L^1(\mathbb{R})$, we have

A.3.2. *Proposition.* The closure of $\mathcal{S}_o(\mathbb{R})$ in $L^2_{1,1}(\mathbb{R})$ is $\{f \in L^2_{1,1}(\mathbb{R}) : \hat{f}(0) = 0\}$.

A.3.3. *Remark.* $L^2_{1,1}$ is a Hilbert space so that its dual is isomorphic to $L^2_{1,1}$. As in the case of L^2_m, this isomorphism is complicated and the continuous linear functions on $L^2_{1,1}$ have an alternate explicit representation. For example, the proof of Proposition A.3.1 shows that the constants are elements of $(L^2_{1,1})'$, and so, noting that the elements of $L^2_{1,1}(\mathbb{R})$ are locally absolutely continuous, we see that there is $g \in L^2_{1,1}(\mathbb{R})'$ so that $[f,g] = f(0)$ for each $f \in L^2_{1,1}(\mathbb{R})$.

B. Notation

Besides the usual notation in analysis as found in the books by L. Hörmander, L. Schwartz, and E. Stein and G. Weiss, we use the following conventions and notation.

The integral over $\mathbb{R}^d$ is designated by "$\int$". The Fourier transform of f is $\hat{f}(\gamma) = \int f(t)e^{-2\pi i t\cdot\gamma}dt$, $\gamma \in \hat{\mathbb{R}}^d(=\mathbb{R}^d)$, and $f = (\hat{f})^{\vee}$.

$\mathcal{S}_o(\mathbb{R}^d) = \{f \in \mathcal{S}(\mathbb{R}^d) : \hat{f}(0) = 0\}$ and $\mathcal{S}_{oa}(\mathbb{R}^d) = \{f \in \mathcal{S}(\mathbb{R}^d) : \hat{f}(\gamma_1,\cdots,\gamma_d) = 0$ if some $\gamma_j = 0\}$.

$C_c(\mathbb{R}^d)$ (resp., $C_c^{\infty}(\mathbb{R}^d)$) is the space of continuous (resp., infinitely differentiable) compactly supported functions on $\mathbb{R}^d$. $C_b(\mathbb{R}^d)$ consists of the bounded continuous functions on $\mathbb{R}^d$.

$M(\hat{\mathbb{R}}^d)$ (resp., $M_b(\hat{\mathbb{R}}^d)$, $M_+(\hat{\mathbb{R}}^d)$, $M_{b+}(\hat{\mathbb{R}}^d)$) is the space of Radon measures (resp., bounded, positive, bounded and positive Radon measures) on $\hat{\mathbb{R}}^d$.

$L^p_v(\mathbb{R}^d) = \{f : \|f\|_{p,v} = (\int|f(t)|^p v(t)dt)^{1/p} < \infty\}$ and L^p_E is L^p_v for $v = \chi_E$, the characteristic function of $E \subseteq \mathbb{R}^d$ with Lebesgue measure $|E|$. $\mathcal{L}(X,Y)$ is the space of continuous linear maps between the topological vector spaces X and Y.

σ_{d-1} designates surface measure on $\hat{\mathbb{R}}^d$. Its restriction to the unit sphere Σ_{d-1} is μ_{d-1}, and $\mu_{d-1}(\Sigma_{d-1}) = \omega_{d-1} = 2\pi^{d/2}/\Gamma(d/2)$. Finally, the ball of radius T centered at the origin $0 \in \mathbb{R}^d$ is $B(0,T) = B(T)$ and if $d = 1$ we write $B(T) = \chi_{(T)}$.

REFERENCES

[Bal] Balian, R. (1981) "Un principe d'incertitude fort en théorie du signal on en mécanique quantique," C. R. Acad. Sci., Paris 292, 1357-1362.

[Ba] Barnes, J. (1970) "Laplace-Fourier transformation, the foundation for quantum information theory and linear physics," Problems in analysis (R. Gunning, ed.), Princeton University Press.

[Bat] Battle, G. (1988) "Heisenberg proof of the Balian-Low theorem," Lett. Math. Phys. 15, 175-177.

[B1] Benedetto, J. (1984) "Some mathematical methods for spectrum estimation," and "Fourier uniqueness criteria and spectrum estimation theorems," Fourier techniques and applications (J. Price, ed.), Plenum Press, N.Y.

[B2] Benedetto, J. (1985) "An inequality associated with the uncertainty principle," Rend. Cir. Mat., Palermo 34, 407-421.

[B3] Benedetto, J., "A multidimensional Wiener-Wintner theorem and spectrum estimation," (submitted).

[BH1] Benedetto, J. and Heinig, H. (1983) "Weighted Hardy spaces and the Laplace transform," Springer Lecture Notes 992, 240-277.

[BH2] Benedetto, J. and Heinig, H., "Fourier transform inequalities with measures weights," Advances in Math. (to appear).

[Bo] Bourgain, J. (1988) "A remark on the uncertainty principle for Hilbertian basis," J. of Functional Anal. 79, 136-143.

[CP] Cowling, M. and Price, J. (1984) "Bandwidth versus time concentration: the Heisenberg-Pauli-Weyl inequality," SIAM J. Math. Anal. 15, 151-165.

[D1] Daubechies, I., "The Wavelet transform, time-frequency localization and signal analysis," IEEE Trans. Inf. Theory (to appear).

[D2] Daubechies, I. (1988) "Orthonormal bases of compactly supported wavelets," Comm. Pure and Appl. Math. 41, 909-996.

[DGM] Daubechies, I., Grossmann, A. and Meyer, Y. (1986) "Painless non-orthogonal expansions," J. Math. Phys. 27, 1271-1283.

[DJ] Daubechies, I. and Janssen, A., "Two theorems on lattice expansions," IEEE Trans. Inf. Theory (submitted).

[De] DeBruijn, N. (1967) "Uncertainty principles in Fourier analysis," Inequalities (O. Shisha, ed.), Academic Press, N.Y. 57-71.

[F] Faris, W. (1978) "Inequalities and uncertainty principles," J. Math. Phys. 19, 461-466.

[FG] Feichtinger, H. and Gröchenig, K., "Banach spaces related to integrable group representations and their atomic decompositions I," J. Functional Anal. (to appear).

[FJ] Frazier, M. and Jawerth, B., "A discrete transform and decomposition of distribution spaces," (preprint).

[G] Gabor, D. (1946) "Theory of communication," J. IEE (London) 93, 429-457.

[Gr] Grenander, U. (1951) "On empirical spectral analysis of stochastic processes," Arkiv för Mat. 1, 503-531.

[GR] Garcia-Cuerva, J. and Rubio de Francia (1985) Weighted norm inequalities and related topics, North-Holland, Amsterdam.

[HW] C. Heil and Walnut, D., "Continuous and discrete wavelet transforms," SIAM Review (to appear).

[HSi] Heinig, H. and Sinnamon, G., "Fourier inequalities and integral representations of functions in weighted Bergman spaces over tube domains," Indiana Math. J. (to appear).

[HS] Heinig, H. and Smith, M. (1986) "Extensions of the Heisenberg-Weyl inequality," International J. Math. and Math. Sci. 9, 185-192.

[He] Hernández, E., "Factorization and extrapolation of pairs of weights," (preprint).

[HJJ] Høholdt, T., Jensen, H. and Justesen, J. (1988) "Double series representation of bounded signals," IEEE Trans. Inf. Theory 34, 613-624.

[K] Kjellberg, B. (1943) "Ein momentumproblem," Ark. Mat. Fys. Astr. 29, A2.

[L] Landau, H. (1987) "Maximum entropy and the moment problem," Bull. AMS 16, 47-77.

[LPS] Landau, H., Pollak, H., and Slepian, D. (1961) "Prolate spheroidal wave functions, Fourier analysis and, uncertainty," I-V, Bell System Tech. J. 40, 43-64, 40(1961), 65-84, 41(1962), 1295-1336, 43(1964), 3009-3058, 57(1978), 1371-1430.

[M1] Meyer, Y. (Feb. 1986) "Principe d'incertitude, bases Hilbertiennes, et algèbres d'operateurs," Sem. Bourbaki 662.

[M2] Meyer, Y. (1986) "Ondelettes fonctions splines, et analyses graduées," U. of Torino.

[P] Price, J. F. (1987) "Sharp local uncertainty inequalities," Studia Math. 85, 37-45.

[PI] Proc. IEEE (1982) "Special issue on spectrum estimation," 70.

[PS] Price, J. F. and Sitaram, A. (1988) "Local uncertainty inequalities for locally compact groups," Trans. AMS 308, 105-114.

[Si] Sinnamon, G., "A weighted gradient inequality," Proc. Roy. Soc. Edinburgh (to appear).

[W] Wiener, N., Collected works (P. Masani, ed.) MIT Press.

THE GENERALIZED PENCIL-OF-FUNCTION (GPOF) METHOD FOR EXTRACTING POLES FROM TRANSIENT RESPONSES

Yingbo Hua, Tapan K. Sarkar, and Fengdo Hu

Department of Electrical and Computer Engineering
Syracuse University
Syracuse, NY 13244-1240

ABSTRACT

The generalized pencil-of-function (GPOF) method for extracting the poles of EM (electromagnetic) systems from their transient responses is presented. The GPOF method solves a generalized eigenvalue problem to find the poles. This is in contrast to the conventional Prony and Pencil-of-Function (POF) methods which yield the solution in two steps, namely, the solution of an ill-conditioned matrix equation and finding the roots of a polynomial. To optimize the performance of the GPOF method, the subspace decomposition approach is used. The GPOF method has advantages over the Prony method in both computation and noise sensitivity, and it approaches the Cramer-Rao bound when SNR is above threshold. To further lower the threshold of the GPOF method, a circular weighting matrix is proposed. An application of the GPOF method to a thin-wire target is also presented.

The work was supported in part by the Office of Naval Research under Contract N00014-79-C-0598.

----

1. INTRODUCTION

It is known [1-8] that in target identification, extracting various features of the target from its transient response is desired. The target poles that contain the information of the decaying factors and the resonant frequencies should be estimated with high accuracy while the residues can be computed by solving a linear least squares problem [1] after the poles are obtained.

The Prony method [1-3] has been a very popular technique for pole retrieval. There are also many versions of the Prony method, which include the least-square (LS) Prony method, the total-least-square (TLS) Prony method and the SVD Prony method.

An alternative method is the pencil-of-function (POF) method [4, 9]. Very recently, the idea of the POF method

J. S. Byrnes and J. F. Byrnes (eds.), Recent Advances in Fourier Analysis and Its Applications, 183–199.

has been explored along with ESPRIT [10], and this has resulted in improved and generalized versions [11-14]. This paper, which is a result of this exploration, presents a generalized pencil-of-function (GPOF) method. The GPOF method finds poles by directly solving a generalized eigenvalue problem instead of the conventional two steps process where the first step involves the solution of a matrix equation and the second step is to find the roots of a polynomial, as is required by the Prony method. We develop the GPOF method and discuss its computational aspects in Section 2. The noise sensitivity of the GPOF method is addressed in Section 3. In Section 4, a circular weighting matrix is applied to the GPOF to further lower the threshold. In Section 5, an application of the GPOF method to a thin-wire target is presented.

2. GENERALIZED PENCIL-OF-FUNCTION METHOD

It is known that the EM transient signal can be described by the samples y_k, defined as

$$y_k = \sum_{i=1}^{M} b_i \exp(s_i k\delta t) \tag{2.1}$$

where k=0, 1, ..., N-1, b_i's are the complex residues, s_i's are the complex poles, and δt is the sampling interval. We let $z_i = \exp(s_i \delta t) = \exp(\alpha_i + jw_i)$ which are the poles in the Z-plane. It is clear that b_i's and s_i's should each be in complex conjugate pairs for real valued y_k.

Following the idea of the pencil-of-function method, we consider the following set of "information" [9] vectors:

$\underline{y}_0, \underline{y}_1, \ldots, \underline{y}_L$

where

$$\underline{y}_i = [y_i, y_{i+1}, \ldots, y_{i+N-L-1}]^T \tag{2.2}$$

The superscript "T" denotes transpose of a matrix. Based on these vectors, we define the matrices, Y_1 and Y_2, as

$$Y_1 = [\underline{y}_0, \underline{y}_1, \ldots, \underline{y}_{L-1}] \tag{2.3}$$

$$Y_2 = [\underline{y}_1, \underline{y}_2, \ldots, \underline{y}_L] \tag{2.4}$$

To look into the underlying structure of the two matrices, one can write

$$Y_1 = Z_1 \; B \; Z_2 \tag{2.5}$$

(2.6) $Y_2 = Z_1 \ B \ Z_0 \ Z_2$

where

(2.7) $$Z_1 = \begin{bmatrix} 1 & 1 & \dots & 1 \\ z_1 & z_2 & \dots & z_M \\ \cdot & \cdot & \dots & \cdot \\ z_1^{N-L-1} & z_2^{N-L-1} & \dots & z_M^{N-L-1} \end{bmatrix}$$

(2.8) $$Z_2 = \begin{bmatrix} 1 & z_1 & \dots & z_1^{L-1} \\ & & \dots & \\ 1 & z_M & \dots & z_M^{L-1} \end{bmatrix}$$

(2.9) $Z_0 = \mathrm{diag}[z_1 , z_2 , \dots, z_M]$

(2.10) $B = \mathrm{diag}[b_1 , b_2 , \dots, b_M]$

where diag[ ] denotes a diagonal matrix. Based on the above decomposition of Y_1 and Y_2 , one can show that if $M \leq L \leq N-M$, the poles $\{z_i ; i=1, \dots, M\}$ are the generalized eigenvalues [10, 17] of the matrix pencil $Y_2 - zY_1$. Namely, if $M \leq L \leq N-M$, $z=z_i$ is a rank reducing number of $Y_2 - zY_1$. If $L=M$, this method is the same as the basic POF method as in [12] and equivalent to a version of Ibrahim time-domain (ITD) method [16]. However, in the GPOF method, we are more interested in different values of L for $M < L < N-M$. The significance of this will be shown later.

To develope, and to illustrate the use of, an algorithm for computing the generalized eigenvalues of matrix pencil, we write

(2.11) $$\begin{aligned} & Y_1^{+} Y_2 \\ & = Z_2^{+} B^{-1} Z_1^{+} Z_1 B Z_0 Z_2 \\ & = Z_2^{+} Z_0 Z_2 \end{aligned}$$

where the superscript "+" denotes the (Moore-Penrose) pseudoinverse [17], whereas we use "-1" for the (regular) inverse. It can be seen from (2.11) that there exist vectors $\{\mathbf{p}_i ; i=1, \dots, M\}$ such that

(2.12) $Y_1^{+} Y_1 \mathbf{p}_i = \mathbf{p}_i$ and

(2.13) $Y_1^{+} Y_2 \mathbf{p}_i = z_i \mathbf{p}_i$

The $\mathbf{p}_i$'s are called the generalized eigenvectors of $Y_1 - zY_2$
To compute the pseudoinverse Y_1^{+} , one can use the Singular Value Decomposition (SVD) [17] of Y_1 as follows.

$$(2.14)\quad Y_1 = \Sigma_{i=1,M}\ \sigma_i\ \underline{u}_i\ \underline{v}_i^{\,H} = U\,D\,V^H$$

$$(2.15)\quad Y_1^{\,+} = V\,D^{-1}\,U^H$$

where $U=[\underline{u}_1, \ldots, \underline{u}_M]$, $V=[\underline{v}_1, \ldots, \underline{v}_M]$, and $D=\mathrm{diag}[\sigma_1, \ldots, \sigma_M]$. The superscript "H" denotes the conjugate transpose of a matrix. U and V are matrices of left and right singular vectors, respectively. Note that for noisy data y_k, one should choose $\sigma_1, \ldots, \sigma_M$ to be the M largest singular values of Y_1, and the resulting $Y_1^{\,+}$ is called the truncated pseudoinverse of Y_1. Since $Y_1^{\,+} Y_1 = VV^H$ and $V^H V=I$, substituting (2.15) into (2.13) and left multiplying (2.13) by V^H yields

$$(2.16)\quad (T - z_i I)\,\underline{f}_i = 0$$

where i=1, ..., M, and

$$(2.17)\quad T = D^{-1}\,U^H\,Y_2\,V\text{, and}$$

$$(2.18)\quad \underline{f}_i = V^H\,\underline{p}_i$$

Note that T is an MxM matrix, and z_i's and $\underline{f}_i$'s are, respectively, eigenvalues and eigenvectors of T. Now we have completed the description of an algorithm of the GPOF method. In summary, from the data one constructs the two rectangular matrices Y_1 and Y_2 as described by (2.3) and (2.4). Then the SVD of Y_1 is computed (e.g., using the IMSL software package) as in (2.14). Then the MxM matrix T is constructed according to (2.17). Finally, the poles z_i's are the eigenvalues of T.

It is important to mention that the number, M, of poles can be estimated from the singular values, $\sigma_1 \geq \sigma_2 \geq \ldots \geq \sigma_M \geq \ldots \geq \sigma_{\min(N-L,L)}$, since $\sigma_{M+1} = \ldots = \sigma_{\min(N-L,L)} = 0$ for noiseless data. Namely, M can be chosen to be the number of dominant singular values of Y_1.

If L=M, the SVD of Y_1 is not required, and $\{z_i;\ i=1, \ldots, M\}$ are the eigenvalues of the MxM matrix $(Y_1^{\,H} Y_1)^{-1} Y_1^{\,H} Y_2$ which is obtained by substituting $Y_1^{\,+} = (Y_1^{\,H} Y_1)^{-1} Y_1^{\,H}$ into (2.13) for L=M. Furthermore, one ca verify that with or without noise,

$$(2.19)\quad (Y_1^{\,H} Y_1)^{-1} Y_1^{\,H} Y_2 = \begin{bmatrix} 0 & \cdots & 0 & -c_M \\ 1 & & 0 & -c_{M-1} \\ & \ddots & & \vdots \\ & & 1 & -c_1 \end{bmatrix}$$

which is the companion matrix of the polynomial

(2.20) $1 + \Sigma_{i=1,M} \; c_i \, z^{-i} = 0,$

and

(2.21) $$\begin{bmatrix} c_M \\ . \\ . \\ c_1 \end{bmatrix} = -(Y_1^H Y_1)^{-1} Y_1^H \underline{y}_M$$

is the solution of the least-square (LS) Prony method. Hence, for L=M, the GPOF method is equivalent to the LS Prony method [1].

The total-least-square (TLS) Prony method is to compute the polynomial coefficients as follows. Let $Y=[\underline{y}_0 , \underline{y}_1 , ..., \underline{y}_M]$ and $\underline{c}=[c_M , ..., c_1 , c_0]^T$. Then the polynomial coefficient vector $\underline{c}$ is computed by nimimizing

(2.22) $\underline{c}^H Y^H Y\underline{c}$ subject to $||\underline{c}||_2 =1$

A perturbation analysis [12] has shown that the LS Prony method and the TLS Prony method are equivalent to the first order perturbation approximation.

The SVD Prony method (for L > M) is to compute $\underline{c}_L =[c_{L-1} , ..., c_0]^T$ by

(2.23) $\underline{c}_L = -(Y_2^H Y_2)^+ Y_2^H \underline{y}_0 = -Y_2^+ \underline{y}_0$

and detect the M signal roots, i.e., $\{z_i^{-1} ; i=1, ..., M\}$, from the L roots of the L-degree polynomial $1 + \Sigma_{i=1,L} \; c_{L-i} z^{-i}$. But the detection is guaranteed (without noise) to be successful only when all z_i 's (z_i^{-1} 's) are inside (outside) or on the unit circle [21]. Computationally, solving for the roots of an L-degree polynomial is also a disadvantage of the SVD Prony method, comparing to solving M eigenvalues of an MxM matrix for the GPOF method.

3. NOISE SENSITIVITY OF GPOF METHOD

In this section, we illustrate the noise sensitivity of the GPOF method through some numerical examples. Specifically, we assume a real data sequence

(3.1) $y_k = \Sigma_{i=1,J} \; a_i \sin(w_i k+\varphi_i)\exp(\alpha_i k)$

where $k=0, 1, ..., N-1$, $N=30$, $J=2$, $a_1 =a_2 =1$, $w_1 =0.2\pi$, $w_2 =0.35\pi$, $\varphi_1 =\varphi_2 =0$, $\alpha_1 =-0.02\pi$, and $\alpha_2 =-0.035\pi$. Note that

$M=2J=4$, $b_i=(1/2j)a_i\exp(j\varphi_i)$ for $i=1,2$. It is also important to note that α_i's and w_i's are, respectively, damping factors and resonant frequencies normalized by the sampling frequency $f_s=1/\delta t$. (There is no need to specify the sampling frequency for our numerical simulations. The sequence as in (3.1) is the only sampled data sequence used in this section for illustration purposes.) The first order perturbation analysis of the GPOF method is outlined in Appendix. For analysis, it is assumed that the additive noise in y_k's is white and sufficiently weak so that the first order approximation can be carried out through our derivation. Figure 1 shows the inverted perturbation variance in dB of w_1 (imaginary part of $s_i\,\delta t$) versus the pencil parameter L. The Cramer-Rao bound provides the "absolute" best result that any technique can achieve under the present "noisy" environment. The GPOF method approximately reached the Cramer-Rao bound. This implies that "no" other theretical technique can do any better! Figure 2 shows the same thing for α_1 (real part of $s_i\,\delta t$). The plots for w_2 and α_2 are similar to the above two figures, and hence omitted from this paper. As one observes, the optimal choice of L is around $L=N/2$. Intuitively, this phenomenon can be explained as follows.

The noiseless Y_1 (or Y_2) has a column subspace of dimension M. (Note that the GPOF method requires $L\geq M$ and $N-L\geq M$.) This subspace is called signal subspace, denoted by S_S. But the noisy Y_1 (or Y_2), which consists of signal and noise, spans a column subspace, denoted by S_{S+N}, of dimension equal to min(N-L, L). It is clear that we can write $S_{S+N}=S_S+S_N$, where S_S contains all signal component and noise component projected onto the signal subspace, and S_N contains noise only. As one sees, in the GPOF method, we do subspace decomposition of Y_1 as in (2.14) and throw away the noise component in S_N which is spanned by $\underline{u}_i$ for $i>M$. It seems, therefore, that the larger the noise subspace S_N is, the more noise we can filter out by the GPOF method. The dimension of S_N is the largest when $N-L=L$, i.e., $L=N/2$. This is consistent with our perturbation analysis. Note that around $L=N/2$ (practically between N/3 and 2N/3), the performance of the GPOF method is very close to the optimal bound, i.e., the Cramer-Rao bound [14, 19]. A different interpretation of a similar phenomenon with ITD method was made in [16].

With the choice $L=N/2=15$, simulation results for the GPOF method are shown in Figures 3-4. The (Monte-Carlo) simulation was conducted with 200 runs. During each run, we computed the estimated α_i's and w_i's from the data contaminated by (pseudo) white Gaussian noise. (Note that $\alpha_i=\mathrm{Re}\{\log[z_i]\}$ and $w_i=\mathrm{Im}\{\log[z_i]\}$.) The noise used in each run is independent of those used in others. Figures 3-4 show the inverted sample variances in dB (denoted by the plus signs "+") of the estimated w_1 and α_1 versus SNR (Signal to Noise Ratio) which is defined by

$$\text{(3.2)} \quad SNR = \Sigma_{k=0,N-1} \, |y_k|^2 /NP \quad \text{and}$$

$$\text{(3.3)} \quad SNR(dB) = 10\log_{10} (SNR).$$

where P is the noise power. The straight lines are obtained from the perturbation analysis. As one observes, for high SNR, the simulation results agree with the analytical results. As a reference, the simulation results of the LS Prony method are also shown in the plots. The noisy data used for the LS Prony method were the same as those used for the GPOF method. The detailed perturbation analysis of the Prony method is available in [14-15]. We should mention that the SVD Prony method as represented by (2.21) performs better than the LS Prony method and the TLS Prony method [21]. In fact, the SVD Prony method performs almost as well as the GPOF method for this particular example. However, it can be shown [12-14] both from a theoretical and numerical point that the GPOF method is less sensitive to noise than the SVD Prony method. This is in addition to the fact that GPOF is computationally more efficient than the SVD Prony method.

4. GPOF with Circular Weighting Matrix

In the previous section, we have emphasized that the subspace decomposition plays an important role in reducing the noise effect. The subspace filtering technique has the advantage that it does not require a priori information about the signal parameters (damping factors and frequencies).

In this section, we show that the estimation accuracy of signal parameters can be further improved by using a circular weighting matrix. Specifically, we apply a (N-L)x(N-L) circular matrix H to the set of "information" vectors $\{\underline{y}_i \; ; \; i=0,1,...,L\}$ to yield

$$\text{(4.1)} \quad \underline{g}_i = H\underline{y}_i$$

for i=0,1,...,L, where the first row of H is

$$\text{(4.2)} \quad \underline{h}^T = [h_0 \;\; h_1 \;\; ... \;\; h_{N-L-1}]$$

It is clear that (4.1) is simply the circular convolution between $\underline{h}$ and $\underline{y}_i$, i.e.,

$$\text{(4.3)} \quad \underline{g}_i = \underline{h} * \underline{y}_i$$

Based on some initial frequency estimates obtained by FFT or the GPOF (without weighting), one can design the FIR filter $\underline{h}$ to amplify the signal frequency component and suppress the noise component. If $\underline{h}$ is properly chosen, one expects that the weighted vectors $\{\underline{g}_i \; ; \; i=0,1,...L\}$ are "cleaner" than the unweighted vectors $\{\underline{y}_i \; ; \; i=0,1,...L\}$. Therefore, applying the GPOF to $\{\underline{g}_i \; ; \; i=0,1,...,L\}$ should yield better estimates of signal parameters.

It is important to note that as long as H has rank equal to or larger tham M, $\{\underline{g}_i \; ; \; i=0,1,...,L\}$ lead to exact pole estimates in noiseless case. This is in contrast to conventional filtering approaches.

To show an example of using the circular weighting matrix, we consider the same data given in (3.1). Note that the two resonant frequencies are $w_1 = 0.2\pi = 0.628$ and $w_2 = 0.35\pi = 1.1$. We chose the FIR sequence to be

$$(4.4) \qquad h(k) = h_d(k)\, w(k)$$

where $h_d(k)$ is an ideal bandpass filter with lower cutoff frequency 0.5 and upper cutoff frequency 1.2, and $w(k)$ is the Kaiser window [23] of length=19 and $\beta=5.658$.

In the GPOF method, we let L=10 and N-L=20. Padding a zero to the 19-element sequence h(k), we obtain the first row of the 20x20 circular matrix H.

Figures 5-6 show the results of 200-run simulation. The stars (*) are inverted sample variances in dB for the GPOF without weighting. The pluses (+) are inverted sample variances in dB for the GPOF with weighting. (The dash lines are inverted mean square errors in dB.) As one can see, above the threshold of the GPOF, the weighting has little effect. But at low SNR, the weighted GPOF is more robust to noise.

5. AN APPLICATION

Consider that a 2-meter dipole antenna of radius 0.001m is illuminated by a short EM pulse of the form

$$E^{inc} = \eta/(\pi^{1/2} ce) \exp[-(t-t_0)^2 / e^2]$$

from broadside where $\eta=377$ ohms, $c=3\times10^8$ m/s, $e=0.5/c$, and $t_0=6e$. The current response at the center point of the dipole was computed by the approach in [20] with sampling time $\delta t=0.5$ ns, and is shown in Figure 7. To get the intrinsic poles of the dipole itself, we considered a segment of the current for t=8.4 through 25.5 light-meters, which consists of 114 samples. Note that for t>8.4 light-meters the EM pulse E^{inc} is almost zero. Figure 8 shows the FFT amplitude spectrum of the current. From the spectrum, four resonant components are detected at 0.0684, 0.203, 0.391, and 0.414 GHz. The spectrum was computed with resolution equal to f_s /1024=0.00195 GHz.

Applying the GPOF method (without weighting) with L=N/2=57 to the 114 sampled data, we observed that the ten largest singular values of the data matrix Y_1 are 9.3, 7.7, 0.45, 0.42, 0.057, 0.056, 0.039, 0.0388, 0.0052, and 0.0049. With M=8 (since there is a large drop from the 8th singular value to the 9th), the GPOF method yielded the following poles (i.e., $s_i \delta t=\log[z_i]=\alpha_i+jw_i$):

$-0.0204\pm j0.218$ $-0.0266\pm j0.642$ $-0.0046\pm j1.17$ $0.0039\pm j1.27$

By least squares fitting [1], the corresponding residues (absolute values) were computed to be

0.3835×10^{0} 0.2456×10^{-1} 0.9379×10^{-3} 0.7506×10^{-3}

From the first six stable poles, three estimated resonant frequencies ($f_i=w_i/2\pi\delta t$) are obtained to be 0.0694, 0.204, and 0.372 GHz. The unstable estimated poles with the frequency 0.404 GHz appear to correspond to the "fourth" resonant component as is shown in the FFT spectrum. Applying the LS Prony method to the same data samples and using the assumption M=8, we found the following poles:

$-0.0203\pm j0.218$ $-0.0261\pm j0.645$ $0.3\times10^{-3}\pm j1.251$ $0.611\pm j2.244$

It is seen that the first two pairs of the poles are close to the corresponding pairs obtained by the GPOF method while the next two pairs differ from those obtained by the GPOF method. In the following table, resonant frequencies estimated by different approaches are compared for the identical data records.

Frequency(GHz)	f_1	f_2	f_3	f_4
By $f_i=c(2i-1)/2L$	0.0750	0.225	0.375	*
By GPOF	0.0694	0.204	0.372	0.404
By FFT	0.0684	0.203	0.391	0.414
By LS Prony	0.0694	0.205	0.398	0.714

where L is the length of the dipole and c is the light velocity. It does seem that for the same data the GPOF method provides a stable solution.

6. CONCLUSION

We have presented the GPOF method which solves a generalized eigenvalue problem to estimate poles of EM systems. In this new approach, one finds the poles directly by solving a generalized eigenvalue problem. This is in contrast to the contemporary methods like Prony and Pencil-of-Function methods, which are generally two step processes. In the first step one solves a matrix equation followed by finding the

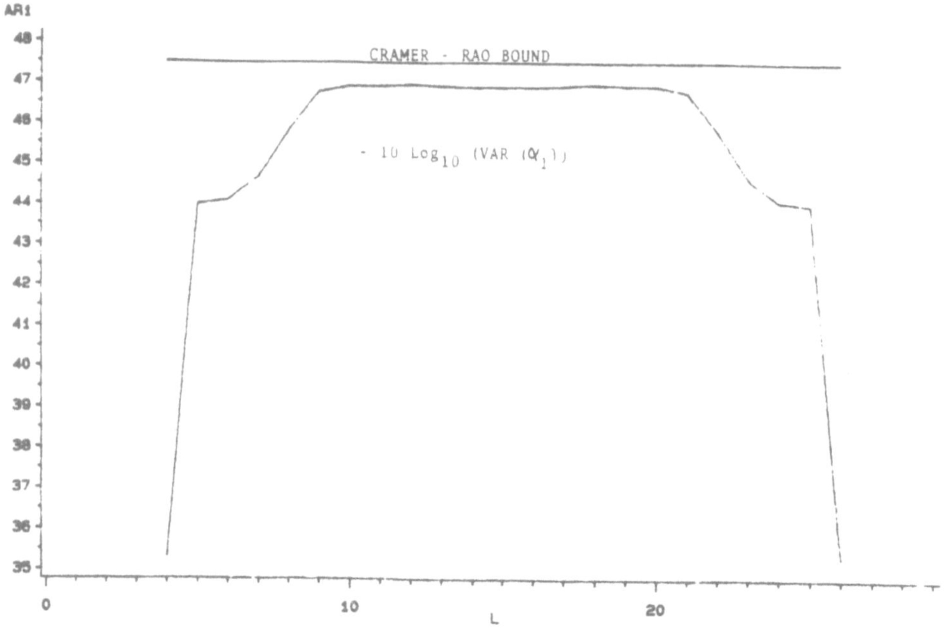

1. Inverted perturbation variances in dB of the estimated α_1 versus the pencil parameter L.

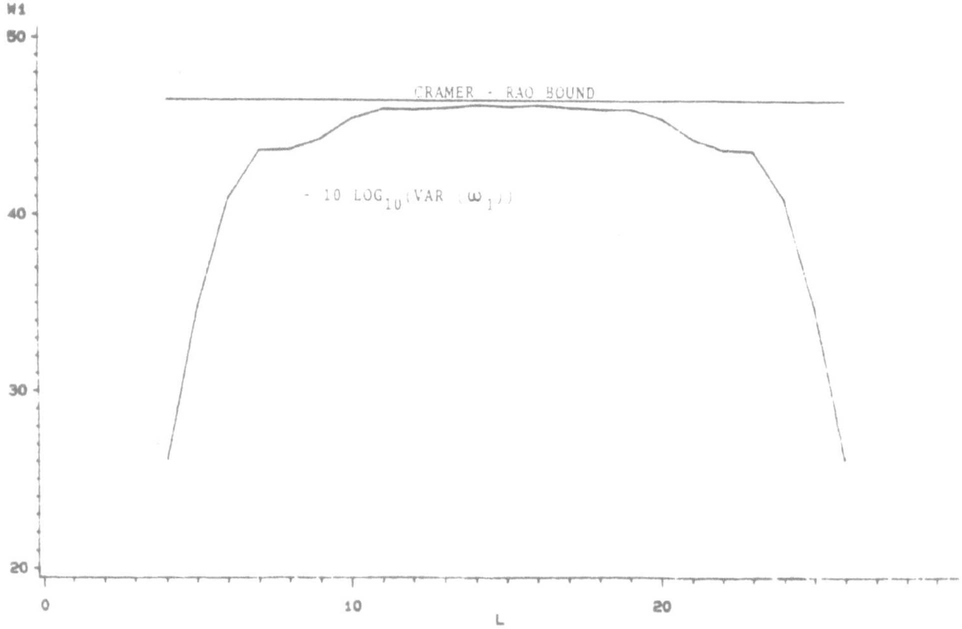

2. Inverted perturbation variances in dB of the estimated w_1 versus the pencil parameter L.

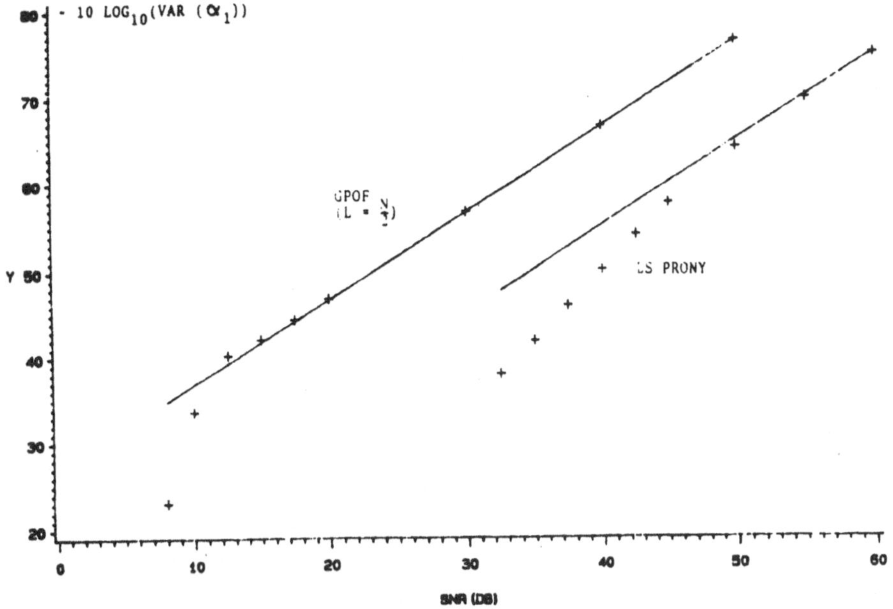

3. Simulation results (denoted by "+") of the inverted sample variance in dB of α_1 estimated by the GPOF method and the LS Prony method. The straight lines are obtained from the perturbation analysis.

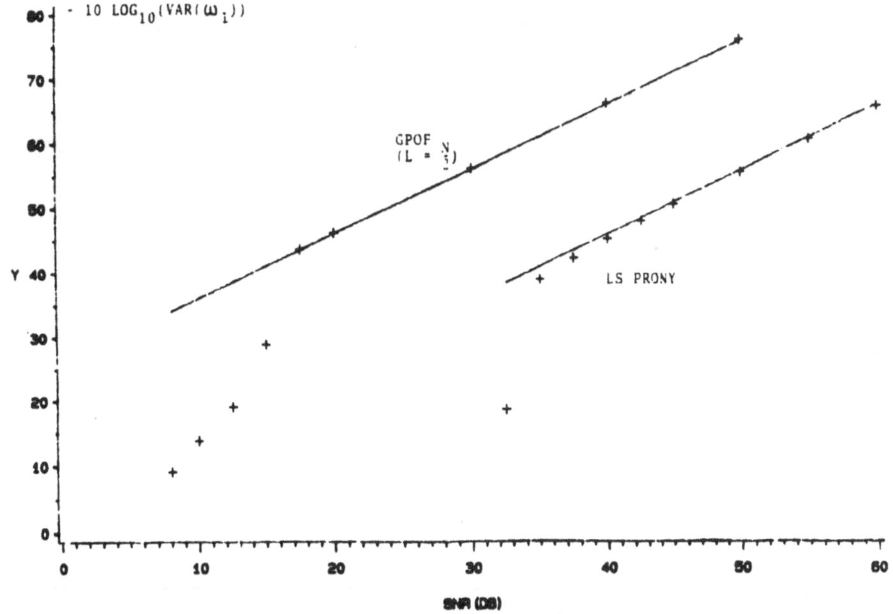

4. Simulation results (denoted by "+") of the inverted sample variance in dB of w_1 estimated by the GPOF method and the LS Prony method. The straight lines are obtained from the perturbation analysis.

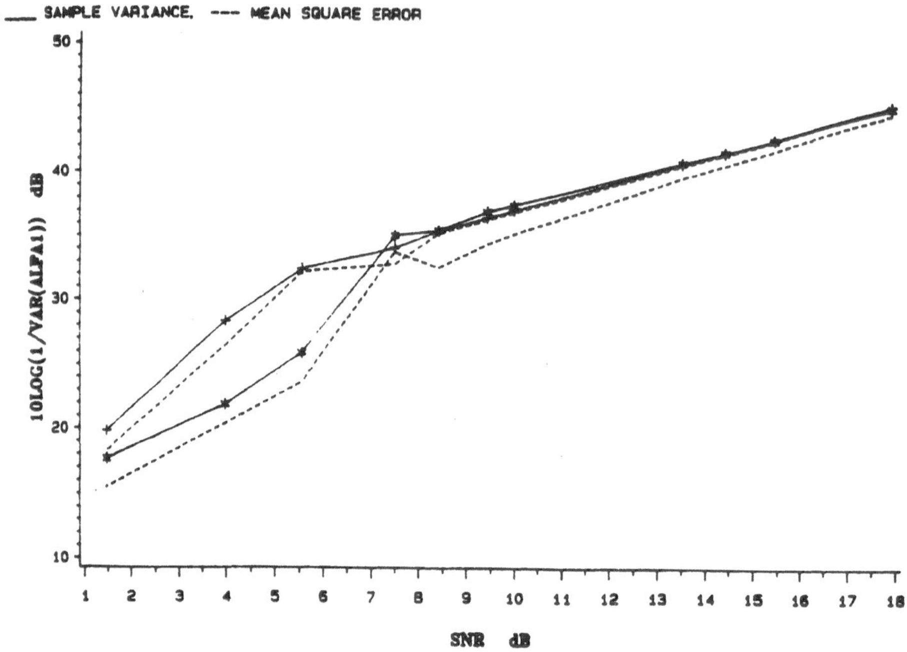

5. For α_1 . Comparison of weighted GPOF and unweighted GPOF. The Pluses (+) and Stars (*) are inverted sample variances in dB for weighted GPOF and unweighted GPOF, respectively. The dash lines are inverted mean square errors in dB.

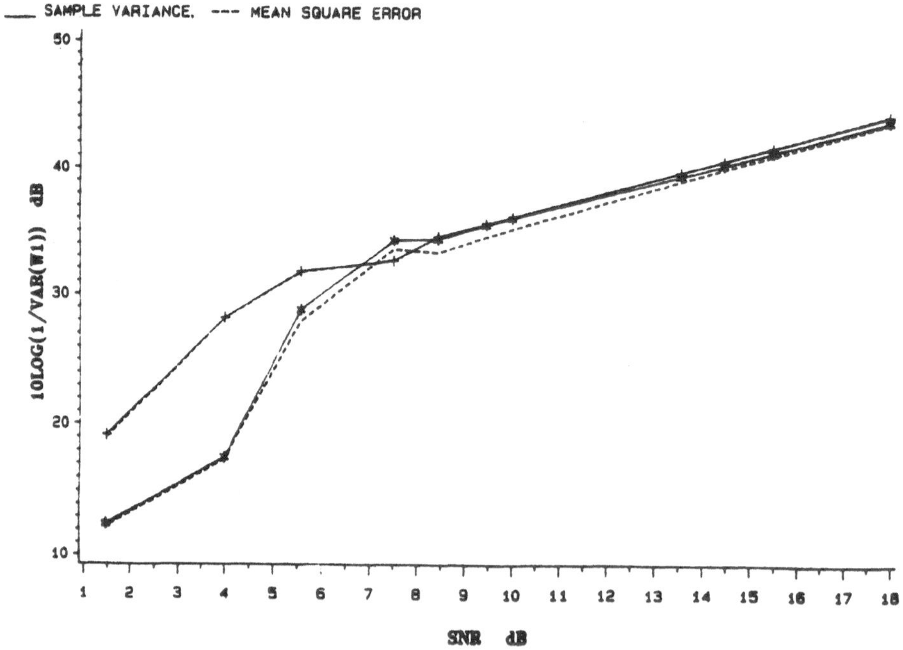

6. For w_1 . Same as in Figure 5.

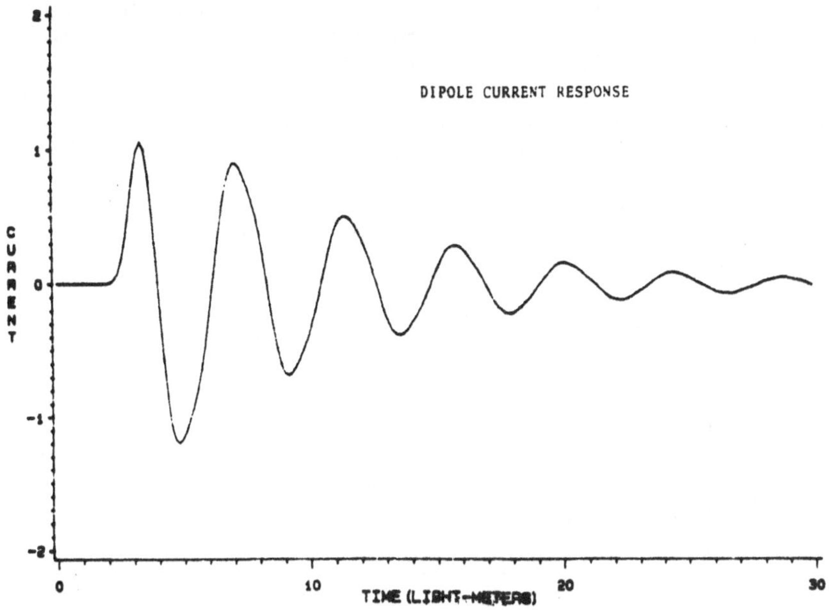

7. Current response at the center of the 2-meter dipole illuminated by a short pulse.

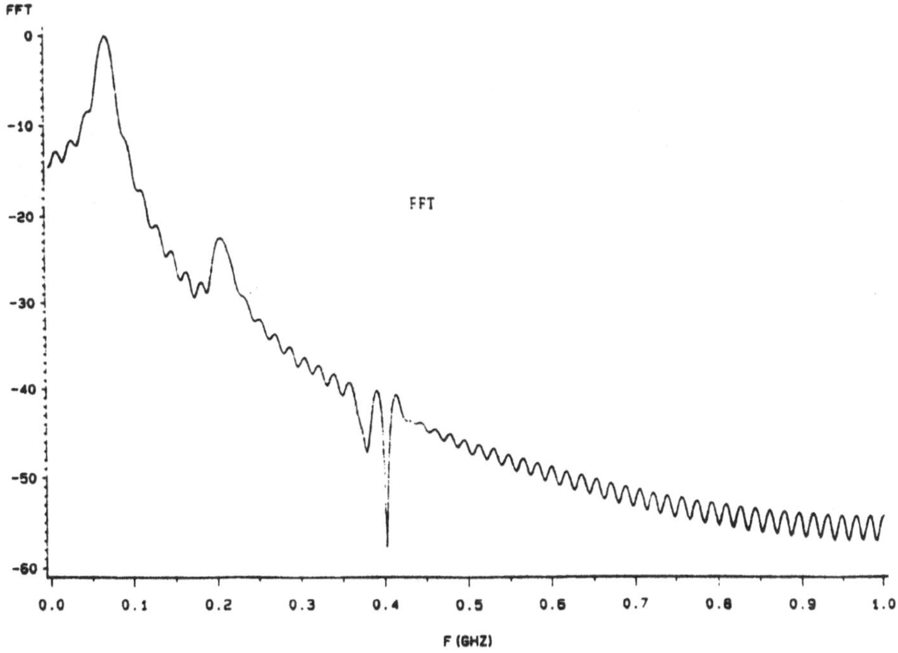

8. FFT amplitude spectrum of the current.

roots of a polynomial. Compared to the SVD Prony method, the GPOF method is less restricted because it does not require that all system poles must be either stable or nonstable, and computationally more efficient because it does not solve an L-degree polynomial. Compared to the LS or TLS Prony method, the GPOF method is more robust to noise as has been shown. A weighting approach has been presented, which increases the noise robustness of the GPOF. An application of the GPOF method to a synthesized dipole has been shown, with comparison to the FFT and the LS Prony method.

APPENDIX

In this section, we give an outline of the first order perturbation analysis of the GPOF method. For detailed discussion, see [14].

First, it can be verified that the eigenvalues of T are the same as the nonzero eigenvalues of $Y_1^+ Y_2$ whether or not the data is noisy. Then, it is known [18] that the first order variation in the poles is given by

$$\text{(A.1)} \qquad \delta z_i = \mathbf{q}_i^H \, \delta(Y_1^+ Y_2)\mathbf{p}_i / \mathbf{q}_i^H \mathbf{p}_i$$

where "δ" is the first order differential operator, and $\mathbf{p}_i$ and $\mathbf{q}_i$ are, respectively, the right and the left eigenvectors of the noiseless $Y_1^+ Y_2$. So $\mathbf{p}_i$ (see (2.11)) is the ith column of Z_2^+, and $\mathbf{q}_i$ is the ith column of Z_2^H. Therefore, $\mathbf{q}_i^H \mathbf{p}_i = 1$.

Secondly, it can be shown [13-14] that

$$\text{(A.2)} \qquad \mathbf{q}_i^H \, \delta Y_1^+ Y_2 \mathbf{p}_i = -\mathbf{q}_i^H Y_1^+ \, \delta Y_1 \, Y_1^+ Y_2 \mathbf{p}_i$$

Note that Y_1^+ is a truncated pseudoinverse of Y_1 in noisy data case. From (A.1) and (A.2), after some algebraic manipulations one can obtain the following.

$$\text{(A.3)} \qquad \delta z_i = (1/b_i)\mathbf{r}_i^H (\delta Y_2 - z_i \, \delta Y_1)\mathbf{p}_i$$

where $\mathbf{r}_i^H$ is the ith row of Z_1^+, and

$$\text{(A.4)} \qquad \delta Y_1 = \begin{bmatrix} n_0 & n_1 & \dots & n_{L-1} \\ & & \dots & \\ n_{N-L-1} & n_{N-L} & \dots & n_{N-2} \end{bmatrix}$$

$$\text{(A.5)} \qquad \delta Y_2 = \begin{bmatrix} n_1 & n_2 & \dots & n_L \\ & & \dots & \\ n_{N-L} & n_{N-L+1} & \dots & n_{N-1} \end{bmatrix}$$

Finally, it can be shown from (A.3-5) that if $\{n_i ; i=0, \ldots, N-1\}$ are white with variance equal to P, then the first order perturbation variances are

(A.6) $$E\{|\delta z_i|^2\} = [P/|b_i|^2] \Sigma_{k=1,N} |C_{i,k} - z_i C_{i,k+1}|^2$$

(A.7) $$E\{|\delta \alpha_i|^2\} = P \Sigma_{k=1,N} Re^2 \{(C_{i,k} - z_i C_{i,k+1})/b_i z_i\}$$

(A.8) $$E\{|\delta w_i|^2\} = P \Sigma_{k=1,N} Im^2 \{(C_{i,k} - z_i C_{i,k+1})/b_i z_i\}$$

where

(A.9) $$C_{i,k} = \begin{cases} \Sigma_{t=1,k-1} f_{i,k,t}, & 2 \leq k \leq \min(L,N-L) \\ \Sigma_{t=k-L,k-1} f_{i,k,t}, & L \leq N-L \text{ and } L+1 \leq k \leq N-L+1 \\ \Sigma_{t=1,N-L} f_{i,k,t}, & N-L \leq L \text{ and } N-L+1 \leq k \leq L+1 \\ \Sigma_{t=k-L,N-L} f_{i,k,t}, & \max(L,N-L)+2 \leq k \leq N \end{cases}$$

(A.10) $$f_{i,k,t} = r_{i,t}{}^* p_{i,k-t}$$

in which $r_{i,t}{}^*$ is the tth element of the row vector $\underline{r}_i{}^H$ and $p_{i,t}$ is the tth element of the column vector $\underline{p}_i$.

REFERENCES

[1] M. L. Van Blaricum and R. Mittra, "A technique for extracting the poles and residues of a system directly from its transient response," IEEE Trans. on Ant. and Prop., Vol. 23, No. 6, pp. 777-781, Nov. 1975.

[2] A. J. Poggio, et al., "Evaluation of a processing technique for transient data," IEEE Trans. on Ant. and Prop., Vol. 26, No. 1, pp. 165-173, Jan. 1978

[3] M. L. Van Blaricum and R. Mittra, "Problem and solutions associated with Prony's method for processing transient data," IEEE Trans. on Ant. and Prop., Vol. 26, No. 1, pp. 174-182, Jan. 1978.

[4] T. K. Sarkar, et al., "Sub-optimal approximation /identification of transient waveforms from electromagnetic system by pencil-of-function method," IEEE Trans. on Ant. and Prop., Vol. 28, No. 6, pp. 928-933, Nov. 1980.

[5] F. I. Tseng and T. K. Sarkar, "Experimental determination of resonant frequencies by transient scattering from conducting spheres and cylinders," IEEE Trans. Ant. and Prop., Vol. 32, No. 9, pp. 914-918, Sept. 1984.

[6] F. I. Tseng and T. K. Sarkar, "Deconvolution of the impulse response of a conducting sphere by the conjugate

gradient method," IEEE Trans. on Ant. and Prop., Vol. 35, No. 1, pp. 105-109, Jan. 1987.

[7] E. M. Kennaugh and D. L. Moffatt, "Transient and impulse response approximations," Proc. IEEE, Vol. 53, pp. 893-901, Aug. 1965.

[8] D. L. Moffatt and R. K. Mains, "Detection and discrimination of radar targets," IEEE Trans. on Ant. and Prop., Vol. 23, No. 3, pp. 358-367, May 1975.

[9] V. K. Jain, T. K. Sarkar, and D. D. Weiner, "Rational modeling by pencil-of-function method," IEEE Trans. on Acous., Speech, and Signal Proc., Vol. 31, No. 3, pp. 564-573, June 1983.

[10] A. Paulraj, R. Roy, and T. Kailath, "Estimation of signal parameters via rotational invariance techniques-ESPRIT," in Proc. 19th Asilomar Conf. Circuit, Syst. Comp., Asilomar, CA, Nov. 1985.

[11] A. J. Mackay and A. McCowen, "An improved pencil-of-function method and comparisons with traditional methods of pole extraction," IEEE Trans. on Ant. and Prop., Vol. 35, No. 4, pp. 435-441, April 1987.

[12] Y. Hua and T. K. Sarkar, "Further analysis of three modern techniques for pole retrieval from data sequence," in Proc. 30th Midwest Symposium on Circuits and Syst., Syracuse, NY, Aug. 1987.

[13] Y. Hua and T. K. Sarkar, "Matrix pencil method and its performance," in Proc. ICASSP-88, New York, NY, April 1988.

[14] Y. Hua, "On techniques for estimating parameters of exponentially damped/undamped sinusoids in noise," Ph.D. Dissertation, Syracuse University, 1988.

[15] Y. Hua and T. K. Sarkar, "Perturbation analysis of TK method for harmonic retrieval problem," IEEE Trans. on Acous. Speech, and Signal Proc., Vol. 36, No. 2, pp. 228-240, Feb. 1988.

[16] S. R. Ibrahim and R. S. Pappa, "Large model survey testing using the Ibrahim time domain identification technique," J. Spacecraft, Vol. 19, No. 5, pp. 459-465, Sept.-Oct. 1982.

[17] G. H. Golub, C. F. Van Loan, Matrix Computations, John Hopkins Baltimore, Maryland, 1983.

[18] J. H. Wilkinson, The Algebraic Eigenvalue Problem, Clarendon Press, Oxford, 1965.

[19] C. R. Rao, Linear Statistical Inference and Its Applications, 2nd ed., New York, John Wiley and Son., 1973.

[20] S. M. Rao, T. K. Sarkar, and S. A. Dianat, "The application of the conjugate gradient method to the solution of transient electromagnetic scattering from thin wires," Radio Science, Vol. 19, No. 5, pp. 1319-1326, Sept.-Oct. 1984.

[21] R. Kumaresan, "Estimating the parameters of exponentially damped or undamped sinusoidal signals in noise," Ph.D. dissertation, University of Rhode Island, 1982.

[22] Y. Hua and T. K. Sarkar, "Generalized pencil-of-function method for extracting poles of an EM system from its transient response," IEEE Trans. on Antennas and Propagation, Vol. 37, No. 2, pp. 229-234, Feb. 1989.

[23] L. Rabiner and B. Gold, Theory and Application of Digital Signal Processing, Prentice-Hall, Inc., Englewood Cliffs, NJ, pp. 93-94, 103, 1975.

APPLICATION OF CONJUGATE GRADIENT METHOD AND FFT TO ELECTROMAGNETICS AND SIGNAL PROCESSING PROBLEMS

Saila Ponnapalli
Tapan K. Sarkar
Ercument Arvas

Department of Electrical Engineering
Syracuse University
Syracuse, New York 13244-1240

ABSTRACT

The conjugate gradient method (CGM) has found a wide variety of applications in electromagnetics and in signal processing as an efficient method for solving matrix equations. In addition, CGM when used in conjunction with FFT (CGFFT) is extremely efficient for solving Hankel and Teoplitz or block Toeplitz matrix systems which frequently arise in both electromagnetics and signal processing applications. The FFT may be utilized because of the convolutional nature of the matrix. CGM has also been used in adaptive spectral estimation. The objective of this paper is the describe CGM and CGFFT, outline some applications and compare their performance with other existing techniques.

I. INTRODUCTION

The conjugate gradient method was simultaneously developed by M. Hestenes and E. Steifel and was first presented in a joint paper [1]. The method solves a matrix equation in an iterative fashion and is guaranteed to converge in a finite number of steps. Theoretically, for an NxN matrix, CG converges to the solution in M steps where M is the number of independent eigenvalues of the matrix. It was initially presented in electromagnetics as an efficient means of solving matrix equations [2].

In electromagnetics, the integral equation for scattering problems is of the form,

$$AJ = Y \tag{1}$$

J. S. Byrnes and J. F. Byrnes (eds.), Recent Advances in Fourier Analysis and Its Applications, 201–220.

where A is an integro-differential operator. The original problem, which has an infinite number of unknowns, is projected onto a finite dimensional space, yielding a matrix equation. In many applications, the matrix is Toeplitz or block Toeplitz, since the operation of A on J involves a convolution. In such cases, the FFT may be incorporated into CGM to perform the convolution as a multiplication in the frequency domain. CGFFT has been used to solve scattering problems for thin wires, and conducting strips or plates using a surface formulation [3-5], or arbitrarily shaped dielectric bodies using a volume formulation [6-7]. The use of CGFFT reduces the order of complexity of the problem from N^2 to $O(N)$.

In many signal processing applications, it is required to make a rational function approximation of data, or to make an all-pole model of data. This gives rise to a matrix equation of the form,

$$AW = 0 \tag{2}$$

where A is a Hankel or Toeplitz matrix which may represent the covariance or autocorrelation matrix, and W are filter parameters. If one of the w's is set to 1, an equation of the form Aw=y results, and CGFFT may be applied to efficiently solve this system [8]. CGM has also been applied to adaptive spectral estimation, where the spectral component of the signal may be adaptively found by solving a generalized eigenvalue problem using CGM [9]. Many authors have applied CGM to various signal processing problems [10-13].

In section II, two widely used CG algorithms are detailed. Section III describes the application of CGFFT to Toeplitz and Hankel systems which arise in electromagnetics problems involving thin wires and plates. Finally, section IV outlines the use of CG in adaptive spectral estimation.

II. CONJUGATE GRADIENT METHOD

The conjugate gradient method is a technique which, rather than solve a matrix equation $Ax=y$ directly, minimizes a functional of interest. This minimization is equivalent to solving $A^*Ax = A^*y$ for an arbitrary operator, where * denotes conjugate transpose, or solving $Ax = y$, if the operator is Hermitian positive/negative definite. (If the operator is negative definite the functional is maximized.) Numerous variations of CG exist which treat arbitrary and positive definite matrices. The algorithms may differ in several respects - the functionals which are minimized may vary, or preconditioning matrices or scale factors may be involved. The reader may refer to [14] for a survey of the various algorithms. Another method is the biconjugate gradient method, which is similar to CGM; however, in this method a power norm, rather than a functional is minimized at each iteration, and one solves $Ax = y$ rather than the normal form of this equation[15]. The biconjugate gradient method is not as popular as CGM since it is not known apriori when the method will break down. Two widely used CG algorithms are now described.

Consider an arbitrary operator. The functional which is minimized is

$$F(x)=<R\ ;R> = |R|^2 \tag{3}$$

where R is the residual given by,

$$R = y - Ax \tag{4}$$

Then,

$$F(x) = \bar{y}^t y - \bar{x}^t A^* y - \bar{y}^t Ax + \bar{x}^t A^* Ax \tag{5}$$

Here $^-$ denotes conjugate and * denotes conjugate transpose. This is a quadratic form and has at most one stationary point which occurs at

$$\nabla F(x) = -2A^* y + 2A^* Ax = 0 \tag{6}$$

So that minimizing the functional is equivalent to solving the normal equation,

$$A^* Ax = A^* y \tag{7}$$

The second derivative is positive since $A^* A$ is positive definite and so the stationary point is a minimum.

In any iterative technique an initial guess x_i is made, a search direction P_i is chosen, and the vector x_i is updated so as to minimize the functional along the search direction. For the method of steepest descent, the search direction is chosen along the negative gradient of the functional. However, if the condition number of the matrix is large, the convergence properties of this method are poor [16]. In CG this pitfall is avoided by choosing the search direction at the ith iteration which is closest to the negative gradient of the functional and "A-conjugate" to all the previous search directions.

Let the initial guess be x_i and the search direction be P_i . Then

$$x_{i+1} = x_i + t_i P_i \tag{8}$$

Using (4) and (8),

$$R_{i+1} = R_i - t_i A P_i \tag{9}$$

where t_i is chosen such that $F(x_{i+1})$ is minimized.

Let,

$$t_i = a_i + j\ b_i \tag{10}$$

Then,

$$\frac{\partial F(x_{i+1})}{\partial a_i} = 0 \quad \frac{\partial F(x_{i+1})}{\partial b_i} = 0 \tag{11}$$

where,

$$F(x_{i+1}) = <R_{i+1};R_{i+1}>$$

$$= <R_i;R_i> - \bar{t}_i <R_i;AP_i> - t_i <AP_i;R_i> + |t_i|^2 <AP_i;AP_i> \tag{12}$$

Applying (11) to (12) gives,

$$t_i = \frac{<R_i\ ;\ AP_i>}{<AP_i\ ;\ AP_i>} \tag{13}$$

From (9) and (13) it is clear that,

$$<R_{i+1}\ ;\ AP_i> = 0 \tag{14}$$

The search direction is updated in the direction closest to the negative gradient of the functional. From (6),

$$-\nabla F(x_{i+1})\ \alpha\ A^* y - A^* A x_{i+1} = A^* R_{i+1} \tag{15}$$

Therefore we choose,

$$P_{i+1} = A^* R_{i+1} + q_i P_i \tag{16}$$

It is required that the search directions be A-conjugate to all the previous search directions, and q_i is chosen to ensure this condition. This guarantees that the method converges in a finite number of steps.

$$<AP_i \, ; AP_j> \; = 0 \text{ for } \textit{all } i \neq j \tag{17}$$

Applying (17) to (16) gives,

$$q_i = -\frac{<AA^* R_{i+1} \, ; AP_i>}{<AP_i \, ; AP_i>} \tag{18}$$

From (9), (14) and (16) it is clear that,

$$<R_{i+1} ; AA^* R_i> \; = 0 \tag{19}$$

Using (14) and (16), t_i may be simplified to,

$$t_i = \frac{|A^* R_i|^2}{|AP_i|^2} \tag{20}$$

Using (16), (19) and (20) q_i may be simplified to

$$q_i = \frac{|A^* R_{i+1}|^2}{|A^* R_i|^2} \tag{21}$$

The algorithm then proceeds as follows:

Guess x_0

$$R_0 = y - Ax_o \tag{22}$$

$$P_0 = A^* R_0 \tag{23}$$

$$x_{i+1} = x_i + t_i P_i \tag{24}$$

$$R_{i+1} = R_i - t_i AP_i \tag{25}$$

$$t_i = \frac{|A^* R_i|^2}{|AP_i|^2} \tag{26}$$

$$P_{i+1} = A^* R_{i+1} + q_i P_i \tag{27}$$

$$q_i = \frac{|A^* R_{i+1}|^2}{|A^* R_i|^2} \tag{28}$$

The iterations continue until $F(x)$ reaches the desired minimum.

Next, consider a positive definite operator. The functional which is minimized is the error between the approximate and true solution.

$$F(x) = < A(x - A^{-1}y) \; ; \; x - A^{-1}y >$$

$$= < Ax \; ; \; x > -2 < y \; ; \; x > \tag{29}$$

Then,

$$\nabla F = 2(Ax - y) \tag{30}$$

so that minimizing the functional is equivalent to solving $Ax = y$. The second derivative is positive since A is positive definite. The search direction is chosen closest to the negative gradient of F (R_i for the ith iteration), and A-conjugate to all the previous search directions. The algorithm is then,

Guess x_0

$$R_0 = y - Ax_0 \tag{31}$$

$$P_0 = R_0 \tag{32}$$

$$x_{i+1} = x_i + t_i P_i \tag{33}$$

$$R_{i+1} = R_i - t_i AP_i \tag{34}$$

$$t_i = \frac{|R_i|^2}{< AP_i \, ; P_i >} \tag{35}$$

$$P_{i+1} = R_{i+1} + q_i P_i \tag{36}$$

$$q_i = \frac{|R_{i+1}|^2}{|R_i|^2} \tag{37}$$

The iterations continue until F(x) reaches the desired minimum. In this case $A^* R$ is not formed.

Several points should be noted about the two algorithms. If the operator is not positive or negative definite, the equivalent problem which is solved is $A^* Ax = A^* y$, so that the condition number of the original problem is squared. This reduces the efficiency of CGM, since the number of iterations required to reach a given minimum increases as the condition number of the matrix increases. Also, condition (17) guarantees convergence of the method in at most M steps where M is the number of independent eigenvectors of A, barring round-off error in the computation of AP or $A^* R$, or recursive computation of R_i and P_i . Numerically, it has been found that greater error arises from the former than from recursive computation of R_i or P_i . In practical problems, it is usually found that for large matrices the number of iterations required for convergence is far fewer than M, because the eigenvalues of the matrix tend to cluster together.

III. CGFFT IN ELECTROMAGNETICS PROBLEMS

The application of CGFFT to thin wires and plates is analyzed. Consider a z-directed straight wire of length L and radius a irradiated by an incident field $\vec{E}^i$. Pocklington's equation gives,

$$(k^2 + \frac{d^2}{dz^2}) \int_0^L J(z') G(z,z') dz' = -j\omega 4\pi\epsilon_0 E^i{}_{\tan}(z) \tag{38}$$

where $\vec{E}^i{}_{\tan}$ is the tangential component of the incident field on the wire, J is the unknown current and $G(z,z')$ is the Green's function given by,

$$G(z,z') = \frac{1}{2\pi} \int_0^{2\pi} \frac{e^{-jkR}}{R} d\phi \tag{39}$$

where,

$$R=[(z-z')^2+4a^2\sin^2\phi/2]^{1/2} \tag{40}$$

$$k=2\pi/\lambda \tag{41}$$

where λ is the wavelength. This may be written compactly as,

$$A\ J=Y \tag{42}$$

with A is the integro-differential operator in (38), J is the unknown current and $y=-j\omega 4\pi\epsilon_0 \vec{E}^i{}_{\tan}$. The method of moments (MOM) procedure for solving this problem is to divide the wire into N segments, expand the current on each segment in terms of a basis function, take the inner product of both sides of (38) with respect to a chosen weight function, and solve for the unknown currents. Consider the basis functions to be pulses, the weight functions to be delta functions, and the Green's function to be constant over each segment. Then (38) may be written as,

$$\sum_{i=1}^{N} J(z_i)\Delta z(k^2+\frac{d^2}{dz^2})G(z_m,z_i)=-j\omega 4\pi\epsilon_0 E^i{}_{\tan}(z_m) \tag{43}$$

Denote,

$$\Gamma(m-i)=\Delta z(k^2+\frac{d^2}{dz^2})G(z_m,z_i) \tag{44}$$

$$y(m)=-j\omega 4\pi\epsilon_0 E^i{}_{\tan}(z_m) \tag{45}$$

$$J(m)=J(z_m) \tag{46}$$

Then the matrix equation to be solved is,

$$\begin{bmatrix} \Gamma(0) & \Gamma(1) & \Gamma(2) & \cdots & \Gamma(N) \\ \Gamma(1) & \Gamma(0) & \Gamma(1) & \cdots & \Gamma(N-1) \\ \cdot & \cdot & \cdot & \cdot & \cdot \\ \cdot & \cdot & \cdot & \cdot & \cdot \\ \cdot & \cdot & \cdot & \cdot & \cdot \\ \Gamma(N) & \Gamma(N-1) & \Gamma(N-2) & \cdots & \Gamma(0) \end{bmatrix} \begin{bmatrix} J(0) \\ J(1) \\ \cdot \\ \cdot \\ \cdot \\ J(N) \end{bmatrix} = \begin{bmatrix} y(0) \\ y(1) \\ \cdot \\ \cdot \\ \cdot \\ y(N) \end{bmatrix} \tag{47}$$

The matrix Γ in (47) is Toeplitz, and may be embedded in a circulant matrix. The convolution of Γ with J may be evaluated in the frequency domain as a multiplication. Furthermore, the differential operation may also be transformed to a simple multiplication, thereby reducing the error incurred in making finite difference approximations as is the conventional procedure. Therefore, for an initial guess J_0 in CGM, the Fourier transform of the operator acting on J_0 is,

$$\tilde{AJ_0} = (k^2 - k_z^2)\tilde{J}_0(k_z)\tilde{G}(k_z) \tag{48}$$

where,

$$\tilde{J}_0(k_z) = \int_{-\infty}^{\infty} J_0(z)e^{-jk_z z}\,dz \tag{49}$$

$$\tilde{G}(k_z) = \frac{1}{2\pi}\int_{-\infty}^{\infty} e^{-jk_z z}\,dz \frac{1}{2\pi}\int_0^{2\pi} \frac{e^{-jkR}}{R}\,d\phi \tag{50}$$

with R given by (40). The Fourier transform of the Green's function may be evaluated in closed form.

$$\tilde{\Gamma}(k_z) = \Delta z(k^2 - k_z^2)I_0[\, a\sqrt{(k^2 - k_z^2)}]\, K_0[\, a\sqrt{(k^2 - k_z^2)}] \tag{51}$$

where I_0 and K_0 are the zeroth order modified Bessel functions of the first and second kind respectively. It should be noted that multiplication by $k^2 - k_z^2$ cancels the log singularity in the transform of the Green's function at $k_z = k$. When k_z is sampled and the multiplication in (48) is performed the equivalent matrix multiplication of (47) is carried out in the frequency domain. The CGFFT method then involves the algorithm as described in section II, with the modification that the terms AP and A^*R are evaluated as follows. For a wire of N segments:

1. Consider P or R to be of dimension N. Pad the array to M where M is a composite number. This step is taken to avoid aliasing.

2. Take an FFT of the array. Multiply by the sampled version of $\Gamma(k_z)$ to compute AP or its complex conjugate to compute A^*R .

3. Take an inverse FFT of the result and extract the first N elements.

The iterations in CGM continue until,

$$\frac{|AJ_n - y|}{|y|} < F_{\min} \tag{52}$$

where $F_{\min}$ is the desired minimum.

As a numerical example, consider a 2.5 λ antenna of radius 0.001 λ irradiated by an incident field of 1 V/m, with $F_{\min} = 10^{-5}$, and starting with an initial guess of 0 in CGFFT. Three different methods all produce comparable results; however, the order of complexity for each method is significantly different. For a wire discretized into N segments,

1. If the matrix is computed using MOM procedure and inverted, N^3 operations are required to invert the matrix and N^2 operations are required to find the solution.

$$\# \; of \; computations = N^3 + N^2 = O(N^3) \tag{53}$$

If Trench's or Levinson's algorithm for Toeplitz matrices is used the order of complexity is reduced to $O(N^2)$.

2. If the matrix is computed using MOM procedure and solved using CGM $2N^2$ operations are required per iteration to compute AP and $A^* R$.

$$\# \; of \; computations = p2N^2 = O(N^2) \tag{54}$$

where p is the number of iterations.

3. If operator equation is solved using CGFFT, 4 FFT's and 2(2N) multiplications are required per iteration to compute AP and $A^* R$.

$$\# \; of \; computations = p8N[log(2N) + \frac{1}{4}] = O(N) \tag{55}$$

For this thin wire example, the CPU time versus the number of unknowns using CGFFT with the conjugate gradient algorithm described in section II is shown in Figure 1. The CPU time is linearly proportional to the matrix size.

Another problem to which CGFFT has been applied is scattering from a conducting plate. Consider a flat plate of dimensions $L_1 \times L_2$, located in the x-y plane irradiated by an incident field of intensity $\vec{E}^i$. Applying the boundary condition that $E_{\tan} = 0$ on the plate gives,

$$k^2 \left[\hat{x} \int_0^{L_1} dx' \int_0^{L_2} dy' \, J_x(x',y') G(x,y) + \hat{y} \int_0^{L_1} dx' \int_0^{L_2} dy' \, J_y(x',y') G(x,y) \right.$$

$$+ \left(\hat{x}\frac{\partial}{\partial x} + \hat{y}\frac{\partial}{\partial y}\right) \int_0^{L_1} dx' \int_0^{L_2} dy' \left(\frac{\partial J_x}{\partial x'}(x',y') + \frac{\partial J_y}{\partial y'}(x',y') \right) G(x,y)$$

$$= -j\omega 4\pi\epsilon_0 \left[\hat{x}E_x{}^i(x,y) + \hat{y}E_y{}^i(x,y) \right] \tag{56}$$

where the Green's function is,

$$G(x,y) = \frac{e^{-jkR}}{R} \tag{57}$$

with,

$$R = ((x - x')^2 + (y - y')^2))^{1/2} \tag{58}$$

$E_x{}^i$ and $E_y{}^i$ denote the x and y components of the incident field. If the problem is solved using a MOM procedure, a block Toeplitz matrix results. However, the solution method can be considerable improved by using CGFFT. The transform of the Green's function is,

$$\tilde{G}(k_x,k_y) = \int_{-\infty}^{\infty} e^{-jk_x x} \, dx \int_{-\infty}^{\infty} e^{-jk_y y} \, dy G(x,y)$$

$$= \frac{2\pi}{\sqrt{k_x{}^2 + k_y{}^2 - k^2}} \tag{59}$$

Then the transform of the operator acting on J_0 is given by,

$$\tilde{AJ_0} = k^2[\hat{x}\tilde{J}_{x_0}\tilde{G} + \hat{y}J_{y_0}\tilde{G}] + [\hat{x}jk_x + \hat{y}jk_y][jk_x\tilde{J}_{x_0}\tilde{G} + jk_y\tilde{J}_{y_0}\tilde{G}] \tag{60}$$

All the Fourier transforms are two dimensional and AJ_0 is the two dimensional inverse Fourier transform of $\tilde{AJ_0}$.

$$AJ_0 = \int_{-\infty}^{\infty} e^{jk_x x} \, dx \int_{-\infty}^{\infty} e^{jk_y y} \, dy \tilde{AJ_0} \tag{61}$$

AP and A^*R are computed in a similar manner as for the thin wire problem.

As a numerical example, consider a 1.0 λ square plate irradiated by a normally incident field of intensity 377 V/m, with $F_{\min} = 10^{-4}$. There are nine unknowns per wavelength, and the total number of unknowns is 162. The FFT is 64x64. Figure 2 shows good agreement between the results using CGFFT and Rao's results using MOM with triangular expansion functions [17]. A 0.5 λ square plate with 30 unknowns per wavelength, (a total of 450 unknowns), and $F_{\min} = 10^{-5}$ is analyzed. The CPU time versus the number of unknowns using the biconjugate gradient method and FFT is shown in Figure 3 for this problem. The figure illustrates the O(N) complexity; the ordinary conjugate gradient method is also of O(N), however, the slope of the curve is much steeper. This is due to an increase in the number of iterations because of the squaring of the condition number of the matrix which occurs in CGM.

Hankel matrix systems which arise in signal processing problems may be solved in a similar manner using CGFFT. However, in such applications, the analytic Fourier transform of the data is not known, and the transform of the matrix is found by using the FFT.

IV. ADAPTIVE SPECTRAL ESTIMATION USING CONJUGATE GRADIENT METHOD

In signal processing problems, data which consists of sinusoids in noise is frequently encountered. It is desired to adaptively find the spectral component of the data. Consider a data sequence,

$$x(0),x(1),\ldots,x(k),..$$

$$=s(0)+n(0),s(1)+n(1),\ldots,s(k)+n(k).. \tag{62}$$

where, s(k), n(k) are samples of the signal and zero-mean noise at time k, respectively. The covariance matrix of the observed data is,

$$R_{xx} = \begin{bmatrix} r_{xx}(0) & r_{xx}(-1) & \cdots & r_{xx}(-N) \\ r_{xx}(1) & r_{xx}(0) & \cdots & r_{xx}(-N-1) \\ \cdot & \cdot & \cdot & \cdot \\ \cdot & \cdot & \cdot & \cdot \\ r_{xx}(N) & r_{xx}(N-1) & \cdots & r_{xx}(0) \end{bmatrix} \tag{63}$$

where the autocorrelation function is,

$$r_{xx}(k) = E[x(i+k)x(i)^*]$$

$$= E[s(i+k)s^*(i)] + E[n(i+k)n^*(i)]$$

$$= s_{xx}(k) + n_{xx}(k) \tag{64}$$

S_{xx} is the covariance matrix of the signal and N_{xx} is that of the noise in the data. These are,

$$S_{xx} = \begin{bmatrix} s_{xx}(0) & s_{xx}(-1) & \cdots & s_{xx}(-N) \\ s_{xx}(1) & s_{xx}(0) & \cdots & s_{xx}(-N-1) \\ \cdot & \cdot & \cdot & \cdot \\ \cdot & \cdot & \cdot & \cdot \\ s_{xx}(N) & s_{xx}(N-1) & \cdots & s_{xx}(0) \end{bmatrix} \tag{65}$$

$$N_{xx} = \begin{bmatrix} n_{xx}(0) & n_{xx}(-1) & \cdots & n_{xx}(-N) \\ n_{xx}(1) & n_{xx}(0) & \cdots & n_{xx}(-N-1) \\ \cdot & \cdot & \cdot & \cdot \\ \cdot & \cdot & \cdot & \cdot \\ n_{xx}(N) & n_{xx}(N-1) & \cdots & n_{xx}(0) \end{bmatrix} \tag{66}$$

The signal space and noise space are orthogonal; the noise parameters may be related to the signal parameters by,

$$S_{xx}\ W = 0 \tag{67}$$

or,

$$(R_{xx} - N_{xx})\ W = 0 \tag{68}$$

If the covariance matrix of the noise is not known, an estimate Z_{xx} is used, which results in a generalized eigensystem,

$$R_{xx}\ W = \lambda Z_{xx}\ W \tag{69}$$

where Z_{zz} is the eigenvector corresponding to the minimum generalized eigenvalue. If the noise is white, $N_{zz} = \sigma^2 I$ where σ^2 is the variance of the noise and I is the identity matrix.

Then,

$$R_{zz}\ W = \sigma^2\ W \tag{70}$$

In this case σ^2 is the minimum generalized eigenvalue of R_{zz} when R_{zz} is $2N+1 \times 2N+1$. In either case CGM can be used to efficiently compute the eigenvector corresponding to the minimum eigenvalue of a generalized eigensystem. The algorithm, called MEVMCG encorporates some modifications into the previously outlined CGM and is found in [9]. Many other methods exist for this problem; amongst them are LMS (LMSALG) [18], the unit-norm constrained algorithm (UNCALG) [19], the γ - LMS algorithm (GAMLMS) and the least squares type algorithm (LSTALG) [20]. Once the noise parameters are ascertained, the power spectral density of the signal can be found from,

$$P(f) = \frac{1}{\left|\sum_{0}^{N} w_i\ Z^{N-i}\ \right|_{Z=\exp(j2\pi f)}} \tag{71}$$

As an example, consider two equiamplitude sinusoids of normalized frequencies 0.18 and 0.38 at 0dB SNR. The results of five different algorithms are presented in Figure 4. The plot shows that LMSALG cannot resolve the two sinusoids with 5 weight even after 5000 iterations. UNCALG and GAMLMS obtain almost the same results and use almost the same CPU time. LSTALG and MEVMCG obtain the best results for this example; however, MEVMCG uses approximately $\frac{1}{45}$ the CPU time as the former. In all the cases, the length of the data equals the number of iterations, except MEVMCG. In this case, the length of the data is 89 and the number of iterations is 142.

V. CONCLUSION

In this paper, the conjugate gradient method and CGFFT are described. The application of CGFFT to several electromagnetics problems is discussed. The computational complexity of CGFFT is compared with other methods, and CGFFT is found to reduce the order of complexity of the problem from $O(N^2)$ to $O(N)$. The use of CGM in adaptive signal processing is outlined and an example is presented which illustrates that MEVMCG performs considerably better than existing techniques.

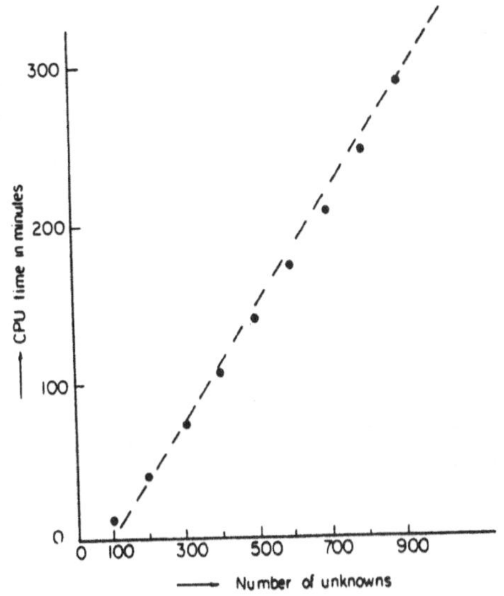

Figure 1. CPU time vs. the number of unknowns using CGM for thin wire.

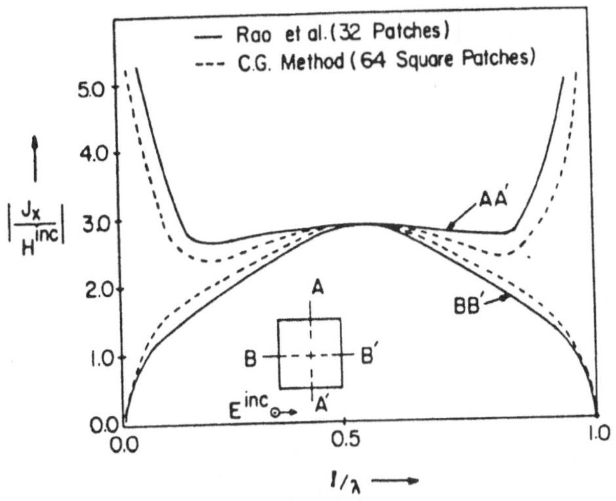

Figure 2. Comparison of surface electric current distribution in a 1.0λ square plate illuminated by a normally incident plane wire.

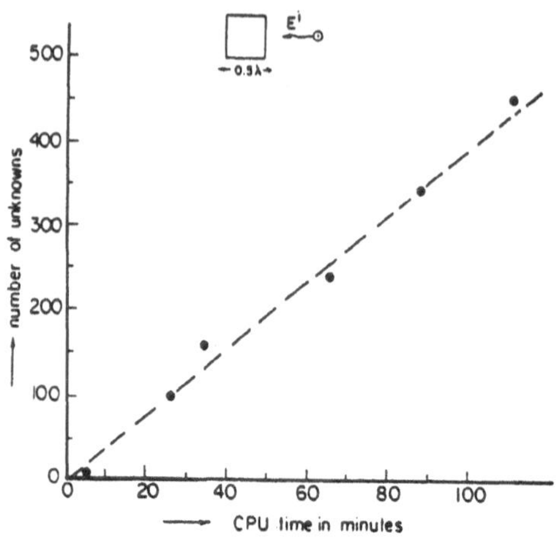

Figure 3. CPU time vs number of unknowns using Biconjugate Gradient Method for a conducting plate.

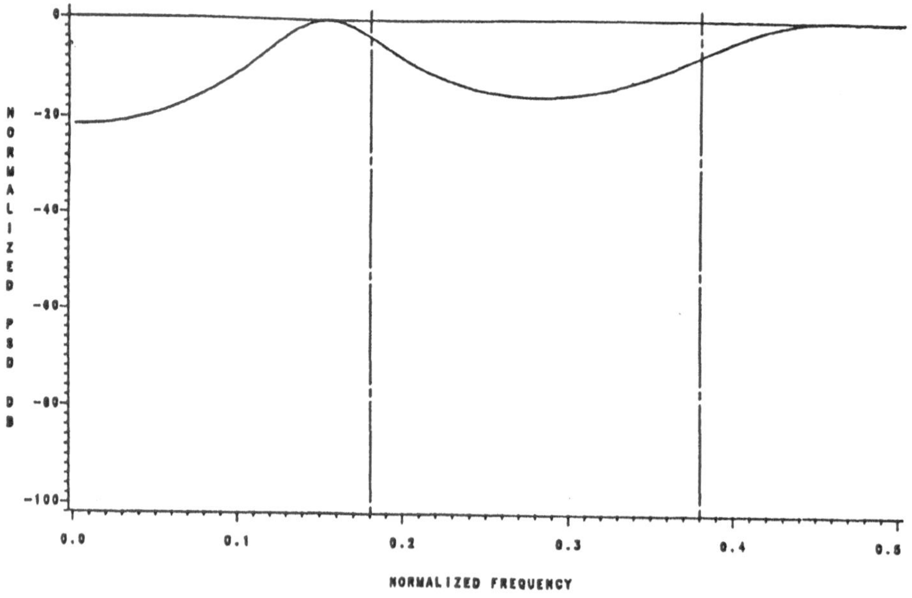

Figure 4(a). LMSALG spectral estimates with white noise. F1 = 0.18, F2 = 0.38, ks = -0.0075, SNR = 0 db. Number of iterations is 5000, and CPU time is 50.15s.

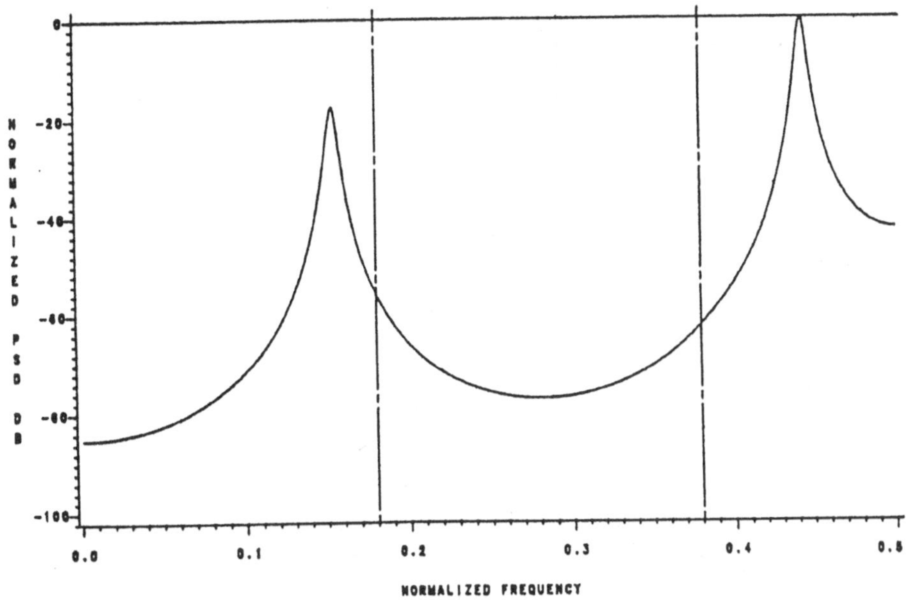

Figure 4 (cont.)(b).UNCALG spectral estimates with white noise. F1 = 0.18, F2 = 0.38, MU = 0.015, SNR = 0dB. Number of iterations is 5000, and CPU time is 50.08 s.

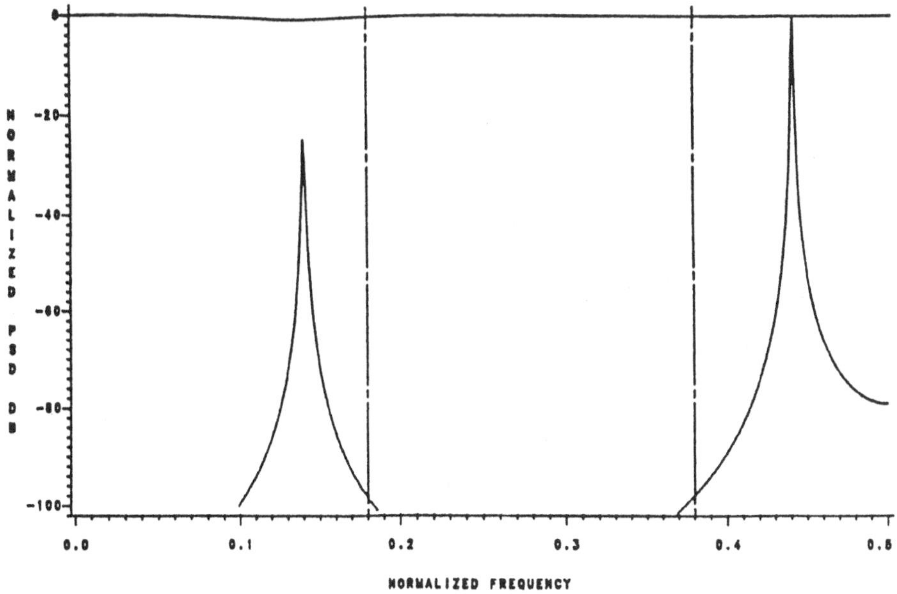

Figure 4(c). GAMLMS spectral estimates with white noise. F1 = 0.18, F2 = 0.38, MU = 0.015, SNR = 0dB. Number of iterations is 5000, and CPU time is 50.73 s.

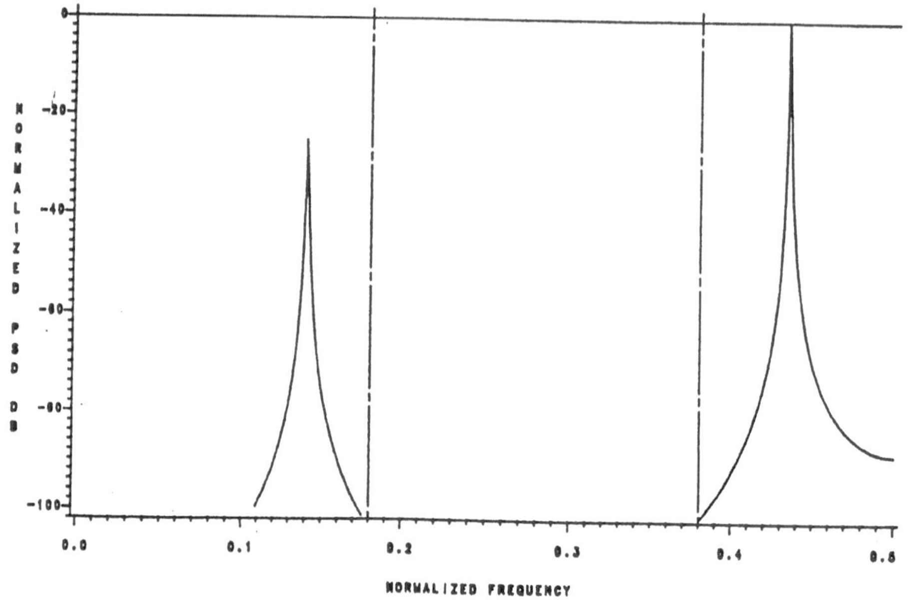

Figure 4(cont.)(d). LSTALG spectral estimates with white noise. F1 = 0.18, F2 = 0.38, PZ = 100, SNR = 0 dB. Number of iterations is 5000, and CPU time is 62.17 s.

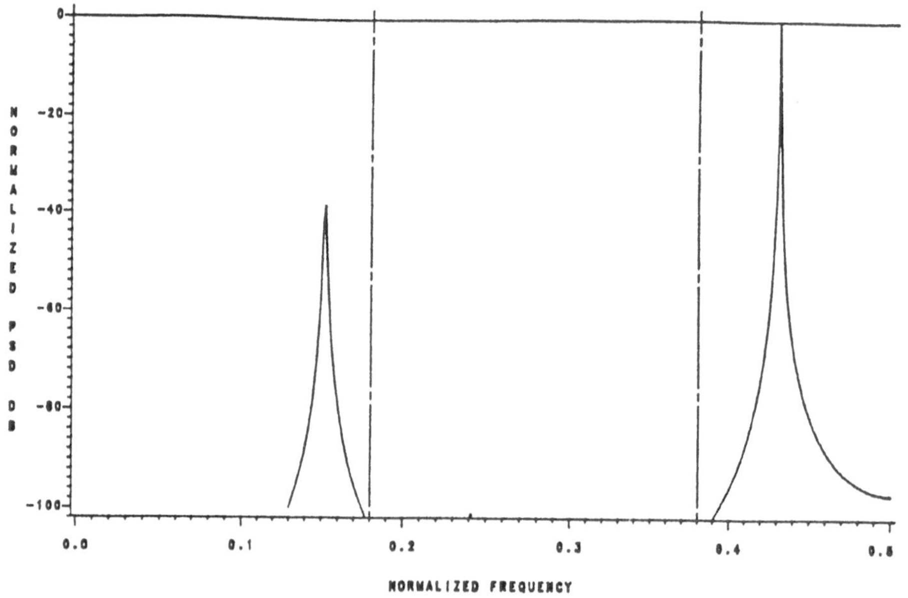

Figure 4(e). MEVMCG spectral estimates with white noise. F1 = 0.18, F2 = 0.38, RC = 0.001, SNR = 0dB. Number of iterations is 124, CPU time is 1.38 s. Number of updating data is 80.

REFERENCES

[1]. Hestenes, M. and Stiefel, E. (1952) 'Method of Conjugate Gradients for Solving Linear Systems', J. Res. Nat. Bur. Stand., Vol. 49, 409-436.

[2]. Sarkar, T. K., Siarkiewicz, and Stratton, R. (1981) 'Survey of Numerical Methods for Solution of Large Systems of Linear Equations for Electromagnetic Field Problems', IEEE Trans. Antennas Propagat. 29, 847-856.

[3]. Sarkar, T.K., Arvas, E., and Rao, S.M. (1986), 'Application of FFT and the Conjugate Gradient Method for the Solution of Electromagnetic Radiation for Electrically Large and Small Conducting Bodies', IEEE Trans. Antennas and Propagat. 34, 635-640.

[4]. Sarkar, T.K., (1987) 'On the Application of the Generalized Biconjugate Gradient Method', J. Electromagn. Waves and Appls. 1, 223-242.

[5]. Zohar, S. (1974) 'The Solution of Toeplitz Sets of Linear Equations', J. ACM 21, 272-276.

[6]. Su, C. C. (1987) 'Calculation of Electromagnetic Scattering from a Dielectric Cylinder Using the Conjugate Gradient Method and FFT', IEEE Trans. Antennas and Propogat., vol. AP-37, 1418-1425.

[7]. Su, C. C. (1989) 'Electromagnetic Scattering by a Dielectric Body with Arbitrary Inhomogeneity and Anistropy', IEEE Trans. Antennas and Propogat., vol. AP-37, 384-389.

[8]. Sarkar, T.K. and Yang, X. (1987) 'Efficient Solution of Hankel Systems Utilizing FFTs and the Conjugate Gradient Method', Proc. of International Conf. on Acoustics, Speech and Signal Processing (ICASSP 86), Dallas, TX, 1835-1838.

[9]. Chen, H., Sarkar, T.K., Dianat, S.A and Brule, J.D. (1986) 'Adaptive Spectral Estimation by the Conjugate Gradient Method', IEEE Trans. on Speech, Acoustics and Signal Processing, vol ASSP-34, 272-283.

[10]. Fienup, J.R. (1982) 'Phase Retrieval Algorithms: A Comparison', Applied Optics, 2758-2769.

[11]. Koehler, F. and Taner, M. (1985) 'The Use of the Conjugate Gradient Algorithm in the Computation of the Predictive Deconvolution Operators', Geophysics, vol. 50, 2752-2758.

[12]. Kaneata, S. and Nalcioglu, O. (1985) 'Constrained Iterative Reconstruction by the Conjugate Gradient Method, IEEE Trans. on Medical Imaging, vol. MI-4, 66-71.

[13]. Friel, J. O. (1975) "Adaptive Predication of Stationary Time Series by Modified Conjugate Direction Methods', Comm. in Statistics, vol. 4, 19-32.

[14]. Sarkar, T.K, Yang, X. and Arvas, E. (1988) 'A Limited Survey of Various Conjugate Gradient Methods for Solving Complex Matrix Equations Arising in Electromagnetic Wave Interactions', Wave Motion 10, 527-546.

[15]. Sarkar, T. K. (1987) 'On the Application of the Generalized Biconjugate Gradient Method', Journal of Electromagnetic Waves and Applications, vol. 1, No. 3, 223-242.

[16]. Golub, G. and Van Loan, C. F. (1983) Matrix Computations, Johns Hopkins University Press.

[17]. Rao, S. M. et al. (1982) 'Electromagnetic Scattering by Surfaces of Arbitrary Shape', IEEE Trans. on Antennas Propagat., vol. Ap-30, 409-418.

[18]. Widrow, B. (1971) Adaptive Filters, Holt, Rhinehart, Winston, 563-587.

[19]. Thompson, P. A. (1979) 'An Adaptive Spectral Analysis Technique for Unbiased Frequency Estimation in the Presence of White Noise', Proc. of 13th Asilomar Conf. on Circuits, Syst., Comput., 529-533.

[20]. Reddy, V.V., Edgardt, B., and Kauath T. (1982) "Least Squares Type Algorithm for Adaptive Implementation of Pisarenko's Harmonic Retrieval Method', IEEE Trans. Acoustics, Speech, Signal Processing, vol. ASSP-30, 399-405.

ON THE APPLICATION OF FOURIER TRANSFORMATION IN ANTENNA MEASUREMENTS.

K. VAN 'T KLOOSTER,
ESA/ESTEC,
P.O.Box 299,
2200 AG, Noordwijk,
The Netherlands.

Abstract. Determination of antenna radiation characteristics can be done today with a number of methods. Far-field (FF), Compact Antenna Test Ranges (CATR), intermediate and near-field (NF) methods are applied. Processing of measured data is usually required to calculate desired parameters from the measured data. The Fourier transformation takes an important place and efficient algorithms are available to perform data reduction tasks (NF), in which use of the FFT is exploited. An outline description is given of a number of antenna test methods and some examples are described of application of measurement techniques involving datareduction in planar NF scanning, FF, intermediate and CATR testing.

1. Introduction.

The usual parameters of interest for an antenna are its far field parameters. The straightforward method to determine the parameters is testing under far-field conditions. Such conditions are realised if sufficient range distance is kept between the antenna under test and the measurement location. That condition leads in general to outdoor testranges, often several hundred meters in length, as is known for more than forty years [1].

Interest in indoor testing has stimulated work on different testmethods and today two interesting methods are commonly available: Compact Antenna Test Range and Near-Field antenna testing. The NF techniques use with advantage FFT-algorithm in the data-reduction. The processing techniques are studied further to improve on results obtained from CATR, FF and intermediate (=Fresnel type) techniques. Further work is done on new methods for indoor testing. The modulated scattering technique [2] is an example with promising results.

This paper is limited to some examples encountered in the more standard NF, FF, intermediate and CATR techniques.

J. S. Byrnes and J. F. Byrnes (eds.), Recent Advances in Fourier Analysis and Its Applications, 221–236.

2. Testing geometry, principles and some history.

2.1 CATR

The compact range exploits a focussing device, which can be one or more reflectors. A spherical wave is converted into a plane wave. The plane wave is used, as on a FF-range to illuminate the antenna under test. The use of collimation for testing was reported several years ago [3]. The use of a reflector was explored in the late sixties [10] resulting in a single off-set reflector configuration. Lateron a pair of two parabolic cylindrical reflectors was introduced [11]. Further work led to an introduction of a CATR with double curved reflectors, among which [12]. All CATR configurations have a very short range length, in the order of a few meters between antenna under test and CATR reflector. Different configurations are available or under study today [13].
Fig.1 shows the CATR at ESTEC with two cylindrical reflectors [11]. The feed of the CATR is hidden in the corner on the right. It illuminates the first reflector (left), which has its focal line perpendicular to the focal line of the second ('main') reflector on the right (fig.1). The edge illumination of the reflectors is high. Continuous edges would cause strong diffraction effects disturbing the quality of the testzone. To lower such effects, the edges are serrated, causing less stray radiation into the testzone [50].

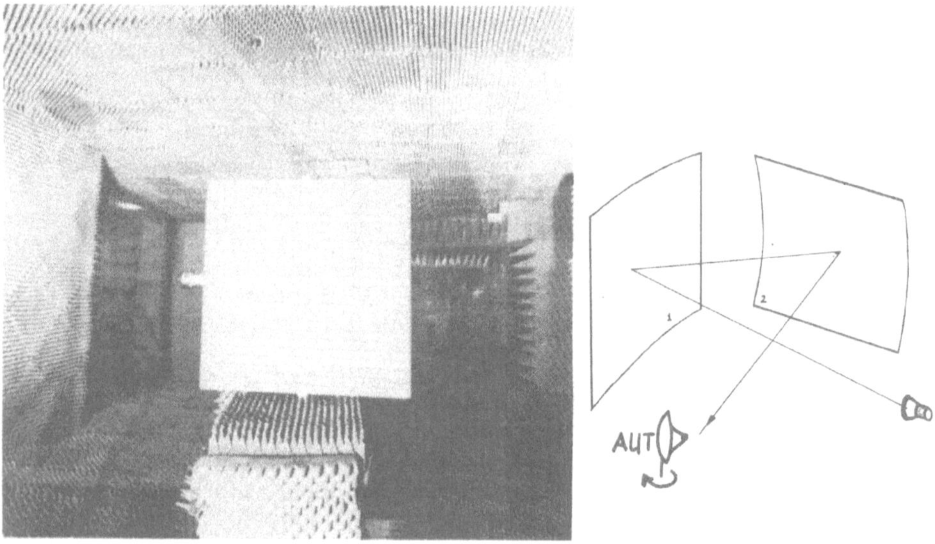

Fig.1. Compact Antenna Test Range at ESTEC. The antenna under test is a slotted waveguide antenna of 1*1 meter. (See also fig.5).

Fig.2. Analog computer for calculation of discrete Fourier series with 41 complex coefficients [4].

2.2 NF testing.

In NF antenna testing, one calculates the FF radiation performances from the NF data. In the late fifties an analog computer was used to perform the necessary calculations. An example of an analog computer as used in those days is shown in fig.2. The calculation of Fourier series with 41 complex coefficients was possible on this machine. The modification of one single coefficient caused a direct observable impact (oscilloscope) in the final result. The analog computer was used for calculation of patterns from NF data and for other type of antenna related calculations like pattern synthesis and root-finding of complex polynomials [4,6].
The analysis of series with N>41 was done by using the computer several times, as reported in [4]. The latter technique is comparable to a segmented transformation, which is sometimes seen as new today with FFT applications. Analog computers became commercially available for applications in antenna testing and analysis [5].

An overview of NF testing techniques is given in [22]. The planar NF test method (fig.3) has thoroughly been investigated at NBS and Georgia Institute of Technology. The work of Kerns [14] must be mentioned. He applied the scattering matrix description for the antenna and considered a measurement problem as ´two-antenna´ problem, in which one antenna (in NF scanning the probe) is known [8]. The formulation has a general application in ´two-antenna´ problems [29].
Important planar ´pilot´ facilities for planar NF scanning

are at the mentioned institutes [15,16].

Analysis of NF measurements on a cylindrical contour was carried out [7] based on expansions in Hankel-functions. It included the effect of directive properties of the probe. It was noticed, that radiation patterns can be completely specified using measurement data taken at angular intervals equal to half-power beamwidth ('sampled representation'). The work was extended to two dimensions and probe correction was described [17]. An approach based on a scattering matrix description for the antenna and probe was given [9]. Software written for this approach is available from NBS. Fig.4 shows a cylindrical facility at MBB [34].

Analysis for spherical NF measurements was first described in [18], using a transmission formula based on Lorentz reciprocity theorem. A formulation based on a scattering matrix description was given by Wacker [19]. Further elaboration led to a fast computer program (SNIFT) [20], in which probe correction is applied. This software is used today in various test facilities. The spherical NF range of Technical University of Denmark is shown in fig.5, with the same 1*1m antenna on the range as in fig.1. A recent book treats spherical NF scanning in detail [21].

Work on plane-polar scanning was done, see [23,24]. As found [25,33], an interpolation from polar to a rectangular grid and subsequent processing as for rectangular NF testing can be used. Comparison of algorithms, among which a Fast Hankel transform, showed, that the interpolation combiwith rectangular 2D FFT is the fastest method [25].

Fig.3. Phased array under test in planar NF facility [27] of TNO, The Hague.

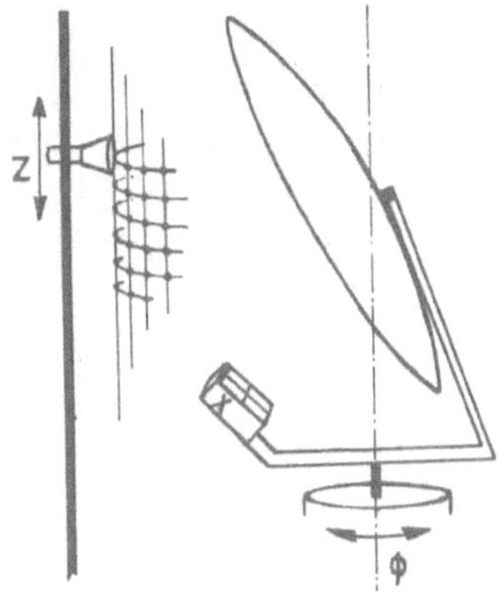

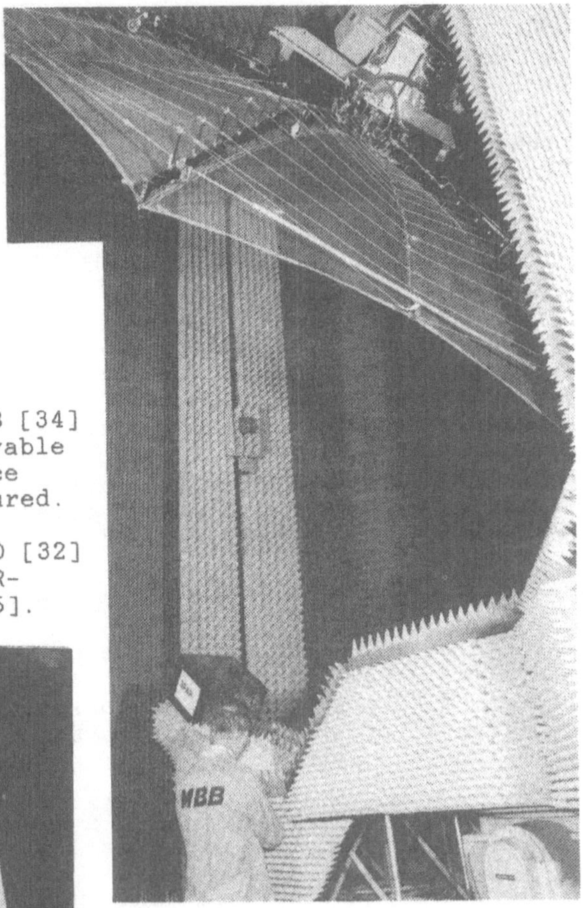

Fig.4. NF facility MBB [34] Cylindrical. A deployable mesh antenna for space applications is measured.

Fig.5. NF facility TUD [32] Spherical [21]. A SAR-panel is measured [35]. (shown below).

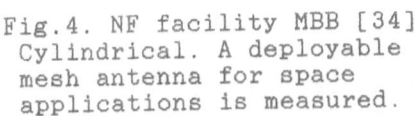

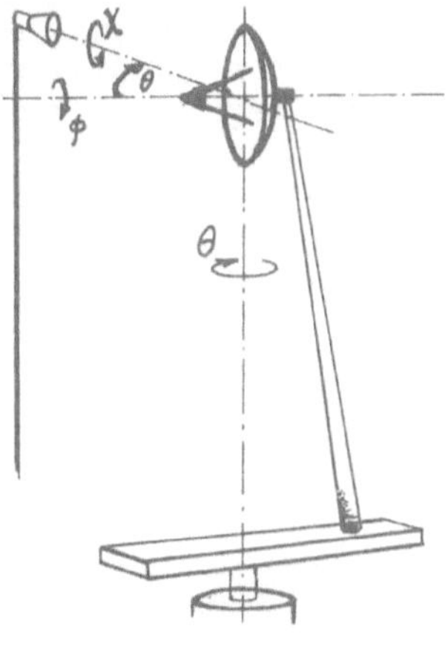

3. Planar scanning

In the scattering matrix description the antenna is seen as multiport: one (or a few) port(s) on the 'guided wave' side and many virtual ports in free space, each related to a single TE or TM-mode in the mode spectrum. The latter spectrum contains an infinite number of modes, of which only a finite set is propagating, the number being related to the antenna dimensions.

In planar scanning (and in all other type of NF scanning), one neglects multiple scattering between probe and antenna.

In planar scanning, a 2D Fourier transformation of measured data is directly proportional to the product of transmit and receive spectrum of antenna under test and probe. The probe behaviour is known from a separate calibration and the spectrum of the antenna under test can be derived. An asymptotic relation between spectrum of the antenna and its far-field provides the FF radiation performances of the antenna.

With $\bar{S}(\bar{k})$ as vectorial antenna spectrum and $\bar{R}(\bar{k})$ as probe vectorial spectrum, we obtain after Fourier transformation of data b(x,y):

$$\vec{S}(\vec{k})\cdot\vec{R}(\vec{k}) = C_1 e^{j\gamma d} \int_{-\infty}^{+\infty}\int_{-\infty}^{+\infty} b(x,y,d)\, e^{j(k_x x + k_y y)}\, dx\, dy. \qquad (1)$$

The distance between scan plane and antenna reference is 'd' and C1 is a complex calibration constant. Gamma is the component of $\bar{k}$ perpendicular to the scanplane. A second dataset provides after transformation a second equation as needed to solve for two spectral components. The far-field follows as:

$$\bar{E}_f(\bar{r}) = jk\cos(\vartheta)\, C_2 \frac{e^{-jkr}}{r} \vec{S}(\vec{k}). \qquad (2)$$

C2 is a complex constant. The output is complete within the region of validity, including polarisation, FF phase, etc. Complex constants are important for abs. gain measurement.

An antenna of 2*1 m (slotted waveguide panel) was used for some experiments. From $\bar{S}(\bar{k})$ an equivalent field was calculated in a tomographic fashion at planes parallel to the scan plane. Examples are shown in fig.6. The raw measured data (amplitude) are shown in fig.6a with contours at 5 dB intervals. Five slots were covered with metal patches. Transformation back to the aperture plane showed the patch locations. (fig.6b). An equivalent field distribution in a plane transverse to the aperture is shown in fig.6c. The line AA' is in the aperture and BB' in the measurement plane. The solid and dotted curves in fig.6d are relative (equivalent) amplitude and phase along the lines AA' and BB' respectively.

The back-transformation was used on an antenna panel of the ERS-1 SAR antenna. Some deviating results were found [35]. An X-ray inspection revealed a small crack in a waveguide.

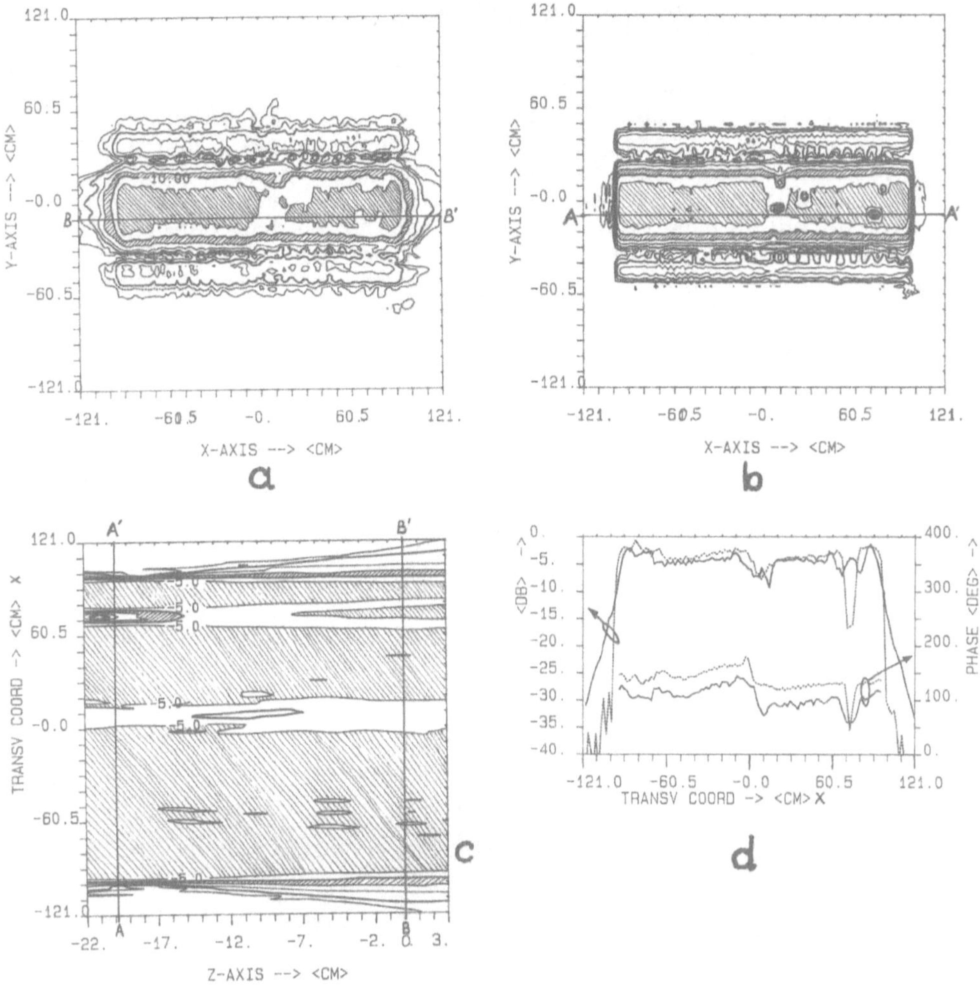

Fig.6a Measured amplitude distribution of panel with five small aluminium patches.

Fig.6b Equivalent aperture distribution with the location of the patches.

Fig.6c Transversal distribution of amplitude.

Fig.6d Relative amp. and phase along AA´ (dotted) and BB .

Planar NF scanning is also studied at TNO, The Hague, NL. A number of antennas was tested [27], among which a phased array (fig.3) [26], a shaped beam (cosec-2 antenna) and a slotted waveguide antenna. A multiplexing of frequencies was carried out during the scanning, thus providing acquisition of datasets for different frequencies during one scan. The depth of a tracking null of the phased array antenna was derived using Fourier interpolation. The retrieval of a faulty element was done with back-transformation.

The slotted waveguide antenna had larger length than could be accomodated by the NF scanner. To have the antenna tested it was put on a sledge parallel to the scanner. Two datasets were taken, one with the antenna shifted to the left and one with the antenna shifted to the right. A larger dataset was constructed on the computer by matching in the central overlapping zone [27]. The latter matching was relatively simple due to the saw-tooth like NF phase behaviour of this travelling wave slotted waveguide antenna.

The polarisation purity a circular waveguide with chokes as a probe was exploited. Ludwigs third definition provides a coordinate set for decomposition of the vector-field. With a good alignment one can measure co- and cross-polarisation characteristics of the antenna separately for all directions within the forward hemisphere in a scalar fashion, provided that the probe has 'zero' cross-polarisation. Fig.8 shows a probe with low Xpol, developed at TNO [28,51].

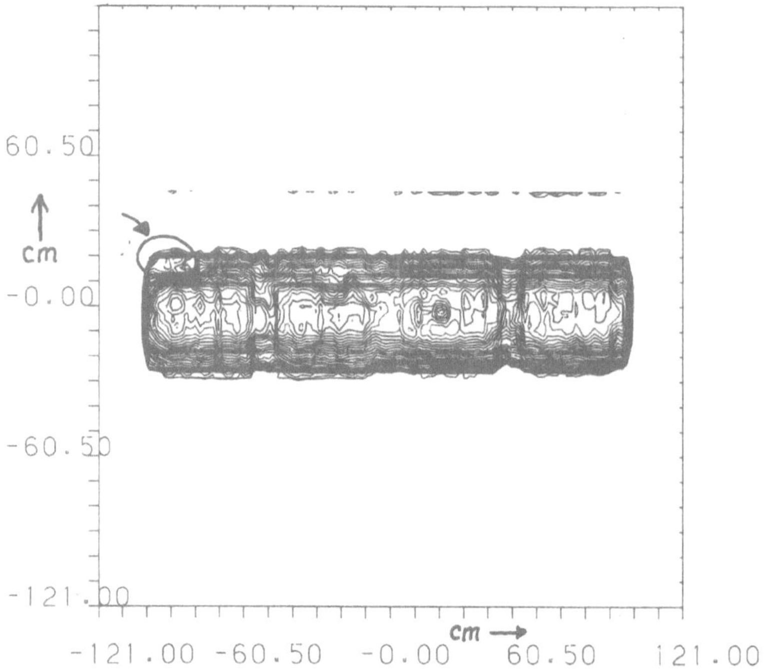

Fig.7. Equivalent aperture distribution (amp) of CFRP slotted waveguide panel. The increments are at .5 dB. A crack was found with X-ray investigation in the indicated area.

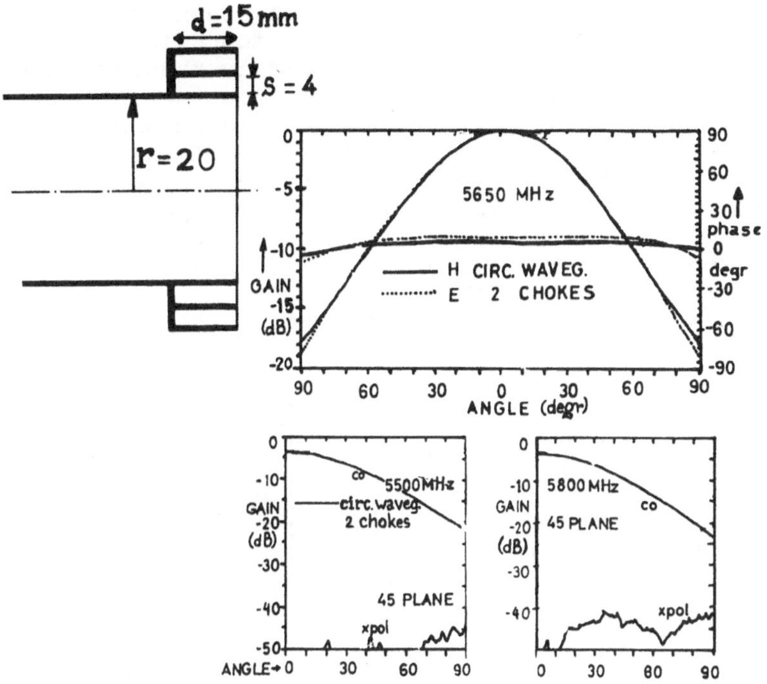

Fig.8. Circular waveguide with chokes as a probe.

The probe is less sensitive to stray radiation from the back than an open-ended waveguide as probe. The latter has a less directive pattern than as shown in fig.8. It also implies that the circular probe attenuates the far-out components in the plane wave spectrum of the antenna under test more, with as a consequence lower S/N, thus an increased error level in the very far-out sidelobe prediction (>say 60 deg.).

Interesting recent results in planar scanning were obtained with the measurement of the ERS-1 SAR-antenna on a large (5m by 12m scan-plane) scanner [30,37]. Phase-tuning of the feed network of the SAR-antenna was also done on this scanner.
A segmented NF-scanning at the electrical panel level was used to obtain an equal phase setting between the 10 electrical sub-panels within the 10 meter long SAR antenna.

The gain-evaluation involves accurate mismatch evaluation and use of correctly normalised complex constants in (1,2).

The verification of measurement performance was done by a number of comparison tests, using a know antenna [35].

For an extensive treatment of spherical scanning, see [21]. Examples of spherical NF facilities are found in [30,32,13]. Just some result is shown here. The highly shaped, linearly polarised slotted waveguide wind-scatterometer antennas for ERS-1 were tested on a spherical range. The antennas operate at 5.3 GHz. Fig.9 shows the pattern of one of them (MID-WSA)

The strong shaping in this antenna pattern is of interest, also within the context of this workshop, where methods with 'constrained coefficients' are discussed. Pattern realisation involves synthesis, but it is only part of the problem. Physical limitations, as present in couplers, lead to constraints in the excitations. Not all solutions from the synthesis are realisable.

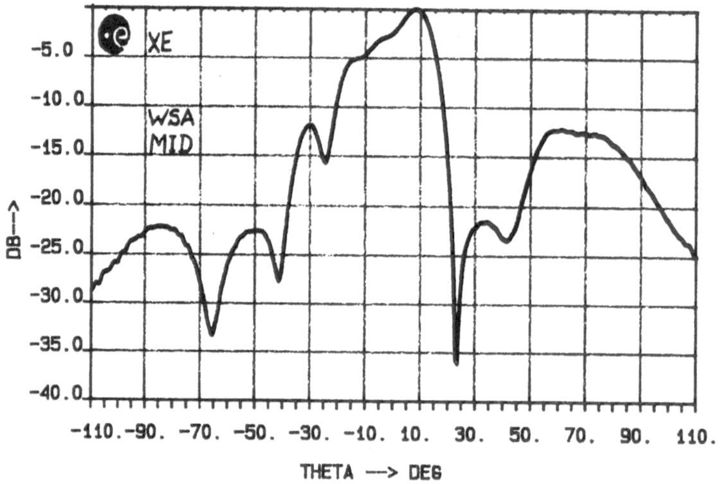

Fig.9 WSA-mid FF elevation pattern from spherical NF data.

4. Example of data processing on a FF or intermediate range.

As example the evaluation of performances of large reflector antennas in the mm-wave region is discussed. Such antennas are used today in radio-astronomy. New methods have been implemented to adjust the reflector panels of such antennas [38,40]. Fourier relationships are exploited again between equivalent current distribution in the aperture-plane and the far-field [41]. Sampling the FF in k and the use of an inverse Fourier transformation yields an equivalent current distribution in the aperture plane. An equispaced sampling allows the use of the FFT algorithm. Ray-tracing interrelates the phase distribution in the aperture to reflector properties. Deviation in the feed pattern can be taken into account (astigmatism). The phase can be related to an accurate position of reflector panels, which can then be tuned for an optimum performance [39,40,43]. For FF-sampling a source at or beyond the Rayleigh distance is needed. This can be well beyond 100 km. Celestial sources [39] or satellite beacons are used [40].

A Fresnel region sampling has also been used. It requires an additional correction, but it can be used at shorter distances. Then the source can be controlled [47,48].
The field in the aperture can be expressed in a form nearly similar to the inverse of (1) but with a quadratic term:

$$\vec{E}_{ap}(x,y) = -j\lambda R \left(\frac{2}{1+\cos(\theta)}\right) e^{jkR} . e^{jk(x^2+y^2)/2R} \; F^{-1}\left[\vec{E}_f(u,v)\right] \quad (3)$$

R is the distance at which the (u,v)-space (or $\bar{k}$-space) is sampled. The minimum R is still large for mm-wave antennas. A value of 1/20 * Rayleigh distance is quoted, still in the order of kilometers [47]. In both cases the measurement of the amplitude and phase is required (holography = ´writing down all´, thus amplitude and phase).

The measurement of an accurate phase is difficult in the mm-wave region. Therefore so-called phase retrieval methods based on amplitude-only data are explored. The phase is here retrieved by a series of iterations, using two distinct FF amplitude sets, which are interrelated in a predictable way.

One measures the FF amplitude pattern twice: once with the feed in focus and once with the feed defocussed. The latter situation yields is related to the first one in a systematic predictable fashion.
The process starts with an inverse Fourier transformation of FF amplitude data for the focussed configuration. A phase pattern is assigned. A complex distribution results. It is multiplied by a phase correction to predict the result for the de-focussed configuration. Fourier transformation gives a complex distribution, in which the amplitude is replaced by the measured FF amplitude for the de-focussed situation. An inverse Fourier transformation and phase correction (conjugate to first one) is now applied to go from the de-focussed to focussed ´aperture-plane´. After Fourier transformation a complex distribution results, in which the amplitude is replaced by the measured FF data for the focussed situation. The phase distribution is kept all the time. After a series of iterations, the process converges. A particular phase distribution results for the aperture plane. From this distribution one derives as before information about reflector properties or radiation performances.
Radio telescopes have been perfectioned with this method. RMS errors down to 65 micrometer resulted at 86 GHz [42,44].

The Fourier relationship is used above. The cylindrical and spherical NF-FF algorithms can be used as well [45] in an iterative fashion, using only amplitude data (2 NF datasets, 2 FF data sets and perhaps even 1 FF and 1 NF dataset) Uniqueness and convergence seems realised with the first algorithm. It is not yet guaranteed [45] always.

Phase retrieval from one single FF amplitude set is reported in [46]. The zero-field condition outside the aperture is used in the iteration. Initial tests show convergence for this method for a particular application, but more work is needed to assess accuracies. Results were discussed for use of amplitude-only (1 dataset) in planar NF scanning [49]. An important problem also here is the uniqueness [45].

5. Data-processing using 'hardware'

5.1 Range gating for indoor testranges.

Measurement equipment (HP8510) have hardware installed to perform fourier transformation back or forward and filtering (or gating) in between. It is used with advantage on antenna testranges to eliminate undesired spurious signals.
An example is shown in fig.10 of gating in the time domain. The result is a recording of a fourier transformed frequency respons of an antenna on an indoor test configuration for a particular angular position (freq. band was around 1.6 GHz). The level at t=0 (centre) is of interest. A range-gate is centred around this peak, corresponding to the actual location of antenna (time//range). The peak at 32 nanosec is related to reflection from the backwall, 5 m farther away.

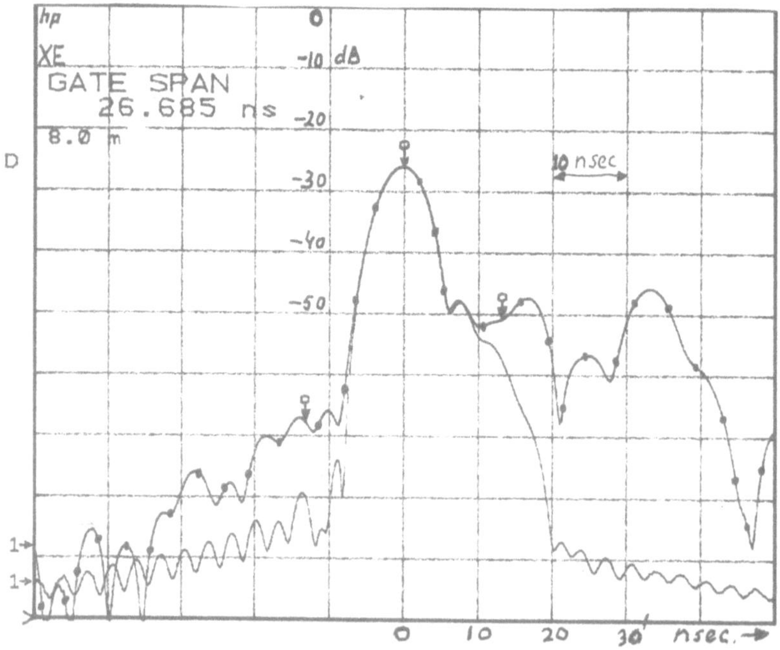

Fig.10. Recording of respons (time domain) of antenna under test. Filtered respons (solid) has the same value at t=0.

After gating (solid line) and inverse transformation, the filtered signal is obtained. This action is performed during the testing for all angles of observation.
Attention has to be spent, when the antenna has narrow bandwidth. Then the time domain respons is more spread-out and gating will lead to errors.

5.2 Absolute gain measurement on CATR with RCS measurements.

Absolute gain of a 1*1 m slotted waveguide antenna was measured on the CATR in ESTEC using gating with the HP8510 [31].

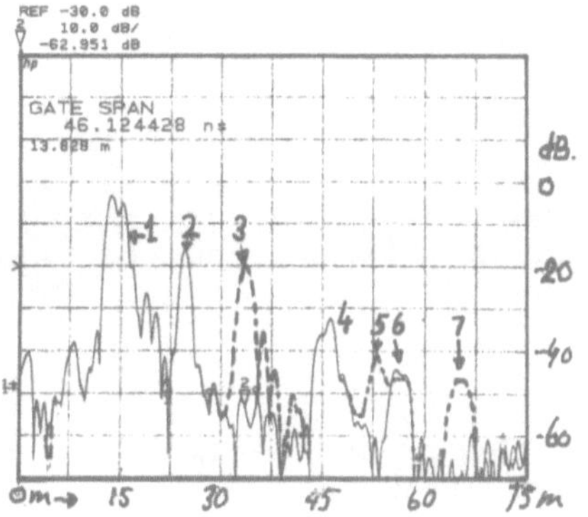

Fig.11. Time domain curves
Explanation see text.

The technique resembles known methods based on radar cross section (RCS) measurements.
The use of gating is new [31]. The absolute scale for backscattering is set with a standard object (flat plate). Then the backscatter from the AUT is measured for two terminations at its input port. One is a coax cable of particular length with a load. The other termination is a cable with a short. All reflections not coming from the short are separated out with gating.
Measurement of the reflection from the short on the cable at the antenna input allows the determination of S12 of the antenna, provided that the coax cable is calibrated. Discrimination is checked using a load on the coax cable (solid line, marker 2 in fig.11).
The S12 describes transfer through the antenna and relates to the gain (similar formulation as encountered in planar NF scanning). It is illustrative to discuss the backscattering in the time domain (fig.11):

Peak 1: Reflected signal from the aperture (at boresight). The peak is broadened due to dispersion and resonances in the antenna.
Peak 2: Reflection from the backwall in the CATR room, about 5.5 meter behind the antenna under test.
Peak 3: Peak of interest: short on coax cable (solid line is related to load on cable). Peak 3 is gated out.
Peak 4: Multiple reflection aperture-feed CATR-aperture.
Peak 5: Multiple reflection short-antenna internal.
Peak 6: Multiple reflection backwall-feed CATR-aperture.

The gate is selected around the third peak to provide the desired information. The length of the cable is carefully selected, so that peak '3' related to the short occurs in a 'low-level' region of the curve.

Table I shows the gain values in comparison with the values measured at Technical University of Denmark (TUD).

Frequency (GHz)	Gain (dB) ESTEC	TUD
5.290	30.40	30.52
5.292	30.39	30.47
5.300	30.47	30.53
5.308	30.54	30.53
5.310	30.57	30.75

Table I, Comparison gain of 1*1 meter slotted wvg panel.

6. Conclusions

Several type of antenna measurement techniques are in use and a number of methods has been outlined. Processing of data plays an important role, in which the Fourier transform is exploited. Examples were given of software and hardware type applications.

7. References:

[1] Silver, S. 'Microwave antenna theory and design', McGrawHill, MIT series, Vol 12, 1949.
[2] Fine, M.G. 'Methode de correction de sonde en mesures de champs proches', Dr.Thesis, Univ. Paris-Sud, July 1989.
[3] Woonton,G. 'Indoor measurements of microwave antenna radiation patterns by means of a metal lens',Journ.Appl. Physics, Vol.21,pp428,1950.
[4] Hammer,J.A.'The evaluation of diffraction problems in aerial theory by means of an analyser for complex Fourier series', Proc. Conf. Circuits and Antennas, Paris,Oct. 1957. L'Onde Electr.38e Ann,376, Tome I,p8.
[5] Clayton,L.,Hollis,J.R.'Calculation of microwave antenna radiation patterns by the Fourier integral method', The Essay, Scient. Atlanta Inc. March 1960.
[6] Goldbohm,E.'An automatic phase plotter for use in the near-field of microwacve aerials',Proc.Conf. Circuits and Antennas, Paris,oct.1957,p804.
[7] Brown,J., 'The prediction of aerial radiation patterns from near-field measurements', Proc. Inst.Electr.Eng. p635, Nov.1961.
[8] Kerns,D.M. 'Plane wave scattering matrix theory of antennas and antenna-antenna interactions', NBS Monograph 162, Boulder, Colorado,1981.
[9] Yaghjian,A.'Near-field measurements on a cylindrical surface',NBS Tech.Note 696,sept.1977.
[10]Johnson,R. 'Compact range techniques and measurements , IEEE TAP Vol AP-17, No 5, 1969.

[11]Vokurka,V. 'New compact range with cylindr. reflectors and high efficiency',Proc.Electr.Conf.Muenchen, 1976.

[12]Dudok,E. Fasold,D.Steiner,H. 'Development of optimised compact antenna test range', 11th ESTEC Workshop on Antenna Measurements, ESA WPP-001,June 1988.

[13] 11th ESTEC Antenna Workshop on Antenna Measurements, ESA WPP-001,June 1988.

[14]Kerns,D. 'Plane wave scattering matrix theory of antennas and antenna-antenna interactions', Journ.of Research.,NBS,Vol.80B,No.1,Jan.'76.

[15]Wacker,P. 'Description of NBS scanning facilities', in ESA SP127, p115, June 1977, Noordwijk,NL.

[16]Joy,E.e.a.'Applications of probe-compensated near-field measurements',IEEE-TAP,Vol.AP-26,May,1978.

[17]Leach,W.,Paris,D. 'Probe compensated near-field measurements on a cylinder', IEEE-TAP,Vol.AP-21,July,1973.

[18]Jensen,F. 'Electromagnetic near-field far-field correlations', PhD Thesis LD-15 , 1970,Techn.Univ.Denmark.

[19]Wacker,P. 'Non-planar near-field measurements: spherical scanning', Report NBSIR 75-809,Boulder 1975.

[20]Larsen,F. 'Probe corrected spherical near-field measurements', PhD Thesis LD-36 Techn.Univ.Denmark.

[21]Hansen,J. 'Spherical near-field antenna measurements', IEE Electr. Waves Nr.26, P.Peregrinus,Ltd,London,1988.

[22]Yaghjian, 'An overview of near-field measurements', IEEE-TAP,Vol.AP-34, Jan.1986.

[23]Rahmat Samii,Y. Galindo Israel,Mittra,R. 'A plane-polar approach for FF construction from NF measurements', IEEE-TAP Vol.AP-28, March 1980.

[24]Rahmat Samii,Y. Gatti,M. 'Far-field patterns of spaceborne antennas from plane-polar NF measurements', IEEE-TAP,Vol.AP-33, June 1985.

[25]Cao,C, V.Lil,E., V.d.Capelle,A.,V.'t Klooster, 'Comparison of plane-polar NF to FF transformation techniques' ESTEC 11th Workshop antenna measurements, ESA WPP-001.

[26]V.Schaik,H. 'Theory and performance of space-fed planar phased array' PhD Thesis, Delft June 1979.

[27]V.'t Klooster,K.'Planar near-field measurements' (dutch) PHL-1984-22,Physics Lab.TNO,The Hague,NL.

[28]V.'t Klooster,K.'Co- and crosspolar patterns of cosec-2 antenna from planar NF measurements', Conf.Precision Electr.Meas. Delft,1984.

[29]Yaghjian,A. 'Efficient computation of antenna coupling and fields within NF region', IEEE-TAP V-30,Jan.1982.

[30]'The ESA-Ericsson Near-Field Test Facility'. ESA Brochure ESA BR-38, Noordwijk, NL.

[31]Hammer,J. 'Absolute gain calibration of an antenna by comparing with reflection of a flat plate', 11th ESTEC Workshop on Antenna Measurements,ESA WPP-001,June 1988.

[32]'The TUD-ESA NF antenna test facility', ESA Brochure ESA BR-19, Noordwijk NL.

[33]Gatti,M., Rahmat-Samii,Y., 'FFT applications to plane-polar NF antenna measurements',IEEE-TAP,V.36 June,1988.
[34]Fischer,C. 'A cylindrical near-field test facility for large satellite antennas',Proc.Eu.Microw.Conf.1983,p829
[35]Lemanczyk,G.,v.'t Klooster,K. 'An interrange comparison in suppport of the evaluation of testranges for ERS-1', 11th ESTEC Ant.Workshop Antenna Meas.,ESA WPP-001,1988.
[36]v.'t Klooster,K. 'Radiation characteristics of SAR panel of ERS-1 from planar NF scanning',17th EuMC 1987, Rome.
[37]v.'t Klooster,K.,Petersson,R., Kallas,E., 'Radiation performance of the ERS-1 SAR EM antenna', IEEE-AP Symp. Syracuse,1988.
[38]Bennett,J., Anderson,, McInnes,'Microwave holographic metrology of large refl. antennas' IEEE-TAP,V.24,1976.
[39]Morris,D. Baars,J. Hein,H. e.a.,'Radio-holographic refl. measurement of 30-m millimeter radio telescope at 22GHz with a cosmic signal source',Astr.Astroph.203,'88,p399.
[40]Godwin,M.Schoessow,E.Grahl,B.,'Improvement 100m Effelsberg telescope based on holographic reflector surface measurement',Astr.Astroph.167,1986,p390.
[41]Rahmat-Samii,Y.'Antenna diagnosis by microwave holographic metrology', Il Ciocco, Adv.Study Inst. 1987.
[42]Morris,D. Baars,J. Hein,H.Steppe,H. 'Phase retrieval radio holography in the Fresnel region: tests on 30-m telescope at 86 GHz.' IEE proc.135,pt.H,Febr.1988.
[43]Baars,J. Morris,D.'Millimeter wave metrology of millimeter radio telescopes', 11th ESTEC Ant. Workshop on Antenna Measurements, ESA WPP-001,June 1988.
[44]Hills,R.,Lasenby,A. 'Millimeter wave metrology of the James Clark Maxwell telescope', 11th ESTEC Ant. Workshop on Antenna Measurements, ESA WPP-001,June 1988.
[45]Bucci,O.M.,Pierri,R.,Leone,G.,Elia,'Far-field pattern determination by amplitude only NF measurements', 11th ESTEC Ant. Workshop Ant. Meas.,ESA WPP-001,1988.
[46]McCormack,J.,Anderson,A.P.,'Phase retrieval microwave metrology of refl. antennas from single data sets. 11th ESTEC Ant. Workshop Ant. Meas.,ESA WPP-001,1988.
[47]Morris,D. 'Telescope testing by radio holography', Proc.URSI Symp. MM and sub-MM Wave Radio Astronomy, Granada, Spain, Sept.1984,p29.
[48]Orta,R.'Microwave holographic measurement of reflector surface accuracy',ESA Journal,No.3,1985,p329.
[49]Anderson,A.,Junkin,G.,McCormack,J.,'NF-FF predictions from single-intensity-planar-scan phase retrieval', Electr.Letters,13 april 1989,Vol.25,No.8.
[50]Beeckman,P.'Control of FF radiation patterns of microwave reflector antennas by using serrated edges', IEE Proc.Pt.H,No.3,June 1987.
[51]V.Spaendonck,R.A.C.M.,'NF component evaluation', Rep.PHL 1983-57, (dutch), Phys.Lab.TNO, The Hague,NL.

--oOo--

COMPUTATIONS IN X-RAY CRYSTALLOGRAPHY

Richard Tolimieri and Myoung An
Center for large scale computation
25 West 43rd Street, Suite 400
New York, NY 10036

1. FOURIER TRANSFORM IN X-RAY CRYSTALLOGRAPHY

In normalized coordinates, the electron density F of a crystal is a function of $r \in R^3$, the Euclidean 3-space, that satisfies the *periodicity condition*,

$$F(r+n) = F(x), \quad r \in R^3, \; n \in Z^3. \tag{1}$$

We can write $F(r)$ as a Fourier series

$$F(r) = \sum_{n \in Z^3} \rho(n) exp(2\pi i n \cdot r), \tag{2}$$

where $n = (n_1, n_2, n_3)$, $r = (r_1, r_2, r_3)$ and

$$n \cdot x = n_1 r_1 + n_2 r_2 + n_3 r_3. \tag{3}$$

The complex constants

$$\rho(n), \quad n \in Z^3, \tag{4}$$

are called ***structure factors***. The method of X-ray diffraction determines thousands and even millions of intensities $| \rho(n) |^2$, but not the structure factors themselves. Computing the phases of these structure factors, the ***phase problem***, is a major problem in X-ray crystallography and has been successfully solved for small crystals largely due to the efforts of Hauptman and Karl [1,2,3,4]. For proteins and virus', theses methods, so far, appear in general to be unsuccessful. Bricogne have introduced the methods of noncrystallographic symmetry [5], and maximal entropy [6] which offer new tools in the *Direct Method* approach for solving large structures. These ideas are beyond the scope of this work, but by their nature require many backward and forward Fourier transform (FT) computations which justfies the need for highly optimized code on large machines.

In this paper, we describe several algorithms for computing finite Fourier transform (FFT) for X-ray crystallography. Many efficient FFT algorithms exist– even specialized algorithms suitable for use in crystallography. The need for specialized FFT algorithms for crystallography comes from the fact that crystallographic

J. S. Byrnes and J. F. Byrnes (eds.), Recent Advances in Fourier Analysis and Its Applications, 237–250.

data contains considerable redundancy. This fact has been known to crystallographers since the beginning stages of the X-ray method, but has never been used systematically, especially in conjunction with highly sophisticated and fast computers until recently. In 1973, L. Ten Eyck [7] introduced a method of incorporating a class of crystal symmetries into existing efficient FFT algorithms.

The algorithms we are about to describe here all have an algebraic nature as a common feature. Instead of viewing symmetries merely as a device of identifying redundancies in the data, an algebraic scheme coming from the group theoretic properties of the symmetry groups is developed. The algebraic scheme sets up a way of organizing the algorithms in the way crystallographic groups are organized, as well as to control the flow of computations in each of the algorithms.

We will describe two classes of algorithms: *Orbital Exchange* method and *Orbital Collapse* method. Orbital exchange method provides a way of introducing symmetry group into existing FFT algorithms that are decomposition algorithms. Orbital collapse, on the other hand, generates blocks of matrices that are computationally advantageous. We will begin the process with a discussion on the sampling theory underlying the general FFT.

2. FFT

Take an integer $N > 0$ and consider the subgroup of Z,

$$NZ = \{Nr \mid r \in Z\}. \tag{5}$$

Form the subgroup of Z^3

$$\Delta = NZ \times NZ \times NZ, \tag{6}$$

and the quotient group

$$Z^3/\Delta = Z/NZ \times Z/NZ \times Z/NZ, \tag{7}$$

which can be identified with the set of points $n = (n_1, n_2, n_3)$ satisfying $0 \leq n_1, n_2, n_3 < N$. Define $g(m)$, $m \in Z^3$, by

$$g(m) = F(\frac{m_1}{N}, \frac{m_2}{N}, \frac{m_3}{N}), \quad m = (m_1, m_2, m_3), \tag{8}$$

and observe that by the periodicity condition (1)

$$g(m) = g(m + r), \quad m \in Z^3, \; r \in \Delta. \tag{9}$$

The sample function $g(m)$, $m \in Z$, can be viewed as a function on Z^3/Δ.

Define $f(n)$, $n \in Z^3$, by

$$f(n) = \sum_{r \in \Delta} \rho(n + r), \tag{10}$$

where $\rho(n)$ are the structure factors. The *periodized* structure factors $f(n)$, $n \in Z$, satisfy

$$f(n) = f(n+r), \quad n \in Z^3, \ r \in \Delta, \tag{11}$$

and can be viewed as a function on Z^3/Δ. The basic sampling theory result (Poisson summation formula) relates the sample values of the electron density map to the periodized structure factors by

$$g(m) = \sum_{n \in Z^3/\Delta} f(n) exp(-2\pi i m \cdot n), \quad m \in Z^3/\Delta. \tag{12}$$

In matrix form,

$$\underline{g} = F_3(N)\underline{f}, \tag{13}$$

where $F_3(N)$ is the 3-dimensional $N \times N \times N$ FFT matrix

$$F_3(N) = [w^{m \cdot n}], \quad m, \ n \in Z^3/\Delta, \ w = exp(\frac{-2\pi i}{N}) \tag{14}$$

and $\underline{g}$ and $\underline{f}$ are the vectors of size N^3 given by

$$\underline{g} = [g(m)]_{m \in Z^3/\Delta},$$

$$\underline{f} = [f(n)]_{n \in Z^3/\Delta},$$

with the lexicographic ordering of Z^3/Δ.

In practice, only a finite number, however large, of structure constants are used in the computation. The periodization of structure constants is replaced essentially by the assumption that the structure constants vanish outside a bounded sphere. The control of *aliasing* erros resulting from this assumption is a part of the art of crystallographic computations. It should, however, be observed that the size of this aliasing error depends to a large extent on the size of high frequency structure factors. If the electron density map is viewed as the sum of delta functions then the high frequency structure factors are large and resulting aliasing errors are large.

3. ALGORITHMS TO COMPUTE MULTIDIMENSIONAL FFT

There are several standard techniques for computing multidimensional FFT's. The language of tensor product [8] will be convinient. The 3-dimensional $N \times N \times N$ FT matrix can be written as

$$F_3(N) = F(N) \otimes F(N) \otimes F(N), \tag{15}$$

where $F(N)$ is the N-point FFT matrix. Let $P \equiv P(N^3, N)$ denote the N-point stride matrix on N^3 data points. Implicit use of this permutation can be found repeatedly in many algorithms [9]. The reason for distinguishing this permutation is that on many vector computations (and DSP chips), it can be carried out by the machine instruction

$$\text{LOAD} - \text{STRIDE}$$

This fact has ranging consequences on the simplicity of code writing especially as related to the complex addressing required for data flow through the computation and to the matching of the computation to a given machine architecture. The size of the computation both globally and locally can be fitted to machine characteristics such as the size and the number of vector registers and communication links.

The permutation P can be viewed as follows. Assume first that the data is arranged in a 3-dimensional array

$$a(k,m,n),\ \ 0 \leq k,\ m,\ n < N.$$

We can order the array linearly by ordering (k,m,n) lexicographically with n the fastest running variable, followed by m and k the slowest running variable. Applying the permutation P to the linear array is equivalent to making m the fast running variable, followed by k with n becoming the slowest running variable. In terms of tensor product language,

$$\begin{aligned} F_3(N) &= (F(N) \otimes I_{N^2})P(F(N) \otimes I_{N^2})P^{-1}P^2(F(N) \otimes I_{N^2})P^{-2} \\ &= (F(N) \otimes I_{N^2})P(F(N) \otimes I_{N^2})P(F(N) \otimes I_{N^2})P. \end{aligned}$$

Up to the stride permutation P, the computation of $F_3(N)$ can be carried out as three vector computations $F(N) \otimes I_{N^2}$. In effect, the permutation P sets up the data in the appropriate order for the next stage of the computation. In each stage, an N-point FFT is applied on N^2 vectors of size N. The Cooley-Tukey FFT can be appealed to implement each stage in $N^3 logN$ computations. This algorithm will be referred to as the row-column method.

A second method, the Good-Thomas Prime Factor algorithm can also be useful and in fact is closely related to the Orbital Exchange method introduced by J. Cooley and M. An [10] for dealing with symmetrized data. We require that N be composite having the form $N = PQ$, with P and Q relatively prime. In this case, we can write

$$F(N) = R(F'(P) \otimes F'(Q))R^{-1}$$

where R is a permutation matrix given by the Chinese remainder theorem and $F'(P)$ and $F'(Q)$ are modified FFT matrices that can be implemented using any FFT algorithms. We can now write

$$F_3(N) = F(N) \otimes F(N) \otimes F(N)$$

$$= R(F'(P) \otimes F'(Q))R^{-1} \otimes R(F'(P) \otimes F'(Q))R^{-1} \otimes R(F'(P) \otimes F'(Q))R^{-1}$$
$$= (R \otimes R \otimes R)(F'(P) \otimes F'(Q))(R \otimes R \otimes R)$$

The large computation $F_3(N)$ is broken up into the smaller computations $F'(P) \otimes F'(Q)$. Continuing in this way, we can decompose the initial computation into 'small' 3-dimensional FFT's gien by the prime power factors of N. In case N is a product of distinct primes, Winograd-type multiplicative algorithms can be applied [9].

A third approach is due to Winograd [11] and referred to as *Winograd's multiplicatve* FFT algorithm. For a prime P, let α be a generator of the multiplicative units of the ring Z/P. Represent α as a 3×3 matrix

$$u = \begin{bmatrix} \alpha & 0 & 0 \\ 0 & \alpha & 0 \\ 0 & 0 & \alpha \end{bmatrix}. \tag{16}$$

Denote by U_P the group generated by u. U_P acting on $(Z/P)^3$ decomposes $(Z/P)^3$ into distinct orbits. Let

$$O_0,\ O_1,\ \cdots,\ O_l \tag{17}$$

be the distinct U_P-orbits in $(Z/P)^3$, with $O_0 = \{(0,0,0)\}$. Order O_i, $1 \le i \le l$ by the cyclic power of u. We can order $(Z/P)^3$ linearly by ordering the O_i's, beginning with O_0. FFT matrix with this ordering consists of blocks of $(P-1) \times (P-1)$ matrices of a special form. The matrix associated with the orbits O_i and O_j is

$$M_{ij} = [w^{m \cdot n}]_{m \in O_i, n \in O_j}. \tag{18}$$

Let $x = (x_1, x_2, x_3)$, $y = (y_1, y_2, y_3) \in (Z/P)^3$ be the first elements of O_i and O_j. The exponent of the kl-th entry of M_{ij} is

$$(\alpha^{k-1}x_1, \alpha^{k-1}x_2, \alpha^{k-1}x_3) \cdot (\alpha^{l-1}y_1, \alpha^{l-1}y_2, \alpha^{l-1}y_3)$$
$$= \alpha^{k+l-2}x_1y_1 + \alpha^{k+l-2}x_2y_2 + \alpha^{k+l-2}x_3y_3)$$
$$= \alpha^{k+l-2}x \cdot y$$

Setting $m(k) = w^{\alpha^k x \cdot y}$, $0 \le k \le P-1$ we can rewrite

$$M_{ij} = \begin{bmatrix} m(0) & m(1) & m(2) & \cdots & m(P-1) \\ m(1) & m(2) & m(3) & \cdots & m(0) \\ m(2) & m(3) & m(4) & \cdots & m(1) \\ \cdot & \cdot & \cdot & \cdots & \cdot \\ \cdot & \cdot & \cdot & \cdots & \cdot \\ \cdot & \cdot & \cdot & \cdots & \cdot \\ m(P-1) & m(0) & m(1) & \cdots & m(P-2) \end{bmatrix} \tag{19}$$

Matrices of the form ( ) are called *skew circulant.* There are algorithms for computing with skew circulant matrices. In particular, Winograd's multiplicative algorithm computes with such matrix as *polynomial multiplication* modulo $x^P - 1$.

4. SYMMETRIZED FFT

We will see how crystal symmetry enters into the computation of the FFT [10,14]. Consider any subgroup G of $SL(3, Z/N)$ and a function $f(\underline{n})$ on $(Z/N)^3$. For $g \in G$, we define the g-translate of $f(\underline{n})$ by

$$f_g(\underline{n}) = f(g\underline{n}), \ \underline{n} \in (Z/N)^3, \tag{20}$$

where $g\underline{n}$ denotes the action of the matrix of g on the point $\underline{n}$. Straightforward computation leads to the following results.

$$1. \ (F_3(N)f)(g^*\underline{m}) = (F_3(N)f_g)(\underline{m}),$$

where $g^* = (g^{-1})^t$.

$$2. \ \text{If} f = f_g \text{then } F(f) = F(f)_{g^*}. \tag{21}$$

It follows that if f is G-invariant then $F(f)$ is G^*-invariant. Assume now that f is G-invariant for the remainder of this section. Then F is completely determined by its actions on a G-fundamental region according to the following definition. For $\underline{n} \in (Z/n)^3$,

$$G(\underline{n}) = \{g\underline{n} \mid g \in G\} \tag{22}$$

is called the G-orbit of $\underline{n}$ and

$$Iso_G(\underline{n}) = \{g \in G \mid g\underline{n} = \underline{n}\} \tag{23}$$

is called the isotropy subgroup in G at $\underline{n}$. A G-fundamental region in $(Z/N))^3$ is any subset X satisfying the following two conditions.

$$(Z/N)^3 = \cup_{\underline{n} \in X} G(\underline{n})$$

$$G(\underline{n}) \cap G(\underline{n}') = \phi \text{ (empty set)}. \tag{24}$$

Every G-orbit intersects a fundamental region X in exactly one point. Since the G-invariant function f takes the same value at each point of a G-orbit, it is uniquely determined by its values on the fundamental region X. Suppose X is a G-fundamental region and Y is a G^*-fundamental region. Denote the elements of X by

$$\underline{x}_1, \cdots \underline{x}_r. \tag{25}$$

We will now see how to compute $F(N)f$ on Y soley from the values of f on X. Set

$$Gx_j = \{\underline{x}_{j1}, \cdots, \underline{x}_{jm}\}. \tag{26}$$

Then

$$(F(N)f)(\underline{y}) = \sum_{j=1}^{r} \sum_{l=1}^{jm} f(\underline{x}_{jl}) exp(< \underline{y}, \underline{x}_{jl} >)$$

$$= \sum_{j=1}^{r} f(\underline{x}_j) \sum_{l=1}^{jm} exp(< \underline{y}, \underline{x}_{jl} >). \tag{27}$$

Set

$$c_{kj} = \sum_{l=1}^{jm} exp(< \underline{y}_k, \underline{x}_{jl} >). \tag{28}$$

From (24), we can compute $F(N)f(\underline{y})$ on Y by the matrix multiplication

$$\begin{bmatrix} (F(N)f)(\underline{y}_1) \\ \cdot \\ \cdot \\ \cdot \\ (F(N)f)(\underline{y}_r) \end{bmatrix} = C \begin{bmatrix} f(\underline{x}_1) \\ \cdot \\ \cdot \\ \cdot \\ f(\underline{x}_r) \end{bmatrix} \tag{29}$$

where C is the $r \times r$ matrix

$$C = [c_{kj}]. \tag{30}$$

Computing $F(N)f$ by (25) removes all redundant calculations. However, straightforward application of 1-dimensional FFT algorithms is lost in the process. The effort in the next four sections have as their goal the designing of fast FT algorithms which operate soley on data given on fundamental regions.

5. CRYSTALLOGRAPHIC FFT ALGORITHMS

Orbital exchange method is an algebraic way of finding fundamental regions, thereby yielding readily to algebraic manipulations and computer implementations. As in the row-column method and the prime factor algorithm, decomposition algorithms contain permutation steps. Orbital exchange sets up fundamental regions that can be algebraically manipulated to replace the permutation step.

5.1. Orbital Exchange: Row-Column Algorithm

The primitive crystallographic groups in monoclinic and the orthorhombic crystal classes act 'diagonally' on the 3-space [12,13]. We will call these groups the diagonal groups. These are the classes of symmetries analyzed in Ten Eyck approach [7]. In homogeneous coordinate system, an element d of such a group is represented by the following matrix.

$$d = \begin{bmatrix} g_1 & 0 & 0 & t_1 \\ 0 & g_2 & 0 & t_2 \\ 0 & 0 & g_3 & t_3 \\ 0 & 0 & 0 & 1 \end{bmatrix}, \tag{31}$$

where $g_i = \pm 1$ and $t_i = \frac{N_i}{2}$, for some positive even integer N_i. A point $n = (n_1, n_2, n_3) \in Z/N_1 \times Z/N_2 \times Z/N_3$ is identified with the matrix

$$\underline{n} = \begin{bmatrix} n_1 \\ n_2 \\ n_3 \\ 1 \end{bmatrix},$$

and the action of d on $\underline{n}$ is defined as the matrix multiplication modulo the respective numbers. Note that N_1, N_2 and N_3 may be different. This means that the sampling rate along each coordinate maybe different. (This is often the desired case.) Thus the computation we will consider is

$$[F(N_1) \otimes F(N_2) \otimes F(N_3)]\underline{f}, \tag{32}$$

where $\underline{f}$ is lexicographically ordered as before. This computation can be obtained in the following 3 steps.

$$\begin{aligned} 1.\ [I_{N_1} \otimes I_{N_2} \otimes F(N_3)]\underline{f} &= \underline{f}_3 \\ 2.\ [I_{N_1} \otimes F(N_2) \otimes I_{N_3}]\underline{f}_3 &= \underline{f}_2 \\ 3.\ [F(N_1) \otimes I_{N_2} \otimes I_{N_3}]\underline{f}_2 &= \underline{f}_1 = \hat{\underline{f}} \end{aligned} \tag{33}$$

Assumue for the remainder of this section that f is invariant under a diagonal group D. To save notation, we will use X, Y and Z respectively for Z/N_1, Z/N_2 and Z/N_3. Obtain D_3, D_2 and D_1 by setting $t_3 = 0$, $t_2 = 0$ and $t_1 = 0$ successively. Then we have the following symmetry conditions upto phase factors. For $k = 1,\ 2,\ 3,$

$$\underline{f}_k \text{ is invariant under } D_k. \tag{34}$$

The above follows from straightforward computations. We will prove the case $k = 2$. Let

$$d_2 = \begin{bmatrix} g_1 & 0 & 0 & t_1 \\ 0 & g_2 & 0 & 0 \\ 0 & 0 & g_3 & 0 \\ 0 & 0 & 0 & 1 \end{bmatrix}$$

be an element of D_2. For $(x, y, z) \in X \times Y \times Z$,

$$\underline{f}_2(d_2(x, y, z)) = \underline{f}_2(g_1 x + t_1, g_2 y, g_3 z)$$

$$= \sum_{v \in Y} \underline{f}_3(g_1x + t_1, v, g_3z)exp(\frac{2\pi i}{N_2} g_2yv)$$

$$= \sum_{(g_2v+t_2) \in Y} \underline{f}_3(g_1x, g_2v + t_2, g_3z)exp(\frac{2\pi i}{N_2} g_2y(g_2v + t_2))$$

$$= \sum_{(g_2v+t_2) \in Y} \underline{f}_3(d_3(x, v, z)exp(\frac{2\pi i}{N_2} g_2y(g_2v + t_2))$$

$$= \sum_{(g_2v+t_2) \in Y} \underline{f}_3(x, v, z)exp(\frac{2\pi i}{N_2} yv)exp(\frac{2\pi i}{N_2} g_2yt_2)$$

$$= exp(\frac{2\pi i}{N_2} g_2yt_2)\underline{f}_2(x, y, z).$$

The factor $exp(\frac{2\pi i}{N_2} g_2yt_2)$ is the phase factor referred to earlier. The value of this factor is 1 or -1, depending on t_2 and y. (Recall that $t_2 = 0$ or $\frac{N_2}{2}$ and $g_2 = \pm 1$.)

The three computational steps can now be reduced to computations on each of the respective fundamental regions. We will now illustrate the use of orbital exchange.

For a diagonal group D, consider $D \mid_X$, the restriction of D to X. We will denote a $D \mid_X$-fundamental region by X/D. For each $x \in X/D$, consider $Iso(x) \mid_Y$, the action of the subgroup $Iso(x)$ restricted to Y. Denoting a fundamental region of this action by $Y/Iso(x)$, note that

$$(X \times Y)/D = \cup_{x \in X/D}(\{x\} \times Y/Iso(x)) \tag{35}$$

is a fundamental region of $D \mid_{(X \times Y)}$. Continuing in the same way, we can find

$$(X \times Y \times Z)/D = \cup_{(x,y) \in (X \times Y)/D}(\{(x, y)\} \times Z/Iso(x, y)). \tag{36}$$

$(X \times Y \times Z)/D$ is a D-fundamental region in $X \times Y \times Z$. $Iso(x, y)$ acting on a point $(x, y, z) \in (X \times Y \times Z)/D$ moves the point only along the Z-coordinate. This follows from the fact that the function f is invariant under $Iso(x, y)$. Thus we can compute $\underline{f}_3$ for each $(x, y) \in (X \times Y)/D$ on the subsets

$$\{(x, y)\} \times Z/Iso(x, y)\} \tag{37}$$

of $(X \times Y \times Z)/D$ using 1-dimensional $Iso(x, y)$-symmetrized FFT routine. Thus the computation of $F(N_3)$ can be replaced by an $Iso(x, y)$-symmetrized N_3-point FT for each $(x, y) \in (X \times Y)/D$. In this way, we have the values of $\underline{f}_3$ on a D_3-fundamental region. For the next stage of the computation, we need a D_3-fundamental region that prepares the Y-coordinate for symmetrized FFT computation in the way $(X \times Y \times Z)/D$ prepares the Z-coordinate. We can obtain $(X \times Y \times Z)/D_3$ which is of the desired form by repeating the steps for finding

$(X \times Y \times Z)/D$ with D_3 replacing D. Since $(X \times Y \times Z)/D_3$ is different from the D_3-fundamental region on which f_3 is defined, we need a mapping between the two D_3-fundamental regions. Such a mapping replaces an element in a fundamental region by an element in its orbit. Thus, this mapping is an adjunction of elements of the diagonal group actiong on subsets of a fundamental region. We will call this mapping an *orbit exchange map.* In practice, rather than constructing $(X \times Y \times Z)/D_3$, orbit exchange map is defined from the preimage to have such a D_3-fundamental region as its image. (See [13] for the details.) Observe that orbit exchange map is a permutation or a reindexing of the elements of f_k and replaces the stride permutation in the row-column algorithm.

5.2. Orbital exchange; Prime Factor Algorithm

In the defining formula

$$(F(N)f)(y) = \sum_x f(x) w^{<x,y>},$$

we can replace the bilinear form $< x, y >$ by any nondegenerate bilinear form

$$x^t By$$

where B is any 3×3 nonsingualr matrix over Z/N. The resulting computation

$$(F_B(N)f)(y) = \sum_x f(x) w^{x^t By} \tag{38}$$

is related to the standard FFT by

$$(F(N)f)(y) = (F_B(N)f)(B^{-1}y). \tag{39}$$

For simplicity, we will assume that G is a subgroup of $SL(3, Z/N)$ such that

$$g^t Bg = g, \quad g \in G \tag{40}$$

for some 3×3 nonsingular matrix B over Z/N. The Weyl unitary trick guarantees the existence of a symmetric B over Z for any finite subgroup G of $SL(3, Z)$ but in general a nonsingular B over Z/N does not exist. See [10] for a complete discussion. The value of such a B is that for any G-invariant f,

$$(F_B f)(y) = (F_B(f))(gy), \quad g \in G, \; y \in (Z/N)^3$$

which means that $F_B(f)$ is also G-invariant. In this case, the entries of the nonredundant symmetrized FFT matrix are

$$c_{kj} = \sum_{l=1}^{jm} w^{y_k^t B g_l x_j}.$$

Take $N = PQ$ where P and Q are relatively prime integers. By the Good-Thomas algorithm, up to input and output permutation, we can replace the computation of $F_3(N)$ by the following problem. Let $f(a,b)$ be a function on $(Z/P)^3 \times (Z/Q)^3$ and define

$$(F(P,Q)f)(c,d) = \sum_{a\in(Z/P)^3} \sum_{b\in(Z/Q)^3} f(a,b) w_P^{a^t Bc} w_q^{b^t Bd} \tag{41}$$

where B in the expression $w_P^{a^t Bc}$ is taken *mod* P and in $w_Q^{a^t Bc}$ is taken *mod* Q. We can carry out this computation in two stages.

1. Compute for all $a \in (Z/P)^3$

$$f_1(a,d) = \sum_{b\in(Z/Q)^3} f(a,b) w_Q^{b^t Bd} \tag{42}$$

Setting $f_a(b) = f(a,b)$,

$$f_1(a,d) = (F_B(Q)f_a)(d). \tag{43}$$

2. Compute for all $d \in (Z/Q)^3$

$$f_2(c,d) = \sum_{a\in(Z/P)^3} f(a,d) w_P^{a^t Bd} \tag{44}$$

Setting $f_1(d) = f_1(a,d)$,

$$f_2(c,d) = (F_B(P)(f_1)_d(c). \tag{45}$$

This completes the computation, since $f_2(c,d) = (F_B(P,Q)f)(c,d)$.

The assumption that f is G-invariant implies

$$f_1(ga, gb) = f_1(a,b)$$

$$f_2(gc, gd) = f_2(c,d).$$

Take any $a \in (Z/P)^3$, then

$$f_a(gb) = f_a(b),$$

for all $g \in Iso(a)$ which means that we can compute $F_B(Q)f_a$ by a $Iso(a)$-symmetrized routine. An analogous statement can be made for the computation of $F_B(P)(f_1)_d$.

Computations (42) and (44) state the invariance of $f_1(a,b)$ and $f_2(c,d)$ under the 'diagonal action' of G on the product space $(Z/P)^3 \times (Z/Q)^3$. This situation is similar to that encounterd in the preceeding section. It follows that the functions are completely determined by their values on any G-fundamental region in $(Z/P)^3 \times (Z/Q)^3$. Two types will be described. Let X be a G-fundamental region in $(Z/p)^3$ and Y a G-fundamental region in $(Z/q)^3$. For each $a \in X$, form the the subgroup

$Iso(a)$ and consider the action of $Iso(a)$ on $(Z/Q)^3$. Denote the corresponding fundamental region in $(Z/Q)^3$ by Y_a. Then

$$\mathcal{F}_1 = \cup_{a \in X}(\{a\} \times Y_a) \tag{46}$$

is a fundamental region for the diagonal action of G in $(Z/P)^3 \times (Z/Q)^3$. We can compute f_1 on $\mathcal{F}_1$ by computing, for each $a \in X$, $F_B(Q)f_a$ using an $Iso(a)$-symmetrized FFT algorithm.

In the same way,

$$\mathcal{F}_2 = \cup_{d \in Y}(X_d \times \{d\}) \tag{47}$$

is a G-fundamental region in $(Z/P)^3 \times (Z/Q)^3$, where X_d is a fundamental region in $(Z/P)^3$ corresponding to the action of $Iso(d)$. The function f_2 can be computed on $\mathcal{F}_2$ by computing $F_B(P)(f_1)_d$ by an $Iso(d)$-symmetric algorithm. The data transfer step where from the values of f_1 on $\mathcal{F}_1$ we write down the values of f_1 on $\mathcal{F}_2$ corresponds to the transposition step appearing in nonsymmetric FFT algorithms. As in the preceeding section, we call this the orbit exchange map.

5.3. Orbital Collapse

The orbital collapse method is based on Winograd's multiplicative FFT methods. The generic ring structure that can be placed on $(Z/N)^3$ as a direct product of rings can be used to extend Winograd's methods but in the orbital collapse approach, multiplicative structure is introduced, that is closely related to the group G of crystal symmetries. Relative to G, a group of matrix actions, the centralizer Γ of G is defined on the data set such that a G-fundamental region can be written as Γ-orbits. These Γ-orbits are linearly parametrized and 1-dimensional FFT's or cyclic convolutions are taken. In this discussion, however, we will only look at the simplest case.

We will use the notation of the previous sections. For a crystallographic group G, note that $u \in U_P$ commutes with every element of $g \in G$. For $n \in (Z/P)^3$,

$$gU_P(n) = U_P(gn). \tag{48}$$

Hence $g \in G$ maps a U_P-orbit onto a U_P-robit, and G acts on the orbit space of orbits of $(Z/P)^3$. In this way, G *collapses* U_p-orbits. This means that the symmetrized FFT matrix consists of blocks of sums of skew circulant matrices. Since sums of skew circulant matrices are again skew circulant, this property is preserved in introducing symmetry groups.

This work, first considered in [14] and in the case of 120 degree rotational symmetry in [15], has reached the point where code has been written to handle data admitting any space group symmetry on many data sizes [16]. Many prime size codes have been efficiently programmed and work is in progress to create composite size codes using orbital exchange. See also [17], where a special case of the approach was carried out in detail.

ACKNOWLEDGMENTS

This research was supported by the Advanced Research Projects Agency of the Department of Defense and was monitored by the Air Force Office of Scientific Research under Contract No. F49620-89-0-0020. The United States Government is authorized to reproduce and distribute reprints for governmental purposes notwithstanding any copyright notation hereon.

REFERENCES

1. H. Hauptman, J. Karle, 'Solution of the Phase Problem', 1953 ACA Monograph NO. 3, Pittsburg: Polycrystal Book Service.
2. H. Hauptman, J. Karle, 'Solution of the Phase Problem', 1956 Acta Cryst. 9, 45-55.
3. J. Karle, 'In Crystallographic Computing', p. 25, 1970 Copenhagen; Munksgaard.
4. H. Hauptman, J. Karle, 'Solution of the Phase Problem', 1957 Acta Cryst. 10, 515.
5. G. Bricogne, 'Acta Crystallogr' A32, 832-847.
6. G. Bricogne, 'Maximum Entropy Method', Acta Crystallography, 1984.
7. L. F. Ten Eyck, 'Crystallographic Fast Fourier Transforms', Acta. Cryst., 183–191, 1973.
8. J. Johnson, R. Johnson, D. Rodriguez, R. Tolimieri, 'A Methodology for Designing, Modifying, and Implementing Fourier Transform Algorithms on Various Architectures', 1989, Accepted for publication in Circuits, Signals and system.
9. R.E. Blahut, Fast Algorithms for Digital Signal Processing, Addison Wesley Publishing Company, 1984.
10. M. An, J. Cooley, R. Tolimieri, 'Factorization Method for Crystallographic FT', 1989, Accepted for publication in Advances in Applied Mathematics.

11. S. Winograd, 'Arithmetic Complexity of computations', CBMS-NSF Conference Series in Applied Math. V33, SIAM 1980.

12. M. An, L. Auslander M. Cook, 'Method of FT Computation for Monoclinic and Orthorhombic classes', Accepted for publication in Acta Cryst., 1989.

13. M. An, L. Auslander M. Cook, 'Programming Methods in Crystallography', to appear in the Proceedings of 1989 SPIE Conference.

14. L. Auslander and M. An, 'Fourier Transforms That Respect Crystallographic Symmetries'. IBM Journal of research and development, 31(2):213–223, March 1987.

15. M. An, R. Tolimieri, 'Notes on FT Computation for P3 Symmetry', Working paper, 1988. Preprint available at Center for Large Scale Computation.

16. R. Johnson, 'Evaluating FFT's That Respect Crystallographic Symmetries, II', Submitted for publication.

17. G. Bricogne, R. Tolimieri, 'Symmetrized FFT Algorithms', IMA Springer-Verlag, to appear.

A New Approach to Irregular Sampling of Band-Limited Functions

Karlheinz Gröchenig*

1 Introduction

This is a report on joint work in progress [10,11,12] on irregular sampling by H. G. Feichtinger, University of Vienna, and the author.

In the irregular sampling problem one is asked to reconstruct a band-limited function f from its values $f(x_n)$ at irregularly distributed points x_n. The case where the $x_n = \alpha n$ are equally spaced has been studied extensively and has created a huge literature on the classical Shannon-Whittacker-Kotel'nikov sampling theorem and its variations, cf. [4,16,15,21].

Irregular sampling, however, has been investigated in less detail. For a review of the available engineering literature we refer to [20]. Usually irregular sampling is treated in the context of the jitter error, e.g. [22]. By these methods only an approximation of the original function can be obtained.

Nevertheless is it known that "error-free recovery of signals from irregularly spaced samples" is possible [3] and can be easily obtained from gap and density theorems on Fourier series [19]. In the mathematical literature [1,2,18] one finds strong uniqueness theorems for band-limited functions. Using a heavy machinery of complex function theory general conditions on the sampling set $x_i, i \in I$, are given under which a band-limited function f is uniquely determined by the values $f(x_i)$. The question of how to reconstruct f from the $f(x_i)$ is not touched. This may be the reason why these results are hardly ever referred to in the engineering literature.

[0]Department of Mathematics U-9, University of Connecticut,
Storrs, Connecticut 06269-3009, USA
This work was supported by a faculty research grant of The University of Connecticut.

J. S. Byrnes and J. F. Byrnes (eds.), Recent Advances in Fourier Analysis and Its Applications, 251–260.

For practical purposes the mere uniqueness is insufficient. What is required is a constructive method to recover a band-limited function from its samples. Such methods can be obtained by means of perturbation arguments and lead to interesting results in non-harmonic Fourier series [6,17,24].

Since in the applications only a finite number of (irregular) sampling values is available, "bunched sampling" [21], Ch. 6, is often an adequate method for the reconstruction.

In most solutions mentioned of the irregular sampling problem certain limitations have to be accepted. The uniqueness theorems, though quite profound, are not constructive and thus rarely applicable. The reconstructions by means of perturbation theory and iteration are based on Hilbert space techniques. All of them lack locality, and consequently many samples are required to obtain a good approximation of a function at a given point, see [4] for similar problems with the cardinal series. In some instances [23] one even faces instabilities.

The situation is even worse for multivariate irregular sampling. Although this would be of interest in image processing, acoustics, geophysics, there are no multivariate versions of irregular sampling known that go beyond the obvious generalizations of the one-dimensional case.

In this article we present a new approach to irregular sampling. It is also based on a perturbation argument, but the starting point is now the observation that band-limited functions satisfy a reproducing formula, in contrast to the references mentioned which start with the classical Shannon-Whittacker sampling theorem. We will indicate how to derive an irregular sampling theorem that satisfies the following requirements.

1. The original function should be recovered from the irregularly sampled values by a constructive method.
2. The method should have good localization properties.
3. The reconstruction should be possible in stronger norms than just L^2, at least in weighted L^p-spaces.
4. Stability of the reconstruction under errors
5. Arbitrary dimension

2 Reviewing the Classical Sampling Theorem

Let us briefly take a glance at the usual proof of the classical sampling theorem and some of its variations. We assume that f is band-limited with band-width 1 and of finite energy, i.e. $\text{supp}\hat{f} \subseteq [-\pi, \pi]$ and $\int |f(t)|^2\, dt < \infty$.

A regular sampling theorem provides a reconstruction of f from its sampled values $f(\beta n), n \in Z$, in the form of a series

$$f(t) = \beta \sum_{n \in Z} f(\beta n)\, L_{\beta n} h \tag{1}$$

Here $L_y f(t) = f(t-y)$ denotes the translation by y, and h and β are to be chosen later. After taking the Fourier transform on both sides and applying Poisson's formula, (1) is equivalent to

$$\hat{f}(\omega) = (\beta \sum f(\beta n) e^{-2\pi i \beta n \omega}) \hat{h}(\omega) = \tag{2}$$

$$= (\sum \hat{f}(\omega + \frac{2\pi k}{\beta}))\, \hat{h}(\omega)$$

The easiest way to obtain such a representation is to assume that all translates of $\hat{f}$ have a support disjoint of supp $\hat{h}$, i.e. $\hat{f}(\omega + \frac{2\pi k}{\beta})\, \hat{h}(\omega) \equiv 0$ for $k \neq 0$. In this case (2) reduces to the reproducing formula $\hat{f} = \hat{f}\hat{h}$. Consequently *a reproducing formula $f = f * h$ will imply a sampling theorem.* This seems to be a fundamental principle and also occurs in other context of harmonic analysis, see [14]. A reproducing formula is easily obtained: if we take h to be band-limited and

$$\hat{h}(\omega) = 1 \text{ for } |\omega| \leq \pi \text{ and}$$

$$\hat{h}(\omega) = 0 \text{ for } |\omega| \geq \alpha,$$

we may choose β such that $\frac{2\pi}{\beta} \geq \alpha + \pi$. This yields the desired sampling theorem (1) with sampling density $\beta \leq 2\pi/(\alpha + \pi)$.

The special case $\hat{h}(\omega) = \chi_{[-\pi,\pi]}, h(t) = \frac{\sin \pi t}{\pi t}$ and $\beta = 1$ is the famous sampling theorem due to Shannon, Whittacker, Kotel'nikov and many others. The reconstruction of f is by means of the cardinal series

$$f(t) = \sum_{n \in Z} f(n)\, \frac{\sin \pi(t-n)}{\pi(t-n)}. \tag{3}$$

The sampling rate $\beta = 1$, the so-called Nyquist rate, is actually the minimal rate required for a complete reconstruction. If $\beta > 1$, then other terms $\hat{f}(\omega + \frac{2\pi}{\beta})\hat{h}(\omega)$ will necessarily contribute in (2). Consequently $\hat{f}$ cannot be isolated and a reconstruction is impossible in this case.

Note that $\frac{\sin \pi t}{\pi t} = O(\frac{1}{t})$ is not absolutely integrable, therefore the cardinal series cannot converge in the L^1-norm or in certain weighted L^p-norms. This can be avoided if supp$\hat{h} \subseteq [-\alpha, \alpha], \alpha > \pi$. Then $h \in L^1$ or even in the Schwartz class is possible. Thus the reconstruction (1) displays much better convergence properties, however at the price of oversampling by a rate $\beta \leq \frac{2\pi}{\alpha+\pi} < 1$, cf. [4], Thm. 4.3.

3 A Change of Perspective

It is clear that in irregular sampling Poisson's formula is not available and that a different approach is required. Section 2 gives some hints where to start. The essential part of the derivation was a reproducing formula $f = f*h$ for band-limited functions f. The reconstruction on the right side of (1) can be considered as a Riemann sum for the integral $\int f(t)\, L_t h\, dt = f * h = f$. In this description of (1) Poisson's formula does not play a part.

From this point of view it is clear that other types of Riemann sums could be used, especially sums that involve the irregularly sampled values of f! We can no longer expect that such general Riemann sums equal the convolution as in the regular case, but at least they should be good approximations of the integral. In such a situation, it is tempting to form a remainder term and apply the same argument once more. If we end up in the right spaces and if all estimates fit together, then this should lead to a convergent iteration scheme which recovers f from its sampled values. To carry out this idea, two steps have to be investigated.

STEP I (Approximation of Convolutions): We have to find good and appropriate approximations to the convolution C_h on spaces of band-limited functions.

STEP II (Iteration): We have to set up an iteration procedure and prove its convergence.

Similar ideas have been used for some time in other areas of harmonic analysis to obtain simple decompositions of functions, so-called wavelet and Gabor type expansions, see [5,13,8,9,14]. As an approach to irregular sampling theorems it seems to be new.

4 Approximation of Convolutions

To be precise from now on, we shall work with the weighted L^p-spaces $L^p_\alpha(R^n) = \{f \in L^1_{loc} : \int_{R^n} |f(t)|^p (1+|t|)^{\alpha p}\, dt := \|f\|^p_{p,\alpha} < \infty\}$, where $1 \le p < \infty$ and $\alpha \ge 0$. We shall frequently make use of the convolution inequality

$$\|f * h\|_{p,\alpha} \le \|f\|_{p,\alpha} \|h\|_{1,\alpha} \tag{4}$$

or $L^p_\alpha * L^1_\alpha \subseteq L^p_\alpha$ respectively.

Working in several dimensions, we consider spaces of band-limited functions with fixed spectrum in a compact set Ω:

$$B^p_\alpha(\Omega) = \{f \in L^p_\alpha,\ \operatorname{supp} \hat{f} \subseteq \Omega\}$$

Note that $f \in B^p_\alpha(\Omega)$ automatically implies some decay on f:

$$|f(x)| \le C(1+|x|)^{-\alpha}. \tag{5}$$

Our goal is to derive an irregular sampling theorem for $B^p_\alpha(Q)$, the band-limited functions in L^p_α with spectrum in the cube $Q = [-\pi, \pi]^n$. The case of functions with general compact spectrum can easily be reduced to this case by applying (possibly inhomogeneous) dilations.

For the following, h will be a fixed band-limited function in L^1_α, such that $\hat{h}(\omega) = 1$ for $\omega \in Q$, $\mathrm{supp}\hat{h} \subseteq \Omega \supseteq Q$. Then the reproducing formula

$$f = f * h \tag{6}$$

holds for $f \in B^p_\alpha(Q)$. Since (4) is available, the convolution operator $C_h f = f * h$ is bounded from L^p_α into $B^p_\alpha(\Omega)$ and acts as the identity on $B^p_\alpha(Q)$.

We will consider sampling sets $X = (x_i)_{i \in I} \subseteq R^n$ which are δ-dense, i.e. $\bigcup_{i \in I} B_\delta(x_i) = R^n$ and separated, i.e. $|x_i - x_j| \geq \delta_0 > 0$ for all $i \neq j$ for some δ_0. Given X, we take any partition of unity $\Psi = (\psi_i)$ associated to X with the following properties:
(1) supp $\psi_i \subseteq B_\delta(x_i)$, (2) $0 \leq \psi_i \leq 1$ measurable, and (3) $\sum_{i \in I} \psi_i \equiv 1$.

In R^n one could take the characteristic functions of the Voronoi regions $V_i := \{x \in R^n : |x - x_i| \leq |x - x_j| \ \forall j \neq i\}$.

Then the following approximations of the convolution C_h can be considered to replace the regular Riemann sum (1). They depend on the sampling set X and involve only the irregularly sampled function f.

$$Af := (\sum_{i \in I} f(x_i)\psi_i) * h \tag{7}$$

or

$$A_1 f := \sum_{i \in I} f(x_i)(\int \psi_i)\, L_{x_i} h \tag{8}$$

In the sequel we shall use the approximation operator A, because the arguments are slightly simpler than for A_1, but it should be understood that similar conclusions also hold for A_1, see [11]. For numerical purposes A_1 is preferable, because it avoids convolutions and uses translations instead.

It is easy to see that A is well-defined on $B^p_\alpha(\Omega_0)$ for any compact spectrum Ω_0, see [10] for the general case. If $f \in B^p_\alpha$, then $\sum_{i \in I} |f(x_i)|^p\,(1+|x_i|)^{p\alpha} < \infty$ and thus $\sum_i f(x_i)\psi_i \in L^p_\alpha$. Because supp $(Af)\hat{} \subseteq$ supp $\hat{h} \subseteq \Omega$ and 4, A maps any $B^p_\alpha(\Omega_0)$ into $B^p_\alpha(\Omega)$.

In the following proposition it is shown how good an approximation of C_h the operator A is.

Proposition 1 *Given $h \in B^1_\alpha(\Omega)$, there exists a minimal sampling density $\delta_0 > 0$, such that on $B^p_\alpha(\Omega)$*

$$|||C_h - A||| < 1 \tag{9}$$

holds for <u>*any*</u> *δ_0-dense and separated set $X = (x_i)$, i.e.*

$$||f * h - Af||_{p,\alpha} \leq C||f||_{p,\alpha} \tag{10}$$

holds for all $f \in B^p_\alpha(\Omega)$ with a constant $C < 1$.

Sketch of the Proof: We introduce the δ *-oscillation* of f:

$$\operatorname{osc}_\delta f(t) := \sup_{u \in B_\delta(t)} |f(t+u) - f(t)| \tag{11}$$

With this the pointwise estimate

$$|f * h(t) - Af(t)| = |\int \sum_{i \in I} (f(u) - f(x_i))\, \psi_i(u)\, L_u h(t)\, du|$$

$$\leq \int \sum_i \operatorname{osc}_\delta f(u)\, \psi_i(u) L_u |h|(t)\, du = \operatorname{osc}_\delta f * |h|$$

is easily obtained. To get an estimate for $\operatorname{osc}_\delta f$, we write $f = f * p$ for some reproducing function $p \in L^1_\alpha, \hat{p}(\omega) \equiv 1$ on Ω. Then it follows from (11) that

$$\operatorname{osc}_\delta f = \operatorname{osc}_\delta (f * p) \leq |f| * \operatorname{osc}_\delta p. \tag{12}$$

Consequently

$$||f * h - Af||_{p,\alpha} \leq ||\, |f| * \operatorname{osc}_\delta p * |h|\, ||_{p,\alpha} \leq ||f||_{p,\alpha}\, ||\operatorname{osc}_\delta p||_{1,\alpha}\, ||h||_{1,\alpha}$$

Since $||\operatorname{osc}_\delta p||_{1,\alpha} \to 0$ as $\delta \to 0$ for appropriate p, the proposition is proved.

Remark: By the mean value theorem $||\operatorname{osc}_\delta p||_{1,\alpha} = O(\delta)$, if p is band-limited and δ can be maximized by a suitable choice of p. More precisely, δ is determined by the condition

$$\inf_{p:\hat{p}=1 \text{ on } \Omega} ||\operatorname{osc}_\delta p * |h|\, ||_{1,\alpha} = \inf_p ||\operatorname{osc}_\delta p||_{1,\alpha} ||h||_{1,\alpha} < 1.$$

This leads to some classical questions of Fourier analysis, such as the determination of local Fourier norms.

5 The Iteration

Once we have found good approximations to the convolution operator C_h, we would like to start an iteration. This is not entirely trivial, because starting with $f \in B^p_\alpha(Q)$ already in the first step we leave this space and end up in the larger space $B^p_\alpha(\Omega)$. This can be corrected by using another auxiliary function, see [7]. Instead of that rather tricky argument we will present an abstract argument which makes the iteration entirely transparent. It leads to the factorization of a convolution operator and might be of independent interest.

Proposition 2 *Let $g, h \in B^p_\alpha(\Omega)$ be given and assume that $h * g = g$ ($\Longleftrightarrow$ $C_h C_g = C_g$). Let A be a good approximation of C_h on $B^p_\alpha(\Omega)$ in the sense that*

$$|||C_h - A||| < 1 \tag{13}$$

Then the operator $D = \sum_{n=1}^{\infty}(C_h - A)^n$ is well-defined on $B^p_\alpha(\Omega)$ and is factorized over A as follows

$$C_g = D\,A\,C_g \tag{14}$$

Proof:Denote by $R = C_h - A$ the remainder term. By the assumptions $|||R^{n+1}||| \to 0$ as $n \to \infty$, and the geometric series is therefore well-defined on $B^p_\alpha(\Omega)$.

The factorization follows from

$$C_g = C_h C_g \quad \text{and} \quad C_h\, C_g = RC_g + AC_g \tag{15}$$

by induction.

$$C_h\, C_g = RC_g + AC_g = RC_h\, C_g + AC_g =$$

$$R(RC_g + AC_g) + AC_g = R^2 C_h\, C_g + R^1 AC_g + R^0 AC_g =$$

$$= \ldots = R^{n+1}C_g + (\sum_{k=0}^{n} R^k)AC_g$$

Letting $n \to \infty$, the factorization $C_g = DAC_g$ is obtained.

6 The Irregular Sampling Theorem

We have seen in Section 4 that good approximation operators A are easily constructed. Thus for appropriate choices of g, h the irregular sampling theorem is contained in (14). For this we chose g such that $\hat{g}(\omega) = 1$ for $\omega \in Q$ and $\hat{h}(\omega) = 1$ for $\omega \in \operatorname{supp} \hat{g}$ and $h \in B^p_\alpha(\Omega)$, where $Q \subseteq \operatorname{supp} \hat{g} \subseteq \Omega$.

If the spectrum of f is contained in the cube Q, $f \in B^p_\alpha(Q)$, then these properties and (14) imply

$$f = f * g = C_g f = DAC_g f = DAf =$$

$$= D(\sum_{i \in I} f(x_i)\, \psi_i * h) = \sum_{i \in I} f(x_i) D(\psi_i * h) \tag{16}$$

Note that the precise form of g is not important as long as $f * g = f$, because it drops out in (16) for $f \in B^p_\alpha(Q)$. Denoting the expanding functions

$$e_i = D(\psi_i * h) \tag{17}$$

we have proved the following

Theorem 3 *For $Q \subseteq R^n$ compact and $h \in B^p_\alpha(\Omega)$, $\Omega \supseteq Q, \hat{h}(\omega) = 1$ for $\omega \in Q$, there exists a density $\delta > 0$, depending only on Q and h, not on p, such that any $f \in B^p_\alpha(Q)$, $1 \leq p < \infty$, $\alpha \geq 0$ can be reconstructed from <u>any</u> δ-dense and separated sampling set $X = (x_i)_{i \in I}$ as a series*

$$f(t) = \sum_{i \in I} f(x_i) e_i \tag{18}$$

where the series converges in L^p_α.

Remarks: (1) It can be shown that $e_i = D(\psi_i * h) \in B^1_\alpha(\Omega)$ and consequently $e_i(t) = O((1+|t|)^{-\alpha})$. This implies better localization and convergence properties of the reconstruction (16) as compared to the cardinal series (3) even for irregular sampling. As a consequence of the favorable localization properties one may use very general norms in (16), cf. [10,11]

(2) The factorization (14) implies that $f \in B^p_\alpha(Q)$ can be recovered from the sampled values $f(x_i)$ by the following iterative procedure: starting with $f(x_i)$, set

$$\phi_0 = \sum_{i \in I} f(x_i)\, \psi_i * h \tag{19}$$

the first approximation of f. The n -th correction term is defined inductively by

$$\phi_n = \phi_{n-1} * h - \sum_{i \in I} \phi_{n-1}(x_i) \psi_i * h. \tag{20}$$

Then finally

$$f = \sum_{n=0}^{\infty} \phi_n \tag{21}$$

The iteration steps are easier if the approximation operator A_1 of (8) is used. Then only pointwise evaluations of ϕ_n and shifts of h have to be used for a reconstruction of f.

(3) The reconstruction (19),(20) and (21) is stable with respect to the usual input errors such as round off errors, jitter errors, truncation errors and aliasing errors. While this is not at all clear from the iteration method, the factorization (14) gives us a convenient tool to carry out an error analysis for the irregular sampling theorem. In contrast to usual error estimates, which are always in the L^∞-norm, from (14) can also be obtained L^p_α -estimates [12].

(4) Utilizing other types of approximation operators for C_h, one may obtain quite different theorems on band-limited functions. For instance, if instead of A the operators $B = \sum_{i \in I}(\int f(t)\psi_i(t)dt) L_{x_i} h$ are used, then by arguments similar to those discussed in Sections 4 and 5 one obtains series expansions of the form

$$f = \sum_{i \in I} \lambda_i L_{x_i} g$$

for $f \in B^p_\alpha(Q)$ with coefficients λ_i depending linearly on f and $\sum_{i\in I} |\lambda_i|^p(1+|x_i|)^\alpha < \infty$ and equivalent to f. For a detailed discussion of series expansion of band-limited functions related to irregular sampling we refer to [10,11].

References

[1] A. Beurling: Local harmonic analysis with some applications to differential operators. In "Some Recent Advances in the Basic Sciences", p. 109–125, Belfer Grad. School of Science, Annual Science Conference Proc., A. Gelbart, ed. Vol. I, 1962–64.

[2] A. Beurling, P. Malliavin: On the closure of characters and the zeros of entire functions. Acta Math. **118** (1967), 79–95.

[3] F.J.Beutler: Error-free recovery of signals from irregularly spaced samples. SIAM Review **8** (1966), 328–335.

[4] P.L.Butzer, W. Splettstößer, R.Stens: The sampling theorem and linear prediction in signal analysis. Jber.d. Dt. Math.-Verein. **90** (1987), 1–70.

[5] R.R. Coifman and R. Rochberg Representation theorems for holomorphic and harmonic functions in L^p, Asterisque 77 (1980),11-66.

[6] R. Duffin, A. Schaeffer: A class of nonharmonic Fourier series. Trans. Amer. Math. Soc. **72**(1952), 341–366.

[7] H.G. Feichtinger: Discretization of convolution and reconstruction of band–limited functions. (To appear in J. Approx. Th.)

[8] H.G. Feichtinger, K. Gröchenig: A unified approach to atomic decompositions via integrable group representations. Proc. Conf. Lund 1986, "Function Spaces and Applications", Lecture Notes in Math. **1302** (1988), 52–73.

[9] H.G. Feichtinger, K. Gröchenig: Banach spaces related to integrable group representations and their atomic decompositions I. (To appear in J. Functional Anal.)

[10] H.G. Feichtinger, K.Gröchenig: Reconstruction of Band-Limited Functions from Irregular Sampling Values. (Preprint)

[11] H.G. Feichtinger, K. Gröchenig: Irregular sampling theorems and series expansions for band-limited functions. (Preprint)

[12] H.G. Feichtinger, K. Gröchenig: Error analysis in irregular sampling theory. (in preparation)

[13] M. Frazier, B. Jawerth: The ϕ-transform and applications to distribution spaces. Proc. Conf. Lund 1986, "Function Spaces and Applications", Lecture Notes in Math. 1302 (1988), 223–246.

[14] K.Gröchenig. Describing functions: atomic decompositions versus frames. (Submitted)

[15] J.R. Higgins: Five short stories about the cardinal series. Bull. Amer. Math. Soc. **12** (1985), 45–89.

[16] A.J. Jerri: The Shannon sampling theorem — its various extensions and applications. A tutorial review. Proc. IEEE **65** (1977), 1565–1596.

[17] S. Jaffard: A density criterium for frames of complex exponentials. Thèse de Doctorat de l'Ecole Polytechnique (1989).

[18] H. Landau: Necessary density conditions for sampling and interpolation of certain entire functions. Acta Math. **117** (1967), 37–52.

[19] N. Levinson: Gap and Density Theorems. Coll. Publ. 26, Amer. Math. Soc., New York, 1940.

[20] F.A. Marvasti: A unified approach to zero-crossing and nonuniform sampling of single and multidimensional systems. Nonuniform, P.O.Box 1505, Oak Park, IL 60304. (1987)

[21] A. Papoulis: Signal Analysis. McGraw-Hill. New York. 1977.

[22] A. Papoulis: Error analysis in sampling theory. Proc. IEEE **54/7** (1966), 957–955.

[23] W. Splettspößer: Unregelmäßige Abtastungen determinierter und zufälliger Signale. In Kolloquium DFG-Schwerpunktprogramm Digitale Signalverarbeitung , H.G. Zimmer, V. Neuhoff, Eds., Göttingen, 1981, p. 1–4.

[24] R. Young: An Introduction to Nonharmonic Fourier Series. Academic Press, New York. 1988.

AUTO–CROSS CORRELATION FUNCTIONS OF THE EVEN AND THE ODD PARTS OF FOURIER SERIES

Önder TÜZÜNALP

SUMMARY: *In the present work, the auto and cross correlation functions of the even and the odd parts of simple and complex Fourier series are computed and consequent theorems with relative properties are given.*

Such correlation functions are applied to some characteristic functions, in order to give some insight into the resulting correlograms.

The work concludes by the implementation of such correlograms by using AEON parallel array processor.

FOURİER SERİLERİNİN TEK VE ÇİFT KISIMLARININ OTO VE ÇAPRAZ KORRELASYON FONKSİYONLARI

ÖZET: *Basit ve kompleks Fourier serilerinin tek ve çift bileşenlerinin oto ve kros korrelasyon fonksiyonları ile ilgili teoremler sonuçlanan özelliklerin incelenmesi ile beraber sunulmuştur.*

Adı geçen korrelasyon fonksiyonları bazı karakteristik fonksiyonlara uygulanmış, sonuçlanan korrelogramların ODTÜ–Fizik Bölümünde tasarımı yapılmış, AEON paralel dizi prosesörü ile gerçekleştirilmeleri verilmiştir.

— * —

INTRODUCTION

In reference (1), it is shown that a time function can be split into its even and odd components by virtue of a new real–time parallel porcessing method, based upon the parametric time delaying and time reversal, experimentally implemented by means of the AEON processor, (3).

In this presentation, the auto and cross correlation functions of the even and the odd parts are computed for simple and complex Fourier series and the resulting theorems are given together with some typical correlogram applications.

A realized implementation scheme is also given for the real time computations of the correlograms.

Geliş Tarihi: 14/7/1985

(*) *University of Ankara — Physics Engineering Department, Ankara*

J. S. Byrnes and J. F. Byrnes (eds.), Recent Advances in Fourier Analysis and Its Applications, 261–272.

THEORETICAL DEVELOPMENT

Given simple Fourier series expansion of a periodic time function $(f(t))$, obeying the Dirichlet's conditions, as,

$$f(t) = \frac{1}{2} q_0 + \sum_{n=1}^{\infty} (a_n \cos \frac{2\pi}{T} nt + b_n \sin \frac{2\pi}{T} nt) \tag{1}$$

where the (a_0) term represents the static and the expression under the summation operator corresponds to the fluctuating compenents respectively. Then the even and the odd parts of above series can be separated as follows:

$$f_e(t) = \frac{1}{2} a_0 + \sum_{n=1}^{\infty} a_n \cos \frac{2\pi}{T} nt.$$

$$f_o(t) = \sum_{n=1}^{\infty} b_n \sin \frac{2\pi}{T} nt.$$

The reflection of $(f(t))$ is then obtained by inverting the sign of the odd part, i.e., the phase inversion:

$$f(-T) = \frac{1}{2} a_0 + \sum_{n=1}^{\infty} a_n \cos \frac{2\pi}{T} nt - b_n \sin \frac{2\pi}{T} nt.$$

$$\theta_n = \text{arc tg} \frac{b_n}{a_n}$$

then by definition,

$$f_e(t) = \frac{1}{2} [f(t) + f(-t)] \tag{6}$$

$$f_o(t) = \frac{1}{2} [f(t) - f(-t)] \tag{7}$$

The Fourier Coefficients:

$$a_n = \frac{2}{T} \int_{-T/2}^{T/2} f(t) \cos \frac{2\pi}{T} nt = 0 \tag{8}$$

for a fully odd function and,

$$b_n = \frac{2}{T} \int_{-T/2}^{T/2} f(t) \sin \frac{2\pi}{T} nt = 0 \tag{9}$$

for a fully even function. Note that, by the change of origin certain functions can be converted from fully odd into fully even and the series assume then special forms.

Time Shift with Respect to Time Origin:

Now, introduce a negative shift (ν) to $(f(-t))$ with respect to origin:

$$f(-t + \nu) = \frac{1}{2} a_0 + \sum_{n=1}^{\infty} a_n \cos \frac{2\pi}{T} n(t + \nu) - b_n \sin \frac{2\pi}{T} n(t + \nu) \tag{10}$$

then $f(-t+\nu)$ can be redefined by the new coefficients:

$$f(-t+\nu) = \frac{1}{2}a_0 + \sum_{n=1}^{\infty} a'_n \cos\frac{2\pi}{T}nt + b'_n \sin\frac{2\pi}{T}nt. \tag{11}$$

taking the mirror symmetry (reflection) of (11) with respect to origin:

$$= \frac{1}{2}a_0 + \sum_{n=1}^{\infty} a'_n \cos\frac{2\pi}{T}nt + b'_n \sin\frac{2\pi}{T}nt. \tag{12}$$

is obtained.

Similarly, introducing an equal amount of time shift to (f(t)) in the positive direction:

$$f(t-\nu) = \frac{1}{2}a_0 + \sum_{n=1}^{\infty} a_n \cos\frac{2\pi}{T}n(t-\nu) + b_n \sin\frac{2\pi}{T}n(t-\nu) \tag{13}$$

and redefining new coefficients:

$$f(t-\nu) = \frac{1}{2}a_0 + \sum_{n=1}^{\infty} a''_n \cos\frac{2\pi}{T}nt + b''_n \sin\frac{2\pi}{T}nt. \tag{14}$$

and comparing the new coefficients of (12) and (14), the following identities are obtained:

$$a'_n = a''_n \tag{15}$$

$$b'_n = b''_n \tag{16}$$

Therefore, the reflection of the shifted function (12) with respect to the time origin is identical with ($f(t-\nu)$ of equation (13), See (Fig. 1).

THEOREM I.

If (f(−t)) is shifted in time by an amount (ν) with respect to (f(t)), then (f(t)) shifts from the origin by the same amount, in the opposite direction of time.

By means of this theorem, the shift of (f(t)) from the origin is ensured in order to get both (a_n) and (b_n) coefficients, irrespective of the fact that (f(t)) is fully even or fully odd. This fact, which escaped from the eyes up to now, to our knowledge, will later be shown to be of prime interest from experimental point of views.

Autocorrelation Functions for the Even and the Odd Components of Fourier Series:

For a periodic real function (f(t)) the general expression for the autocorrelation function is

$$\varphi_{11}(z) = \frac{1}{2}\int_{-T/2}^{T/2} f(t)\,.\,f(t+z)dt. \tag{17}$$

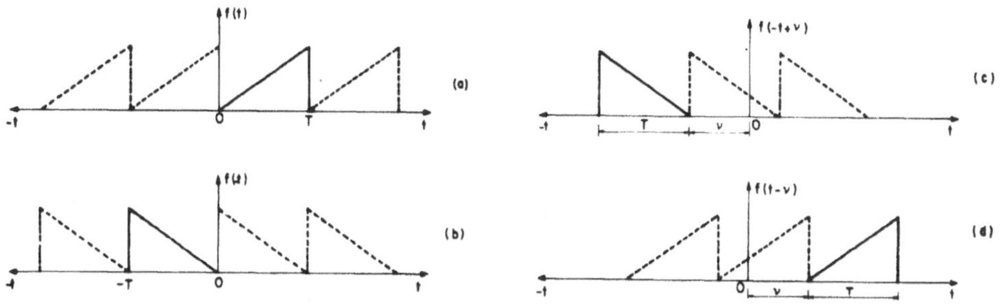

Figure 1. Graphical interpretation of theorem I for a ramp function.

then as the even and the odd parts of the Fourier series are separately available, one can prove, the autocorrelation functions ($\varphi_{ee}(z)$) and ($\varphi_{oo}(z)$) for both parts:

$$\varphi_{ee}(z) = \frac{1}{2} \int_{-T/2}^{T/2} \frac{a_0}{2} + [\sum_{n=1}^{\infty} a^*_n \cos \frac{2\pi}{T} nt] [\sum_{m=1}^{\infty} a^*_m \cos \frac{2\pi}{T} m(t+z)] dt. \tag{18}$$

$$\varphi_{oo}(z) = \frac{1}{2} \int_{-T/z}^{T/z} [\sum_{n=1}^{\infty} b^*_n \sin \frac{2\pi}{T} nt] [\sum_{m=1}^{\infty} b^*_m \sin \frac{2\pi}{T} m(t+z)] dt. \tag{19}$$

the solutions of above integrals lead to the following theorem:

THEOREM II.

The autocorrelation functions of the even and the odd parts of simple Fourier series are cosine series associated with the squares of respective Fourier coefficients as shown as

II.a. $$\varphi_{ee}(z) = \frac{a_0^2}{4} + \frac{1}{2} \sum_{n=1}^{\infty} a_n^{*2} . \cos \frac{2\pi}{T} nz \tag{20}$$

II.b. $$\varphi_{oo}(z) = \frac{1}{2} \sum_{n=1}^{\infty} b_n^{*2} . \cos \frac{2\pi}{T} nz \tag{21}$$

These new correlation series are uniformly and quickly convergent, term by term differentiable and integrable as with the same mathematical rigor of the original Fourier series. In addition, they include the following main properties:

Both ($\varphi_{ee}(z)$) and ($\varphi_{oo}(z)$) are even functions of (z) with even symmetry with respect to (z) origin.

($\varphi_{oo}(z) = \varphi_{oo}(-z)$), ($\varphi_{ee}(z) = \varphi_{ee}(-z)$). for all ($z$), than is both have maximae at ($z = 0$).

($\varphi_{ee}(0), \varphi_{oo}(0) \geqslant | \varphi_{ee}(z), \varphi_{oo}(z) |$) for all ($z$), than is both have maximae at $z = 0$

For fully even functions ($b_n = 0$), ($\varphi_{oo}(z) = 0$) and for fully odd functions ($a_n = 0$), ($\varphi_{ee}(z) = a_0^2/4$), fluctuating part of ($\varphi_{ee}(z) = 0$).

$$(\varphi_{ee}(z) + \frac{a_0^2}{4} + \frac{1}{2} \sum_{n=1}^{\infty} (a_n^{*2} + b_n^{*2}) \cos \frac{2\pi}{T} nz = \varphi_{11}(z)).$$

This sum connects with the Parcevals' theorem for ($z = 0$) as,

$$\varphi_{11}(z) = \frac{a_0^2}{4} + \frac{1}{2} \sum_{n=1}^{\infty} (a_n^{*2} + b_n^{*2}) = \frac{1}{T} \int_{-T/2}^{T/2} [f(t)]^2 dt.$$

expressing the maximum power. If ($z \neq 0$) than, it expresses socalled "partialized powers" of the even and odd parts by ($\cos \frac{2\pi}{T} nz$).

Cross Correlation Function of the Even and the Odd Components of Fourier Series:

The general expression for the cross correlation of a periodic real function of time is:

$$\varphi_{12}(z) = \frac{1}{T} \int_{-T/z}^{T/z} f_1(t) . f_2(t+z) dt \tag{22}$$

and,

$$\varphi_{eo}(z) = \frac{1}{T} \int_{-T/2}^{T/2} f_e(t) \,.\, f_o(t + z)dt. \tag{23}$$

is evaluated. In this computation, the cross terms are zero due to orthogonality and the result forms a sine series, associated with both Fourier coefficients.

THEOREM III.

The cross correlation of the even and the odd parts of simple Fourier Series is a sine series associated with the product of the even and the odd Fourier coefficients:

$$\varphi_{eo}(z) = \frac{1}{a} \sum_{n=1}^{\infty} a^*{}_n \, b^*{}_n \sin \frac{2\pi}{T} nz \tag{24}$$

These new cross correlation series are uniformly and quickly convergent, term by term differentiable and integrable in the same mathematical rigor of the original Fourier Series. They further include the following outstanding properties:

They are odd functions with odd symmetric with respect to the origin of (z), $(\varphi_{ee}(z) = -\varphi_{eo}(-z))$.

They cancel (a_0), the static part of the Fourier Series and respond only to fluctuating components of the Fourier Series.

For fully even and fully odd functions, that is, $(a_n = 0; b_n = 0)$, $(\varphi_{eo}(z) = 0)$.

For $(z = 0)$, $(\varphi_{eo}(z) = 0)$.

Interchanging the sense of cross correlation: $(\varphi_{eo}(z) = -\varphi_{oe}(z))$, $(\varphi_{eo}(-z) = \varphi_{oe}(z))$.

They include the phase information inherently with the Fourier coefficients:

$$\varphi_{eo}(z) = -\frac{1}{2} \sum_{n=1}^{\infty} a_n^{*2} \operatorname{tg} \theta_n \sin \frac{2\pi}{T} nz = -\frac{1}{2} \sum_{n=1}^{\infty} b_n^{*2} \frac{1}{\operatorname{tg}\theta_n} \sin \frac{2\pi}{T} nz$$

$(\varphi_{eo}(z))$ contains only those frequencies common to both the even and the odd components of $(f(t))$.

$(\varphi_{eo}(z))$ have the same generality with the convolution: $(f_et)*(f_o(t))$ but, reversed in time (retrocorrelation).

$(\varphi_{eo}(z))$ of an $(f(t))$ is equivalent to passage through a filter having an amplitude-frequency response of the same form as the amplitude spectrum of the original function and a phase–frequency response of the same form of the phase spectrum of the original function but reversed is sign.

$$(|\varphi_{eo}(z)|^2 \leqslant \varphi_{ee}(0) \,.\, \varphi_{oo}(0))$$

$$(|\varphi_{eo}(z)| \leqslant \frac{1}{2} [\varphi_{ee}(0) + \varphi_{oo}(0)])$$

$(\varphi_{eo}(z))$ is equal to the sum of the areas of (n) number of triangles, having sides of (a_n, b_n) and the angle between them (nz), therefore, each harmonic (n) is represented by the area of such a triangle.

From the sum of $(\varphi_{ee}(z) \pm \varphi_{eo}(z))$, the original function $(f(t))$ can easily be rocevered performing a successive series expansion over the sum.

Generalization to Correlation of Complex Fourier Series:

Representing $(f(t))$ by complex Fourier Series:

$$f(t) = \sum_{n=-\infty}^{\infty} C_n \,.\, e^{j \frac{2\pi}{T} nt} \tag{25}$$

where the complex Fourier coefficients are:

$$C_n = \frac{1}{2} \int_{-T/2}^{T/2} f(t) \, . \, e^{-j \frac{2\pi}{T} nt.} \tag{26}$$

Now, defining again $(f(t))$ as:

$$f(t) = f_e(t) + f_0(t) \tag{27}$$

where this time, $(f_e(t))$ and $(f_0(t))$ are in general complex, in the sense of an analytic signal:

$$f_c(t) = \underbrace{\text{Re } f_e(t) + J \text{ Im } f_e(t)}_{\text{even part}} + \underbrace{\text{Re } f_0(t) + J \text{ Im } f_0(t)}_{\text{odd part}} \tag{28}$$

Then, the even and the odd parts $(f_c(t))$:

$$f_{ce}(t) = \sum_{n=-\infty}^{\infty} C'_n \, . \cos \frac{2\pi}{T} nt; \; f_{co}(t) = J \sum_{n=-\infty}^{\infty} C'_n \sin \frac{2\pi}{T} nt. \tag{29}$$

and calculating the autocorrelation function of the even part:

$$\varphi_{cee}(z) = \frac{1}{T} \int_{-T/2}^{T/2} [C'_n \, . \cos \frac{2\pi}{T} nt] \, [\sum_{m=-\infty}^{\infty} C'_m \sin \frac{2\pi}{T} n(t+z)]dt. \tag{30}$$

and that of corresponding to the odd part is:

$$\varphi_{coo}(z) = -\frac{1}{T} \int_{-T/2}^{T/2} [\sum_{n=-\infty}^{\infty} C'_n \sin \frac{2\pi}{T} nt] \, [\sum_{m=-\infty}^{\infty} C'_m \sin \frac{2\pi}{T} m(n+z)]dt. \tag{31}$$

Then again by orthogonality, $(n = m)$ in the solutions of integrals (30, 31) and the following theorems can be proved.

THEOREM V.

The complex autocorrelation function of the even and odd parts of a complex time function $(f_c(t))$ forms cosine series, respectively, as follow:

$$\varphi_{cee}(z) = \frac{1}{2} \sum_{n=-\infty}^{\infty} C'^2_n \cos \frac{2\pi}{T} nz \tag{32}$$

and,

$$\varphi_{coo}(z) = \frac{1}{2} \sum_{n=-\infty}^{\infty} C'^2_n \cos \frac{2\pi}{T} nz \tag{33}$$

Now, the complex correlation between the even and the odd parts:

$$\varphi_{ceo}(z) = \frac{1}{T} \int_{-T/2}^{T/2} [\sum_{n=-\infty}^{\infty} C'_n \cos \frac{2\pi}{T} nt] \, [\sum_{m=-\infty}^{\infty} C'_m \sin \frac{2\pi}{T} m(t+z)]dt. \tag{34}$$

and again letting $(n = m)$ by orthogonality, the following theorem can be formulated:

THEOREM VI

The complex cross correlation function between the even and the odd parts of complex Fourier Series is a sine series, such that:

$$\varphi_{ceo}(z) = \frac{j}{2} \sum_{n=-\infty}^{\infty} c_n^2 \sin \frac{2\pi}{T} nz \tag{35}$$

where the complex coefficients and phases are related to the simple series by:

$$c_n = \frac{1}{2} \sqrt{a_n^2 + b_n^2} \,.\, e^{-j\phi_n} \tag{36}$$

and,

$$\phi_n = \text{arc tg} \frac{-b_n^*}{a_n^*} \tag{37}$$

As far as the properties of complex correlation functions, in addition to all properties valid for simple series correlation functions, one can say that, they are cosine and sine series associated with complex Fourier coefficients.

The special kinds of symmetry properties possessed by $(f(t))$ play important roles namely such as Hermitian functions. In this respect the real and imaginary parts of $(f_c(t))$ correspond to the even and odd parts and therefore, several combinations of particular correlation functions can be defined.

APPLICATIONS

In this section, the theorems introduced in the theoretical development part will be applied to some characteristic real time functions namely chosen as a sine–wave, symmetric and non–symmetric square waves and a narrow band pseudo–random periodic wave, all parametrically shifted from the time origin.

The properties of the resulting correlation functions will briefly be examined in order to give some insight for their interpretations. For convenience, in the following calculations the period is taken at $(T = 2\pi)$, i.e, $(\omega = 1)$.

Sine Wave

Taking a sine wave shifted from the origin, in the negative direction of time, by an amount (ν), (see Fig. 2) as:

$$f(t + \nu) = \sin \left(\frac{2\pi}{T} t + \nu\right) \tag{38}$$

The Fourier coefficients for such a function are:

$$a_1^* = \sin \nu, \qquad b_1^* = \cos \nu \tag{39}$$

and with linearly changing phases with (ν):

$$\theta_1 = k \frac{\pi}{2} \nu; \; k = 1, 3, 5, 7, \ldots \tag{40}$$

then the cross correlation function of the even and the odd parts are computed according to Theorem III. as:

$$\varphi_{eo}(z) = \frac{1}{2} \sin z\nu \,.\, \sin z \tag{41}$$

This function is plotted in (Fig. 3) as a function of (z) and (ν). This figure shows that the result is again a sine wave as expected out of the general properties of the correlation func-

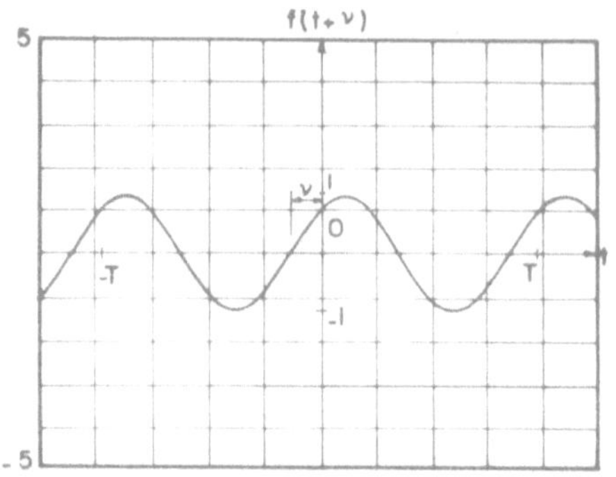

Figure 2. A sine wave shifted with respect to origin.

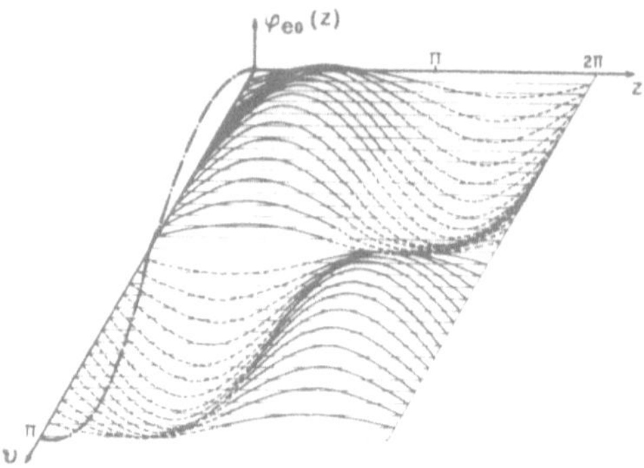

Figure 3. Cross correlation of even and odd parts of a sine wave shifted from the origin.

tions, in (z) domain with a period of (2π). Sinusoidal correlation exists up to (π) value and lack of correlation from (π) to (2π).

The projection of the correlation function into the $(\varphi_{eo}(z, \nu)$ plane forms again a sine wave, this time with a period of (π). Note that the phases of $(\varphi_{eo}(z))$ are zero, in (z) domain and only the amplitudes of $(\varphi_{eo}(z))$ change sinusoidally.

Finally, it is observed that:

$$\varphi_{eo}(z) = 0 \quad \text{for} \quad \nu = \frac{k\pi}{2}\ ; k = 0, 2, 4, 6, \ldots \tag{42}$$

Square Wave

Define a square wave shifted from the origin by an amount (ν) with the duty cycle parameters as indicated in (Fig. 4). Such a square wave is expressed analytically as follows:

$$f(t+\nu) = \left[\begin{array}{ll} 1 & (Tn-\nu) < t < (Tn-\nu-\frac{T}{a}) \\ \text{for} & \\ 0 & (Tn-\nu+\frac{T}{a}) < t < (Tn-\nu+T) \end{array}\right. \tag{43}$$

for which the coefficients are;

$$a^*_n = \frac{1}{\pi n}\cos\frac{\pi n}{T}\left(\frac{T}{a} - 2\nu\right).\ \sin\frac{\pi n}{a} \tag{44}$$

$$b^*_n = \frac{1}{\pi n}\sin\frac{\pi n}{T}\left(\frac{T}{a} - 2\nu\right).\ \sin\frac{\pi n}{a} \tag{45}$$

According to Theorem III., the cross correlation for such a square wave is computed as:

$$\varphi_{eo}(z) = \frac{1}{4\pi^2}\sum_{n=1}^{\infty}\frac{1}{n^2}\sin\frac{2\pi}{T}n\left(\frac{T}{a} - 2\nu\right)\left(\sin\frac{\pi n}{a}\right)^2 \cdot \sin\frac{2\pi}{T}nz. \tag{46}$$

Now, consider the above formulated square wave for: (a = 2) and for (a = 8) cases.

– Symmetric Case (a = 2).

The $(\varphi_{eo}(z))$ for symmetric case is plotted in (Fig. 5). It can directly be revealed that, $(\varphi_{eo}(z))$ consists out of positive and negative correlation and lack of correlation volumes

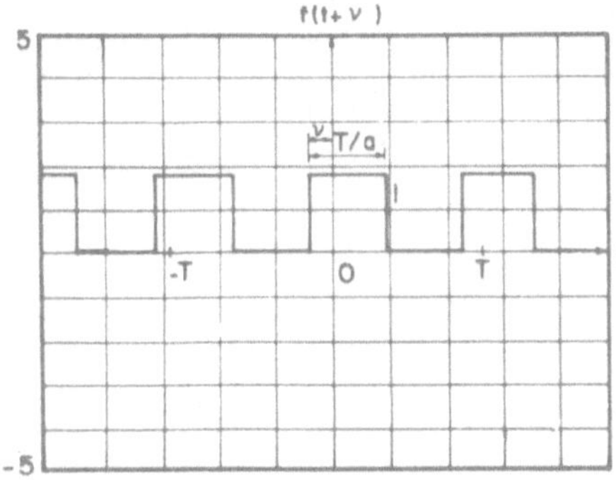

Figure 4. A square wave shifted with respect to origin.

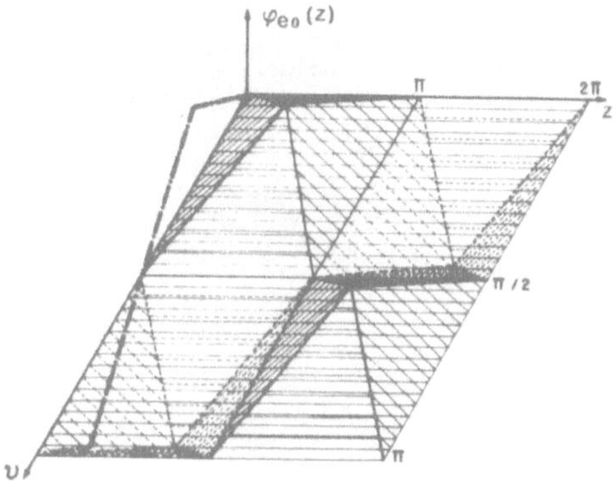

Figure 5. Cross correlation of even and odd parts of a symmetric square wave shifted from the origin.

–pyramids– while (ν) is changing, and odd symmetry with respect to (z) axis. The correlation exists up to (π) value followed by a back of correlation from (π) to (2π).

The projection on the ($\varphi_{eo}(z, \nu)$) plane is again a triangular function with a period of (π) in (ν) axis.

– Non–Symmetric Case (a = 8).

The cross correlation function for a non– symmetric square wave is shown in (Fig. 6). This is an interesting case, including the cross behavior of the correlation and the back of correlation regions in (z) and (ν) plane consisting of trapezoids, with a trapezoidal projection into ($\varphi_{eo}(z, \nu)$) plane. In this figure, the correlation reflections are clearly observable.

A Narrow Band Pseudo Random Wave

Define a narrow band pseudo random periodic wave with the following coefficients representing the time function plotted in (Fig. 7), having only five haronics given as follow:

(n)	(a_n)	(b_n)
0	2.620	0
1	−1.570	−0.392
2	−1.098	0.663
3	−0.458	0.916
4	0.671	0.158
5	−0.063	−0.291

The even and odd parts composing this function are plotted in (Fig. 8) and in (Fig. 9) respectively for the full interval (O–2π).

The autocorrelation function ($\varphi_{11}(z)$) is also shown as in (Fig.10), again for the full interval which clearly puts forward the symmetry at (π) where it folds.

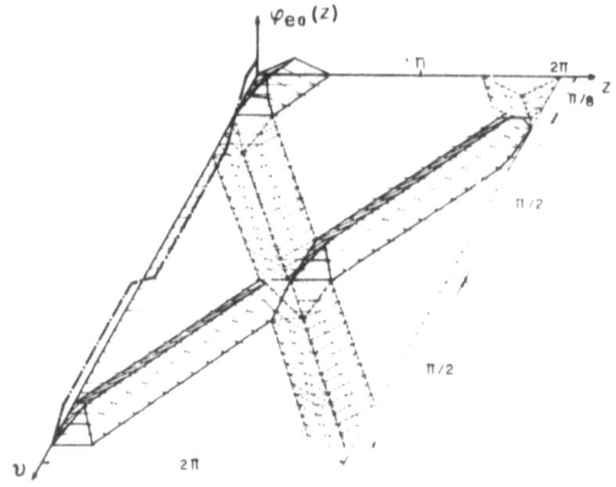

Figure 6. Cross correlation of even and odd parts of a non symmetric square wave shifted from the origin.

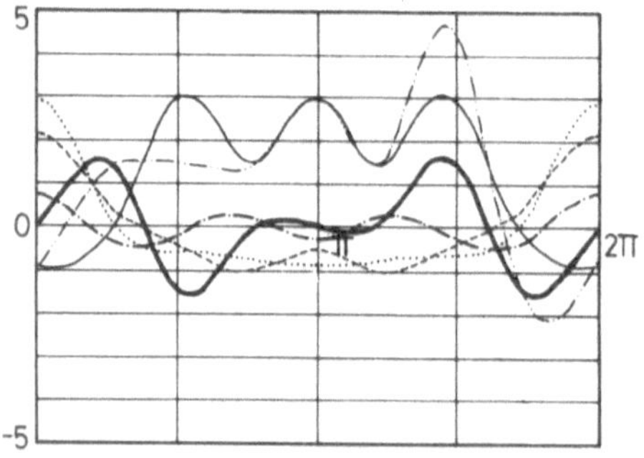

Figure 7. A narrow band pseudo random wave.

Figure 8. Even part of the pseudo random wave.

Figure 9. Odd part of the pseudo random wave.

Figure 10. Auto—correlation function of the pseudo random wave.

Figure 11. Auto—correlation function for the even part, $\varphi_{ee}(z)$.

Figure 12. Auto—correlation function for the odd part, $\varphi_{oo}(z)$).

In (Fig. 11) and (Fig. 12), the autocorrelation functions for the even part ($\varphi_{ee}(z)$) and for the odd part ($\varphi_{oo}(z)$) are seperately indicated for the full intervals, respectively. It is interesting to note, how *the random component of (f(t)) is proportionally shared among the autocorrelation functions of the even and odd components* in correspondance of (z = 0). It is also important to note that the *odd part emphasizes better the periodic component of the original function (f(t)).*

Now, introduce a shift to (f(t)) from the origin by an amount (ν) measured in the opposite direction of time. Then, new coefficients are needed as (a_n^*, b_n^*) corresponding to the shifted function:

$$a'_n = a_n \cos n\nu + b_n \sin n\nu \tag{47}$$

$$b'_n = -a_n \sin n\nu + b_n \cos n\nu \tag{48}$$

Then the shifted function:

$$f(t + \nu) = \frac{1}{2} a_0 + \sum_{n=1}^{\infty} a_n^* \cos \frac{2\pi}{T} nt + b_n^* \sin \frac{2\pi}{T} nt. \tag{49}$$

and the phases of the shifted function (θ'_n):

$$\theta'_n = n\nu + \theta_n \tag{50}$$

that is, the original phase spectrum (θ_n) added together with the phase shift due to (ν), by $(n\nu)$.

Calculating the cross correlation function:

$$\varphi_{eo}(z) = \frac{1}{2} \sum_{n=1}^{5} a_n^* \cdot b_n^* \sin \frac{2\pi}{T} nz \tag{51}$$

is obtained. This function is plotted in (Fig.13) as a function of (z) and (ν). One surprisingly observes from this three dimensional plot that is $(\varphi_{eo}(z, \nu)$ is projected over a vertical plane, diagonally intersecting (ν) and (z) plane with an intersection line expressed as:

$$\nu = -\frac{\nu_0}{z_0} \cdot z + 1 \tag{52}$$

and if this plane is scanned radially by changing (ν_0), around the center $(z = 2\pi)$, the time reversed version of the original time function (f(t)) is seemingly decorrelated (or deconvolved) and recovered out of $(\varphi_{eo}(z, \nu)$ with frequency doubling, (Sec. Fig. 14). The projection of $(\varphi_{eo}(z))$ over (ν, z) diagonal plane is also analytically calculated and illustrated in (Fig. 15).

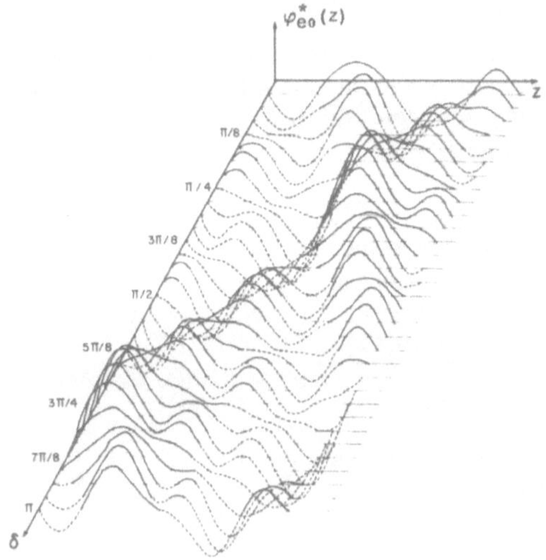

Figure 13. Cross correlation of even and odd parts of the pseudo random wave.

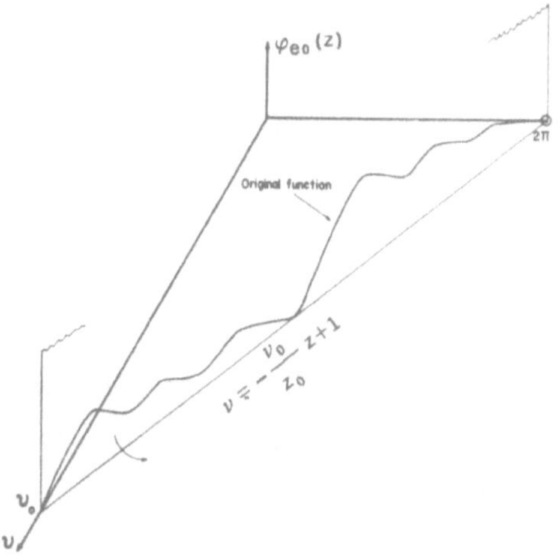

Figure 14. Recovery of the original function from the cross correlation function $\varphi_{eo}(z)$.

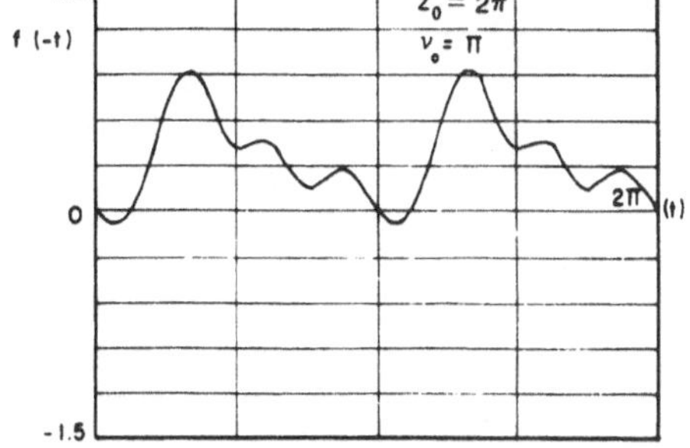

Figure 15. Recovered reflection function from the cross correlation.

IMPLEMENTATION

The AEON parallel array processor developped in References (1), (2), (3), is very suitably applicable for the real–time computations of such correlograms. In this respect an implementation scheme is given in (Fig. 16). It consists of a delay and time reversal processor followed by two summing nodes in order to form the even and the odd components, then another processor to compute the correlation functions by parametric scanning in (ν).

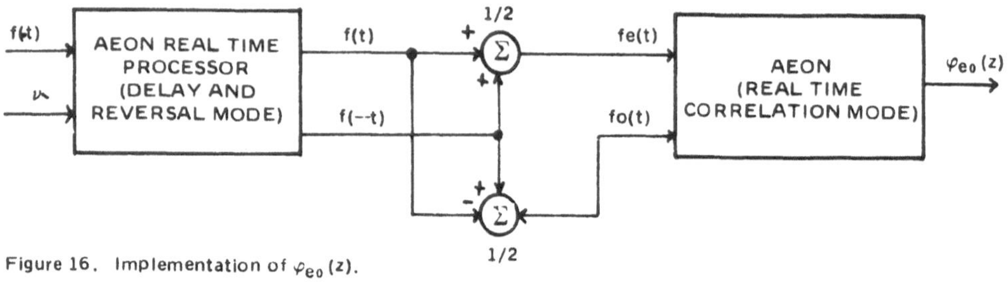

Figure 16. Implementation of $\varphi_{eo}(z)$.

CONCLUSION

In above presented preliminary approach of deriving new correlation series by making use of the eveness and oddness concepts is believed to lead to important consequences mostly for identification problems. The resulting properties of the proved theorems are still wide open to many further developments and applications, such as the recovery of the original wave form from the correlation functions.

Author believes that, such correlograms– called here as "oddevenograms" would contribute very positively to the problems of wave phenomena involved in various fields of physics namely as, optics, seismology, acoustics, communication, image processing etc.,

Finally, the real time electronic implementation of presented correlograms by means of AEON processor is another outstanding feature of the present work.

ACKNOWLEDGEMENT

Author extends his gratitute to Şükrü Akçay and Aysun Sinanoğlu for their help for the analytic developments, to Fatih Güler for the exhaustive computer aid.

— * —

REFERENCES

1. *Tüzünalp, Ö.*, A Discrete Convolution Array and Its Applications, *TUBITAK–The Turkish Technical and Scientific Research Council–Report, 1972 Ankara, Turkey.*

2. *Tüzünalp, Ö., Adaptive Methods in Underwater Acoustics, ed. by H.G. Urban, NATO ASI Series,* Math. and Phy. Sciences, *Vol. 151, 1985.*

3. *Tüzünalp, Ö., Real–Time Dissasociation of a Time Function into Its Odd and Even Components,* METU Journal of Pure and Applied Sciences, *17, 3, 253–267, 1984.*

— * —

FOURIER METHODS APPLICABLE IN THE NUMERICAL SOLUTION OF ELECTROMAGNETIC TIME-DOMAIN SCATTERING PROBLEMS

ANTON G. TIJHUIS
Laboratory of Electromagnetic Research
Department of Electrical Engineering
Delft University of Technology
P.O. Box 5031, 2600 GA Delft, The Netherlands
Electronic Mail TIJHUIS@ET.TUDELFT.NL

ABSTRACT. In transient electromagnetic scattering problems, Fourier or Laplace transformations are employed to attain a spectral decomposition in terms of solutions of reduced problems of a lower dimensionality. The resulting inversion integrals are evaluated either by contour deformation into the complex frequency or order plane, or by the direct evaluation of one or more Fourier or Bromwich inversion integrals. The modal representations resulting from the former approach provide a detailed physical insight into the scattering process; the latter approach is more suitable for an efficient computation of the desired time-domain solution. Both approaches are illustrated for the "tutorial" scattering problem of transient scattering by a radially inhomogeneous, lossy dielectric circular cylinder embedded in a homogeneous, lossless dielectric. Numerical results are presented and discussed.

1. Introduction

Transient electromagnetic scattering problems have, in recent ycens, become increasingly more important in metrology and inverse scattering, parametric modeling of electromagnetic systems, remote probing of the earth, processing of transient data in EMP and EMC studies, and in many other applications. Mathematically, time-domain electromagnetic fields are described by Maxwell's equations. For most practical situations, these equations can only be solved numerically. To this end, a variety of techniques is available, including finite-difference, finite-element and integral-equation methods.

In this context, Fourier transformations have been employed in two different ways. One possibility is to subject the relevant space-time domain integral equations to a spatial Fourier transformation, and to solve the result by marching on in time [1]. A possible difficulty associated with this approach is that Gibbs' phenomenon complicates the computation of those field components that are discontinuous across interfaces between media differing in their electromagnetic properties. Moreover, any time-marching approach inherently suffers from the accumulation of discretization errors as the computation progresses. The fact that the time-domain results obtained are stable does not exclude the occurrence of such an error accumulation.

J. S. Byrnes and J. F. Byrnes (eds.), Recent Advances in Fourier Analysis and Its Applications, 273–309.

In the present chapter, we review a second possibility, in which Fourier or Laplace transformations are carried out with respect to only those space and time coordinates along which the configuration is translationally invariant, while the incident field (forcing function) is not. The effect of such transformations is that the solution of the original space-time problem is reduced to the repeated solution of a transform-domain problem of a lower dimensionality, and the evaluation of one or more Fourier or Bromwich inversion integrals. The most obvious example is a transformation with respect to time. In the time window of interest, the scattering configuration may be regarded as stationary, while the incident field is pulsed. A second example is a spatial Fourier transformation in cases where the configuration consists of a horizontally or cylindrically stratified material excited by a localized source.

For the evaluation of the inversion integrals, two different approaches can be followed. First, these integrals can be determined indirectly by contour deformation in the relevant complex plane(s). This results in modal representations for the electromagnetic fields excited. The principal issues in this approach are the validity (closing-contour contribution) and convergence (truncation) of the representations obtained. Second, these inversion integrals can be computed directly with the aid of an FFT algorithm. In such a "brute-force" approach, the key issue is the discretization of the relevant lower-dimensional integral equations.

In this chapter, both approaches are discussed for the "tutorial" example of plane-wave scattering by a radially inhomogeneous, lossy dielectric circular cylinder embedded in a homogeneous, lossless dielectric. Special attention is devoted to recent developments indicating that it may be more convenient to analyze some of the convergence and discretization aspects referred to above in the space-time domain, rather than in the transform domain. Illustrative numerical results are presented and discussed.

The presentation is organized as follows. In Section 2, we review the basic theory of transient electromagnetic fields. Section 3 contains the mathematical formulation of our "tutorial" scattering problem and of the pertinent Fourier transformations. Sections 4 and 5 describe the evaluation of the temporal and spatial inversion integrals by contour deformation in the relevant complex planes, while Section 6 is devoted to the "brute-force" approach. Our conclusions, finally, are stated in Section 7.

2. Basic Theory

In an interdisciplinary textbook like the present volume, it is unlikely that the reader will be familiar with the theory of every particular physical phenomenon that is addressed. In this section, therefore, a brief summary of the theory of transient electromagnetic fields is presented. With the applications in mind, we restrict the discussion to the case of an inhomogeneous, linear, isotropic, locally and instantaneously reacting, time-invariant medium.

2.1. ELECTROMAGNETIC FIELD EQUATIONS

The equations describing the behavior of transient electromagnetic fields in matter as specified above can be written as

$$\begin{aligned} -\nabla\times\mathcal{H}(\mathbf{r},t)+\epsilon(\mathbf{r})\partial_t\mathcal{E}(\mathbf{r},t)+\sigma(\mathbf{r})\mathcal{E}(\mathbf{r},t) &= -\mathcal{J}^e(\mathbf{r},t), \\ \nabla\times\mathcal{E}(\mathbf{r},t)+\mu(\mathbf{r})\partial_t\mathcal{H}(\mathbf{r},t) &= -\mathcal{J}^m(\mathbf{r},t). \end{aligned} \tag{1}$$

In this form, the electromagnetic-field equations are known as *Maxwell's equations* in a dielectric. In (1), we encounter the physical quantities

$$\begin{aligned} \mathbf{r} &= \text{Cartesian position vector (m)}, \\ t &= \text{time coordinate (s)}, \\ \mathcal{E}(\mathbf{r},t) &= \text{electric-field strength (V/m)}, \\ \mathcal{H}(\mathbf{r},t) &= \text{magnetic-field strength (A/m)}, \\ \mathcal{J}^e(\mathbf{r},t) &= \text{volume density of external electric current (A/m}^2), \\ \mathcal{J}^m(\mathbf{r},t) &= \text{volume density of external magnetic current (V/m}^2), \\ \sigma(\mathbf{r}) &= \text{conductivity (S/m)}, \\ \epsilon(\mathbf{r}) &= \text{permittivity (F/m)}, \\ \mu(\mathbf{r}) &= \text{permeability (H/m)}, \end{aligned}$$

expressed according to the International System of Units (Système International d'Unités), abbreviated SI.

In the right-hand sides of (1), the external or source current densities $\mathcal{J}^e(\mathbf{r},t)$ and $\mathcal{J}^m(\mathbf{r},t)$ constitute the active parts, which describe the action of the sources that generate the field. For the case of plane-wave incidence considered in the present chapter, these current densities may be regarded as being located at infinity. In that case, $\mathcal{J}^e(\mathbf{r},t)$ and $\mathcal{J}^m(\mathbf{r},t)$ may be set to zero for any finite value of the position vector $\mathbf{r}$.

For the constitutive parameters $\sigma(\mathbf{r})$, $\epsilon(\mathbf{r})$ and $\mu(\mathbf{r})$, which describe the physical properties of the material at hand, we have the inequalities

$$\sigma(\mathbf{r})\geq 0, \quad \epsilon(\mathbf{r})\geq\epsilon_0, \quad \mu(\mathbf{r})\geq\mu_0, \tag{2}$$

where ϵ_0 and μ_0 denote the permittivity and permeability of free space (i.e. vacuum), respectively. The value of μ_0 is fixed by SI as $\mu_0 = 4\pi\times10^{-7}$ H/m; the value of ϵ_0 follows from $\epsilon_0 = 1/\mu_0 c_0^2$, where $c_0 = 299792458$ m/s is the speed of electromagnetic waves in free space.

2.2. BOUNDARY CONDITIONS

At the interface between two media differing in their electromagnetic properties, the electric- and magnetic-field strengths $\mathcal{E}(\mathbf{r},t)$ and $\mathcal{H}(\mathbf{r},t)$ will in general vary discontinuously. Due to this discontinuous behavior, the electric-field equations given in (1) lose their significance at the interface. Hence, these equations must be supplemented by conditions that connect the field values on both sides of the interface, the so-called *boundary conditions*.

To formulate these conditions, we consider an interface S between two domains $\mathcal{D}_1$ and $\mathcal{D}_2$ that contain matter with different material properties. The properties of the media in both domains and the location of the interface S are assumed to be time invariant. Moreover, it is assumed that S has everywhere a unique tangent plane, and, accordingly, a unit vector $\mathbf{n}(\mathbf{r})$ pointing into $\mathcal{D}_1$. Now, let $\mathbf{r}$ be the position vector of some point on S, and let $\{\mathcal{E}_1(\mathbf{r},t), \mathcal{H}_1(\mathbf{r},t)\}$ and $\{\mathcal{E}_2(\mathbf{r},t), \mathcal{H}_2(\mathbf{r},t)\}$ represent the limiting values upon approaching $\mathbf{r}$ from $\mathcal{D}_1$ and $\mathcal{D}_2$, respectively.

As mentioned above, we only consider the case where both an electric and a magnetic field are present in the entire space $\mathbb{R}^3$. Consistency of the integral form of the field equations then requires that the components of both fields that are tangential to the interface be continuous upon crossing the interface, i.e.

$$\begin{aligned} \mathbf{n}(\mathbf{r}) \times [\mathcal{H}_1(\mathbf{r},t) - \mathcal{H}_2(\mathbf{r},t)] &= \mathbf{0}, \\ \mathbf{n}(\mathbf{r}) \times [\mathcal{E}_1(\mathbf{r},t) - \mathcal{E}_2(\mathbf{r},t)] &= \mathbf{0}. \end{aligned} \tag{3}$$

These two equalities are the fundamental boundary conditions at the interface of two penetrable media.

2.3. Electromagnetic Energy and Power

The relation that describes the local energy balance of transient electromagnetic fields is obtained directly by taking the vector inner product of the first and second lines of (1) with $\mathcal{E}(\mathbf{r},t)$ and $\mathcal{H}(\mathbf{r},t)$, respectively, and adding the results. Using the vector identity

$$\nabla \cdot [\mathcal{E}(\mathbf{r},t) \times \mathcal{H}(\mathbf{r},t)] = -\mathcal{E}(\mathbf{r},t) \cdot [\nabla \times \mathcal{H}(\mathbf{r},t)] + \mathcal{H}(\mathbf{r},t) \cdot [\nabla \times \mathcal{E}(\mathbf{r},t)],$$

we arrive at

$$\begin{aligned} \nabla \cdot [\mathcal{E}(\mathbf{r},t) \times \mathcal{H}(\mathbf{r},t)] = & -\sigma(\mathbf{r})\mathcal{E}^2(\mathbf{r},t) \\ & - \partial_t \left[\tfrac{1}{2}\epsilon(\mathbf{r})\mathcal{E}^2(\mathbf{r},t) + \tfrac{1}{2}\mu(\mathbf{r})\mathcal{H}^2(\mathbf{r},t)\right] \\ & + \mathcal{E}(\mathbf{r},t) \cdot \mathcal{J}^e(\mathbf{r},t) + \mathcal{H}(\mathbf{r},t) \cdot \mathcal{J}^m(\mathbf{r},t). \end{aligned} \tag{4}$$

This identity is *Poynting's theorem*; it represents the basic electromagnetic energy relation.

Each term in (4) has a distinct physical interpretation in terms of some form of energy or power. The relevant interpretations and dimensions are

$\mathcal{E}(\mathbf{r},t) \times \mathcal{H}(\mathbf{r},t)$ = area density of electromagnetic power flow (W/m^2),
$\sigma(\mathbf{r})\mathcal{E}^2(\mathbf{r},t)$ = volume density of dissipated power (W/m^3),
$\tfrac{1}{2}\epsilon(\mathbf{r})\mathcal{E}^2(\mathbf{r},t)$ = volume density of electric energy (J/m^3),
$\tfrac{1}{2}\mu(\mathbf{r})\mathcal{H}^2(\mathbf{r},t)$ = volume density of magnetic energy (J/m^3),
$\mathcal{E}(\mathbf{r},t) \cdot \mathcal{J}^e(\mathbf{r},t)$ = volume density of power delivered by the external electric current (W/m^3),
$\mathcal{H}(\mathbf{r},t) \cdot \mathcal{J}^m(\mathbf{r},t)$ = volume density of power delivered by the external magnetic current (W/m^3).

The vector quantity $\mathcal{S}(\mathbf{r},t) \stackrel{\text{def}}{=} \mathcal{E}(\mathbf{r},t) \times \mathcal{H}(\mathbf{r},t)$ is known as *Poynting's vector.* From (3), it is observed directly that its normal component $\mathbf{n}(\mathbf{r}) \cdot \mathcal{S}(\mathbf{r},t)$ is continuous across an interface between two different penetrable media.

The relation (4) is the *local* form of the electromagnetic power balance. Corresponding *global* forms can be obtained by integrating (4) over a finite domain $\mathcal{D}$, and applying Gauss' theorem. If the domain $\mathcal{D}$ is crossed by one or more interfaces between media differing in their electromagnetic properties, the integration must be carried out over subdomains in which these properties vary continuously. The continuity of $\mathbf{n}(\mathbf{r}) \cdot \mathcal{S}(\mathbf{r},t)$ can then be invoked to cancel out the portions of the resulting surface integrals along opposite sides of each interface.

2.4. UNIQUENESS

With a suitable initial condition, the electric-field equations (1) and the boundary conditions (3) uniquely determine the electromagnetic field $\{\mathcal{E}(\mathbf{r},t), \mathcal{H}(\mathbf{r},t)\}$. To close this section, we briefly outline one way of establishing this uniqueness.

Let $\{\mathcal{E}_A(\mathbf{r},t), \mathcal{H}_A(\mathbf{r},t)\}$ and $\{\mathcal{E}_B(\mathbf{r},t), \mathcal{H}_B(\mathbf{r},t)\}$ be two sufficiently well-behaved solutions of (1) and (3), square integrable over $\mathbb{R}^3$ at each instant t, such that

$$\begin{aligned} \mathcal{E}_A(\mathbf{r},t_0) &= \mathcal{E}_B(\mathbf{r},t_0), \\ \mathcal{H}_A(\mathbf{r},t_0) &= \mathcal{H}_B(\mathbf{r},t_0). \end{aligned} \tag{5}$$

Further, let the field $\{\Delta\mathcal{E}(\mathbf{r},t), \Delta\mathcal{H}(\mathbf{r},t)\}$ be the difference between both solutions, i.e.

$$\begin{aligned} \Delta\mathcal{E}(\mathbf{r},t) &\stackrel{\text{def}}{=} \mathcal{E}_A(\mathbf{r},t) - \mathcal{E}_B(\mathbf{r},t), \\ \Delta\mathcal{H}(\mathbf{r},t) &\stackrel{\text{def}}{=} \mathcal{H}_A(\mathbf{r},t) - \mathcal{H}_B(\mathbf{r},t). \end{aligned} \tag{6}$$

Then, the difference $\{\Delta\mathcal{E}(\mathbf{r},t), \Delta\mathcal{H}(\mathbf{r},t)\}$ satisifies the field equations (1) with zero right-hand sides, as well as the boundary conditions (3).

Consequently, Poynting's theorem as given in (4) also applies to this difference field. A spatial integration over $\mathbb{R}^3$ and a temporal integration over the time interval $t_0 < t < t_1$ then result in

$$\begin{aligned} \int_{\mathbb{R}^3} dV(\mathbf{r}) \left[\tfrac{1}{2}\epsilon(\mathbf{r})\, \Delta\mathcal{E}^2(\mathbf{r},t_1) + \tfrac{1}{2}\mu(\mathbf{r})\, \Delta\mathcal{H}^2(\mathbf{r},t_1)\right] &\leq \\ \int_{\mathbb{R}^3} dV(\mathbf{r}) \left[\tfrac{1}{2}\epsilon(\mathbf{r})\, \Delta\mathcal{E}^2(\mathbf{r},t_0) + \tfrac{1}{2}\mu(\mathbf{r})\, \Delta\mathcal{H}^2(\mathbf{r},t_0)\right] &= 0, \end{aligned} \tag{7}$$

for any $t_1 \geq t_0$. In deriving (7), the fact was used that $\sigma(\mathbf{r}) \geq 0$, as specified (2). With the inequalities for $\epsilon(\mathbf{r})$ and $\mu(\mathbf{r})$ specified in that equation, it follows immediately from (7) that

$$\begin{aligned} \Delta\mathcal{E}(\mathbf{r},t_1) &= \mathcal{E}_A(\mathbf{r},t_1) - \mathcal{E}_B(\mathbf{r},t_1) = 0, \\ \Delta\mathcal{H}(\mathbf{r},t_1) &= \mathcal{H}_A(\mathbf{r},t_1) - \mathcal{H}_B(\mathbf{r},t_1) = 0, \end{aligned} \tag{8}$$

for any $t_1 \geq t_0$, which is what we set out to prove.

The electromagnetic-field equations discussed above hold for a wide class of dielectric materials, particularly when the electromagnetic fields do not vary too rapidly in time. For most practical dielectric materials, the analysis can be simplified to the case where $\mu(\mathbf{r}) = \mu_0$ for $\mathbf{r} \in \mathbb{R}^3$. When the time variation does become rapid, relaxation effects must be taken into account. This means that the terms in (1) containing the material parameters $\sigma(\mathbf{r})$, $\epsilon(\mathbf{r})$ and $\mu(\mathbf{r})$ are replaced with causal convolution integrals involving field values at previous instants. The remainder of the theory given in this section can readily be modified to incorporate such a change.

3. Formulation of the Problem

Now that the basic theory of transient electromagnetic fields in penetrable media is known, we turn our attention to the "tutorial" scattering problem for wich the various Fourier techniques will be demonstrated.

3.1. The Configuration

We consider a radially inhomogeneous, lossy dielectric circular cylinder with radius a embedded in a homogeneous, lossless dielectric (see Figure 1). A cylindrical coordinate system is introduced, with the z-axis coinciding with the symmetry axis, and ρ and ϕ as polar coordinates in the plane perpendicular to it. As indicated, the configuration consists of an exterior domain $\mathcal{D}_1$ and an interior domain $\mathcal{D}_2$. In $\mathcal{D}_1$, we have the permittivity $\epsilon_1(\mathbf{r}) = \epsilon_{1r}\epsilon_0$, the conductivity $\sigma_1(\mathbf{r}) = 0$, and the permeability $\mu_1(\mathbf{r}) = \mu_0$. In $\mathcal{D}_2$, we have $\epsilon_2(\mathbf{r}) = \epsilon_{2r}(\rho)\epsilon_0$, $\sigma_2(\mathbf{r}) = \sigma_2(\rho)$, and $\mu_2(\mathbf{r}) = \mu_0$, respectively. In all cases, the values of these parameters are chosen in accordance with the inequalities given in (2).

From $\mathcal{D}_1$, an electrically polarized pulse of finite duration is normally incident on the cylinder from the direction $\phi = \pi$. We write the electric-field intensity and the magnetic-field intensity of the incident field as

$$\begin{aligned} \mathcal{E}^i(\mathbf{r},t) &\stackrel{\text{def}}{=} \mathcal{F}(t - [\rho\cos\phi + a]/c_1)\,\mathbf{i}_z, \\ \mathcal{H}^i(\mathbf{r},t) &\stackrel{\text{def}}{=} -Y_1\,\mathcal{F}(t - [\rho\cos\phi + a]/c_1)\,(\sin\phi\,\mathbf{i}_\rho + \cos\phi\,\mathbf{i}_\phi), \end{aligned} \tag{9}$$

where $Y_1 \stackrel{\text{def}}{=} (\epsilon_1/\mu_0)^{\frac{1}{2}}$ is the wave admittance in $\mathcal{D}_1$ and $c_1 \stackrel{\text{def}}{=} (\epsilon_1\mu_0)^{-\frac{1}{2}}$ the speed of electromagnetic waves in $\mathcal{D}_1$, and where $\mathcal{F}(t)$ is a piecewise continuous function of bounded variation in t that vanishes outside the interval $0 < t < T$.

Then, it follows from the source-free version of the electromagnetic-field equations (1) that the problem is two-dimensional, and that the field intensities are functions of ρ, ϕ and t only:

$$\begin{aligned} \mathcal{E}(\mathbf{r},t) &= \mathcal{E}(\rho,\phi,t)\,\mathbf{i}_z, \\ \mathcal{H}(\mathbf{r},t) &= \mathcal{H}(\rho,\phi,t)\,\mathbf{i}_\phi - \frac{1}{\rho\mu_0}\int_{-\infty}^{t} dt'\,\partial_\phi\mathcal{E}(\rho,\phi,t')\,\mathbf{i}_\rho. \end{aligned} \tag{10}$$

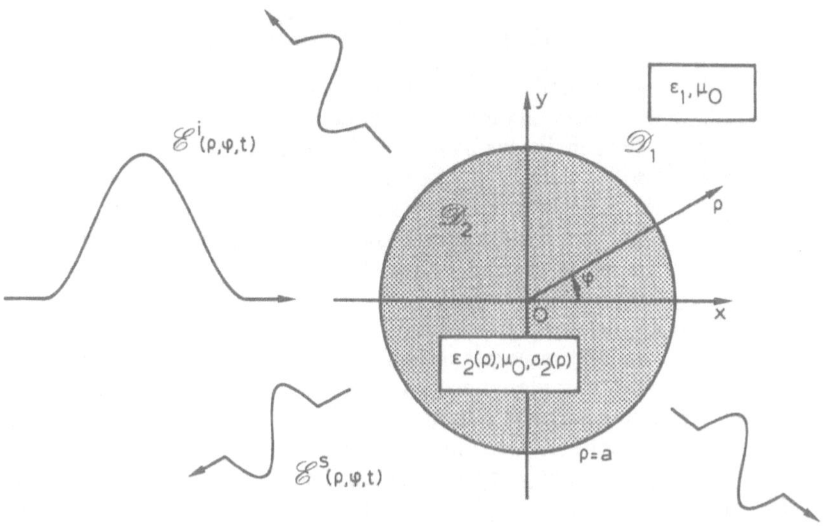

Figure 1: Scattering of an electrically polarized, pulsed electromagnetic plane wave by a radially inhomogeneous, lossy dielectric circular cylinder embedded in a homogeneous, lossless dielectric.

In (10), we have restricted the unknowns to those field quantities that, according to (3), are continuous at any discontinuity in $\epsilon(\rho)$ and $\sigma(\rho)$. The scattered field is defined by

$$\begin{aligned} \mathcal{E}^s(\rho,\phi,t) &\stackrel{\text{def}}{=} \mathcal{E}(\rho,\phi,t) - \mathcal{E}^i(\rho,\phi,t), \\ \mathcal{H}^s(\rho,\phi,t) &\stackrel{\text{def}}{=} \mathcal{H}(\rho,\phi,t) - \mathcal{H}^i(\rho,\phi,t). \end{aligned} \tag{11}$$

3.2. Fourier Transformations

As remarked in the introduction, it is convenient to carry out Fourier transformations with respect to those space and time coordinates along which the scattering configuration is translationally invariant, while the incident field is not. For the present configuration, this concerns the time coordinate t and the angular coordinate ϕ.

Let us start with the angular transformation. Since $\mathcal{F}(t)$ is piecewise continuous and of bounded variation in t, it follows from the field equations (1) that, for ρ and t fixed, $\mathcal{E}(\rho,\phi,t)$ and $\mathcal{H}(\rho,\phi,t)$ also exhibit these properties as a function of ϕ. Hence, these quantities can, as far as their dependence on ϕ is concerned, be represented in terms of a Fourier series, e.g.

$$\mathcal{E}(\rho,\phi,t) = \sum_{m=-\infty}^{\infty} \exp(im\phi)\, \mathcal{E}_m(\rho,t), \tag{12}$$

with

$$\mathcal{E}_m(\rho,t) \stackrel{\text{def}}{=} \frac{1}{2\pi}\int_{-\pi}^{\pi} d\phi \, \exp(-im\phi)\, \mathcal{E}(\rho,\phi,t), \tag{13}$$

which converges in mean square to the relevant quantity [2].

Furthermore, it follows from the energy balance (4) that, for the scattered field in $\mathcal{D}_1$ and for the total field in $\mathcal{D}_2$, we have as well mean-square convergence on the entire relevant spatial domain uniformly in t. The proof, which is rather lengthy, is deferred to Appendix A. This global convergence implies that the Fourier series (12), when truncated at $|m| = M$, approximates the relevant field quantities with the same global accuracy in the entire space-time domain. This is essentially different from the usual frequency-domain result (see e.g. [3]), where, for an increasing frequency, M has to be increased to maintain the accuracy. Consequently, we are left with the calculation of the corresponding Fourier coefficients for $|m| \leq M$, where M is determined by the global accuracy desired.

Next, we turn our attention to the temporal behavior of the Fourier coefficients $\{\mathcal{E}_m(\rho,t), \mathcal{H}_m(\rho,t)\}$. Since the incident field reaches the cylinder at $t = 0$, causality ensures that both $\{\mathcal{E}^s(\rho,\phi,t), \mathcal{H}^s(\rho,\phi,t)\}$ in $\mathcal{D}_1$ and $\{\mathcal{E}(\rho,\phi,t), \mathcal{H}(\rho,\phi,t)\}$ in $\mathcal{D}_2$ vanish in the time interval $-\infty < t < 0$. From the definition of the Fourier coefficients given in (13), it follows that the Laplace transforms of these field quantities are identical to the corresponding one-sided Laplace transforms, e.g.

$$E_m(\rho,s) \stackrel{\text{def}}{=} \int_0^{\infty} dt \, \exp(-st)\, \mathcal{E}_m(\rho,t), \tag{14}$$

which are regular for $\mathrm{Re}(s) \geq 0$.

The corresponding coefficients for the incident field follow from applying an angular Fourier transformation and the shift rule plus a one-sided temporal Laplace transformation to (9) as

$$E^i_m(\rho,s) = (-1)^m \exp(-sa/c_1)\, I_m(s\rho/c_1)\, F(s), \tag{15}$$

and a related expression for $H^i_m(\rho,s)$. In (15), we have

$$F(s) \stackrel{\text{def}}{=} \int_0^{T} dt \, \exp(-st)\, \mathcal{F}(t),$$

while $I_m(z)$ denotes the modified Bessel function of the first kind of order m. Both $E^i_m(\rho,s)$ and $H^i_m(\rho,s)$ are regular for all complex s.

When the Laplace-transformed Fourier coefficients are known for all m and s, the space-time response follows from the inversion formula

$$\mathcal{E}(\rho,\phi,t) = \frac{1}{2\pi i}\lim_{M\to\infty}\left\{\sum_{m=-M}^{M} \exp(im\phi) \int_{\mathcal{L}_\beta} ds\, \exp(st)\, E_m(\rho,s)\right\}, \tag{16}$$

and the same operation for $\mathcal{H}(\rho,\phi,t)$, where $\mathcal{L}_\beta$ is the Bromwich contour

$$\mathcal{L}_\beta \stackrel{\text{def}}{=} \{s \mid \mathrm{Re}(s) = \beta,\ \beta \geq 0\}. \tag{17}$$

3.3. Dimensionless Coordinates

In view of subsequent calculations, we carry out a normalization to dimensionless quantities that are more natural to the problem at hand. We normalize all distances with respect to the radius a and all times with respect to the free-space travel time from the boundary of the cylinder to its center, i.e. a/c_0; e.g.

$$\bar{\rho} = \rho/a, \quad \bar{t} = c_0 t/a.$$

In addition, we introduce the dimensionless quantities

$$\bar{s} = sa/c_0, \quad \bar{\sigma}(\bar{\rho}) = Z_0\sigma(\rho)a, \quad \bar{\epsilon}(\bar{\rho}) = \epsilon_r(\rho),$$

and the normalized field quantities

$$\bar{E}_m(\bar{\rho}, \bar{s}) = E_m(\rho, s), \quad \bar{H}_m(\bar{\rho}, \bar{s}) = Z_0\, H_m(\rho, s),$$

where $Z_0 \stackrel{\text{def}}{=} (\mu_0/\epsilon_0)^{\frac{1}{2}}$ is the wave impedance of free space. Henceforth, we will use the normalized quantities, omitting the bars.

The main advantage of the above normalization it that it leads to a simplification of almost all mathematical expressions in the remainder of the discussion. Further, the dimensionless quantities are needed for use as variables in the computer programs.

3.4. Reduced Equations

The Fourier and Laplace transformed, normalized, source-free version of the electromagnetic field equations (1) is given by

$$\left[\partial_\rho + \rho^{-1}\right] H_m(\rho, s) = s\left[\epsilon_r(\rho) + \sigma(\rho)/s + m^2/s^2\rho^2\right] E_m(\rho, s), \tag{18}$$

$$\partial_\rho E_m(\rho, s) = s\, H_m(\rho, s). \tag{19}$$

Elimination of $H_m(\rho, s)$ leads to the second-order differential equation

$$\left[\partial_\rho^2 + \rho^{-1}\partial_\rho - \left(m^2/\rho^2 + s^2\epsilon_{1r}\right)\right] E_m(\rho, s) = s^2 C_1(\rho, s)\, E_m(\rho, s). \tag{20}$$

In (20), the *contrast function* $C_1(\rho, s)$, defined as

$$C_1(\rho, s) \stackrel{\text{def}}{=} \epsilon_r(\rho) - \epsilon_{1r} + \sigma(\rho)/s, \tag{21}$$

represents the deviation in material parameters with respect to the homogeneous, lossless dielectric in region $\mathcal{D}_1$. Accordingly, $C_1(\rho, s)$ differs from zero only when $0 \le \rho \le 1$.

With the aid of the Green's function for the operator working on $E_m(\rho, s)$ in the left-hand side of (20), i.e.

$$G_m(\rho, \rho'; s) \stackrel{\text{def}}{=} \rho'\, K_m(s\epsilon_{1r}^{\frac{1}{2}}\rho_>)\, I_m(s\,\epsilon_{1r}^{\frac{1}{2}}\rho_<), \tag{22}$$

where $K_m(z)$ denotes the modified Bessel function of the second kind of order m, and where $\rho_> \stackrel{\text{def}}{=} \max\{\rho, \rho'\}$ and $\rho_< \stackrel{\text{def}}{=} \min\{\rho, \rho'\}$, the following integral relation is derived, which is equivalent to (20):

$$E_m(\rho, s) = E^i_m(\rho, s) - s^2 \int_0^1 d\rho' \, C_1(\rho', s) \, G_m(\rho, \rho'; s) \, E_m(\rho', s). \tag{23}$$

When $0 \leq \rho \leq 1$, (23) constitutes an integral equation of the second kind in $E_m(\rho, s)$.

For a general, radially inhomogeneous cylinder, either the system of first-order differential equations (18)–(19), the second-order differential equation (20), or the integral equation (23) must be solved numerically to obtain the Laplace-transformed Fourier coefficients $\{E_m(\rho, s)\}$. All three equations are one-dimensional, with ρ as the dependent variable, and m and s as parameters. Two possible solution schemes are discussed in Appendix B.

Solving either (18)–(19), (20) or (23) leaves us with carrying out the inversion procedure prescribed in (16). In the upcoming sections, three different ways of doing this will be reviewed, each of which highlights one particular aspect of the scattering problem at hand; the choice between them should be made based on the relative importance of those aspects.

4. Contour Deformation in the Complex Frequency Plane

One way of evaluating (16) is, for each angular order m, to continue $E_m(\rho, s)$ analytically into the domain $\text{Re}(s) < 0$, supplement $\mathcal{L}_\beta$ by circular arcs at infinity in the complex s-plane, and apply Cauchy's theorem. In case the integrals along the supplementary arcs vanish, only the singularities in the analytic continuation of $E_m(\rho, s)$ contribute to the result. In the literature on transient wave propagation, this approach is therefore known as the *singularity expansion method*, abbreviated SEM.

When $\mathcal{L}_\beta$ can be closed to the right, the field represented by the inversion integral in (16) is zero, since the Laplace transform of a causal signal is regular in the right half of the complex s-plane. Upon closure to the left, we end up with a field representation in terms of contributions from the poles, and, possibly, branch points in the complex s-plane. Each pole corresponds with a natural freqency of the scatterer; the corresponding natural-mode field distribution constitutes, along with the corresponding magnetic field, a nontrivial solution to the electromagnetic-field equations. Branch cuts, when present, can be identified with a continuum of natural modes.

Natural-mode representations have been the subject of investigation for about two decades. An extensive bibliography was compiled by Michalski [4]. Insight into the development and the applications of the SEM can be obtained from the reviews authored by Baum [5,6,7], and from the special issues edited by Pearson and Marin [8] and by Langenberg [9].

One question which has remained unresolved for a long time is that of the vanishing or nonvanishing contributions from the closing contours at infinity. Only in recent years [10]–[17], the inadequacy of field descriptions solely in terms of modal contributions has been emphasized widely in the literature. In order to determine whether the contribution from the closing contour vanishes in the left or the right half of the complex-frequency plane, at least a closed-form first-order high-frequency approximation to the Laplace-transformed electromagnetic field $\{\mathbf{E}(\mathbf{r},s),\mathbf{H}(\mathbf{r},s)\}$ is needed, which holds over the entire range of $\arg(s)$. For most practical scattering configurations, such an approximation is not immediately available.

In this section, we briefly summarize the results of analyzing all aspects of the SEM for our "tutorial" scattering problem of plane-wave scattering by a lossy, radially inhomogeneous dielectric circular cylinder. For more details, the reader is referred to [18] and [19, Section 2.5].

4.1. Formal Expression for the Fourier Coefficients

Formally, the normalized, Laplace-transformed Fourier coefficient $E_m(\rho,s)$ can be written as

$$E_m(\rho,s) = \begin{cases} (-1)^m \exp(-s\epsilon_{1r}^{\frac{1}{2}})\, F(s)\, [I_m(s\epsilon_{1r}^{\frac{1}{2}}\rho) + b_m(s)\, K_m(s\epsilon_{1r}^{\frac{1}{2}}\rho)] & \text{in } \mathcal{D}_1, \\ (-1)^m \exp(-s\epsilon_{1r}^{\frac{1}{2}})\, F(s)\, c_m(s)\, e_m(\rho,s) & \text{in } \mathcal{D}_2. \end{cases} \tag{24}$$

The corresponding coefficient for the magnetic field assumes the form

$$H_m(\rho,s) = \begin{cases} (-1)^m \exp(-s\epsilon_{1r}^{\frac{1}{2}})\, F(s)\, \epsilon_{1r}^{\frac{1}{2}}\, [I_m'(s\epsilon_{1r}^{\frac{1}{2}}\rho) + b_m(s)\, K_m'(s\epsilon_{1r}^{\frac{1}{2}}\rho)], \\ (-1)^m \exp(-s\epsilon_{1r}^{\frac{1}{2}}) F(s)\, c_m(s)\, h_m(\rho,s), \end{cases} \tag{25}$$

for $0 \le \rho \le 1$ and for $1 \le \rho < \infty$, respectively, where the primes indicate a differentation of the modified Bessel functions $I_m(z)$ and $K_m(z)$ with respect to their arguments. In (24) and (25), the normalized pair of solutions $\{e_m(\rho,s), h_m(\rho,s)\}$ is defined by

$$\begin{aligned} \left[\partial_\rho^2 + \rho^{-1}\partial_\rho - s^2\epsilon_r(\rho) - s\sigma(\rho) - m^2/\rho^2\right] e_m(\rho,s) &= 0, \\ \lim_{\rho\downarrow 0} \rho^{-|m|} e_m(\rho,s) &= 1, \\ h_m(\rho,s) &= s^{-1}\partial_\rho e_m(\rho,s). \end{aligned} \tag{26}$$

The first equation of (26) is identical to (20), while the last one follows from (19). The values of the coefficients $b_m(s)$ and $c_m(s)$ follow from invoking the boundary conditions (3) at the interface between regions $\mathcal{D}_1$ and $\mathcal{D}_2$. Enforcing the continuity of $e_m(\rho,s)$ and $h_m(\rho,s)$ at $\rho = 1$ results in

$$\begin{aligned} b_m(s) &= \left[\epsilon_{1r}^{\frac{1}{2}}\, I_m'(s\epsilon_{1r}^{\frac{1}{2}})\, e_m(1,s) - I_m(s\epsilon_{1r}^{\frac{1}{2}})\, h_m(1,s)\right] / g_m(s), \\ c_m(s) &= 1/s g_m(s), \end{aligned} \tag{27}$$

with

$$g_m(s) \stackrel{\text{def}}{=} K_m(s\epsilon_{1r}^{\frac{1}{2}})\, h_m(1,s) - \epsilon_{1r}^{\frac{1}{2}}\, K'(s\epsilon_{1r}^{\frac{1}{2}})\, e_m(1,s)$$

being the characteristic cylinder denominator.

4.2. Contributions from the Singularities

Since the field distributions $e_m(\rho, s)$ and $h_m(\rho, s)$ defined by (26) are entire functions of s, we observe from (24) and (27) that the singularities in the complex s-plane comprise the poles $\{s_{mn}\}$, which are found as zeros of the cylinder denominator $g_m(s)$:

$$g_m(s_{mn}) = g_m(s^\star_{mn}) = 0, \tag{28}$$

with $n = 0, 1, \ldots, \infty$, as well as a branch point at $s = 0$ due to the presence of the modified Bessel function $K_m(z)$. The corresponding branch cut is chosen along the negative real s-axis.

By applying Cauchy's formula to a contour circumscribing the whole complex s-plane excepting the singularities and the branch cut, we obtain the generalized Mittag-Leffler representation

$$E_m(\rho, s) = \sum_{n=0}^{\infty} \left\{ \frac{r_{mn}(\rho)}{s - s_{mn}} + \frac{r^\star_{mn}(\rho)}{s - s^\star_{mn}} \right\} + \int_0^\infty d\nu \left\{ \frac{r_{m\nu}(\rho)}{s + \nu} \right\} + E^e_m(\rho, s). \tag{29}$$

In (29), $E^e_m(\rho, s)$ is an entire function and $r_{mn}(\rho)$ denotes the residual contribution from a pole at $s = s_{mn}$. $r_{m\nu}(\rho)$ is proportional to the jump across the branch cut:

$$r_{m\nu}(\rho) \stackrel{\text{def}}{=} \frac{i}{2\pi} \lim_{\delta \downarrow 0} \left[E_m(\rho, -\nu + i\delta) - E_m(\rho, -\nu - i\delta) \right], \tag{30}$$

and can be interpreted as the residual contribution from the point $s = -\nu$ on the branch cut.

With (16), the contribution from the singular part of (29) to the space-time domain field is directly obtained as

$$\mathcal{E}_{\text{singularities}}(\rho, \phi, t) = \lim_{M \to \infty} \sum_{m=-M}^{M} \exp(im\phi) \left\{ 2\text{Re} \sum_{n=0}^{\infty} r_{mn}(\rho) \exp(s_{mn} t) + \int_0^\infty d\nu \, r_{m\nu}(\rho) \exp(-\nu t) \right\}. \tag{31}$$

In (31), the term between braces is real-valued and invariant under a change of sign in m. Hence, the complex exponential $\exp(im\phi)$ can be replaced by $\cos(m\phi)$, and the summation over m can be reduced to one over $m \geq 0$.

4.2.1. Residual Contributions from the Poles The residues $r_{mn}(\rho)$ can be determined by following a procedure first outlined by Marin for perfectly conducting scatterers [20], and generalized to dielectric obstacles in [11,18]. Starting point is the formulation of (23) in operator form:

$$L_m(s) E_m(\rho, s) = E^i_m(\rho, s), \tag{32}$$

in which

$$L_m(s) f(\rho) \stackrel{\text{def}}{=} \int_0^\infty d\rho' \left[\delta(\rho - \rho') + s^2 C_1(\rho', s) G_m(\rho, \rho'; s) \right] f(\rho'). \tag{33}$$

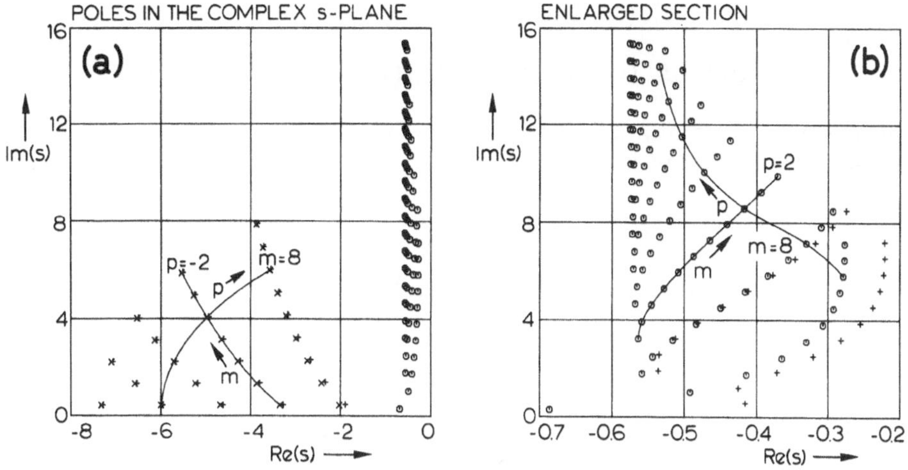

Figure 2: Natural frequencies s_{mn} with $0 \le m \le 10$ and $0 \le n \le 10$ for a dielectric cylinder with a parabolic permittivity profile $\epsilon_{2r}(\rho) = 6.25 - 4\rho^2$ and a constant conductivity $\sigma = 2$ embedded in free space. The solid lines connect families of poles with $m = 8$ and with $p = \pm 2$, respectively. Poles corresponding to whispering-gallery modes are indicated by o's; creeping-wave poles by x's; the +'s are asymptotic approximations. In Figure 2.b, the real s-axis was stretched to bring out the ordering in the poles with $p \ge 0$.

Substituting (29) in (32) and using the analytic properties of $L_m(s)$, $E^i_m(\rho, s)$ and $E^e_m(\rho, s)$ in the vicinity of a pole, we obtain

$$r_{mn}(\rho) = \lambda_{mn} E_{mn}(\rho), \tag{34}$$

where λ_{mn} is an as yet unknown constant and where

$$E_{mn}(\rho) \stackrel{\text{def}}{=} \begin{cases} e_m(\rho, s_{mn}), & \text{for } 0 \le \rho \le 1, \\ K_m(s_{mn}\, \epsilon_{1r}^{\frac{1}{2}}\, \rho) e_m(1, s_{mn}) / K_m(s_{mn}\, \epsilon_{1r}^{\frac{1}{2}}), & \text{for } 1 \le \rho < \infty, \end{cases} \tag{35}$$

denotes the natural-mode field distribution. Note that, with (26) and (28), it follows directly that $E_{mn}(\rho)$ satisfies the source-free integral equation

$$L_m(s_{mn}) E_{mn}(\rho) = 0. \tag{36}$$

Next, we define the *object product* of two functions $f(\rho, s)$ and $g(\rho, s)$ as

$$\langle f(\rho, s_1) \mid g(\rho, s_2) \rangle = \int_0^1 \rho d\rho\, f(\rho, s_1) C_1(\rho, s_2) g(\rho, s_2) \tag{37}$$

The kernel in this object product is the contrast function $C_1(\rho, s)$ for the complex frequency $s = s_2$ that also occurs in the function $g(\rho, s)$. Using (33), we find that the operator $L_m(s)$ has the following property:

$$\langle f(\rho, s_1) \mid L_m(s_2) g(\rho, s_2) \rangle = \langle L_m(s_2) f(\rho, s_1) \mid g(\rho, s_2) \rangle. \tag{38}$$

Finally, combining (32), (36) and (38), we find

$$\langle \{L_m(s) - L_m(s_{mn})\} E_m(\rho, s) \mid E_{mn}(\rho)\rangle = \langle E^i_m(\rho, s) \mid E_{mn}(\rho)\rangle. \tag{39}$$

The unknown constant λ_{mn} is determined by substituting in (39) the formulas (29) and (34) and the expansion

$$L_m(s) = L_m(s_{mn}) + L'_m(s_{mn})(s - s_{mn}) + \mathcal{O}\left[(s - s_{mn})^2\right], \tag{40}$$

where $L'_m(s)$ is the derivative operator of $L(s)$:

$$L'_m(s) f(\rho) \stackrel{\text{def}}{=} \int_0^1 d\rho' \, \partial_s \left[s^2 C_1(\rho', s) G_m(\rho, \rho', s)\right] f(\rho'). \tag{41}$$

Taking the limit $s \to s_{mn}$ in (39), we then determine λ_{mn} as

$$\lambda_{mn} = F(s_{mn}) D_{mn}, \tag{42}$$

where the coupling coefficient D_{mn} is given by

$$D_{mn} \stackrel{\text{def}}{=} (-1)^m \exp(-s_{mn}\epsilon_{1r}^{\frac{1}{2}}) \frac{\langle I_m(s_{mn}\, \epsilon_{1r}^{\frac{1}{2}} \rho) \mid E_{mn}(\rho)\rangle}{\langle L'_{mn}(s_{mn}) E_{mn}(\rho) \mid E_{mn}(\rho)\rangle}. \tag{43}$$

D_{mn} depends on the configuration only. In its actual computation from $E_{mn}(\rho)$, both a single and a double space integral have to be evaluated.

4.2.2. Contribution from the Branch Cut The residual contribution from the branch cut follows directly by substituting (24) and (27) in (30). For real-valued positive ν, we have a jump discontinuity in $K_m(-\nu\, \epsilon_{1r}^{\frac{1}{2}})$ [21, formula 9.6.31], while $e_m(\rho, s)$, $h_m(\rho, s)$ and $I_m(s\, \epsilon_{1r}^{\frac{1}{2}} \rho)$ are continuous at $s = -\nu$. Using these properties directly yields a closed-form expression for $r_{m\nu}(\rho)$.

More important is that, from (15), (30) and (32), it follows that $r_{m\nu}(\rho)$ satisfies the homogeneous equation

$$L_m(-\nu) r_{m\nu}(\rho) = 0. \tag{44}$$

This confirms the statement in the introduction of this section, that the branch-cut contribution can be regarded as a continuous superposition of natural-mode contributions.

4.3. Contribution from the Closing Contours

In the total time-domain response, the entire function $E^e_m(\rho, s)$ in (29) yields, in the time domain, the contribution of the supplementary arc in either the left or the right half of the s-plane. The role of the entire function can be analyzed by constructing a first-order WKB approximation for m fixed and $|s| \to \infty$, and applying Jordan's lemma. Details can be found in [18] and [19, Subsection 2.5.3]. We find that the contributions along the closing contours vanish as indicated in Table 1.

Table 1: Time intervals for vanishing contributions from closing contours.

closure in half-plane	total field inside cylinder $(0 \leq \rho \leq 1)$	scattered field outside cylinder $(1 \leq \rho < \infty)$	total field outside cylinder $(1 \leq \rho < \infty)$
right	$-\infty < t < \tau(\rho)$	$-\infty < t < \tau(\rho)$	$-\infty < t < -\tau(\rho)$
left	$T - \tau(\rho) < t < \infty$	$T + \tau(\rho) + 2 < t < \infty$	$T + \tau(\rho) < t < \infty$

In this table, $\tau(\rho)$ denotes the time it takes for a wave to travel in the radial direction from $\rho = 1$ to the point of observation:

$$\tau(\rho) \stackrel{\text{def}}{=} \begin{cases} \int_\rho^1 d\rho' \, \epsilon_{2r}^{\frac{1}{2}}(\rho'), & \text{for } 0 \leq \rho \leq 1, \\ (\rho - 1)\epsilon_{1r}^{\frac{1}{2}}, & \text{for } 1 \leq \rho < \infty. \end{cases} \tag{45}$$

From Table 1, we observe that for $T < 2\tau(1)$, there is some ρ_1 in $\mathcal{D}_2$ such that $T - \tau(\rho_1) = \tau(\rho_1)$. For $0 \leq \rho < \rho_1$, the electric field can always be determined by closing the Bromwich contour such that the integral along the supplementary arc vanishes. For $\rho > \rho_1$, however, there is always a time interval in which the possible contribution from the supplementary arcs remains to be investigated. When $T > 2\tau(1)$, such an interval is present for all ρ.

From the last line of Table 1, we also see that, for $1 \leq \rho < \infty$, there is a time interval of length 2 where the contour can be closed to the left for the total field, while it cannot be closed for the scattered field. Physically, this corresponds to the phenomenon that, for these space-time points, the part of $\mathcal{E}_m^i(\rho, t)$ that represents a wave traveling away from the cylinder is at least partially canceled by the scattered field. In that space-time region, special care is needed in the numerical evaluation of the angular Fourier series (12), because this series only converges globally for the scattered field.

From Table 1, it is observed that the closing conditions do not depend on the azimuthal angle of observation. This is caused by the fact that, in (12), we have decomposed the problem into a series of one-dimensional subproblems, each of which describes a wave that travels in the radial direction only, and whose angular field distribution is $\cos(m\phi)$, independently of ρ and t. The true causal nature of the field (e.g. the turn-on time of the field at a given space point) only follows when all the azimuthal constituents $\{\mathcal{E}_m(\rho, t)\}$ are added up according to (12). This is the price we have to pay for obtaining a globally converging representation and, hence, for being able to analyze the role of the entire function as outlined above.

4.4. Results

The theory presented in this section has been implemented numerically. In order to obtain accurate values of the natural-mode parameters, the normalized field distributions $\{e_m(\rho, s), h_m(\rho, s)\}$ were computed with the Runge-Kutta-Verner scheme explained in Appendix B.1. In view of the symmetry observed in (28) and (31), the computation was restricted to $m \geq 0$ and $\text{Im}(s) \geq 0$. The approximate location of

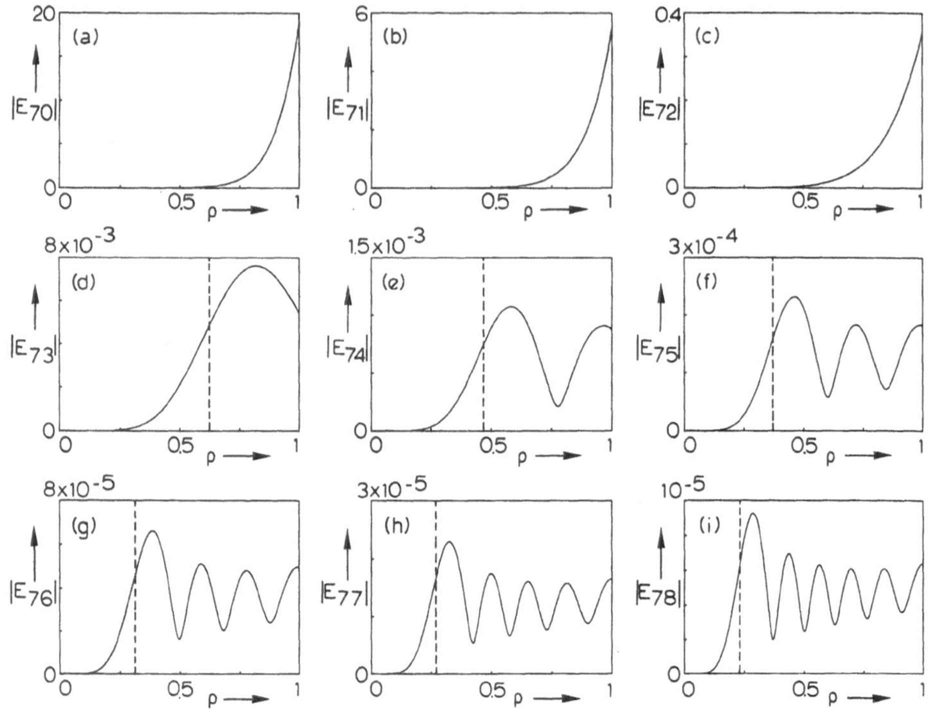

Figure 3: The magnitude of the natural-mode field distributions $E_{mn}(\rho)$ with $m = 7$, $0 \leq n \leq 8$, for the configuration specified in Figure 2 plotted for $0 \leq \rho \leq 1$. The dashed lines indicate the locations of the transition points referred to in Subsection 5.2.

the zeros of $g_m(s)$ was determined from argument-type integrals [22]. Subsequently, the estimates obtained were improved by Muller's method [23]. The integrals in (43) were evaluated by applying a single and a twice-repeated trapezoidal rule.

In Figures 2–5, results are shown for a lossy, radially inhomogeneous dielectric cylinder with a quadratic permittivity profile $\epsilon_{2r}(\rho) = 6.25 - 4\rho^2$ and a constant conductivity $\sigma = 2$, embedded in free space. In Figure 2.a, the natural frequencies are shown. In Figure 2.b, we have enlarged the portion of the complex plane near the imaginary axis. In both these figures, two types of orderings are observed. As expected, the frequencies can be grouped according to their angular index m. From the results shown in Figure 2, it is observed that the natural frequencies can also be grouped into families with $n - \text{ent}(m/2) = p$ being a constant, integer number. For $p < 0$, their locations stretch out into the negative half of the complex s-plane; for $p \geq 0$, they remain near the imaginary s-axis.

In Figure 3, the first few natural-mode distributions have been plotted in absolute value for $0 \leq \rho \leq 1$ and $m = 7$. Note that, for modes with $p < 0$, the magnitude of the field increases exponentially with increasing ρ (see Figures 3.a,b,c) while, for modes with $p \geq 0$, p minima are observed in the interval (see Figures 3.d–i).

In Figure 4, the electric field at the center of the cylinder is given as a function of time. Because $E_m(0, s) = 0$ for $m \neq 0$, only the term with $m = 0$ needs to be

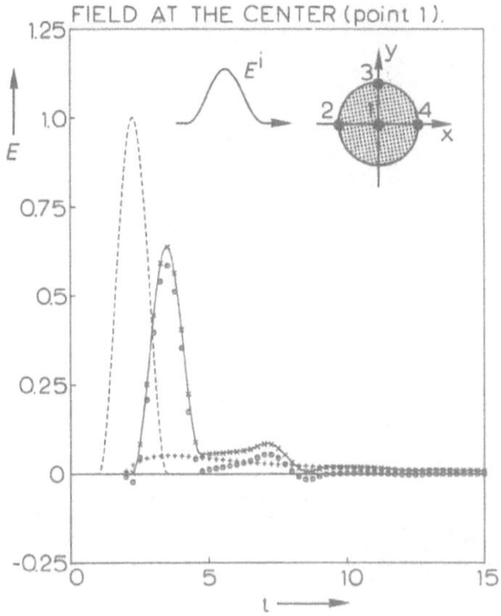

Figure 4: The electric field at the center of the cylinder specified in Figure 2 for the incident-pulse shape $\mathcal{F}(t) = \sin^2(\pi t/T)\text{rect}(t - T/2; T)$ with $T = 2.5$. Dashed line: incident field; solid line: result of FFT; x: total SEM result; o: total pole contribution; +: branch-cut contribution. Inset: the scattering configuration.

taken into account. Figure 4 shows the result of directly evaluating the Bromwich inversion integral in (16) as well as the contributions from the singularities in the complex s-plane and the contributions from the poles and the branch cuts separately. Fields obtained from a similar computation for points on the boundary are shown in Figure 5. The wavefronts observed in Figures 4 and 5 are associated with reflections at or propagation along the cylinder boundary. A comparison with results for different scattering configurations has confirmed this conclusion.

5. Contour Deformation in the Complex Order Plane

From the results presented in Subsection 4.4, it was observed that the natural modes of our "tutorial" scattering problem can be ordered into families in two complementary ways. In addition, a physical interpretation was given for the total residual contribution from a family of poles with constant angular order m. In this section, we analyze families with constant index p.

An analysis of the manner in which natural frequencies are grouped in the complex-frequency plane was first given by Überall and Gaunaurd [24], for the case of radar scattering by a perfectly conducting sphere. For that configuration, it was shown that the poles can be ordered into families such that the contribution from each family of poles constitutes a creeping wave that circumnavigates the sphere during the scattering process. Since then, numerous configurations have

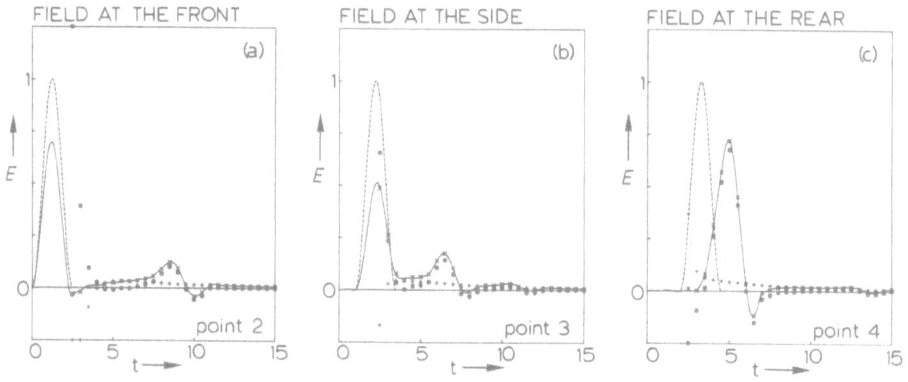

Figure 5: Time-domain electric fields at the points 2, 3 and 4 shown in Figure 4 for the configuration specified in Figures 2 and 4. The summations were truncated at $|m| = 5$ and $n = 10$. Symbols as given in Figure 4.

been analyzed in a similar manner (see e.g. [25,26,27] and references cited therein). As far as electromagnetic problems are concerned, the attention has primarily been devoted to rotationally symmetric, perfectly conducting targets, with or without a dielectric coating. We mention in particular the work of Heyman and Felsen [28,29], who considered a perfectly conducting cylinder. All these applications have in common the feature that the field does not penetrate into the interior of the scattering obstacle; i.e. that only "external" modes are excited.

For a dielectric cylinder, the incident field does cause a response inside the obstacle. In [30] and [19, Section 2.6], it was shown that the modes can be subdivided into whispering-gallery and creeping-wave modes, which propagate on the inside and the outside, respectively, along the cylinder boundary. In this section, we summarize te results. As in the previous section, we restrict ourselves to presenting the most salient aspects of the analysis.

5.1. Watson Transformation

As long as we are not interested in numerical efficiency, we can also carry out the angular Fourier transformation and the temporal Laplace transformation as defined in Subsection 3.2 in reverse order. This leads to

$$\mathcal{E}(\rho,\phi,t) = \frac{1}{2\pi i}\int_{\mathcal{L}_\beta} ds\, \exp(st - s\,\epsilon_{1r}^{\frac{1}{2}})\, F(s)\hat{E}(\rho,\phi,s), \tag{46}$$

where

$$\hat{E}(\rho,\phi,s) = \sum_{m=-\infty}^{\infty} (-1)^m \exp(im\phi)\hat{E}_m(\rho,s), \tag{47}$$

with

$$\hat{E}_m(\rho,s) \stackrel{\text{def}}{=} \begin{cases} I_m(s\,\epsilon_{1r}^{\frac{1}{2}}\rho) + b_m(s)K_m(s\,\epsilon_{1r}^{\frac{1}{2}}\rho), & \text{in } \mathcal{D}_1, \\ c_m(s)e_m(\rho,s), & \text{in } \mathcal{D}_2. \end{cases} \tag{48}$$

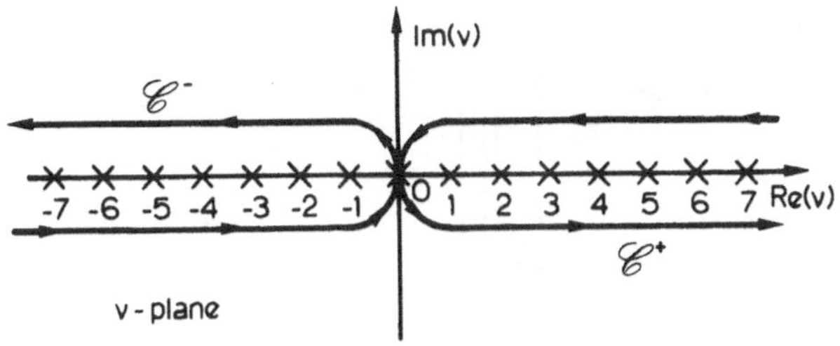

Figure 6: Integration contour for the Watson transformation (49).

In (48), the normalized field distributions $\{e_m(\rho, s), h_m(\rho, s)\}$ are defined according to (26), and the coefficients $b_m(s)$ and $c_m(s)$ according to (27).

Compared with (16), we have taken out the factor $\exp(-s\,\epsilon_{1r}^{\frac{1}{2}})\,F(s)$, which represents the spectrum of the incident pulse. $\hat{E}(\rho, \phi, s)$ reduces to the customary unit-amplitude frequency-domain response when s is replaced with $i\omega$. Apart from the complex time factor $\exp(i\omega t)$, $\hat{E}(\rho, \phi, i\omega)$ represents the field excited by an electrically polarized, monochromatic plane wave with angular frequency ω, incident from the direction $\phi = \pi$.

As is well known from the literature (see e.g. [3,31]), the convergence of (47) slows down as s increases in magnitude. This was the reason for carrying out the two transformations in their original order in Subsection 3.2. To resolve this problem, we convert the summation over m into an integral in the complex ν-plane by the *Watson transformation*:

$$\hat{E}(\rho, \phi, s) = \frac{1}{2i} \int_C \frac{d\nu}{\sin(\nu\pi)} \exp(i\nu\phi)\, \hat{E}_\nu(\rho, s), \tag{49}$$

where $\hat{E}_\nu(\rho, s)$ is obtained from $\hat{E}_m(\rho, s)$ by replacing $I_m(z)$, $K_m(z)$, $e_m(\rho, s)$ and $h_m(\rho, s)$ with $I_{\sqrt{\nu^2}}(z)$, $K_\nu(z)$, $e_\nu(\rho, s)$ and $h_\nu(\rho, s)$, respectively. The pair of normalized solutions $\{e_\nu(\rho, s), h_\nu(\rho, s)\}$ is defined by

$$\begin{aligned} &\left[\partial_\rho^2 + \rho^{-1}\partial_\rho - s^2\epsilon_r(\rho) - s\sigma(\rho) - \nu^2/\rho^2\right] e_\nu(\rho, s) = 0, \\ &\lim_{\rho\downarrow 0} \rho^{-\sqrt{\nu^2}} e_\nu(\rho, s) = 1, \\ &h_\nu(\rho, s) = s^{-1}\partial_\rho e_\nu(\rho, s). \end{aligned} \tag{50}$$

In $I_{\sqrt{\nu^2}}(z)$ and in (50), the square root is chosen according to its principal value, i.e. $\sqrt{\nu^2} = \nu$ when $\mathrm{Re}(\nu) > 0$, and $\sqrt{\nu^2} = -\nu$ when $\mathrm{Re}(\nu) < 0$.

The integration contour consists of two contours C^+ and C^-, which encircle the positive and the negative half of the real ν-axis, respectively as shown in Figure 6. Both integrals are defined according to their Cauchy principal value at $\nu = 0$; i.e. half the residue at $\nu = 0$ is attributed to either integral.

Finally, it should be noted that, in (49) and (50), we have represented the total field in terms of constituents that remain bounded near the origin. From the

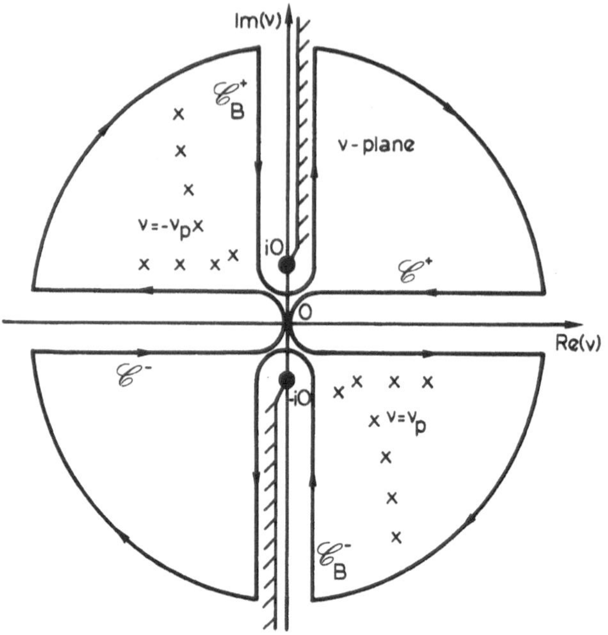

Figure 7: Contour deformation and singularities in the complex ν-plane (the pole locations apply to $s = i\omega$ with ω positive real).

definition (50), it is observed that $e_\nu(\rho, s)$ satisfies the symmetry relations

$$\begin{aligned} e_{-\nu}(\rho, s) &= e_\nu(\rho, s), \\ e_{\nu^*}(\rho, s^*) &= [e_\nu(\rho, s)]^* . \end{aligned} \tag{51}$$

Moreover, these relations also hold for $h_\nu(\rho, s)$, $K_\nu(s\,\epsilon_{1r}^{\frac{1}{2}}\,\rho)$ and $I_{\sqrt{\nu^2}}(s\,\epsilon_{1r}^{\frac{1}{2}}\,\rho)$ and, hence, by (27), (48) and (50), for $\hat{E}_\nu(\rho, s)$ as well.

At this stage in the discussion, it may be helpful to note that the integral in (49) is in itself a Fourier representation of the harmonic response $\hat{E}(\rho, \phi, s)$. This representation can also be obtained in a more direct manner, by applying a continuous spatial Fourier transformation to the periodic extension of $\hat{E}(\rho, \phi, s)$ into the infinite domain $-\infty < \phi < \infty$ (see [32] or [33, Section 6.7]). As such, this representation constitutes an alternative starting point for the modal analysis of our "tutorial" scattering problem.

In the literature on wave propagation, the representation (49) is known as an *angular transmission representation.* As argued in [33, Subsection 6.2.b], the contour C encloses all singularities in the complex ν-plane that correspond to *radially* propagating waves but no others. Equation (46) is then the discrete representation of $\hat{E}(\rho, \phi, s)$ in terms of such waves. This is in agreement with the analytical and numerical results obtained in Section 4.

A form equivalent to (46) in terms of *angularly* propagating waves is obtained by evaluating the contour integral in (49) by contour deformation in the complex ν-plane away from the real ν-axis. To obtain this representation, we supplement C^+ and C^- by circular arcs at infinity in the complex ν-plane, and we apply Cauchy's theorem. Analyzing the asymptotic behavior of $\hat{E}_\nu(\rho, s)$ shows that, unless $\arg(\nu) = \pm\pi/2$, $\hat{E}_\nu(\rho, s)$ decays faster than exponentially as $|\nu| \to \infty$ (see [18], [19, Section 2.6]). This means that the integrals along the closing contours vanish and that only the singularities of $\hat{E}_\nu(\rho, s)$ contribute to the result. These singularities comprise a denumerable set of poles and a pair of branch cuts along the negative and positive imaginary axes, as shown in Figure 7 ($\sqrt{\nu^2}$ may be thought of as $[(i\nu)(-i\nu)]^{\frac{1}{2}}$ with branch points at $\nu = \pm i0$; see also [34]).

5.2. Residual Contribution from the Poles

Let us consider the poles in more detail. Since $\hat{E}_\nu(\rho, s)$ satisfies (51), it follows that these poles occur in pairs $\nu = \pm\nu_p(s)$, with $\mathrm{Re}(\nu_p) \geq 0$. The $\{\nu_p(s)\}$ are found by searching the right half of the complex ν-plane for the roots of the characteristic equation

$$g_\nu(s) \stackrel{\mathrm{def}}{=} K_\nu(s\,\epsilon_{1r}^{\frac{1}{2}})\,h_\nu(1, s) - \epsilon_{1r}^{\frac{1}{2}}\,K_\nu'(s\,\epsilon_{1r}^{\frac{1}{2}})\,e_\nu(1, s) = 0, \tag{52}$$

with $g_\nu(s)$ being the analytic continuation of the characteristic cylinder denominator $g_m(s)$ into the complex ν-plane, excepting the imaginary axis. From the same symmetry, it also follows that

$$\mathrm{res}\;\hat{E}_\nu(\rho, s)\Big|_{\nu=-\nu_p} = -\,\mathrm{res}\;\hat{E}_\nu(\rho, s)\Big|_{\nu=\nu_p} \stackrel{\mathrm{def}}{=} -\,r_p(\rho, s). \tag{53}$$

With this relation, the residual contribution from a pole pair $\nu = \pm\nu_p(s)$ to $\hat{E}(\rho, \phi, s)$ takes the form:

$$(-4\pi i)\,r_p(\rho, s)\,\frac{\cos[\nu_p(s)\phi]}{2i\sin[\nu_p(s)\pi]} = \tag{54}$$

$$(-2\pi i)\,r_p(\rho, s)\sum_{n=0}^{\infty}\{\exp[i\nu_p(s)\,(\phi - \pi - 2\pi n)] + \exp[i\nu_p(s)\,(-\phi - \pi - 2\pi n)]\},$$

where, in the second line, it has been assumed that $\mathrm{Im}(\nu_p(s)) < 0$. The residue $r_p(\rho, s)$ can be determined according to the same procedure that was used in Subsection 4.2 to determine the residues of $E_m(\rho, s)$ at poles in the complex s-plane. In this manner, the numerical differentiation of the numerically obtained denominator $g_\nu(s)$ is avoided. Details can be found in [19, Appendix 2.6.B]. As in (34), the residue $r_p(\rho, s)$ is, for $0 \leq \rho \leq 1$, linearly proportional to the normalized electric-field distribution $e_{\nu_p}(\rho, s)$.

5.2.1. Angular Propagation The physical interpretation of the residual contribution (54) follows when we take $s = i\omega$ with ω real and positive, and combine the second line of (54) with the complex time factor $\exp(i\omega t)$. Then, it is observed that (54) represents this contribution in terms of traveling waves that originate from $\phi = \pi + 2\pi n$ and $\phi = -\pi - 2\pi n$, with $n = 0, 1, \ldots, \infty$, and propagate in

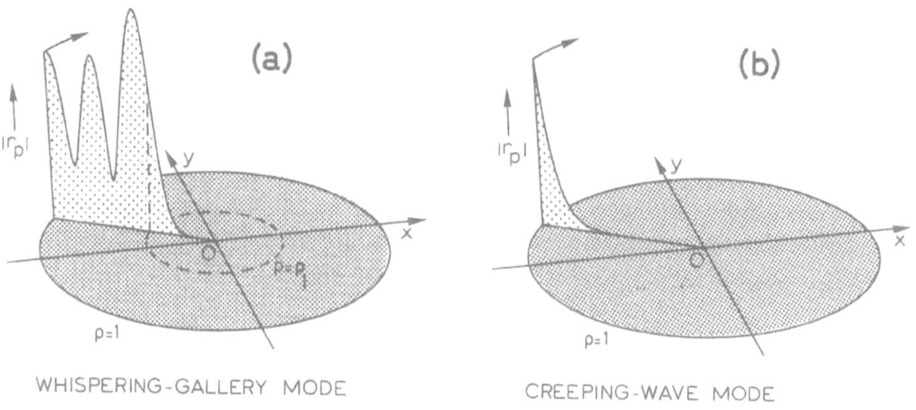

Figure 8: Behavior of a natural mode inside the cylinder. (a) Whispering-gallery mode. (b) Surface-wave part of creeping wave mode. In both cases, a similar mode moves in the opposite ϕ-direction.

the negative and positive ϕ-direction, respectively. Consequently, the terms with index n represent traveling waves that circumnavigate the central axis n times. The angular behavior is illustrated graphically in Figure 8.

This interpretation also explains why in Figure 7 and in the second line of (54) the $\{\nu_p(s)\}$ have been restricted to the fourth quadrant of the complex ν-plane; a pole in the first quadrant would give rise to a similar expansion as in (54) involving noncausal terms.

5.2.2. Radial Distribution For a further interpretation of the poles in the complex ν-plane and the corresponding residual contributions, information is needed on the behavior of $\hat{e}_\nu(\rho, s)$ as a function of ρ and ν. Since this distribution can, in general, only be determined numerically, we use asymptotic expressions resulting from a generalized WKB asymptotics. Our asymptotic theory is applicable when the magnitude of one or both of the parameters s and ν is large. Details can be found in [30] and in [19, Appendix 2.6.A]. For s large, this theory yields an expression for $e_\nu(\rho, s)$ that is valid for each complex ν.

The radial behavior of $e_\nu(\rho, s)$ will be different according to whether or not a transition point $\rho = \rho_1$ is present on the interval $0 < \rho < 1$. When $\arg(\nu) = \arg(s) - \pi/2 + \mathcal{O}(\alpha^{-1})$, with $\alpha \overset{\text{def}}{=} \max\{|\nu|, |s|\}$, a transition point is present. In that case $e_\nu(\rho, s)$ increases exponentially in magnitude from $\rho = 0$ to $\rho = \rho_1$. Between the transition point $\rho = \rho_1$ and the boundary $\rho = 1$, $e_\nu(\rho, s)$ has the form of a standing wave with p minima in its magnitude. This leads to the visualization in Figure 8.a, which suggests that a pole of this type corresponds to a *whispering-gallery mode*. The standing wave in $\rho_1 < \rho < 1$ is then identified as being caused by a partial reflection at the cylinder boundary, and a total reflection at the surface $\rho = \rho_1$ (see also [35]). The exponentially increasing part of the modal field distribution up to $\rho = \rho_1$ can be understood as being the broken wave caused by the total reflection at $\rho = \rho_1$. This interpretation was confirmed by considering the case where the outer medium becomes impenetrable, i.e. by taking the limit $\epsilon_{1r}^{\frac{1}{2}} \to \infty$. In that limit, our theory reproduces the results obtained previously by Wasylkiwskyj [36].

In the absence of a transition point, i.e. for all combinations of s and ν that are not covered by the preceding paragraph, $e_\nu(\rho, s)$ increases exponentially in magnitude across the entire interval $0 \leq \rho \leq 1$. In combination with (54), this leads to a residual contribution which in $\mathcal{D}_2$ behaves as a surface wave propagating along the cylinder boundary, as visualized in Figure 8.b. Like the part of the whispering-gallery mode up to $\rho = \rho_1$, such a wave can be understood as being a broken wave caused by the total reflection of a part of the incident field at the surface. The dominant part of the propagation takes place in $\mathcal{D}_1$, where this reflection gives rise to a *creeping wave*, i.e. a wave that creeps along the outer boundary of the cylinder, gradually radiating away its energy. It therefore seems appropriate to designate a mode of the present type as a *creeping-wave mode*. An additional verification of this interpretation can be obtained by letting the inner medium become impenetrable, i.e. by taking the limit for $\epsilon_{2r}(1) \to \infty$. In that limit, our results are in agreement with those obtained by Heyman and Felsen [28].

5.3. TRANSIENT RESPONSE

The transient response caused by the incident pulse specified in (9) now follows by evaluating the integral along the Bromwich contour $\mathcal{L}_\beta$ in (46). The SEM representation (31) is obtained when this evaluation is effectuated by deforming $\mathcal{L}_\beta$ into the negative half of the complex s-plane. Obviously, the modal representation only holds when the contribution from the closing contour vanishes, i.e. when the closing conditions listed in Table 1 are met.

The alternative derivation of the SEM representation follows when $\hat{E}(\rho, \phi, s)$ in (46) is replaced by the result of evaluating the contour integral in (49) by contour deformation in the complex ν-plane. As discussed in Subsection 5.1, this procedure leads to a representation of $\hat{E}(\rho, \phi, s)$ in terms of a series of residues as given in the first line of (54) and a contour integral around the branch cuts. Since the factor $\exp(st - s\,\epsilon_{1r}^{\frac{1}{2}})\,F(s)$ is analytic in the entire complex s-plane, the poles in that plane must, in this derivation, correspond to the zeros of the factor $\sin(\nu_p(s)\pi)$ in the denominator in (54). This leads to the characteristic equation

$$\nu_p(s) = m, \qquad m \in \mathbb{Z}. \tag{55}$$

In the terminology of the SEM approach, the solutions $s = s_{mp}$ of (55) are then the complex natural frequencies.

The pair of equations (52) and (55) provides a description of these natural frequencies which is dual to the description given in (28). In Section 4, the angular order was immediately restricted to $\nu = m$, with $m \in \mathbb{Z}$, which led to the definition of the natural frequencies as being zeros of the characteristic cylinder denominator $g_m(s)$ in the complex s-plane.

This connection between the poles in the ν-plane and those in the s-plane provides a physical interpretation of the total residual contribution from a family of the latter with constant p. By (46) and (49), that contribution equals an integral of the type (46) with $\beta = 0$ and with $\hat{E}(\rho, \phi, s)$ replaced by the second line of (54) for $\text{Im}(s) > 0$

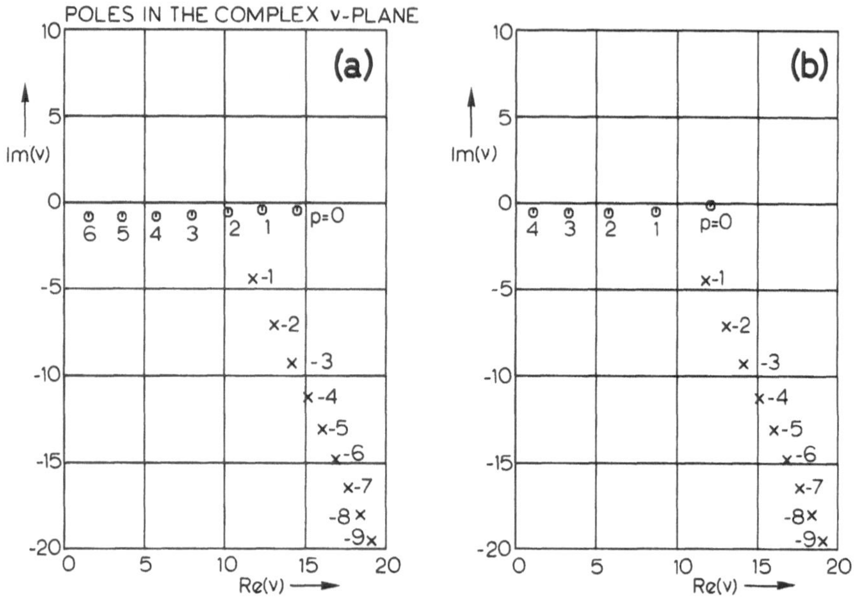

Figure 9: Poles in the complex ν-plane for $s = 10i$ for two configurations with $\epsilon_{1r} = 1$ and $\epsilon_{2r}(1) = 2.25$. (a) $\epsilon_{2r}(\rho) = 6.25 - 4\rho^2$, $\sigma_2(\rho) = 2$; (b) $\epsilon_{2r}(\rho) = 2.25$, $\sigma_2(\rho) = 0$; Poles corresponding to whispering-gallery modes are indicated by o's; creeping-wave poles by x's.

and by an equivalent form for $\mathrm{Im}(s) < 0$. In view of the interpretation of (54) given in Subsection 5.2, it is clear that this integral, and hence also the equivalent residual contribution, represents a pair of angularly propagating waves. The nature of these waves depends on the index p: they are of the whispering-gallery and of the creeping-wave type for $p \geq 0$ and $p < 0$, respectively.

5.4. Results

As in the case of the SEM, the theory presented in this section has been verified numerically. The normalized field distributions $\{e_\nu(\rho, s), h_\nu(\rho, s)\}$ were computed with a generalized version of the Runge-Kutta-Verner scheme explained in Appendix B.1. In view of the symmetry relations (51), the computation was restricted to $\mathrm{Re}(\nu) \geq 0$ and $\mathrm{Im}(s) \geq 0$. The zeros in the complex ν-plane were determined with the same procedure that was used to search the zeros of $g_m(s)$ in the complex s-plane.

As an illustration, Figure 9 shows the result for a real-valued frequency $\omega = 10$ ($s = 10i$) for two configurations with the same permittivity at the cylinder boundary and a vacuum outside the cylinder. The permittivity and the conductivity in $\mathcal{D}_2$ are different for both configurations. In Figure 9, only the exact poles have been plotted. On the scale of Figure 9, the poles resulting from our asymptotic theory cannot be distinguished from the exact ones. From Figure 9, it is observed that the creeping-wave poles for both configurations are nearly identical, while the poles corresponding to whispering-gallery modes differ considerably. This is in agreement with the physical interpretation given above: the creeping waves only encounter the boundary of the cylinder, while the whispering-gallery modes also depend on its interior properties. Similar computations were carried out for complex natural frequencies $s = s_{mp}$, available from the SEM analysis of Section 4. The results were in agreement with (55).

Further, the dashed lines in Figure 3 indicate the location of the transition points $\rho = \rho_1$ figuring in the asymptotic theory referred to in Subsection 5.2, while the +'s in Figure 2 indicate the location of the asymptotic natural frequencies resulting from this theory. Both approximations turn out to be surprisingly accurate.

6. Bromwich Inversion

While modal representations as discussed in Sections 4–5 provide a detailed insight into the physical aspects of the scattering process, they may not always constitute the most convenient vehicle for determining the time-domain response of a scattering obstacle. For one, the SEM representation is not valid at early times and for special pulse shapes such as a Gaussian pulse, which has an infinite duration. If we are merely interested in determining the time-domain response, it may actually be more efficient to carry out a "brute-force" computation.

As mentioned in the introduction, one possibility is to perform a marching-on-in-time computation, solving either a discretized space-time integral equation [37,38] or a spatially transformed equivalent [1]. Such an approach, however, may produce unstable results, and will certainly lead to the accumulation of discretization errors at late times (for an error analysis of time-marching methods, see [39] and [19, Section 3.2]).

An alternative is to directly carry out an inversion procedure of the form (16). Traditionally, the frequency-domain constituents are, in such an approach, computed with the same relative accuracy for all pertinent complex frequencies $\{s = \beta + i\omega \mid \omega \in \mathbb{R}\}$. The major drawback of doing this is that, for large angular frequencies ω, the computational effort required to determine a single constituent becomes proportional to some positive, integer power of $|\omega|$. As a consequence, the application of "brute-force" Fourier techniques has, in the last few years, mainly been restricted to special problems for which at least part of the frequency-domain solution is known in closed form. Typical examples can be found in [40,41,42].

Recently [43], it was observed that aiming for a constant relative accuracy in the frequency-domain constituents is not really necessary. Instead, based on the success of time-marching methods, it was proposed to discretize the space-time integral equation in space only, and to solve the resulting system of linear time-domain differential equations of a *fixed* dimension indirectly, by transforming to the time-Laplace domain. As a consequence, the computational effort needed to determine the approximate frequency-domain constituents becomes independent of ω.

One of the examples considered in [43] is our "tutorial" scattering problem of plane-wave scattering by a lossy, radially inhomogeneous dielectric circular cylinder. In this section, we put the results obtained for that configuration into the perspective of Fourier theory.

6.1. Determination of the Fourier Coefficients $\{\mathcal{E}_m(\rho, t)\}$

As argued in Subsection 3.2, the global convergence of the angular Fourier representation (12) can be utilized to truncate the summation over m at some $|m| = M$. This means that we only need to evaluate $M+1$ Fourier coefficients $\{\mathcal{E}_m(\rho, t) \mid m = 0, 1, \ldots, M\}$ for $0 \leq t < \infty$.

To this end, the relevant part of the inversion formula (16) is written in the form

$$\mathcal{E}_m(\rho, t) = \frac{1}{2\pi} \exp(\beta t) \int_{-\infty}^{\infty} d\omega \exp(i\omega t)\, E_m(\rho, \beta + i\omega), \tag{56}$$

with β being a real-valued, nonnegative parameter. Typically, the integral in (56) is evaluated numerically with the aid of a Fast Fourier Transformation (FFT). It should be stressed that β need not be identical to zero. On the other hand, β cannot be chosen arbitrarily large in view of the multiplication of the FFT result by the exponentially increasing factor of $\exp(\beta t)$, which enhances the numerical error in this result for large t.

In Appendix B, two possible procedures are described for determining the Laplace-transformed Fourier coefficients $\{E_m(\rho, \beta + i\omega)\}$. The Runge-Kutta-Verner integration of the first-order differential equations (18)–(19) was already employed implicitly in computing the time-domain results shown in Figures 4–5. The solid lines displayed in those figures were computed by evaluating the inversion integral in (56) for $\beta = 0$. The Runge-Kutta-Verner integration is an adaptive procedure, which consumes a fixed amount of computation time for each oscillation in the field distributions $\{e_m(\rho, \beta + i\omega), h_m(\rho, \beta + i\omega)\}$. Especially for short incident pulses, this may make the technique rather time-consuming.

An alternative is to solve the integral equation (23) with the aid of the recursive procedure proposed in Appendix B.2. As remarked in that appendix, this necessitates a choice of $\beta > 0$. Considered from a frequency-domain point of view, the idea is to require only a constant *absolute* error in $E_m(\rho, \beta + i\omega)$. When the pulse shape $\mathcal{F}(t)$ is two times piecewise continuously differentiable, as most practical pulse shapes are, application of the Riemann-Lebesgue theorem yields the asymptotic estimate

$$F(\beta + i\omega) = \mathcal{O}(|\omega|^{-2}), \tag{57}$$

as $|\omega| \to \infty$. Substituting this result in the error estimate (B.14), and eliminating a factor of ω in view of the oscillating behavior of $E_m(\rho, \beta + i\omega)$ directly shows that, for a fixed h, the error $\Delta \tilde{E}_m(n, \beta + i\omega)$ will at most remain constant in magnitude as ω becomes large. This means that we can use the same *fixed* discretization for all values of ω. The major improvement in this approach compared with "conventional" frequency-domain techniques is that the spectrum of the incident pulse is now accounted for. As a consequence, less computational effort is spent on computing overly accurate frequency-domain solutions for large ω.

Finally, it follows from (B.14) that the error in the time-domain solution obtained will be of $\mathcal{O}(h^2) = \mathcal{O}(N^{-2})$ as the number of space steps N increases.

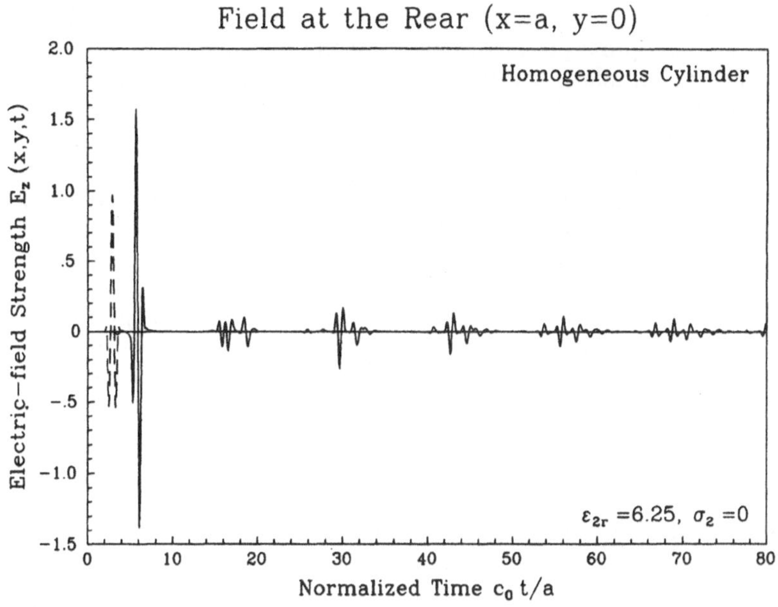

Figure 10: Time-domain response at the rear (point 4 as shown in Figure 4) of a homogeneous, lossless cylinder with $\epsilon_{2r} = 6.25$ in free space. The pulse shape is given by a raised-cosine pulse $\mathcal{F}(t) = \cos(7t/T)[1 + \cos(7t/2T)]\text{rect}(t; 4\pi T/7)$ with $T = 1$. The angular Fourier series was truncated at $|m| = 20$, the trapezoidal rule in (B.9) was taken with $N = 200$, and the Bromwich inversion (56) was carried out for $\beta = 0.05$ with an FFT of order 1500 over the time interval $-5 < t < 95$.

6.2. Results

Numerical results were obtained for a variety of configurations. We restrict ourselves to presenting a single time-domain result that confirms the physical interpretation given in Section 5.

In Figure 10, we consider plane-wave scattering by a homogeneous, lossless cylinder with a refractive index of 2.5 embedded in free space. The incident pulse was chosen such that it contains sufficient high-frequency information to excite the angularly propagating waves. For this configuration, the Laplace-transformed Fourier component $E_m(\rho, s)$ can be expressed in terms of modified Bessel functions. Hence, we can verify the results of our hybrid scheme independently.

Figure 10 displays the field at the rear end of the cylinder. In this figure, three interesting physical effects can be observed. In the first place, we can just distinguish a small precursor arriving at approximately $t = 4$. This precursor is not a numerical artefact. It is also observed in the "analytical" reference solution, even when both the number of Fourier components in the truncated spectral representation (12) and the order of the FFT in the temporal Fourier inversion (56) are doubled. Because of the value of the refractive index, a pulse traveling through the cylinder cannot arrive before $t = 5$. This precursor, therefore, can only be generated by creeping waves.

Second, we notice the arrival of a large pulse at $t = 5$. This arrival time is in accordance with the propagation through a homogeneous medium with refractive index 2.5. The amplitude is considerably larger than the amplitude of the corresponding incident pulse. This focusing effect can be attributed to the fact that the cylinder acts as a lens.

Third, we observe the repeated arrival of a more and more dispersed wavelet. This phenomenon can be explained from the fact that the cylinder supports whispering-gallery modes. For angular propagation along the interior of the cylinder boundary, the travel time would be given by $\tau_\phi = 5\pi \approx 15.7$. The intervals in Figure 10 are a little shorter, which indicates that the propagation takes place just inside that boundary. The dispersion effects are caused by the presence of several angularly propating waves with slightly different travel times. Additional confirmation of this interpretation has been obtained by comparing the time signal shown in Figure 10 with time signals at other points on the cylinder boundary.

As hinted above, we compared for this configuration all the results that were obtained with the integral-equation method with "reference" results obtained by expressing $E_m(\rho, s)$ in terms of modified Bessel functions. The corresponding results turned out to be graphically indistinguishable.

7. Conclusions

In this chapter, we have reviewed a number of different Fourier techniques for solving electromagnetic time-domain scattering problems. In particular, the attention has been focused on applying temporal and/or spatial Fourier or Laplace transformations with respect to those coordinates along which the scattering configuration is translationally invariant, while the incident field is not.

As far as the *singularity expansion method* is concerned, the principal advantage is that its application provides insight into the physical aspects of the scattering problem. In this context, the ordering of the poles in the complex-frequency plane is of particular importance. To obtain a complete understanding of this ordering, a *Watson transformation* in the complex-order plane is employed. This tranformation is also useful in computing the conventional frequency-domain response for large values of the angular frequency ω.

The principal disadvantage of modal methods is that the field representations obtained are not valid at early times and for special pulse shapes. In addition, the repeated solution of individual frequency-domain problems up to a constant *relative* accuracy needed in this approach consumes a considerable amount of computation time.

The result of the *Bromwich inversion* is valid at all times, and for all integrable pulse shapes, but provides little physical insight. The procedure turns out to be relatively efficient, especially when an integral-equation approach as described in Appendix B.2 and Subsection 6.1 is employed to evaluate the integrand up to a constant *absolute* accuracy. Whether such an approach is applicable to more complicated scattering geometries remains to be investigated. The deciding factor should be whether results with a given accuracy can be obtained more rapidly than with other available methods.

A. Mean Square Convergence of the Angular Fourier Series

In Subsection 3.2., it was argued that, for a suitable choice of the incident-pulse shape $\mathcal{F}(t)$, the electromagnetic field intensities $\mathcal{E}(\rho,\phi,t)$ and $\mathcal{H}(\rho,\phi,t)$ can, as far as their dependence on ϕ is concerned, be represented in terms of a Fourier series of the type (12). For a fixed ρ and t, the series obtained converges in mean square to the relevant quantity. In this appendix, we investigate the convergence as a function of ρ and t.

First, we consider the total field $\{\mathcal{E}(\rho,\phi,t),\mathcal{H}(\rho,\phi,t)\}$ inside the cylinder. Integrating the local form of the electromagnetic power balance (4) over the space-time domain $0<\rho<a$, $-\pi<\phi<\pi$, $z_0<z<z_0+1$, $0<t'<t$, results in

$$\frac{1}{2}\int_0^a \rho d\rho \int_{-\pi}^{\pi} d\phi\,[\epsilon(\rho)\boldsymbol{\mathcal{E}}\cdot\boldsymbol{\mathcal{E}}(\rho,\phi,t)+\mu_0\boldsymbol{\mathcal{H}}\cdot\boldsymbol{\mathcal{H}}(\rho,\phi,t)]$$
$$+\int_0^t dt'\int_0^a \rho d\rho\int_{-\pi}^{\pi} d\phi\,[\sigma(\rho)\boldsymbol{\mathcal{E}}\cdot\boldsymbol{\mathcal{E}}(\rho,\phi,t')]=$$
$$a\int_0^t dt'\int_{-\pi}^{\pi} d\phi\,[\mathcal{E}(a,\phi,t')\mathcal{H}(a,\phi,t')], \tag{A.1}$$

where $\mathcal{E}(\rho,\phi,t)$ and $\mathcal{H}(\rho,\phi,t)$ denote the tangential field components as specified in (10). Using the properties of $\epsilon(\rho)$ and $\sigma(\rho)$ given in (2), we arrive at the inequality

$$\int_0^a \rho d\rho\int_{-\pi}^{\pi} d\phi\,[\epsilon_0\mathcal{E}^2(\rho,\phi,t)+\mu_0\mathcal{H}^2(\rho,\phi,t)]$$
$$\leq 2a\int_0^t dt'\int_{-\pi}^{\pi} d\phi\,[\mathcal{E}(a,\phi,t')\mathcal{H}(a,\phi,t')]. \tag{A.2}$$

Now, we consider the difference

$$\Delta\mathcal{E}_M(\rho,\phi,t)\stackrel{\text{def}}{=}\mathcal{E}(\rho,\phi,t)-\sum_{m=-M}^{M}\exp(im\phi)\,\mathcal{E}_m(\rho,t), \tag{A.3}$$

and the similarly defined difference $\Delta\mathcal{H}_M(\rho,\phi,t)$. $\mathcal{E}(\rho,\phi,t)$ and each linear combination

$$\exp(im\phi)\,\mathcal{E}_m(\rho,t)+\exp(-im\phi)\,\mathcal{E}_{-m}(\rho,t), \tag{A.4}$$

along with the corresponding magnetic field, constitutes a real-valued solution of Maxwell's equations (1).

By virtue of the superposition principle, the linear combination on the right-hand side of (A.3) and the corresponding magnetic field also meet these conditions and can therefore be substituted in (A.2). Employing the Cauchy-Schwarz inequality to estimate the right-hand side of (A.2), we arrive at

$$\int_0^a \rho d\rho\int_{-\pi}^{\pi} d\phi\,[\epsilon_0\Delta\mathcal{E}_M^2(\rho,\phi,t)+\mu_0\Delta\mathcal{H}_M^2(\rho,\phi,t)]$$
$$\leq 2a\int_0^t dt'\left\{\int_{-\pi}^{\pi} d\phi\,\Delta\mathcal{E}_M^2(a,\phi,t')\right\}^{\frac{1}{2}}\left\{\int_{-\pi}^{\pi} d\phi\,\Delta\mathcal{H}_M^2(a,\phi,t')\right\}^{\frac{1}{2}}. \tag{A.5}$$

Except for a normalization factor, the left-hand side of (A.5) represents the sum of the integrated squared errors in approximating the total fields $\mathcal{E}(\rho,\phi,t)$ and $\mathcal{H}(\rho,\phi,t)$ inside the cylinder at time t by a Fourier series truncated at $|m| = M$. The integrals between braces on the right-hand side of (A.5) are the integrated squared errors in approximating $\mathcal{E}(a,\phi,t')$ and $\mathcal{H}(a,\phi,t')$ by a truncated Fourier series. Since, for a fixed ρ and t', the Fourier series representations of these field quantities converge in mean square, these integrals vanish for each t' as $M \to \infty$. Moreover, since $\mathcal{E}(a,\phi,t')$ and $\mathcal{H}(a,\phi,t')$ are of bounded variation in t' by virtue of the choice of the incident field, the right-hand side of (A.5) is continous in t for each fixed value of M. Consequently, this term assumes its maximum on any closed interval $0 \le t \le t_0$ of a finite length. From this fact and (A.5), it follows directly that the Fourier series representations of the total electromagnetic field intensities $\mathcal{E}(\rho,\phi,t)$ and $\mathcal{H}(\rho,\phi,t)$ converge in mean square on $\mathcal{D}_2$, uniformly in t for $0 \le t \le t_0$.

Second, we consider the electromagnetic field outside the cylinder. Clearly, there is no global convergence for the incident pulsed plane wave. Therefore, we restrict ourselves to the scattered field $\{\mathcal{E}^s(\rho,\phi,t), \mathcal{H}^s(\rho,\phi,t)\}$, which, in $\mathcal{D}_1$, is by itself a valid solution to the source-free electromagnetic-field equations (1) in the exterior medium. Integrating the local form of the electromagnetic power balance (4) for this field over the space-time domain $a < \rho < R$, $-\pi < \phi < \pi$, $z_0 < z < z_0 + 1$, $0 < t' < t$, results in

$$\frac{1}{2}\int_a^R \rho d\rho \int_{-\pi}^{\pi} d\phi \left[\epsilon_1 \boldsymbol{\mathcal{E}}^s \cdot \boldsymbol{\mathcal{E}}^s(\rho,\phi,t) + \mu_0 \boldsymbol{\mathcal{H}}^s \cdot \boldsymbol{\mathcal{H}}^s(\rho,\phi,t)\right]$$
$$- R\int_0^t dt' \int_{-\pi}^{\pi} d\phi \left[\mathcal{E}^s(R,\phi,t')\mathcal{H}^s(R,\phi,t')\right] =$$
$$- a\int_0^t dt' \int_{-\pi}^{\pi} d\phi \left[\mathcal{E}^s(a,\phi,t')\mathcal{H}^s(a,\phi,t')\right]. \qquad \text{(A.6)}$$

From the principle of causality, it follows that, for $\mathbf{r} \in \mathcal{D}_1$, the scattered field $\{\mathcal{E}^s(\rho,\phi,t), \mathcal{H}^s(\rho,\phi,t)\}$ vanishes for $\rho > a + c_1 t$, with $c_1 = (\epsilon_1\mu_0)^{-\frac{1}{2}}$, as defined in connection with (9). Taking in (A.6) the limit for $R \to \infty$, therefore, we obtain

$$\int_a^\infty \rho d\rho \int_{-\pi}^{\pi} d\phi \left\{\epsilon_1 \left[\mathcal{E}^s(\rho,\phi,t)\right]^2 + \mu_0 \left[\mathcal{H}^s(\rho,\phi,t)\right]^2\right\}$$
$$\le -2a \int_0^t dt' \int_{-\pi}^{\pi} d\phi \left[\mathcal{E}^s(a,\phi,t')\mathcal{H}^s(a,\phi,t')\right]. \qquad \text{(A.7)}$$

By arguments similar to those given above for the total field inside the cylinder, it follows that the Fourier series representation of the scattered field intensities $\mathcal{E}^s(\rho,\phi,t)$ and $\mathcal{H}^s(\rho,\phi,t)$ converge in mean square on $\mathcal{D}_1$, uniformly in any closed time interval $0 \le t \le t_0$.

Third, we make use of the finite duration of the incident-pulse shape $\mathcal{F}(t)$. From the definition of the incident field in (9) and the definition of the Fourier coefficients in (13), it is observed that the incident field $\{\mathcal{E}^i(\rho,\phi,t), \mathcal{H}^i(\rho,\phi,t)\}$ as well as the corresponding Fourier coefficients $\{\mathcal{E}^i_m(\rho,t), \mathcal{H}^i_m(\rho,t)\}$ vanish for $\mathbf{r} \in \mathcal{D}_2$ and $t > T + 2a/c_1$. This implies that, at those space-time points, the total field and the corresponding Fourier coefficients for the total field are identical to their counterparts for the scattered fields.

Considering the field difference $\{\Delta\mathcal{E}^s_M(\rho,\phi,t), \Delta\mathcal{H}^s_M(\rho,\phi,t)\}$, therefore, we integrate the local form of the power balance (4) the over the space-time domain $0<\rho<\infty$, $-\pi<\phi<\pi$, $z_0<z<z_0+1$, $t_0<t'<t$, with $t_0>T+2a/c_1$. Using (A.1), (A.6) and the Cauchy-Schwarz inequality then result in

$$\begin{aligned}
&\int_0^\infty \rho d\rho \int_{-\pi}^{\pi} d\phi\, \left\{\epsilon_0 \left[\Delta\mathcal{E}^s_M(\rho,\phi,t)\right]^2 + \mu_0\left[\Delta\mathcal{H}^s_M(\rho,\phi,t)\right]^2\right\} \\
&\quad \le 2a \int_0^{t_0} dt' \left\{\int_{-\pi}^{\pi} d\phi\, \Delta\mathcal{E}^2_M(a,\phi,t')\right\}^{\frac{1}{2}} \left\{\int_{-\pi}^{\pi} d\phi\, \Delta\mathcal{H}^2_M(a,\phi,t')\right\}^{\frac{1}{2}} \\
&\quad + 2a \int_0^{t_0} dt' \left\{\int_{-\pi}^{\pi} d\phi\, \left[\Delta\mathcal{E}^s_M(a,\phi,t')\right]^2\right\}^{\frac{1}{2}} \left\{\int_{-\pi}^{\pi} d\phi\, \left[\Delta\mathcal{H}^s_M(a,\phi,t')\right]^2\right\}^{\frac{1}{2}}. \qquad \text{(A.8)}
\end{aligned}$$

Combining (A.8) with the convergence results obtained above for finite time intervals $0\le t\le t_0$, finally, we conclude that, for the scattered field in $\mathcal{D}_1$ and for the total field in $\mathcal{D}_2$, the Fourier series representation (12) converges in mean square on the entire relevant spatial domain, uniformly in $0<t<\infty$.

B. Numerical Determination of Field Distributions in $\mathcal{D}_2$

For a general, radially inhomogeneous cylinder, the radial behavior of $E_m(\rho,s)$ or $\hat{E}_\nu(\rho,s)$ must be determined numerically. In this appendix, we outline two possible approaches for doing this. Since the structure of the computations is almost identical for integer and for complex values of the angular order, we restrict ourselves to nonnegative integer values $m\ge 0$.

B.1. First-Order Differential Equations

The first way of determining field distributions inside a radially inhomogeneous cylinder is to solve the system of first-order differential equations (18)–(19). Since the normalization condition in (26) provides a starting value at $\rho=0^+$, it seems logical to determine $e_m(\rho,s)$ and $h_m(\rho,s)$ by integrating these equations from $\rho=0^+$ to $\rho=1$. However, the coefficients in (18)–(19) are singular at $\rho=0$. Furthermore, $e_m(\rho,s)$ and $h_m(\rho,s)$ vanish as $\rho\downarrow 0$ for $m>0$ and for all m, respectively.

For $m>0$, we circumvent these difficulties by introducing the vector

$$\mathsf{f}_m(\rho,s) = \begin{bmatrix} f^e_m(\rho,s) \\ f^h_m(\rho,s)\end{bmatrix} \stackrel{\text{def}}{=} \begin{bmatrix} \rho^{-m}\, e_m(\rho,s) \\ \rho^{-m+1}\, h_m(\rho,s)\end{bmatrix} \qquad \text{(B.1)}$$

(see also [44]). In terms of this vector, (18)–(19) reduces to the matrix equation

$$\partial_\rho \mathsf{f}_m(\rho,s) = \rho^{-1}\,\mathsf{A}_m(\rho,s)\,\mathsf{f}_m(\rho,s), \qquad \text{(B.2)}$$

with

$$\mathsf{A}_m(\rho,s) \stackrel{\text{def}}{=} \begin{bmatrix} -m & s \\ \rho^2 s[\epsilon_{2r}(\rho)+\sigma_2(\rho)/s]+m^2/s & -m\end{bmatrix}. \qquad \text{(B.3)}$$

The singular coefficients are avoided by substituting in (B.2) the first-order Taylor expansions at $\rho = 0$:

$$\begin{aligned} \mathbf{f}_m(\rho, s) &= \mathbf{f}_m(0, s) + \rho\, \mathbf{f}'_m(0, s) + \mathcal{O}(\rho^2), \\ \mathsf{A}_m(\rho, s) &= \mathsf{A}_m(0, s) + \rho\, \mathsf{A}'_m(0, s) + \mathcal{O}(\rho^2), \end{aligned} \tag{B.4}$$

where the primes denote differentiations with respect to ρ. Comparing the terms containing factors ρ^{-1} and ρ^0, respectively, we find

$$\mathbf{f}_m(0) = \begin{bmatrix} 1 \\ m/s \end{bmatrix}, \quad \mathbf{f}'_m(0) = \mathbf{0}. \tag{B.5}$$

Next, we compute, with the aid of (B.4)–(B.5), a starting vector $\mathbf{f}_m(\rho_0)$ for a small, positive value of ρ_0. The matrix equation (B.2) is then integrated numerically by a Runge-Kutta-Verner fifth- and sixth-order method [45, Subsection 5.1.1].

For $m = 0$, we follow the same procedure, starting with the introduction of the vector

$$\mathbf{f}_0(\rho, s) = \begin{bmatrix} f_0^e(\rho, s) \\ f_0^h(\rho, s) \end{bmatrix} \stackrel{\text{def}}{=} \begin{bmatrix} e_0(\rho, s) \\ \rho^{-1} f_0(\rho, s) \end{bmatrix}. \tag{B.6}$$

A possible disadvantage of the procedure outlined above is that the Runge-Kutta-Verner procedure loses accuracy as the angular order m increases in magnitude. This is probably caused by the factor of ρ^{-1} on the right-hand side of (B.2).

B.2. Integral Equation

An alternative way to determine the transformed Fourier coefficients $\{E_m(\rho, s)\}$ is to solve the integral equation (23). With (15), this equation can be rewritten as

$$\begin{aligned} E_m(\rho, s) &= A_m(s)\, I_m(s\, \epsilon_{1r}^{\frac{1}{2}} \rho) \\ &\quad - s^2 \int_0^1 \rho' d\rho'\, C_1(\rho', s) I_m(s\, \epsilon_{1r}^{\frac{1}{2}} \rho_<)\, K_m(s\, \epsilon_{1r}^{\frac{1}{2}} \rho_>)\, E_m(\rho', s), \end{aligned} \tag{B.7}$$

for $0 < \rho < 1$. In (B.7), the amplitude $A_m(s)$ is given by

$$A_m(s) \stackrel{\text{def}}{=} (-1)^m \exp(-s\, \epsilon_{1r}^{\frac{1}{2}})\, F(s). \tag{B.8}$$

To discretize the integral over ρ' in (B.7), we introduce a spatial grid $\rho_n = nh$ with $n = 0, 1, \ldots, N$, and $h = N^{-1}$. Restricting the observation points to this grid, we can approximate the space integral in (B.7) by a repeated trapezoidal rule. The result can be written as

$$\tilde{E}(n, s) + k_n \sum_{n'=1}^{n} C_{n'} i_{n'} \tilde{E}(n', s) + i_n \sum_{n'=n+1}^{N} C_{n'} k_{n'} \tilde{E}(n', s) = i_n A_m(s), \tag{B.9}$$

where $\tilde{E}(n, s)$ is the *exact* solution of the discretized integral equation (B.9), and where the sums are understood to vanish when the lower index is smaller than the upper one. In (B.9), we have

$$
\begin{aligned}
i_n &\stackrel{\text{def}}{=} I_m(s\,\epsilon_{1r}^{\frac{1}{2}}\,nh) && \text{for } 0 \le n \le N,\\
k_n &\stackrel{\text{def}}{=} K_m(s\,\epsilon_{1r}^{\frac{1}{2}}\,nh) && \text{for } 0 < n \le N,\\
C_n &\stackrel{\text{def}}{=} s^2\,w_n\,\tfrac{n}{N}\,C_1(nh,s) && \text{for } 0 < n \le N,
\end{aligned}
\tag{B.10}
$$

where the weighting factor w_n is given by

$$
w_n = \begin{cases} N^{-1} & \text{for } 0 < n < N,\\ (2N)^{-1} & \text{for } n = 0, N. \end{cases}
\tag{B.11}
$$

Because of the special Toeplitz-like structure of the matrix equation (B.9), most of its inversion can be carried out in closed form. First, we reduce (B.9) to a suitable, tridiagonal form. Let row(n) denote the n'th row of (B.9). Then, the steps taken to reduce the matrix to this form can be formulated as follows. First, we remove all but the first superdiagonals with the aid of the row operation

$$
\text{row}(n) \to \text{row}(n) - \frac{i_n}{i_{n+1}}\,\text{row}(n+1)
$$

for $n = 0, 1, \ldots, N-1$. This reduces the matrix equation (B.9) to

$$
\tilde{E}(n,s) + k_n' \sum_{n'=1}^{n} C_{n'} i_{n'} \tilde{E}(n',s) - \frac{i_n}{i_{n+1}} \tilde{E}(n+1,s) = i_N A_m(s)\delta_{n,N}
\tag{B.12}
$$

for $n = 0, 1, \ldots, N$. In (B.12), $\delta_{n',n}$ denotes the Kronecker symbol. For $n = N$, the last term on the left-hand side is understood to vanish. Further, we have

$$
k_n' \stackrel{\text{def}}{=} \begin{cases} -\frac{i_0}{i_1} k_1 & \text{for } n = 0,\\ k_n - \frac{i_n}{i_{n+1}} k_{n+1} & \text{for } 0 < n < N,\\ k_N & \text{for } n = N. \end{cases}
\tag{B.13}
$$

Note that the reduction from (B.9) to (B.12) involves the division by $i_n = I_m(s\,\epsilon_{1r}^{\frac{1}{2}}\,nh)$ with $n = 1, 2, \ldots, N$, which may be zero for purely imaginary s. Hence, we can no longer choose $\beta = 0$ in the Bromwich contour (17).

Next, we remove all but the first subdiagonals from (B.12) by carrying out the row operation

$$
\text{row}(n) \to \text{row}(n) - \frac{k_n'}{k_{n-1}'}\,\text{row}(n-1)
$$

for $n = 2, 3, \ldots, N$. This operation does not affect the right-hand side and the first superdiagonal in (B.12).

The row operations indicated above result in a tridiagonal matrix equation that can be inverted recurrently. The solution scheme consists of two parts. In the forward part, we eliminate the subdiagonal by taking suitable linear combinations of row(n) and row$(n+1)$ for $n = 2, 3, \ldots, N$. This leads to a matrix equation with only a diagonal and a single superdiagonal. The backward part of our scheme consists of solving $\tilde{E}(n,s)$ from this matrix equation by interpreting it as a system of two-term recurrence relations. Using this procedure, we can invert (B.9) in $\mathcal{O}(N)$ operations provided that $\mathrm{Re}(s) > 0$.

One aspect of the numerical implementation that needs to be explained is the evaluation of the Bessel-function values $\{i_n\}$ and $\{k_n\}$. To avoid numerical overflow and underflow problems, elements of the tridiagonal matrix equation mentioned above are expressed in terms of $i_n k_n$, i_n/i_{n+1} and k_{n+1}/k_n only. These quantities, in turn, are obtained simultaneously for all orders m. To this end, we determine $I_0(z)$, $I_1(z)$, $K_0(z)$ and $K_1(z)$ from the usual series representations [21, Formulas 9.6.10, 9.6.11, 9.2.5–9.2.10 and 9.7.2], and evaluate the quotients $I_{m+1}(z)/I_m(z)$ and $K_{m+1}(z)/K_m(z)$ by backward and forward recurrence in m, respectively [46].

The error in the numerically obtained frequency-domain solution $\tilde{E}(m,\beta+i\omega)$ is governed by the choice of the space step h. Since this solution can be obtained from the recurrence scheme derived above, it seems reasonable to assume that its error will be of the same order of magnitude as the one made in replacing the integral in the Laplace-tranformed integral equation by the sum in (B.9). For that error, an estimate is available from the literature (see e.g. [21, Formulas 25.4.1, 25.4.2]). Using the second-order partial differential equation (20) to eliminate the factor of $\partial_\rho^2 E_m(\rho,\beta+i\omega)$ figuring in this estimate, we end up with

$$\Delta\tilde{E}_m(n,\beta+i\omega) \propto h^2\omega^3 F(\beta+i\omega), \tag{B.14}$$

as $h \to 0$ and $|\omega| \to \infty$.

The usual reasoning behind the choice for the discretization step h is that one wants to attain a constant *relative* error over the entire range of ω. With $h \propto |\omega|^{-1}$, the factor of h^2 in (B.14) cancels out two factors of ω. The remaining factor of ω can be eliminated by realizing that (B.14) is a worst-case estimate based on the observation that the errors caused by repeating the trapezoidal rule may accumulate. For an oscillatory solution $E_m(\rho,\beta+i\omega)$, it is more reasonable to assume that these errors average out. The amplitude factor $F(\beta+i\omega)$ need not be considered in this context since it is also present in $\tilde{E}_m(n,\beta+i\omega)$.

Finally, it should be mentioned that the Laplace-transformed Fourier coefficient $H_m(\rho,s)$ should not be determined by numerically performing the space differentiation in (19). Instead, this coefficient can be found by computing the integral in the representation

$$H_m(\rho,s) = H_m^i(\rho,s) - s\int_0^1 d\rho'\, C_1(\rho',s)\, \partial_\rho G_m(\rho,\rho';s)\, E_m(\rho',s), \tag{B.15}$$

which is obtained by combining (19) with (23). Carrying out a numerical differentiation causes the loss of an order of h in accuracy compared with the numerically obtained $E_m(\rho,s)$, while evaluating the integral in (B.15) preserves the accuracy of $E_m(\rho,s)$.

Acknowledgments

The research presented in this chapter was mainly carried out at the Delft University of Technology, and has benefited considerably from continual discussions with Professor Hans Blok of that University. The research described in Section 6 forms part of work performed while the author was on leave at the University of Colorado at Boulder with the financial support of the Netherlands Organization for the Advancement of Scientific Research (NWO). The author gratefully acknowledges this support and the hospitality received from the members of the Electromagnetics Laboratory at Boulder. In particular, he would like to thank his host, Professor Edward F. Kuester, both for his hospitality and for many stimulating discussions.

References

[1] Bojarski, N.N. (1982) "The *k*-space formulation of the scattering problem in the time domain", *J. Acoust. Soc. Am.* *72*, 570–584 (This paper includes a comprehensive review of Bojarski's work).

[2] Titchmarsh, E.C. (1950) *The Theory of Functions*, Oxford University Press, London, second edition, Chapter 13.

[3] Franz, W. (1957) *Theorie der Beugung elektromagnetischer Wellen*, Springer Verlag, Berlin, Chapter 11.

[4] Michalski, K.A. (1981) "Bibliography of the singularity expansion method and related topics", *Electromagnetics 1*, 493–511.

[5] Baum, C.E. (1976) "The singularity expansion method", in L.B. Felsen (ed.), *Transient Electromagnetic Fields*, Spinger Verlag, Berlin, Chapter 3.

[6] Baum, C.E. (1976) "Emerging technology for transient and broad-band analysis and synthesis of antennas and scatterers", *Proc. IEEE 64*, 1598–1616.

[7] Baum, C.E. (1978) "Toward an engineering theory of scattering: the singularity and eigenmode expansion methods", in P.L.E. Uslenghi (ed.), *Electromagnetic Scattering*, Academic Press, New York, Chapter 15.

[8] Pearson, L.W. and Marin, L. (eds.) (1981) "Special issue on the singularity expansion method", *Electromagnetics 1*.

[9] Langenberg, K.-J. (ed.) (1983) "Special issue on transient fields", *Wave Motion 5*.

[10] Tijhuis, A.G. and Blok, H. (1984) "SEM approach to the transient scattering by an inhomogeneous, lossless dielectric slab; Part 1: the homogeneous case", *Wave Motion 6*, 61–78.

[11] Tijhuis, A.G. and Blok, H. (1984) "SEM approach to the transient scattering by an inhomogeneous, lossless dielectric slab; Part 2: the inhomogeneous case", *Wave Motion 6*, 167–182.

[12] Morgan, M.A. (1984) "Singularity expansion representations of fields and currents in transient scattering", *IEEE Trans. Antennas Propagat. 32*, 466–473.

[13] Morgan, M.A. (1985) "Response to comments regarding SEM representation", *IEEE Trans. Antennas Propagat. 33*, 120.

[14] Pearson, L.W. (1984) "A note on the representation of scattered fields as a singularity expansion", *IEEE Trans. Antennas Propagat. 32*, 520–524.

[15] Felsen, L.B. (1984) "Progressing and oscillatory waves for hybrid synthesis of source excited propagation and diffraction", *IEEE Trans. Antennas Propagat. 32*, 775–796.

[16] Felsen, L.B. (1985) "Comments on early time SEM", *IEEE Trans. Antennas Propagat. 33*, 118–119.

[17] Dudley, D.G. (1985) "Comments on SEM and the parametric inverse problem", *IEEE Trans. Antennas Propagat.* *33*, 119–120.

[18] Tijhuis, A.G. and Van der Weiden, R.M. (1986) "SEM approach to transient scattering by a lossy, radially inhomogeneous dielectric circular cylinder", *Wave Motion 5*, 43–63.

[19] Tijhuis, A.G. (1987) *Electromagnetic Inverse Profiling: Theory and Numerical Implementation*, VNU Science Press, Utrecht, The Netherlands.

[20] Marin, L. (1973) "Natural-mode representation of transient fields", *IEEE Trans. Antennas Propagat.* *21*, 809–818.

[21] Abramowitz, M.A. and Stegun, I.A. (1972) *Handbook of Mathemathical Functions*, Ninth Dover printing, Dover Publications, New York.

[22] Singaraju, B.K., Giri, D.V. and Baum, C.E. (1976) "Further developments in the application of contour integration to the evaluation of the zeros of analytic functions and relevant computer programs", *Air Force Weapons Laboratory Mathematics Notes 42*, Air Force Weapons Laboratory, Albuquerque, New Mexico.

[23] Traub, J.F. (1964) *Iterative Methods for the Solution of Equations*, Prentice-Hall, Englewood Cliffs, New Jersey, Chapter 10.

[24] Überall, H. and Gaunaurd, G.C. (1981) "The physical content of the singularity expansion method", *Appl. Phys. Lett.* *39*, 362–364.

[25] Howell, W.E. and Überall, H. (1984) "Complex frequency poles of radar scattering from coated conducting spheres", *IEEE Trans. Antennas Propagat.* *32*, 624–627.

[26] Überall, H. and Gaunaurd, G.C. (1984) "Relation between the ringing of resonances and surface waves in radar scattering", *IEEE Trans. Antennas Propagat.* *32*, 1071–1079.

[27] Gaunaurd, G.C. and Werby, M.F. (1985) "Resonance responses of submerged, acoustically excited thick and thin shells", *J. Acoust. Soc. Am.* *77*, 2081–2093.

[28] Heyman, E. and Felsen, L.B. (1983) "Creeping waves and resonances in transient scattering by smooth convex objects", *IEEE Trans. Antennas Propagat.* *31*, 426–437.

[29] Heyman, E. and Felsen, L.B. (1985) "Wavefront interpretation of SEM resonances, turn-on times and entire functions", in L.B. Felsen (ed.), *Proceedings of the Nato Advanced Workshop on Hybrid Formulation of Wave Propagation and Scattering, IAFE, Castel Gandolfo, Italy, August 30–September 3, 1983*, Martinus Nijhoff Publishers, Dordrecht, The Netherlands.

[30] Tijhuis, A.G. (1986) "Angularly propagating waves in a radially inhomogeneous, lossy dielectric circular cylinder and their connection with the natural modes", *IEEE Trans. Antennas Propagat.* *34*, 813–824.

[31] Miyazaki, Y. (1981) "Scattering and diffraction of electromagnetic waves by inhomogeneous circular cylinder", *Radio Science 16*, 1009–1014.

[32] Friedlander, F.G. (1958) *Sound Pulses*, Cambridge University Press, Cambridge, pp. 149–166.

[33] Felsen, L.B. and Marcuvitz, N. (1973) *Radiation and Scattering of Waves*, Prentice-Hall, Englewood Cliffs, New Jersey.

[34] Lewin, L., Chang, D.C. and Kuester, E.F. (1977), *Electromagnetic Waves and Curved Structures*, Peter Peregrinus Ltd., Stevenage, United Kingdom.

[35] Wait, J.R. (1967) "Electromagnetic whispering gallery modes in a dielectric rod", *Radio Science 2*, 1005–1007.

[36] Wasylkiwskyj, W. (1975) "Diffraction by a concave perfectly conducting circular cylinder", *IEEE Trans. Antennas Propagat.* *23*, 480–492.

[37] Mitzner, K.M. (1967) "Numerical solution for transient scattering from a hard surface of arbitrary shape - Retarded potential technique," *J. Acoust. Soc. Am.* *42*, 391–397.

[38] Bennett, C.L. and Weeks, W.L. (1970) "Transient scattering from conducting cylinders," *IEEE Trans. Antennas Propagat.* *18*, 627–633.

[39] Tijhuis, A.G. (1984) "Toward a stable marching-on-in-time method for two-dimensional transient electromagnetic scattering problems," *Radio Science 19*, 1311–1317.

[40] Sezginer, A., and Kong, J.A. (1984) "Transient response of line source excitation in cylindrical geometry," *Electromagnetics 4*, 35–54.

[41] Moffatt, D.L., Lai, C.Y. and Lee, T. (1984) "Time-domain electromagnetic scattering by open ended circular waveguide and related structure," *Wave Motion 6*, 363–387.

[42] Pearson, L.W. (1986) "A construction of the fields radiated by z-directed point sources of current in the presence of a cylindrically layered obstacle," *Radio Science 21*, 559–569.

[43] Tijhuis, A.G., Wiemans R. and Kuester, E.F. (1989) "A hybrid method for solving time-domain integral equations in transient scattering", *Journal of Electromagnetic Waves and Applications 3*, 485–511.

[44] Dil, J.G. and Blok, H. (1972) "Propagation of electromagnetic surface waves in a radially inhomogeneous optical waveguide", *Optoelectronics 5*, 415–428.

[45] IMSL (1987) *Math/Library User's Manual; FORTRAN Subroutines for Mathematical Applications*, Version 1.0, Houston, Texas.

[46] Lindberg, R.C. (1966), "Bessel, a subroutine for the generation of Bessel functions with real or complex arguments," *Air Force Weapons Laboratory Mathematics Notes 1*, Air Force Weapons Laboratory, Albuquerque, New Mexico.

SOME RESULTS ON THE ABSOLUTELY CONVERGENT SERIES OF FUNCTIONS AND OF DISTRIBUTIONS

by
Dr Nicolas K. Artemiadis
Emeritus University Professor
Regular member of the Academy of Athens
169, Megalou Alexandrou St.,
136 71 Thrakomakedones
Greece

Introduction

Let $\mathscr{A}$ be the class of all Lebesgue integrable complex-valued functions on the circle T (the additive group of the reals modulo 2π) with absolutely convergent Fourier series and let A be the collection of all continuous functions in $\mathscr{A}$.

The study of the class $\mathscr{A}$ is one of the primary objectives of Harmonic Analysis. An approach to studying $\mathscr{A}$ (or A) has concentrated attention on seeking conditions on a function $f \in L^1(T)$, which are necessary and/or sufficient, that ensure that $f \in \mathscr{A}$.

A similar problem arises if f is a function of $L^1(\mathbb{R})$ or if f is a distribution defined on T.

The present paper is organized as follows:

Part 1 contains some preliminaries and notations. In Part 2 we give the statements of a few theorems (without proofs) concerning functions in $L^1(T)$ and $L^1(\mathbb{R})$. More precisely, Th.1 constitutes a generalization of the well known fact: If f is continuous on T and has nonnegative Fourier coefficients then $f \in A$. Th. 2 provides a necessary and sufficient condition for a function to be in $\mathscr{A}$. Th. 3 provides a necessary anf sufficient condition, so that the Fourier transform of a function in $L^1(\mathbb{R})$ to be in $L^1(\mathbb{R})$. The proofs of all these statements can be found in [1] and [2].

It seems that one of the useful aspects of the method used in [1] and [2] is that it can be extended to the case where f is a distribution, and this is exactly what we do in Part 3 of the paper.

J. S. Byrnes and J. F. Byrnes (eds.), Recent Advances in Fourier Analysis and Its Applications, 311–316.

PART 1.

$\mathbb{R}$ is the real line. $L^1(T)$ and $L^1(\mathbb{R})$ are the spaces of all Lebesgue integrable complex-valued functions on T and $\mathbb{R}$ respectively.

For $f \in L^1(T)$ the numbers

$$\hat{f}(n) = \frac{1}{2\pi} \int_T f(t) e^{-int} dt \quad , \quad (n \in Z)$$

are the Fourier coefficients of f, where Z is the group of integers.

As usual $\sum_{n \in Z} \hat{f}(n) e^{int}$ is the Fourier series of f.

For $f \in L^1(\mathbb{R})$ the Fourier transform of f is

$$\hat{f}(t) = \int_{\mathbb{R}} f(x) e^{ixt} dx \quad , \quad (t \in \mathbb{R}).$$

Let f be a measurable function in an interval $[a, b] \in \mathbb{R}$. A point $x \in (a, b)$ is said to be a Lebesgue point for f if and only if

$$\lim_{h \to 0} \frac{1}{h} \int_0^h |f(x+t) + f(x-t) - 2f(x)| dt = 0.$$

With C^∞ we denote the set of all 2π-periodic infinitely differentiable functions.

The space of all infinite sequences $<c_n>$ of complex numbers such that $\sum_{n \in Z} |c_n| < +\infty$ is denoted by l^1.

Let S be a distribution defined on T, which is equal to a function of the class L^∞ (of essentially bounded functions) on an open interval I of T containing the point $\alpha \in \mathbb{R}$. For each $\varphi \in C^\infty$ we set $<S_\alpha, \varphi> = <S, \varphi_\alpha>$ where $\varphi_\alpha(t) = \varphi(t - \alpha)$, $\alpha \in \mathbb{R}$. Clearly S_α is also a distribution which is equal to a function of L^∞ in an open interval containing the origin.

By S_0 is meant the distribution S.

The Fourier coefficients of the distribution S_α are:

$$\hat{S}_\alpha(n) = <S_\alpha, e^{-itn}> = <S, e^{-in(t-\alpha)}> = e^{in\alpha} <S, e^{-int}> = e^{in\alpha} \hat{S}(n).$$

For $b \in \mathbb{R}$ we define, as usual, $b^+ = max(b, 0)$, $b^- = max(-b, 0)$. $R_e z$, $J_m z$, mean the real and imaginary part of z, respectively.

PART 2 ([1], [2]).

Theorem 1. Let f be a continuous complex-valued function on T with the property:

«There is $\alpha \in \mathbb{R}$ such that $\alpha \leq \arg \hat{f}(n) < \alpha + (\pi/2)$, $n \in Z$».

Then $f \in A$, and every $f \in A$ is a linear combination of continuous functions on T with the above property.

Theorem 2. Let $f \in L^1(T)$. Then $f \in \mathscr{A}$, if and only if, there is a Lebesgue point α for f such that the following two sequences
(*) $<(R_e \hat{f}_\alpha(n))^->_{n \in Z}$, $<(J_m \hat{f}_\alpha(n))^->_{n \in Z}$, belong to l^1.

Corollary. A function $f \in L^1(T)$ is equal almost everywhere to a linear combination of positive definite functions if, and only if, there is a Lebesgue point α for f such that condition(*) is satisfied.

Note. Theorem 2 and the Corollary remain true if the sentence: "*there is a Lebesgue point α for f*" is replace by the sentence: "*f is essentially bounded in a neighborhood of some real number α*".

Theorem 3. Let $f \in L^1(\mathbb{R})$. Then $\hat{f} \in L^1(\mathbb{R})$, if, and only if there is a Lebesgue point α for f such that $(R_e \hat{f}_\alpha)^-$, $(J_m \hat{f}_\alpha)^-$ belong to $L^1(\mathbb{R})$. (Here $f_\alpha(t) = f(t+\alpha)$).

A consequence of Th. 3 and of the inversion formula is that: "A function $f \in L^1(\mathbb{R})$ is equal, a.e., to a finite linear combination of positive definite functions if, and only if, there is a Lebesgue point α for f such that both functions $(R_e \hat{f}_\alpha)^-$, $(J_m \hat{f}_\alpha)^-$ belong to $L^1(\mathbb{R})$.

Now the criterion given by Th. 2 is to be compared with the following two well known criteria.

Criterion of M. Riesz([4]). $f \in A$ if, and only if, it can be expressed in the form $f = u * v$ with $u, v \in L^2(T)$.

This criterion is very difficult to apply in any specific case that is not alredy decidable on more evident grounds.

Criterion of Stečkin ([4]). $f \in A$ if, and only if,

$$\sum_{n=1}^{\infty} \frac{1}{\sqrt{n}} e_n(f) < +\infty$$

where

$$e_n(f) = inf\left\{\frac{1}{2\pi} \int_{-\pi}^{\pi} |f(x) - P(x)|^2 dx\right\}^{1/2}$$

the infimum being taken with respect to all choices of the not necessarily harmonic trigonometric polynomials P of degree at most n, that is the functions of the type

$$P(x) = \sum_{|m| \leqq n} c_m \cdot \exp(i\lambda_m x)$$

where the c_m are complex numbers and the λ_m are distinct real numbers satisfying $\lambda_{-m} = -\lambda_m$ for $|m| \leq n$.

The main drawback with this criterion is the extreme difficulty encountered in estimating the numbers $e_n(f)$ for a given function f.

PART 3 ([3]).

Theorem 4. Let S be a distribution (in the sense of L. Schwartz) defined on T. Suppose that on an open interval I of T, S is equal to a function of L^∞. Then the Fourier series of S is absolutely convergent if and only if the sequences

(1) $\langle(R_e\,\hat{S}_\alpha(n))^-\rangle_{n=1}^\infty$, $\langle(J_m\,\hat{S}_\alpha(n))^-\rangle_{n=1}^\infty$, both belong to l^1.

Proof. Suppose (1) is satisfied. We first consider the case $\alpha = 0$. Let $\varphi \in C^\infty$ with support in the interval $(-\pi, \pi)$ and such that $\hat{\varphi}(t) > 0$, $\hat{\varphi}(0) = 1$. Set

$$\varphi_\varepsilon(x) = \frac{1}{\varepsilon}\,\varphi\left(\frac{x}{\varepsilon}\right), \quad x \in \mathbb{R}.$$

For sufficiently small $\varepsilon > 0$ the function φ_ε belongs to C^∞, and its support is contained in $(-\pi, \pi)$. Then φ_ε can be extended to a 2π-periodic function $\widetilde{\varphi}_\varepsilon \in C^\infty$.

Now if F is a distribution and $p \in C^\infty$, then the convolution $F * p$ is defined to be the function for which

$$(F * p)(x) = F(T_x p(-t))$$

where T_x is the translation operator. It is known ([4]) that $F * p \in C^\infty$.

Hence if we put $S * \widetilde{\varphi}_\varepsilon = u_\varepsilon$ we have $u_\varepsilon \in C^\infty$, so that u_ε equals the sum of its Fourier series. We find

$$u_\varepsilon(0) = \sum_n \hat{\varphi}(\varepsilon n)\,\hat{S}(n) = \lim_{N\to\infty} \sum_{n=-N}^{N} \hat{\varphi}(\varepsilon n)\,[Re(\hat{S}(n)) + iJ_m(\hat{S}(n))]$$

Set

$$\sigma_{N,\varepsilon} = \sum_{n=-N}^{N} \hat{\varphi}(\varepsilon n)\,\hat{S}(n) = \sum_{n=-N}^{N} \hat{\varphi}(\varepsilon n)\,[R_e(\hat{S}(n)) + iJ_m(\hat{S}(n))]$$

$$= \sum_{n=-N}^{N} \hat{\varphi}(\varepsilon n)\,[(R_e\hat{S}(n))^+ - (R_e\hat{S}(n))^- + i(J_m\,\hat{S}(n))^+ - i(J_m\,\hat{S}(n))^-]$$

$$= \sum_{n=-N}^{N} \hat{\varphi}(\varepsilon n)\,(R_e\hat{S}(n))^+ + i\sum_{n=-N}^{N} \hat{\varphi}(\varepsilon n)\,(J_m\,\hat{S}(n))^+$$

$$- \sum_{n=-N}^{N} \hat{\varphi}(en)\,(R_e\check{S}(n))^- - i\sum_{n=-N}^{N} \hat{\varphi}(\varepsilon n)\,(J_m\,\hat{S}(n))^-.$$

Next observe that, due to the hypothesis that S equals a function of L^∞ in I, $u_\varepsilon(0)$ remains uniformly bounded when $\varepsilon \longrightarrow 0$. Furthermore since, by assumption, the sequences in (1) belong to l^1, it follows that the expressions

$$\sum_{n=-N}^{N} \hat{\varphi}(\varepsilon n)\,(R_e\,\hat{S}(n))^-, \quad \sum_{n=-N}^{N} \hat{\varphi}(\varepsilon n)\,(J_m\,\hat{S}(n))^-$$

remain uniformly bounded for all ε and N. Set

$$A_{m,n} = \hat{\varphi}\left(\frac{n}{m}\right)(R_e\,\hat{S}(n))^+,$$

where m, n are natural numbers.

It follows from what we have just proved that the double series $\sum_{m,n} A_{m,n}$ is convergent since $A_{m,n} \geq 0$. We have

139.2

$$\sum_{m,n} A_{m,n} = \lim_{N\to\infty}\left(\lim_{m\to\infty}\sum_{n=-N}^{N}\hat{\varphi}\left(\frac{n}{m}\right)(R_e\,\hat{S}(n))^+\right)$$

$$= \lim_{N\to\infty}\sum_{n=-N}^{N}(R_e\,\hat{S}(n))^+ = \sum_{n\in Z}(R_e\,\hat{S}(n))^+ < +\infty$$

In a similar way we prove that

$$\sum_{n\in Z}(J_m\,\hat{S}(n))^+ < +\infty\,.$$

It follows that

$$\sum_{n\in Z}|\hat{S}(n)| < +\infty\,.$$

This proves the theorem in the case $\alpha = 0$.

Next assume $\alpha \neq 0$ and consider the distribution S_α. As we have noticed before, S_α is a distribution which is equal to a function of L^∞ in an interval containing the origin, so that the above result holds for S_α. We have

$$\sum_{n\in Z}|\hat{S}_\alpha(n)| = \sum_{n\in Z}|e^{ina}\,\hat{S}(n)|\sum_{n\in Z}|\hat{S}(n)| < +\infty.$$

This proves the one half of the theorem.

To prove the converse, let S be a distribution on T such that $\sum_{n\in Z}|\hat{S}(n)| < +\infty$. The sequence $\langle\hat{S}(n)\rangle_{n\in Z}$ being tempered (see [4] p.65) the distributions $S_N = \sum_{|n|\leq N}\hat{S}(n)e_n$ (where e_n is the function $x \longrightarrow e^{inx}$) converge in the space of distributions, as $N \longrightarrow \infty$, to a distribution F such that $\hat{F}(n) = \hat{S}(n)$, $(n\in Z)$. Hence $F = S$. Now the function $f(x) = \sum_{n\in Z}\hat{S}(n)e^{inx}$ can be considered as a distribution. It is easily seen that, due to the uniform convergence of the last series, we have, for each $\varphi \in C^\infty$

$$<S, \varphi> = \lim_{N\to\infty} <S_N, \varphi> = \frac{1}{2\pi}\int f(x)\,\varphi(x)dx = <f, \varphi>,$$

which shows that S is an L^∞ function on T. The theorem is proved.

One might find the above result interesting, also because of the following remark.

Remark. Call a numerical series $\sum(\alpha_n + ib_n)$, $(\alpha_n, b_n \in \mathbb{R})$, "one sidedly absolutely convergent" (O.A.C.), if and only if:
(at least one of $\sum\alpha_n^+$, $\sum\alpha_n^-$) and (at least one of $\sum b_n^+$, $\sum b_n^-$) is finite. Now it is possible that a series $\sum(\alpha_n + ib_n)$ is not O.A.C., while the series $\sum(\alpha_n + ib_n)e^{i\lambda}$ is O.A.C. In other words a non O.A.C. series can, in some cases, be converted to an O.A.C. series by just multiplying each term by a factor of the form $e^{i\lambda}$ (λ = some constant) or perhaps in some other way.

Example. Let $c_n = \alpha_n + ib_n$, where $c_{2n} = 1 + i$, $c_{2n+1} = 1 - i$, $n = 0, 1, 2, \ldots$, and $\lambda = \pi/4$. Then it is easily seen that $\sum c_n$ is not O.A.C. while $\sum c_n e^{i\lambda}$ is.

The theorem we have just proved, essentially says that: "The Fourier series of a distribution converges absolutely if, and only if, $\sum \hat{S}_\alpha(n)$ is O.A.C. for some $\alpha \in \mathbb{R}$.

REFERENCES

[1] N. ARTEMIADIS, "Criteria for absolute convergence of Fourier Series" Proc. Amer. Math. Soc. 50 (1975), 179-183.

[2] N. ARTEMIADIS, "Quelques résultats sur les transformées de Fourier avec applications" Bull. Sc. Math., 2e serie, 97, 1973, 177-191.

[3] N. ARTEMIADIS, "Absolutely convergent Fourier series of distributions" Proc. Amer. Math. Soc. 83 (1981), 276-278.

[4] R.E. EDWARDS, "Fourier Series" vols. I, II, Holt, Rinehart and Winston, New York, 1967.

[5] J.-P. KAHANE, "Séries de Fourier absolument convergentes" Ergebnisse der Mathematik und ihrer Grenzgebiete, Band 50, Springer-Verlag, Berlin and New York, 1970, MR 43 # 801.

Invariance Techniques in Spectral Estimation and Signal Processing

RICHARD ROY
Information Systems Laboratory
Stanford University
Stanford, CA 94305

October 2, 1989

Abstract

High-resolution signal parameter estimation is a problem of significance in many signal processing applications. Such applications include *direction-of-arrival* (DOA) estimation, system identification, and time series analysis. A novel approach to the general problem of signal parameter estimation is described. Though discussed in the context of direction-of-arrival estimation, ***ESPRIT*** can be applied to a wide variety of problems. It exploits an underlying *rotational invariance* among signal subspaces induced by an array of sensors with a *translational invariance* structure. A few simulation results are presented including the application of the algorithm to estimation of dispersion relations of a large molcular chain.

Contents

J. S. Byrnes and J. F. Byrnes (eds.), Recent Advances in Fourier Analysis and Its Applications, 317–365.

1. Introduction

IN MANY PRACTICAL SIGNAL PROCESSING PROBLEMS, the objective is to estimate, using measurements from a sensor array, a set of *constant* parameters upon which the signals being received depend. Problems of this type include high resolution direction-of-arrival (DOA) estimation in sensor systems such as radar, sonar, and electronic surveillance, and high resolution spectral analysis of time series. In such problems, the functional form of the underlying signals can often be assumed to be known (*e.g.*, narrowband plane waves, cisoids). The quantities to be estimated are parameters (*e.g.*, frequencies and DOAs of plane waves, cisoid frequencies) upon which the sensor outputs depend, and these parameters are assumed to be *constant.*[1]

There have been several approaches to such problems including the so-called maximum likelihood (ML) method of Capon (1969) [1] and Burg's (1967) [2] maximum entropy (ME) method. Though often successful and widely used, these methods have certain fundamental limitations (*esp.* bias and sensitivity in the parameter estimates), largely because they do not explicitly use the actual underlying model of the measurements. Pisarenko (1973) [3] was one of the first to exploit the structure of the data model, doing so in the context of estimation of parameters of cisoids in additive noise using a covariance approach. Schmidt (1977) [4] and independently Bienvenu (1979) [5] were the first to correctly exploit the measurement model in the case of sensor arrays of arbitrary form. Schmidt, in particular, accomplished this by first deriving a complete geometric solution in the absence of noise, then cleverly extending the geometric concepts to obtain a *reasonable* approximate solution in the presence of noise. The resulting algorithm was called MUSIC (MUltiple SIgnal Classification) and has been widely studied. In a detailed evaluation based on thousands of simulations, MIT's Lincoln Laboratory concluded that, among currently accepted high-resolution algorithms, MUSIC was the most promising and a leading candidate for further study and actual hardware implementation. However, though the performance advantages of MUSIC are substantial, they are achieved at a considerable cost in computation (searching over parameter space) and storage (of array calibration data, *cf.* Section 3).

In the sequel. a recently developed algorithm (***ESPRIT***) that dramatically reduces the aforementioned computation and storage costs is described. In the context of DOA estimation, the reductions are achieved by requiring that the sensor array possess a *displacement invariance*, *i.e.*, that sensors occur in matched pairs with identical *displacement* vectors (*cf.* Figure 3). Fortunately, there are many practical problems in which these conditions are or can easily be satisfied. Linear arrays of equi-spaced identical sensors are commonplace in sonar applications, as are regular rectangular arrays of identical elements (*e.g.*, phased-arrays) in radar applications. In addition to obtaining signal parameter estimates efficiently, *optimal signal copy vectors* for reconstructing the signals are also provided by ***ESPRIT***. ***ESPRIT*** is also manifestly less sensitive to array imperfections than previous techniques including MUSIC [6], a desirable property when system implementation is considered.

---

[1] Extensions to situations in which the parameters may be time varying can be made, however they rely on an inherent *time-scale* or *eigenvalue separation* between the parameter dynamics and the dynamics of the signal process. Fundamentally, the assumption is made that over time intervals long enough to collect sufficient information from which to obtain accurate parameter estimates, the parameters have not changed significantly.

1.1. Notation and Mathematical Preliminaries

To make the presentation as clear as possible, an attempt is made to adhere to a somewhat standard notational convention. Lower case **boldface** characters will generally refer to vectors. Upper case **BOLDFACE** characters will generally refer to matrices. For either real or complex-valued matrices, $(\cdot)^*$ will be used to denote the Hermitian conjugate (or complex-conjugate transpose) operation. Eigenvalues of square Hermitian matrices are assumed to be ordered in decreasing magnitude, as are the singular values of non-square matrices. Knowledge of the fundamental theorems of matrix algebra dealing with eigendecompositions and singular value decompositions (SVD) is assumed (*cf.* [7]).

2. The Data Model

Though ***ESPRIT*** is generally applicable to a wide variety of problems, for illustrative purposes the discussions herein focus on direction-of-arrival (DOA) estimation. In many practical signal processing applications, data from an array of sensors are collected and the objective is to locate point sources assumed to be radiating energy that is detectable by the sensors as depicted in Figure 1. Mathematically, such problems are quite simply, though abstractly, modeled using *Green's functions* for the particular differential operator that describes the physics of radiation propagation from the sources to the sensors. For the intended applications however, a few *reasonable* assumptions can be invoked to simplify the model.

The transmission medium is assumed to be isotropic and non-dispersive so that the radiation propagates in *straight lines*, and the sources are assumed to be in the *far-field* of the array. Consequently, the radiation impinging on the array is in the form of a sum of *plane waves*. For simplicity, it is assumed that the problem is planar and that all sources lie in a half-space, thus reducing the location parameter space to a single-dimensional subset of $\Re$, *i.e.*, $\theta_i \in [0, \pi]$, where θ_i is the direction-of-arrival (DOA) of the ith source.

The signals are assumed to be *narrowband* processes, and can be considered to be sample functions of a stationary stochastic process or deterministic functions of time. In terms of estimating the signals themselves, these two assumptions lead to entirely different estimation algorithms. However, as far as estimation of signal parameters such as DOA is concerned, both assumptions lead to the same (suboptimal) algorithm under certain assumptions (*e.g.*, *persistent excitation*).

Since the narrowband signals are assumed to have the same *known* center frequency[2] (ω_0), the ith signal can be written as

$$\tilde{s}_i(t) = u_i(t)\cos(\omega_0 t + v_i(t)), \tag{1}$$

where $u_i(t)$ and $v_i(t)$ are slowly varying[3] functions of time that define the amplitude and phase of $\tilde{s}_i(t)$ respectively. For such signals, it is often more convenient to use the so-called *complex envelope* representation [8] in which $\tilde{s}_i(t) = \mathrm{Re}\{s(t)\}$, where $s(t) = u(t)e^{j(\omega_0 t + v(t))}$

[2] In the multiple signal environment, this condition is often termed *co-channel interference* in communication applications. If the center frequencies are not the same, the problem can be greatly simplified by first separating the signals in the frequency domain.

[3] The definition of slowly varying is taken to mean that the approximation $\tilde{s}_i(t - \tau_k(\theta_i)) \approx u_i(t)\cos(\omega_0(t - \tau_k(\theta_i)) + v_i(t))$, is valid, *i.e.*, the amplitude and phase variations as functions of spatial position for fixed t are *negligible* over the extent of the array.

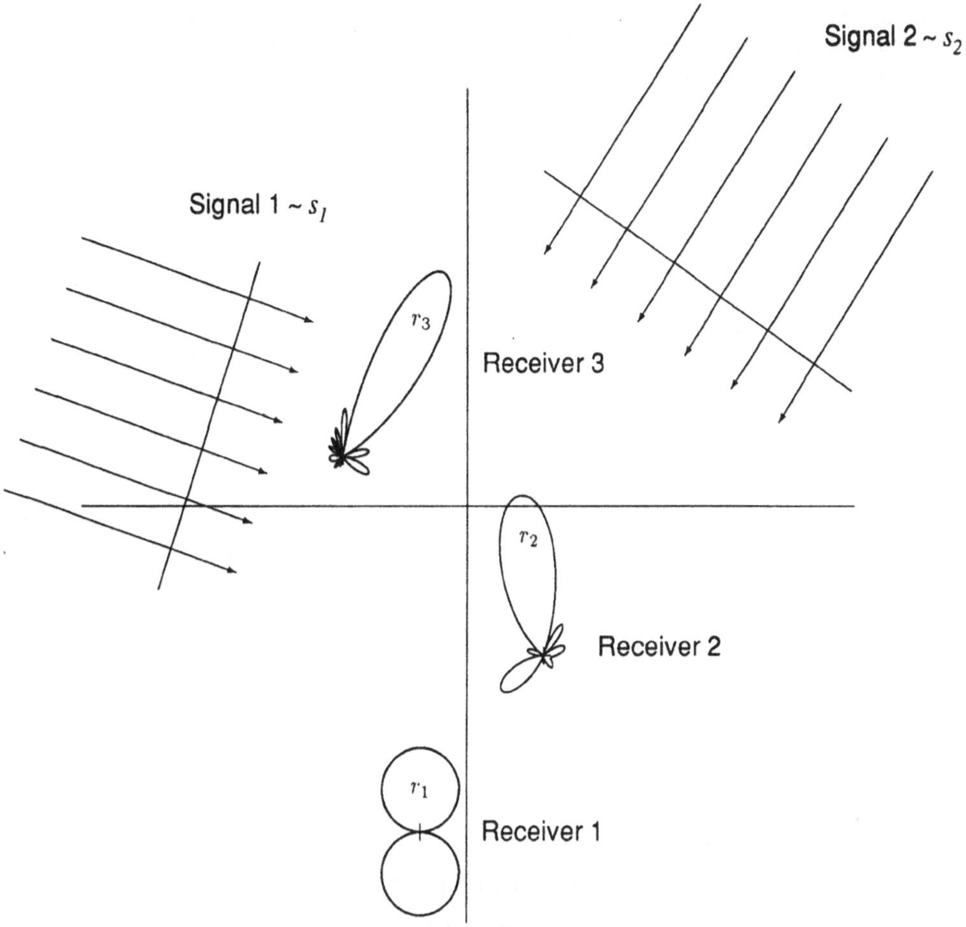

Figure 1: Passive Sensor Array Geometry

Noting that the narrowband assumption implies $u(t) \approx u(t-\tau)$ and $v(t) \approx v(t-\tau)$ for all possible propagation delays τ, the effect of a time-delay on the received waveforms is simply a phase shift, *i.e.*,

$$s(t-\tau) \approx s(t)e^{-j\omega_0\tau} \tag{2}$$

The result is that $x_k(t)$, the complex signal output of the kth sensor at time t, can be written as

$$x_k(t) = \sum_{i=1}^{d} a_k(\theta_i)s_i(t-\tau_k(\theta_i)) = \sum_{i=1}^{d} a_k(\theta_i)s_i(t)e^{-j\omega_0\tau_k(\theta_i)} , \tag{3}$$

where $\tau_k(\theta_i)$ is the propagation delay between a reference point and the kth sensor for the ith wavefront impinging on the array from direction θ_i, $a_k(\theta_i)$ is the corresponding sensor element complex response (gain and phase) at frequency ω_0, and there are assumed to be d point sources present.

Employing vector notation for the outputs of the m sensors, the *data model* becomes

$$\mathbf{x}(t) = \sum_{i=1}^{d} \mathbf{a}(\theta_i)s_i(t) , \tag{4}$$

where

$$\mathbf{a}(\theta_i) = [a_1(\theta_i)e^{-j\omega_0\tau_1(\theta_i)}, \ldots, a_m(\theta_i)e^{-j\omega_0\tau_m(\theta_i)}]^T , \tag{5}$$

often termed the *array steering vector* for direction θ_i. Setting

$$\mathbf{A}(\boldsymbol{\theta}) = [\mathbf{a}(\theta_1) \vdots \cdots \vdots \mathbf{a}(\theta_d)], \tag{6}$$

$$\mathbf{s}(t) = [s_1(t), \ldots, s_d(t)]^T , \tag{7}$$

and adding measurement noise $\mathbf{n}(t)$, the *measurement model* for the passive sensor array narrowband signal processing problem becomes

$$\mathbf{z}(t) = \mathbf{A}(\boldsymbol{\theta})\mathbf{s}(t) + \mathbf{n}(t) , \tag{8}$$

where $\mathbf{z}(t), \mathbf{n}(t) \in \mathcal{C}^m$, $\mathbf{s}(t) \in \mathcal{C}^d$, and $\mathbf{A}(\boldsymbol{\theta}) \in \mathcal{C}^{m \times d}$, and it is assumed that $d < m$.

By employing the simplifying assumptions above, the DOA estimation problem, which is in general a complex problem in two independent variables, space and time, is reduced to a parameter estimation problem in one independent variable, time. The resulting measurement model is equivalent to one commonly employed in time series analysis. Therein, given a sequence of samples (indexed by a single real independent variable, usually time) from a series presumed to be composed of a superposition of sinusoids with possibly growing or decaying envelopes, the objective is to obtain estimates of the sinusoidal frequencies, the envelopes, and the relative strengths (powers) of all the components. Mathematically, the measurement model can be expressed as follows:

$$z(t) = \sum_{i=1}^{d} a_i e^{\sigma_i t} \cos(\omega_i t + \phi_i) + n(t) , \tag{9}$$

where a_i is the residue at the ith pole $\gamma_i = \sigma_i + j\omega_i$, and ϕ_i is the relative phase of the ith cisoidal component. The term $n(t)$ represents *additive* noise present in the measurements and accounts for such sources as thermal noise in amplifiers and quantization noise introduced in analog-to-digital (A/D) conversion. In a manner entirely analogous to the development of the DOA data model, the complex representation can be employed to simplify the notation. Assuming uniform sampling, any vector of m consecutive samples can be written as

$$\tilde{\mathbf{z}}(t) = \sum_{i=1}^{d} \mathbf{a}(\gamma_i)\tilde{s}_i(t) + \tilde{\mathbf{n}}(t) , \tag{10}$$

where $\tilde{s}_i(t) = a_i e^{(\gamma_i t + j\phi_i)}$ and $\mathbf{a}(\gamma_i) = [1 , e^{-\gamma_i D} , \ldots , e^{-\gamma_i (m-1)D}]$, a Vandermonde vector. Combining the column vectors, $\mathbf{a}(\gamma_i)$, into a matrix $\mathbf{A}(\boldsymbol{\gamma})$,

$$\tilde{\mathbf{z}}(t) = \mathbf{A}(\boldsymbol{\gamma})\tilde{\mathbf{s}}(t) + \tilde{\mathbf{n}}(t), \tag{11}$$

which is exactly the same form as equation (8). Recall that in the DOA estimation application, however, the columns of $\mathbf{A}$ are arbitrary array steering vectors. They are Vandermonde vectors only in the special case of a uniform linear array of sensors.

3. The Geometric Approach

In 1977, Schmidt [4] developed the **MUSIC** (**MU**ltiple **SI**gnal **C**lassification) algorithm by taking a geometric view of the signal parameter estimation problem. As aforementioned, one of the major breakthroughs afforded by the **MUSIC** algorithm was the ability to handle arbitrary arrays of sensors. Until the mid-1970's, direction finding techniques required knowledge of the array directional sensitivity pattern in analytical form, and the task of the antenna designer was to build an array of antennae with a pre-specified sensitivity pattern. Inevitable imperfections and interelement interactions make this a difficult task. The work of Schmidt essentially relieved the designer from such constraints by exploiting the reduction in analytical complexity that could be achieved by *calibrating* the array. Thus, the highly nonlinear problem of calculating the array response to a signal from a given direction was reduced to that of measuring and storing the response. Although **MUSIC** did not mitigate the computational complexity of solution to the DOA estimation problem, it did extend the applicability of high-resolution DOA estimation to *arbitrary* arrays of sensors.

3.1. Array Manifolds and Signal Subspaces

To introduce the concepts of the *array manifold* and the *signal subspace*, recall the noise-free data model (*cf.* equation (4)):

$$\mathbf{x}(t) = \mathbf{A}(\boldsymbol{\theta})\mathbf{s}(t).$$

The vectors $\mathbf{a}(\theta_i) \in \mathbb{C}^m$, the columns of $\mathbf{A}(\boldsymbol{\theta})$, are elements of a set (not a subspace), termed the **array manifold**[4] ($\mathcal{A}$), composed of all array response (*steering*) vectors obtained as θ ranges over the entire parameter space. $\mathcal{A}$ is completely determined by the sensor directivity patterns and the array geometry, and can sometimes be computed analytically. For example, as discussed in Section 2, in DOA estimation using uniform linear arrays as well

[4]Technically, a *k-dimensional manifold* in $\mathbb{C}^m$ is a subset of points in $\mathbb{C}^m$ satisfying certain local continuity and differentiability conditions. The physics of sensor arrays guarantee the continuity and differentiability properties will be satisfied. Associated with each point on the manifold is a vector to that point from the origin in $\mathbb{C}^m$ using the standard basis.

as in time series analysis using a uniform tapped-delay line, the elements of $\mathcal{A}$ are Vandermonde vectors of the form

$$\mathbf{a}(\omega) = [1, \lambda, \lambda^2, \ldots, \lambda^{m-1}].$$

where the parameter λ is a (nonlinear) function of the signal parameters of interest (DOAs or cisoid frequencies). However, for complex arrays that defy analytical description, $\mathcal{A}$ can be obtained by calibration (*i.e.*, physical measurements). For azimuth only DOA estimation, the array manifold is a one-parameter manifold that can be viewed as a *rope* weaving through $\mathbb{C}^m$. For azimuth and elevation DOA estimation, the manifold is a *sheet* in $\mathbb{C}^m$. Note that to avoid ambiguities, it is necessary to assume that the map from θ to $\mathbf{a}(\theta)$ is one-to-one, a property that can be ensured by proper array design.[5]

The key observation is that if $\mathbf{x}(t) = \mathbf{a}(\theta)s_\theta(t)$ is an appropriate data model for a single signal, the data are confined to a *one-dimensional subspace* of $\mathbb{C}^m$ characterized by the vector $\mathbf{a}(\theta)$. For d signals, the observed data vectors $\mathbf{x}(t) = \mathbf{A}(\boldsymbol{\theta})\mathbf{s}(t)$ are constrained to the d-dimensional subspace of $\mathbb{C}^m$, termed the **signal subspace ($\mathcal{S}_Z$)**, that is spanned by the d vectors $\mathbf{a}(\theta_i)$, the columns of $\mathbf{A}(\boldsymbol{\theta})$.[6]

3.2. Intersections as Solutions

The concepts of an observed signal subspace and a calibrated array manifold permit an immediate visualization of the solution. In the absence of noise, the outputs of the sensor array lie in a d-dimensional subspace of $\mathbb{C}^m$, the *signal subspace* ($\mathcal{S}_Z$) spanned by the columns of $\mathbf{A}(\boldsymbol{\theta})$. Once d independent vectors have been observed,[7] $\mathcal{S}_Z$ is known, and intersections between the observed subspace and the array manifold yield the set of vectors from the array manifold that span the observed signal subspace. A three-sensor, two-source example is graphically depicted in Figure 2. Assuming that the sensor array has been designed such that the map from parameters to array manifold vectors is unique, the parameters are immediately determined.

Problems arise when only noisy measurements $\mathbf{z}(t) = \mathbf{A}(\boldsymbol{\theta})\mathbf{s}(t) + \mathbf{n}(t)$ of the array output are available, since $\mathcal{S}_Z$ must be estimated. Imposing the constraint that the estimate $\widehat{\mathcal{S}}_Z$ be spanned by elements from $\mathcal{A}$ and assuming unknown deterministic signals and Gaussian noise, a maximum-likelihood (ML) estimator can be formulated as described in Appendix D. However, the ML solution (of obtaining a set of vectors from the array manifold that *best fits* all the measurements) is computationally prohibitive in most practical applications.

Schmidt's idea was to employ a suboptimal *two-step* procedure instead. First, an *unconstrained* set of d vectors that best fits all the measurements is found. Then points of closest approach of the space spanned by those vectors to the array manifold are sought. This procedure, though clearly suboptimal, retains some of the key properties of the ML solution, including the fact that the *exact* answer is obtained asymptotically as the number of measurements goes to infinity.

---

[5]This condition is necessary, but not sufficient, to guarantee unique solutions in the multiple source environment. Higher order ambiguities can exist as discussed in detail in [6].

[6]It is also convenient to define the noise subspace ($\mathcal{S}_X^\perp$) as the orthogonal complement of the signal subspace, $\mathcal{S}_Z$, in $\mathbb{C}^m$.

[7]The problem of degenerate signal spaces, *i.e.*, fully correlated signals, is discussed in [6]. For the purposes of this discussion, it is assumed that the signals are not fully correlated.

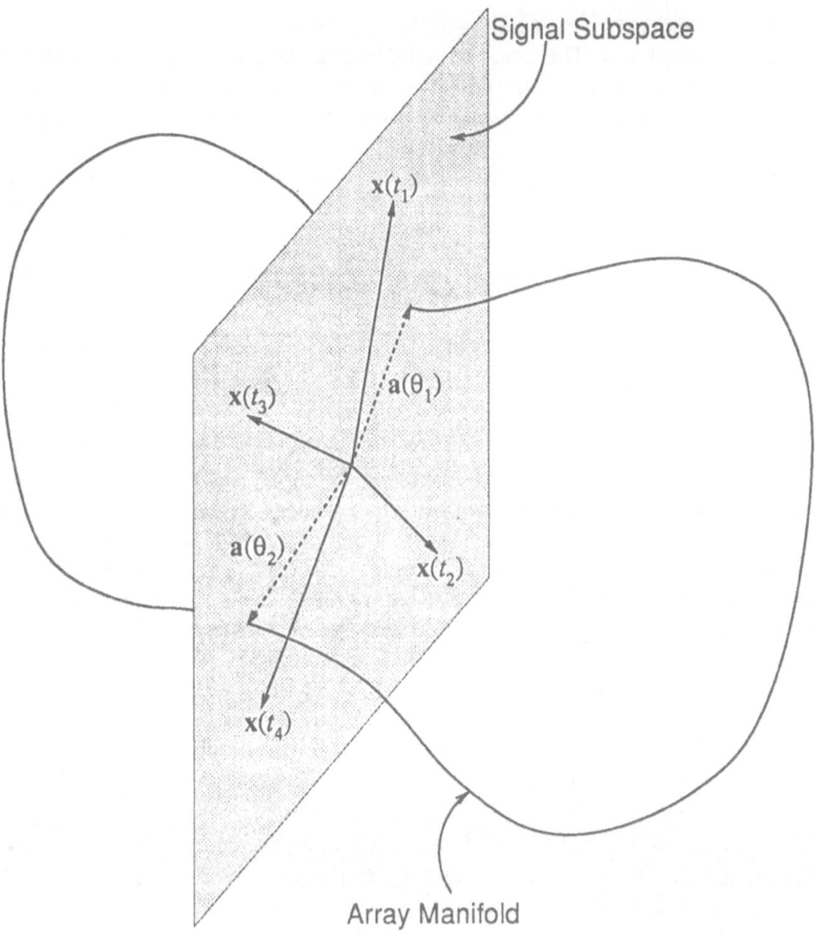

Figure 2: The Geometry of MUSIC for a Three-Sensor/Two-Source Example (No Noise)

3.3. Estimating the Signal Subspace

To obtain an unconstrained estimate of the signal subspace, the ***least-squares*** (LS) criterion is most often employed. The idea is to find a set of d vectors that span a subspace of $\mathcal{C}^m$ that ***best fits***, in a LS sense, the observed data. Assuming the signals and noise are zero-mean, a method can be derived by first examining the covariance[8] matrix of the measurements. If the signals are modeled as stationary stochastic processes, they are assumed to be uncorrelated with the noise and possess a positive definite covariance matrix $\mathbf{R}_{SS} > 0$. If, on the other hand, a deterministic signal model is chosen, a persistent excitation condition, *i.e.*,

$$\mathbf{R}_{SS} \stackrel{\text{def}}{=} \lim_{N\to\infty} \frac{1}{N} \sum_{t=1}^{N} \mathbf{s}(t)\mathbf{s}^*(t) > 0,$$

is assumed. Without loss of generality, the covariance matrix of the measurement noise[9] ($\mathbf{R}_{NN}$) can be assumed to be $\sigma^2\mathbf{I}$, where σ may be unknown.

Under the conditions given above, the covariance matrix of the measurements is given by

$$\mathbf{R}_{ZZ} \stackrel{\text{def}}{=} E\{\mathbf{z}\mathbf{z}^*\} = \mathbf{A}\mathbf{R}_{SS}\mathbf{A}^* + \sigma^2\mathbf{I}. \tag{12}$$

The objective is to find a set of d linearly independent vectors that is contained in $\mathcal{S}_Z = \mathcal{R}\{\mathbf{A}\}$, the subspace spanned[10] by the columns of $\mathbf{A}$. One such set of vectors is easily obtained as the set of eigenvectors $\{\mathbf{e}_i, i = 1, \ldots, d\}$ of $\mathbf{R}_{ZZ}$ corresponding to the d largest eigenvalues, a fact that follows from elementary linear algebra as follows. First, any $m \times m$ unitary[11] matrix $\mathbf{E}$ can be taken as a matrix of eigenvectors for $\sigma^2\mathbf{I}$ since the identity matrix commutes with any matrix, *i.e.*, $\mathbf{E}\mathbf{I}\mathbf{E}^* = \mathbf{I}$ for unitary $\mathbf{E}$. Thus, any set of eigenvectors of $\mathbf{A}\mathbf{R}_{SS}\mathbf{A}^*$ is also a set[12] of eigenvectors of $\mathbf{A}\mathbf{R}_{SS}\mathbf{A}^* + \sigma^2\mathbf{I}$. Since $\mathbf{A}\mathbf{R}_{SS}\mathbf{A}^*$ is rank d and positive semi-definite by construction, there are d positive eigenvalues and $m - d$ zero eigenvalues. Thus, the d largest eigenvalues of $\mathbf{R}_{ZZ}$ are simply the d non-zero eigenvalues of $\mathbf{A}\mathbf{R}_{SS}\mathbf{A}^*$ augmented by σ^2, and the associated eigenvectors lie in the subspace spanned by the columns of $\mathbf{A}$. Note also that the $m - d$ smallest eigenvalues of $\mathbf{R}_{ZZ}$ are all equal to σ^2, a fact that can be used to find d and σ. Thus, $\mathcal{R}\{\mathbf{E}_Z\} = \mathcal{S}_Z$ where $\mathbf{E}_Z \stackrel{\text{def}}{=} [\mathbf{e}_1 \,\vdots\, \cdots \,\vdots\, \mathbf{e}_d]$. Similarly defining $\mathbf{E}_n \stackrel{\text{def}}{=} [\mathbf{e}_{d+1} \,\vdots\, \cdots \,\vdots\, \mathbf{e}_m]$, $\mathcal{R}\{\mathbf{E}_n\} = \mathcal{S}_X^{\perp}$.

In most situations, the covariance matrices required above are not known and must be estimated from measurements. Practically, the noise covariance can be estimated from measurements made when signals of interest are not present. Measurements made when signals of interest are present are used to estimate $\mathbf{R}_{ZZ}$. The standard estimate

$$\hat{\mathbf{R}}_{ZZ} = \frac{1}{N} \sum_{t=1}^{N} \mathbf{z}(t)\mathbf{z}^*(t),$$

---

[8] The sample covariance is easily seen to be a *sufficient statistic* for this least-squares estimation problem.

[9] The more general situation where the noise covariance $\mathbf{R}_{NN} = \sigma^2\mathbf{\Sigma}_n$, where $\mathbf{\Sigma}_n > 0$, leads naturally to generalized eigenproblems that can be avoided by *pre-whitening* the measurements by first filtering the data using $\mathbf{\Sigma}_n^{-\frac{1}{2}}$.

[10] This subspace is also referred to as the *range* of $\mathbf{A}$.

[11] A matrix $\mathbf{E} \in \mathcal{C}^{m\times m}$ is unitary if and only if $\mathbf{E}\mathbf{E}^* = \mathbf{I}$.

[12] Since there are eigenvalues with multiplicity greater than one, the eigenvectors are not unique. However, the invariant subspaces spanned by eigenvectors corresponding to distinct eigenvalues are unique, and this is all that is required.

is most often employed, and in terms of the $m \times N$ data matrix $\mathbf{Z}$,

$$\hat{\mathbf{R}}_{ZZ} = \frac{1}{N}\mathbf{Z}\mathbf{Z}^*.$$

Note that this estimate can be scaled by $N/(N-m)$ to obtain an unbiased estimate, but the scaling only effects the magnitudes of the eigenvalues, not the eigenvectors, so the subspace estimates remain unaffected.

An alternative to first forming the measurement covariance matrix and then performing an eigendecomposition is to operate directly on the measurements using the singular value decomposition. In addition to avoiding *squaring* the data, this approach has a nice geometric interpretation. Letting $\mathbf{Z} \in \mathbb{C}^{m \times N}$ denote the data matrix, the objective is to obtain a set of vectors spanning the column space of the rank d matrix, $\hat{\mathbf{Z}} \in \mathbb{C}^{m \times N}$, that best approximates $\mathbf{Z}$ in a least-squares sense. The solution is given by the d left singular vectors of $\mathbf{Z}$ corresponding to the d largest singular values.[13]

It is easily demonstrated that the eigendecomposition and the SVD yield the same subspace estimate. If the SVD of $\mathbf{Z}/\sqrt{N}$ is given by $\mathbf{U}\mathbf{\Sigma}\mathbf{V}^*$,

$$\frac{1}{N}\mathbf{Z}\mathbf{Z}^* = \mathbf{U}\mathbf{\Sigma}^2\mathbf{U}^* = \hat{\mathbf{R}}_{ZZ},$$

since $\mathbf{\Sigma}$ is diagonal and real, and $\mathbf{U}$ and $\mathbf{V}$ are unitary. Thus, the left singular vectors, $\mathbf{U}$, of $\mathbf{Z}$ are the right eigenvectors of $\hat{\mathbf{R}}_{ZZ}$, the sample covariance matrix. Thus, in the absence of finite precision effects in computing the decompositions, the subspace estimates from both techniques are identical.

There are, however, significant computational differences. The computation of the full SVD of $\mathbf{Z}$ is of order mN^2 which can be significantly larger than the $O(m^3)$ operations required for an eigendecomposition of $\hat{\mathbf{R}}_{ZZ}$. The increase in computation is due to the fact that the full SVD is also obtaining a set of d vectors, the first d columns of $\mathbf{V} \in \mathbb{C}^{N \times N}$, that span the d-dimensional subspace of N-dimensional space spanned by the d signal vectors, vectors in $\mathbb{C}^N$ whose components are the samples of the underlying signals.[14] If the information of the full SVD is not required, *partial* SVD algorithms that compute only the left singular vectors and singular values can be employed resulting in substantial computational savings [9].

If N is the number of measurements to be processed, and m is the number of elements in each sample vector, forming the sample covariance matrix requires on the order of Nm^2 operations. Eigendecompositions of matrices in $\Re^{m \times m}$ require on the order of $10m^3$ operations, whereas a standard SVD of $\mathbf{Z} \in \Re^{m \times N}$ requires $\approx 2Nm^2 + 4m^3$ (*cf.* [7, page 175]) if only the singular values and left singular vectors are required. The computational effort is therefore on the same order for both the SVD and eigenvector approaches to reducing the measurements to the statistic $\mathbf{E}_Z$ used by ***ESPRIT*** and MUSIC. Again, the major advantage of the SVD over the eigendecomposition is that the measurements are processed directly without *squaring* them. Thus, numerical problems associated with ill-conditioned matrices are mitigated to some extent by using the SVD.

[13] A discussion of this problem can be found in [7, pp. 19-20], though therein the 2-norm is employed. The extension to the Frobenius norm (*cf.* total least-squared error) is straightforward.

[14] In the presence of noise, these d vectors are biased estimates due to the fact that for each measurement taken, another signal vector must be estimated. Thus, there is no *averaging of the noise*.

When the covariances must be estimated from finite data matrices (or the data matrices used directly), the $m-d$ smallest eigenvalues (singular values) are clustered around, but not all equal to σ^2 (σ). In this case, special statistical techniques based on likelihood ratio (LR) tests (*cf.* Appendix E) can be used to obtain an estimate $\hat{d}$ of the signal subspace dimension. In any case, in the presence of a finite amount of noisy measurements, $\mathcal{R}\{\mathbf{E}_z\} = \hat{\mathcal{S}}_Z$ and $\mathcal{R}\{\mathbf{E}_n\} = \hat{\mathcal{S}}_Z^{\perp}$, the *estimated signal and noise subspaces.*

3.4. Estimating the Signal Parameters

In the absence of noise, parameter estimates can be obtained by finding *intersections* of $\mathcal{A}$ with $\hat{\mathcal{S}}_Z$, or equivalently finding elements of $\mathcal{A}$ that are orthogonal to $\hat{\mathcal{S}}_Z^{\perp}$. At this point in the suboptimal signal subspace algorithms, the real computational effort[15] begins. Even with perfect knowledge of the signal subspace, searching the array manifold for d intersections with $\mathcal{S}_Z$ can be quite costly, especially for multi-dimensional parameter spaces (*e.g.*, azimuth, elevation, and range).

The problem is further complicated in the presence of noise since, with probability one, $\hat{\mathcal{S}}_Z \cap \mathcal{A} = \emptyset$; there are no intersections. Consequently there are no elements of $\mathcal{A}$ that are orthogonal to $\hat{\mathcal{S}}_Z^{\perp}$. Referring to Figure 2, it would seem that intersections could almost always be found. In this respect, the figure is somewhat misleading. For three sensors and two sources, the signal subspace is a two-dimensional *complex* subspace of three-dimensional *complex* space. In the real field, the estimated signal subspace is actually a four-dimensional subspace of six-dimensional space, and it need not intersect the one-dimensional manifold at all. Obviously, elements of $\mathcal{A}$ that are *closest* to $\hat{\mathcal{S}}_Z$ should be considered as potential solutions, but the issue of an appropriate measure of *closeness* remains.

Schmidt [4] proposed the following function as one possible measure[16] of the *closeness* of an element of $\mathcal{A}$ to $\hat{\mathcal{S}}_Z$:

$$P_M(\theta) = \frac{\mathbf{a}^*(\theta)\mathbf{a}(\theta)}{\mathbf{a}^*(\theta)\mathbf{E}_N\mathbf{E}_N^*\mathbf{a}(\theta)}. \tag{13}$$

In the absence of noise, this measure, termed the **MUSIC** *spectrum*, is infinite for elements of $\mathcal{A}$ belonging to $\mathcal{S}_Z$. In the presence of noise, this measure is clearly peaked near points of closest approach of $\mathcal{A}$ to $\hat{\mathcal{S}}_Z$, points where $\mathcal{A}$ is nearly orthogonal to $\hat{\mathcal{S}}_Z^{\perp}$. This property is used to obtain parameter estimates as those parameters, or parameter vectors in the case of multi-dimensional array manifolds, that yield the d largest *peaks in the spectrum.*

Though conceptually simple, the *one-dimensional*[17] **MUSIC** measure has several drawbacks. Primarily, problems in the finite measurement case arise from the fact that since d signals are known to be present, d parameter estimates, $\{\theta_1, \ldots, \theta_d\}$, should be sought *simultaneously* by maximizing an appropriate functional (*cf.* [6]) rather than obtaining estimates one at a time as is done in the search over $P_M(\theta)$. However, multi-dimensional searches are exponentially more expensive than one-dimensional searches. The price paid

[15]Only for extremely large arrays such as phased array radars with hundreds of elements is the eigendecomposition of the measurement covariance matrix computationally significant in comparison to the search over $\mathcal{A}$.

[16]In [4], the numerator was not explicitly included in the measure since it was assumed that the array manifold vectors were all *normalized in some suitable fashion.*

[17]The conventional **MUSIC** measure is herein referred to as a *one-dimensional* measure, though the search over $\mathcal{A}$ is potentially a multi-dimensional one, the dimension being that of the parameter vector (*e.g.*, three for a parameter vector consisting of range, azimuth and elevation).

for the computational reduction achieved by employing a *one-dimensional search* for d parameters is that the method is finite-sample-biased in the multiple source environment (*cf.* [6]).

Problems also arise in low SNR scenarios, and in situations where even small sensor array errors are present. The ability of the conventional **MUSIC** *spectrum* to *resolve* closely spaced sources (*i.e.*, observe multiple peaks in the measure) is severely degraded (*cf.* [6]). Fundamentally, the problem is the same as the cisoidal frequency resolution problem in time series analysis using DFTs. As discussed in detail in [6], this *lack of resolution* is a consequence of the one-dimensional nature of the measure. As with the finite-sample-bias, the problem is mitigated by proper choice of a *multi-dimensional measure* of closeness of d elements of $\mathcal{A}$ to $\widehat{\mathcal{S}}_Z$. Nevertheless, it should be emphasized that in spite of these drawbacks, **MUSIC** has been shown to outperform previous techniques (*cf.* [10]).

Finally, as indicated above, **MUSIC** asymptotically yields unbiased parameter estimates since as the amount of data becomes infinite, errors in the estimate $\widehat{\mathcal{S}}_Z$ vanish. If the noise is spatially Gaussian and temporally independent, the distribution of the eigenvectors of the sample covariance is asymptotically Gaussian, with mean equal to the true eigenvectors (assuming distinct eigenvalues) and covariances that go to zero [11, 12]. Thus, the estimated signal subspace *converges in mean-square* to the true signal subspace, and the parameter estimates converge to the true values as well.

3.5. Summary of the MUSIC Algorithm

The following is a summary of the **MUSIC** algorithm based on the covariance[18] approach described above.

1. Collect the data and estimate $\mathbf{R}_{ZZ} = E\{\mathbf{zz}^*\} = \mathbf{AR}_{SS}\mathbf{A}^* + \sigma^2\mathbf{I}$ denoting the estimate $\widehat{\mathbf{R}}_{ZZ}$.

2. Solve for the eigensystem;
$$\widehat{\mathbf{R}}_{ZZ}\bar{\mathbf{E}} = \bar{\mathbf{E}}\mathbf{\Lambda},$$
where $\mathbf{\Lambda} = \operatorname{diag}\{\lambda_1, \ldots, \lambda_m\}$, $\lambda_1 \geq \cdots \geq \lambda_m$, and $\bar{\mathbf{E}} \overset{\text{def}}{=} [\mathbf{e}_1 \,\vdots\, \cdots \,\vdots\, \mathbf{e}_m]$.

3. Estimate the multiplicity of $\lambda_{\min}$, *i.e.*, estimate $m - d$.

4. Evaluate
$$P_M(\theta) = \frac{\mathbf{a}^*(\theta)\mathbf{a}(\theta)}{\mathbf{a}^*(\theta)\mathbf{E}_N\mathbf{E}_N^*\mathbf{a}(\theta)},$$
where $\mathbf{E}_n = [\mathbf{e}_{d+1} \,\vdots\, \cdots \,\vdots\, \mathbf{e}_m]$.

5. Find the d (largest) peaks of $\mathbf{P}(\theta)$ to obtain estimates of the parameters.

---

[18] Alternatively, a signal subspace estimate can be obtained by performing a (generalized) SVD of the data matrix $\mathbf{Z} = [\mathbf{z}(t_1) \,\vdots\, \cdots \,\vdots\, \mathbf{z}(t_N)]$. The resulting subspace estimate is the same; the choice of which approach to use is made based on external factors such as numerical and hardware implementation considerations.

4. *ESPRIT*

Though MUSIC was the first of the high-resolution algorithms to correctly exploit the underlying data model of narrowband signals in additive noise, the algorithm has several limitations including the fact that complete knowledge of the array manifold is required, and that the search over parameter space is computationally very expensive. In this section, an approach (*ESPRIT*) to the signal parameter estimation problem that exploits sensor array invariances is described.[19] *ESPRIT* is similar to MUSIC in that it correctly exploits the underlying data model, while manifesting significant advantages over MUSIC as described in Section 1.

4.1. Array Geometry

ESPRIT retains most of the essential features of the *arbitrary* array of sensors, but achieves a significant reduction in computational complexity by imposing a constraint on the structure of the sensor array, a constraint most easily described by an example. Consider a planar array of arbitrary geometry composed of m sensor *doublets* as shown in Figure 3. The elements in each doublet have identical sensitivity patterns and are translationally separated by a known constant displacement vector $\mathbf{\Delta}$. Other than the obvious requirement that each sensor have non-zero sensitivity in all directions of interest, the gain, phase, and polarization sensitivity of the elements in the doublet are arbitrary. Furthermore, there is no requirement that any of the doublets possess the same sensitivity patterns, though as discussed in [6, 14], there are advantages to employing arrays with such characteristics.

4.2. The Data Model

Assume that there are $d \leq m$ narrowband sources[20] centered at frequency ω_0, and that the sources are located sufficiently far from the array such that in homogeneous isotropic transmission media, the wavefronts impinging on the array are planar. As before, the sources may be assumed to be stationary zero-mean random processes or deterministic signals. Additive noise is present at all $2m$ sensors and is assumed to be a stationary zero-mean random process with a *spatial* covariance $\sigma^2\mathbf{I}$. There is no loss of generality in comparison with the assumption that the covariance is $\sigma^2\Sigma_n$. As long as Σ_n is known, *pre-whitening* of the measurement noise can be performed.

To describe mathematically the effect of the *translational invariance* of the sensor array, it is convenient to describe the array as being comprised of two subarrays, Z_X and Z_Y, identical in every respect although physically displaced (not rotated) from each other by a known displacement vector $\mathbf{\Delta}$. The signals received at the ith doublet can then be expressed as

$$\begin{aligned} x_i(t) &= \sum_{k=1}^{d} s_k(t)a_i(\theta_k) + n_{x_i}(t)\,, \\ y_i(t) &= \sum_{k=1}^{d} s_k(t)e^{j\omega_0\Delta\sin\theta_k/c}a_i(\theta_k) + n_{y_i}(t)\,, \end{aligned} \qquad (14)$$

[19] A patent has been issued on the sensor array design and concepts embodied in *ESPRIT* [13].

[20] MUSIC imposes the requirement $d < 2m$, and can therefore can handle roughly twice as many sources as *ESPRIT* in general. For uniform linear arrays, however, *ESPRIT* can handle as many sources as MUSIC [6] by employing *overlapping* subarrays.

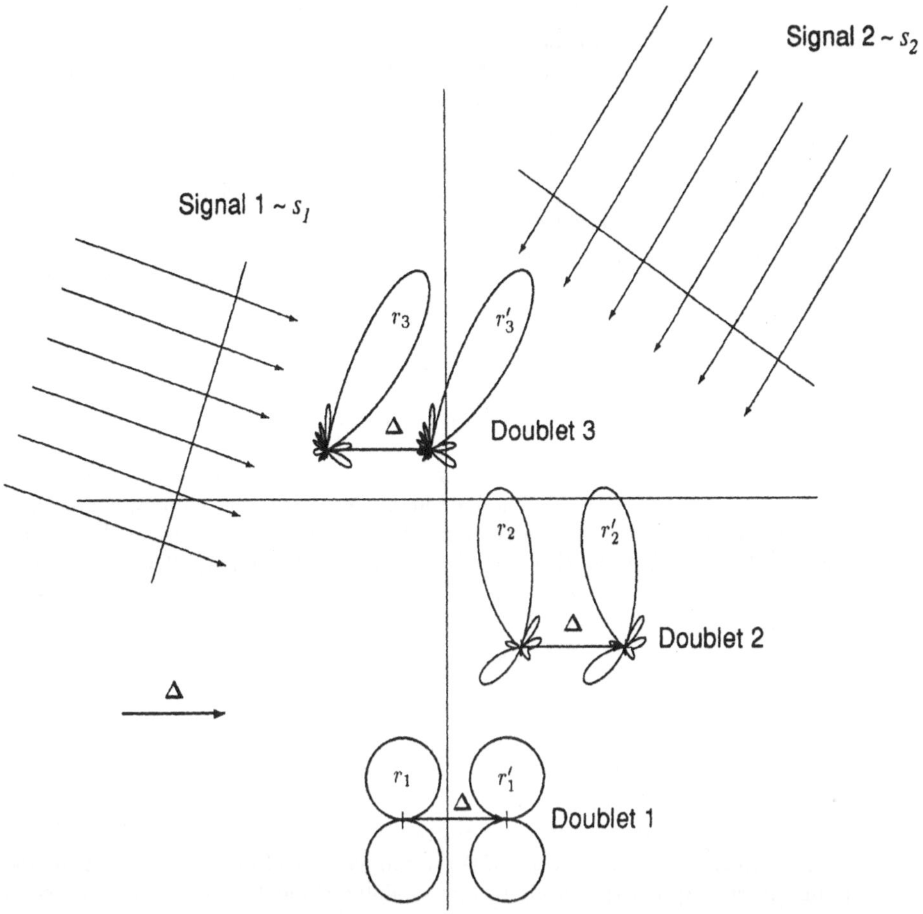

Figure 3: Sensor Array Geometry for Multiple Source DOA Estimation Using ***ESPRIT***

where θ_k is now the direction of arrival of the kth source relative to the direction of the translational displacement vector $\boldsymbol{\Delta}$.

Since the sensor gain and phase patterns are arbitrary and since ***ESPRIT*** does not require any knowledge of the sensitivities, the subarray displacement vector $\boldsymbol{\Delta}$ sets not only the scale for the problem, but the *reference direction* as well. The DOA estimates obtained are angles-of-arrival with respect to the direction of the vector $\boldsymbol{\Delta}$, and they are obtained as nonlinear functions of the ***ESPRIT*** *electrical* angles, angles that are linear functions of $\Delta = \|\boldsymbol{\Delta}\|_2$. A natural consequence of this fact is the necessity for a corresponding displacement vector for each *dimension* in which parameter estimates are desired.

Combining the outputs of each of the sensors in the two subarrays, the received data vectors can be written as follows:

$$\mathbf{x}(t) = \mathbf{A}\mathbf{s}(t) + \mathbf{n}_x(t)\,, \tag{15}$$

$$\mathbf{y}(t) = \mathbf{A}\boldsymbol{\Phi}\mathbf{s}(t) + \mathbf{n}_y(t)\,, \tag{16}$$

where the vector $\mathbf{s}(t)$ is the $d \times 1$ vector of impinging signals (wavefronts) as observed at the reference sensor of subarray X. The individual signals can be temporally correlated, *i.e.*, $E\{s_i(t)s_j^*(t)\} \neq 0,\ i \neq j$, though the case of coherent (or fully correlated) sources is not considered herein (*cf.* [6, Section 7.11] for a complete discussion). The matrix $\boldsymbol{\Phi}$ is a diagonal $d \times d$ matrix of the phase delays between the doublet sensors for the d wavefronts, and is given by

$$\boldsymbol{\Phi} = \mathrm{diag}[\, e^{j\gamma_1}, \ldots, e^{j\gamma_d}\,]\,, \tag{17}$$

where

$$\gamma_k = \omega_0 \Delta \sin\theta_k / c\,. \tag{18}$$

$\boldsymbol{\Phi}$ is a unitary matrix (operator) that relates the measurements from subarray Z_X to those from subarray Z_Y. In the complex field, $\boldsymbol{\Phi}$ is a simple scaling operator. However, it is isomorphic to the real two-dimensional rotation operator and is herein referred to as a *rotation*[21] operator. The unitary nature of $\boldsymbol{\Phi}$ is a consequence of the narrowband planewave assumption, an assumption that leads to unit-modulus cisoidal *signals* in the spatial domain. In time series analysis, the diagonal elements of $\boldsymbol{\Phi}$ are potentially arbitrary complex numbers in which case $\boldsymbol{\Phi}$ could be an *expansive* or *contractive* operator.

Defining the total array output vector as $\mathbf{z}(t)$, the subarray outputs can be combined to yield

$$\mathbf{z}(t) = \begin{bmatrix} \mathbf{x}(t) \\ \mathbf{y}(t) \end{bmatrix} = \bar{\mathbf{A}}\mathbf{s}(t) + \mathbf{n}_z(t)\,, \tag{19}$$

$$\bar{\mathbf{A}} = \begin{bmatrix} \mathbf{A} \\ \mathbf{A}\boldsymbol{\Phi} \end{bmatrix}, \quad \mathbf{n}_z(t) = \begin{bmatrix} \mathbf{n}_x(t) \\ \mathbf{n}_y(t) \end{bmatrix} \tag{20}$$

It is the structure of $\bar{\mathbf{A}}$ that is exploited to obtain estimates of the diagonal elements of $\boldsymbol{\Phi}$ *without having to know* $\mathbf{A}$.

[21] This is the origin of the term *rotational* in the acronym ***ESPRIT***.

4.3. Scale Invariance

From equation (19), it is easily seen that the estimation problem posed is ***scale-invariant*** in the sense that absolute signal powers are not observable. For any nonsingular diagonal matrix, $\mathbf{D}$, the data model is invariant with respect to the transformations $\mathbf{s}(t) \to \mathbf{D}^{-1}\mathbf{s}(t)$ and $\bar{\mathbf{A}} \to \bar{\mathbf{A}}\mathbf{D}$. Thus, estimates of the signals and the associated array manifold vectors derived herein are to be interpreted modulo an arbitrary scale factor unless knowledge of the gain pattern of *one* of the sensors is available.

4.4. *ESPRIT* – The Invariance Approach

The basic idea behind ***ESPRIT*** is to exploit the *rotational* invariance of the underlying signal subspace induced by the translational invariance of the sensor array. The relevant signal subspace is the one that contains the outputs from the two subarrays described above, Z_X and Z_Y. Simultaneous sampling of the output of the arrays leads to two sets of vectors, $\mathbf{E}_X$ and $\mathbf{E}_Y$, that span the same signal subspace (ideally, that spanned by the columns of $\mathbf{A}$).

The ***ESPRIT*** algorithm is based on the following results for the case in which the underlying $2m$-dimensional signal subspace containing the entire array output is known. In the absence of noise, the signal subspace can be obtained as before by collecting a sufficient number of measurements and finding *any set of d linearly independent vectors* that span the same column space, the subspace spanned by $\bar{\mathbf{A}}$. The signal subspace can also be obtained from knowledge of the covariance of the measurements $\mathbf{R}_{ZZ} = \bar{\mathbf{A}}\mathbf{R}_{SS}\bar{\mathbf{A}}^* + \sigma^2\mathbf{I}$. As shown in Section 3.3, if $d \leq m$ (an assumption required in later developments), the $2m-d$ smallest eigenvalues of $\mathbf{R}_{ZZ}$ are equal to σ^2. The d eigenvectors ($\mathbf{E}_Z$) corresponding to the d largest eigenvalues satisfy $\mathcal{R}\{\mathbf{E}_Z\} = \mathcal{R}\{\bar{\mathbf{A}}\}$, the signal subspace. Also as shown in Section 3.3, the *same* set of vectors $\mathbf{E}_Z$ can be obtained by performing a singular value decomposition (SVD) of the data matrix $\mathbf{Z}$.

Since $\mathcal{R}\{\mathbf{E}_Z\} = \mathcal{R}\{\bar{\mathbf{A}}\}$, there must exist a unique (recall $d \leq m$), nonsingular $\mathbf{T}$ such that

$$\mathbf{E}_Z = \bar{\mathbf{A}}\mathbf{T}, \tag{21}$$

Furthermore, the invariance structure of the array implies $\mathbf{E}_Z$ can be decomposed into $\mathbf{E}_X \in \mathbb{C}^{m\times d}$ and $\mathbf{E}_Y \in \mathbb{C}^{m\times d}$ (*cf.* X and Y subarrays) such that

$$\mathbf{E}_X = \mathbf{A}\mathbf{T}, \tag{22}$$

$$\mathbf{E}_Y = \mathbf{A}\mathbf{\Phi}\mathbf{T}. \tag{23}$$

which implies that

$$\mathcal{R}\{\mathbf{E}_X\} = \mathcal{R}\{\mathbf{E}_Y\} = \mathcal{R}\{\mathbf{A}\}.$$

Since $\mathbf{E}_X$ and $\mathbf{E}_Y$ share a common column space, the rank of

$$\mathbf{E}_{XY} \stackrel{\text{def}}{=} [\mathbf{E}_X \,\vdots\, \mathbf{E}_Y] \tag{24}$$

is d, which implies there exists a unique (recall $d \leq m$) rank d matrix[22] $\mathbf{F} \in \mathbb{C}^{2d\times d}$ such that

$$\mathbf{0} = [\mathbf{E}_X \,\vdots\, \mathbf{E}_Y]\mathbf{F} = \mathbf{E}_X\mathbf{F}_X + \mathbf{E}_Y\mathbf{F}_Y, \tag{25}$$

[22]This derivation, though somewhat more lengthy than at first glance seems necessary, will prove useful when noisy estimates of $\mathbf{E}_X$ and $\mathbf{E}_Y$ are available.

$$= \mathbf{ATF}_X + \mathbf{A\Phi TF}_Y . \tag{26}$$

$\mathbf{F}$ spans the *null-space* of $[\mathbf{E}_X \,\vdots\, \mathbf{E}_Y]$. Defining

$$\Psi \stackrel{\text{def}}{=} -\mathbf{F}_X[\mathbf{F}_Y]^{-1} , \tag{27}$$

equation (26) can be rearranged to yield[23]

$$\mathbf{AT}\Psi = \mathbf{A\Phi T} \Longrightarrow \mathbf{AT}\Psi\mathbf{T}^{-1} = \mathbf{A\Phi} . \tag{28}$$

Assuming $\mathbf{A}$ to be full rank implies

$$\boxed{\mathbf{T}\Psi\mathbf{T}^{-1} = \Phi .} \tag{29}$$

Therefore, the eigenvalues of Ψ must be equal to the diagonal elements of Φ and the columns of $\mathbf{T}$ are the eigenvectors of Ψ. This is the key relationship in the development of TLS ***ESPRIT*** and its properties. The signal parameters are obtained as nonlinear functions of the eigenvalues of the operator Ψ that maps (rotates) one set of vectors ($\mathbf{E}_X$) that span an m-dimensional signal subspace into another ($\mathbf{E}_Y$).

4.5. Estimating the Subspace Rotation Operator

In practical situations where only a finite number of noisy measurements are available, $\mathbf{E}_Z$ is estimated from the covariance matrices of the measurements, $\mathbf{R}_{ZZ}$, or equivalently from the data matrix $\mathbf{Z}$. The result is that $\mathcal{R}\{\mathbf{E}_Z\}$ is only an estimate of $\mathcal{S}_Z$, and with probability one, $\mathcal{R}\{\mathbf{E}_Z\} \neq \mathcal{R}\{\bar{\mathbf{A}}\}$. Furthermore, $\mathcal{R}\{\mathbf{E}_X\} \neq \mathcal{R}\{\mathbf{E}_Y\}$. Thus the objective of finding a Ψ such that $\mathbf{E}_X\Psi = \mathbf{E}_Y$ is no longer achievable. A criterion for obtaining a suitable estimate must be formulated. The most commonly employed criterion for problems of this nature is the *least-squares* (LS) criterion.

The standard LS criterion applied to the model $\mathbf{AX} = \mathbf{B}$ to obtain an estimate of $\mathbf{X}$ assumes $\mathbf{A}$ is known and the error is to be attributed to $\mathbf{B}$. Assuming the set of equations is overdetermined, the columns of $\mathbf{A}$ are linearly independent, and the noise in $\mathbf{B}$ is zero mean with covariance proportional to $\mathbf{I}$, the LS solution is

$$\hat{\mathrm{X}} = [\mathrm{A}^*\mathrm{A}]^{-1}\mathrm{A}^*\mathrm{B}.$$

It is easily verified that the estimate is unbiased and minimum variance. If both $\mathbf{A}$ and $\mathbf{B}$ are *noisy* however, the LS solution is known to be biased.

Since it is not difficult to argue that the estimates $\mathbf{E}_X$ and $\mathbf{E}_Y$ are *equally* noisy, the LS criterion is clearly inappropriate. A criterion that takes into account noise on both $\mathbf{A}$ and $\mathbf{B}$ is the *total least-squares* (TLS) criterion. The TLS criterion can be stated [7] as finding *residual* matrices $\mathbf{R_A}$ and $\mathbf{R_B}$ of minimum Frobenius norm, and $\hat{\mathbf{X}}$ such that

$$[\mathrm{A} + \mathrm{R_A}]\hat{\mathrm{X}} = \mathrm{B} + \mathrm{R_B}. \tag{30}$$

[23] The same argument used in deriving equation (21) can be used to derive equation (28) directly. However, the derivation only insures the existence and uniqueness of such a full-rank Ψ. The advantage of the preceding derivation is that implicitly a prescription for obtaining Ψ is given. Note that the existence and uniqueness of a full-rank Ψ guarantees the invertibility of $\mathbf{F}_Y$.

This criterion is equivalent (*cf.* Appendix A) to replacing the zero matrix in equation (25) by a matrix of errors, the Frobenius norm (*i.e.*, *total* least-squared error) of which is to be minimized. Appending a nontriviality constraint $\mathbf{F}^*\mathbf{F} = \mathbf{I}$ to eliminate the zero solution and applying standard Lagrange techniques (*cf.* Appendix B) leads to a solution for $\mathbf{F}$ given by the eigenvectors corresponding to the d smallest eigenvalues of $\mathbf{E}_{XY}^*\mathbf{E}_{XY}$. The eigenvalues of $\mathbf{\Psi}$ as defined above and calculated from the estimates $\mathbf{F}_X$ and $\mathbf{F}_Y$ are taken as estimates of the diagonal elements of $\mathbf{\Phi}$.

4.6. Summary of the TLS *ESPRIT* Covariance Algorithm

The TLS ***ESPRIT*** algorithm based on a covariance formulation can be summarized as follows.

1. Obtain an estimate of $\mathbf{R}_{ZZ}$, denoted $\hat{\mathbf{R}}_{ZZ}$, from the measurements $\mathbf{Z}$.
2. Compute the eigen-decomposition of $\hat{\mathbf{R}}_{ZZ}$
$$\hat{\mathbf{R}}_{ZZ}\bar{\mathbf{E}} = \bar{\mathbf{E}}\mathbf{\Lambda}\,,$$
where $\mathbf{\Lambda} = \mathrm{diag}\{\lambda_1, \ldots, \lambda_{2m}\}$, $\lambda_1 \geq \cdots \geq \lambda_{2m}$, and $\bar{\mathbf{E}} = [\mathbf{e}_1 \mid \cdots \mid \mathbf{e}_{2m}]$.
3. Estimate the number of sources $\hat{d}$ (*cf.* Appendix E).
4. Obtain the signal subspace estimate $\hat{\mathcal{S}}_Z = \mathcal{R}\{\mathbf{E}_Z\}$ where
$$\mathbf{E}_Z \stackrel{\mathrm{def}}{=} [\mathbf{e}_1 \mid \cdots \mid \mathbf{e}_{\hat{d}}] = \begin{bmatrix} \mathbf{E}_X \\ \mathbf{E}_Y \end{bmatrix}.$$
5. Compute the eigendecomposition ($\lambda_1 > \ldots > \lambda_{2\hat{d}}$),
$$\mathbf{E}_{XY}^*\mathbf{E}_{XY} \stackrel{\mathrm{def}}{=} \begin{bmatrix} \mathbf{E}_X^* \\ \mathbf{E}_Y^* \end{bmatrix} [\mathbf{E}_X \mid \mathbf{E}_Y] = \mathbf{E}\mathbf{\Lambda}\mathbf{E}^*,$$
and partition $\mathbf{E}$ into $\hat{d} \times \hat{d}$ submatrices,
$$\mathbf{E} \stackrel{\mathrm{def}}{=} \begin{bmatrix} \mathbf{E}_{11} & \mathbf{E}_{12} \\ \mathbf{E}_{21} & \mathbf{E}_{22} \end{bmatrix}.$$
6. Calculate the eigenvalues of $\mathbf{\Psi} = -\mathbf{E}_{12}\mathbf{E}_{22}^{-1}$,
$$\hat{\phi}_k = \lambda_k(-\mathbf{E}_{12}\mathbf{E}_{22}^{-1}), \;\; \forall\, k = 1, \ldots, \hat{d}.$$
7. Estimate $\hat{\theta}_k = f^{-1}(\hat{\phi}_k)$; *e.g.*, for DOA estimation, $\hat{\theta}_k = \sin^{-1}\{c\arg(\hat{\phi}_k)/(\omega_0\Delta)\}$.

As alluded to earlier, in many instances it is preferable to avoid forming covariance matrices, and instead to operate directly on the *data.* This approach leads to singular value decompositions (SVDs) of data matrices, and an SVD variant of ***ESPRIT*** summarized in Appendix C.

From the key relation, equation (29), several other quite striking results can be derived. For example, not only is knowledge of the array manifold not required, but the elements thereof associated with the estimated signal parameters (DOAs) can be estimated if desired. The same is true of the source correlation matrix, knowledge of which is not needed in ***ESPRIT***.

4.7. Multiple Invariance ***ESPRIT***

As mentioned earlier, the popularity of the **MUSIC** algorithm is due in part to its applicability to arbitrary, but known (*i.e.*, calibrated) antenna arrays. Its principal disadvantages are the heavy computational and memory requirements inherent in the calibration and spectral search procedures. For the special case of a uniform linear array (ULA), however, the resulting array manifold has a simple analytical form that can be exploited to mitigate these requirements (see, for example, the **Root-MUSIC** procedure [4, 6]. The ***ESPRIT*** algorithm was the first to demonstrate that similar computational savings could also be achieved for more general antenna array configurations. However, as currently formulated, ***ESPRIT*** cannot exploit additional structure in the sensor array beyond a single displacement invariance. As a result, for many arrays there exist several possible choices of subarray pairs which satisfy the displacement invariance requirement, and there is no general criterion for choosing one pair over another.

Though the single invariance ***ESPRIT*** algorithm may be applied to arrays with more than one displacement invariance (via overlapping subarrays), it is expected that an algorithm that fully exploits information about the physical array should yield superior performance. The desire to incorporate all such information *and* eliminate the need for the calibration procedure are the primary motivations for developing an algorithm for the multiple invariance problem. Though important in practice, ULAs are only a subset of the set of arrays possessing multiple invariances. Commonly oriented subarrays each consisting of a ULA are easily envisaged in spaceborne as well as underwater acoustic applications. The problem formulation as discussed in Appendix A provides the framework for generalizing the *subspace-fitting* interpretation of ***ESPRIT*** to such cases.

Assume the array is composed of p identical subarrays of m sensors, each with its own displacement vector $\mathbf{\Delta}_i$ relative to the reference subarray ($p_{\mathrm{ref}} = 0$). The displacement vectors are assumed to be colinear and on the same line. The total number of sensors M need not be equal to mp since overlapping subarrays are allowed ($M < mp$ in such cases). Let $\mathbf{J}_i$ be the $m \times M$ matrix that selects the output of the ith subarray from each snapshot,and define

$$\mathbf{J} = \begin{bmatrix} \mathbf{J}_0 \\ \mathbf{J}_1 \\ \vdots \\ \mathbf{J}_{p-1} \end{bmatrix}.$$

The structure inherent in $\mathbf{E}_S$ may be described by the following equation:

$$\mathbf{J}\mathbf{E}_s \stackrel{\mathrm{def}}{=} \begin{bmatrix} \mathbf{E}_0 \\ \mathbf{E}_1 \\ \mathbf{E}_2 \\ \vdots \\ \mathbf{E}_{p-1} \end{bmatrix} = \begin{bmatrix} \mathbf{A} \\ \mathbf{A}\mathbf{\Phi} \\ \mathbf{A}\mathbf{\Phi}^{\delta_2} \\ \vdots \\ \mathbf{A}\mathbf{\Phi}^{\delta_{p-1}} \end{bmatrix} \mathbf{T}. \tag{31}$$

where $\delta_i \stackrel{\mathrm{def}}{=} |\mathbf{\Delta}_i|/|\mathbf{\Delta}_1|$. The multiple invariance ***ESPRIT*** problem can now be stated as follows:

Given p subspace estimates $\mathbf{E}_0, \mathbf{E}_1, \ldots, \mathbf{E}_{p-1} \in \mathbb{C}^{m\times d}$, *find an operator* $\mathbf{\Psi} \in \mathbb{C}^{d\times d}$ *and a matrix* $\mathbf{B} \in \mathbb{C}^{m\times d}$ *to minimize*

$$V = \left\| \begin{bmatrix} \mathbf{E}_0 \\ \mathbf{E}_1 \\ \mathbf{E}_2 \\ \vdots \\ \mathbf{E}_{p-1} \end{bmatrix} \mathbf{W}^{1/2} - \begin{bmatrix} \mathbf{B} \\ \mathbf{B\Psi} \\ \mathbf{B\Psi}^{\delta_2} \\ \vdots \\ \mathbf{B\Psi}^{\delta_{p-1}} \end{bmatrix} \right\|_F^2 .$$

Note that the $\mathbf{\Delta}_i$'s are collinear, though not necessarily of equal magnitude. Also note that a weighting matrix $\mathbf{W} > 0$ has been included in the cost function V. The asymptotically optimal weighting matrix has been shown to be given by [15]

$$\mathbf{W}_{\text{opt}} = (\hat{\mathbf{\Lambda}}_s - \hat{\sigma}^2\mathbf{I})^2 \hat{\mathbf{\Lambda}}_s^{-1} , \tag{32}$$

where $\hat{\mathbf{\Lambda}}_s$ is the diagonal eigenvalue matrix associated with the signal subspace eigenvectors. $\mathbf{B} = \mathbf{AT}$ and the estimates of the parameters of interest are again obtained from the eigenvalues of the operator $\mathbf{\Psi} = \mathbf{T}^{-1}\mathbf{\Phi}\mathbf{T}$. For the special case of a uniform linear array, $\mathbf{\Delta}_i = i\mathbf{\Delta}_1$ and $\delta_i = i$, and the cost function becomes

$$V = \left\| \begin{bmatrix} \mathbf{E}_0 \\ \mathbf{E}_1 \\ \mathbf{E}_2 \\ \vdots \\ \mathbf{E}_{p-1} \end{bmatrix} \mathbf{W}^{1/2} - \begin{bmatrix} \mathbf{B} \\ \mathbf{B\Psi} \\ \mathbf{B\Psi}^{2} \\ \vdots \\ \mathbf{B\Psi}^{p-1} \end{bmatrix} \right\|_F^2 .$$

Note that in this case p is the number of sensors and $m = 1$.

If we define the following quantities,

$$\bar{\mathbf{E}}_W \stackrel{\text{def}}{=} \left[\mathbf{E}_0\mathbf{W}^{1/2} \,\vdots\, \mathbf{E}_1\mathbf{W}^{1/2} \,\vdots\, \cdots \,\vdots\, \mathbf{E}_{p-1}\mathbf{W}^{1/2}\right]$$

$$\bar{\mathbf{\Psi}} \stackrel{\text{def}}{=} \left[\mathbf{I} \,\vdots\, \mathbf{\Psi} \,\vdots\, \mathbf{\Psi}^{\delta_2} \,\vdots\, \cdots \,\vdots\, \mathbf{\Psi}^{\delta_{p-1}}\right],$$

standard properties of the *trace* operator can be applied to simplify the expression for V:

$$V = \text{Tr}\left\{[\bar{\mathbf{E}}_W - \mathbf{B}\bar{\mathbf{\Psi}}]^* [\bar{\mathbf{E}}_W - \mathbf{B}\bar{\mathbf{\Psi}}]\right\} . \tag{33}$$

This form for the cost function is familiar to separable least-squares minimization problems, so $\mathbf{B}$ can be eliminated from equation (33) by solving $\partial_{\mathbf{B}} V = \mathbf{0}$ and substituting back into (33). The resulting expression for $\mathbf{B}$ is

$$\mathbf{B} = \bar{\mathbf{E}}_W \bar{\mathbf{\Psi}}^* \left[\bar{\mathbf{\Psi}}\bar{\mathbf{\Psi}}^*\right]^{-1} , \tag{34}$$

which, when substituted into (33) gives

$$V = \text{Tr}\left\{\bar{\mathbf{E}}_W^* \bar{\mathbf{E}}_W (\mathbf{I} - \bar{\mathbf{\Psi}}^*[\bar{\mathbf{\Psi}}\bar{\mathbf{\Psi}}^*]^{-1}\bar{\mathbf{\Psi}})\right\} . \tag{35}$$

4.8. Algorithm Implementation

The original MI-***ESPRIT*** problem statement (4.7) is now formulated as

$$\hat{\Psi} = \arg\min_{\Psi} V = \arg\min_{\Psi} \mathrm{Tr}\{\bar{\mathbf{E}}_W^* \bar{\mathbf{E}}_W (\mathbf{I} - \bar{\Psi}^*[\bar{\Psi}\bar{\Psi}^*]^{-1}\bar{\Psi})\} \,. \tag{36}$$

The minimization of V over all $\Psi \in \mathbb{C}^{d \times d}$ is a nonlinear, computationally complex problem, and there appears to be no easily formulated solution as there is for the single invariance case. Though some type of search technique such as a Newton-Raphson procedure is required to achieve $\partial_{\Psi} V = \mathbf{0}$, the fact that the single invariance version of ***ESPRIT*** can be applied to produce an excellent initial estimate of Ψ is of immense practical importance. ***ESPRIT*** can be applied in different ways to obtain an initial estimate, though in general, the most favorable configuration is obtained with $\mathbf{J}_x = [\mathbf{J}_0^T \cdots \mathbf{J}_{p-2}^T]^T$ and $\mathbf{J}_y = [\mathbf{J}_1^T \cdots \mathbf{J}_{p-1}^T]^T$, often referred to as a *maximum overlap* configuration.

5. Simulation Results

To demonstrate the capabilities of the ***ESPRIT*** algorithm, simulations results for several simulated DOA estiamtion problems are presented. Following these, application of the technique to the estimation of dispersion relations of a large polyethylene (backbone) molecule is briefly described.

5.1. DOA Estimation Simulations

Many simulations have been conducted exploring different aspects of ***ESPRIT*** and making comparisons with other techniques (*cf.* [6]). Herein, only one of the scenarios, but one that addresses several issues that arise in a practical implementation of ***ESPRIT***, is presented. Thus, sensor gain and phase errors, as well as sensor spacing errors are included.[24] Finally, unequal source powers and a high degree of source correlation are assumed.

The array chosen was a ten element array with doublet spacing $\lambda/4$ and the five doublets randomly spaced on a line resulting in an aperture of approximately 4λ. Two sources were located at 24° and 28° and were of unequal powers, 20 dB and 15 dB respectively. Sensor response errors were introduced by including zero-mean normal random additive errors, independent of angle, with sigmas of 0.1 dB in amplitude and 2° in phase. Additive sensor position errors were similarly included with sigmas of 0.5% of the nominal separation in each doublet. An identity noise covariance was used, *i.e.*, $\mathbf{R}_{NN} = \mathbf{I}$.

The sources were initially 50% temporally correlated. The results are given in Figures 4 through 6. The indicated failure rate for MUSIC of 6% indicates the percentage of the number of trials in which the conventional MUSIC spectrum either exhibited less than two peaks, or either of the final DOA estimates was outside the interval [20°, 32°]. The sample means and variances of the ***ESPRIT*** and MUSIC estimates (excluding the failures) are given in Table 1. Notice in Figure 6 that there is clearly a small overlap in the distributions of the ***ESPRIT*** parameter estimates. Since a simple *nearest-to-the-true-location* assignment scheme was employed in the calculation of the sample statistics,[25] the resulting mean and

[24] While the array is linear, the elements are not uniformly spaced rendering root-MUSIC inapplicable.

[25] This does not imply that the statistics are compiled by *splitting the histogram* down the middle and computing the center of mass and second moments of the truncated distributions.

Estimator	Source DOA Estimates	
	$\hat{\theta}_1$ $(\theta_1 = 24°)$	$\hat{\theta}_2$ $(\theta_2 = 28°)$
MUSIC*	$24.25° \pm 0.23°$	$27.69° \pm 0.27°$
ESPRIT	$23.97° \pm 0.89°$	$28.01° \pm 1.01°$

* Based on 4713 successes out of 5000 trials.

Table 1: ***ESPRIT*** and MUSIC DOA Sample Means and Sigmas — Random 10 Element Linear Array, SNR [20,15] dB, Source Correlation 50%, $\Delta = \lambda/4$.

variance estimates are biased, though for this scenario the effect is not large. Ideally, a 5-parameter (two means, two variances, and a correlation coefficient) fit of a two-dimensional Gaussian distribution to the estimates should be performed.

The results indicate the presence of a finite sample bias in the conventional MUSIC estimates, the source of which is described in detail in [6, pp. 98-100]. They also indicate that knowledge of the array manifold is certainly valuable, at least in situations where the ***ESPRIT*** subarray separation is small. As the subarray separation increases, however, the relative value of the knowledge of the subarray manifold decreases as is clearly demonstrated in the third of the three scenarios to be described herein.

To investigate the sensitivity to source correlation, that parameter was increased from 50% to 90% while holding all other parameters fixed. The results are shown in Figures 7 through 9. The means and sigmas of the DOA estimates are given in Table 2. The results

Estimator	Source DOA Estimates	
	$\hat{\theta}_1$ $(\theta_1 = 24°)$	$\hat{\theta}_2$ $(\theta_2 = 28°)$
MUSIC*	$24.35° \pm 0.28°$	$27.48° \pm 0.38°$
ESPRIT	$23.93° \pm 1.07°$	$28.06° \pm 1.37°$

* Based on 3175 successes out of 5000 trials.

Table 2: ***ESPRIT*** and MUSIC DOA Sample Means and Sigmas — Random 10 Element Linear Array, Source Powers [20,15] dB, Source Correlation 90%, $\Delta = \lambda/4$.

indicate the relative lack of sensitivity of ***ESPRIT*** to the dramatic increase in correlation, while MUSIC degrades significantly. The MUSIC failure rate has increased to 37% as manifest in the spectra shown in Figure 7, and the finite sample bias in the estimate of the source DOA for the lower power source at 28° has increased significantly. Note that in computing the statistics, only the trials that resulted in *successes* are included. Note that as in the previous scenario, the ***ESPRIT*** mean and variance estimates are approximate as a

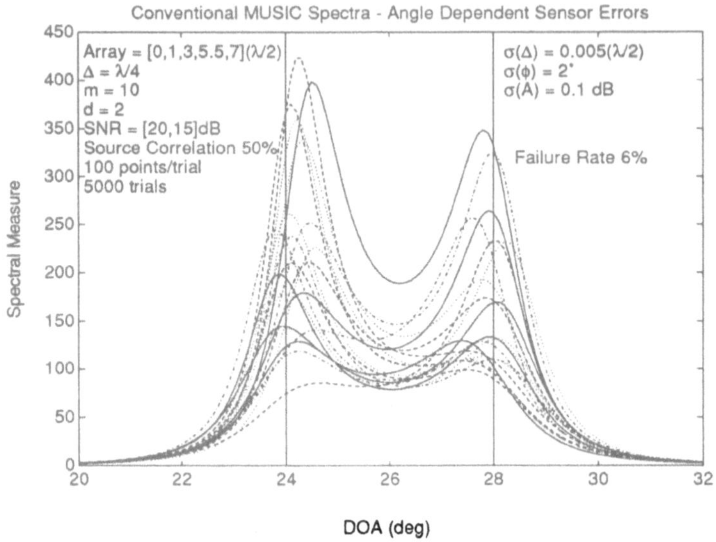

Figure 4: Conventional **MUSIC** Spectra — Random 10 Element Linear Array, Source Correlation 50%, Small Array Aperture.

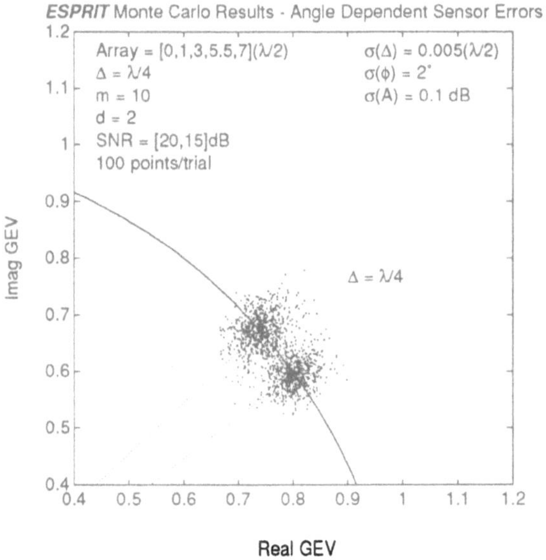

Figure 5: ***ESPRIT*** GEVs — Random 10 Element Linear Array, Source Correlation 50%, Small Array Aperture ($\Delta = \lambda/4$).

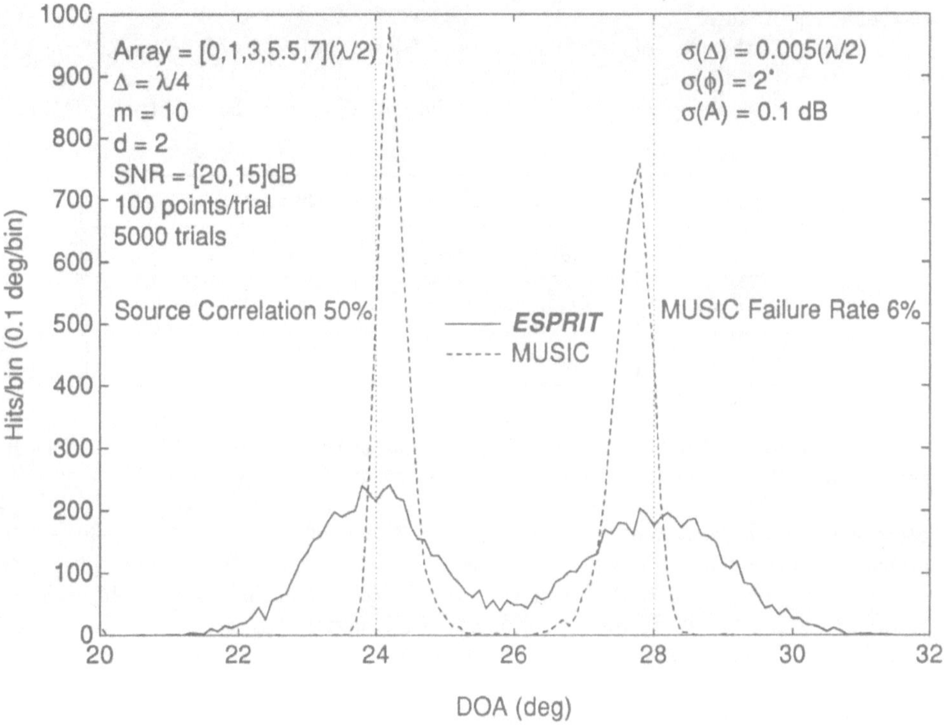

Figure 6: Histogram of MUSIC and ***ESPRIT*** Results — Random 10 Element Linear Array, Source Correlation 50%, Small Array Aperture ($\Delta = \lambda/4$).

two-dimensional Gaussian distribution fit was eschewed in favor of a simple angle ordering assignment procedure.

To investigate the effect of increasing the subarray separation, Δ was increased from $\lambda/4$ to 4λ. The results are given in Figures 10 through 12, and the sample means and sigmas in Table 3. The conventional **MUSIC** algorithm did not fail once in the 1000 trials. This is a consequence of the increase in the magnitude of the gradient of the array manifold vectors evaluated at the nominal angles resulting from the increase in array aperture. As alluded to earlier, as the subarray separation increases, knowledge of the subarray manifold becomes relatively less important. In this scenario, ***ESPRIT*** is performing as well as, if not better than **MUSIC**, and at a fraction of the computational cost!

Estimator	*Source DOA Estimates*	
	$\hat{\theta}_1 \quad (\theta_1 = 24°)$	$\hat{\theta}_2 \quad (\theta_2 = 28°)$
MUSIC	$24.011° \pm 0.056°$	$27.986° \pm 0.078°$
ESPRIT	$24.003° \pm 0.062°$	$28.002° \pm 0.089°$

Table 3: ***ESPRIT*** and MUSIC DOA Sample Means and Sigmas — Random 10 Element Linear Array, Source Powers [20,15] dB, Source Correlation 90%, $\Delta = 4\lambda$.

5.2. Dispersion Relation Estimation

In this section, a comparison of the results obtained in applying both **MUSIC** and ***ESPRIT*** to the problem of estimating dispersion relations (frequency versus wave number) for nonlinear simulations of a 500-element carbon-carbon backbone of a polyethylene chain are presented. Using an appropriate Hamiltonian, Hamilton's equations of motion were solved using a ODE solver on a CRAY XMP with an estimated absolute error of 10^{-11} per integration step. From the trajectories of the atoms in the chain, a *particle density correlation function*, ρ, was calculated:

$$\rho(K,t) = \sum_{i=1}^{500} e^{iKq_i(t)}. \tag{37}$$

The Fourier transform of $\rho(K,t)$ yields the dynamic structure factor

$$S(K,\omega) = \int_{-\infty}^{+\infty} e^{i\omega t} \rho(K,t)\, dt, \tag{38}$$

and the location of the major peaks of $S(K,\omega)$ leads to dispersion relations $\omega(K)$.

Unfortunately, calculation of the Fourier transform in equation (38) is not practical due to the rather large time interval over which the particle density correlation function is required. In similar situations involving spectral analysis of time series, a discretely sampled version over a finite time interval of the function under consideration, in this case the particle density correlation function $\rho(K,t)$, is often substituted in place of the continuous version and a discrete Fourier transform (DFT) used instead.

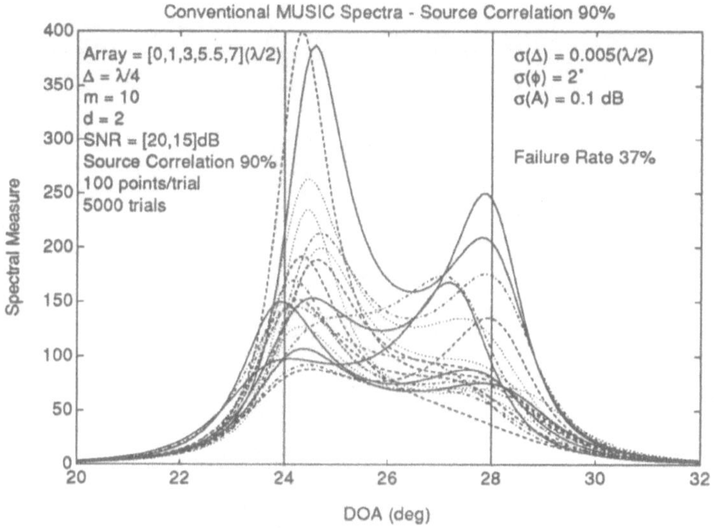

Figure 7: Conventional **MUSIC** Spectra — Random 10 Element Linear Array, Source Correlation 90%, Small Array Aperture.

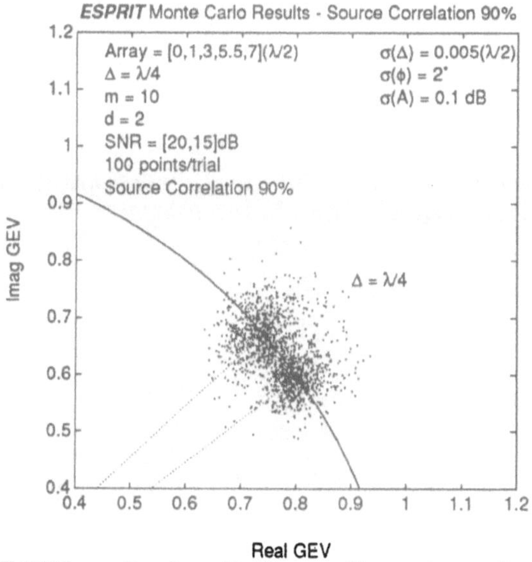

Figure 8: ***ESPRIT*** GEVs — Random 10 Element Linear Array, Source Correlation 90%, Small Array Aperture ($\Delta = \lambda/4$).

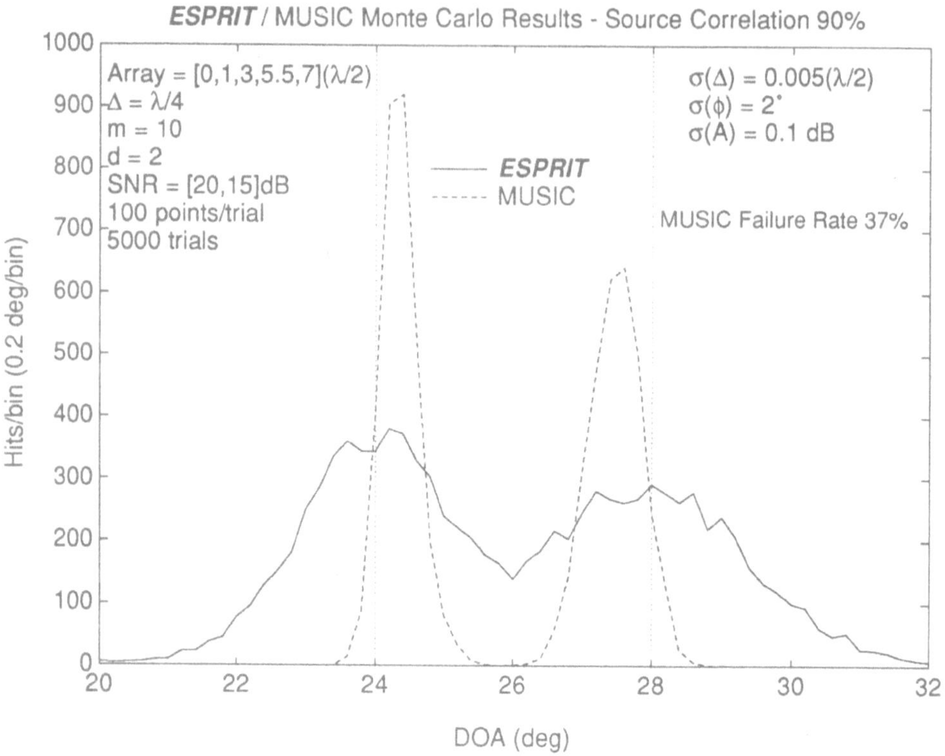

Figure 9: Histogram of MUSIC and ***ESPRIT*** Results — Random 10 Element Linear Array, Source Correlation 90%, Small Array Aperture ($\Delta = \lambda/4$).

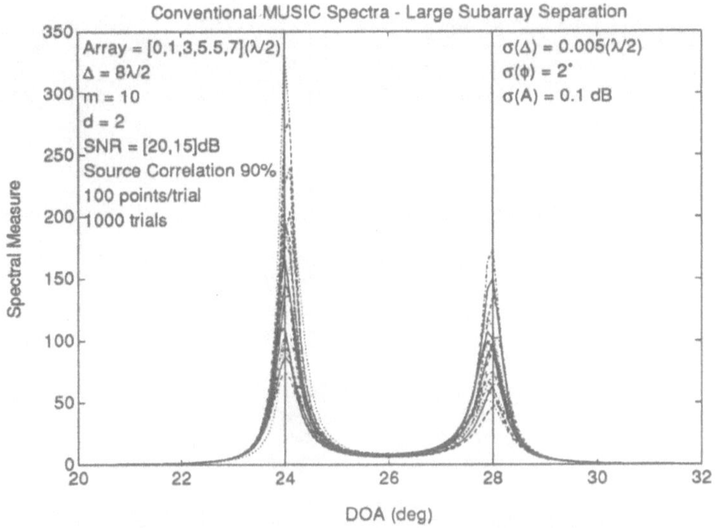

Figure 10: Conventional **MUSIC** Spectra — Random 10 Element Linear Array, Source Correlation 90%, Large Array Aperture.

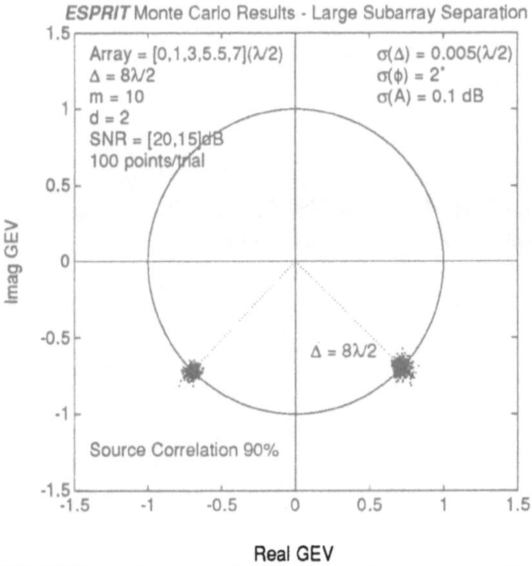

Figure 11: ***ESPRIT*** GEVs — Random 10 Element Linear Array, Source Correlation 90%, Large Array Aperture ($\Delta = 4\lambda$).

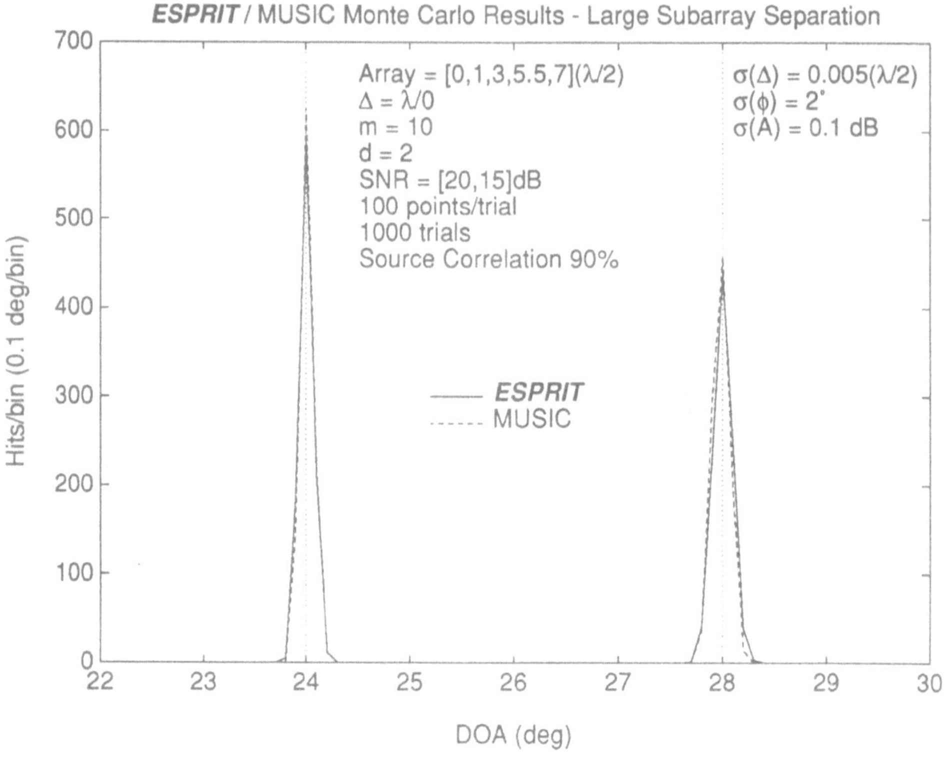

Figure 12: Histogram of MUSIC and ***ESPRIT*** Results — Random 10 Element Linear Array, Source Correlation 90%, Large Array Aperture ($\Delta = 4\lambda$).

There are many problems associated with this approach, however. For the problems considered herein, the most significant is the fact that the frequency resolution is limited by the inverse of the time interval available. For molecular dynamics simulations, obtaining the particle density correlation function over time intervals large enough ($\approx 33\,ps$) to produce the often desired frequency resolution ($\approx 1\,cm^{-1}$) is quite costly. The second issue involves detecting the damping coefficient associated with the various modes of oscillation (*i.e.*, detection of the location of the characteristic frequency in the complex plane and not necessarily on the unit circle). Calculation of the DFT does not directly provide this information even in the absence of noise or nonlinear perturbation effects. Extracting damping coefficients from Fourier coefficients requires further calculations, calculations that are quite sensitive to errors in the Fourier coefficients.

In situations where the data can be adequately modeled as a finite sum of cisoids (damped, growing, or constant modulus exponential signals), the limited resolution associated with small time intervals and the classical Fourier approach can be overcome. The basic idea is to exploit the data model in the estimation procedure, thereby compressing the available information into a few parameters instead of using an invertible (information preserving) transformation that *spreads* the information over a parameter space that is much too large.

The particle density correlation function $\rho(K,t)$ was calculated for approximately $0.27\,ps$ (64 samples at a sampling rate of 240.04 THz) and for 10 values of K ranging from $0.0497\,a_0^{-1}$ to $2.2843\,a_0^{-1}$ in $0.2483\,a_0^{-1}$ increments. The correlation function for a few values of K is shown in Figures 13 and 14. Though the similarity to the artificially generated correlation

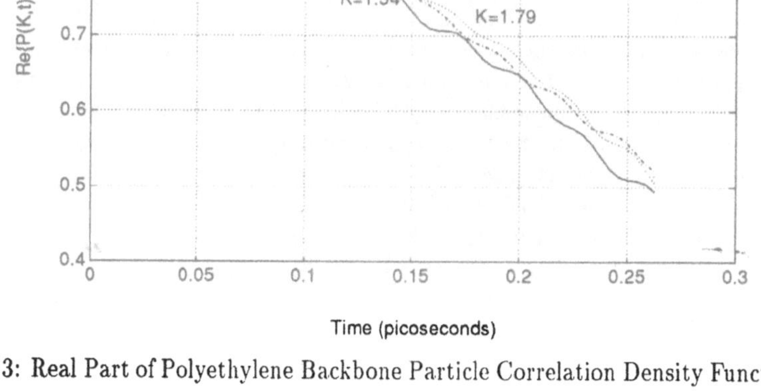

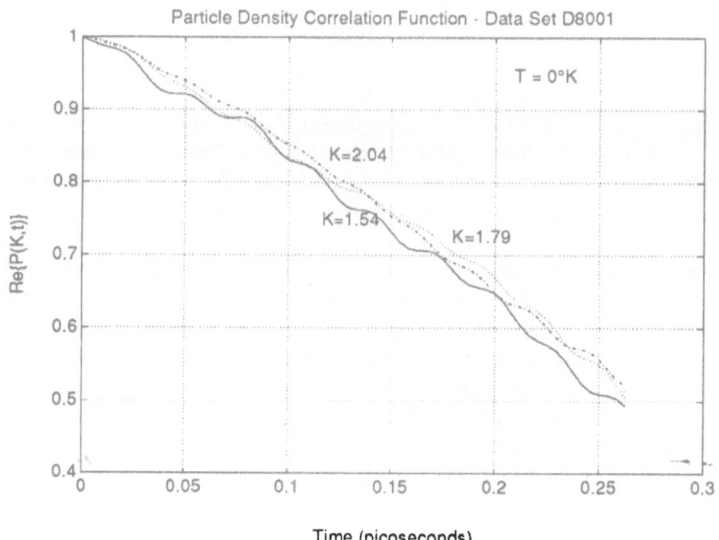

Figure 13: Real Part of Polyethylene Backbone Particle Correlation Density Function versus Time

function of the previous section is obvious, the effects of the nonlinearities in the molecular

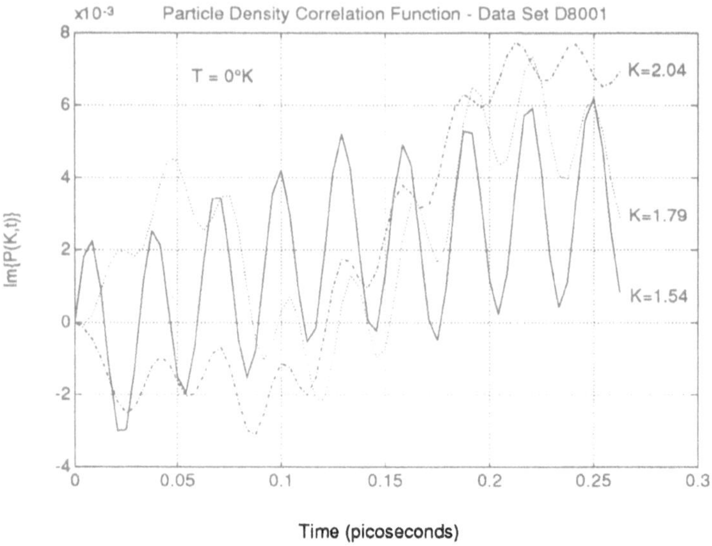

Figure 14: Imaginary Part of Polyethylene Backbone Particle Correlation Density Function versus Time

dynamics are discernible. It should be noted that these *nonlinear* components represent errors in the underlying data model of a sum of cisoids in additive noise.

The inadequacy of the classical FFT approach to power spectrum estimation with only a limited amount of available data is again manifest in Figure 15 where the magnitude of the FFT of the data shown in Figures 13 and 14 is given.

To obtain the necessary *snapshots* from a single 64-sample time series, a sliding window of length 30 was used to generate a 30 × 35 data matrix $\mathbf{Z}$. The preliminary step of the estimation of the number of cisoidal components present in the data in order to obtain an appropriate estimate of the dimension of the signal subspace was performed by inspection of the eigenvalues of the correlation matrix associated with $\mathbf{Z}$, *i.e.*, $\mathbf{R}_{ZZ} \propto \mathbf{Z}\mathbf{Z}^*$. It is easily shown that these eigenvalues are the squares of the singular values of the data matrix $\mathbf{Z}$. The eigenvalues (in decreasing order) of the correlation matrices for all ten values of K are shown in Figure 16. For all values of K, there is a noticeable decrease in the magnitude of the eigenvalues between λ_{10} and λ_{15} where the magnitude drops below 10^{-10}. Since the absolute integration error was approximately 10^{-11} per integration step, and generation of $\rho(K,t)$ required in excess of 10^6 steps, numerical errors on the order of 10^{-5} can be expected. Thus, eigenvalues (which are the squares of the singular values) less than 10^{-10} and the corresponding eigenvectors are associated with the *numerical noise subspace*. Though more sophisticated statistical tests could be employed, most are all based on the additive independent noise model which is clearly invalid in this problem. Based on these arguments, a signal subspace dimension $d = 10$ was selected as appropriate.

Results of applying **MUSIC** with $m = 30$ and $d = 10$ to $\rho(K,t)$ for all values of K are shown in Figures 17 and 18. The phenomenon of *obscured peaks* evident in the simulated

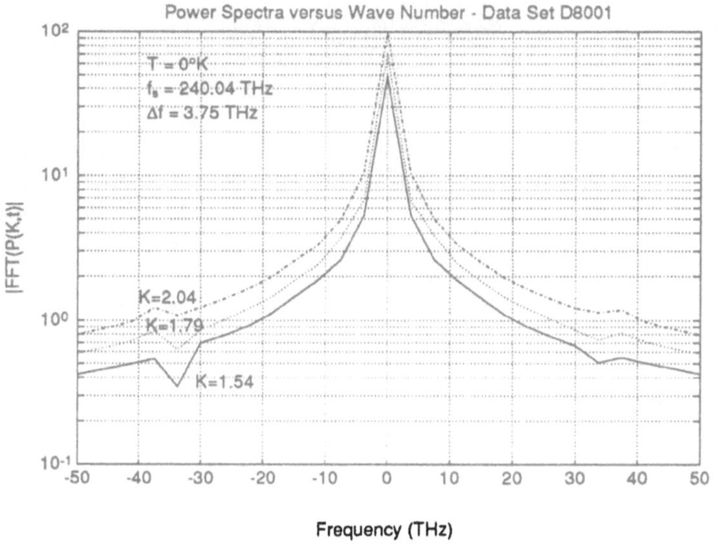

Figure 15: Magnitude of the FFT of the Polyethylene Backbone Particle Correlation Density Function

results of the previous section are clearly manifest in Figure 17 as well. The peaks located outside the interval $(-40, 40)$ *THz* are clearly *outliers.*

The results of applying ***ESPRIT*** to the same data are summarized in Figures 19 and 20. The dispersion relation estimates in Figure 19 are for a signal subspace dimension $d = 8$ as indicated and three values of $\Delta = 1, 2, 3$ where Δ is the *subarray separation* in units of taps in the TDL. Maximally overlapping subarrays were chosen as such structures have been shown to incorporate the maximum amount of information present in the measurements from a TDL. Consistency of the estimates as a function of Δ for the results shown in Figure 20 for a subspace dimension $d = 10$ versus the variation present in the estimates for $d = 8$ shown in Figure 19 provide further justification for the choice $d = 10$. The *stability* of the estimates versus Δ in the *optical branch* above 30 *THz* are most illustrative.

Also note that as indicated in the figures, ***ESPRIT*** required an order of magnitude less computation than did MUSIC for $d = 10$, with another factor of 2 resulting from a decrease in subspace dimension to $d = 8$. Increasingly ordered ***ESPRIT*** frequency (pole angle) estimates and the corresponding magnitude of the pole estimates are given in Tables 4 and 5 respectively. Though complex magnitudes of each of the cisoidal components can be calculated, they are of limited value without knowledge of the initial conditions. Since the initial conditions were randomly chosen subject to a zero center-of-mass momentum condition, the excitation of the different modes was random as well.

This example demonstrates the relative computational efficiency and improved resolution capability of ***ESPRIT*** over MUSIC for problems of this type. ***ESPRIT*** achieves its resolution superiority by effectively performing a d-dimensional parameter search in the complex plane for cisoid poles that best fit the observations. Practically, MUSIC is constrained

$K\,(a_0^{-1})$	***ESPRIT*** Frequency Estimates (THz)									
0.0497	-34.13	-30.61	-17.15	-1.10	0.00	0.95	1.10	17.11	30.52	34.14
0.2980	-33.82	-30.79	-17.01	-7.04	-0.22	0.22	7.04	17.56	30.89	33.84
0.5462	-33.20	-30.81	-16.38	-12.19	-0.32	0.35	12.20	16.94	30.90	33.21
0.7945	-32.78	-30.05	-17.00	-15.73	-0.33	0.37	13.82	15.69	29.16	32.13
1.0428	-33.44	-29.57	-16.07	-15.95	-0.30	0.36	14.86	16.13	29.99	33.38
1.2911	-34.35	-30.39	-15.72	-13.98	-0.28	0.40	13.63	15.10	30.41	34.49
1.5394	-34.00	-30.93	-15.57	-9.43	-0.44	0.40	9.50	15.78	30.84	34.14
1.7877	-35.23	-29.42	-15.39	-4.43	-0.03	0.84	4.14	15.77	29.70	35.23
2.0360	-36.00	-29.35	-15.57	-6.87	-0.27	0.71	8.52	15.68	30.75	36.28
2.2843	-36.62	-30.20	-15.50	-1.72	-0.14	0.57	9.21	16.63	31.31	35.77

Table 4: ***ESPRIT*** Frequency Estimates versus K for $\Delta = 1$, $m = 30$, $d = 10$

$K\,(a_0^{-1})$	Magnitude of ***ESPRIT*** Complex Pole Estimates									
0.0497	0.9900	0.9640	1.0048	0.9501	0.9797	1.0175	1.0183	0.9803	0.9497	1.0048
0.2980	0.9989	1.0003	0.9473	0.9919	1.0164	1.0140	0.9890	0.9514	1.0002	0.9990
0.5462	0.9979	0.9989	0.8975	0.9980	0.9945	0.9937	1.0013	0.9151	0.9990	0.9969
0.7945	0.9980	0.8741	1.0037	1.0275	0.9958	0.9979	1.0186	0.8937	1.0018	0.9968
1.0428	0.9993	0.8920	1.0026	0.9933	1.0342	1.0241	1.0204	0.9005	1.0169	0.9975
1.2911	1.0011	0.9939	0.9254	0.9946	1.0297	0.9876	0.9895	0.8808	0.9973	0.9975
1.5394	0.9972	1.0091	0.9451	0.9881	1.0039	1.0107	0.9682	0.9420	1.0105	0.9982
1.7877	0.9204	1.0082	0.9666	1.0182	1.0247	1.0153	0.9802	0.9594	1.0290	0.9879
2.0360	0.9804	0.8953	0.9800	1.0358	1.0201	1.0178	1.0048	0.9689	1.0233	0.9852
2.2843	1.0175	0.9752	0.9982	1.0241	1.0104	1.0098	1.0336	0.9823	1.0491	1.0017

Table 5: Magnitude of ***ESPRIT*** Complex Pole Estimates versus K for $\Delta = 1$, $m = 30$, $d = 10$

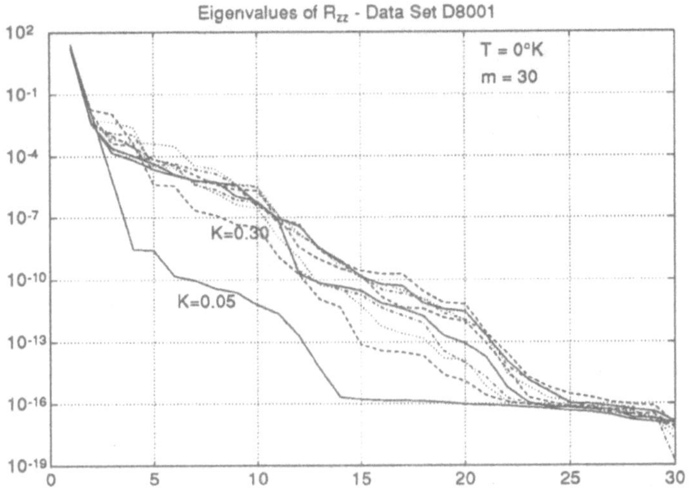

Figure 16: Eigenvalues of the Correlation Matrix of the Polyethylene Backbone Particle Correlation Density Function for All Values of K

to a one-dimensional search over the unit-circle for d frequencies (angles) that best fit the data. In addition to being computationally more efficient, ***ESPRIT*** provides additional information in the form of the magnitude of the pole estimate (*cf.* damping). In many cases, estimates of damping coefficients are important.

6. Discussion

Herein, a new technique (TLS ***ESPRIT***) for signal parameter estimation has been described. The effectiveness of the algorithm has been demonstrated in the several examples presented.

6.1. Numerical Concerns

The earlier versions of ***ESPRIT*** described in [16, 17] sought generalized eigenvalues of a pair of matrices that ideally shared a common $(2m - d)$-dimensional nullspace, a fact that leads to some concern over potential numerical difficulties in solving the generalized eigenproblem. A preliminary solution was to find a least-squares estimate of an $m \times m$ operator whose action was restricted to a d-dimensional subspace. Imposing the subspace restriction (*e.g.,* by subspace *deflation*) as a constraint prior to solving the generalized eigenproblem mitigates these numerical concerns; however, the least-squares property of the estimate is retained [6, pp. 133-145]. In many cases, the difference between the LS and TLS parameter estimates is insignificant, the difference becoming notable at very low SNRs. The LS estimates, as predicted, are biased, while the TLS estimates are not. The major advantage of the TLS approach at higher SNRs is clearly numerical. By properly formulating the esti-

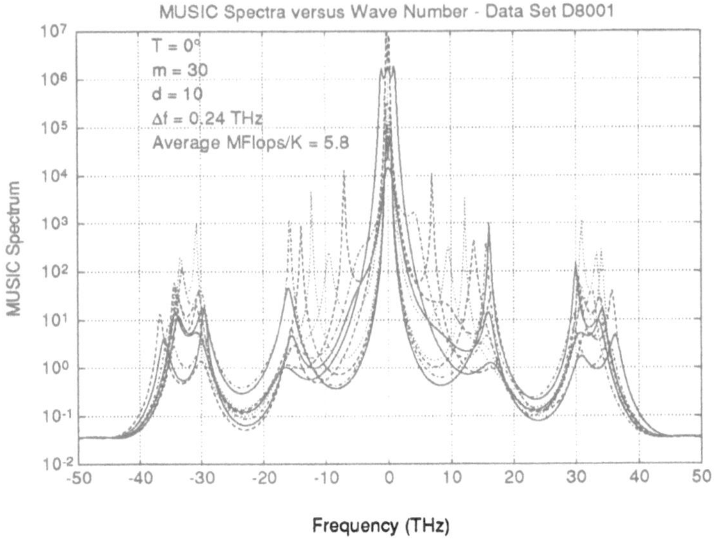

Figure 17: Polyethylene Backbone Particle Correlation Density Function **MUSIC** *Spectra*

mation problem in a subspace of the appropriate dimension (determined by the dimension of the signal subspace where the observed signal *power* is maximum), numerical problems associated with subspace deflation are mitigated altogether.

6.2. *ESPRIT* AND NULL-STEERING

Many of the previous high-resolution parameter estimation techniques are based on *steering beams* toward signal directions and, in some cases, simultaneously attempting to otherwise minimize the power in some weighted combination of sensor outputs. Parameter estimates are associated with DOAs at which peaks in the power output occur. Though intuitively appealing at first glance, there is a much more powerful alternative philosophy, that of *null-steering*. It is well-known that deep, sharp notches in directivity patterns and filter gain functions are much easier to achieve than sharp peaks. Interferometers exploit this fact to obtain accurate estimates of source parameters by finding the relative phase required to cancel signal components in two channels. In this context, ***ESPRIT*** can be interpreted as a *multi-dimensional null-steering* parameter estimation algorithm. Calculation of the eigenvalues of the (rotation) operator Ψ, which are the roots of its characteristic polynomial, can be interpreted as multi-dimensional null-steering. Instead of steering broad beams, ***ESPRIT*** steers sharp nulls at all sources simultaneously and does so without relying on knowledge of the array manifold!

6.3. COMPUTATIONAL ADVANTAGES OF *ESPRIT*

The primary computational advantage of ***ESPRIT*** is that it eliminates the search procedure inherent in all previous methods (ML, ME, MUSIC). ***ESPRIT*** produces signal parameter

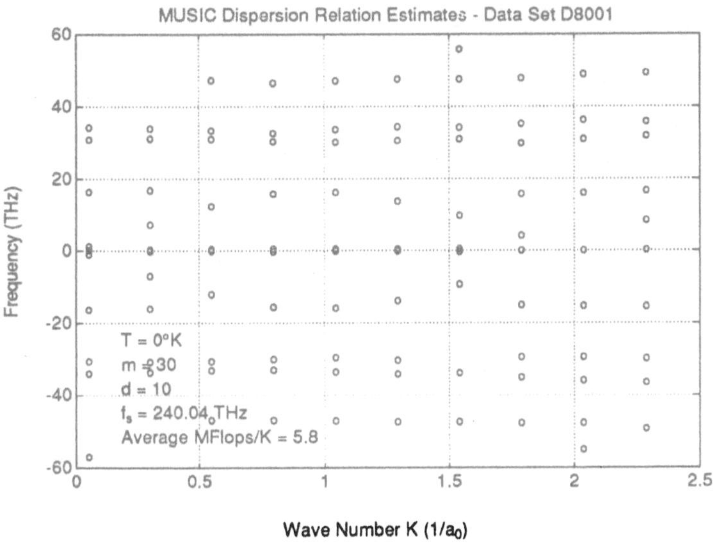

Figure 18: Polyethylene Backbone Particle Correlation Density Function **MUSIC** Dispersion Relation Estimates

estimates directly in terms of (generalized) eigenvalues. As noted previously, this involves computations of the order d^3. On the other hand, **MUSIC** and the other high resolution techniques require a search over $\mathcal{A}$, and it is this search that is computationally expensive.[26]

A simple example serves to demonstrate the notable advantage of ***ESPRIT***. Consider a DOA estimation application with a sensor array composed of twenty (20) elements. Assume that the parameter space is one dimensional, *i.e.*, that azimuth estimates are to be obtained, and that two signals are present. Further assume the aperture covers an arc of 2 radians and that a one-half milliradian resolution in azimuth is required. Neglecting comparison operations in searching for the four largest peaks in the **MUSIC** *spectrum*, a search of the entire array manifold requires on the order of $20 \times 4000 \approx 10^5$ multiplications and additions (operations). The resulting ***ESPRIT*** eigenproblems (SVDs) require on the order of $10 \times 2^3 \approx 10^2$ operations. The resulting computational advantage of ***ESPRIT*** is on the order of 10^3 over **MUSIC**.[27] Note that an exhaustive search of $\mathcal{A}$ is assumed for the search required. **MUSIC** would certainly benefit from a more sophisticated search technique, but the advantage of ***ESPRIT*** is still overwhelming.

[26]The significant computational advantage of ***ESPRIT*** becomes even more pronounced in multi-dimensional parameter estimation where the computational load grows linearly with dimension in ***ESPRIT***, while that of **MUSIC** grows exponentially. If r_l is the resolution (*i.e.*, number of vectors) required in the calibration of $\mathcal{A}$ for the lth dimension in Θ, the computation required to search over L dimensions for d parameter vectors is proportional to $\prod_{l=1}^{L} r_l$. For $r_l = r$, the computational load is r^L.

[27]For a two-dimensional manifold, *e.g.*, azimuth and elevation, the advantage improves to a remarkable factor of 10^7.

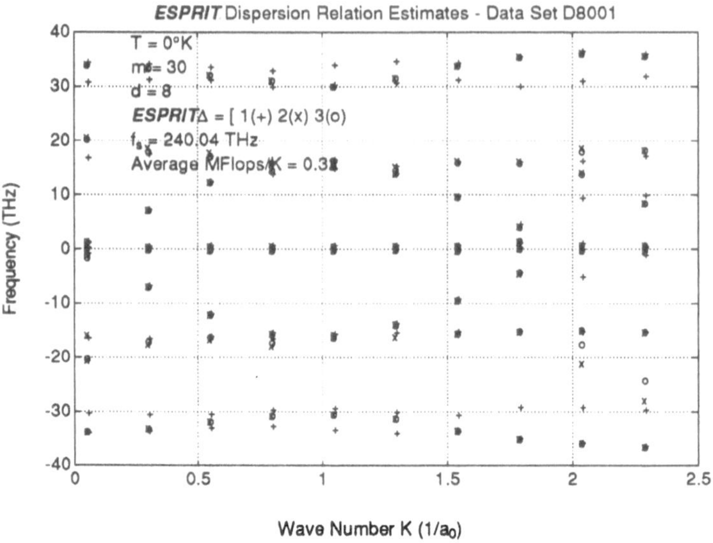

Figure 19: Polyethylene Backbone Particle Correlation Density Function ***ESPRIT*** Dispersion Relation Estimates

7. Directions of Current and Future Research

7.1. Multidimensional ***ESPRIT***

The planar array of sensors in Figure 3 has, as far as ***ESPRIT*** is concerned, DOA sensitivity in one dimension only, *i.e.*, azimuth sensitivity. If the sources are actually parameterized by an azimuth and an elevation angle, *i.e.*, a DOA estimation problem in three-dimensional space is being addressed, another sensor array of doublets with a displacement vector independent of the first is required. If these two arrays have no elements in common and are sampled independently, then independent processing of the array outputs is *nearly optimal.* The first sensor array yields *azimuth* DOA estimates which locate the sources on *cones of ambiguity* in three-space. Assuming that the displacement vector of the second array is orthogonal to the first (though it need not be), the *elevation* DOA estimates from the second array reduce the ambiguity to that of a pair of DOA {azimuth,elevation} tuple estimates that are symmetric with respect to the plane defined by $\boldsymbol{\Delta}_{az} \wedge \boldsymbol{\Delta}_{el}$. Practically, this ambiguity can be resolved in many situations by physical reasoning. The important point to be made is that for each *dimension in parameter space* in which an estimate is desired, a *displacement invariance* is required in that dimension. This invariance manifests itself in the data model as a complex scaling of the elements of $\mathcal{A}$, a fact that is the key to ***ESPRIT***.

There is an important issue to be addressed when employing the decoupled procedure described above in the presence of multiple sources. The estimates from the two subarrays must be *paired* to obtain a tuple associated with a particular source. One possible approach is to perform signal copy in both azimuth and elevation separately, and correlate signal esti-

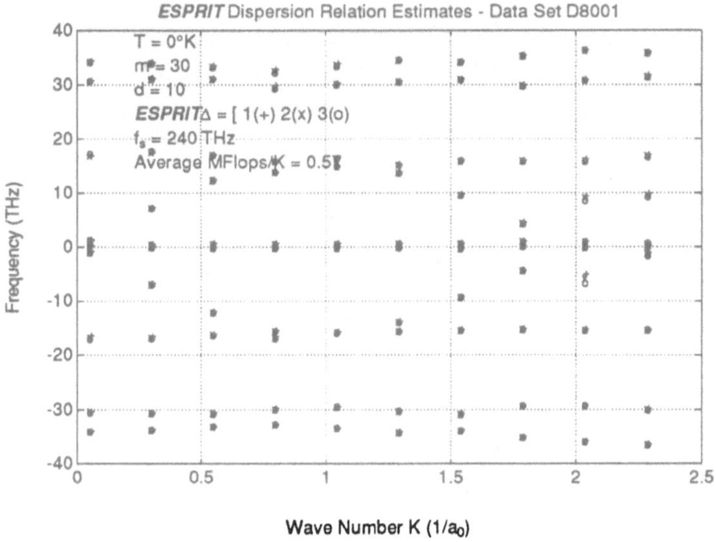

Figure 20: Polyethylene Backbone Particle Correlation Density Function ***ESPRIT*** Dispersion Relation Estimates

mates. This suboptimal strategy can be employed only if the arrays are sampled such that the signal components are highly correlated (recall the assumption of temporally independent snapshots). Such a correlation results if the arrays are *nearly* simultaneously sampled in the sense that the distance travelled by the signals between azimuth and elevation array sampling times is small compared to the extent of the entire array.

Note that practically, azimuth and elevation estimates can be obtained from an array of *triplets* of sensors, a generalization of the commonly used rectangular array of sensors (*cf.* phased-arrays) for source location in two dimensions. If frequency estimates are desired as well, tapped delay lines with pairs of taps separated by a fixed *time increment behind each sensor* can be used to obtain the necessary information. Practically this might be implemented by a sampler that at regular or irregular intervals obtains a pair of samples separated by a fixed time increment. In such practical implementations, decoupled estimation is clearly suboptimal.

For the purposes of illustration, consider a regular rectangular array of identical sensors (commonly referred to as a phased-array). The array clearly possesses many *invariances* which can be exploited. Consider for the moment however, only the three subarrays depicted in Figure 21. Though it is certainly possible to choose subarrays that do not share any elements in common, the subarrays shown are maximally *overlapping*, a condition that has some interesting associated mathematical consequences as discussed later on. By combining subarrays 0 and 1 and subarrays 0 and 2, two arrays with orthogonal displacement vectors ($\boldsymbol{\Delta}_1 \perp \boldsymbol{\Delta}_2$) result. Furthermore, the two arrays share a common subarray, subarray 0, by construction, so that by simultaneously sampling the output of each sensor in the entire

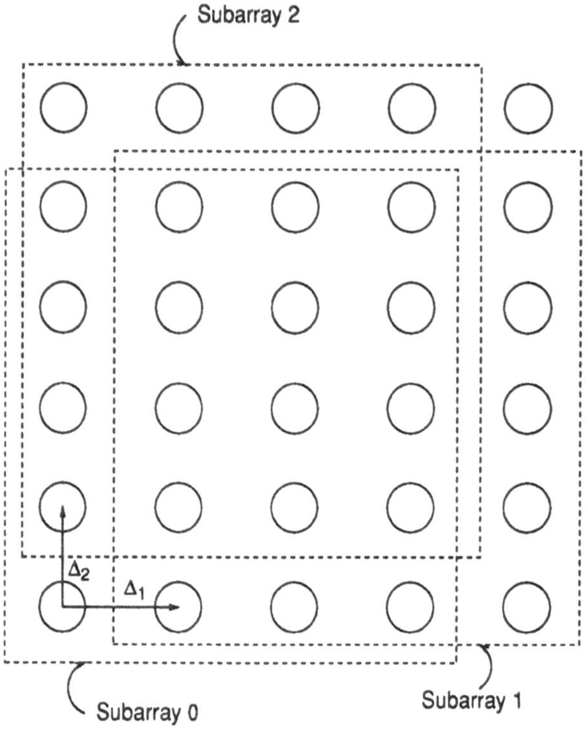

Figure 21: Possible Subarray Structure of a Regular Rectangular Array

array, the following *data model/signal subspace relationships* are easily derived.

$$\begin{bmatrix} \mathbf{E}_0 \\ \mathbf{E}_1 \\ \mathbf{E}_2 \end{bmatrix} = \begin{bmatrix} \mathbf{B} \\ \mathbf{B}\mathbf{\Psi}_1 \\ \mathbf{B}\mathbf{\Psi}_2 \end{bmatrix} = \begin{bmatrix} \mathbf{AT} \\ \mathbf{ATT}^{-1}\mathbf{\Phi}_1\mathbf{T} \\ \mathbf{ATT}^{-1}\mathbf{\Phi}_2\mathbf{T} \end{bmatrix}. \tag{39}$$

$\mathbf{E}_i$ is an $m \times d$ matrix that spans the signal subspace observed by subarray i.

In the presence of noisy data, equation (39) does not hold since $\mathbf{E}$ must be estimated. As in the one-dimensional case, a TLS approach is adopted and the objective is to find

$$\min_{\mathbf{B},\mathbf{\Psi}_1,\mathbf{\Psi}_2} \left\| \begin{bmatrix} \mathbf{E}_0 \\ \mathbf{E}_1 \\ \mathbf{E}_2 \end{bmatrix} - \begin{bmatrix} \mathbf{B} \\ \mathbf{B}\mathbf{\Psi}_1 \\ \mathbf{B}\mathbf{\Psi}_2 \end{bmatrix} \right\|_F^2, \tag{40}$$

subject to the constraint that

$$[\mathbf{\Psi}_1, \mathbf{\Psi}_2] \stackrel{\text{def}}{=} \mathbf{\Psi}_1\mathbf{\Psi}_2 - \mathbf{\Psi}_2\mathbf{\Psi}_1 = \mathbf{0},$$

that is, $\mathbf{\Psi}_1$ commutes with $\mathbf{\Psi}_2$, a constraint that is essentially equivalent to the constraint that they share a common set of eigenvectors, $\mathbf{T}$. A computationally efficient solution to this optimization problem is an area of current investigation. Several iterative algorithms have been derived, but the unavoidable issues of local minima, numerical stability, and convergence rates make these approaches unattractive (unsuitable) for real-time autonomous implementation. The objective is derive an algorithm in the spirit of ***ESPRIT***, involving SVDs and eigendecompositions as the fundamental operations.

7.2. Alternatives to the SVD and Eigendecompositions

In estimating the signal subspace, the indicated technique in all current high resolution signal subspace algorithms is to perform an SVD of the data matrix or an eigendecomposition of the sample covariance matrix. Once the measurements are reduced to yield $\mathbf{E}_Z$, unless d is exceedingly large, the majority of the computational effort required to obtain parameter estimates is complete. The bulk of the remaining effort as far as ***ESPRIT*** is concerned involves an eigendecomposition of a matrix in $\mathbb{C}^{2d \times 2d}$, where d is the estimated number of signals present (*i.e.*, the number of columns of $\mathbf{E}_Z$).

As should be clear from the developments and discussions to this point, eigenvectors or singular vectors are *not required* in the implementation of ***ESPRIT***, or any of the other signal subspace algorithms for that matter, in obtaining estimates of the signal subsapce. They can be replaced by *any* set of vectors (not necessarily orthonormal) that span the signal subspace, *i.e.*, any set of vectors $\bar{\mathbf{E}}_Z$ that satisfy

$$\bar{\mathbf{E}}_Z = \mathbf{E}_Z \bar{\mathbf{T}}, \tag{41}$$

for *any* nonsingular $\bar{\mathbf{T}}$ is informationally to $\mathbf{E}_Z$ as far as ***ESPRIT*** is concerned.[28] Thus, any efficient algorithm for obtaining any set of linearly independent vectors spanning the (estimated) signal subspace can be employed. An area for further research involves deriving a computationally efficient (*i.e.*, more efficient than a full eigendecomposition) algorithm for obtaining such a set of vectors. It might also prove fruitful to investigate applications of multilinear algebra to this problem, since an object that quite succinctly characterizes the signal subspace is the *wedge* or *Grassmann* product of any set of vectors that span the subspace!

Appendices

A. Equivalent TLS Formulations

In this appendix, the equivalence between two formulations of the TLS ***ESPRIT*** solution is demonstrated. Though certainly of pedagogical interest, the equivalence also yields a method for obtaining TLS estimates of the array manifold vectors, as well as a method for extending ***ESPRIT*** to arrays with multiple invariances.

The most common statement of the TLS linear parameter estimation problem as stated, for example, in [7] is as follows. Given $\mathbf{A}, \mathbf{B} \in \mathbb{C}^{m \times d}$, $m > d$, find $\mathbf{R_A}$ and $\mathbf{R_B}$ of minimum

[28]The structure of $\mathcal{A}$ being exploited by ***ESPRIT*** is the linear structure manifest in the *row space*, and is independent of *any linear* operations in the *column space*, *i.e.*, linear operations on the *right*.

Frobenius[29] norm, and $\mathbf{X}$ such that

$$[\mathbf{A}+\mathbf{R_A}]\mathbf{X}=\mathbf{B}+\mathbf{R_B}\,. \tag{42}$$

Note that in general $\mathbf{B}$ need not have the same number of columns as $\mathbf{A}$ though for the applications described herein, $\mathbf{A}$ and $\mathbf{B}$ have the same dimensions and $\mathbf{X}$ is square. With reference to ***ESPRIT***, $\mathbf{R}_{\{\mathbf{A},\mathbf{B}\}}$ represent errors in the subspace estimates $\mathbf{E}_X = \mathbf{A}$ and $\mathbf{E}_Y = \mathbf{B}$ respectively, and $\mathbf{X}$ is the operator Ψ whose eigenvalues are parameters of interest. Thus, the TLS ***ESPRIT*** estimation problem can be stated as follows:

Given subspace estimates $\mathbf{E}_X$ and $\mathbf{E}_Y$, find an operator Ψ and residual matrices $\mathbf{R}_{E_X}, \mathbf{R}_{E_Y}$ to minimize

$$J = \|[\mathbf{R}_{E_X} \mid \mathbf{R}_{E_Y}]\|_F^2 \tag{43}$$

subject to

$$[\mathbf{E}_X+\mathbf{R}_{E_X}]\,\Psi = \mathbf{E}_Y+\mathbf{R}_{E_Y}\,. \tag{44}$$

The objective is to show that the resulting estimate, Ψ, is equivalent to that obtained from the constrained minimization problem posed in Section 4, specifically:

Given subspace estimates $\mathbf{E}_X$ and $\mathbf{E}_Y$, find a matrix $\mathbf{F} \in \mathbb{C}^{2d\times d}$ to minimize

$$J = \|[\mathbf{E}_X \mid \mathbf{E}_Y]\,\mathbf{F}\|_F^2 \tag{45}$$

subject to

$$\mathbf{F}^*\mathbf{F} = \mathbf{I}\,, \tag{46}$$

where

$$\mathbf{F} \stackrel{\text{def}}{=} \begin{bmatrix} \mathbf{F}_X \\ \mathbf{F}_Y \end{bmatrix}, \tag{47}$$

$$\Psi = -\mathbf{F}_X[\mathbf{F}_Y]^{-1}. \tag{48}$$

Starting with the TLS formulation in equations (43) and (44), define $\mathbf{B} \stackrel{\text{def}}{=} \mathbf{E}_X + \mathbf{R}_{E_X}$. Substituting into equation (44) gives $\mathbf{E}_Y + \mathbf{R}_{E_Y} = \mathbf{B}\Psi$. Rearranging these equations gives

$$\mathbf{R}_{E_X} = \mathbf{E}_X - \mathbf{B}\,, \tag{49}$$

$$\mathbf{R}_{E_Y} = \mathbf{E}_Y - \mathbf{B}\Psi\,, \tag{50}$$

so the minimization becomes

$$\min_{\{\mathbf{B},\Psi\}} J = \min_{\{\mathbf{B},\Psi\}} \|[\mathbf{E}_X - \mathbf{B} \mid \mathbf{E}_Y - \mathbf{B}\Psi]\|_F^2\,, \tag{51}$$

which using standard properties of the trace operator (Tr) can be written as follows:

$$\min_{\{\mathbf{B},\Psi\}} J = \min_{\{\mathbf{B},\Psi\}} Tr\left\{[\mathbf{E}_{XY} - \mathbf{B}\bar{\Psi}]^*\,[\mathbf{E}_{XY} - \mathbf{B}\bar{\Psi}]\right\}\,, \tag{52}$$

[29] If knowledge of the covariance of the errors is available (in general in the form of a 4-tensor), a weighted Frobenius norm can be employed. Since for the problems considered herein it can be argued that the errors are independent and identically distributed (at least asymptotically), the *weighting matrix* or metric is proportional to the identity.

where

$$\mathbf{E}_{XY} \stackrel{\text{def}}{=} [\mathbf{E}_X \,\vdots\, \mathbf{E}_Y] \, , \tag{53}$$

$$\bar{\mathbf{\Psi}} \stackrel{\text{def}}{=} [\mathbf{I} \,\vdots\, \mathbf{\Psi}] \, . \tag{54}$$

The minimization problem can be solved by setting[30] $\partial_{\mathbf{B}} J = \mathbf{0}$ and solving for $\mathbf{B}$ in terms of $\bar{\mathbf{\Psi}}$ which yields[31]

$$\mathbf{B} = \mathbf{E}_{XY}\bar{\mathbf{\Psi}}^* \left[\bar{\mathbf{\Psi}}\bar{\mathbf{\Psi}}^*\right]^{-1} . \tag{55}$$

Substituting this expression for $\mathbf{B}$ back into equation (52), and employing the cyclic property of the trace operator gives

$$\min_{\mathbf{\Psi}} J = \min_{\mathbf{\Psi}} Tr\left\{\mathbf{E}_{XY}^*\mathbf{E}_{XY}\left(\mathbf{I} - \bar{\mathbf{\Psi}}^* \left[\bar{\mathbf{\Psi}}\bar{\mathbf{\Psi}}^*\right]^{-1} \bar{\mathbf{\Psi}}\right)\right\} . \tag{56}$$

A similar sequence of manipulations on equation (45) yields

$$\min_{\mathbf{F}} J = \min_{\mathbf{F}} Tr\left\{\mathbf{E}_{XY}^*\mathbf{E}_{XY}\left(\mathbf{F}\left[\mathbf{F}^*\mathbf{F}\right]^{-1}\mathbf{F}^*\right)\right\}, \tag{57}$$

where an identity derived from the constraint equation (46) has been inserted to demonstrate that an equivalent constraint is $\mathbf{F}$ be full-rank d so that $\mathbf{F}^*\mathbf{F}$ is invertible. Since $\bar{\mathbf{\Psi}}$ satisfies this condition by construction, $\mathbf{I} - \bar{\mathbf{\Psi}}^* \left[\bar{\mathbf{\Psi}}\bar{\mathbf{\Psi}}^*\right]^{-1} \bar{\mathbf{\Psi}} \in \mathbb{C}^{2d \times 2d}$ is rank d as well, a fact easily proven using eigendecompositions. In general, for $\mathbf{A} \in \mathbb{C}^{m \times n}$ and $m > n$,

$$P_{\mathbf{A}} \stackrel{\text{def}}{=} \mathbf{A}\left[\mathbf{A}^*\mathbf{A}\right]^{-1}\mathbf{A}^* \tag{58}$$

defines a *projection operator* $P_{\mathbf{A}}$ on $\mathbb{C}^m$, mapping elements of $\mathbb{C}^m$ to the subspace of $\mathbb{C}^m$ spanned by the columns of $\mathbf{A}$, and $\mathbf{I} - P_{\mathbf{A}}$ defines a projection operator $P_{\mathbf{A}}^{\perp}$ into the orthogonal complement of the subspace spanned by the columns of $\mathbf{A}$. Employing this notation convention, the minimization problems are manifestly equivalent,[32] since both are minimizations of the same functional form over rank d projection matrices in $\mathbb{C}^{2d \times 2d}$. Furthermore, the solution is easily visualized geometrically (and proven algebraically in Appendix B) to be the sum of the d smallest eigenvalues[33] of $\mathbf{E}_{XY}^*\mathbf{E}_{XY}$ (squares of the d smallest singular values of $\mathbf{E}_{XY}$), obtained by choosing to project into the subspace spanned by the d corresponding right eigenvectors (right singular-vectors of $\mathbf{E}_{XY}$). Letting

$$\mathbf{E}_{XY}^*\mathbf{E}_{XY} = [\mathbf{E}_1 \,\vdots\, \mathbf{E}_2]\,\mathbf{\Lambda}\,[\mathbf{E}_1 \,\vdots\, \mathbf{E}_2]^* \, , \tag{59}$$

[30] Since complex conjugation is not analytic, care must be taken in the calculation of the derivative. It turns out, however, that treating $\mathbf{B}^*$ as an independent variable yields the correct equations, though there is certainly nothing mathematically rigorous in this approach.

[31] Note that for any $\delta\mathbf{B}$, $Tr\left\{\delta\mathbf{B}\bar{\mathbf{\Psi}}^*\bar{\mathbf{\Psi}}\delta\mathbf{B}\right\} > 0$, so the stationary point is actually a minimum.

[32] There is a subtle difference in the constraints however. Mathematically speaking, the set of matrices of the form $\bar{\mathbf{\Psi}}^*$ is a subset of all full-rank matrices in $\mathbb{C}^{2d \times d}$ since full-rank matrices whose upper $d \times d$ block is not invertible can not be converted into the form of $\bar{\mathbf{\Psi}}^*$. For the problems considered herein, this set has zero probability of occurring and need not be considered further.

[33] This sum is clearly an achievable lower bound since replacing any of the d smallest eigenvalues with any of the remaining eigenvalues results in an increase in J, unless the dth smallest eigenvalue has multiplicity greater than one. Though mathematically possible, by the problem construction this also will occur with probability zero, and therefore need not be considered.

where $\mathbf{\Lambda} = \text{diag}\{\lambda_1, \ldots, \lambda_{2d}\}$ and $\lambda_1 \geq \lambda_2 \geq \ldots \geq \lambda_{2d}$, $P_{\mathbf{F}} = P^{\perp}_{\bar{\mathbf{\Psi}}^*} = P_{\mathbf{E}_2}$.

Having solved the minimization problem, equivalence of the estimates of $\mathbf{\Psi}$ remains to be shown. Defining

$$[\mathbf{E}_1 \,\vdots\, \mathbf{E}_2] \stackrel{\text{def}}{=} \begin{bmatrix} \mathbf{E}_{11} & \mathbf{E}_{12} \\ \mathbf{E}_{21} & \mathbf{E}_{22} \end{bmatrix}, \tag{60}$$

the TLS ***ESPRIT*** estimate of $\mathbf{\Psi}$ is easily seen to be $-\mathbf{E}_{12}\mathbf{E}_{22}^{-1}$. By orthogonality of the eigenvectors of $\mathbf{E}^*_{XY}\mathbf{E}_{XY}$, $P^{\perp}_{\bar{\mathbf{\Psi}}^*} = P_{\mathbf{E}_2}$ implies $P_{\bar{\mathbf{\Psi}}^*} = P_{\mathbf{E}_1}$, and therefore there exists a non-singular matrix $\mathbf{T}$ such that

$$\begin{bmatrix} \mathbf{I} \\ \mathbf{\Psi}^* \end{bmatrix} = \begin{bmatrix} \mathbf{E}_{11} \\ \mathbf{E}_{21} \end{bmatrix} \mathbf{T}. \tag{61}$$

Thus, the estimate of $\mathbf{\Psi}$ is given by $\mathbf{E}_{11}^{-*}\mathbf{E}^*_{21}$. The orthogonality relation

$$\mathbf{E}^*_{11}\mathbf{E}_{12} + \mathbf{E}^*_{21}\mathbf{E}_{22} = \mathbf{0} \tag{62}$$

now easily establishes the equivalence of the two estimates of $\mathbf{\Psi}$ as was to be shown. Note that substituting the equation for $\mathbf{\Psi}$ into the equation for $\mathbf{B}$ yields

$$\mathbf{B} = \mathbf{E}_{XY}\mathbf{E}_1\mathbf{E}^*_{11} \tag{63}$$

using the easily proven fact that $\bar{\mathbf{\Psi}}\bar{\mathbf{\Psi}}^* = \mathbf{E}_{11}^{-*}\mathbf{E}_{11}^{-1}$. Also note that if the SVD of $\mathbf{E}_{XY}$ is used to obtain $[\mathbf{E}_1 \,\vdots\, \mathbf{E}_2]$ as the matrix of right singular-vectors, defining $[\mathbf{U}_1 \,\vdots\, \mathbf{U}_2]$ as the matrix of left-singular vectors, and $\mathbf{\Sigma}_{XY}$ as the diagonal matrix of ordered singular values,

$$\mathbf{E}_{XY} = [\mathbf{U}_1 \,\vdots\, \mathbf{U}_2]\mathbf{\Sigma}_{XY}[\mathbf{E}_1 \,\vdots\, \mathbf{E}_2]^*. \tag{64}$$

Substituting this expression into equation (63) yields

$$\mathbf{B} = \mathbf{U}_1\mathbf{\Sigma}_1\mathbf{E}^*_{11}, \tag{65}$$

where $\mathbf{\Sigma}_1 = \text{diag}\{\sigma_1, \ldots, \sigma_d\}$, the d *largest* singular values of $\mathbf{E}_{XY}$. Thus, once the SVD of $\mathbf{E}_{XY}$ has been computed, $\mathbf{B}$ can be computed without computing $\mathbf{\Psi}$.

B. TLS Estimation of Ψ

In this appendix, a short derivation of the TLS estimate of the (rotation) operator $\mathbf{\Psi}$ is presented. The objective is to find a matrix $\bar{\mathbf{\Psi}} \in \mathbb{C}^{2dxd}$ satisfying

$$\bar{\mathbf{\Psi}}^*\bar{\mathbf{\Psi}} = \mathbf{I},$$

that minimizes the Frobenius norm of $\mathbf{E}_{XY}\bar{\mathbf{\Psi}}$. Mathematically,

$$\begin{aligned} \bar{\mathbf{\Psi}} &= \arg \min_{\bar{\mathbf{\Psi}}^*\bar{\mathbf{\Psi}}=\mathbf{I}} \|\mathbf{E}_{XY}\bar{\mathbf{\Psi}}\|_F, \\ &= \arg \min_{\bar{\mathbf{\Psi}}^*\bar{\mathbf{\Psi}}=\mathbf{I}} Tr\{\bar{\mathbf{\Psi}}^*\mathbf{E}^*_{XY}\mathbf{E}_{XY}\bar{\mathbf{\Psi}}\}. \end{aligned}$$

Defining the cost function J

$$J \stackrel{\text{def}}{=} Tr\{\bar{\mathbf{\Psi}}^*\mathbf{E}^*_{XY}\mathbf{E}_{XY}\bar{\mathbf{\Psi}}\},$$

and appending the constraint (inside the trace) with a matrix of Lagrange multipliers Λ,

$$J = Tr\{\bar{\Psi}^*\mathbf{E}_{XY}^*\mathbf{E}_{XY}\bar{\Psi} - \Lambda(\bar{\Psi}^*\bar{\Psi} - \mathbf{I})\}.$$

At a minimum of J, the gradient of J with respect $\bar{\Psi}$ is zero, *i.e.*,

$$\frac{\partial J}{\partial \bar{\Psi}} = \mathbf{E}_{XY}^*\mathbf{E}_{XY}\bar{\Psi} - \bar{\Psi}\Lambda = 0. \tag{66}$$

This implies that $\bar{\Psi}$ spans an *invariant subspace* of $\mathbf{E}_{XY}^*\mathbf{E}_{XY}$. Multiplying equation (66) on the left by $\bar{\Psi}^*$ and using the constraint $\bar{\Psi}^*\bar{\Psi} = \mathbf{I}$ gives

$$\Lambda = \bar{\Psi}^*\mathbf{E}_{XY}^*\mathbf{E}_{XY}\bar{\Psi}. \tag{67}$$

Clearly, Λ is a positive semi-definite Hermitian matrix and therefore diagonalizable. Thus, there exists a non-singular matrix $\mathbf{T}$, whose columns are the right eigenvectors of Λ, such that

$$\mathbf{T}^{-1}\Lambda\mathbf{T} = \text{diag}\{\lambda_1, \ldots, \lambda_n\} \stackrel{\text{def}}{=} \Lambda_d. \tag{68}$$

Multiplying equation (66) on the right by $\mathbf{T}$ leads to the following *partial* eigenproblem:

$$\mathbf{E}_{XY}^*\mathbf{E}_{XY}\bar{\Psi}\mathbf{T} = \bar{\Psi}\mathbf{T}\mathbf{T}^{-1}\Lambda\mathbf{T} = \bar{\Psi}\mathbf{T}\Lambda_d. \tag{69}$$

Thus, the columns of $\bar{\Psi}\mathbf{T}$ are eigenvectors of $\mathbf{E}_{XY}^*\mathbf{E}_{XY}$ and the associated eigenvalues are the diagonal elements of Λ_d. Since the objective is to minimize

$$J = Tr\{\bar{\Psi}^*\mathbf{E}_{XY}^*\mathbf{E}_{XY}\bar{\Psi}\} = Tr\,\Lambda_d,$$

Λ_d must contain the *d smallest* eigenvalues of $\mathbf{E}_{XY}^*\mathbf{E}_{XY}$. $\bar{\Psi}$ therefore spans the d-dimensional *invariant subspace* with the *smallest volume*, *i.e.*, the total *least*-squared error. The estimate of the operator Ψ_{XY} is obtained as described in Appendix A.

C. The SVD TLS *ESPRIT* Algorithm

As mentioned earlier, it is often preferable to avoid *squaring* data to obtain covariance matrices, and instead to operate directly on the *data* itself. Benefits accrue from the resulting reduction in matrix condition numbers. With infinite precision computation, the algorithm is theoretically the same whether eigendecompositions or singular value decompositions (SVD) are used. The resulting parameter estimates are identical. Practically, finite precision arithmetic must be employed, and numerical issues can become important. For large arrays composed of hundreds of sensors, and for large amounts of data, round-off and overflow are potential problems to be aware of when forming covariance matrices. Often such problems can be overcome by operating directly on the data. Applying ***ESPRIT*** to data matrices directly requires the use of the SVD. The SVD version of ***ESPRIT*** is obtained by simply replacing all the eigendecompositions in the covariance version with SVDs[34] except for the final eigendecomposition[35] of Ψ. The following is a summary of the algorithm.

[34] See [7] for a discussion of the SVD.

[35] Note that SVD's must sooner or later give way to eigendecompositions since SVD's do not yield phase information in the singular values. The extra degrees of freedom in the SVD allow the positive real singular value constraint without loss of generality.

Summary of the TLS *ESPRIT* SVD Algorithm

1. Form the matrix $\mathbf{Z}$ from the available measurements.

2. Compute the SVD of $\mathbf{Z}$:
$$\bar{\mathbf{E}}^*\mathbf{Z}\bar{\mathbf{V}} = \boldsymbol{\Sigma},$$
where $\boldsymbol{\Sigma} = \text{diag}\{\sigma_1, \ldots, \sigma_{2m}\}$, $\sigma_1 \geq \cdots \geq \sigma_{2m}$, and $\bar{\mathbf{E}} = [\mathbf{e}_1 \mid \cdots \mid \mathbf{e}_{2m}]$.

3. If necessary, estimate the number of sources $\hat{d}$ (*cf.* Appendix E.

4. Obtain the signal subspace estimate $\hat{\mathcal{S}}_Z = \mathcal{R}\{\mathbf{E}_z\}$, where
$$\mathbf{E}_z = [\mathbf{e}_1 \mid \cdots \mid \mathbf{e}_{\hat{d}}] = \begin{bmatrix} \mathbf{E}_X \\ \mathbf{E}_Y \end{bmatrix}.$$

5. Compute the SVD of $\{[\,\mathbf{E}_X \mid \mathbf{E}_Y\,]\} = \mathbf{U}\boldsymbol{\Sigma}\mathbf{V}^*$,
$$\mathbf{U} = [\,\mathbf{U}_X \mid \mathbf{U}_Y\,];\ \boldsymbol{\Sigma} = \text{diag}\{\sigma_1, \ldots, \sigma_{2d}\};\ \mathbf{V} = \begin{bmatrix} \mathbf{V}_{11} & \mathbf{V}_{12} \\ \mathbf{V}_{21} & \mathbf{V}_{22} \end{bmatrix}.$$

6. Calculate the eigenvalues of $-\mathbf{V}_{11}\mathbf{V}_{22}^{-1}$;
$$\hat{\phi}_k = \lambda_k(-\mathbf{V}_{12}\mathbf{V}_{22}^{-1}).$$

7. Estimate the signal parameters $\hat{\theta}_k = f^{-1}(\hat{\phi}_k)$.

Note that if the noise covariance is $\sigma^2\Sigma_n$, a Mahalanobis transformation on the measurements, *i.e.*, $\mathbf{Z} \Longleftarrow \Sigma_n^{-\frac{1}{2}}\mathbf{Z}$, to *whiten the noise* can be used, or the SVD performed to obtain $\mathbf{E}_z$ can be replaced by a generalized SVD of the matrix pair $\{\mathbf{Z}^*, \Sigma_n^{\frac{1}{2}}\}$. Further details on this variant of ***ESPRIT*** can be found in [6]. A thorough discussion of the GSVD is given in [7].

D. The Maximum Likelihood Estimator

For the class of problems considered herein, the maximum likelihood estimator is simple to derive analytically, though in most practical real-time applications, computationally prohibitive. For deterministic (non-random) signals in Gaussian noise with covariance Σ_n, $\mathbf{z}(t) = \mathbf{A}(\boldsymbol{\theta})\mathbf{s}(t) + \mathbf{n}(t)$, the likelihood function is easily written [18, 6]:

$$\mathcal{L}[\mathbf{z}(t)] = -\ln\left[P\{\mathbf{z}(t)|\mathbf{z}(t) = \mathbf{A}(\boldsymbol{\theta})\mathbf{s}(t) + \mathbf{n}\}\right] \tag{70}$$

$$\propto -[\mathbf{z}(t) - \mathbf{A}(\boldsymbol{\theta})\mathbf{s}(t)]^*\Sigma_n^{-1}[\mathbf{z}(t) - \mathbf{A}(\boldsymbol{\theta})\mathbf{s}(t)]. \tag{71}$$

The maximization of $\mathcal{L}$ is over $\{\mathbf{s}(t); t \in [0, N]\} \times \{\boldsymbol{\theta} \in \Theta\}$, and is therefore a nonlinear optimization problem. It belongs, however, to the class of *separable* nonlinear optimization problems. Golub and Pereyra [19] prove that the optimization can be carried out in two steps. A solution for the optimal $\mathbf{s}(t)$ is sought as a function of $\boldsymbol{\theta}$, then the maximization over $\boldsymbol{\theta}$ is performed. Employing this procedure gives $\hat{\mathbf{s}}(t) = \mathbf{w}^*(\boldsymbol{\theta})\mathbf{z}(t)$, where $\mathbf{w}(\boldsymbol{\theta}) =$

$\Sigma_n^{-1}\mathbf{A}(\boldsymbol{\theta})[\mathbf{A}^*(\boldsymbol{\theta})\Sigma_n^{-1}\mathbf{A}(\boldsymbol{\theta})]^{-1}$. Substituting the expression for $\mathbf{s}(t)$ back into (71) and using standard properties of the trace operator,

$$\mathcal{L}(\boldsymbol{\theta}) \propto - Tr\left\{P_{\mathbf{A}^{\perp}(\boldsymbol{\theta})}\mathbf{R}_{zz}\Sigma_n^{-1}\right\}, \tag{72}$$

where $P_{\mathbf{A}^{\perp}(\boldsymbol{\theta})}$ is the *oblique* projection operator onto the complement of the space spanned by $\mathbf{A}(\boldsymbol{\theta})$ (in the metric Σ_n). Maximization of this criterion is equivalent to finding

$$\max_{\boldsymbol{\theta}} Tr\left\{P_{\mathbf{A}(\boldsymbol{\theta})}\mathbf{R}_{zz}\Sigma_n^{-1}\right\}, \tag{73}$$

as can be easily verified.[36] Though easy to describe analytically, the computational burden of actually carrying out the multi-dimensional projection and maximization over $\boldsymbol{\theta}$ is generally prohibitive, resulting in the need for *reasonable* approximate solutions such as MUSIC and ***ESPRIT***.

E. Estimating Subspace Dimensions

The estimation of the dimension of the signal subspace $\mathcal{S}_Z$ is not required as a preliminary step in ***ESPRIT***. In the absence of *a priori* knowledge, the maximum number of signals (equal to the number of doublets m) can be assumed to be present and that many parameters estimated. However, an accurate estimate of the true signal subspace dimension improves parameter estimates in the sense of reducing the root-mean-square (RMS) error of the parameter estimates.

If the signal subspace dimension is to be estimated, the likelihood ratio techniques described in detail in [6] can be applied directly. These techniques rely on the fact that when the covariances of the measurements and noise are known, the minimum repeated generalized eigenvalue is the noise power. When the covariances must be estimated from a finite number of measurements, a sequence of hypotheses can be tested against a common null hypothesis to obtain an estimate of d. If the noise covariance is completely known, likelihood threshold tests can be devised for a given level of significance.

If the noise power is not known, more sophisticated statistical criteria such as An Information Criterion (AIC) and the Minimum Description Length (MDL) criteria can be applied as well. These criteria are basically generalized likelihood ratios (GLRs) for testing the equality of the k smallest generalized eigenvalues augmented by terms that attempt to account for the increasing *complexity of the model* with increasing number of signals. Since the GLRs are generally monotonically decreasing functions of the hypothesized number of signals, augmenting them with terms that are monotonically increasing functions of the number of signals generally yields a convex function with a minimum in $(1, d_{\max})$ though the minimum can be attained at one of the boundary points in some cases.

[36] This development assumes that the number of signals is known. When the number of signals is not known *a priori*, the maximum likelihood estimator must be redefined [18, 6].

For reference, the MDL criterion (under the stochastic signal assumption[37]) is given by

$$\mathrm{MDL}(k) = -\log\left\{\frac{\prod_{i=k+1}^{m}\hat{\lambda}_i^{\frac{1}{m-k}}}{\frac{1}{m-k}\sum_{i=k+1}^{m}\hat{\lambda}_i}\right\}^{(m-k)N} + \frac{k}{2}(2m-k)\log N\,; \tag{74}$$

where $\hat{\lambda}_i$ are the generalized eigenvalues of $\{\mathbf{R}_{ZZ}, \mathbf{\Sigma}_n\}$. The use of this criterion was first proposed by Wax and Kailath (1985) [20]. The estimate $(\hat{d})$ of the number of sources d is given by the value of k for which the MDL function is minimized.

References

[1] J. Capon, High resolution frequency wave number spectrum analysis, *Proc. IEEE*, **57**:1408–1418, 1969.

[2] J. P. Burg, Maximum entropy spectral analysis, In *Proceedings of the* 37th *Annual International SEG Meeting*, Oklahoma City, OK., 1967.

[3] V. F. Pisarenko, The retrieval of harmonics from a covariance function, *Geophys. J. Royal Astronomical Soc.*, **33**:347–366, 1973.

[4] R. O. Schmidt, *A Signal Subspace Approach to Multiple Emitter Location and Spectral Estimation*, PhD thesis, Stanford University, Stanford, CA., 1981.

[5] G. Bienvenu and L. Kopp, Adaptivity to background noise spatial coherence for high resolution passive methods, In *Proc. IEEE ICASSP*, pages 307–310, Denver, CO., 1980.

[6] R. H. Roy, ***ESPRIT*** - *Estimation of Signal Parameters via Rotational Invariance Techniques*, PhD thesis, Stanford University, Stanford, CA., 1987.

[7] G. H. Golub and C. F. Van Loan, *Matrix Comptutations*, Johns Hopkins University Press, Baltimore, MD., 1984.

[8] H. L. Van Trees, *Detection, Estimation, and Modulation Theory*, John Wiley and Sons, New York, 1971.

[9] J.J. Dongarra, J.R. Bunch, C.B. Moler, and G.W. Stewart, *LINPACK Users' Guide*, SIAM, Philadelphia, PA, 1979.

[10] A. J. Barabell, J. Capon, D. F. Delong, J. R. Johnson, and K. Senne, Performance comparison of superresolution array processing algorithms, Technical Report TST-72, Lincoln Laboratory, M.I.T., 1984.

[11] T. W. Anderson, Asymptotic theory for principal component analysis, *Ann. Math. Statist.*, **34**:122–148, 1963.

[37] The likelihood ratios obtained when the signals are assumed to be parameters to be estimated are different from those obtained when the signals are modeled as stochastic processes whose covariance is to be estimated (or is known). Since little is currently known about even the asymptotic properties of the MDL or AIC criteria under the deterministic signal assumption, the stochastic formulation is presented herein.

[12] T. W. Anderson, *An Introduction to Multivariate Statistics Analysis*, John Wiley and Sons, Inc., New York, 2nd edition, 1985.

[13] A. Paulraj, R. Roy, and T. Kailath, Patent Application: *Methods and Means for Signal Reception and Parameter Estimation*, Stanford University, Stanford, CA, 1985.

[14] R. Roy, B. Ottersten, A. L. Swindlehurst, and T. Kailath, Multiple invariance **ESPRIT**, In *Proc.* 22nd *Asilomar Conference of Signals, Systems, and Computers*, Asilomar, CA., November 1988.

[15] B. Ottersten and M. Viberg, Analysis of subspace fitting based methods for sensor array processing, In *Proc. ICASSP 89*, Glasgow, Scotland, May 1989.

[16] R. Roy, A. Paulraj, and T. Kailath, **ESPRIT** - A subspace rotation approach to estimation of parameters of cisoids in noise, *IEEE Trans. on ASSP*, **34**(4):1340–1342, October 1986.

[17] A. Paulraj, R. Roy, and T. Kailath, A subspace rotation approach to signal parameter estimation, *Proceedings of the IEEE*, pages 1044–1045, July 1986.

[18] M. Wax, *Detection and Estimation of Superimposed Signals*, PhD thesis, Stanford University, Stanford, CA., 1985.

[19] G. H. Golub and A. Pereyra, The differentiation of pseudo-inverses and nonlinear least squares problems whose variables separate, *Siam J. Numer. Anal.*, **10**:413–432, 1973.

[20] M. Wax and T. Kailath, Detection of signals by information theoretic criteria, *IEEE Trans. on ASSP*, **33**(2):387–392, April 1985.

Fourier Analysis in Reflector Antenna Synthesis

Dr. R.C.Brown
Department of Electrical and Electronic Engineering
Queen Mary College
Mile End Road
London

ABSTRACT. The degree to which it is possible to shape microwave beams for telecommunication is ultimately limited by the Fourier relation between the near and far field. Synthesis of an appropriately shaped reflector antenna is a search for a balance or geometry and diffraction. The upper limit to shaping is the required phase space measured in Nyquist units. Unfortunately, as a simple example shows, specifications are often incomplete, and a variety of results can emerge from synthesis.

1. Introduction

A convenient way of directing microwaves from a satellite to a given region on Earth is shown in figure 1.

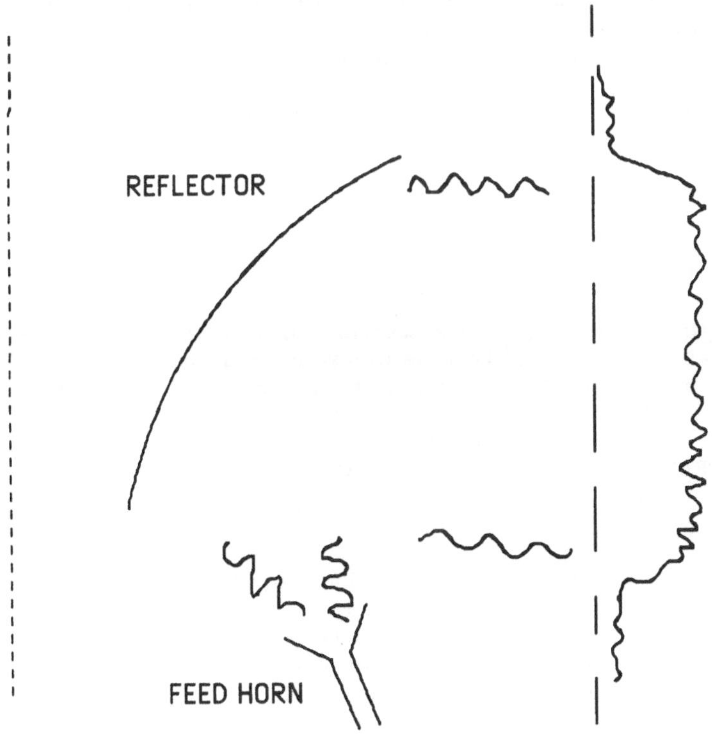

FIG.1 REFLECTOR ANTENNA GEOMETRY

J. S. Byrnes and J. F. Byrnes (eds.), Recent Advances in Fourier Analysis and Its Applications, 367–376.

The shaped reflector focusses microwave energy from a small feed horn into a shaped beam. The angular pattern of radiation is determined by a Fourier transform of the field in a plane near the antenna, such as the dashed line in the figure. The field $\psi(x,y)$ in this plane is mostly contained in the area projected by geometrical rays from the reflector. Local change in the reflector shape produces local change in the aperture field.

If we measure distances in wavelengths, the field far from the antenna takes the simple form

$$\psi(\mathbf{R}) = \psi(\alpha,\beta)\ \exp(-j2\pi R)/R \qquad [1]$$

where f is the frequency. An $\exp(j2\pi ft)$ dependence is assumed. Alpha and beta are direction cosines for the vector distance $\mathbf{R}=R\hat{\mathbf{R}}$ with $\hat{\mathbf{R}} = (\alpha,\beta,\gamma)$, and $\alpha^2 + \beta^2 + \gamma^2 = 1$. The angular field pattern $\psi(\alpha,\beta)$ is the sum of contributions from the field in the aperture plane,

$$\psi(\alpha,\beta) = \int_{-\infty}^{\infty} dx \int_{-\infty}^{\infty} dy\ \psi(x,y)\ \exp(+j2\pi[\alpha x+\beta y]) \qquad [2]$$

Although this is a scalar relation, it applies equally to electric or magnetic field components by including the inner product of different transverse polarisation bases for $\psi(\alpha,\beta)$ and $\psi(x,y)$.

$$\psi_v(\alpha,\beta) = \sum_{p=1}^{2} \int_{-\infty}^{\infty} dx \int_{-\infty}^{\infty} dy\ \psi_p(x,y)\ v^*.p\ \exp(+j2\pi[\alpha x+\beta y]) \qquad [3]$$

The inverse transform is,

$$\psi_p(x,y) = \sum_{v=1}^{2} \int_{-\infty}^{\infty} dx \int_{-\infty}^{\infty} dy\ \psi_v(\alpha,\beta)\ p^*.v\ \exp(-j2\pi[\alpha x+\beta y]) \qquad [4]$$

Antenna engineers prefer to write equation [3] as a transform between electric field vectors with the same Cartesian polarisation basis in near (E) and far (F) field. The v*.p bits emerge as an operator to keep the far field transverse to its direction of propagation

$$\mathbf{F}(\alpha,\beta) = [\ \gamma\mathbf{1}-\hat{\mathbf{z}}\hat{\mathbf{R}}\ ] \int_{-\infty}^{\infty} dx \int_{-\infty}^{\infty} dy\ \mathbf{E}(x,y)\ \exp(+j2\pi[\alpha x+\beta y]) \qquad [5]$$

The use f, α,β, and γ (instead of angular frequency ω and wavevector $\mathbf{k}$) seems appropriate for antenna problems. It also means we don't have to worry about normalising the Fourier transforms by sharing out a $1/2\pi$ factor between them. From here on all integrals will range from $-\infty$ to $+\infty$ and all sums from 1 to 2 unless stated otherwise

Equation [3] (or [5]) only describes the field in the 'forward' hemisphere - to the right in figure 1. An aperture plane on the other side of the antenna would be required to express the backward field. The dotted line in figure 1 shows a possible choice.

The angular distribution and total power in the forward hemisphere are

$$\frac{d^2P_v}{d\alpha d\beta} = \frac{1}{2}\,|\psi_v(\alpha,\beta)|^2 \quad \text{and} \quad P = \frac{1}{2}\sum_v \int d\alpha \int d\beta\, |\psi(\alpha,\beta)|^2 \qquad [6]$$

Usually this distribution is discussed in terms of directivity Ð or gain G. Directivity is defined as the ratio of the radiated power per solid angle to an isotropic distribution of the total radiated power, viz

$$Ð_v(\alpha,\beta) = \frac{4\pi}{P}\,\frac{d^2P_v}{d\alpha d\beta} \qquad [7]$$

Gain is similar but has a power denominator which includes non radiative losses. Both are normally quoted logarithmically in "units" of dBi - decibels over isotropic.

2. Diffraction

The shape of any antenna pattern is limited by diffraction. It is not possible to establish a map from x,y to α,β below the limit set by the uncertainty principle intrinsic to the Fourier transform. There are different ways of expressing this principle. One version (Heisenberg's) applies to r.m.s. widths for the field in conjugate domains [1]

$$\sigma_x\,\sigma_\alpha \geq 1/4\pi \qquad [8]$$

In antenna work this is not as useful as the version associated with the name of Nyquist. If equivalent widths are defined for polarisations p and v, as

$$w_x = \left|\frac{\int dx \int dy\, \psi_p(x,y)\; p^*\!\cdot v}{\psi_p(x{=}0,y{=}0)}\right| \quad \text{and} \quad w_\alpha = \left|\frac{\int d\alpha \int d\beta\, \psi_v(\alpha,\beta)\; v^*\!\cdot p}{\psi_v(\alpha{=}0,\beta{=}0)}\right| \qquad [9]$$

it is not difficult to see that

$$w_x\,w_\alpha = 1 \qquad [10]$$

This relation can also be seen as a variation on Bragg's law for interference between radiation from the edges of an aperture of size D : D sin(θ) = n (λ). The angle of the first minimum is given by D sin(θ_o) = 1/2, i.e. eq.[10] with spatial and angular widths D and 2sin(θ_o). Similar relations apply between y and β and (although not needed here) between z and γ and f and t).

This limit to simultaneous localisation in position and direction space has profound implications (quantum mechanics). In geometrical optics the phase space (x,y,α,β) is populated by infinitely many points - the geometrical optics rays that map spatial points (x,y) to directions (α,β). Equation [10] blurs this picture and defines a quantum of information - the Nyquist cell. Despite this limit, smaller details are often specified in specifications for satellite broadcast beams. Figure 2 shows a simple example [2].

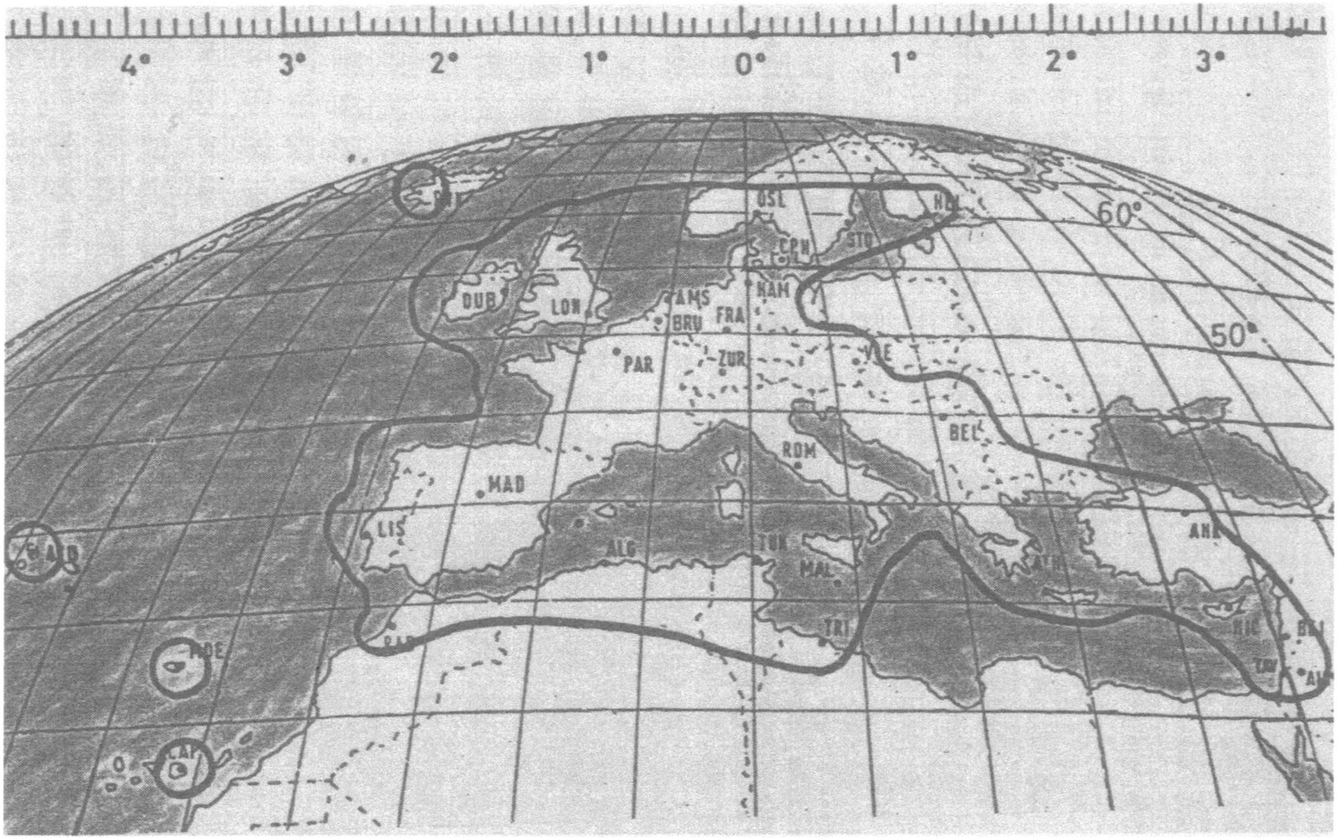

FIG.2 A TYPICAL SPECIFICATION :
28 dBi inside coverage, -2dBi outside
Maximum Reflector Diameter D = 55 λ (1/D ≃ 1°)

This pattern of gain "islands" contains detail which cannot be achieved with the maximum allowed reflector diameter of 55 wavelengths. Its realisation is both limited and determined by diffraction.

3. Synthesis

As pattern specifications are only notional and contain no phase information, reflector synthesis must proceed by optimisation techniques. A shaped reflector can be generated indirectly by finding an aperture field whose transform approximates the specification, and by a further optimisation to find the reflector surface. However in practice it is simpler to take a more direct route [3]. Currents J on the reflector surface z(x,y) give the far field via a Fourier-like transform.

$$F(\alpha,\beta) = [\ 1-\hat{R}\hat{R}\ ] \int dx \int dy\ J(x,y)\ \exp(+j2\pi[\alpha x+\beta y+\gamma z]) \qquad [11]$$

The phase in equation [5] has received an extra γz term, and the integral ranges over the now finite surface. Figure 3 suggests how this surface "skin" can be stretched to cover equivalent problems, linking equation [11] and the "forward" and "backward" aperture forms of equation [5]. In all cases the surface encloses the model field sources,

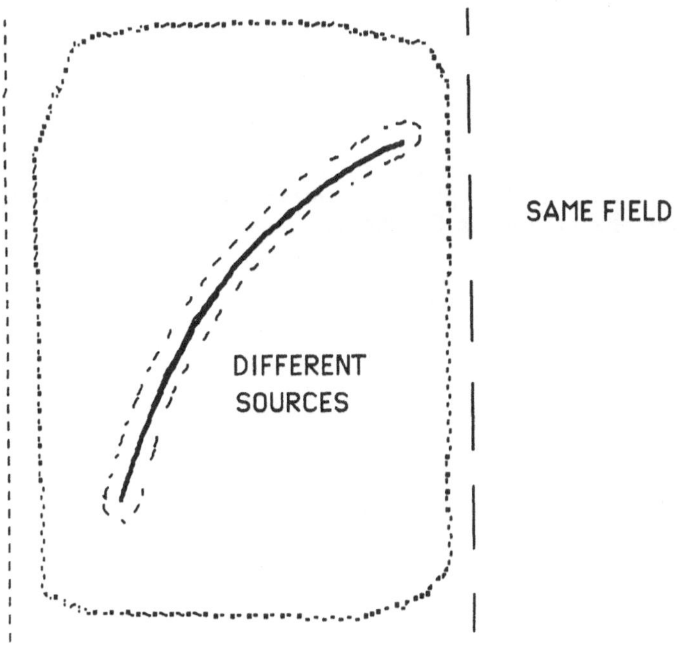

FIG.3 EQUIVALENT PROBLEMS

Surface currents are calculated in the physical optics approximation, i.e. $2\hat{n} \times H$, where $\hat{n}$ is an outward normal to the surface, and H is the field incident on the reflector from its feed. This is essentially a geometrical approximation, and current on the "shadow" side of the reflector is assumed to be zero. If the feed horn is small enough so that only the main lobe of its radiation pattern (one Nyquist unit) illuminates the reflector, its field can make little contribution to the final beam shape. A suitable feed model is a vector version of equation [1]

$$E(R) = [\, E_\theta \cos(\varphi)\ \hat{\theta} - E_\varphi \sin(\varphi)\ \hat{\varphi}]\ \exp(-j2\pi R)/R \qquad [12]$$

where R, θ and φ are polar coordinates from the feed "phase" centre, and $\hat{\theta}$ and $\hat{\varphi}$ associated unit vectors. In the examples shown later the two parameters E_θ and E_φ are both equal and given by a power of cos(θ) which gives 15 dB taper at the edge of the reflector.is used makes little contribution to field shape.

To shape the reflector surface z(x,y), we describe it by a polynomial plus a series of Fourier harmonics,

$$z(x,y) = ax + by + cx^2 + dy^2 + exy + \sum_{1}^{N_x}{}_r \sum_{1}^{N_y}{}_s C_{rs}\, f_r(x) f_s(y)$$

where $f_r(x)$ = 1, cos(x), sin(x), cos(2x), sin(2x) etc
for r = 1 2 3 4 5

[13]

where x and y are scaled here to lie in the range [-π/2,+π/2] over the projected reflector area. The polynomial in x and y defines a basic focussing shape on which to "play" the Fourier harmonics. All coefficients in equation [13] are varied to synthesise the reflector shape. A quasi Newton optimiser attempts to reduce a measure of the difference of computed and specified patterns. We assume all the power is radiated,(so gain = directivity,) and calculate the sum of squares F given by

$$F = \frac{1}{NDF} \sum_{1}^{N}{}_i \left[\frac{G^{(i)}_{calc} - G^{(i)}_{spec}}{\delta G^{(i)}} \right]^2 \qquad [14]$$

in N directions $(\alpha,\beta)_i$. The δG^i are chosen tolerances, and NDF is discussed below.

4. Information.

The number of Fourier terms in eq.[13] must match the information in the specification. An upper limit is set by the the number (N) of Nyquist cells in the required volume of phase space. For a solid angle Ω and reflector area A this is $(A/\lambda^2)\Omega$ (double this if you include polarisation). This is the information capacity, not the content, and the number required may be less. When counting the terms in equation [13], the number of terms is more than the $N_x N_y$ terms in the series. Some allowance must be made for the polynomial terms. Our "guesstimate" is $NU \simeq N_x N_y + 2$

If all directions in a specified coverage Ω are equally important, it is clear that $N \gg NU$ comparison points should be uniformly distributed over Ω and all tolerances δG_i should be equal. If some directions are more important than others, we can sample in a non uniform manner, and/or use different tolerances δG_i.

As the search for a minimum proceeds, the objective function decreases in an exponential fashion resembling a cooling curve. In general no minimum is found. One exception was a test case where we used a computed radiation pattern as a specification [4]. The spurious detail in a real specification may act like an effective noise and make convergence an approach to thermal equilibrium.

A lack of a minimum is no problem in practice. The search process can be stopped once the objective function falls below one. At this point the residual gain differences are statistically consistent with the tolerances δG^i, i.e. the sum of squares has a statistical expectation value equal to the residual degrees of freedom in the data, NDF=N-NU+1.

In each iteration the optimiser requires typically NU+1 analyses, each calculating N pattern points. Typically the computation takes a few hundred seconds for vectorised code on a Cray 1. "Ok for some !" you may think. However our job turn-around time is a day or two, and this is just the sort of time it would take on a fast micro/mini such as a Sun 4. The nervous can monitor progress by breaking the task into a one or two shorter jobs.

5.Examples.

The contours from figure 2 are reproduced in figures 4 to 7 together with the results of different syntheses. All used tolerances δG of 10% inside coverage, and 35% outside the coverage regions. The reflector was described with a 5x5 series, so that NU≃27. Plot axes are direction cosines times the reflector diameter in wavelengths, i.e. Nyquist units.

In figures 4, 5 and 6 the comparison points were a regular grid of Nyquist spacing. Figure 4 shows the result achieved if contributions to eq.[16] are only taken if gain fell below the requirement inside coverage. There were N=15 comparison points within the coverage regions, so we may be overshaping a little here. Figure 5 and 6 show the effect of including contributions when gain rose above the specification outside . The difference between them is that figure 5 ignores a two cell margin around coverage having N=55, whilst figure 6 uses the almost the whole picture, N=88. The wide difference in results reflect these different assumptions as to the area over which the specification applies. Results can be "massaged" by playing around with the distribution of comparison points. Figure 7 shows an example where some 30 comparison points were used inside and on the edge of coverage. At first sight the beam shape looks better, but there are one or two sidelobes as high as -22 dBi in mid Atlantic. Whether such results are really better or not depends on the real pattern of tolerances $\delta G(\alpha,\beta)$ required by the underlying application.This is after all what we really should be optimising !

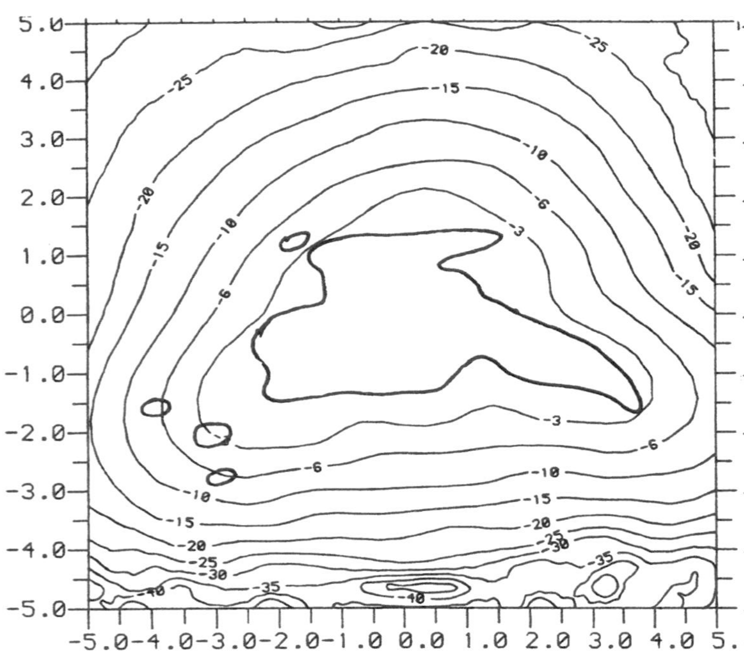

FIG.4 SYNTHESISED RADIATION PATTERN
Peak Gain 32.2 dBi. Axes are Dα and Dβ, D=55 λ
Contours at peak - 3, -6, -10, -20, -25, -30, -35 dB

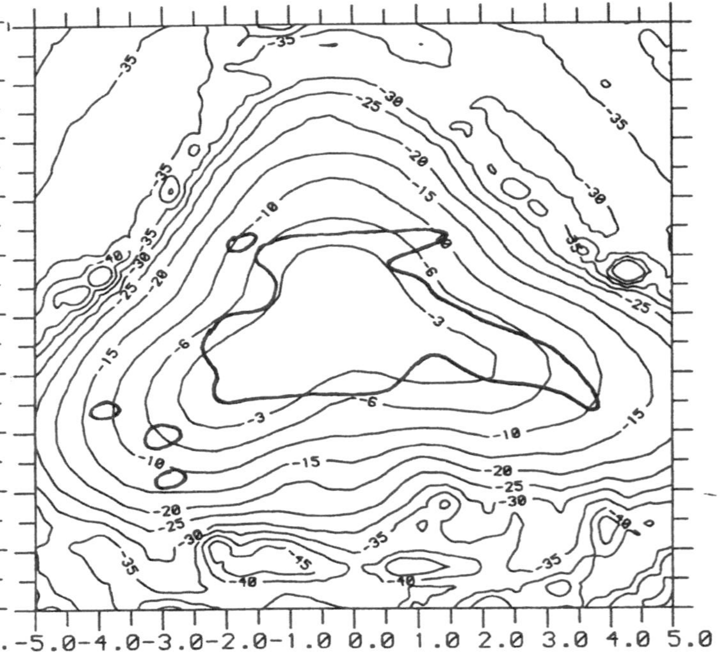

FIG.5 SYNTHESISED RADIATION PATTERN
Peak Gain 35.4 dBi.

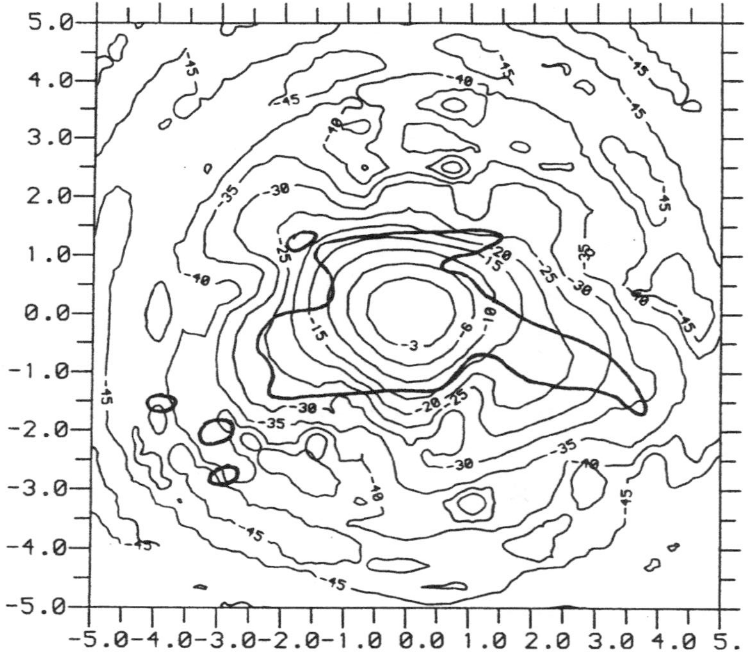

FIG.6 SYNTHESISED RADIATION PATTERN
PeaK Gain 42.5 dBi.

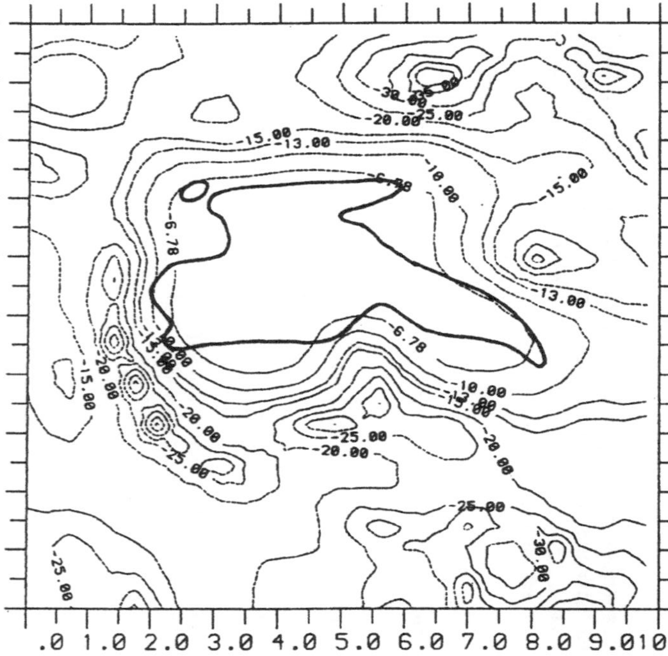

FIG.7 SYNTHESISED RADIATION PATTERN
Peak Gain 34.8 dBi.
Contours at peak -6.8, -10, -13, -15, -20,
-25, -35 dB

6.Conclusions

It is clear that the success of shaped beam synthesis depends on two things, the size of the antenna and the completeness of the specification. We need a real goal such as signal to noise. Antenna design would become an integral part of the system design process.

References

[i] Bracewell "The Fourier Transform and its Applications", Mcgraw Hill, N.Y., 1986, 2nd edition.

[2] kindly supplied by Dr.G.Crone of ESTEC,Noordvjik.

[3] Bergmann,J.R., Brown,R.C., Clarricoats,P.J.B. and Zhou,H., "Synthesis of Shaped Beam Reflector Antenna Patterns", Proc.IEE, part H, 1988,1 135,

[4] Brown R.C., Clarricoats P.J.B. and Zhou H., "Optimum Shaping of Reflector Antennas for Specified Radiation Patterns", Elec.Lett.,1985,21,pp.1164-1165

BISPECTRA AND SYSTEM IDENTIFICATION

ATHANASIOS PAPOULIS
Polytechnic University
Department of Electrical Engineering
Route 110
Farmingdale, NY 11735

ABSTRACT. The concept of bispectrum is introduced and is used to determine the amplitude and the phase of a linear system driven by non-Gaussian white noise, in terms of the third order moments of the input.

1. INTRODUCTION

The Fourier integral is a central concept in signal analysis. One of its many applications is the identification of the system function

$$H(\omega) = A(\omega)e^{j\phi(\omega)} \tag{1}$$

of a linear system in terms of the moments of its input x(t) and output y(t). The amplitude $A(\omega)$ of $H(\omega)$ is uniquely determined if only second order moments are known. However, for the determination of $\phi(\omega)$ higher order moments are required. In this paper we develop the underlying theory concentrating on the use of Fourier transforms and bispectra in the solution of the phase problem. In the last section we present a simple explanation of the spectral representation of a stochastic process based on the formal properties of the impulse function, and use this representation to reexamine the earlier results in terms of Fourier transforms. As a preparation, we review briefly the second order theory of stochastic processes.

Consider a real process x(t) with zero mean and autocorrelation

$$R_{xx}(t_1,t_2) = E\{x(t_1)\, x(t_2)\} \tag{2}$$

If x(t) is the input to a linear system, the resulting output is a stochastic process given by

$$y(t) = L_t[x(t)] = \int_{-\infty}^{\infty} x(t-\alpha)h(\alpha)d\alpha \tag{3}$$

In the above, L_t indicates a linear operator operating on the variable t, and h(t) is the impulse response of the system (Green's Function) which we assume deterministic and real. The integral in (3) is the convolution of x(t) with h(t), written symbolically in the form

J. S. Byrnes and J. F. Byrnes (eds.), Recent Advances in Fourier Analysis and Its Applications, 377–386.

$$y(t) = x(t) * h(t) \tag{4}$$

From (3) and the linearity of expected values it follows that

$$E\{y(t)\} = E\{L_t[x(t)]\} = L_t\,[E\{x(t)\}] \tag{5}$$

This is the basis of the relationship between input and output moments of any order.
We introduce the second order moments

$$R_{xy}(t_1,t_2) = E\{x(t_1)y(t_2)\} = E\{x(t_1)\;L_2\,[x(t_2)]\}$$

$$R_{yy}(t_1,t_2) = E\{y(t_1)y(t_2)\} = E\{L_1\,[x(t_1)]\;y(t_2)\}$$

where L_1 and L_2 indicate the operator L operating on t_1 and t_2 respectively as in (3). Applying (5) we obtain

$$R_{xy}(t_1,t_2) = \int_{-\infty}^{\infty} R_{xx}(t_1,t_2-\beta)h(\beta)d\beta$$

$$R_{yy}(t_1,t_2) = \int_{-\infty}^{\infty} R_{xy}(t_1-\alpha,t_2)h(\alpha)d\alpha \tag{6}$$

1.1. Stationary Processes

Suppose now that x(t) is a stationary process. In this case,

$$R_{xx}(t_1,t_2) = R_{xx}(\tau) = E\{x(t+\tau)x(t)\} \qquad \tau = t_1-t_2$$

and (6) yields

$$R_{xy}(\tau) = E\{x(t+\tau)y(t)\} = \int_{-\infty}^{\infty} R_{xx}(\tau+\beta)h(\beta)d\beta$$

$$R_{yy}(\tau) = E\{y(t+\tau)y(t)\} = \int_{-\infty}^{\infty} R_{xy}(\tau-\alpha)h(\alpha)d\alpha$$

or, equivalently,

$$R_{xy}(\tau) = R_{xx}(\tau)*h(-\tau) \qquad\qquad R_{yy}(\tau) = R_{xy}(\tau)*h(\tau) \tag{7}$$

1.1.1. *Spectra.* The power spectrum $S_{xx}(\omega)$ of a stationary process x(t) is the Fourier transform of its autocorrelation:

$$S_{xx}(\omega) = \int_{-\infty}^{\infty} R_{xx}(\tau)e^{-j\omega\tau}d\tau.$$

The spectra $S_{xy}(\omega)$ and $S_{yy}(\omega)$ are the Fourier transforms of $R_{xy}(\tau)$ and $R_{yy}(\tau)$ respectively. From (7) and the convolution theorem it follows that

$$S_{xy}(\omega) = S_{xx}(\omega)H(-\omega) \qquad S_{yy}(\omega) = S_{xy}(\omega)H(\omega) \tag{8}$$

where

$$H(\omega) = \int_{-\infty}^{\infty} h(t)e^{-j\omega t}dt. \tag{9}$$

Since h(t) is real, $H(-\omega) = H^*(\omega)$ and (8) yields

$$S_{yy}(\omega) = S_{xx}(\omega) \,|\, H(\omega) \,|^2 \tag{10}$$

1.1.2. *White Noise.* If the samples $x(t_i)$ of the process x(t) are independent, then x(t) is called a white noise process. In this case,

$$R_{xx}(t_1,t_2) = p(t_1)\ \delta(t_1-t_2) \tag{11}$$

where $\delta(t)$ is the impulse function specified formally in terms of the identity

$$\int_{-\infty}^{\infty} \delta(\tau)\phi(t-\tau)d\tau = \phi(t)$$

Thus, $\delta(t)$ can be defined as the identity element in the operation of convolution.

If x(t) is stationary white noise, then p(t) = p is a constant and

$$R_{xx}(\tau) = p\delta(\tau) \qquad S_{xx}(\omega) = p \int_{-\infty}^{\infty} \delta(\tau)e^{-j\omega\tau}d\tau = p \tag{12}$$

In this case, (10) yields

$$S_{yy}(\omega) = p \,|\, H(\omega) \,|^2 \tag{13}$$

Thus, the magnitude $A(\omega)$ of $H(\omega)$ is determined within a factor in terms of the second order moments of y(t). To determine the phase $\phi(\omega)$ we turn to third order moments.

2. BISPECTRA

The function

$$R(t_1, t_2, t_3) = E\{x(t_1)x(t_2)x(t_3)\} \tag{14}$$

is the third order moment of x(t). For stationary processes, this function depends only on the differences

$$t_1-t_3 = \mu, \qquad t_2-t_3 = \nu.$$

With $t_3=t$ this yields

$$R(t_1,t_2,t_3) = R(\mu,\nu) = E\{x(t+\mu)\ x(t+\nu)\ x(t)\} \tag{15}$$

The Fourier transform S(u,v) of $R(\mu,\nu)$ is the bispectrum of x(t):

$$S(u,v) = \iint_{-\infty}^{\infty} R(\mu,\nu)e^{-j(u\mu+v\nu)}d\mu d\nu \tag{16}$$

If x(t) is white noise, then

$$R(t_1,t_2,t_3) = q(t_1)\delta(t_1-t_3)\delta(t_2-t_3) \tag{17}$$

If x(t) is stationary white noise, then $q(t_1) = q$ and

$$R(\mu,\nu) = q\delta(\mu)\delta(\nu) \qquad\qquad S(u,v) = q \tag{18}$$

2.1. Symmetries

As we see from (14), the function $R(t_i,t_j,t_k)$ has the same value if (t_i,t_j,t_k) is any one of the six permutations of (t_1,t_2,t_3). For stationary processes,

$$t_1-t_3 = \mu \qquad t_2-t_3 = \nu \qquad t_1-t_2 = \mu-\nu$$

This yields the following table for the corresponding symmetries of $R(\mu,\nu)$. The last column will be explained presently:

TABLE 1

1	t_1,t_2,t_3	μ , ν	u , v
2	t_2,t_1,t_3	ν , μ	v , u
3	t_3,t_1,t_2	$-\nu$, $\mu-\nu$	-u-v , u
4	t_3,t_2,t_1	$-\mu$, $-\mu+\nu$	-u-v , u
5	t_2,t_3,t_1	$-\mu+\nu$, $-\mu$	v , -u-v
6	t_1,t_3,t_2	$\mu-\nu$, $-\nu$	u , -u-v

Thus,

$$R(\mu,\nu)= R(\nu,\mu)= R(-\nu,\mu-\nu)= R(-\mu,-\mu+\nu)= R(-\mu+\nu,-\mu)= R(\mu-\nu,-\nu) \tag{19}$$

This shows that if we know the function $R(\mu,\nu)$ in any one of the six regions of Fig. 1, we can determine it everywhere.

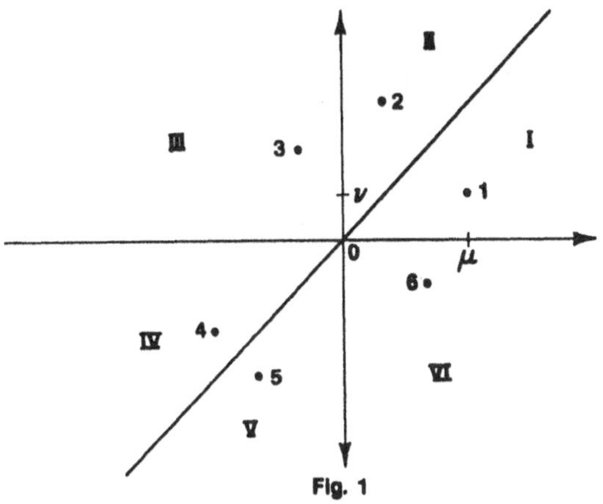

Fig. 1

Inserting (19) into (16), we obtain the symmetries:

$$S(u,v) = S(v,u) = S(-u-v,u) = S(-u-v,v) = S(v,-u-v) = S(u,-u-v) \qquad (20)$$

of the last column of Table 1. Since $R(\mu,\nu)$ is real, we also have

$$S^*(u,v) = S(-u,-v) \qquad (21)$$

This yields six additional conjugate symmetries.

The six symmetries in (20) show that if we know the bispectrum S(u,v) in one of the six regions I, II, ..., VI of Fig. 2 (shaded) we can determine it in any one of the others. The conjugate condition (21) generates the six regions I*, II*, ... VI*. Thus, if S(u,v) is known in any one of the 12 regions of Fig. 2, it can be determined everywhere.

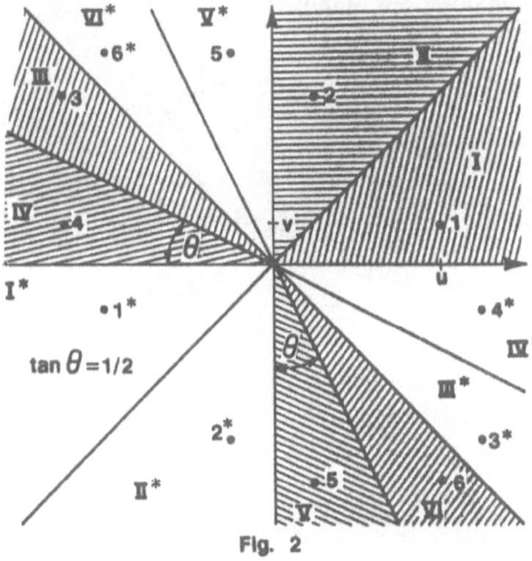

Fig. 2

2.2. Linear Systems

We shall identify the mixed third order moments between the input x(t) and the output y(t) of a linear system by appropriate subscripts. For example,

$$R_{xxy}(t_1,t_2,t_3) = E\{x(t_1)x(t_2)y(t_3)\}$$

Proceeding as in (5), we obtain

$$R_{xxy}(t_1,t_2,t_3) = \int_{-\infty}^{\infty} R_{xxx}(t_1,t_2,t_3-\gamma)h(\gamma)d\gamma$$

Repeated application of this yields

$$R_{yyy}(t_1,t_2,t_3) = \iiint_{-\infty}^{\infty} R_{xxx}(t_1-\alpha,t_2-\beta,t_3-\gamma)h(\alpha)h(\beta)h(\gamma)d\alpha d\beta d\gamma \qquad (22)$$

For stationary processes,

$$R_{xxx}(\mu,\nu) = R_{xxx}(t_1,t_2,t_3), \qquad \mu = t_1-t_3, \qquad \nu = t_2-t_3$$

hence,

$$R_{xxx}(t_1-\alpha,t_2-\beta,t_3-\gamma) = R_{xxx}(\mu+\gamma-\alpha,\ \nu+\gamma-\beta)$$

and (22) yields

$$R_{yyy}(\mu,\nu) = \iiint_{-\infty}^{\infty} R_{xxx}(\mu+\gamma-\alpha,\ \nu+\gamma-\beta)h(\alpha)h(\beta)h(\gamma)d\alpha d\beta d\gamma \qquad (23)$$

Taking transforms of both sides, we obtain

$$S_{yyy}(u,v) = \iint_{-\infty}^{\infty} R_{yyy}(\mu,\nu)e^{-j(u\mu+v\nu)}d\mu d\nu$$

$$= \iiint_{-\infty}^{\infty} S_{xxx}(u,v)e^{j[u(\gamma-\alpha)+v(\gamma-\beta)]}h(\alpha)h(\beta)h(\gamma)d\alpha d\beta d\gamma$$

And since

$$H(\omega) = \int_{-\infty}^{\infty} h(t)e^{-j\omega t}dt$$

the above yields

$$S_{yyy}(u,v) = S_{xxx}(u,v)H(u)H(v)H^*(u+v) \tag{24}$$

This relationship will be used to determine the amplitude $A(\omega)$ and phase $\phi(\omega)$ of $H(\omega)$.

3. THE IDENTIFICATION PROBLEM

We assume next that x(t) is white noise with bispectrum

$$S_{xxx}(u,v) = q \neq 0 \tag{25}$$

For normal processes, $S_{xxx}(u,v) = 0$, hence, the identification problem cannot be so solved if the input process x(t) is normal.

Inserting (25) into (24) we obtain

$$S_{yyy}(u,v) = qH(u)H(v)H^*(u+v) \tag{26}$$

With

$$S_{yyy}(u,v) = B(u,v)e^{j\theta(u,v)}$$

$$H(\omega) = A(\omega)e^{j\phi(\omega)}$$

(polar form) we conclude from (26) equating amplitudes and phases that

$$B(u,v) = qA(u)A(v)A(u+v)$$

$$\theta(u,v) = \phi(u) + \phi(v) - \phi(u+v) \tag{27}$$

We shall use these relationships to determine the amplitude $A(\omega)$ and phase $\phi(\omega)$ of $H(\omega)$ in terms of the amplitude $B(u,v)$ and phase $\theta(u,v)$ of the bispectrum $S_{yyy}(u,v)$ of y(t).

If q is unknown, $A(\omega)$ can be determined only within a factor. If $\phi(\omega)$ satisfies (27), then so does the sum $\phi(\omega)+b\omega$ for every b. We can assume therefore that

$$q = 1 \qquad \phi'(0) = 0 \tag{28}$$

<u>Theorem</u>

$$A^2(u) = \frac{2}{A(0)} \int_0^u B_v(\gamma,0)d\gamma$$

$$\phi(u) = -\int_0^u \theta_v(\gamma,0)d\gamma \tag{29}$$

where

$$B_v(u,0) = \left.\frac{\partial B(u,v)}{\partial v}\right|_{v=0} \qquad \theta_v(u,0) = \left.\frac{\partial \theta(u,v)}{\partial v}\right|_{v=0}$$

Proof The function $A(\omega)$ is even because $h(t)$ is real, hence, if $A(\omega)$ is differentiable at the origin, $A'(0) = 0$. Differentiating each equation in (27) with respect to v and setting $v = 0$, we obtain

$$B_v(u,0) = A(0)A(u)A'(u) = \frac{1}{2} A(0) \frac{d}{du} A^2(u)$$

$$\theta_v(u,0) = - \phi'(u)$$

and (29) results because, under the stated assumptions, $B_v(0,0) = 0$ and $\phi(0) = 0$.

The preceding theorem shows that, if the input to a system is non-Gaussian white noise the bispectrum $S_{yy}(u,v)$ of its output is uniquely determined in terms of its values near the u-axis. If $S_{yy}(u,v)$ is known for every (u,v), this information is redundant. In a numerical solution of the problem the redundancy can be used to reduce the variance of the estimate of $A(\omega)$ and $\phi(\omega)$. The numerical solution to the problem, however, will not be discussed.

4. SPECTRAL REPRESENTATION

In the preceding discussions, we used Fourier transforms of deterministic signals only. We now introduce the Fourier transform of a random process x(t) and use it to rederive our earlier results. In the usual approach to this topic all Fourier transforms are interpreted as Fourier-Stieltjes integrals. We show that the analysis can be simplified with the introduction of the impulse function.

4.1. Spectra

The Fourier transform

$$X(\omega) = \int_{-\infty}^{\infty} x(t)e^{-j\omega t}dt$$

of a real stochastic process x(t) is a complex process with zero mean and autocorrelation

$$E\{X(u)X^*(v)\} = \iint_{-\infty}^{\infty} E\{x(t_1)\, x(t_2)\}e^{j(ut_1-vt_2)}dt_1dt_2$$

With

$$\Gamma(u,v) = \iint_{-\infty}^{\infty} R(t_1, t_2)e^{-j(ut_1+vt_2)}dt_1dt_2 \tag{30}$$

the two-dimensional Fourier transform of the autocorrelation of x(t), it follows that

$$E\{X(u)X^*(v)\} = \Gamma(u,-v) \tag{31}$$

If the process x(t) is stationary with power spectrum $S(\omega)$, then

$$\Gamma(u,v) = \iint_{-\infty}^{\infty} R(t_1-t_2)e^{-j(ut_1+vt_2)}dt_1dt_2$$

$$= \int_{-\infty}^{\infty} R(\tau)e^{-ju\tau}d\tau \int_{-\infty}^{\infty} e^{-j(u+v)t}dt = 2\pi S(u)\delta(u+v)$$

because

$$\int_{-\infty}^{\infty} e^{-j\omega t}dt = 2\pi\delta(\omega)$$

Hence,

$$E\{X(u)X^*(v)\} = 2\pi S(u)\delta(u-v) \tag{32}$$

Thus, the second moment of $X(\omega)$ consists of line masses on the line u=v of the uv plane.

If y(t) is the output of a linear system with input x(t) as in (3), then

$$Y(\omega) = H(\omega)\ X(\omega)$$

$$E\{Y(u)Y^*(v)\} = H(u)H^*(v)E\{X(u)X^*(v)\}$$

$$= 2\pi H(u)H^*(v)S(u)\delta(u-v)$$

And since

$$H^*(v)\delta(u-v) = H^*(u)\delta(u-v)$$

we conclude that

$$E\{Y(u)Y^*(v)\} = 2\pi\,|H(u)|^2S(u)\delta(u-v)$$

This shows that

$$S_{yy}(\omega) = S_{xx}(\omega)\ |H(\omega)|^2$$

as in (10).

4.2. Bispectra

Suppose that x(t) is a stationary process with third order moment $R(\mu,\nu)$ and bispectrum $S(u,v)$. We introduce the third order moment

$$E\{X(u)X(v)X^*(w)\} = \iiint_{-\infty}^{\infty} E\{x(t_1)x(t_2)x(t_3)\}e^{-j(ut_1+vt_2-wt_3)}dt_1dt_2dt_3$$

$$= \iint_{-\infty}^{\infty} R(\mu,\nu)e^{-j(u\mu+v\nu)}d\mu d\nu \int_{-\infty}^{\infty} e^{-j(u+v-w)t}dt$$

$$= 2\pi S(u,v)\delta(u+v-w) \tag{33}$$

Thus, this moment consists of surface masses on the plane

$$w = u+v$$

of the uvw space with mass density $2\pi S(u,v)$. We shall use this result to rederive (22). For this purpose, we form the third order moment of $Y(\omega)$:

$$E\{Y(u)Y(v)Y^*(w)\} = H(u)H(v)H^*(w)E\{X(u)X(v)X^*(w)\}$$

$$= 2\pi H(u)H(v)H^*(w)S_{xx}(u,v)\delta(u+v-w)$$

$$= 2\pi H(u)H(v)H^*(u+v)S_{xx}(u,v)\delta(u+v-w)$$

This shows that [see (33)]

$$S_{yyy}(u,v) = S_{xxx}(u,v)H(u)H(v)H^*(u+v)$$

Thus, the formal properties of Fourier transforms of stochastic processes can be simply expressed in terms of appropriate singularity functions.

5. REFERENCES

1. D. R. Brillinger (1981) Times Series, Data Analysis and Theory, Holden-Day, San Francisco, CA.

2. K. S. Lii and M. Rosenblatt (1982) "Deconvolution and Estimation of Transfer Function Phase and Coefficient for Non-Gaussian Linear Processes," The Annals of Statistics, Vol. 10.

3. A. Papoulis (1984) Probability, Random Variables, and Stochastic Processes, (2nd Ed.), McGraw-Hill, NY.

COMBINED LAPLACE AND FOURIER TRANSFORMATIONS IN THE THEORY OF TRANSIENT WAVE PROPAGATION IN LAYERED MEDIA

Adrianus T. de Hoop and Sytze M. de Vries
Laboratory of Electromagnetic Research
Faculty of Electrical Engineering
Delft University of Technology
P.O. Box 5031
2600 GA Delft
The Netherlands

ABSTRACT. A combination of a one-sided Laplace transformation with respect to time and spatial Fourier transformations can adequately be used to analyse wave propagation problems in configurations that are time invariant and spatially partially shift-invariant. In this category the modified Cagniard method plays an important role. In it causality is employed to restrict the time Laplace-transform parameter to positive real values. After appropriate contour deformations in the complex spatial Fourier-transform domain inversion back to the space time domain is carried out with the aid of Lerch's theorem. To illustrate the method, the acoustic wave motion generated by an impulsive point source in a layered fluid medium is calculated. The wave motion is represented as a superposition of generalized-ray constituents, that follow from a recursive solution of the transform-domain wave amplitudes.

1. INTRODUCTION

One of the methods to analyze the excitation and propagation of transient waves in layered media is the modified Cagniard method. This method, originally developed by Cagniard (1939) and subsequently modified by De Hoop (1960) (see also Achenbach, 1973; Miklowitz, 1978; Aki and Richards, 1980), employs a specific scheme of integral transformations with respect to those coordinates in space-time in which the configuration is shift invariant. In a time-invariant, horizontally layered medium these coordinates are: the time coordinate t and the horizontal space coordinates, which are chosen to be x_1 and x_2. In the vertical space coordinate x_3 the medium varies in its properties. Because of the difference in properties in the (x_1,x_2)-directions on one hand and the x_3-direction on the other hand, it is for such configurations advantageous to introduce a separate subscript notation

J. S. Byrnes and J. F. Byrnes (eds.), Recent Advances in Fourier Analysis and Its Applications, 387–425.

(with corresponding summation convention) applying to the (x_1,x_2)-directions. In the following, lower-case Greek subscripts are used for this purpose.

As far as the time coordinate is concerned, a one-sided Laplace transformation is carried out over the semi-infinite interval whose lower limit is the instant at which the sources that generate the wave field are switched on, assuming zero initial values of the wave field. In this manner, the wave motion that is causally related to the action of the sources is accounted for. Let $t=t_0$ be the relevant instant, and let $u=u(\mathbf{x},t)$ denote any space-time wave function representative of the generated wave motion, then the one-sided Laplace transform $\hat{u} = \hat{u}(\mathbf{x},s)$ of $u = u(\mathbf{x},t)$ is defined as

$$(1.1) \qquad \hat{u}(\mathbf{x},s) = \int_{t\epsilon T} \exp(-st)u(\mathbf{x},t)dt,$$

where T is the set

$$(1.2) \qquad T = \{t\epsilon T; t>t_0\}$$

and s is the time-Laplace-transform parameter. Following Cagniard, s is taken to be REAL and POSITIVE. Then, the integral at the right-hand side of (1.1) is convergent for all bounded wave functions with T as support (i.e., the set in time on which the wave function at most differs from zero). Next, a Fourier transformation is carried out with respect to the horizontal space coordinates x_1 and x_2. For the wave function $\hat{u} = \hat{u}(\mathbf{x},s)$ $= \hat{u}(x_\mu,x_3,s)$ the corresponding transformation is taken as

$$(1.3) \qquad \tilde{u}(j\alpha_\mu,x_3,s) = \int_{x_\mu \epsilon R^2} \exp(js\alpha_\mu x_\mu)\hat{u}(x_\mu,x_3,s)dx_1dx_2.$$

In (1.3), the spatial Fourier-transform parameters are $s\alpha_\mu$ with $\alpha_\mu \epsilon R^2$. For the convergence of the integral at the right-hand side of (1.3) it is sufficient that $|\hat{u}(x_\mu,x_3,s)|\to 0$ as $x_\mu x_\mu \to \infty$. The transformation inverse to (1.3) is given by (note that $s\alpha_\mu$ are the actual Fourier transformation parameters)

$$(1.4) \qquad \hat{u}(x_\mu,x_3,s) = (s/2\pi)^2 \int_{\alpha_\mu \epsilon R^2} \exp(-js\alpha_\mu x_\mu)\tilde{u}(j\alpha_\mu,x_3,s)d\alpha_1 d\alpha_2.$$

Upon applying the transformations (1.1) and (1.3) to the partial differential equations that govern any type of wave propagation, the properties of (1.1) and the vanishing of the initial values of the wave function ensure that $\partial_t \to s$, while the properties of (1.3) and the continuity of the wave functions in x_μ ensure that $\partial_\mu \to -js\alpha_\mu$.

After having carried out the indicated integral transformations, a system of "transform-domain" differential equations results, in which x_3 occurs as the independent variable and the "transform-domain" wave function $\hat{u} = \hat{u}(\alpha_\mu, x_3, s)$ as dependent variable, while the constitutive parameters that determine the vertical profile of the configuration occur as coefficients. Once the solution, for given source distributions, of the transform-domain differential equations has somehow been obtained, the relevant values of $\hat{u}$ are substituted in the right-hand side of (1.4) and the transformation back to space-time can take place.

The general outline for the steps to be taken in this procedure will be given for point-source excitation, where the source is located at $x_1=0$, $x_2=0$, $x_3=z_S$. (Since extended source distributions can be regarded as the superposition of point sources, and the configuration is horizontally shift invariant, the choice $x_1=0$, $x_2=0$ for the horizontal space coordinates of the source does not restrict the generatity of the method.) Once this has been done, the orientation of the horizontal axes of the chosen Cartesian reference frame (which was up to now arbitrary) is taken such that one of the axes, the axis along which x_1 is measured, say, joins the projection of the source point on the plane $x_3=0$ with the projection of the point of observation (where some receiver is placed) on the plane $x_3=0$. Let $r>0$ be the "horizontal offset" (or "epicentral distance") from source to receiver, then at the receiver $x_1=r$, $x_2=0$. Substitution of these values in the exponential function at the right-hand side of (1.4) yields

$$\hat{u} = (s/2\pi)^2 \int_{\alpha_1=-\infty}^{\infty} d\alpha_1 \int_{\alpha_2=-\infty}^{\infty} \exp(-js\alpha_1 r) \tilde{u}(j\alpha_1, j\alpha_2, x_3, s, z_S)\, d\alpha_2, \tag{1.5}$$

where the dependence of $\hat{u}$ on z_S has been indicated in its argument. Next, $j\alpha_1$ is replaced by p, α_2 by q, and the order of integration is interchanged. This results into

$$\hat{u} = (s^2/4\pi^2 j) \int_{q=-\infty}^{\infty} dq \int_{p\in I} \exp(-spr)\, \tilde{u}(p, jq, x_3, s, z_S)\, dp, \tag{1.6}$$

where I denotes the imaginary axis in the complex p-plane. The integrals with respect to p and q are now rearranged as follows:

$$(1.7)\qquad \int_{q=-\infty}^{\infty} dq \int_{p\in I} \exp(-spr)\, \tilde{u}(p,jq,x_3,s,z_S)dp$$

$$= \left(\int_{q=-\infty}^{0} dq + \int_{q=0}^{\infty} dq\right) \int_{p\in I} \exp(-spr)\, \tilde{u}(p,jq,x_3,s,z_S)dp$$

$$= -\int_{q=0}^{\infty} dq \int_{p^*\in I^*} \exp(-sp^*r)\, \tilde{u}(p^*,-jq,x_3,s,z_S)dp^*$$

$$+ \int_{q=0}^{\infty} dq \int_{p\in I} \exp(-spr)\, \tilde{u}(p,jq,x_3,s,z_S)dp$$

$$= 2j\ \mathrm{Im}\left[\int_{q=0}^{\infty} dq \int_{p\in I} \exp(-spr)\, \tilde{u}(p,jq,x_3,s,z_S)dp\right].$$

In (1.7) we have used the properties that $I^*=I$, $dp^*=-dp$, and that

$$(1.8)\qquad \begin{aligned} &\exp(-sp^*d)\tilde{u}(p^*,-jq,x_3,s,z_S) \\ &= [\exp(-spr)\tilde{u}(p,jq,x_3 s,z_S)]^*, \end{aligned}$$

where * denotes "complex conjugate". Equation (1.8) follows from the fact that the relevant transform-domain wave function results from carrying out the Laplace and Fourier transformations under consideration on partial differential equations with real-valued coefficients, and the fact that s has been chosen to be a real-valued transform parameter. Using (1.7) in (1.6), we obtain

$$(1.10)\qquad \hat{u} = (s^2/2\pi^2)\,\mathrm{Im}\left[\int_{q=0}^{\infty} dq \int_{p\in I} \exp(-spr)\, \tilde{u}(p,jq,x_3,s,z_S)dp\right].$$

Now, the wavelike nature of the phenomenon of which $u=u(\mathbf{x},t)$ is descriptive entails a decomposition of $\tilde{u}$ that at any level of reception $x_3=z_R$ can be written as

$$(1.11)\qquad \tilde{u} = \hat{Q}_S(s)\textstyle\sum_N \Pi_N(p,jq)\exp[-s\Phi_N(p,jq)],$$

in which $\hat{Q}_S$ is related to the one-sided Laplace transform of the "signature" (i.e., the time shape) of the strength of the source, Φ_N is the transform-domain complex s-independent, vertical travel time of the relevant wave constituent that can be envisaged as to travel from the source level to the receiver level via a number of single or multiple, transmissions and reflections at interfaces in the layered configuration, and Π_N is the product of the corresponding s-independent excitation coefficient at the source level and the transmission and reflection coefficients at the interfaces that the wave constituent has encountered. The decomposition of the type (1.11) is the so-called transform-domain generalized-ray representation of the wave motion (Wiggins and Helmberger, 1974; Helmberger, 1968; Spencer, 1960).

The first step in the transformation back to the space-time domain consists of applying (1.4) to (1.10). Upon using (1.11), the result is written as

$$(1.12)\qquad \hat{u}(\mathbf{x},s) = \textstyle\sum_N \hat{u}_N(\mathbf{x},s),$$

with

$$(1.13)\qquad \hat{u}_N(\mathbf{x},s) = s^2\hat{Q}_S(s)\,\hat{G}_N(\mathbf{x},s),$$

in which

$$(1.14)\qquad \hat{G}_N(\mathbf{x},s) = (2\pi^2)^{-1}\,\mathrm{Im}\Big[\int_{q=0}^{\infty} dq \int_{p\in I} \Pi_N(p,jq) \exp\{-s[pr + \Phi_N(p,jq)]\}dp\Big],$$

is by definition the s-domain Green's function of the generalized ray. The modified Cagniard method now aims at rewriting, somehow, (1.14) as

$$(1.15)\qquad \hat{G}_N(\mathbf{x},s) = \int_{\tau=T_N}^{\infty} \exp(-s\tau)G_N(\mathbf{x},\tau)d\tau,$$

where τ is a real variable of integration, $T_N>0$, while $G_N=G_N(\mathbf{x},\tau)$ does not depend on s. Suppose that this has been achieved, then the uniqueness theorem of the Laplace transformation with real, positive, transform parameter (Lerch's theorem, see Widder, 1946) ensures that

$$(1.16)\qquad G_N(\mathbf{x},t) = \{0,g_N(\mathbf{x},t)\} \text{ for } \{t<T_N, t>T_N\}.$$

Additional properties of the Laplace transformation then lead to the final result

$$(1.17)\quad u(\mathbf{x},t) = \sum_N u_N(\mathbf{x},t),$$

with

$$(1.18)\quad u_N(\mathbf{x},t) = \{0,\ \partial_t^2 \int_{\tau=T_N} Q_S(t-\tau)G_N(\mathbf{x},\tau)d\tau\} \quad \text{for } \{t<T_N, t>T_N\}.$$

Equation (1.18) shows that T_N is the arrival time of the generalized-ray constituent at the point of observation.

To achieve the transformation from (1.14) to (1.15), the integrand in the integration with respect to p in (1.14) is continued analytically into the complex p-plane away from the imaginary axis, while q is kept real. In this process the singularities in Π_N and Φ_N are encountered. These may consist of branch points and poles. Since in the transformation Cauchy's theorem of complex function theory is applied, the analytic continuation must be made single-valued. To this end, branch cuts are introduced such that the relevant branches of the multi-valued functions that occur in Φ_N lead to causal time-domain results.

In the applications that we shall discuss later on (wave propagation in isotropic media), the integrand in (1.14) takes on complex conjugate values in complex conjugate points of the p-plane (Schwarz's principle of reflection), while all singularities of Π_N and Φ_N are located on the real p-axis. The required causality in time then necessitates the introduction of branch cuts that join the relevant branch points via infinity.

Next, the integration along the imaginary axis in the complex p-plane is replaced by one along a path defined by

$$(1.19)\quad pr + \Phi_N(p,jq) = \tau,$$

where τ is real and positive. Such a path is denoted as a modified Cagniard path. Equation (1.19) is satisfied for values on the part of the real p-axis where no branch cuts are located, and on a curve in the right half of the complex p-plane that is located symmetrically with respect to the real p-axis and extends to infinity. Now, as $|p|\to\infty$, Φ_N has the property

$$(1.20)\quad \Phi_N \sim \mp ipH \quad \text{as } |p|\to\infty,$$

where H is a real, non-negative constant that is independent of q, and where the upper and lower signs apply to the first and the fourth quadrants of the complex p-plane, respectively. Since τ goes to infinity at both ends of the complex modified Cagniard path, it reaches a minimum $\tau=T_N(q)$ which follows upon differentiation of (1.19) with respect to p from

$$(1.21) \quad r + \partial_p\Phi_N(p,jq) = 0.$$

Let $p=p_N(q)$ be the solution of (1.21) that for the applications to be considered turns out to be a single value on the real p-axis, then substitution of this value in (1.19) yields

$$(1.22) \quad \tau_N = p_N r + \Phi(p_N,jq).$$

Hence, the complex part of the modified Cagniard path has the asymptotic representation

$$(1.23) \quad p \sim \tau/(r \mp iH) \quad \text{as } \tau\to\infty.$$

Equation (1.23) is representative for two straight half-lines starting at the origin of the complex p-plane, one in the first quadrant (upper sign) and one in the fourth quadrant (lower sign). The parametric representation of the part of complex modified Cagniard path that is located in the first quadrant of the complex p-plane is now written as

$$(1.24) \quad p = p_N^B(q,\tau) \quad \text{for } \tau_N(q) \le\tau<0,$$

the parametric representation of its part in the fourth quadrant is then given by

$$(1.25) \quad p = p_N^{B*}(q,\tau) \quad \text{for } \tau_N(q)\le\tau<0,$$

where * denotes complex conjugate and Schwarz's reflection principle has been used.

In a number of cases the domain in the complex p-plane in between the imaginary axis and the modified Cagniard path defined by (1.24) and (1.25) is free from singularities. Further, Π_N is at most of algebraic growth as $|p|\to\infty$. On account of Jordan's lemma (cf. Whittaker and Watson, 1950) the contribution from circular arcs at infinity joining the two paths vanishes and the integration along the imaginary p-axis can, in view of Cauchy's theorem, be replaced by an integration along the modified Cagniard path. Combining the contributions from the parts (1.24) and (1.25) and again using Schwarz's reflection principle, we arrive at

$$\hat{G}_N(\mathbf{x},s) = \pi^{-2}\{\int_{q=0}^{\infty} dq \int_{\tau=\tau_N(q)}^{\infty} \tag{1.26}$$
$$\exp(-s\tau)\,\mathrm{Im}[\Pi_N(p_N^B,jq)(\partial p_N^B/\partial\tau)]d\tau\},$$

in which $\partial p_N^B/\partial\tau$ is the one-dimensional Jacobian that transforms the variable of integration from p into τ and where it has been taken into account that q, τ and s are real-valued.

Again for the cases we are going to consider (wave propagation in isotropic media) the function $\Phi_N = \Phi_N(p,jq)$ is an even function of q with the additional property that $\tau=\tau_N(q)$ is a monotonically increasing function of q in the interval $q\epsilon(0,\infty)$. From (1.19) it then follows that $\tau=\tau_N(q)$ has a single minimum at q=0. Let $q=q_N(\tau)$ be the (unique) inverse of $\tau=\tau_N(q)$, then an interchange in the order of integration in (1.26) yields

$$\hat{G}_N(\mathbf{x},s) = \pi^{-2}\int_{\tau=\tau_N(0)}^{\infty} \exp(-s\tau)d\tau \tag{1.27}$$
$$\int_{q=0}^{q_N(\tau)} \mathrm{Im}[\Pi_N(p_N^B,jq)(\partial p_N^B/\partial\tau)]dq.$$

Application of Lerch's theorem finally leads to (1.16) with

$$T_N = \tau_N(0) \tag{1.28}$$

and

$$g_N(\mathbf{x},\tau) = \pi^{-2}\int_{q=0}^{q_N(\tau)} \mathrm{Im}[\Pi_N(p_N^B,jq)(\partial p_N^B/\partial\tau)]dq, \tag{1.29}$$

with which the time-domain Green's function for the generalized-ray constituent has been determined, at least as far as the contribution from the complex part of the modified Cagniard path - the so-called body-wave contribution - is concerned. Under certain circumstances, the body-wave contribution has to be supplemented by a head-wave contribution; this will be investigated in the specific examples that follow.

In some cases, especially in the ones where the acoustic radiation is due to moving sources, it is advantageous to choose one of the horizontal axes in the direction of the motion of the source and not along some line passing through the projection of the point of observation on the plane $x_3=0$. In that case, we employ the polar-coordinate representation $\{r,\theta\}$ of the point of observation with respect to the fixed origin. These are defined through

$$(1.30) \quad x_1 = r\cos(\theta),\ x_2 = r\sin(\theta),$$

with $0 \le r < \infty$, $0 \le \theta < 2\pi$. Now, the variables of integration in the α_1, α_2-plane are replaced by p and q that follows from

$$(1.31) \quad j\alpha_1 = p\cos(\theta) - jq\sin(\theta),$$

$$(1.32) \quad j\alpha_2 = p\sin(\theta) + jq\cos(\theta),$$

with $p \in I$ and $q \in R$, which makes $d\alpha_1 d\alpha_2 = j^{-1}dpdq$, $\alpha_1^2 + \alpha_2^2 = q^2 - p^2$ and, using (1.30),

$$(1.33) \quad j\alpha_1 x_1 + j\alpha_2 x_2 = pr.$$

From here on, the steps in the modified Cagniard method proceed as earlier in this section. The transformation (1.31) - (1.32) was introduced by De Hoop (1960)

2. THE POINT-SOURCE SOLUTION TO THE THREE-DIMENSIONAL SCALAR WAVE EQUATION

In this section we shall apply the modified Cagniard method as outlined in Section 1 to construct the point-source solution to the three-dimensional scalar wave equation Let $u = u(\mathbf{x},t)$ be the relevant wave function and let $Q_S = Q_S(t)$ denote the source strength as a function of time of a point source located at $\{0,0,0\}$. Then, u satisfies the equation

$$(2.1) \quad \partial_1^2 u + \partial_2^2 u + \partial_3^2 u - c^{-2}\partial_t^2 u = -Q_S(t)\delta(x_1,x_2,x_3),$$

where c is the wave speed of the medium in which the wave motion occurs. Upon carrying out the integral transformations (1.1) and (1.3), the transform-domain equivalent of (2.1) is obtained as

$$(2.2) \quad \partial_3^2\tilde{u} - s^2\gamma^2\tilde{u} = -\hat{Q}_S(s)\delta(x_3),$$

in which

(2.3) $\qquad \gamma^2 = \alpha_1^2 + \alpha_2^2 + 1/c^2$. When $x_3 \neq 0$, the right-hand side of (2.2) vanishes and $\tilde{u}$ is a linear combination of $\exp(s\gamma x_3)$ and $\exp(-s\gamma x_3)$, where

$$\gamma = (\alpha_1^2 + \alpha_2^2 + 1/c^2)^{1/2} > 0. \tag{2.4}$$

In view of the condition of boundedness of $\tilde{u}$ as $|x_3| \to \infty$, we have

$$\tilde{u} = A^- \exp(s\gamma x_3) \quad \text{when } x_3 < 0 \tag{2.5}$$

and

$$\tilde{u} = A^+ \exp(-s\gamma x_3) \quad \text{when } x_3 > 0, \tag{2.6}$$

where A^- and A^+ are constants. The values of the latter follow from the source conditions at $x_3 = 0$, where $\partial_3 \tilde{u}$ must make a step at $x_3 = 0$, while $\tilde{u}$ must be continuous at $x_3 = 0$. Hence,

$$\lim_{x_3 \downarrow 0} \tilde{u} - \lim_{x_3 \uparrow 0} \tilde{u} = 0 \tag{2.7}$$

and

$$\lim_{x_3 \downarrow 0} \partial_3 \tilde{u} - \lim_{x_3 \uparrow 0} \partial_3 \tilde{u} = -\hat{Q}_S. \tag{2.8}$$

Substitution of (2.5) and (2.6) into (2.7) and (2.8) leads to

$$A^+ - A^- = 0, \tag{2.9}$$

$$-s\gamma A^+ - s\gamma A^- = -\hat{Q}_S, \tag{2.10}$$

from which it follows that

$$A^+ = A^- = \hat{Q}_S / 2s\gamma. \tag{2.11}$$

Consequently, $\tilde{u}$ is obtained as

(2.12) $\tilde{u} = (\hat{Q}_S/2s\gamma)\exp(-s\gamma|x_3|)$.

Substituting this result into (1.5) and carrying out the subsequent transformations, the s-domain wave function is obtained as (cf. (1.10) and (1.12))

(2.13) $\hat{u}(\mathbf{x},s) = s\hat{Q}_s(s)\,\hat{G}(\mathbf{x},s)$,

in which

(2.14) $\hat{G}(\mathbf{x},s) = (4\pi^2)^{-1}\,\mathrm{Im}\{\int_{q=0}^{\infty} dq \int_{p\varepsilon I} \exp[-s(pr + \gamma|x_3|)]\gamma^{-1}dp\}$.

The modified Cagniard path in the complex p-plane follows from (cf. (1.19))

(2.15) $pr + \gamma|x_3| = \tau$,

where τ is real and positive and where

(2.16) $\gamma = [S(q)^2 - p^2]^{1/2}$ with $\mathrm{Re}(\gamma)\geq 0$ for all $p\varepsilon C$,

in which

(2.17) $S = (q^2 + 1/c^2)^{1/2} > 0$.

Obviously, γ has branch points at $p=-S(q)$ and $p=S(q)$ and the choice $\mathrm{Re}(\gamma)\geq 0$ for all $p\varepsilon C$ implies that branch cuts are introduced along $\{p\varepsilon C; |\mathrm{Re}(p)|>1/c,\ \mathrm{Im}(p) = 0\}$.

Solving Eq. (2.15) for p, the parametric representation of the modified Cagniard path is obtained as

(2.18) $p = p^B(q,\tau)$,

and

(2.19) $p = p^{B*}(q,\tau)$,

where

(2.20) $p^B = \dfrac{\tau r + j|x_3|[\tau^2 - \tau^B(q)^2]^{1/2}}{r^2 + x_3^2}$ for $\tau^B(q)\leq\tau<\infty$,

with

$$(2.21)\qquad \tau^B(q) = S(q)(r^2 + x_3^2)^{1/2}.$$

From (2.20), the Jacobian of the transformation from p^B to τ follows as

$$(2.22)\qquad \partial p^B/\partial\tau = \frac{r + j|x_3|\tau[\tau^2 - \tau^B(q)^2]^{-1/2}}{r^2 + x_3^2} \quad \text{for } \tau^B(q)\le\tau<\infty,$$

while upon using (2.20) in (2.15) it is found that

$$(2.23)\qquad \gamma(p^B,q) = \frac{\tau|x_3| - jr[\tau^2 - \tau^B(q)^2]^{1/2}}{r^2 + x_3^2} \quad \text{for } \tau^B(q)\le\tau<\infty.$$

Hence

$$(2.24)\qquad \gamma^{-1}(p^B,q)(\partial p^B/\partial\tau) = j[\tau^2 - \tau^B(q)^2]^{-1/2} \quad \text{for } \tau^B(q)\le\tau<\infty.$$

Equation (2.20) defines a hyperbola in the right half of the complex p-plane that is symmetrical with respect to the real p-axis and has the asymptotes $p\sim\tau/(r - j|x_3|)$ as $\tau\to\infty$. Using (2.15) - (2.23) in (2.14) and observing that

$$(2.25)\qquad \mathrm{Im}[\gamma^{-1}(p^B,q)(\partial p^B/\partial\tau)] = [\tau^2 - \tau^B(q)^2]^{-1/2} \quad \text{for } \tau^B(q)\le\tau<\infty,$$

the expression (2.14) for the s-domain Green's function changes into

$$(2.26)\qquad \hat{G}(\mathbf{x},s) = (2\pi^2)^{-1}\int_{q=0}^{\infty} dq \int_{\tau=\tau^B(q)}^{\infty} \exp(-s\tau)[\tau^2 - \tau^B(q)]^{-1/2} d\tau.$$

Interchanging the order of the integrations we arrive at

$$(2.27)\qquad \hat{G}(\mathbf{x},s) = (2\pi^2)^{-1}\int_{\tau=\tau^B(0)}^{\infty} \exp(-s\tau)d\tau \int_{q=0}^{Q^B(\tau)} [\tau^2 - \tau^B(q)^2]^{-1/2} dq,$$

in which (cf. (2.17) and (2.21))

(2.28) $$\tau^B(0) = (r^2 + x_3^2)^{1/2}/c$$

and

(2.29) $$Q^B(\tau) = [\tau^2 - (r^2 + x_3^2)/c^2]^{1/2}(r^2 + x_3^2)^{-1/2}.$$

To evaluate the integral with respect to q, we introduce the variable of integration ψ defined through

(2.30) $$q = Q^B(\tau)\sin(\psi) \quad \text{with } 0 \le \psi \le \pi/2.$$

With this,

(2.31) $$\int_{q=0}^{Q^B(\tau)} [\tau^2 - \tau^B(q)^2]^{-1/2} dq$$

$$= (r^2 + x_3^2)^{-1/2} \int_{q=0}^{Q^B(\tau)} [Q^B(\tau)^2 - q^2]^{-1/2} dq$$

$$= (r^2 + x_3^2)^{-1/2} \int_{\psi=0}^{\pi/2} d\psi$$

$$= (\pi/2)(r^2 + x_3^2)^{-1/2}.$$

Since, further,

(2.32) $$\int_{\tau=\tau^B(0)}^{\infty} \exp(-s\tau)d\tau = s^{-1}\exp[-s\tau^B(0)] \quad \text{for } s>0,$$

the s-domain Green's function (cf. (2.27) reduces to

(2.33) $$\hat{G}(\mathbf{x},s) = s^{-1}\exp[-s(r^2 + x_3^2)^{1/2}/c]/4\pi(r^2 + x_3^2)^{1/2}$$

and hence (cf. (2.13))

(2.34) $$\hat{u} = \hat{Q}_S(s)\exp[-s(r^2 + x_3^2)^{1/2}/c]/4\pi(r^2 + x_3^2)^{1/2}.$$

The corresponding time-domain result is

$$(2.35)\qquad u = \frac{Q_S[t - (r^2 + x_3^2)^{1/2}/c]}{4\pi(r^2 + x_3^2)^{1/2}},$$

which is a well-known result. Obviously,

$$(2.36)\qquad T^B = (r^2 + x_3^2)^{1/2}/c$$

is the travel time from the source to the point of observation. With (2.36, the result (2.35) can be rewritten as

$$(2.37)\qquad u = \frac{Q_S(t - T^B)}{4\pi(r^2 + x_3^2)^{1/2}} .$$

In the simple case under consideration, the integral in (2.31) could be evaluated analytically, as well as the integral in (2.32). After that, elementary rules of the one-sided time Laplace transformation led to the direct expression (2.35) for the generated wave function. In more complicated cases, this is not possible. Then, we have to fall back on Lerch's theorem and we obtain (cf. (1.16))

$$(2.38)\qquad G(\mathbf{x},\tau) = \{0, g^B(\mathbf{x},t)\} \quad \text{for } \{t<T^B, t>T^B\},$$

in which (cf. (1.29))

$$(2.39)\qquad \begin{aligned} g^B(\mathbf{x},\tau) &= (2\pi^2)^{-1} \int_{q=0}^{q^B(\tau)} [\tau^2 - \tau^B(q)^2]^{-1/2} dq \\ &= 1/4\pi(r^2 + x_3^2)^{1/2}, \end{aligned}$$

and the wave function is given by (cf. (2.13))

$$(2.40)\qquad u(\mathbf{x},t) = \{0,\ \partial_t \int_{\tau=T^B}^{t} Q_S(t-\tau) g^B(\mathbf{x},\tau) d\tau\} \quad \text{for } \{t<T^B,\ t<T^B\}.$$

Equation (2.40) directly leads to (2.37).

EXERCISES

Exercise 2-1. Determine with the aid of the modified Cagniard method the impulsive line-source solution to the two-dimensional scalar wave equation

(2.41) $\partial_1^2 u + \partial_3^2 u - c^{-2}\partial_t^2 u = -Q_S(t)\delta(x_1,x_3)$,

where $u = u(x_1,x_3,t)$. (Hint: Carry out the one-sided Laplace transformation (1.1) with respect to t, and a single Fourier transformation of the type (1.3) with respect to x_1. Show (a) that the transform-domain wave function $\tilde{u} = \tilde{u}(j\alpha_1,x_3,s)$ satisfies the differential equation

(2.42) $\partial_3^2\tilde{u} - s^2\gamma^2\tilde{u} = -\hat{Q}_S(s)\delta(x_3)$,

in which now

(2.43) $\gamma = (\alpha_1^2 + 1/c^2)^{1/2} > 0$.

Show (b) that $\tilde{u}$ follows from (2.42) as

(2.44) $\tilde{u} = (\hat{Q}_S/2s\gamma)\exp(-s\gamma|x_3|)$.

(c) Write the spatial Fourier inversion integral as

(2.45) $\hat{u}(\mathbf{x},s) = \hat{Q}_S(s)\hat{G}(\mathbf{x},s)$,

in which

(2.46) $\hat{G}(\mathbf{x},s) = (4\pi j)^{-1}\int_{p\varepsilon I}\exp[-s(px_1 + \gamma|x_3|)]\gamma^{-1}dp$,

where $j\alpha_1$ has been replaced by p. (d) By replacing the integral along $p\varepsilon I$ by one along the modified Cagniard path

(2.48) $px_1 + \gamma|x_3| = \tau$,

show that

(2.49) $G(\mathbf{x},t) = \{0, g^B(\mathbf{x},t)\}$ for $\{t<T^B, t>T^B\}$,

where

(2.50) $g^B(\mathbf{x},\tau) = 1/2\pi[\tau^2 - (T^B)^2]^{1/2}$,

with

$$(2.51)\quad T^B = (x_1^2 + x_3^2)^{1/2}/c.$$

(e) Show that the final result is obtained as

$$(2.52)\quad u(\mathbf{x},t) = \{0, \int_{\tau=T^B}^{t} Q_S(t-\tau)g^B(\mathbf{x},\tau)d\tau\} \quad \text{for } \{t<T^B, t>T^B\}.)$$

3. ACOUSTIC WAVES IN GEOPHYSICAL PROSPECTING

Acoustic waves are one of the standard diagnostic tools to probe the subsurface structure of the earth in the search for fossile energy resources. In the acquisition of land seismic data we distinguish between surface seismics, vertical seismic profiling, and cross-borehole seismics. In surface seismics, the acoustic source is located either on the earth's surface (for a mechanical vibrator) or somewhat below, but close to, it (for an explosion source), while the resulting acoustic wave motion is picked up by a number of acoustic receivers (geophones) placed at the surface of the earth. In this kind of seismics, the offset between source and receivers is predominantly horizontal, and the interpretation of the data is mainly based on the acoustic wave reflection against interfaces between the successive layers out of which the earth is composed. In Vertical Seismic Profiling (VSP), again an acoustic source of the type as used in surface seismics is employed, while the acoustic receivers are now placed in a borehole. In this kind of seismics, the offset between source and receivers is predominantly vertical, and the interpretation of the data is mainly based on the acoustic wave transmission across the successive layers of the earth. In cross-borehole seismics, the acoustic source (usually of an explosion type or a piezo-electric transducer) is placed in one borehole, while the acoustic receivers are placed in an adjacent one. In this kind of seismics, there is no predominant offset, and the interpretation of the data is based on both the reflection and the transmission properties of the successive layers of the earth.

In the acquisition of marine seismic data, both the acoustic source (usually an airgun) and the acoustic receivers (hydrophones) are situated somewhat below, but close to, the surface of the sea; they are towed by a surface vessel. Here, too, the offset between source and receiver is predominantly horizontal.

The problem of seismic prospecting is, in fact, an inverse one: given the measured data, one is to reconstruct the physical properties of the probed structure. In seismic practice, most of this inversion is still carried out by assuming that the earth's structure changes in its mechanical properties (volume density of mass, elastic compliance or stiffness) much more rapidly in the (downward) vertical direction than in the horizontal direction. Accordingly, one takes a horizontally layered model of the earth as point of departure. Using this model, the thicknesses of the layers and their mechanical parameters are, roughly speaking, inferred by comparing the measured data with the theoretically determined response of a number of model configurations of the indicated type. Hence, there is a need for a versatile and efficient computational method to generate so-called synthetic seismograms for a wide variety of structures of the indicated kind. The modified Cagniard method, the general ideas of which have been sketched in Section 1, provides such a method

The present chapter discusses this method for the case of acoustic waves generated by an impulsive point source located in a layered fluid medium. The results are also of importance to the case of a layered isotropic solid if the acoustic wave motion in the solid is predominantly of the compressional type (as is the case in fluids), which amounts to neglecting, in the first instance, the influence of shear waves. It is observed, however, that the modified Cagniard method can without difficulty also incorprate the presence of shear waves (cf. De Hoop and Van der Hijden, 1983, 1984, 1985), while with the unavoidable increase in the degree of complexity it can also be applied to arbitrarily anisotropic solid media (Van der Hijden, 1987). Through the method, the total acoustic wave motion is expressed as a time convolution of the strength of the source as a function of time ("source signature") and a properly defined space-time Green's function that is characteristic for the type of source (volume-injection source, force source) and for the configuration in which the acoustic radiation takes place. The space-time Green's function is expressed as the superposition of generalized rays. This representation is exact, and within a finite time interval the number of contributing generalized rays is finite. The expressions for the acoustic pressure and the particle velocity of each generalized ray consist of a single integral over a real, finite range, the integrand of which is an algebraic function of a certain complex ray parameter. The so-called modified Cagniard path maps this complex ray parameter in a one-to-one way to the real time parameter. For large horizontal and vertical offsets, simplified asymptotic representations for the space-time Green's function can be derived (see Roever, Vining, and Strick, 1959; Wiggins and Helmberger, 1974).

4. DESCRIPTION OF THE CONFIGURATION

We investigate theoretically the acoustic wave motion in a layered fluid (or fluid equivalent of a solid) the mechanical properties of which vary in a single rectilinear direction in space only. This direction is taken to be the vertical one. To specify position in the configuration we employ, as before, the coordinates $\{x_1,x_2,x_3\}$ with respecto to a fixed, orthogonal, Cartesian reference frame with the origin 0 and the three mutually perpendicular base vectors $\{\mathbf{i}_1,\mathbf{i}_2,\mathbf{i}_3\}$ of unit length each. In the indicated order, the base vectors form a right-handed system. In accordance with the geophysical conventions, $\mathbf{i}_3$ points vertically downwards. The subscript notation for Cartesian vectors and tensors is used. Lower-case Latin subscripts are used for this purpose; they are to be assigned the values 1,2 and 3. Further, the summation convention applies to repeated subscripts. Whenever appropriate, the position is also specified by the position vector $\mathbf{x}=x_p\mathbf{i}_p$, with $\mathbf{x}\in R^3$. The time coordinate is denoted by t, with $t\in R$. Partial differentiation with respect to x_p is denoted by ∂_p; the symbol ∂_t is reserved for partial differentiation with respect to t.

The acoustic properties of the fluid are characterized by its volume density of mass ρ (which is characteristic for the inertia properties of the fluid) and its compressibility κ (which is characteristic for the elastic compliance of the fluid). Both ρ and κ are functions of x_3. They are independent of x_1, x_2, or t, which makes the configuration shift invariant in the horizontal x_1,x_2-plane, and time invariant as well. The real-valued functions $\rho=\rho(x_3)$ and $\kappa=\kappa(x_3)$ are taken to be positive and piecewise constant. The acoustic wave speed associated with ρ and κ is given by $c=(\rho\kappa)^{-1/2}$. Their values in the different subdomains are listed in Table 4.1.

Table 4.1. Nomenclature of the configuration.

Domain	Vertical coordinate	Volume density of mass	Compress-ibility	Wave speed
D_1	$-\infty<x_3<z_1$	ρ_1	κ_1	c_1
...	...	...	...	...
D_S	$z_{S-1}<x_3<z_S$	ρ_S	κ_S	c_S
...	...	...	...	...
D_R	$z_{R-1}<x_3<z_R$	ρ_R	κ_R	c_R
...	...	...	...	...
D_{ND}	$z_{ND-1}<x_3<\infty$	ρ_{ND}	κ_{ND}	c_{ND}

An impulsive point source, located at $\{0,0,z_S\}$, generates the acoustic waves. A point receiver is placed at $\{x_1,x_2,z_R\}$ to pick up the generated wave motion. For our further analysis it turns out to be advantageous to introduce interfaces at the levels $x_3=z_S$ and $x_3=z_R$, even if the source and/or the receiver are located in the interior of one of the domains. The relevant levels have been included already in the nomenclature listed in Table 4.1. The source starts to act at the instant t=0. To determine the wave motion that is causally related to the action of the source, we put the state quantities describing this wave motion equal to zero in the time interval t<0 (initial condition). Figure 4.1 shows schematically the configuration and the location of the source.

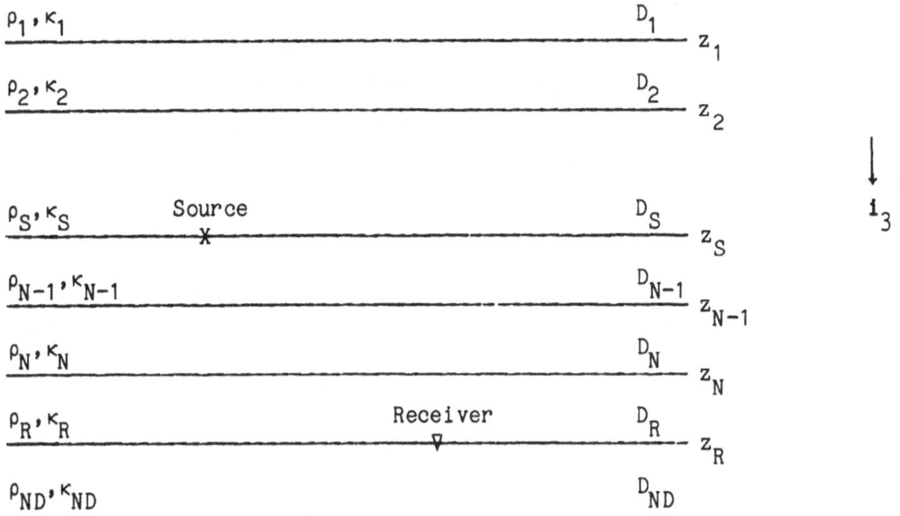

Figure 4.1. Horizontally layered fluid medium in which acoustic waves are generated by an impulsive point source and picked up by a point receiver.

5. BASIC EQUATIONS FOR THE ACOUSTIC WAVE MOTION

The state quantities that characterize the acoustic wave motion are the acoustic pressure p and the particle velocity v_k. In a subdomain of the configuration where ρ and κ vary continuously with position, p and v_k are continuously differentiable functions of $\mathbf{x}$ and t, and they satisfy the acoustic wave equations

(5.1) $$\partial_k p + \rho\, \partial_t v_k = f_k,$$

(5.2) $$\partial_r v_r + \kappa\, \partial_t p = q,$$

in which f_k is the volume source density of force and q is the volume source density of injection rate. The source densities are assumed to be piecewise continuous functions of $\mathbf{x}$ and t. At the interfaces where ρ and/or κ show a jump discontinuity, the pressure and the normal component of the particle ,elocity are to be continuous.

For the specific case of a point source located at $\{0,0,z_S\}$ we have

$$(5.3)\qquad \{q,f_k\} = \{Q_S(t),F_{S;k}(t)\}\delta(x_1,\ x_2,\ x_3 - z_S),$$

where $\{Q_S(t),F_{S;k}(t)\}$ characterize the strength of the source as a function of time and δ denotes the three-dimensional unit pulse (Dirac distribution). It is understood that $\{Q_S(t),F_{S;k}(t)\}=0$ when $t<0$. When $Q_S\neq0$ and $F_{S;k}=0$, the source is of the volume-injection type; such a source is also denoted as an acoustic monopole source. When $Q_S=0$ and $F_{S;k}\neq0$, the source is of the force type; such a source is also denoted as an acoustic dipole source.

To solve Eqs.(5.1) - (5.3) and their accompanying boundary conditions in the configuration discussed in Section 3, we subject them to the integral transformations that are compatible with the time invariance of the configuration and its shift invariance in the horizontal plane and that have been discussed in Section 1.

6. THE TRANSFORM-DOMAIN ACOUSTIC WAVE EQUATIONS

First, the acoustic state quantities and the volume source densities are subjected to a one-sided Laplace transformation with respect to time over the range $t>0$. To show the notation we give the transformation of the acoustic presssure:

$$(6.1)\qquad \hat{p}(\mathbf{x},s) = \int_{t=0}^{\infty} \exp(-st)p(\mathbf{x},t)dt.$$

It is recalled that in (6.1), the Laplace-transform parameter s is taken to be real and sufficiently large positive. Taking $\{Q_S(t),F_{S;k}(t)\}$ to be at most of Order[$\exp(s_0t)$] as $t\to\infty$, it can be shown that also $\{p(\mathbf{x},t),v_k(\mathbf{x},t)\}$ are at most of this order, and the corresponding right-hand sides of (6.1) exist for $s>s_0$. Considering (6.1) as an integral equation for the space-time quantities, given their Laplace transforms, the solution of this integral equation is unique and vanishes in the interval $t<0$ on account of Lerch's theorem (Widder, 1946). Owing to this property, the selection of the causal quantities in the $(\mathbf{x},t)$-domain is brought down to selecting their bounded counterparts in the $(\mathbf{x},s)$-domain.

Next, we carry out a Fourier transformation with respect to the horizontal space coordinates over their entire range $(x_1,x_2)\in R^2$. Again, to show the notation, we give the expression for the acoustic pressure:

$$\tilde{p}(\alpha_1,\alpha_2,x_3,s) = \int_{x_2=-\infty}^{\infty} dx_2 \tag{6.2}$$

$$\int_{x_1=-\infty}^{\infty} \exp[js(\alpha_1 x_1 + \alpha_2 x_2)]\, \hat{p}(x_1,x_2,x_3,s)dx_1 .$$

In (6.2), the Fourier-transform parameters $\{\alpha_1,\alpha_2\}$ are taken to be real, and j is the imaginary unit. Note that in terms of the standard Fourier transformation, $\{s\alpha_1,s\alpha_2\}$ are the Fourier-transform parameters. As a consequence, the inverse transformation is given by

$$\hat{p}(x_1,x_2,x_3,s) = (s/2\pi)^2 \int_{\alpha_1=-\infty}^{\infty} d\alpha_1 \tag{6.3}$$

$$\int_{\alpha_2=-\infty}^{\infty} \exp[-js(x_1\alpha_1 + x_2\alpha_2)]\tilde{p}(\alpha_1,\alpha_2,x_3,s)d\alpha_2 .$$

Applying the transformations (6.1) and (6.2) to Eqs. (5.1) and (5.2), taking into account that $\partial_1 \to -js\alpha_1$, $\partial_2 \to -js\alpha_2$, and $\partial_t \to s$, and eliminating $\tilde{v}_1$ and $\tilde{v}_2$ from the resulting equations, we end up with

$$\partial_3 \tilde{p} + s\gamma Y^{-1}\tilde{v}_3 = \tilde{f}', \tag{6.4}$$

$$s\gamma Y\tilde{p} + \partial_3 \tilde{v}_3 = \tilde{q}', \tag{6.5}$$

in which

$$\gamma = (1/c^2 + \alpha_1^2 + \alpha_2^2)^{1/2} > 0 \tag{6.6}$$

is the vertical slowness,

$$c = (\kappa\rho)^{-1/2} > 0 \tag{6.7}$$

is the acoustic wave speed,

$$Y = \gamma/\rho \tag{6.8}$$

is the vertical acoustic wave admittance, and

$$\tilde{f}' = \tilde{f}_3, \tag{6.9}$$

$$\tilde{q}' = \tilde{q} + \rho^{-1}(j\alpha_1\tilde{f}_1 + j\alpha_2\tilde{f}_2), \tag{6.10}$$

are the transform-domain "notional source distributions" of force and volume injection, respectively.

Equations (6.4) and (6.5) are the transform-domain acoustic wave equations. In a homogeneous, source-free subdomain of the configuration they constitute a system of homogeneous, ordinary differential equations with x_3 as independent variable and $\{\tilde{p},\tilde{v}_3\}$ as dependent variables. At those interfaces between two adjacent domains with different mechanical properties that are source-free, the quantities $\tilde{p}$ and $\tilde{v}_3$ are to be continuous. Since, further, the transform-domain source distributions follow from Eq. (5.3) as

$$\{\tilde{q},\tilde{f}_k\} = \{\hat{Q}_S,\hat{F}_{S;k}\}\delta(x_3 - z_S), \tag{6.11}$$

the boundary conditions at the interfaces $x_3 = z_S$ are on account of (6.4) and (6.5)

$$\lim_{x_3 \downarrow z_S} \tilde{p} - \lim_{x_3 \uparrow z_S} \tilde{p} = \tilde{F}', \tag{6.12}$$

$$\lim_{x_3 \downarrow z_S} \tilde{v}_3 - \lim_{x_3 \uparrow z_S} \tilde{v}_3 = \tilde{Q}', \tag{6.13}$$

in which (cf. (6.9) - (6.11))

$$\tilde{F}' = \hat{F}_{S;3}, \tag{6.14}$$

$$\tilde{Q}' = \hat{Q}_S + \rho^{-1}(j\alpha_1\hat{F}_{S;1} + j\alpha_2\hat{F}_{S;2}), \tag{6.15}$$

are the transform-domain "notional strengths" of the point source at $\{0,0,z_S\}$.

Equations (6.4), (6.5), (6.12) and (6.13) will be solved by expanding, in each homogeneous source-free subdomain of the configuration, $\{\tilde{p},\tilde{v}_3\}$ in terms of transform-domain wave constituents that in space-time correspond to down- and upgoing waves.

7. THE TRANSFORM-DOMAIN ACOUSTIC WAVE MOTION

In the homogeneous source-free subdomain D_N of the configuration, we write

$$\{\tilde{p},\tilde{v}_3\} = \{\tilde{p}_N,\tilde{v}_{3;N}\} = \{\tilde{p}_N^+ + \tilde{p}_N^-, \tilde{v}_{3;N}^+ + \tilde{v}_{3;N}^-\} \tag{7.1}$$

$$\text{with } N\epsilon(1,\ldots,ND\},$$

where (cf. Eqs. (6-4) and (6.5))

$$\tilde{p}_N^+ = W_N^+ (2Y_N)^{-1/2}\exp[-s\gamma_N(x_3 - z_{N-1})], \tag{7.2}$$

$$\tilde{p}_N^- = W_N^- (2Y_N)^{-1/2}\exp[-s\gamma_N(z_N - x_3)], \tag{7.3}$$

$$\tilde{v}_{3;N}^{\pm} = \pm Y_N \tilde{p}_N^{\pm}, \tag{7.4}$$

with (cf. Eqs. (6.6) - (6.8))

$$\gamma_N = (1/c_N^2 + \alpha_1^2 + \alpha_2^2)^{1/2} > 0, \tag{7.5}$$

$$c_N = (\kappa_N\rho_N)^{-1/2} > 0, \tag{7.6}$$

$$Y_N = \gamma_N/\rho_N. \tag{7.7}$$

The wave $\{\tilde{p}_N^+,\tilde{v}_{3;N}^+\}$ propagates away from the interface $x_3 = z_{N-1}$ in the direction of increasing x_3 (and hence downwardly), the wave $\{\tilde{p}_N^-,\tilde{v}_{3;N}^-\}$ propagates away from the interface $x_3 = z_N$ in the direction of decreasing x_3 (and hence upwardly). Since both $s>0$ and $\gamma_N>0$, the right-hand side of (7.2) stays bounded as $x_3\to\infty$ and the right-hand side of (7.3) stays bounded as $x_3\to-\infty$. Hence, the causality condition is satisfied. The quantities $\{W_N^+,W_N^-\}$ are the transform-domain wave amplitudes of the down- and upgoing waves, respectively. The normalization has been taken such that reciprocity is satisfied. In order to satisfy the causality condition in the outer half-spaces, we have $W_1^+=0$ and $W_{ND}^-=0$.

The different wave amplitudes are interrelated by the boundary conditions at the interfaces. At a source-free interface we express this relationship via a scattering discription, whereby the amplitudes of the waves propagating away from the interface are expressed in terms of the amplitudes of the waves propagating toward the interface through the scattering matrix. At the interfaces where sources are located, the scattering description is supplemented by excitation terms that are related to the notional source strengths. Let, accordingly, for any $N \varepsilon \{1,\dots,ND-1\}$,

$$(7.8) \qquad W_N^- = S_N^{-,+} \bar{W}_N^+ + S_N^{-,-} \bar{W}_{N+1}^- + X_N^-,$$

$$(7.9) \qquad W_{N+1}^+ = S_N^{+,+} \bar{W}_N^+ + S_N^{+,-} \bar{W}_{N+1}^- + X_{N+1}^+,$$

where

$$(7.10) \qquad \bar{W}_N^\pm = W_N^\pm \exp(-s\gamma_N d_N) \quad \text{for any } N\varepsilon\{2,\dots,ND-1\},$$

$$(7.11) \qquad \bar{W}_1^+ = 0,$$

$$(7.12) \qquad \bar{W}_{ND}^- = 0,$$

denote the so-called modified wave amplitudes, and

$$(7.13) \qquad d_N = z_N - z_{N-1} \quad \text{with } N\varepsilon\{2,\dots,ND-1\}$$

is the thickness of the layer occupying the domain D_N. For our single source located at $\{0,0,z_S\}$, only X_S^- and X_{S+1}^+ differ from zero. Substitution of (7.8) and (7.9) in the boundary conditions (cf. (6.12) and (6.13))

$$(7.14) \qquad \lim_{x_3 \downarrow z_N} \tilde{p}_{N+1} - \lim_{x_3 \uparrow z_N} \tilde{p}_N = \tilde{F}'_N,$$

$$(7.15) \qquad \lim_{x_3 \downarrow z_N} \tilde{v}_{3;N+1} - \lim_{x_3 \uparrow z_N} \tilde{v}_{3;N} = \tilde{Q}'_N,$$

leads, with the aid of (7.2) - (7.4) to

$$(7.16) \qquad S_N^{-,+} = (Y_N - Y_{N+1})/(Y_N + Y_{N+1}),$$

$$(7.17)\quad S_N^{-,-} = 2(Y_N Y_{N+1})^{1/2}/(Y_N + Y_{N+1}),$$

$$(7.18)\quad S_N^{+,+} = 2(Y_N Y_{N+1})^{1/2}/(Y_N + Y_{N+1}),$$

$$(7.19)\quad S_N^{+,-} = (Y_{N+1} - Y_N)/(Y_N + Y_{N+1}),$$

and

$$(7.20)\quad X_N^{-} = (2Y_N)^{1/2}(\tilde{Q}_N' - Y_{N+1}\tilde{F}_N')/(Y_N + Y_{N+1}),$$

$$(7.21)\quad X_N^{+} = (2Y_{N+1})^{1/2}(\tilde{Q}_N' + Y_N\tilde{F}_N')/(Y_N + Y_{N+1}).$$

The expressions (7.16) and (7.19) are reflection coefficients and the expressions (7.17) and (7.18) are transmission coefficients at the interface $x_3 = z_N$.

To describe the overall behavior of the structure, we introduce the modified scattering coefficients as

$$(7.22)\quad \bar{S}_N^{-,+} = S_N^{-,+} \exp(-s\gamma_N d_N),$$

$$(7.23)\quad \bar{S}_N^{-,-} = S_N^{-,-} \exp(-s\gamma_{N+1} d_{N+1}),$$

$$(7.24)\quad \bar{S}_N^{+,+} = S_N^{+,+} \exp(-s\gamma_N d_N),$$

$$(7.25)\quad \bar{S}_N^{+,-} = S_N^{+,-} \exp(-s\gamma_{N+1} d_{N+1}).$$

Next, (7.8) and (7.9) are rewritten as

$$(7.26)\quad W_N^{-} - \bar{S}_N^{-,-}\, W_{N+1}^{-} = \bar{S}_N^{-,+}\, W_N^{+} + X_N^{-},$$

$$(7.27)\quad W_{N+1}^{+} - \bar{S}_N^{+,+}\, W_N^{+} = \bar{S}_N^{+,-}\, W_{N+1}^{-} + X_{N+1}^{+}.$$

These equations are solved by Neumann iteration where the iteration takes place in increasing orders of the reflection coefficients $\bar{S}_N^{-,+}$ and $\bar{S}_N^{+,-}$. Equation (7.26) is solved by backward recursion starting from $N=ND-1$ and $W_{ND}^{-}=0$; Equation (7.27) is solved by forward recursion starting from N=1 and $W_1^+=0$. In the resulting expressions for the wave amplitudes each term contains products of exponential factors of the type occurring in (7.22) - (7.25). In the $(\mathbf{x},t)$-domain these factors correspond to a non-vanishing time delay. Now, in practice, one only observes the acoustic wave motion in some finite time interval, the so-called time window of observation. From a certain number of factors in the products onward, the terms yield only contributions with a time delay that exceeds the duration of the time window of observation, and the relevant contribution can be ignored. Each product Υ exponential functions contains Υ interactions at the different interfaces; the value $\Upsilon=0$ yields the direct excitation of the acoustic wave at the source levels. The exponential functions contain the Laplace-transform parameter s only through the exponential functions occurring in (7.22) - (7.25). Further, the excitation factors contain s only via the source strengths. These properties play a fundamental role in the transformation back to the $(\mathbf{x},t)$-domain via the modified Cagniard method.

After the wave amplitudes have been determined, the transform-domain expressions for $\tilde{p}_N$ and $\tilde{v}_{3;N}$ follow from Eqs. (7.2)-(7.4), while $\tilde{v}_{1;N}$ and $\tilde{v}_{2;N}$ are, on account of (5.1), (6.1) and (6.2), expressed in terms of $\tilde{p}_N$ as

$$(7.28)\qquad \tilde{v}_{1;N} = \rho_N^{-1}\, j\alpha_1\, \tilde{p}_N,$$

$$(7.29)\qquad \tilde{v}_{2;N} = \rho_N^{-1}\, j\alpha_2\, \tilde{p}_N.$$

With this, the determination of the transform-domain acoustic wave quantities has been completed. Through the wave-amplitude formalism, they are expressed as a superposition of wave constituents that are generated by the sources and then undergo successive reflections and transmissions at the interfaces of the layered structure. Each element of the resulting expressions is denoted as a generalized-ray constituent Now, the transformation back to the space-time domain with the aid of the modified Cagniard method typically applies to the generalized-ray wave constituents. The relevant steps will be discussed in subsequent sections. As the general form of a generalized-ray constituent we take

$$(7.30)\qquad \tilde{w} = \hat{Q}_S(s)\Pi(\alpha_1,\alpha_2)\exp[-s \textstyle\sum_{\lambda\varepsilon\Lambda} \gamma_\lambda(\alpha_1,\alpha_2)h_\lambda],$$

where $\hat{Q}_S = \hat{Q}_S(s)$ is representative for the signature of the source, $\Pi = \Pi(\alpha_1,\alpha_2)$ is the (s-independent) factor that describes the coupling of the generalized-ray constituent to the sources as well as the reflections and transmissions it has undergone at the interfaces, h_λ is the total (possible multiple) vertical path length that the generalized ray has traversed in the domain D_λ, and γ_λ is the vertical slowness in D_λ, and $\Lambda \varepsilon \{1,\ldots,ND\}$.

8. THE TRANSFORMATION BACK TO THE SPACE-TIME DOMAIN

The first step in the transformation back to the space-time domain consists of applying (6.3) to (7.30). This yields

$$\hat{w}(\mathbf{x},s) = (s/2\pi)^2 \hat{Q}_S(s) \int_{\alpha_1=-\infty}^{\infty} d\alpha_1$$

$$\int_{\alpha_2=-\infty}^{\infty} \Pi(\alpha_1,\alpha_2)\exp[-s(j\alpha_1 x_1 + j\alpha_2 x_2 + \textstyle\sum_{\lambda\varepsilon\Lambda} \gamma_\lambda(\alpha_1,\alpha_2)h_\lambda)]d\alpha_2. \tag{8.1}$$

Next, we rewrite (8.1) as

$$\hat{w}(\mathbf{x},s) = s^2 \hat{Q}_S(s)\hat{G}(\mathbf{x},s), \tag{8.2}$$

where

$$\hat{G}(\mathbf{x},s) = (2\pi)^{-2} \int_{\alpha_1=-\infty}^{\infty} d\alpha_1$$

$$\int_{\alpha_2=-\infty}^{\infty} \Pi(\alpha_1,\alpha_2)\exp[-s(j\alpha_1 x_1 + j\alpha_2 x_2 + \textstyle\sum_{\lambda\varepsilon\Lambda} \gamma_\lambda(\alpha_1,\alpha_2)h_\lambda)]d\alpha_2, \tag{8.3}$$

is by definition the Green's function of the generalized ray. The modified Cagniard method now aims at rewriting, somehow, (8.3) as

$$\hat{G}(\mathbf{x},s) = \int_{\tau=T}^{\infty} \exp(-s\tau)w^G(\mathbf{x},\tau)d\tau, \tag{8.4}$$

where τ is a real variable of integration, $T>0$, and where $w^G(\mathbf{x},\tau)$ does not depend on s. Suppose that this has been achieved, then the uniqueness theorem of the Laplace transformation with real, positive, transform parameter (Lerch's theorem; see WIDDER, 1946) ensures that

$$G(\mathbf{x},t) = \begin{cases} 0 & \text{when } t<T, \\ w^G(\mathbf{x},t) & \text{when } t>T. \end{cases} \tag{8.5}$$

Additional theorems of the Laplace transformation then lead to the final space-time result

$$w(\mathbf{x},t) = \begin{cases} 0 & \text{when } t<T. \\ \partial_t^2 \int_{\tau=T}^{t} Q_S(t-\tau) w^G(\mathbf{x},\tau)d\tau & \text{when } t>T. \end{cases} \tag{8.6}$$

Equation (8.6) shows that T is the arrival time of the generalized-ray constituent at the point of observation.

In the rewriting of (8.3) in the form (8.4) several steps are carried out; they are characteristic for the modified Cagniard method. First, the variables of integration $\{\alpha_1,\alpha_2\}$ in (8.3) are changed into $\{p,q\}$ via the substitution

$$\alpha_1 = -jp\cos(\phi) - q\sin(\phi), \tag{8.7}$$

$$\alpha_2 = -jp\sin(\phi) + q\cos(\phi), \tag{8.8}$$

where ϕ follows from the polar-coordinate specification of the point of observation in the horizontal plane, i.e., from

$$x_1 = r\cos(\phi), \quad x_2 = r\sin(\phi), \tag{8.9}$$

with $r\geq 0$ and $0\leq\phi<2\pi$, $p\varepsilon I$ and $q\varepsilon R$. Since $d\alpha_1 d\alpha_2 = j^{-1}dpdq$ and $j\alpha_1x_1 + j\alpha_2x_2 = pr$, we have

$$\hat{G}(\mathbf{x},s) = (4\pi^2 j)^{-1}\int_{q=-\infty}^{\infty} dq \int_{p=-j\infty}^{j\infty} \bar{\Pi}(p,q) \exp[-s(pr + \textstyle\sum_{\lambda\varepsilon\Lambda} \bar{\gamma}_\lambda(p,q)h_\lambda)]dp, \tag{8.10}$$

where $\bar{\Pi}$ results from Π and $\bar{\gamma}_\lambda$ from γ_λ under the substitution (8.7)-(8.8). In the integration with respect to p we next continue the integrand analytically into the complex p-plane, away from the imaginary axis. In this process we encounter the singularities in $\bar{\Pi}$ and $\bar{\gamma}_\lambda$. In view of (7.16)-(7.21) and (7.5), these are the branch points of $\bar{\gamma}_\lambda$ i.e. $p = \pm S_\lambda(q)$, where

$$(8.11) \qquad S_\lambda = (1/c_\lambda^2 + q^2)^{1/2} > 0.$$

To make the analytic continuation single-valued, we introduce branch cuts along $\mathrm{Im}(p) = 0$, $|\mathrm{Re}(p)| \geq S_\lambda$. In the cut p-plane, we then have $\mathrm{Re}(\gamma_\lambda) > 0$. Further, it can be shown that the denominators in (7.16)-(7.21) never vanish in the finite part of the p-plane; hence, no poles occur.

Keeping q real, the integration in the complex p-plane is now carried out along a path along which

$$(8.12) \qquad pr + \sum_{\lambda\in\Lambda} \bar{\gamma}_\lambda(p,q)h_\lambda = \tau,$$

where τ is real and positive; such a path is denoted as a modified Cagniard path. Since $r \geq 0$, and $\mathrm{Im}(\bar{\gamma}_\lambda) < 0$ and > 0 in the upper and lower halves of the complex p-plane, respectively, the modified Cagniard paths are located in the right-half of the p-plane. It is clear that the part of the real p-axis $0 \leq \mathrm{Re}(p) < \min_{\lambda\in\Lambda}[S_\lambda(q)]$, $\mathrm{Im}(p) = 0$ satisfies (8.12). Further, there is a complex path that satisfies (8.12); it has the asymptotic representation

$$(8.13) \qquad p \sim \tau/(r \mp j\sum_{\lambda\in\Lambda} h_\lambda) \quad \text{as } \tau \to \infty,$$

in the upper/lower half of the p-plane. This part is denoted as the body-wave path; its representation in the first quadrant of the p-plane will be denoted as $p = p^B(\tau,q)$. Since the left-hand side of (8.12) satisfies Schwarz's reflection principle, the representation in the fourth quadrant is then $p = p^{B*}(\tau,q)$, where * denotes complex conjugate. The point of intersection of $p = p^B$ with the real p-axis follows from the consideration that at that point τ attains its minimum value. Let $p = p_0(q)$ denote the relevant (real) value of p, then we have

$$(8.14) \qquad r - p \sum_{\lambda\in\Lambda}(h_\lambda/\bar{\gamma}_\lambda) = 0 \quad \text{at } p = p_0(q).$$

Since $p = p_0(q)$ is necessarily located in between $p = 0$ and each of the branch points $p = S_\lambda(q)$ (all $\bar{\gamma}_\lambda$ with $\lambda\epsilon\Lambda$ must be real at p_0), we can write

$$(8.15)\qquad p_0(q) = S_\lambda(q)\sin[\theta_\lambda(q)] \quad \text{for all } \lambda\epsilon\Lambda,$$

with $0\le\theta_\lambda(q)\le\pi/2$. Since, then,

$$(8.16)\qquad \bar{\gamma}_\lambda = S_\lambda(q)\cos[\theta_\lambda(q)] \quad \text{at } p = p_0,$$

equation (8.14) can be rewritten as

$$(8.17)\qquad r - \textstyle\sum_{\lambda\epsilon\Lambda} h_\lambda \tan[\theta_\lambda(q)] = 0 \quad \text{at } p = p_0(q).$$

Let $\tau = \tau^B(q)$ at $p = p_0(q)$, then τ^B follows from (8.12) as

$$(8.18)\qquad \tau^B(q) = \textstyle\sum_{\lambda\epsilon\Lambda}\{S_\lambda(q)h_\lambda/\cos[\theta_\lambda(q)]\}.$$

In the first instance we replace the integration along $p\epsilon I$ in (8.10) by an integration along $p = p^B(\tau,q)$ and $p = p^{B*}(\tau,q)$. In view of Cauchy's theorem, Jordan's lemma, and the properties of $p = p^B(\tau,q)$, this is admissible provided that $\bar{\Pi} = \bar{\Pi}(p,q)$ has either no other singularities than $\{p = S_\lambda(q),\ \lambda\epsilon\Lambda\}$ or additional branch points (due to reflection against a layer in which the generalized-ray constituent under consideration does not propagate) that are outside the range $0\le p\le\min_{\lambda\epsilon\Lambda}[S_\lambda(q)]$. The generalized-ray constituent under consideration then only contains a body-wave part. If $\bar{\Pi} = \bar{\Pi}(p,q)$ does have an additional branch point in the indicated range, the body-wave part of the modified Cagniard path must, depending on the location of the point of observation with respect to the source, be supplemented by a loop integral around the branch cut belonging to this branch point and joining the points $p = p_0(q) - j0$ and $p = p_0(q) + j0$, where $p = p^{B*}(\tau,q)$ and $p = p^B(\tau,q)$, respectively, were tempted to cross the real p-axis. This part of the path of integration is denoted as the head-wave part, and its contribution to the generalized-ray constituent is called the head-wave contribution. The body-wave and the head-wave contributions to the generalized-ray constituent will be investigated in Sections 9 and 10, respectively.

9. BODY-WAVE CONTRIBUTION TO A GENERALIZED-RAY CONSTITUENT

The body-wave contribution to a generalized-ray Green's function follows from (8.10) and (8.12) as

$$\hat{G}^{B}(\mathbf{x},s) = (2\pi^2)^{-1} \int_{q=-\infty}^{\infty} dq \int_{\tau=\tau^{B}(q)}^{\infty} \exp(-s\tau)\,\mathrm{Im}[\bar{\Pi}(p^{B},q)(\partial p^{B}/\partial\tau)]\,d\tau, \tag{9.1}$$

where the parts along $p = p^{B}(\tau,q)$ and $p = p^{B*}(\tau,q)$ have been taken together, Schwarz's reflection principle (that also applies to $\bar{\Pi}$) has been used, and where the Jacobian of the transformation from p^{B} to τ follows from (8.12) as

$$\partial p^{B}/\partial\tau = [r - p^{B}\textstyle\sum_{\lambda\in\Lambda}(h_{\lambda}/\bar{\gamma}_{\lambda})]^{-1}. \tag{9.2}$$

Next, we interchange in the right-hand side of (9.1) the order of the integrations. This yields

$$\hat{G}^{B}(\mathbf{x},s) = \int_{\tau=\tau^{B}(0)}^{\infty} \exp(-s\tau)\,d\tau \; (2\pi^2)^{-1} \int_{q=-q^{B}(\tau)}^{q^{B}(\tau)} \mathrm{Im}[\bar{\Pi}(p^{B},q)(\partial p^{B}/\partial\tau)]\,dq, \tag{9.3}$$

in which $q = q^{B}(\tau)$ is the inversion of the mapping $\tau = \tau^{B}(q)$ for $q \geq 0$. (Note that, in view of (8.12), $\tau^{B}(q)$ is an even function of q, while differentiation of the left-hand side with respect to q shows that $\partial_q \tau^{B}(q) > 0$ if $q > 0$). Now, (9.3) is of the form of (8.4) and hence

$$w^{G;B}(\mathbf{x},t) = \begin{cases} 0 & \text{when } t<\tau^{B}(0), \\ (2\pi^2)^{-1} \displaystyle\int_{q=-q^{B}(\tau)}^{q^{B}(\tau)} \mathrm{Im}[\bar{\Pi}(p^{B},q)(\partial p^{B}/\partial\tau)]\,dq & \text{when } t>\tau^{B}(0). \end{cases} \tag{9.4}$$

With this, the body-wave contribution to the generalized-ray constituent follows as

$$(9.5)\qquad w^{B}(\mathbf{x},t) = \begin{cases} 0 & \text{when } t<\tau^{B}(0), \\ \partial_t^2 \int_{\tau=\tau^{B}(0)}^{t} Q_S(t-\tau)w^{G;B}(\mathbf{x},\tau)d\tau & \text{when } t>\tau^{B}(0). \end{cases}$$

Obviously, $T^{B}=\tau^{B}(0)$ is the arrival time of the body-wave contribution to the generalized-ray constituent.

10. HEAD-WAVE CONTRIBUTION TO A GENERALIZED-RAY CONSTITUENT

Let us consider the head-wave contribution that is due to the occurrence of $\bar{\gamma}_\mu = \bar{\gamma}_\mu(p,q)$ in $\bar{\Pi} = \bar{\Pi}(p,q)$, where $\mu\varepsilon\{1,\ldots,ND\}$, but $\mu\notin\Lambda$. Then, the integral along the body-wave contour must be supplemented by a loop integral around the branch cut associated with $\bar{\gamma}_\mu$ and joining the points $p_0(q) - j0$ and $p_0(q) + j0$ where $p = p^{B*}(\tau,q)$ and $p = p^{B}(\tau,q)$, respectively, were tempted to cross the real axis. Along this loop, too, the parametrization (8.12) has to be carried out. The relevant values of p in the upper and lower halves of the p-plane will be denoted by $p = p^{H}(\tau,q)$ and $p = p^{H*}(\tau,q)$, respectively. Let $\tau = \tau^{H}(q)$ denote the value of τ at $p = S_\mu(q)$, then (8.12) leads to

$$(10.1)\qquad \tau^{H}(q) = S_\mu(q)r + \textstyle\sum_{\lambda\varepsilon\Lambda}[S_\lambda^2(q) - S_\mu^2(q)]^{1/2}h_\lambda .$$

Now, for a head-wave contribution to occur, we must have $S_\mu(q)<p_0(q)$, or, using (8.15),

$$(10.2)\qquad S_\mu(q) < S_\lambda(q)\sin[\theta_\lambda(q)] \quad \text{with } \lambda\varepsilon\Lambda.$$

For those values of q where (10.2) is satisfied, we introduce the "critical angles" $\{\theta_\lambda^\mu(q);\lambda\varepsilon\Lambda\}$, with $0\le\theta_\lambda^\mu(q)\le\pi/2$, through

$$(10.3)\qquad \sin[\theta_\lambda^\mu(q)] = S_\mu(q)/S_\lambda(q).$$

Then, (10.2) implies

(10.4) $\sin[\theta_\lambda(q)] > \sin[\theta_\lambda^\mu(q)]$,

which in its turn implies

(10.5) $\tan[\theta_\lambda(q)] > \tan[\theta_\lambda^\mu(q)]$.

Using (10.5) in combination with (8.17), the condition $S_\mu(q)<p_0(q)$ is then equivalent to

(10.6) $r > \sum_{\lambda\epsilon\Lambda} h_\lambda \tan[\theta_\lambda^\mu(q)]$,

or

(10.7) $$r > \sum_{\lambda\epsilon\Lambda} h_\lambda \frac{S_\mu(q)/S_\lambda(q)}{[1-S_\mu^2(q)/S_\lambda^2(q)]^{1/2}} .$$

When q = 0, (10.7) reduces to

(10.8) $$r > \sum_{\lambda\epsilon\Lambda} h_\lambda \frac{c_\lambda/c_\mu}{(1-c_\lambda^2/c_\mu^2)^{1/2}},$$

which is the condition for "total reflection" against an interface of the layer D_μ in accordance with Snell's law of refraction at the other interfaces. Further, since

(10.9) $S_{\lambda,\mu}(q) \sim |q| + O(q^{-1})$ as $|q|\to\infty$,

equation (8.15) leads to

(10.10) $\theta_\lambda(q) \sim \theta + O(q^{-1})$ as $|q|\to\infty$,

and hence

(10.11) $p_0(q) \sim |q| \sin(\theta) + O(q^{-1})$ as $|q|\to\infty$.

Since $\sin(\theta)<1$, we always have $p_0(q)<S_\mu(q)$ from a certain value of $|q|$ onward, and hence (10.2) is, beyond this value, no longer satisfied. Let (10.2) be satisfied in the range $-Q_\mu<q<Q_\mu$, then the head-wave contribution to the generalized-ray Green's function can be written as

$$\hat{G}^H(\mathbf{x},s) = (2\pi^2)^{-1} \int_{q=-Q_\mu}^{Q_\mu} dq \tag{10.12}$$
$$\int_{\tau=\tau^H(q)}^{\tau^B(q)} \exp(-s\tau)\,\mathrm{Im}[\bar{\Pi}(p^H,q)(\partial p^H/\partial\tau)]d\tau,$$

where the parts along $p = p^H(\tau,q)$ and $p = p^{H*}(\tau,q)$ have been taken together, Schwarz's reflection principle has been used, and where the Jacobian of the transformation from p^H to τ follows from (8.12) as

$$\partial p^H/\partial\tau = [r - p^H \textstyle\sum_{\lambda\in\Lambda}(h_\lambda/\bar{\gamma}_\lambda)]^{-1}. \tag{10.13}$$

Next, we interchange in the right-hand side of (10.12) the order of the integrations. This yields

$$\hat{G}^H(\mathbf{x},s) = \int_{\tau=\tau^H(0)}^{\tau^B(0)} \exp(-s\tau)d\tau \tag{10.14}$$
$$(2\pi^2)^{-1} \int_{q=-q^H(\tau)}^{q^H(\tau)} \mathrm{Im}[\bar{\Pi}(p^H,q)(\partial p^H/\partial\tau)]dq$$
$$+ \int_{\tau=\tau^B(0)}^{T_\mu} \exp(-s\tau)d\tau$$
$$(2\pi^2)^{-1}\left[\int_{q=-q^H(\tau)}^{-q^B(\tau)} + \int_{q=q^B(\tau)}^{q^H(\tau)}\right]$$
$$\mathrm{Im}[\bar{\Pi}(p^H,q)(\partial p^H/\partial\tau)]dq,$$

in which $q = q^H(\tau)$ is the inversion of the mapping $\tau = \tau^H(q)$ for $q\geq 0$. (Note that, in view of (8.12), $\tau^H(q)$ is an even function of q, while differentiation of the left-hand side with respect to q shows that $\partial_q\tau^H(q)>0$ if $q>0$.) In the τ,q-plane, the curves $\tau = \tau^B(q)$ and $\tau = \tau^H(q)$ have the point $q = Q_\mu$, $\tau = T_\mu$ in common. This follows from the definition of Q_μ and T_μ. At this point they have, however, also a common tangent, as follows by differentiation of (8.12) with respect to q and taking (8.14) into account. Now, (10.14) is of the form of (8.4) and hence

(10.15)
$$w^{G;H}(\mathbf{x},t) = \begin{cases} 0 & \text{when } t<\tau^{H}(0), \\ (2\pi^2)^{-1}\displaystyle\int_{q=-q^H(\tau)}^{q^H(\tau)} \mathrm{Im}[\bar{\Pi}(p^H,q)(\partial p^H/\partial\tau)]dq & \text{when } \tau^H(0)<t<\tau^B(0), \\ (2\pi^2)^{-1}\Big[\displaystyle\int_{q=-q^H(\tau)}^{-q^B(\tau)} + \int_{q=q^B(\tau)}^{q^H(\tau)}\Big] & \\ \mathrm{Im}[\bar{\Pi}(p^H,q)(\partial p^H/\partial\tau)]dq & \text{when } T^B(0)<\tau<T_\mu. \end{cases}$$

With this, the head-wave contribution to the generalized-ray constituent follows as

(10.16)
$$w^{H}(\mathbf{x},t) = \begin{cases} 0 & \text{when } t<\tau^{H}(0), \\ \partial_t^2 \displaystyle\int_{\tau=\tau^H(0)}^{t} Q_S(t-\tau)w^{G;H}(\mathbf{x},\tau)d\tau & \text{when } t>\tau^H(0). \end{cases}$$

Note that, for (10.16) to occur, (10.8) must be satisfied, i.e. the horizontal offset between source and receiver must be large enough. Obviously, $T^H=\tau^H(0)$ is the arrival time of the head-wave contribution to the generalized-ray constituent.

11. NUMERICAL IMPLEMENTATION

Except for the simplest case where the generalized-ray constituent contains only a single exponential factor, the modified Cagniard path must be determined with the aid of numerical methods. First, for each q and given r and $\{h_\lambda;\lambda\in\Lambda\}$, (8.14) is solved for p_0. Using (8.12), the mappings $\tau = \tau^B(q)$ and $q = q^B(\tau)$ follow. Next, for each q and given r and $\{h_\lambda;\lambda\epsilon\Lambda\}$, (8.12) is solved for p^B in the range $\tau>\tau^B(q)$, and this value is used in the body-wave contributions. If head-wave contributions are present, (10.1) is used to obtain the mappings $\tau = \tau^H(q)$ and $q = q^H(\tau)$. Subsequently, (8.12) is used to construct the value of p^H that is to be used in the head-wave contribution. For the evaluation of the generalized-ray Green's functions, the integrations occurring in (9.4) and (10.15) have to be carried out numerically. This has to be done carefully because of the inverse square root singularity, due to $\partial p/\partial\tau$, at the end points of the interval of integration. The application of a local stretching procedure circumvents this difficulty. The final evaluation of the convolution integrals (9.5) and (10.16) presents usually no difficulty. Numerical results along these lines can be found in De Hoop and Van der Hijden (1983, 1984, 1985), Van der Hijden (1987), and Drijkoningen and FOKKEMA (1987)

APPENDIX

A. THE MODIFIED CAGNIARD METHOD FOR A GENERALIZED RAY THAT PROPAGATES IN A SINGLE MEDIUM ONLY

For a generalized-ray constituent that propagates in a single medium only, there is only a single exponential factor. The modified Cagniard path then follows from an expression of the form

(A.1) $$pr + [S(q)^2 - p^2]^{1/2}\, h = \tau,$$

in which

(A.2) $$S(q) = (1/c^2 + q^2)^{1/2} > 0.$$

The value of $p_0 = p_0(q)$ then follows from (cf.(8.14))

(A.3) $$r - p[S(q)^2 - p^2]^{-1/2}\, h = 0.$$

Equation (A.3) leads to

(A.4) $$p_0 = r\, S(q)/(r^2 + h^2)^{1/2}.$$

Substitution of (A.4) in (A.1) yields

(A.5) $$T^B = \tau^B(q) = S(q)(r^2 + h^2)^{1/2}.$$

Using (A.2), it follows from (A.5) that

$$\gamma^B = \gamma^B(\tau) = [\tau^2/(r^2 + h^2) - 1/c^2]^{1/2}. \tag{A.6}$$

Solving p from (A.1), we obtain

$$p^B = \frac{\tau r + jh[\tau^2 - T^B(q)^2]^{1/2}}{r^2 + h^2}, \tag{A.7}$$

from which the Jacobian of the transformation from p^B to τ is found as

$$\partial p^B/\partial\tau = \frac{r + jh\tau[\tau^2 - T^B(q)^2]^{-1/2}}{r^2 + h^2}. \tag{A.8}$$

Let, further, the presence of γ_μ in one of the reflection coefficients be responsible for the occurrence of a head wave. Then (cf.(10.1)),

$$T^H = \tau^H(q) = S_\mu(q)r + (1/c^2 - 1/c_\mu^2)^{1/2}\ h, \tag{A.9}$$

in which

$$S_\mu = (1/c_\mu^2 + q^2)^{1/2}. \tag{A.10}$$

Using (A.9) and (A.10), it follows that

$$q^H = q^H(\tau) = \{[\tau/r - (1/c^2 - 1/c_\mu^2)^{1/2}\ h/r]^2 - 1/c_\mu^2\}^{1/2}. \tag{A.11}$$

Solving p from (A.1), we now obtain

$$p^H = \frac{\tau r - h[T^B(q)^2 - \tau^2]^{1/2}}{r^2 + h^2}, \tag{A.12}$$

from which the Jacobian of the transformation from p^H to τ is found as

$$\partial p^H/\partial\tau = \frac{r + h\tau[T^B(q)^2 - \tau^2]^{-1/2}}{r^2 + h^2}. \tag{A.13}$$

The end point of the q-interval in which a head-wave contribution is present, follows as

(A.14) $Q_\mu = [(1/c^2 - 1/c_\mu^2)r^2/h^2 - 1/c_\mu^2]^{1/2}$,

which leads to

(A.15) $T_\mu = (r^2 + h^2)(1/c^2 - 1/c_\mu^2)^{1/2}/h.$

REFERENCES

Achenbach, J.D., 1973, Wave propagation in elastic solids, Amsterdam, North-Holland, p. 298.

Aki, K., and P.G. Richards, 1980, Quantative seismology, San Francisco, Freeman, p. 224.

Cagniard, L., 1939, Réflection et réfraction des ondes séismiques progressives, Paris, Gauthier-Villars, 255 pp. (Translated and revised by FLINN, E.A. and C.H. DIX, Reflection and refraction of progressive seismic waves, New York, McGraw-Hill, 1962, 282 pp.)

De Hoop, A.T., 1960, "A modification of Cagniard's method for solving seismic pulse problems", Applied Scientific Research, B8, 349-356.

De Hoop, A.T., 1961, "Theoretical determination of the surface motion of a uniform elastic half-space produced by a dilatational, impulsive, point source", Proceedings Colloque International C.N.R.S. No. 111, Marseille, 21-31 (in English).

De Hoop, A.T., and Van der Hijden, J.H.M.T., 1983, "Generation of acoustic waves by an impulsive line source in a fluid/solid configuration with a plane boundary", Journal of the Acoustical Society of America, 74, 333-342.

De Hoop, A.T., and Van der Hijden, J.H.M.T., 1984, "Generation of acoustic waves by an impulsive point source in a fluid/solid configuration with a plane boundary", J. Acoust. Soc. Am., 75, 1709-1715.

De Hoop, A.T., and Van der Hijden, J.H.M.T., 1985, "Seismic waves generated by an impulsive point source in a solid/fluid configuration with a plane boundary", Geophysics, 50, 1083-1090.

Drijkoningen, G.G., and J.T. Fokkema, 1987, "The exact seismic response of an ocean and a N-layer configuration", Geophysical Prospecting, 35, 33-61.

Helmberger, D.V., "The crust-mantle transition in the Bering Sea", Bulletin of the Seismological Society of America, Vol. 58, 1968, pp. 179-214

Miklowitz, J., 1978, "The theory of elastic waves and waveguides", Amsterdam, North Holland, p. 302.

Roever, W.L., T.F. Vining, and E. Strick, 1959, "Propagation of elastic wave motion from an impulsive source along a fluid/solid interface", Philsophical Transactions of the Royal Society of London, Series A, No. 1000, Vol. 251, 455-523.

Spencer, T.W., "The method of generalized reflection and transmission coefficients", Geophysics, Vol. 25, 1960, pp. 625-641.

Van der Hijden, J.H.M.T., 1987, Propagation of transient elastic waves in stratified anisotropic media, Amsterdam, North-Holland Publishing Company.

Whittaker, E.T. and G.N. Watson, "A course of modern analysis", Cambridge, Cambridge University Press, 1950, 4th ed., p. 115

Widder, D.V., 1946, The Laplace transform, Princeton, Princeton University Press, p. 63.

Wiggins, R.A., and D.V. Helmberger, 1974, "Synthetic seismogram computation by expansion in generalized rays", Geophysical Journal of the Royal Astronomical Society, 37, 73-90.

COHERENT FRAMES AND IRREGULAR SAMPLING

Hans G. Feichtinger

Department of Mathematics
University of MARYLAND, College Park [1]
MD, 20742 , U S A

Abstract. It is the purpose of this short note to highlight perhaps unexpected connections between the topics described in the papers [G2] *and* [HW1] *in this volume, and at the same time between various papers within a series of joint publications* [FG1-8] *with K.Groechenig on the two topics indicated in the title: Algorithms that allow to recover a function (or tempered distribution) f from a suitable family of coefficients, which arise as integrals of f against a countable coherent family of functions (such as Heisenberg or affine frames) and the problem of reconstructing a band-limited function from a (sufficiently rich) set of irregularly taken sampling values. As will be pointed out in detail below the basic observation, which may be taken as an explanation of most common results for these two settings, concerns properties of functions which arise as convolution products with nice, integrable functions on a locally compact group.*

INTRODUCTION

The paper is going to describe in which sense the constructions of frames (as described in [DGM],[HW1,2],...) can be seen as a result related to the classical Shannon sampling problem. In both cases the "classical" approach puts its main emphasis on the use of Hilbert spaces methods (such as orthonormal bases, frames, positive operators), using various formulas (such as Poisson's formula, expansion of functions into Fourier series, ..), based on the fact that the sampling points or the parameter sets involved in describing these frames form a regular lattice.

[1] Permanent address: Dept.Math.,Univ.Vienna,Strudlhofg.4. A-1090 Wien, AUSTRIA

J. S. Byrnes and J. F. Byrnes (eds.), Recent Advances in Fourier Analysis and Its Applications, 427–440.

As will be pointed out a *group theoretical approach*, using convolution relations between suitable function spaces on the Euclidean space $\mathbb{R}^m$ or suitable locally compact groups naturally associated with these frames gives a much more flexible tool, which allows much more freedom in the choice of "mother wavelets" and sampling coefficients than one might expect from the "regular" methods. Also, these methods (which are iterative ones) are not restricted to the Hilbert space setting (thus a number of other function spaces, including many classical ones, such as Besov spaces,...) can be covered in the discussion. Finally, the methods allow error and stability analysis with respect to the "right" norms, i.e. whenever a function belongs to a reasonable normed and invariant function space, that the error with respect to the norm of this space can be estimated .

As giving background information and pointing out connections is the main concern of this note we only state few of the main results of the two topics under discussions, in a non-technical (hence not most general) form. Thus we shall restrict the discussion here to spaces related to (unweighted) L^p-spaces, because in this setting the relevant theory is already available in the literature (cf. [FG1,5]) and does not require complicated technical explanations.

We recall that for $1 \le p < \infty$ the space $L^p(\mathbb{R}^m)$ consists of all measurable functions F on $\mathbb{R}^m$ such that $\|F\|_p := [\int_{\mathbb{R}^m} |F(x)|^p dx\,]^{1/p} < \infty$. We use the symbol T_y for the translation operator, given by $T_yF(z) := F(z-y)$). We under-stand the Fourier transform $\mathcal{F}F=\hat{F}$ of an integrable functions $F \in L^1(\mathbb{R}^m)$ is given by the usual integral formula. It is well known that the Fourier transform $\mathcal{F}$ may be extended to a mapping from $L^p(\mathbb{R}^m)$ for $1 \le p \le 2$, whereas one has use distribution theory (tempered distributions) in order to define the Fourier transform for $p>2$ (cf. [F4] for an alternative) .

Thus the spectrum of a function is well defined as the support of the Fourier transform: $\mathrm{spec}(F) := \mathrm{supp}(\hat{F})$. A function in L^p is called band-limited (to some bounded set Ω) if $\mathrm{spec}(F) \subseteq \Omega$. Thus $\mathrm{spec}(F) \subseteq \Omega$ if $\hat{F}$ vanishes on the complement of Ω. It is also well known that band-limited functions are continuous (by the Paley Wiener theorem in fact analytic), so pointwise evaluations make perfect sense for these functions.

We start with the description of the most simple version of our problem, the so-called (regular) sampling problem. We first recall it in a slightly more general form than usually presented in textbooks (cf.[G2] for more details, and see [BSS],[H],[J] for surveys of this topic and its relevance).

REGULAR SAMPLING

<u>Theorem :</u> There exists some positive constant C_o (only depending on the definition of the Fourier transform involved) such that for any $C > 0$ the following is true: Given any $\gamma \le C_o C^{-1}$ (the sampling rate), such that any band-limited function $F \in L^p(\mathbb{R}^m)$, for $1 \le p < \infty$, with $\hat{F}(t) = 0$ For $|t| \ge C$ can be completely recovered From the sampling values $(F(\gamma\cdot n))_{n\in\mathbb{Z}^m}$. Actually, there is an explicit formula, telling us that

$$F = \sum_{n\in\mathbb{Z}^m} F(\gamma\cdot n)\, T_{\gamma\cdot n}G \quad (*)$$

whenever G is a band-limited and integrable function, satisfying $\hat{G}(t) \equiv 1$ on Ω, with $\hat{G}$ having sufficiently small support. Moreover, the series converges in the pointwise sense (uniformly over bounded sets) as well as in the L^p-sense.

Proof. Although this result can be easily obtained as a minor modification of the usual Shannon sampling theorem we have no simple reference available. Thus we indicate the main arguments (which actually work also in the setting of weighted L^p-spaces) to be used (besides Poisson's formula, of course):

First we observe that for band-limited functions (with bounded spectrum) in L^p the sampling values $(F(\gamma\cdot n))_{n\in\mathbb{Z}^m}$ are p-summable, in fact $\sum_{n\in\mathbb{Z}^m}|F(\gamma\cdot n)|^{p\ 1/p} \le C\cdot\|F\|_p$ for all $f\in L^p(\mathbb{R}^m)$ with fixed compact spectrum (due to Theorem 3.7.i) and Proposition 3.4. in [FG7], cf. also Remark 2.5 in [FG5]). Using now some notations from [FG5] we may interpret the series as a convolution of the discrete measure $\sum_{n\in\mathbb{Z}^m} F(\gamma\cdot n)\,\delta_{\gamma\cdot n}$ (which is convergent in ML^p) with the function G. Since we have assumed $G \in L^1$ to be band-limited, the local maximal function $G^\#$ also belongs to L^1 and thus the convolution product in (*) is convergent in CL^p, in particular in L^p and locally uniform over compact sets (by Theorem 2.1.iii) and ii) of [FG5]).

Finally we mention, that the dependency of the necessary sampling density from the band-width (size of the spectrum of f = support of f) is easily obtained by proving the result first for Ω being the unit cube and by reducing the problem to that case by means of dilations in the general case.

Remark: For irregular (we should call them quasi-regular here) lattices, which can be obtained from a regular one by the application of a regular matrix, such as rotation or (anisotropic) dilation the same arguments go through.

IRREGULAR SAMPLING

The proof of this so-called Whittaker-Shannon Sampling theorem on regular sampling is usually based on Poisson's formula (in a more or less disguised way). Since there is no substitute for Poisson's formula for irregular sets (this has been shown recently by Cordoba) it is clear that one has to use a completely different approach in the irregular setting. In our papers [FG5-8] we were able to show that full reconstruction is still possible (even if the sampling points are completely irregular)supposing that the sampling density (which is defined in the most natural way) is high enough (in relationship to the size of the spectrum). Several variants of a possible constructive approach are described in more detail in [G2] and [FG5-8]) have been found. One version of the resulting theorems reads as follows:

<u>Theorem :</u> Let Ω be some compact subset of $\mathbb{R}^m$. Then there exists an open neighborhood U (only dependent of Ω) , such that for any U-dense, discrete family $\Lambda = (\lambda_i)_{i\in I}$ of points in $\mathbb{R}^m$ (which means that the system $(\lambda_i + U)_{i\in I}$ covers $\mathbb{R}^m$) any band-limited function $F \in L^p(\mathbb{R}^m)$ with $\mathrm{supp}(\hat{F}) \subseteq \Omega$, can be completely reconstructed from the irregular sampling values $(F(x)_i)_{i\in I}$.

As a starting point for our arguments we took the following observation: A function $F \in L^p(\mathbb{R}^m)$ is band-limited if and only if there exists some other function $G \in L^1(\mathbb{R}^m)$ (which may be assumed to be band-limited itself) such that $G*F = F$. In fact, if we choose $G \in L^1(\mathbb{R}^m)$ such that $\hat{G}$ equals 1 over spec(F) (e.g. some De la Vallée-Poussin kernel), then $\hat{F} = \hat{G}\cdot\hat{F}$. Assume conversely that this holds true for some $G \in L^1(\mathbb{R}^m)$ and some F with unbounded spectrum. Then $\hat{G}$ has to take the value 1 at points arbitrarily far from the origin, which is not possible in view of the Riemann Lebesgue Lemma (stating that the Fourier transform of an integrable functions tends to zero at infinity).

Actually, we were using this reproducing convolution equation for F as our starting point (also not relying on methods from complex analysis). At present there are different methods and algorithms, using this equation in order to recover F from it's sampling values.

The first approach (described in [F3] for weighted L^1-spaces, and in [FG5] for the most general situation, including weighted L^p-spaces) was putting the main emphasis on an iterative approach. The main idea being the following: Use the given information (the sampling values), to approximate the function by some auxiliary function, involving only the sampling values. This can be either (in the most simple case) be a step function (taking the value $F(x_i)$ near x_i (cf. $V_X F$ in [FG6]), or some more sophisticated spline type function ($Sp_\Psi F$, used in [FG5]). Since a band-limited function is quite smooth we may expect that such a function will yield a good approximation to our given function F. But how to recover the remainder? We cannot directly take the remainder term and iterate, because that remainder term does not have any smoothness (at least if we use step functions). In order to circumvent this problem we smooth that first and simple approximation out, in order to get a smooth remainder term, which allows to apply the same procedure over and over and to recapture finally F completely from the sampling values given. In fact, since F itself is not changed by convolution with G we may expect that any approximation to F will still be a good approximation to F after convolving it by G. Altogether, one only has to take care that in this procedure the loss of approximation (due to that additional smoothing) and the win (obtained by having a smoother remainder term for the next step) are in a good balance (so we reduce our approximation quality by investing a small part of it into additional smoothness which in return allows us to keep the machinery running).

The theorem above actually applies to more general situations, based on the same arguments. Thus convergence can be shown to hold true with respect to certain weighted L^p-norms if F belongs to these classes (which simply means that both decay properties and summability properties of f imply better convergence of the series near infinity). In the argument G has to be taken in a suitable weighted L^1-space in that case. The presentation in [FG6] gives the shortest available proof for a (still general) special case. In [FG7] an operator theoretic setting has been choosen to describe the most general version. This is also the basis for the error estimates given in [FG8].

An interesting variant of this result, which might me of great practical use, is the following one (cf. the section on error analysis):

<u>Corollary.</u> In the above situation there exists a bounded family $(E_i)_{i\in I}$ in L^1 (actually uniformly decaying, continuous functions) such that any $F \in L^p(\mathbb{R}^m)$ with $\mathrm{spec}(F) \subseteq \Omega$, $1 \le p < \infty$, can be written as

$$F = \sum_{i\in I} F(x_i)E_i ,$$

with unconditional convergence in $L^p(\mathbb{R}^m)$ and uniformly over compact sets.

FRAMES

Coming to the next topic now we have to speak about <u>coherent frames.</u> Actually, the concept of a frame in a Hilbert space is a classical one (cf. [DS] for a basic account of properties of frames, and [Y] for applications in the theory of non-harmonic Fourier series). It underwent a glorious revival after the discovery of the "wavelet orthonormal system" by Y.Meyer (cf. [M1,2], [LM], [D2].. for details). These are orthonormal systems for $L^2(\mathbb{R}^m)$ having the useful property of being generated from a single function g (called the mother wavelet) by means of elementary operations only, i.e. translations by elements from the lattice $\mathbb{Z}^m \subseteq \mathbb{R}^m$ and dilations (of the argument by powers of 2). Thus the typical basis vector is given by $g_{k,n}(x) := 2^{-m/2}g((x-k)/2^m)$. We consider this family as a <u>coherent</u> one, coherence being understood as the possibility of applying certain elements from a (continuous) group of transformation to one single function in order to obtain the full system. In the wavelet case these transformations are of course unitary transformation induced on $L^2(\mathbb{R}^m)$ induced by the the affine transformations of the argument. We only mention in passing that the notion of "coherent states" is an important concept in quantum mechanics and is the place where this terminology comes from (see [HW1,2] for details).

The observation, that under certain circumstances (e.g.the Heisenberg setting) the requirement of orthonormality turns out to be too strong in companion with coherence, smoothness and decay properties apparently lead to the discussion of frames. First the so-called tight frames came up. These are families $(g_i)_{i\in I}$ in $L^2(\mathbb{R}^m)$ which behave very much like an orthonormal system without being a basis in the usual sense. By definition a <u>tight frame</u> (cf.[DGM]) is identified by the property that $\|f\|_2 = (\sum_{i\in I}|\langle f,g_i\rangle|^2)^{1/2}$. As a simple functional analytic consequence one obtains $f = \sum_{i\in I}\langle f,g_i\rangle g_i$. Thus, in order to expand a given function f one only has to calculate the coefficients $\langle f,g_i\rangle$ and to write the above sum, as if we had an orthonormal bases. As the most simple example of a coherent tight frame one may take as a Hilbert space $\mathbb{R}^2$, and (up to normalization) the unit vectors $(e_1,e_2,-e_1,-e_2)$, which are coherent, because they aris under the action of rotation by multiplies of 90° from e_1. This set is of course no basis (so expansions are not unique), however we are free to require additional properties on the coefficients (in this simple case we could ask for positivity).

An important fact concerning these tight frames is the following. All information about a function is contained in these coefficients, and one can say, if one has an affine tight frame (arising as above by translates and dilates only), that the coefficients $\langle f, g_{k,n} \rangle$ for fixed an (and running k) correspond to the contribution to f at a scaling (or smoothness) level n, also indicating where (depending on the indices k for which large coefficients arise) in the function f relevant contributions at this fixed level prevail. It allows also to find partial sums of the double sum, involving only comparatively few terms and representing nevertheless the function to a variable degree at different places (very precisely on some interval, and only roughly at infinity, for example).

This requirement of storing all information in coefficients (together with the possibility of recovering f "easily" from these coefficients) can also be considerably weakened. The general notation of a frame $(g_i)_{i \in I}$ in the Hilbert space $L^2(\mathbb{R}^m)$ only requires to have constants $A, B > 0$ such that

$$A \cdot \|f\|_2 \le (\Sigma_{i \in I} |\langle f, g_i \rangle|^2)^{1/2} \le B \cdot \|f\|_2 \quad \text{for all} \quad f \in L^2(\mathbb{R}^m) .$$

It is then still possible to recover f by means of an iterative procedure (described in detail in the referred papers).

It was Gröchenig who extended the notion of a frame to a larger class of spaces (so making it a Banach spaces concept as opposed to a Hilbert space notion which is was up to that time). In fact, if we restrict our attention to coherent frames (with few technical restrictions) then it is possible to speak of a frame for a Banach space of functions (actually a coorbit space) by means of an associated Banach space of sequences, if the coefficient mapping allows to embed the Banach space in a complemented way into that sequence space. So, in the case of the Hilbert space the associated Banach space if of course the expected one, namely the space of square summable sequences l^2. Again, there are in this setting norm convergent (iterative) methods to recover the function from its coefficient.

The following turn-around brings us to the heart of the matter, showing that the question of coherent frames (as described in [DGM], [HW], ..) is close related to the irregular sampling problem.

As we mentioned, coherent systems arise from a (continuous) group of operators. We formalize the setting by assuming the we have (unitary) operators T_x, the index x being taken from some group $\mathcal{G}$, fulfilling the rules $T_{x^{-1}} = T_x^*$ (the adjoint corresponds to the inverse group element) and $T_x \circ T_y = T_{x \cdot y}$ (i.e. a group representation). Usually $\mathcal{G}$ can be thought of as a matrix group, so we may use for the wavelets (we describe for simplicity only the case $m=1$) the group of affine transformation of the real line, given by $x \longmapsto ax+b$, $a > 0$, $b \in \mathbb{R}$, and $T_{(a,b)}$ associated with $x=(a,b) \in \mathcal{G}$ by:

$$f \longmapsto T_{(a,b)}f, \text{ with } T_{(a,b)}f(x) := a^{-m/2}g((x-b)/a).$$

Fixing now some "nice" function g_o we can define the following function on the group $\mathcal{G}$ (depending on your background you may call it either a representation coefficient of the above representation, or simply a <u>continuous wavelet transform</u>, also called <u>generalized wavelet transform</u>, if the domain is not only an Hilbert space, but some space of distributions :

$$f \longrightarrow F, \quad \text{with } f(x) := \langle f, T_x g \rangle .$$

In this situation F is a continuous function on the group $\mathcal{G}$. Since any locally compact group carries a unique (left) translation invariant measure, the Haar measure on $\mathcal{G}$, the notion p-integrability with respect to that Haar measure ($F \in L^p(\mathcal{G})$) makes sense, and actually F is square integrable for the elements $f \in L^2(\mathbb{R}^m)$. Also, convolution make sense, being defined by $F*G(x) = \int_{\mathcal{G}} G(y^{-1}x)F(y)dy$ for decent functions F,G. Moreover, as in the case of $\mathbb{R}^m$ we have estimates of the form $\|G*F\|_p \le \|G\|_1 \|F\|_p$, but one has to careful (if the group is not unimodular, i.e. if left and right invariant Haar measures are not the same for $\mathcal{G}$) that one has in general only $\|F*G\|_p \le \|G\|_{1,w}\|F\|_p$, for some weighted L^1-norm $\| \ \|_{1,w}$. Having a "nice" function means now to assume that the function G (i.e. the one associated with g itself by the mapping to induced via g) is integrable over $\mathcal{G}$. In practice this corresponds to decay and/or smoothness or moment conditions on g (cf. [FG1] for details). Moreover, it is possible to recover f from F by some inversion formula (involving again g).

Speaking now about about having a coherent frame of the form $(T_{x_i})_{i \in I}$ turns out to be equivalent (if appropriately defined) to say: f is completely determined by the sampling values of F at the points of $X = (x_i)_{i \in I}$ in $\mathcal{G}$, if $F \in L^p(\mathcal{G})$, for example. Looking at the examples of the wavelet theory and other papers on (tight) frames (cf. [DGM], [FJ], ...) one might get the impression that the sets X arising there have to be lattices in the group (not in the sense of discrete subgroups with compact quotient, but still lattices in a very strict geometric sense). This is actually useful for labeling the points by double indices and allows to use certain formulas, but at the same time restricts the flexibility of choosing X). So what what about sampling information from irregular sets X? Or what can be said about continuous wavelet transforms based on wavelets g satisfying no special properties. The answer turns out to by quite similar to the above one on band-limited functions: Given g (which somehow corresponds to the information of the size of the spectrum in the band-limited case) we can find some neighborhood U of the neutral element in the group (the size of U depends only on certain smoothness and decay properties of G, not on p) such that any U-dense family X (which now means of course that the family $x_i \cdot U$ covers $\mathcal{G}$) gives rise to a coherent frame. In other words, f can be completely be recovered from the coefficients $\langle f, T_{x_i} g\rangle, i\in I$ in this case if $f\in\mathcal{C}o(L^p)=\{f\,|\,F\in L^p(\mathcal{G})\}$.

BASIC CONNECTIONS

In order to reveal now the connection between the two topics and to to stress at the same time the importance of coherence the and of the group theoretic background in this theory we add the following informations (cf.[FG1,2] for details, or [F4]). Under the conditions described (i.e. g to be "nice") we actually have the following facts:
G belongs to the weighted L^1-space $L^1_w(\mathcal{G})$, and the p-integrable functions F arising as continuous wavelet transforms via g can be characterized (among all elements in $L^p(\mathcal{G})$ by the reproducing convolution equation

$$F = F * G .$$

In particular, G is a (symmetric) convolution idempotent, hence the mapping $H \longmapsto H * G$ from $L^p(\mathcal{G}))$ into itself is a projection mapping. This it is not completely unlikely (and as has been shown in fact true) that one can recover F from its sampling values by pretty much the same type of arguments that one can use in the irregular reconstruction problem. Of course, the concept of uniform density of the sampling points X has to be understood in the group theoretical sense now (thus with respect to the hyperbolic metric in the case the upper half plane, identified with the group of affine transformation, the so-called ax+b group), smoothness of F is not described in terms of derivatives (since we do not assume to have a Lie group structure, which however would be available in most examples), but using a suitable family of spaces (the so-called Wiener type spaces) or concepts of local maximal functions and oscillations over groups allows to follow similar patters in both settings at this point (they were introduced in [F1] in full generality, and the underlying method is that of splitting functions into equal parts.

IRREGULAR SAMPLING OF THE SHORT TIME FOURIER TRANSFORM

Although the question of recovering a function from irregular sampling of its STFT (Short Time Fourier Transform) looks more like a problem similar to the sampling problem of band-limited functions at a first sight (Fourier transform being involved), it turns out to be related to the sampling problem of generalized wavelet transforms over the Heisenberg group.

Recall that for a function f (assume for simplicity in $L^2(\mathbb{R})$) the STFT with respect to the window function g (which we will assume to be nice, in the sense of compactly supported and continuous, for example) is defined by $STFT_g(f)(x,s) := \mathcal{F}((T_x g)f)(s)$. A priori, it is understood as a (bounded an continuous) function on $\mathbb{R}^2$ (the so-called <u>time-frequency plane</u>). A reconstruction method, allowing to reconstruct f (given g) from the sampling values of $STFT_g(f)$ at a sufficiently small regular lattice in $\mathbb{R}^2$ was given by [B]. In order to obtain the irregular sampling theorem in this situation we indicate only shortly here how to reinterpret the STFT. In fact, the so-called Heisenberg group acts in a very natural way on $L^2(\mathbb{R})$ by means of translation operators plus modulation operators (cf. [FG1],[G1],[HW1,2]) and (in order to get the group law) multiplications with complex numbers of absolute value one (so the <u>Heisenberg group</u> as a set is just $\mathbb{R}^2 x \mathbb{T}$, where $\mathbb{T}$ denotes the torus group). In an abstract setting this action is defined as the Schrödinger representation of the Heisenberg group. Now looking at the definition of the generalized wavelet transform with respect to this representation (cf. [GF1], Ex.7.1) one can check that for any function f (fixing the analyzing wavelet g) the function $STFT_g(F)$ coincides with the function $F(x,t,1)$. Since $F(x,t,u) = u \cdot F(x,t,1)$ knowledge about the sampling values of $STFT_g(f)$ at sufficiently many points in the time frequency plane gives also sufficient information about F and allows therefore to reconstruct F (and therefrom f) by means of an iterative procedure. Besides the irregularity of the sampling points for $SFTF_g(f)$ (the density depends only on g, not on f !) it is remarkable, that for this reconstruction only minimal requirements on g have to be made.

STABILITY AND ERROR ANALYSIS

The group theoretical approach to coherent frames also allows to cope with questions of stability and to give error estimates in the right norms. Thus, typically one can handle (using the appropriate function spaces on the acting group $\mathcal{G}$) questions of the following type ("jitter error problem"):

To what extent does the reconstruction deliver a wrong variant of f (the difference being measured with a very wide range of norms), if the sampling values (or the coefficients are not picked up by means of the functions $T_{x_i}g$, as assumed, but instead by some "close by" functions such as $T_{y_i}g$, with elements y_i being close to the elements x_i (in the sense that there is a small neighborhood V t of the neutral element in $\mathcal{G}$ such that $y_i \in x_i V$ for $i \in I$).

The stability considerations are of interest in connection with the description of the reconstruction methods in the form $F = \sum_{i\in I} F(x_i)E_i$ (cf. Corollary above; see [F5],Cor.A', [F6],Cor.4.3, or [F7], Theorem1) or $f = \sum_{i\in I}\langle f, T_{x_i}g\rangle e_i$ in the frame setting (cf.[G1],5.10), with suitable families of functions E_i on $\mathcal{G}$ or e_i on $\mathbb{R}^m$. These functions can be precalculated (given only the information about the sampling geometry, i.e. the set of points X, and the reproducing function G), which is an aspect of possible great practical use (in a situation where one has to recover many functions with equal smoothness (in the sense of having the same reproducing function G)from their respective sampling values, taken over the same set of sampling points. In this situation the a priori calculation of these functions E_i could be done on a main-frame, whereas the stored values of the functions E_i could be used to recover F by simple summation (without going into the iterative procedure for each F individually).

In this setting of course the question comes up, to what extent does the particular form of the functions E_i depend on the parameters involved. So, stability in this setting means, that minor changes of the sampling geometry or starting the procedure with a function G_1 only slightly different from a given function G results in an overall system $(E_i^1)_{i\in I}$ of (reconstructing) functions which are in some uniform sense not far from the original functions $(E_i)_{i\in I}$.

LOOKING BACK

In contrast to the presentation given above our approach to both problems (within the last three years) went the other way round. We first worked on atomic decompositions in functions spaces (one should speak perhaps better of non- orthogonal series representations with respect to coherent systems of atoms), which was based on the convolution equation $F = F*G$ and the transfer method available through the generalized wavelet transform (associated with integrable group representations, as pointed out in [FG1-3]). Observing then the connection to the theory of frames, Groechenig pushed the theory of frames to the same general setting (not anymore restricted to Hilbert spaces or regular lattices) which has turned out to be the most general and most natural one for atomic decompositions along coherent systems. [G1]. During these discussions we found the explained analogy with the situation arising in the context of band-limited functions an inspiring background. Working then on the irregular sampling problem we also found that this point of view allows to develop parallel the reconstruction problem methods of expanding band-limited function into a series of functions, each of them obtained from a single function g using only certain of it's translates (cf. [FG6], and [FG7] for the most general version of these results), which again is intimately related to the general atomic decomposition problem solved in [FG2].

Our first hope, that the technical background would be much easier on $\mathbb{R}^m$ turned out to be misleading in one very important point: Whereas one has plenty of integrable idempotents (for convolution) on any group having an integrable and irreducible representation (such as the affine group, the Heisenberg group, but also the $SL(2,\mathbb{R})$ group, mentioned in [FG1]) there are no such function on $\mathbb{R}^m$ ($g \in L^1$ cannot be an idempotent with respect to convolution due to the Riemann Lebesgue Lemma). On the other hand, particular choices of function G satisfying $F = F*G$, such as Schwartz functions G, allow to invoke more elegant estimates in order to verify convergence of the iterative reconstruction method, not available on groups.

SUMMARY

We have tried to point out that from an harmonic analyst's point of view there is a close relationship between previous constructions of coherent (tight) frames or orthonormal wavelet bases and to the regular sampling problem for band-limited function. The same kind of relationship can be found between the results, saying that any nice function can be used to construct a frame as long as the sampling rate on the group is high enough, and the reconstruction of band-limited functions in L^p-spaces from irregular sampling values. The background is in each case a convolution equality and the fact, that it is possible (under various circumstances) to recover functions F satisfying $F = F*G$, for some nice and integrable G, from any family of sampling values, as long as the sampling density of these points is high enough. In the case of coherent frames this strategy (which applies to the continuous wavelet transform) has to be complemented with a transfer method (allowing to transfer questions about f to questions about an appropriate continuous wavelet transform and transferring the result back to f), but this can be done using the machinery of coorbit spaces and the appropriate inversion formulas given in [FG2]. Moreover, since the iterative methods involved in this process work for a large variety of translation invariant function spaces on any locally compact group the range of application of this technique, which also allows good error estimates and the verification of a variety of stability results, is not restricted to the Hilbert space setting.

REFERENCES

[B] Bastiaans, M.J.: Signal description by means of local frequency spectrum SPIE 373, Trans.in Opt.Sign.Proc. (1981), 49-62.

[BSS] Butzer,P. W.Splettstößer and R.L.Stens (1988): The sampling theorem and linear prediction in signal analysis. Jber.d.Dt.Math.-Ver. **90**, 1-70.

[D1] Daubechies, I: The wavelet transform, time-frequency localization and signal analysis.

[D2] - " - : Frames of coherent states, phase space localization, and signal analysis, preprint.

[D3] - " - : Orthonormal bases of compactly supported wavelets, Comm.Pure Appl.Math. 41 (1988), 909-996.

[DG] Daubechies, I. and A.Grossmann: Frames in the Bargmann space of entire functions.

[DGM] Daubechies, I., A. Grossmann and Y.Meyer: Painless on-orthogonal expansions. J.Math. Phys. 27 (1986), 1271-1283.

[DS] Duffin,R; and A.Schaeffer : A class of nonharmonic Fourier series. Trans. Amer.Math.Soc. 72 (1952), 341-366.

[F1] Feichtinger H.G.: Banach convolution algebras of Wiener's type. "Functions, Series, Operators", Proc.Conf., Budapest 1980, Coll. Soc.Janos Bolyai, North Holland, 1983, 509-524.

[F2] - " - : Discretization of convolution and reconstruction of band-limited functions from irregular sampling. J.Approx.Theory, to appear.

[F3] - " - : Nonorthogonal expansions using the Heisenberg group, with applications to signal analysis. Techn.Report, prepared for a talk presented at AT&T Bell Labs.. Murray Hill, April 1988.

[F4] - " - : An elementary approach to the generalized Fourier transform, in "Topics in Mathematical Analysis", World Sci.Publ., Ed. Th.Rassias, Athens, 1989; in honor of Cauchy, 200nd anniversary.

[FG1] Feichtinger, H.G. and K.Gröchenig: A unified approach to atomic characterizations via integrable group representations. Proc. Conf. Lund June 1986, Ed. M.Cwikel et al, Springer Lect.Notes 1302, (1988), 52-73.

[FG2] - " - : Banach spaces related to integrable group representations and their atomic decompositions,I. J.Funct.Anal., to appear, 1989.

[FG3] - " -: Banach spaces related to integrable group representations and their atomic decompositions,II. Monatsh. f. Mathematik, to appear. 1990.

[FG4] - " - : Non-orthogonal expansions of signals and some of their applications. Proc. ECMI conference, Strobl, Mai 1989.

[FG5] - " -: Reconstruction of band-limited functions from irregular sampling values (1989). To appear.

[FG6] - " - :Multidimensional irregular sampling of band-limited functions in L^p-spaces, Proc.Conf. Oberwolfach, Feb. 1989, ISNM, Birkhäuser Publ., 1989.

[FG7] - " - : Irregular sampling theorems and series expansions of band-limited functions I (1989), to appear.

[FG8] - " - : Irregular sampling theorems and series expansions of band-limited functions II. (stability results and error analysis), in preparation.

[FJ1] Frazier,M. and B.Jawerth: Decomposition of Besov spaces. Indiana Univ.Math.J. **34** (1985), 777-799.

[FJ2] - " -: The φ-transform and decompositions of distribution spaces, Proc. Conf. "Functions Spaces and Applications", Lund 1986, Lect.Notes Math. 1302, Springer, Heidelberg, 1988.

[G1] Gröchenig, K.H: Describing functions: Atomic decompositions versus frames, to appear.

[G2] - " - : A new approach to irregular sampling; this volume.

[GMP] Grossmann, A. J.Morlet and T.Paul: Transforms associated to square integrable group representations I, J.Math.Phys. 26 (1985), 2473-2479.

[HW1] Heil, C. and D. Walnut: Continuous and Discrete wavelet transform. To appear in SIAM Rev., 1989.

[HM2] - " - : Gabor and wavelet expansions. This volume.

[H] Higgins, J.R: Five short stories about the cardinal series. Bull. Amer. Math.Soc. 12(1985), 45-89.

[J] Jerri, A.J.: The Shannon sampling theorem - its various extensions and applications, a tutorial review. Proc. IEEE 65 (1977), 45-89.

[LM] Lemarié, P. and Y.Meyer: Ondelettes et bases hilbertiennes, Rev.Mat. Iberoamericana 2 (1986), 1-18.

[M1] Meyer, Y.: De la recherche petroliere a la geometrie des espaces de Banach en passant par les paraproduits. Sem.Equ.Der.Part. 1985/86, Ecole Polytechn. Paris.

[M2] Meyer, Y.: Principe d'incertitude, bases hilbertiennes et algèbres d'opérateurs. Sém. Bourbaki 662 (1985-86).

[P] Papoulis, A: Error analysis in sampling theory. Proc. of the IEEE 54/7 (1966), 947-955.

[Y] Young, R: An Introduction to Nonharmonic Fourier Series. Acad.Press, New York, 1980.

GABOR AND WAVELET EXPANSIONS

Christopher Heil and David Walnut
The MITRE Corporation
7525 Colshire Drive
McLean, Virginia 22102
USA

ABSTRACT. This paper is an examination of techniques for obtaining Fourier series-like expansions of finite-energy signals using so-called Gabor and wavelet expansions. These expansions decompose a given signal into time a frequency localized components. The theory of frames in Hilbert spaces is used as a criteria for determining when such expansions are good representations of the signals. Some results on the existence of Gabor and wavelet frames in the Hilbert space of all finite-energy signals are presented.

0. Introduction.

The frequency analysis of signals by expanding them in terms of a fixed collection of sinusoids, i.e., of representing the signal by means of its Fourier transform, has long been a useful technique in mathematics and engineering. However, the fact that the Fourier transform is not localized in time can make it an unnatural way of representing a signal. For example, music can be thought of as a signal in which the frequency content is changing over time since the combination of notes being received by the ear is constantly changing. One would like to define a transform which reflects this evolutionary nature of the spectrum of a signal. Gabor and wavelet transforms are one means of accomplishing this. They are generalizations of the ordinary Fourier transform defined for periodic functions in the sense that they give Fourier series-like expansions of signals which display both the time and frequency content of the signal.

Gabor-type expansions were introduced in the 1940s by D. Gabor [**Gab**]. He proposed the representation of an arbitrary signal as a sum of translated and modulated Gaussian functions. Gabor's idea can be illustrated by a decades-old technique known as the *Short-time Fourier transform* in which Fourier transforms are taken of short consectuive segments of a given signal. This transform gives an unambiguous representation of the signal and also gives a frequency picture of the

J. S. Byrnes and J. F. Byrnes (eds.), Recent Advances in Fourier Analysis and Its Applications, 441–454.

signal locally in time.

To describe the Gabor transform, first fix a "windowing" function g. In order to analyze an arbitrary signal f, one forms the product of f with a shifted version of g, i.e., $f(t)\overline{g(t-na)}$ (where the bar indicates complex conjugation), then computes the Fourier series coefficients of this product, i.e.,

$$c_{mn}(f) = \int_{\mathbf{R}} f(t)\overline{g(t-na)}\,e^{-2\pi imbt}\,dt,$$

for $m, n \in \mathbf{Z}$, the set of integers. If g is concentrated at 0, then the coefficient $c_{mn}(f)$ should to some extent give the intensity of the frequency mb at time na. It is hoped that, as n and m range over all integers, the coefficients $c_{mn}(f)$ completely determine f and that some sense can be made of the "Fourier series" representation

$$f(t) \sim \sum_{m,n} c_{mn}(f)\, g(t-na)\, e^{2\pi imbt}.$$

The accuracy of the above representation is highly dependent on the function g and the values of the parameters a and b.

The other type of expansion examined here is known as wavelet (or sometimes *affine* wavelet) expansion. The transform associated with this type of expansion was introduced by A. Grossman and J. Morlet as a way of analyzing seismic signals [**GGM**]. There has been a great deal of recent work done using wavelets in this and related areas. For example, they have been used in the analysis of images [**Mal**], sound patterns [**KMG**], and in numerical matrix computation [**BCR**]. As with the Gabor transforms, the point here is to represent a signal in a way which displays both the time and frequency content of the signal. The way this is accomplished in the wavelet case is as follows. Take a fixed function φ (known as the **mother wavelet**) which is concentrated at 0 and consider the function $a^{n/2}\varphi(a^n t)$. If $a > 1$ and $n > 0$ then this function is compressed in time and consequently expanded in frequency. We form the coefficients $c_{mn}(f)$ by integrating f against translated versions of this function, i.e.,

$$c_{mn}(f) = a^{n/2} \int_{\mathbf{R}} f(t)\overline{\varphi(a^n t - mb)}\,dt.$$

For large $n > 0$, the coefficient $c_{mn}(f)$ gives an idea of the high-frequency content of f in a small time neighborhood about $a^{-n}mb$. As before, we want to make sense of the representation

$$f(t) \sim \sum_{m,n} c_{mn}(f)\, a^{n/2}\, \varphi(a^n t - mb).$$

The criterion we use to make sense of these representations comes from the theory of frames in Hilbert spaces. In this we follow the paper [**DGM**], where this connection was first made. The general concept of frames was first introduced by R. J. Duffin and A. C. Schaeffer in connection with non-harmonic Fourier series [**DS**]. The Hilbert space under consideration here will be $L^2(\mathbf{R})$, the space of all finite-energy signals on the real line $\mathbf{R}$. All of the results presented here can be generalized to more than one dimension. A theory in two dimensions is important for image analysis in particular.

The focus of this paper is on presenting results on the existence of Gabor frames and wavelet frames for the Hilbert space $L^2(\mathbf{R})$. The results here are due to others and a more detailed and complete treatment can be found in the expository paper [**HW**]. The first part is an introduction to Hilbert spaces and frames in Hilbert spaces. In Section 2, the Zak transform is introduced and results on the existence of Gabor frames for arbitrary $g \in L^2(\mathbf{R})$ and parameters a, b such that $ab = 1$ are given. In this case, it turns out that necessary and sufficient conditions can be given having to do with the Zak transform of g. In Section 3, we examine what can be done for finer lattices, that is, when $ab < 1$. We present two existence theorems giving simple conditions on g and a which guarantee that we have a frame for all sufficiently small $b > 0$. Section 4 deals with wavelet frames and we show by means of an example the sense in which the wavelet transform can be thought of as a time and frequency localization operator. This example leads to an existence theorem which appears in [**DGM**].

1. Frames in Hilbert Spaces.

In this section we will describe some of the basic properties of frames in Hilbert spaces, showing that they are useful generalizations of orthonormal bases. By a **Hilbert space**, we mean a vector space, H, which possesses an **inner product** $\langle x, y\rangle$ and which is **complete** in the **norm** $\|x\| = \langle x, x\rangle^{1/2}$ (for the precise meaning of these terms we refer the reader to [**GG**]).

The only Hilbert space we will actually use in this paper is $L^2(\mathbf{R})$, the space of all complex-valued signals f defined on the real line, $\mathbf{R}$, which have finite energy, i.e., for which

$$\|f\| = \left(\int_{\mathbf{R}} |f(t)|^2\, dt\right)^{1/2} < \infty.$$

The inner product in this Hilbert space is

$$\langle f, g\rangle = \int_{\mathbf{R}} f(t)\,\overline{g(t)}\, dt,$$

where the bar indicates complex conjugation.

Frames were first introduced in 1952 by R. J. Duffin and A. C. Schaeffer in the paper [**DS**], in connection with nonharmonic Fourier series. Their first use in connection with wavelets was in 1986 in the paper [**DGM**] by I. Daubechies, A. Grossmann, and Y. Meyer. The precise definition is as follows.

DEFINITION 1.1. A **frame** for a Hilbert space H is a set of vectors $\{x_n\}$ for which there exist constants $A, B > 0$ such that

$$A\|x\|^2 \leq \sum_n |\langle x, x_n\rangle|^2 \leq B\|x\|^2$$

for every $x \in H$.

It is well-known that given any Hilbert space H, there always exists an **orthonormal basis**, i.e., a set of vectors $\{e_n\}$ such that

(1) $\langle e_m, e_n\rangle = \begin{cases} 1, & \text{if } m = n; \\ 0, & \text{if } m \neq n; \end{cases}$

(2) $\sum |\langle x, e_n\rangle|^2 = \|x\|^2$ for all $x \in H$.

Every orthonormal basis is clearly a frame with $A = B = 1$. However, frames are much more general than orthonormal bases, for we do not place on them the stringent requirement of orthonormality (condition (1)), and we relax the equality in condition (2) to an inequality. A fundamental property of orthonormal bases is that every element $x \in H$ can be written in terms of the orthonormal basis in a unique way as $x = \sum \langle x, e_n\rangle e_n$. We shall see that frames also give representations of elements of the Hilbert space, although these representations need not be unique. However, they are still computable and under good control.

As a trivial example of a frame which is not an orthonormal basis, consider the following. Let $\{e_1, e_2, \ldots\}$ be an orthonormal basis for a Hilbert space H. This is surely then a frame with bounds $A = B = 1$. However, the set

$$\{e_1/\sqrt{2}, e_1/\sqrt{2}, e_2/\sqrt{2}, e_2/\sqrt{2}, \ldots\}$$

is also a frame with bounds $A = B = 1$, but is not an orthonormal basis.

In this paper we will be interested in constructing two specific types of frames for the Hilbert space $L^2(\mathbf{R})$. In each case, the frame elements have a particularly simple form, for they are functions which are generated from a single fixed function (called the **mother wavelet**) by combinations of the basic operations of translation, modulation, and dilation, defined by:

Translation:	$T_a f(x) = f(x-a)$,	for $a \in \mathbf{R}$;
Modulation:	$E_a f(x) = e^{2\pi i a x} f(x)$,	for $a \in \mathbf{R}$;
Dilation:	$D_a f(x) = a^{-1/2} f(x/a)$,	for $a > 0$.

In Sections 2 and 3 we will construct frames of the form $\{g_{mn}\}_{m,n\in\mathbf{Z}}$, where

$$g_{mn}(t) = e^{2\pi i m b t} g(t - na) = E_{mb} T_{na} g(t)$$

for some fixed function $g \in L^2(\mathbf{R})$ and fixed parameters $a, b > 0$. Such frames will be called **Weyl-Heisenberg**, or **W-H**, frames, and we say that g **generates** the frame. In Section 4 we discuss **affine** frames, which have the form $\{\varphi_{mn}\}_{m,n\in\mathbf{Z}}$, where

$$\varphi_{mn}(t) = a^{n/2}\varphi(a^n t - mb) = D_{a^n}T_{mb}\varphi(t)$$

for some fixed $\varphi \in L^2(\mathbf{R})$ and fixed $a > 1, b > 0$. We will discuss conditions on g and φ under which we can be sure that Weyl-Heisenberg or affine frames will exist. First, however, we list in the remainder of this section some properties of general frames in Hilbert spaces. Most of these general results first appeared in [**DS**]; expanded proofs and statements also appear in [**HW**].

THEOREM 1.2. *If $\{x_n\}$ is a frame then the following hold.*

(1) $Sx = \sum\langle x, x_n\rangle x_n$ *converges for all $x \in H$.*
(2) *S is an isomorphism of H onto itself, i.e., it is bijective and continuous, and has a continuous inverse.*
(3) $x = \sum\langle x, S^{-1}x_n\rangle x_n$ *for all $x \in H$, but not necessarily uniquely.*
(4) *If $x = \sum c_n x_n$ for some scalars $\{c_n\}$ then*

$$\sum_n |c_n|^2 = \sum_n |\langle x, S^{-1}x_n\rangle|^2 + \sum_n |c_n - \langle x, S^{-1}x_n\rangle|^2.$$

Condition (4) says that, while the representation $x = \sum\langle x, S^{-1}x_n\rangle x_n$ may not be unique, it is in fact the "best" way to write x in terms of the $\{x_n\}$.

The following theorem gives a necessary and sufficient condition for uniqueness in frame representations.

THEOREM 1.3. *Given a frame $\{x_n\}$ in a Hilbert space H. Then the representations $x = \sum\langle x, S^{-1}x_n\rangle x_n$ are unique for every $x \in H$ if and only if there exists an orthonormal basis $\{e_n\}$ and an isomorphism U of H such that $x_n = Ue_n$ for all n.*

2. Weyl-Heisenberg frames with lattice size 1.

In this section we consider the problem of determining when an arbitrary $g \in L^2(\mathbf{R})$ will generate a W-H frame if $a, b > 0$ are such that $ab = 1$. Recall that a W-H frame has the form $\{g_{mn}\}_{m,n\in\mathbf{Z}}$, where $g_{mn}(t) = e^{2\pi imbt}g(t - na)$. If one considers the lattice of points $\{(mb, na)\}_{m,n\in\mathbf{Z}}$ in the plane, then the value ab is the area of the rectangles in the plane determined by this lattice. It can be shown that the value $ab = 1$ is a "critical value" for W-H frames in that it is impossible to construct a W-H frame if $ab > 1$. We will see in this section that it is possible to construct W-H frames when $ab = 1$, and, moreover, such W-H frames have the desirable feature that the frame representations are unique. Unfortunately, we will also find that only "bad" functions g can generate W-H frames for this critical value, bad in the sense of either not being smooth or not having good decay.

A crucial tool in this analysis is what has come to be called the **Zak transform**. This transform has been introduced independently by many groups in many different areas of pure and applied mathematics; in fact, a discrete version was used by Gauss [**S**]. J. Zak investigated it for quantum mechanical reasons beginning in the 1960s [**Z**]; recent work includes that of A. J. E. M. Janssen [**J1; J2**]. Proofs of the theorems in this section can be found in [**J1**] and [**HW**].

DEFINITION 2.1. The **Zak transform** of a function $f \in L^2(\mathbf{R})$ is (formally)

$$Zf(t,\omega) = a^{1/2} \sum_{k=-\infty}^{\infty} f(ta+ka)\, e^{2\pi i k\omega}$$

for $t, \omega \in \mathbf{R}$, and where $a > 0$ is a fixed parameter.

The following facts about Z are easily proved.

THEOREM 2.2.

(1) *Zf is* **quasiperiodic**, *i.e.,*

$$Zf(t+1,\omega) = e^{-2\pi i\omega} Zf(t,\omega)$$
$$Zf(t,\omega+1) = Zf(t,\omega)$$

Thus Zf is completely determined by its values on the unit rectangle $Q = [0,1] \times [0,1]$.

(2) *The series defining Zf converges in an L^2-sense on Q. Z is an unitary map of $L^2(\mathbf{R})$ onto $L^2(Q)$, i.e., it is a norm-preserving isomorphism.*

(3) *$Zg_{mn} = E_{mn} \cdot Zg$, where*

$$E_{mn}(t,\omega) = e^{2\pi i m t} e^{2\pi i n\omega}.$$

Now, since Z is unitary, $\{g_{mn}\}$ will form a frame for $L^2(\mathbf{R})$ if and only if $\{Zg_{mn}\}$ forms a frame for $L^2(Q)$. But from Theorem 2.2, $Zg_{mn} = E_{mn} \cdot Zg$, which, combined with the fact that $\{E_{mn}\}$ is an orthonormal basis for $L^2(Q)$, places great restrictions on the function Zg. In particular, we can prove the following theorem.

THEOREM 2.3. *Given $g \in L^2(\mathbf{R})$ and $a, b > 0$ with $ab = 1$. Then $\{g_{mn}\}$ forms a frame if and only if there exist constants A, B such that*

$$0 < A \leq |Zg(t,\omega)|^2 \leq B < \infty$$

for almost every $(t,\omega) \in Q$.

This theorem implies that the frame representations provided by $\{g_{mn}\}$ will be unique. For, if $\{g_{mn}\}$ is a frame then Zg will be essentially constant by Theorem 2.3, so the mapping U defined on $L^2(Q)$ by $UF = F \cdot Zg$ is an isomorphism of $L^2(Q)$. Since $UE_{mn} = E_{mn} \cdot Zg = Zg_{mn}$ and $\{E_{mn}\}$ is an orthonormal basis, this implies by Theorem 1.3 that the frame representations are unique. This, of course, is a desirable feature. Unfortunately, we can show that only "bad" functions can have Zak transforms which are essentially constant in the sense of Theorem 2.3.

THEOREM 2.4. *If Zg is continuous, then it has a zero.*

PROOF: Assume Zg was continuous but nonvanishing. Then we can find (see [**RR**]) a continuous function $\varphi(t,\omega)$ such that

$$Zg(t,\omega) = |Zg(t,\omega)|\, e^{i\varphi(t,\omega)}$$

for $(t,\omega) \in Q$. Using the quasiperiodicity of Zg, it is not difficult to show that there must then exist integers k,l such that

$$\begin{aligned} \varphi(t+1,\omega) &= \varphi(t,\omega) - 2\pi\omega + 2\pi k \\ \varphi(t,\omega+1) &= \varphi(t,\omega) + 2\pi l \end{aligned}$$

for all $(t,\omega) \in Q$. Hence,

$$\begin{aligned} 0 &= \varphi(0,0) - \varphi(1,0) \\ &\quad + \varphi(1,0) - \varphi(1,1) \\ &\quad + \varphi(1,1) - \varphi(0,1) \\ &\quad + \varphi(0,1) - \varphi(0,0) \\ &= -2\pi, \end{aligned}$$

a contradiction. ∎

It is shown in [**H1**] that if g is continuous and satisfies the mild decay condition

$$\sum_{k=-\infty}^{\infty} \operatorname*{ess\,sup}_{t\in[k,k+1]} |g(t)| < \infty,$$

then Zg must be continuous, and therefore g cannot generate a W-H frame. Thus, for example, no smooth function with compact support, in fact, no Schwartz function, can generate a W-H frame at the critical value $ab = 1$. In particular, the Gaussian function $g(t) = e^{-\pi t^2}$ will not generate a frame. A similar criteria is the following.

THEOREM 2.5. *If $g \in L^2(\mathbf{R})$ and*

$$\left(\int_{\mathbf{R}} |t\,g(t)|^2\,dt\right)\left(\int_{\mathbf{R}} |\gamma\,\hat{g}(\gamma)|^2\,d\gamma\right) < \infty,$$

(where $\hat{g}$ is the Fourier transform of g) then g cannot generate a W-H frame when $ab = 1$.

See [**Bal**], [**Bat**], [**L**], [**D3**], [**DJ**], [**BHW**] for discussions and proofs of this theorem.

3. Existence of W-H frames for smaller lattices.

As mentioned in Section 2, the value $ab = 1$ is a critical value for W-H frames, in that there are no W-H frames when $ab > 1$. Moreover, we showed that W-H frames for $ab = 1$ all have unique representations, but unfortunately can only exist when g is either not smooth or has bad decay. In this section, we examine the effect of taking a smaller lattice size, i.e., of considering $ab < 1$. It can be shown that the frame representations in this case will not be unique, but on the other hand we will see that very good functions g will generate W-H frames if we allow ab to be small. Our goal is to prove existence theorems of the following form: given $g \in L^2(\mathbf{R})$ (the mother wavelet) and $a > 0$, find conditions under which there is an interval $(0, b_0)$ such that $\{g_{mn}\}$ is a W-H frame for $L^2(\mathbf{R})$ for every $b \in (0, b_0)$.

The following theorem, which is a straightforward generalization of a theorem of Daubechies ([**D3**]), is a very general existence theorem and will be examined carefully in subsequent pages.

THEOREM 3.1. *Let $g \in L^2(\mathbf{R})$ and $a > 0$ satisfy:*

(3.1.1) *there exist constants $A, B > 0$ such that for almost every $x \in \mathbf{R}$ we have*

$$A \leq \sum_n |g(x - na)|^2 \leq B,$$

(3.1.2) $\lim_{b \to 0} \sum_{j \neq 0} \beta_a(j/b) = 0$, *where*

$$\beta_a(s) = \operatorname*{ess\,sup}_{x \in \mathbf{R}} \sum_n |g(x - na)\, g(x - s - na)|.$$

Then there exists $b_0 > 0$ such that $\{g_{mn}\}$ is a frame for $L^2(\mathbf{R})$ for all $b \in (0, b_0)$.

A proof of this theorem can be found in [**D3**] and [**HW**]. Condition (3.1.1) is necessary in order that $\{g_{mn}\}$ form a frame. This condition can be thought of as an "overlapping" condition on the mother wavelet g. The existence of the lower bound A means that the successive shifts of the function $|g|^2$ cover the entire real line and leave no "gaps". If there were such a gap, then any function supported in the gap would not be recognized by any of the wavelets g_{mn}. That is, we could find a non-zero function f such that $\langle f, g_{mn} \rangle = 0$ for all m and n, which would imply that $\{g_{mn}\}$ was not a frame.

The existence of the bound B says that g must at least be bounded in order for it to generate a frame. The bound B is required to have good control over the size of the wavelet coefficients over all the functions $f \in L^2(\mathbf{R})$.

If g vanishes outside an interval I of length at most $1/b$ then $\beta_a(s) = 0$ if $|s| \geq 1/b$. In this case, the form of the frame operator is particularly simple, in fact,

$$Sf = \sum_{m,n} \langle f, g_{mn} \rangle g_{mn}(x) = f(x) \cdot b^{-1} \sum_n |g(x - na)|^2.$$

Thus, S is an isomorphism if and only if $\sum |g(x - na)|^2$ is bounded above and below, i.e., if and only if (3.1.1) is satisfied. In this case, then, (3.1.1) is both necessary and sufficient in order that $\{g_{mn}\}$ form a frame.

The condition (3.1.2) can be replaced by the simpler condition of the following theorem. The proof can be found in [**HW**].

THEOREM 3.2. *Let $g \in L^2(\mathbf{R})$ and $a > 0$ satisfy condition (3.1.1) and also*

$$\sum_{n=-\infty}^{\infty} \operatorname*{ess\,sup}_{x\in[n,n+1)} |g(x)| \; < \; \infty. \tag{3.2.1}$$

Then there exists $b_0 > 0$ such that $\{g_{mn}\}$ is a frame for $L^2(\mathbf{R})$ for all $b \in (0, b_0)$.

Condition (3.2.1) means that the sequence of maximum values of the function $|g|$ on successive intervals is summable. There are many examples of such functions, such as the Gaussian $g(x) = e^{-\pi x^2}$, and more generally any function g for which there exists a $C > 0$ and an $\epsilon > 0$ such that

$$|g(x)| \; \leq \; C(1 + |x|)^{-(1+\epsilon)}$$

for all $x \in \mathbf{R}$.

Theorem 3.2 is proved by showing that a function which satisfies condition (3.2.1) must also satisfy condition (3.1.2). This makes Theorem 3.2 less general than Theorem 3.1. However, that (3.2.1) implies (3.1.2) shows that condition (3.1.2) is a condition governed by the growth of the function g at ∞. Also, (3.2.1) is a much easier condition to verify than (3.1.2).

4. Affine Systems.

In this section, we give a brief introduction to some aspects of wavelet decompositions of functions in $L^2(\mathbf{R})$. As in Sections 2 and 3, we will address the question of the existence of such decompositions and give an indication of why such decompositions might be useful in signal processing or image analysis by giving a physical interpretation to the expansion coefficients. As usual, the theory of frames in Hilbert spaces will be used as a criterion to determine whether arbitrary signals can be written down in terms of such a collection of functions.

DEFINITION 4.1. Let $\varphi \in L^2(\mathbf{R})$ and $a > 1$, $b > 0$ be given. Then the **(affine) wavelet system** generated by φ, a, and b is the set of functions $\{\varphi_{mn}\}_{m,n\in\mathbf{Z}}$, where

$$\varphi_{mn} \; = \; a^{n/2}\,\varphi(a^n t - mb).$$

We always assume that the function φ, the **mother wavelet** satisfies the condition

$$\int_{\mathbf{R}} \varphi(t)\,dt \; = \; 0.$$

This means that φ (if it is real valued) has as much area above the axis as below, which gives the function a resemblance to a "wave". This is why the term "wavelet" was coined.

The functions making up an affine system all arise from a single function under the action of a collection of norm-preserving transformations, namely, dilations and translations. Recall that the dilation operator D_α is defined by

$$D_\alpha f(t) = \alpha^{1/2} f(\alpha t).$$

The factor $\alpha^{1/2}$ means that D_α preserves the L^2-norm of the functions it acts upon, that is,

$$\int_{\mathbf{R}} |D_\alpha f(t)|^2 \, dt = \int_{\mathbf{R}} |f(t)|^2 \, dt.$$

If $\alpha > 1$ then D_α has the effect of concentrating a function at the origin by making it narrower and (in order to perserve L^2-norm) taller. If $\alpha < 1$ then D_α spreads the function out and decreases the amplitude. If we use the notation T_β to denote translation by β (as defined in Section 1), then we can write our affine system as

$$\varphi_{mn} = D_{a^n} T_{mb} \varphi.$$

In order to give an interpretation of the physical meaning of the set of wavelet coefficients of a particular signal f, let us consider, for a fixed $n > 0$, the collection of coefficients

$$c_{mn}(f) = \langle f, \varphi_{mn} \rangle = a^{n/2} \int f(t) \overline{\varphi(a^n t - mb)} \, dt.$$

The coefficients $c_{mn}(f)$ represent comparisons of f with successive shifts by $a^{-n}mb$ of the function $a^{n/2}\varphi(a^n t)$. Since $n > 0$ and $a > 1$ we have $a^n > 1$ and consequently this dilated function is highly concentrated at 0. Intuitively, then, it should be the case that these comparisons pick out the high frequency behavior or fine detail of the signal f in a small time interval around $a^{-n}mb$. This type of transform is often referred to as a scaling transform because the coefficients $c_{mn}(f)$ where n is fixed, pick out those features of f which exist on a time-scale of about $a^{-n}b$, so that as n becomes large, the coefficients for that n pick out smaller and smaller scale features of the signal. Features on a scale much smaller than $a^{-n}b$ are not noticed because they are averaged out when the coefficients are computed. Features on a scale much larger than $a^{-n}b$ are not noticed because of the shape of the mother wavelet φ, i.e., because $\int \varphi = 0$. That is, if f were nearly constant in a neighborhood of $a^{-n}mb$, then the value of the corresponding coefficient $c_{mn}(f)$ should be nearly zero. In language suggestive of image processing, these scales are often referred to as levels of resolution, and the wavelet transform as a multiresolution transform. Thus it is accurate to say that the coefficients $c_{mn}(f)$ essentially pick out only the high frequency behavior present in the signal f at the given resolution level which was not present at the previous resolution level. For an image, one would say that these coefficients contain only the additional detail in the signal f at this level of resolution which was not detectable at the previous level of resolution.

We shall give a more precise mathematical formulation of the above rather vague statements by looking at the following simple example. Let φ be a square-integrable function (let us for convenience assume integrability as well) such that $\hat{\varphi}$ vanishes outside an interval of the form $[\ell, L]$, where $\ell, L > 0$, and in fact is non-zero everywhere inside the interval. Since $\hat{\varphi}$ is continuous, this means that $\hat{\varphi}$ is bounded and that it is bounded below on subintervals of $[\ell, L]$. Let $b > 0$ be such that $b^{-1} = L - \ell$ and $a > 1$ such that $a < L/\ell$. Let us consider the collection $\{\varphi_{mn}\}$ for this φ and the sum

$$\sum_{m,n} \langle f, \varphi_{mn} \rangle \varphi_{mn}.$$

Notice that under the action of the Fourier transform, dilations on the time side go into dilations in the opposite sense on the frequency side, that is,

$$(D_\alpha \varphi)^\wedge(\gamma) = D_{\alpha^{-1}} \hat{\varphi}(\gamma).$$

Using this fact, the fact that translations on the time side go into modulations on the frequency side, and Parseval's formula, we can prove the following identity (for details, see [**HW**]):

$$\left(\sum_m \langle f, \varphi_{mn} \rangle \varphi_{mn} \right)^\wedge (\gamma) = \hat{f}(\gamma) \cdot b^{-1} \, |\hat{\varphi}(a^{-n}\gamma)|^2.$$

Now, since $\hat{\varphi}$ vanishes outside the interval $[\ell, L]$, the function $|\hat{\varphi}(a^{-n}\gamma)|^2$ is supported in $[a^n \ell, a^n L]$. Thus the sum

$$\sum_m \langle f, \varphi_{mn} \rangle \varphi_{mn}$$

represents a band-filtered version of f where the band is $[a^n\ell, a^n L]$. If $n > 0$ then these contain the high frequencies of f and if $n < 0$, the low frequencies of f.

It should be noted here that since $\hat{\varphi}$ is supported on the right half-line in the frequency domain, only positive frequencies of f will be measured by the wavelet coefficients and hence such a collection of wavelets can never be a frame for $L^2(\mathbf{R})$. There are several ways of handling this situation. If one is dealing with only real-valued signals f (which is usually the case in practice) then it is enough to know $\hat{f}$ on the right (or left) half-line in order to completely determine f. Another possibility is to consider an additional function $\tilde{\varphi}$ which has the same properties as φ only transferred to the left half-line in the frequency domain. For this purpose, it is enough to take $\tilde{\varphi}(t) = \varphi(-t)$. Finally, it is possible to find a function φ whose Fourier transform is supported on both halves of the frequency axis such that the $\{\varphi_{mn}\}$ are a frame for $L^2(\mathbf{R})$. We shall see that the wavelet orthonormal basis of Y. Meyer is of this type. For details on such frame constructions see [**HW**].

The above example also gives some insight into the notion of the frame operator associated to a collection of wavelets as a "deblurring" operator. For fixed M and φ as before, arguing as above gives

$$\left(\sum_{n\leq M}\sum_{m}\langle f,\varphi_{mn}\rangle\varphi_{mn}\right)^{\wedge}(\gamma) = \hat{f}(\gamma)\cdot b^{-1}\sum_{n\leq M}|\hat{\varphi}(a^{-n}\gamma)|^2.$$

The function $\sum_{n\leq M}|\hat{\varphi}(a^{-n}\gamma)|^2$ is supported in $[0, a^M L]$ and in fact does not vanish on this interval. Thus one can see that the above sum is a low-pass filtered or "blurred" version of f. The effect of increasing M to infinity is to add more and more of the high frequency characteristics (one might say "finer and finer details") of f into the sum until finally a complete reconstruction is achieved.

We are now able to state and prove an existence theorem for affine frames for $L^2(\mathbf{R})$. This theorem appears in [**DGM**].

THEOREM 4.2. *Let $\varphi \in L^2(\mathbf{R})$ and $a > 1$, $b > 0$ satisfy*

(4.2.1) *$\hat{\varphi}$ is supported in $[\ell, L]$ where $L > \ell > 0$, $b^{-1} \leq L - \ell$ and $1 < a \leq l/\ell$,*

(4.2.2) *there exist constants $A, B > 0$ such that for almost every $\gamma \in \hat{\mathbf{R}}$ we have*

$$A \leq \sum_{n}|\hat{\varphi}(a^n\gamma)|^2 \leq B.$$

Then $\{\varphi_{mn}, \tilde{\varphi}_{mn}\}$ is a frame for $L^2(\mathbf{R})$, where $\tilde{\varphi}(t) = \varphi(-t)$.

PROOF: The same type of Fourier series argument as before gives

$$\sum_{m,n}|\langle f,\varphi_{mn}\rangle|^2 = \int_0^{\infty}|\hat{f}(\gamma)|^2\, b^{-1}\sum_{n}|\hat{\varphi}(a^n\gamma)|^2\, d\gamma.$$

Therefore,

$$\begin{aligned}\sum_{m,n}&|\langle f,\varphi_{mn}\rangle|^2 + \sum_{m,n}|\langle f,\tilde{\varphi}_{mn}\rangle|^2 \\ &= \int_0^{\infty}|\hat{f}(\gamma)|^2\, b^{-1}\sum_{n}|\hat{\varphi}(a^n\gamma)|^2\, d\gamma + \int_{-\infty}^{0}|\hat{f}(\gamma)|^2\, b^{-1}\sum_{n}|\hat{\tilde{\varphi}}(a^n\gamma)|^2\, d\gamma.\end{aligned}$$

Hence,

$$\begin{aligned}A\int_{-\infty}^{\infty}|f(t)|^2\, dt &= A\int_{-\infty}^{\infty}|\hat{f}(\gamma)|^2\, d\gamma \\ &\leq \sum_{m,n}|\langle f,\varphi_{mn}\rangle|^2 + \sum_{m,n}|\langle f,\tilde{\varphi}_{mn}\rangle|^2 \\ &\leq B\int_{-\infty}^{\infty}|\hat{f}(\gamma)|^2\, d\gamma \\ &= B\int_{-\infty}^{\infty}|f(t)|^2\, dt,\end{aligned}$$

from which the result follows. ∎

Finally, we state some results on the existence of affine systems which are actually orthonormal bases for $L^2(\mathbf{R})$. The differences in the theorems come from the smoothness and decay properties that the mother wavelets can have. In each case we take $a = 2$ and $b = 1$.

THEOREM 4.3. *The Haar system $\{D_{2^n} T_m \varphi\}$ is an orthonormal basis of compactly supported wavelets for $L^2(\mathbf{R})$. The Haar system is defined by taking $a = 2$, $b = 1$, and*

$$\varphi(t) = \begin{cases} 1, & \text{if } 0 \le t < 1/2, \\ -1, & \text{if } 1/2 \le t < 1, \\ 0, & \text{otherwise.} \end{cases}$$

THEOREM 4.4 (MEYER). *There exists a function $\varphi \in L^2(\mathbf{R})$ such that*

(1) *$\hat{\varphi}$ is infinitely differentiable and the support of $\hat{\varphi}$ is contained in the compact set $[-4/3, -1/3] \cup [1/3, 4/3]$ so that in particular, φ is a Schwartz function, that is, it and all of its derivatives are continuous and decay faster than any polynomial.*
(2) *$\int \varphi(t)\, t^k\, dt = 0$ for every integer k.*
(3) *$\{D_{2^n} T_m \varphi\}$ is an orthonormal basis for $L^2(\mathbf{R})$.*

THEOREM 4.5 (LEMARIE). *Given $N > 0$, there is a function $\varphi \in L^2(\mathbf{R})$ such that*

(1) *φ has N continuous derivatives, and*

$$|\varphi(t)| \le C\, e^{-\alpha t}$$

for some C, $\alpha > 0$.
(2) *$\int \varphi(t)\, t^k\, dt = 0$ for $k = 0, 1, 2, \ldots, N+1$.*
(3) *$\{D_{2^n} T_m \varphi\}$ is an orthonormal basis for $L^2(\mathbf{R})$.*

THEOREM 4.6 (DAUBECHIES). *Given a number $N > 0$, there is a function $\varphi \in L^2(\mathbf{R})$ such that*

(1) *φ has N continuous derivatives, and φ has compact support.*
(2) *$\int \varphi(t)\, t^k\, dt = 0$ for $k = 0, 1, 2, \ldots, N-1$.*
(3) *$\{D_{2^n} T_m \varphi\}$ is an orthonormal basis for $L^2(\mathbf{R})$.*

REFERENCES

[**Bal**] R. BALIAN, *Un principe d'incertitude fort en théorie du signal on en mécanique quantique*, C. R. Acad. Sci. Paris **292** (1981), 1357–1362.

[**Bat**] G. BATTLE, *Heisenberg proof of the Balian-Low theorem*, Lett. Math. Phys. **15** (1988), 175–177.

[**Ben**] J. BENEDETTO, "Gabor representations and wavelets," Commutative Harmonic Analysis, D. Colella, ed., Contemp. Math. 19, American Mathematical Society, Providence, 1989, pp. 9–27.

[BHW] J. BENEDETTO, C. HEIL, AND D. WALNUT, *Remarks on the proof of the Balian theorem*, preprint.

[BCR] G. BEYLKIN, R. COIFMAN, AND V. ROKHLIN, *Fast wavelet transforms and numerical algorithms*, I, preprint.

[D1] I. DAUBECHIES, *Orthonormal bases of compactly supported wavelets*, Comm. Pure Appl. Math. **41** (1988), 909–996.

[D2] ———, *Time-frequency localization operators: a geometric phase space approach*, IEEE Trans. Inform. Theory **34** (1988), 605–612.

[D3] ———, *The wavelet transform, time-frequency localization and signal analysis*, IEEE Trans. Inform. Theory, to appear.

[DGM] I. DAUBECHIES, A. GROSSMANN, AND Y. MEYER, *Painless nonorthogonal expansions*, J. Math. Phys. **27** (1986), 1271–1283.

[DJ] I. DAUBECHIES AND A. J. E. M. JANSSEN, *Two theorems on lattice expansions*, IEEE Trans. Inform. Theory, to appear.

[DS] R. J. DUFFIN AND A. C. SCHAEFFER, *A class of nonharmonic Fourier series*, Trans. Amer. Math. Soc. **72** (1952), 341–366.

[Gab] D. GABOR, *Theory of communications*, J. Inst. Elec. Eng. (London) **93** (1946), 429–457.

[GG] I. GOHBERG AND S. GOLDBERG, "Basic Operator Theory," Birkhäuser, Boston, 1981.

[Gro] A. GROSSMANN, "Wavelet transforms and edge detection," Stochastic Processing in Physics and Engineering, S. Albeverio et al., eds., D. Reidel, Dordrecht, the Netherlands, 1988, pp. 149–157.

[GGM] P. GOUPILLAUD, A. GROSSMANN, AND J. MORLET, *Cycle-octave and related transforms in seismic signal analysis*, Geoexploration **23** (1984/85), 85–102.

[GM] A. GROSSMANN AND J. MORLET, *Decomposition of Hardy functions into square integrable wavelets of constant shape*, SIAM J. Math. Anal. **15** (1984), 723–736.

[H1] C. HEIL, *Generalized harmonic analysis in higher dimensions; Weyl–Heisenberg frames and the Zak transform*, Ph.D. thesis, University of Maryland, College Park, MD, 1990.

[H2] ———, *Wavelets and frames*, Proceedings of the Institute for Mathematics and Its Applications, to appear.

[HW] C. HEIL AND D. WALNUT, *Continuous and discrete wavelet transforms*, SIAM Review **34** (1989).

[J1] A. J. E. M. JANSSEN, *Bargmann transform, Zak transform, and coherent states*, J. Math. Phys. **23** (1982), 720–731.

[J2] ———, *The Zak transform: a signal transform for sampled time-continuous signals*, Philips J. Res. **43** (1988), 23–69.

[KMG] R. KRONLAND-MARTINET, J. MORLET, AND A. GROSSMANN, *Analysis of sound patterns through wavelet transforms*, Internat. J. Pattern Recog. Artif. Int. **1** (1987), 273–302.

[LM] P. LEMARIÉ AND Y. MEYER, *Ondelettes et bases hilbertiennes*, Rev. Mat. Iberoamericana **2** (1986), 1–18.

[L] F. LOW, "Complete sets of wave packets," A Passion for Physics—Essays in Honor of Geoffrey Chew, World Scientific, Singapore, 1985, pp. 17–22.

[Mal] S. MALLAT, *A theory for multiresolution signal decomposition: the wavelet representation*, IEEE Trans. on Pattern Anal. and Machine Intel. **11** (1989), 674–693.

[Mey] Y. MEYER, *Principe d'incertitude, bases hilbertiennes et algèbres d'opérateurs*, Séminaire Bourbaki **662** (1985-86).

[RR] T. RADO AND P. REICHELDERFER, "Continuous Transformations in Analysis," Springer-Verlag, Berlin, New York, 1955.

[S] W. SCHEMPP, *Radar ambiguity functions, the Heisenberg group, and holomorphic theta series*, Proc. Amer. Math. Soc. **92** (1984), 103–110.

[W] D. WALNUT, *Weyl–Heisenberg wavelet expansions: existence and stability in weighted spaces*, Ph.D. thesis, University of Maryland, College Park, MD, 1989.

[Y] R. YOUNG, "An Introduction to Nonharmonic Fourier Series," Academic Press, New York, 1980.

[Z] J. ZAK, *Finite translations in solid state physics*, Phys. Rev. Lett. **19** (1967), 1385–1397.

Translation Invariant Multiscale Analysis

W. R. Madych*

Abstract

The notion of multiscale analysis introduced by R. R. Coifman and Y. Meyer is considered and the translation invariant case is characterized.

1 Introduction

Recall that a *dyadic multiscale analysis* of $L^2(R^n)$ is an increasing sequence $\mathcal{V} = \{V_j : j = \ldots, -1, 0, 1, 2, \ldots\}$ of closed subspaces of $L^2(R^n)$ which has the following properties:

(1). $\bigcup_{j=-\infty}^{\infty} V_j$ is dense in $L^2(R^n)$ and $\bigcap_{j=-\infty}^{\infty} V_j = \{0\}$.

(2). $f(x)$ is in V_j if and only if $f(2x)$ is in V_{j+1}.

(3). There is a lattice Γ in R^n such that for every f in V_0 and every γ in Γ the function f_γ is in V_0. Here and in what follows we use the notation $f_\gamma(x) = f(x - \gamma)$.

(4). There are two positive constants $C_2 \geq C_1 > 0$ and a function g in V_0 such that V_0 is the closed linear span of g_γ, $\gamma \in \Gamma$, and

$$C_1^2 \sum_{\gamma\in\Gamma} |a_\gamma|^2 \leq \int_{R^n} \sum_{\gamma\in\Gamma} |a_\gamma g_\gamma(x)|^2 dx \leq C_2^2 \sum_{\gamma\in\Gamma} |a_\gamma|^2 .$$

*Department of Mathematics, University of Connecticut, Storrs, CT 06268. Preliminary report of work partially supported by a grant from the Air Force Office of Scientific Research.

J. S. Byrnes and J. F. Byrnes (eds.), Recent Advances in Fourier Analysis and Its Applications, 455–462.

An introduction to the subject may be found in [1,2]. A basic property of a multiscale analysis $\mathcal{V}$ is the following:

(5). There is a function ϕ in V_0 such that the collection $\{\phi_\gamma\}_{\gamma\in\Gamma}$ is an orthonormal basis in V_0.

This fact may be regarded as a substitute for (4) and plays an important role in what follows.

A dyadic multiscale analysis is *translation invariant* if all the translates of f, $\{f_y : y \in R^n\}$, are in V_0 whenever f is in V_0.

The canonical example of a translation invariant multiscale analysis of $L^2(R)$ is when V_0 is the collection of those functions in $L^2(R)$ whose Fourier transforms are supported in the interval $[-\pi, \pi]$. A natural choice of ϕ in this case is given by

$$\phi(x) = \frac{\sin \pi x}{\pi x}.$$

The point of this paper is to give a characterization of translation invariant multiscale analyses. For the sake of clarity in what follows we will restrict our attention to the case $n = 1$ and $\Gamma = Z$, the lattice of integers. The statements and arguments in the general case are completely analogous to this basic case.

We now briefly digress to list some of the conventions which are used here: The Fourier transform $\hat{f}$ of a function f is defined by

$$\hat{f}(\xi) = \frac{1}{\sqrt{2\pi}} \int_{-\infty}^{\infty} e^{-i\xi x} f(x)dx$$

whenever it makes sense and distributionally otherwise. Basic facts concerning Fourier transforms and distributions will be used without further elaboration in what follows. To avoid the pedantic repetition of "almost everywhere" and other modifying phrases which are inevitably necessary when dealing with functions defined almost everywhere, all equalities between functions and other related notions are interpreted in the distributional sense whenever possible. The term support is also used in the distributional sense; in particular the support of a function f in $L^2(R)$ is a well defined closed set. If W is a collection of tempered distributions then $\widehat{W}$ is the collection of Fourier transforms of elements of W, in other words $\widehat{W} = \{f : f = \hat{g} \text{ for some } g \text{ in } W\}$. For a subset Ω of R and a real number

r the sets $r\Omega$ and $\Omega + r$ are defined by $r\Omega = \{x : x = r\omega \text{ for some } \omega \text{ in } \Omega\}$ and $\Omega + r = \{x : x = \omega + r \text{ for some } \omega \text{ in } \Omega\}$; $L^2(\Omega)$ is the L^2 closure of the subspace of those functions in $L^2(R)$ whose support is contained in Ω. For notational simplicity we use Q to denote the closed interval $[-\pi, \pi]$.

We can now conveniently state our main observation.

Theorem *Suppose $\mathcal{V}$ is a translation invariant dyadic multiscale analysis of $L^2(R)$. Then $\hat{V}_0 = L^2(\Omega)$ where Ω is a closed subset of R which has the following properties:*

(a). $\Omega \subset 2\Omega$.

(b). $\{\Omega + 2\pi j\} \bigcap \{\Omega + 2\pi k\}$ *is a set of Lebesgue measure 0 for any pair of integers such that* $j \neq k$.

(c). $\bigcup_{k=-\infty}^{\infty} \{\Omega + 2\pi k\} = R$.

(d). $\bigcup_{k=1}^{\infty} L^2(2^k\Omega)$ *is dense in* $L^2(R)$.

Conversely, if V_k, $k \in Z$ is defined by $\hat{V}_k = L^2(2^k\Omega)$ where Ω is a closed subset of R which satisfies the properties above then the sequence of subspaces $\{V_k\}$ is a translation invariant multiscale analysis of $L^2(R)$.

Remark 1 In view of the example given above it is very tempting to conjecture that the set Ω in the Theorem must be of the form $\Omega = Q + \alpha$ for some real number α which satisfies $\pi < \alpha < \pi$. Certainly such Ω's satisfy the desired conditions. However the conditions of the Theorem are satisfied by Ω's which need not be connected as the following example due to Rudi Lorentz shows.

$$\Omega = \left[-\frac{5\pi}{4}, -\pi\right] \bigcup \left[-\frac{3\pi}{4}, \frac{3\pi}{4}\right] \bigcup \left[\pi, \frac{5\pi}{4}\right] .$$

Remark 2 Consider

$$\begin{aligned} \Omega_a &= [-1, 1] \bigcup [2\pi + 1, 4\pi - 1] \\ \Omega_b &= [-5, 5] \\ \Omega_c &= [-1, 1] \\ \Omega_d &= [0, 2\pi] . \end{aligned}$$

It is not difficult to verify that each Ω_α listed above is a closed set which fails to satisfy condition (α) but satisfies the remaining conditions in the Theorem. These examples show that conditions *(a)-(d)* are not redundant.

Remark 3 Note that condition *(a)* implies that 0 is contained in Ω. In addition to this it is clear that if Ω contains a neighborhood of the origin then it satisfies condition *(d)*. In view of this it seems reasonable to suspect that subsets Ω which satisfy the conditions of the Theorem must contain an open neighborhood of the origin. That this is not the case can be seen by considering the following example of Ω:

$$\{\bigcup_{k=1}^{\infty}[-(2-2^{-k})2^{-k}\pi, -2^{-k}\pi]\}\bigcup[0,\pi]\{\bigcup_{k=1}^{\infty}[(2-(2-2^{-k})2^{-k})\pi, (2-2^{-k})\pi]\}\,.$$

Remark 4 In view of the examples listed above it may be of some interest to obtain a significantly more lucid description of the set Ω than that given in the Theorem.

A corollary concerning wavelets generated by $\mathcal{V}$ is recorded at the end of Section 2.

2 Details

We begin by establishing a basic lemma. First recall that the indicator function of a set Ω is usually denoted by χ_Ω and satisfies

$$\chi_\Omega(\xi) = \begin{cases} 1 \text{ if } \xi \in \Omega \\ 0 \text{ otherwise.} \end{cases}$$

Lemma *Suppose $\mathcal{V}$ is a translation invariant dyadic multiscale analysis of $L^2(R)$ and ϕ is a function whose existence is guaranteed by (5). Then*

$$|\hat{\phi}| = \chi_\Omega$$

where χ_Ω is the indicator function of a closed set Ω which has properties (a)-(d) in the statement of the Theorem.

Let Ω be the support of ϕ. To prove the lemma we will first show that Ω satisfies property *(b)*.

Recall that (5) implies that for all f in V_0 we may write

$$\hat{f}(\xi) = g(\xi)\hat{\phi}(\xi) \tag{6}$$

where g is 2π periodic and square integrable over Q. In particular, since V_0 is translation invariant, ϕ_y is in V_0 so setting $\alpha = -y$ we may write

$$e^{i\alpha\xi}\hat{\phi}(\xi) = g(\xi)\hat{\phi}(\xi)$$

for some such g. Hence

$$e^{i\alpha(\xi-2\pi m)}\hat{\phi}(\xi - 2\pi m) = g(\xi - 2\pi m)\hat{\phi}(\xi - 2\pi m) = g(\xi)\hat{\phi}(\xi - 2\pi m)$$

which implies that

$$e^{i\alpha(\xi-2\pi m)} = g(\xi)$$

on $\Omega + 2\pi m$. For two different values of m the last equality implies that

$$e^{i\alpha(\xi-2\pi j)} = e^{i\alpha(\xi-2\pi k)}$$

on $\{\Omega + 2\pi j\} \bigcap \{\Omega + 2\pi k\}$. Re-expressing the last relation as

$$e^{i\alpha(\xi-2\pi k)}(e^{i2\pi\alpha(k-j)} - 1) = 0$$

it is clear that either α is an integer, j is equal to k, or $\{\Omega+2\pi j\}\bigcap\{\Omega+2\pi k\}$ is a set of measure zero. Since α may be any real number we conclude that $\{\Omega + 2\pi j\} \bigcap \{\Omega + 2\pi k\}$ has measure zero whenever $j \neq k$.

Now, since $\hat{\phi}_k(\xi) = e^{-ik\xi}\hat{\phi}(\xi)$, $k \in Z$, are orthonormal, we may write

$$\int_R \phi_k(x)\overline{\phi_\ell(x)}dx = \int_R e^{im\xi}|\hat{\phi}(\xi)|^2 d\xi = \tag{7}$$

$$\int_Q e^{im\xi} \sum_{j\in Z} |\hat{\phi}(\xi - 2\pi j)|^2 dx = \begin{cases} 1 \text{ when } m = 0 \\ 0 \text{ otherwise} \end{cases}$$

where $m = \ell - k$. The last equality implies that

$$\sum_{j\in Z} |\hat{\phi}(\xi - 2\pi j)|^2 = \frac{1}{2\pi}$$

on R and since $\{\Omega+2\pi j\}\bigcap\{\Omega+2\pi k\}$ has measure zero whenever $j\neq k$ we may conclude that

$$|\hat{\phi}(\xi)|=\frac{1}{\sqrt{2\pi}}\chi_{\Omega}(\xi)$$

and

$$\bigcup_{k\in Z}\{\Omega+2\pi k\}=R\,.$$

To see (a) observe that (2) and the facts demonstrated above imply that

$$\chi_{\Omega}(\xi)=h(\xi)\chi_{\Omega}(\xi/2)$$

where h is 4π periodic and square integrable over $2Q$. Since $\chi_{2\Omega}(\xi)=\chi_{\Omega}(\xi/2)$, the last equality involving h implies that χ_{Ω} vanishes whenever $\chi_{2\Omega}$ does so $\Omega\subset 2\Omega$.

Finally, the fact that $\bigcup_{k=1}^{\infty}L^2(2^k\Omega)$ is dense in $L^2(R)$ is an immediate consequence of property (1). The proof of the Lemma is complete.

Now, suppose ϕ and Ω are as in the Lemma and its proof. Since V_0 consists of functions f which satisfy (6) it is clear that $\widehat{V_0}$ is contained in $L^2(\Omega)$.

To see that $\widehat{V_0}=L^2(\Omega)$ let f be any element in $L^2(\Omega)$ and let h be defined by

$$h(\xi)=\begin{cases}\overline{\hat{\phi}(\xi)}/|\hat{\phi}(\xi)| & \text{if } \hat{\phi}(\xi)\neq 0\\ 0 & \text{otherwise.}\end{cases}$$

By virtue of the properties of Ω established above it is clear that

$$\frac{1}{\sqrt{2\pi}}\chi_{\Omega}(\xi)=\left(\sum_{j\in Z}h(\xi-2\pi j)\right)\hat{\phi}(\xi)$$

and

$$\hat{f}(\xi)=g(\xi)\hat{\phi}(\xi)$$

where

$$g(\xi)=\sqrt{2\pi}\left(\sum_{j\in Z}\hat{f}(\xi-2\pi j)\right)\left(\sum_{j\in Z}h(\xi-2\pi j)\right)\,.$$

Thus f satisfies (6) and hence we may conclude that $L^2(\Omega)$ is contained in $\widehat{V_0}$.

The Lemma together with the last observation imply the first assertion of the Theorem.

To see the converse, let $\mathcal{V}$ be the sequence of subspaces $\{V_k\}$, $k \in Z$, defined by $\widehat{V}_k = L^2(2^k\Omega)$ where Ω is a closed set which satisfies properties *(a)-(d)* of the Theorem.

The fact that $\mathcal{V}$ is translation invariant and, in particular, satisfies property (3) with $\Gamma = Z$ is an immediate consequence of the definition. Property (2) is also immediate. That $\mathcal{V}$ is an increasing sequence of subspaces and $\bigcup_{k\in Z} V_k$ is dense in $L^2(R)$ are consequences of properties *(a)* and *(d)*.

That $\bigcap_{k\in Z} V_k = \{0\}$ follows from the fact that the measure of Ω is finite. Indeed, its measure is 2π which can be seen from

$$\int_R \chi_\Omega(\xi)d\xi = \sum_{j\in Z}\int_{Q+2\pi j}\chi_\Omega(\xi)d\xi = \int_Q \sum_{j\in Z}\chi_\Omega(\xi - 2\pi j)d\xi = 2\pi$$

by using properties *(b)* and *(c)*.

Finally, to see property (5) take

$$\hat{\phi} = \frac{1}{\sqrt{2\pi}}\chi_\Omega$$

and use properties (b) and (c) to write (7) which shows that $\phi_k(x)$, $k \in Z$, are orthonormal and, for f in V_0,

$$\hat{f}(\xi) = \sqrt{2\pi}\left(\sum_{k\in Z}\hat{f}(\xi - 2\pi k)\right)\hat{\phi}(\xi)$$

or

$$f(x) = \sum_{k\in Z} f(j)\phi(x-k)$$

which shows that they are complete in V_0.

This completes the proof of the Theorem.

Remark 5 Suppose $\mathcal{V}$ is a translation invariant multiscale analysis and Ω is a closed set such that $\widehat{V}_0 = L^2(\Omega)$. Then if W_0 is the orthogonal complement of V_0 in V_1, $\widehat{W}_0 = L^2(\Upsilon)$ where $\Upsilon = 2\Omega \setminus \Omega$. Let ψ be such that the set $\{\psi_k : k \in Z\}$ is an orthonormal basis for W_0. Such a ψ may be referred to as a wavelet. Using reasoning analogous to the proof of the Theorem, it is clear that ψ is a wavelet if and only if

$$|\hat{\psi}| = \frac{1}{\sqrt{2\pi}}\chi_{\Upsilon}\,.$$

Now, an analyzing wavelet in the sense of Meyer, [1], is globally integrable and hence its Fourier Transform must be continuous. Clearly ψ is not such an analyzing wavelet.

Corollary *A translation invariant multiscale analysis cannot give rise to analyzing wavelets in the sense of Meyer.*

References

[1] J. M. Combs, A. Grossman, and Ph. Tchamitchian, eds., *Wavelets*, Proceedings of the International Conference in Marseille, 1987, Springer-Verlag, Berlin, 1989.

[2] I. Daubechies, Orthonormal bases of compactly supported wavelets, *Comm. Pure Appl. Math.*, 41 (1988), 909-996.

HIGH-RESOLUTION SPECTRAL ANALYSIS USING RECURRENCE RELATIONS

J. H. G. LAHR
Institut Supérieur de Technologie
Rue Richard Coudenhove-Kalergi
L-1359 Luxembourg-Kirchberg
Grand-Duché de Luxembourg

ABSTRACT. A new algorithm for the determination of frequencies, amplitudes and phases of a signal is derived from difference equations related to superposed sinusoids. The minimum number of samples for a complete spectral analysis of n sinusoids and a constant term is given by 3n+1. The numerical examples show excellent accuracy.

1. INTRODUCTION

Traditional techniques for spectral estimation have been based on fast Fourier transform (FFT) algorithms. Although computational efficient, these techniques have two main limitations, namely frequency resolution and "leakage". For short data records the FFT based spectral estimation techniques may not provide enough resolution in the frequency domain to distinguish between two different signals. In addition, strong spectral components have sidelobes that may mask adjacent but weaker spectral components [1]. In this paper we present a new spectral determination method based on recurrence relations existing between the sampled values of superposed sinusoids. It is shown that the frequencies may be found by solving a set of linear equations and by determining the roots of an algebraic equation. The amplitudes and phases can be obtained subsequently by few formulae.

2. RECURRENCE RELATIONS IN A SAMPLED SINUSOID

2.1. The fundamental recurrence relation and the parameter of a sampled sinusoid

Suppose that we have the following time-discrete signal:

$$S_n = A \sin(2\pi f T_s n + c) \quad \text{with } n = 0,1,2,3... \tag{1}$$

A is the peak amplitude, f the frequency, T_s the sampling period and c the phase angle.

J. S. Byrnes and J. F. Byrnes (eds.), Recent Advances in Fourier Analysis and Its Applications, 463–482.

Using a modified version of Simson's formulae [2], [4],

$$\sin(an + c) = 2\cos(a)\sin[a(n-1) + c] - \sin[a(n-2) + c],$$
$$\cos(an + c) = 2\cos(a)\cos[a(n-1) + c] - \cos[a(n-2) + c], \tag{2}$$

we have immediately:

$$S_n - P\,S_{n-1} + S_{n-2} = 0, \tag{3}$$

$$\text{with } P = 2\cos(2\pi f T_s). \tag{4}$$

Equation (3) is a second-order linear recursion relation and represents the main relation of a sampled sinusoid. In the further development it plays a dominant part and is named "the fundamental recurrence relation of a sampled sinusoid".

The parameter P also is of central importance in the characterization of a sinusoid and is called in this paper "the parameter of a sampled sinusoid".

The angle between two consecutive samples is constant and denoted by a. We have:

$$a = 2\pi f T_s = \cos^{-1}(P/2). \tag{5}$$

Further we can make the following remarks:

a) The fundamental recurrence relation and the parameter are independent of the peak amplitude.

b) The phase angle c which may admit any positive or negative value gives the fundamental recurrence relation the property of "shift invariance". This property signifies that the sampling positions may be shifted linearly without modifying the fundamental recurrence relation.

2.2. Relations between discrete sinusoids and the polynomials of Morgan-Voyce

The Morgan-Voyce polynomials of first kind and those of second kind have the same difference equation as the fundamental recurrence relation of a sampled sinusoid. They differ only by their initial conditions. We have the following definitions:

Morgan-Voyce polynomials of first kind,

$$B_n - P\,B_{n-1} + B_{n-2} = 0, \text{ with } B_1 = 1;\ B_2 = P. \tag{6}$$

Morgan-Voyce polynomials of second kind,

$$b_n - P\,b_{n-1} + b_{n-2} = 0, \text{ with } b_1 = 1;\ b_2 = P - 1. \tag{7}$$

This definition differs from that chosen by Morgan-Voyce and other authors such as M.N.S. Swamy who start with $B_0 = 1$; $B_1 = P$; $b_0 = 1$ and $b_1 = P - 1$. [5], [6], [7], [9], [10], [11].

If S_1 and S_2 are the initial conditions of a sampled sinusoid we obtain

$$S_n = S_2B_{n-1} - S_1B_{n-2}, \tag{8}$$

and

$$S_n = [(S_3-S_2)b_{n-1} - (S_2-S_1)b_{n-2}]/(P - 2). \tag{9}$$

It is also possible to express the Morgan-Voyce polynomials in terms of a sampled sinusoid.

$$B_n = (S_2S_{n+1} - S_1S_{n+2})/(S_2^2 - S_1S_3), \tag{10}$$

and

$$b_n = [(S_1-S_0)S_{n+1} + (S_1-S_2)S_n]/(S_2^2 - S_1S_3). \tag{11}$$

Since the Morgan-Voyce polynomials are defined for all real and complex values of P, we must restrict the application of the above developed relations to the case $-2<P<2$, if we use real sinusoids. The relations (10) and (11) permit us to establish a large number of identities involving samples of sinusoids. A choice of 14 Morgan-Voyce identities may illustrate these possibilities.

$$B_nB_{r-m} = B_rB_{n-m} - B_mB_{n-r}. \tag{12}$$

$$B_{n-1}B_{n+1} = B_n^2 - 1. \tag{13}$$

$$B_{2n} = B_n(B_{n+1} - B_{n-1}). \tag{14}$$

$$B_{2n-1} = B_n^2 - B_{n-1}^2. \tag{15}$$

$$b_{2n} = B_nb_{n+1} - B_{n-1}b_n. \tag{16}$$

$$b_{2n+1} = B_{n+1}b_{n+1} - B_nb_n. \tag{17}$$

$$b_nB_{r-m} = b_{n-m}B_r - B_mb_{n-r}. \tag{18}$$

$$b_nB_{r-m} = B_{n-m}b_r - b_mB_{n-r}. \tag{19}$$

$$b_nB_n - b_{n+1}B_{n-1} = 1. \tag{20}$$

$$B_nb_{n-1} - b_nB_{n-1} = 1. \tag{21}$$

$$\sum_{r=1}^{n} B_r = (b_{n+1} - 1)/(P - 2). \tag{22}$$

$$\sum_{r=1}^{n} b_r = B_n. \tag{23}$$

$$B_n = \prod_{k=1}^{n-1}\left(P - 2\cos\frac{k\pi}{n}\right). \qquad n\geq 2 \tag{24}$$

$$b_n = \prod_{k=1}^{n-1}\left(P - 2\cos\frac{\pi(2k-1)}{2n-1}\right). \quad n\geq 2 \tag{25}$$

In the relations (10) and (11) the expression $S_2^2 - S_1S_3$ appears. It is easy to demonstrate that it depends only on the amplitude A and the angle a. In general we find:

$$S_n^2 - S_{n-1}S_{n+1} = A^2\sin^2 a = A^2[1 - (P/2)^2]. \tag{26}$$

2.3. The determination of the frequency, the amplitude and the phase of a sampled sinusoid.

The parameter P may also be determined by three consecutive samples.

$$P = (S_n + S_{n-2})/S_{n-1}. \tag{27}$$

If the sampling period T_s is known, the frequency may be computed using relation (4).

$$f = \cos^{-1}(P/2)/(2\pi T_s). \tag{28}$$

It is obvious that any formula for the determination of the frequency should not be dependent on the amplitude. This condition holds in our case.

The amplitude of a discrete sinusoid is immediately given by relation (26).

$$A = 2\sqrt{\frac{S_n^2 - S_{n-1}S_{n+1}}{4 - P^2}}. \tag{29}$$

Since $S_n = A\sin(na + c)$, and the values of a and c are arbitrary, the property of shift invariance is true in the case of the parameter P, the frequency f and the amplitude A.

For the determination of the phase angle c it is useful to consider the fundamental recurrence relation

$$S_n - PS_{n-1} + S_{n-2} = 0.$$

The initial terms of this difference equation may be S_1 and S_2. Since the value of P lies between -2 and +2, the general term of S_n may be given by:

$$S_n = 2\sqrt{\frac{S_1^2+S_2^2-PS_1S_2}{4-P^2}}\cos[n\cos^{-1}(\tfrac{P}{2}) + \tan^{-1}\frac{2S_1+PS_2-P^2S_1}{(S_2-S_1P)\sqrt{4-P^2}}].$$

See [6], page 72. (30)

The general term confirms the formulae for the determination of the amplitude and the frequency; it yields also directly an expression for the phase.

$$c = \tan^{-1}\left[\frac{2S_1 + PS_2 - P^2S_1}{(S_2 - PS_1)\sqrt{4 - P^2}}\right] + \frac{\pi}{2}. \tag{31}$$

After some transformations and allowing the use of sample S_0 we have:

$$c = \tan^{-1}\left[\frac{S_0\sqrt{4 - P^2}}{2S_1 - PS_0}\right]. \tag{32}$$

It lies in the nature of the phase angle that the property of shift invariance is not true in this case. If we substitute in (32) S_0 by S_1 and S_1 by S_2, that signifies an incrementation of the subscripts by one, then we obtain a phase angle c_1, which value equals c + a. By mathematical induction we may verify that the following relation holds:

$$c_n = a\,n + c = \tan^{-1}\left[\frac{S_n\sqrt{4 - P^2}}{2S_{n+1} - PS_n}\right]. \tag{33}$$

Let us now summarize the preliminary results. Given are the samples of a sinusoid; the frequency, the amplitude and the phase being unknown. The following expressions permit their determination.

$$P = \frac{S_n + S_{n-2}}{S_{n-1}}, \qquad \text{for any value of n.}$$

$$f = \frac{\cos^{-1}(P/2)}{2\pi T_s}.$$

$$A = 2\sqrt{\frac{S_n^2 - S_{n-1}S_{n+1}}{4 - P^2}}. \tag{34}$$

$$c_n = \tan^{-1}\left[\frac{S_n\sqrt{4 - P^2}}{2S_{n+1} - PS_n}\right].$$

The main conclusion of this development is:

> A discrete sinusoid is completely determined by three samples. (35)

2.4. Some special cases

2.4.1. The parameter P is indeterminate

Suppose that we have three consecutive samples of a sinusoid, S_1, S_2 and S_3. If $S_2 = 0$, we have also $S_3 = -S_1$. In this case P is indeterminate, and the characterization of the sinusoid is not possible with these samples.

2.4.2. P = -2

This case happens if the frequency equals the so-called Nyquist frequency; that is the limit given by the theorem of C. Shannon.

2.4.3. P = 2

This case happens if the three samples have the same value, or in general, if the three samples constitute an arithmetic progression. In this context it is opportune to mention a family of recurrence relations with coefficients coming from Pascal's triangle (with alternating sign).

$$S_n - 2\,S_{n-1} + S_{n-2} = 0. \tag{36}$$

$$S_n - 3\,S_{n-1} + 3\,S_{n-2} - S_{n-3} = 0. \tag{37}$$

$$S_n - 4\,S_{n-1} + 6\,S_{n-2} - 4\,S_{n-3} + S_{n-4} = 0. \tag{38}$$

If the values of the initial conditions of these difference equations belong to successive terms of an arithmetic progression, then all the terms of these equations belong to an arithmetic progression. (39)

2.5. The case of a sinusoid and a superposed constant term

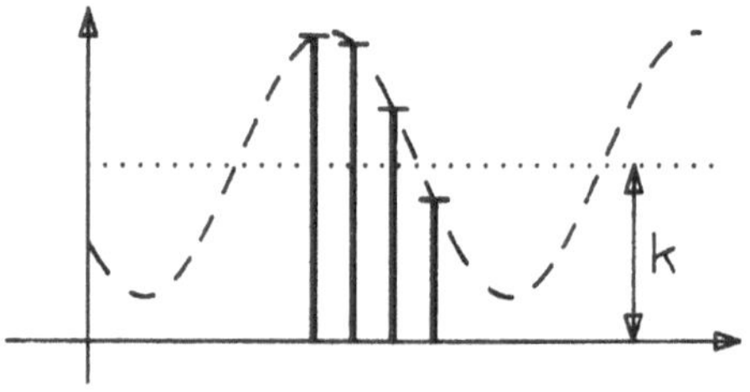

Figure 1. Sinusoid with a constant term

In figure 1 we have represented this case where a constant k is added to all samples of the sinusoid. The determination of the constant may be made by the reflection that the samples number 1,2 and 3 must have the same parameter as the samples number 2,3 and 4. If we subtract the value k from all samples and if we use a general subscript, then we obtain:

$$\frac{S_n + S_{n-2} - 2k}{S_{n-1} - k} = \frac{S_{n-1} + S_{n-3} - 2k}{S_{n-2} - k} \quad .$$

This expression enables us to find the constant k if 4 consecutive samples are known.

$$k = \frac{S_{n-2}(S_n + S_{n-2}) - S_{n-1}(S_{n-1} + S_{n-3})}{S_n - 3S_{n-1} + 3S_{n-2} - S_{n-3}}. \tag{40}$$

After subtracting k from all samples, the determination of f, A and c represents no problem.If for some reason the parameter P is known, then k may be computed in the following way:

$$k = \frac{S_n - S_{n-1}P + S_{n-2}}{(2 - P)}. \tag{41}$$

If the four samples of expression (40) are lying on a straight line, then the denominator gets the value 0, according to the statement (39). The numerator too receives the value 0 in this case. As a matter of fact it is easy to demonstrate that the following theorem holds.

$$\begin{aligned} &S_{n-2}(S_n + S_{n-2}) - S_{n-1}(S_{n-1} + S_{n-3}) = 0, \\ &\text{if } S_n = x + ny; \text{ for any value of x and y.} \end{aligned} \tag{42}$$

Considering these relationships it may be said that in the case of four samples forming an arithmetic progression it is not possible to determine the constant k.

By substituting k in the expression

$$P = \frac{S_n + S_{n-2} - 2k}{S_{n-1} - k},$$

we obtain after some computations:

$$P = \frac{S_n - S_{n-1} + S_{n-2} - S_{n-3}}{S_{n-1} - S_{n-2}}. \tag{43}$$

The fundamental recurrence relation of a sampled sinusoid and a superposed constant term can immediately be established using the preceding equation.

$$S_n - (P+1)S_{n-1} + (P+1)S_{n-2} - S_{n-3} = 0. \tag{44}$$

The relations (43) and (44) may also be found if we substitute in the corresponding expressions without a constant term (relation (27) and (3)) each sample by the difference of two consecutive samples. The following substitutions may be used:

$$S_n \rightarrow S_n - S_{n-1}, \quad \text{or} \quad S_n \rightarrow S_n - S_{n+1}. \tag{45}$$

The main conclusion of this chapter is:

A sinusoid and a constant term are completely determined by four samples.

(46)

3. RECURRENCE RELATIONS IN SUPERPOSED SINUSOIDS

3.1. The fundamental recurrence relations

Let us denoted the samples of a first sinusoid by A_n and those of a second by B_n. The corresponding parameters may be called P and Q. So we have:

$$A_n = P\,A_{n-1} - A_{n-2}. \tag{47}$$

$$B_n = Q\,B_{n-1} - B_{n-2}. \tag{48}$$

The samples of the superposed sinusoid are denoted by S_n, and we obtain:

$$S_n = A_n + B_n. \tag{49}$$

The fundamental recurrence relation is given by:

$$S_n - (P+Q)S_{n-1} + (2+PQ)S_{n-2} - (P+Q)S_{n-3} + S_{n-4} = 0. \tag{50}$$

Several ways are possible to prove this relation. Very useful for this purpose are some theorems established by Maurice d'Ocagne [8],[4]. In the case of three superposed sinusoids we obtain, if we denote by R the parameter of the third sinusoid, the following fundamental recurrence relation:

$$S_n - (P+Q+R)S_{n-1} + (3+PQ+PR+QR)S_{n-2} - [2(P+Q+R)+PQR]S_{n-3} + \\ + (3+PQ+PR+QR)S_{n-4} - (P+Q+R)S_{n-5} + S_{n-6} = 0. \tag{51}$$

In the case of four superposed sinusoids and introducing the parameter T related to the fourth sinusoid we can write:

$$\begin{aligned} S_n &- (P+Q+R+T)S_{n-1} + (4+PQ+PR+PT+QR+QT+RT)S_{n-2} - \\ &- [3(P+Q+R+T)+PQR+PQT+PRT+QRT]S_{n-3} + \\ &+ [6+2(PQ+PR+PT+QR+QT+RT)+PQRT]S_{n-4} - \\ &- [3(P+Q+R+T)+PQR+PQT+PRT+QRT]S_{n-5} + \\ &+ (4+PQ+PR+PT+QR+QT+RT)S_{n-6} - (P+Q+R+T)S_{n-7} + S_{n-8} = 0. \end{aligned} \tag{52}$$

It is possible to determine the coefficients of the fundamental recurrence relation of n superposed sinusoids in a recurrent manner in terms of the coefficients of the fundamental recurrence relation of n-1 sinusoids. For this proposal let us denote by P_i the parameter of the sinusoid number i. A general coefficient is denoted by $T_{i,j}$, where the first subscript i designates the number of the superposed sinusoids and where the second subscript j designates the position of the coefficient in the fundamental recurrence relation. The recurrent determination may then be described by:

$$T_{i,j} = T_{i-1,j-2} - P_i T_{i-1,j-1} + T_{i-1,j}. \tag{53}$$

With $T_{i,j} = 0$ if $j \leq 0$ or $j > 2i+1$, and $T_{0,1} = 1$.

Further we point out the following changes of notations:

$$P_1 = P;\ P_2 = Q;\ P_3 = R;\ P_4 = T;\ \ldots \tag{54}$$

In figure 2 the coefficients are written in tabular form.

i \ j	1	2	3	4	5	6	7	8	9
1	1	-P	1						
2	1	-(P+Q)	2+PQ	-(P+Q)	1				
3	1	-(P+Q +R)	3+PQ+PR +QR	-(P+Q+R) -PQR	3+PQ+PR +QR	-(P+Q+R)	1		
4	1	-(P+Q +R+T)	4+PQ+PR +PT+QR +QT+RT	-3(P+Q+R+T) -PQR-PQT -PRT-QRT	6+2(PQ+PR +PT+QR+QT +RT)+PQRT	-3(P+Q+R+T) -PQR-PQT -PRT-QRT	4+PQ+PR +PT+QR +QT+RT	-(P+Q +R+T)	1

Figure 2. The triangle of the coefficients of the fundamental recurrence relations of superposed sinusoids.

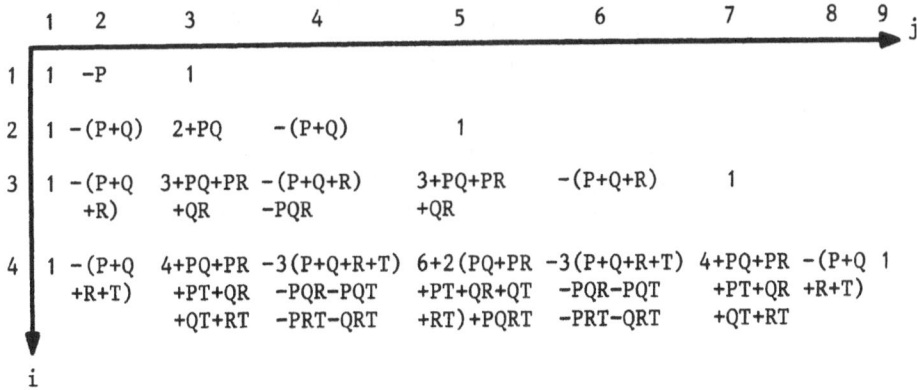

We see that the coefficients of the fundamental recurrence relations are disposed in a symmetrical way and that they appear as sums and products of the parameters P_i. Further we may observe that the following statement is true:

The fundamental recurrence relation of n superposed sinusoids is of the order 2n.	(55)

3.2. The modified fundamental recurrence relations

It may be useful to express the fundamental recurrence relations in terms of the different sums and products of the parameters. So we find: In the case of one sinusoid:

$$S_n + S_{n-2} - PS_{n-1} = 0. \tag{56}$$

In the case of two sinusoids:

$$S_n + 2S_{n-2} + S_{n-4} - (P+Q)(S_{n-1} + S_{n-3}) + PQS_{n-2} = 0. \tag{57}$$

In the case of three sinusoids:

$$\begin{aligned} &S_n+3S_{n-2}+3S_{n-4}+S_{n-6} - (P+Q+R)(S_{n-1}+2S_{n-3}+S_{n-5}) + \\ &\quad + (PQ+PR+QR)(S_{n-2}+S_{n-4}) - PQRS_{n-3} = 0. \end{aligned} \tag{58}$$

And in the case of four sinusoids:

$$\begin{aligned} &S_n+ 4S_{n-2}+6S_{n-4}+4S_{n-6}+S_{n-8} - \\ &\quad - (P+Q+R+T)(S_{n-1}+3S_{n-3}+3S_{n-5}+S_{n-7}) + \\ &\quad + (PQ+PR+PT+QR+QT+RT)(S_{n-2}+2S_{n-4}+S_{n-6}) - \\ &\quad - (PQR+PQT+PRT+QRT)(S_{n-3}+S_{n-5}) + PQRTS_{n-4} = 0. \end{aligned} \tag{59}$$

In this development the coefficients in the different parentheses containing the samples S_{n-i} are given by Pascal's triangle. Further we see that in the parentheses the subscripts are always decremented in steps of two.

Using the relation (53) or the generalization, which may be established on the basis of the modified fundamental recurrence relations using the elements of Pascal's triangle, it is not difficult to proceed to the determination of the fundamental recurrence relation of a superposition of n sinusoids.

3.3. The fundamental recurrence relations of n superposed sinusoids in the presence of a constant term.

We consider first the case of two sinusoids and a constant term k. On the basis of relation (50) we find after some computations, consisting in the elimination of k, the following equation:

$$S_n - S_{n-1} - (P+Q)(S_{n-1} - S_{n-2}) + (2+PQ)(S_{n-2} - S_{n-3}) - (P+Q)(S_{n-3} - S_{n-4}) + S_{n-4} - S_{n-5} = 0. \tag{60}$$

This is the fundamental recurrence relation of two superposed sinusoids and a constant term. This relation is independent of the value of a constant term; it can also be established by the substitution in relation (50) of each sample by the difference of two consecutive samples. See also (44) and (45).

If P and Q are known, the constant term k may be given by:

$$k = \frac{S_n-(P+Q)S_{n-1}+(2+PQ)S_{n-2}-(P+Q)S_{n-3}+S_{n-4}}{(2-P)(2-Q)}. \tag{61}$$

In reality it is sufficient to know the sum and the product of P and Q.

$$k = \frac{S_n-(P+Q)S_{n-1}+(2+PQ)S_{n-2}-(P+Q)S_{n-3}+S_{n-4}}{4-2(P+Q)+PQ}. \tag{62}$$

In the case of three sinusoids and a constant term, the corresponding fundamental recurrence relation is given by:

$$S_n-S_{n-1}-(P+Q+R)(S_{n-1}-S_{n-2})+(3+PQ+PR+QR)(S_{n-2}-S_{n-3}) - [2(P+Q+R)+PQR](S_{n-3}-S_{n-4})+(3+PQ+PR+QR)(S_{n-4}-S_{n-5}) - (P+Q+R)(S_{n-5}-S_{n-6})+S_{n-6}-S_{n-7} = 0. \tag{63}$$

The constant term k is related to the corresponding fundamental recurrence relation by:

$$k = \frac{S_n-(P+Q+R)S_{n-1}+(3+PQ+PR+QR)S_{n-2}-[2(P+Q+R)+PQR]S_{n-3}}{(2-P)(2-Q)(2-R)} + \frac{(3+PQ+PR+QR)S_{n-4}-(P+Q+R)S_{n-5}+S_{n-6}}{(2-P)(2-Q)(2-R)}. \tag{64}$$

The relationships established in this chapter permit the following statements:

The fundamental recurrence relation of n superposed sinusoids and a constant term is of the order 2n+1.	(65)

If we substitute in the fundamental recurrence relation of n superposed sinusoids each sample by the difference of two consecutive samples, then we obtain the fundamental recurrence relation of n superposed sinusoids and a constant term.	(66)

The constant term k and the fundamental recurrence relation of n superposed sinusoids are related by the following equation:

$$k = \frac{\text{fundamental recurrence relation of n sinusoids}}{\prod_{i=1}^{n} (2 - P_i)}. \qquad (67)$$

4. SPECTRAL ANALYSIS

4.1. A special case: Three superposed sinusoids

In order to illustrate the use of the here established recurrence relations for the determination of the spectral components of a signal, let us first consider a set of 9 consecutive samples of a signal composed by three superposed sinusoids.

Let us for this purpose introduce the following notations.

$$\begin{aligned} U_1 &= P + Q + R. \\ U_2 &= PQ + PR + QR. \\ U_3 &= PQR. \end{aligned} \qquad (68)$$

$$\begin{aligned} L_{1,i} &= S_i. \\ L_{2,i} &= S_{i+1} + S_{i-1}. \\ L_{3,i} &= S_{i+2} + 2S_i + S_{i-1}. \\ L_{4,i} &= S_{i+3} + 3S_{i+1} + 3S_{i-1} + S_{i-3}. \end{aligned} \qquad (69)$$

With the modified fundamental recurrence relation (58) we may establish a system of 3 linear equations to determine the unknown variables U_1, U_2 and U_3.

$$L_{1,6}U_3 - L_{2,6}U_2 + L_{3,6}U_1 - L_{4,6} = 0.$$
$$L_{1,5}U_3 - L_{2,5}U_2 + L_{3,5}U_1 - L_{4,5} = 0. \qquad (70)$$
$$L_{1,4}U_3 - L_{2,4}U_2 + L_{3,4}U_1 - L_{4,4} = 0.$$

Using the well-known theorem of Vieta we then solve the following cubic equation to determine the values of P, Q and R.

$$x^3 - U_1x^2 + U_2x - U_3 = 0. \qquad (71)$$

The three roots of this equation are identical with the parameters P, Q and R, and permit the determination of the frequencies of the three sinusoids (with relation (28) for instance).

For the computation of the amplitudes and phases we need two consecutive samples of each involved sinusoid. These samples are denoted by A_1, A_2, B_1, B_2, C_1 and C_2 (C_n is the notation for the samples of the third sinusoid) and may be determined by the following system of linear equations:

A_1	A_2	B_1	B_2	C_1	C_2	
1	0	1	0	1	0	$-S_1$
0	1	0	1	0	1	$-S_2$
-1	P	-1	Q	-1	R	$-S_3$
-P	P^2-1	-Q	Q^2-1	-R	R^2-1	$-S_4$
$-P^2+1$	P^3-2P	$-Q^2+1$	Q^3-2Q	$-R^2+1$	R^3-2R	$-S_5$
$-P^3+2P$	P^4-3P^2+1	$-Q^3+2Q$	Q^4-3Q^2+1	$-R^3+2R$	R^4-3R^2+1	$-S_6$

(72)

It may be mentioned that the Morgan-Voyce polynomials of the first kind appear in the matrix. [5],[6],[7],[9],[10],[11]. This gives us the conviction that the general solution of this linear system could have a very elementary form.

As a matter of fact we obtain:

$$A_n = \frac{S_{n-2} - (Q+R)S_{n-1} + (2+QR)S_n - (Q+R)S_{n+1} + S_{n+2}}{(P-Q)(P-R)}. \qquad (73)$$

$$B_n = \frac{S_{n-2} - (P+R)S_{n-1} + (2+PR)S_n - (P+R)S_{n+1} + S_{n+2}}{(Q-P)(Q-R)}. \qquad (74)$$

$$C_n = \frac{S_{n-2} - (P+Q)S_{n-1} + (2+PQ)S_n - (P+Q)S_{n+1} + S_{n+2}}{(R-P)(R-Q)}. \qquad (75)$$

Two remarks may be made:

The numerator is identical with the fundamental recurrence relation of two superposed sinusoids.

(76)

In the formula for a specific sinusoid related to the parameter P_s, the numerator contains only the two other parameters denoted for this purpose by P_a and P_b. The denominator has the form: $(P_s - P_a)(P_s - P_b)$.

(77)

4.2. The general case of n superposed sinusoids

Suppose that 3n samples of a signal composed by n superposed sinusoids are given, and that the frequency, the amplitude and the phase of each particular sinusoid is wanted.

The parameters are denoted by $P_1, P_2, P_3, \ldots, P_n$. The different sums and products of the parameters are designed by the following auxiliary variables.

$$\begin{aligned}
U_1 &= P_1 + P_2 + P_3 + \ldots + P_n. \\
U_2 &= P_1P_2 + P_1P_3 + P_1P_4 + \ldots + P_{n-1}P_n. \\
U_3 &= P_1P_2P_3 + P_1P_2P_4 + P_1P_2P_5 + \ldots\ldots + P_{n-2}P_{n-1}P_n. \\
&\vdots \\
U_n &= P_1P_2P_3 \ldots P_n.
\end{aligned} \tag{78}$$

The coefficients of the system of linear equations that must be solved are established by using the auxiliary polynomials defined in (69) in which appear the elements of Pascal's triangle. Let us repeat and complete these definitions.

$$\begin{aligned}
L_{1,i} &= S_i. \\
L_{2,i} &= S_{i+1} + S_{i-1}. \\
L_{3,i} &= S_{i+2} + 2S_i + S_{i-2}. \\
L_{4,i} &= S_{i+3} + 3S_{i+1} + 3S_{i-1} + S_{i-3}. \\
L_{5,i} &= S_{i+4} + 4S_{i+2} + 6S_i + 4S_{i-2} + S_{i-4}. \\
&\vdots \\
L_{n+1,i} &= S_{i+n} + nS_{i+n-2} + \ldots + S_{i-n}.
\end{aligned} \tag{79}$$

The matrix of the linear system to determine the unknowns $U_1 \dots U_n$ can be represented in the following manner:

U_n	U_{n-1}	U_{n-2}	$\dots$	
$L_{1,2n}$	$-L_{2,2n}$	$L_{3,2n}$		$(-1)^n L_{n+1,2n}$
$L_{1,2n-1}$	$-L_{2,2n-1}$	$L_{3,2n-1}$		$(-1)^n L_{n+1,2n-1}$
$L_{1,2n-2}$	$-L_{2,2n-2}$	$L_{3,2n-2}$		$(-1)^n L_{n+1,2n-2}$
$\vdots$				
$L_{1,n+1}$	$-L_{2,n+1}$	$L_{3,n+1}$		$(-1)^n L_{n+1,n+1}$

(80)

The determination of the parameters is now possible; we only need to solve the following general algebraic equation:

$$x^n - U_1 x^{n-1} + U_2 x^{n-2} - U_3 x^{n-3} + \dots + (-1)^n U_n = 0. \tag{81}$$

The roots of this algebraic equation are the parameters $P_1 \dots P_n$. The frequencies of the n sinusoids may now be computed using relation (28):

$$f_i = \frac{\cos^{-1}(P_i/2)}{2\pi T_s}, \quad i = 1, 2, 3, \dots, n. \tag{82}$$

The determination of at least two samples of each sinusoid may be executed according to the following algorithm, called "the samples determination algorithm".

<u>Step 1</u> Begin with the sinusoid number n; the corresponding parameter is P_n.

<u>Step 2</u> Establish the coefficients $T_{n-1,j}$ of the fundamental recurrence relation of n-1 superposed sinusoids with the parameters $P_1, P_2, \dots, P_{n-1}$.

$$T_{i,j} = T_{i-1,j-2} - P_i T_{i-1,j-1} + T_{i-1,j}$$

Initial conditions:

$T_{i,j} = 0$ for $j \leq 0$ or for $j > 2i+1$.

$T_{0,1} = 1$.

<u>Step 3</u> The samples N_j of the nth sinusoid are given by: (See (76) and (77).)

$$N_j = \frac{\sum_{k=1}^{2n-1} T_{n-1,k} \cdot S_{j+k-n}}{\prod_{k=1}^{n-1} (P_n - P_k)}. \tag{83}$$

Compute the two samples N_n and N_{n+1} (to avoid the use of samples like $S_0, S_{-1}, \ldots$). If a third sample is needed, then compute: $N_{n+2} = P_n N_{n+1} - N_n$.

<u>Step 4</u> Exchange the values of the parameters according to the following prescription:

$$\begin{aligned} H &= P_n, \\ P_n &= P_{n-1}, \\ P_{n-1} &= P_{n-2}, \\ &\cdots \\ P_2 &= P_1, \\ P_1 &= H. \end{aligned} \tag{84}$$

<u>Step 5</u> Go n times back to step 2.

The steps 2, 3 and 4 are executed n times, and every time the samples of a sinusoid are determined. After these computations the amplitudes and the phases may be obtained using the relations (29) and (31) or (32).

The main characteristics of this new approch for spectral analysis may be enumerated as follows:

> n discrete sinusoids are completely determined by 3n samples. By completely determined we mean the determination of frequencies, amplitudes and phases. (85)

> The main mathematical operations consist of the solution of a system of n linear equations and of the determination of the n roots of an algebraic equation. (86)

From the practical point of view, this method contains two difficulties.

a) If two frequencies are very close together, then the corresponding parameters are almost identical and we are confronted with the problem of numerical instability resulting from the loss of significant digits in formula (83).
b) The resolution of an algebraic equation of degree n could be difficult.

4.3. The general case of n superposed sinusoids in the presence of a constant term.

If we substitute in the fundamental recurrence relation of n superposed sinusoids each sample S_i by the difference S_i-S_{i-1} or by S_i-S_{i+1}, (see relation (45).) then the relation is independent of the constant term and we can determine the parameters of the sinusoids by the same algorithm as in the case without a constant term. After the computation of the frequencies according (82) we determine the constant term k using relation (67). Then we subtract from each sample the constant k. With these modified samples we may use the samples determination algorithm and are then enabled to compute the amplitudes and the phases.

Finally we may say:

n discrete sinusoids and a constant term are completely determined by 3n+1 samples.	(87)
To avoid the effect of a constant term upon the determination of the parameters and the frequencies it is sufficient to replace each sample by the difference of two consecutive samples.	(88)
The effect of a constant term upon the determination of the amplitudes and the phases can not be eliminated only by the substitution of S_i by S_i-S_{i-1}. Indeed after the replacement of each sample of a data record by the difference of two consecutive samples, a Fourier transform of these modified samples yields the correct frequencies (without the frequency 0 of course) but completely false amplitudes and phases.	(89)

5. NUMERICAL EXAMPLES

The numerical examples are chosen under the point of view that there are two fundamental issues in computing spectra of sinusoids:

a) Dynamic range: How great may the difference in amplitudes of sinusoids be and still be identified.

b) Resolution: How closely may two sinusoids be spaced in frequency and still be identified.

5.1. First example

We consider the superposition of three sinusoids consisting of two strong signals and one of about 60 dB lower.
Input data:

sinusoid	amplitude	frequency	phase
1	0 dB	356 Hz	67°
2	0 dB	631 Hz	166°
3	-60.27 dB	533 Hz	37°

The numerical values are identical with those of a publication by R. Küng, where FFT-processors for real-time-signal-processing are presented. In this publication [3] a Fourier transform with a resolution of 1 Hz, a sampling frequency of 2048 Hz and 2048 samples is used. For the weak signal number 3 the FFT-processor gives an amplitude of -59.22 dB and a phase of 33.69°.

Using the method developed in this paper one needs only 9 samples. The corresponding FORTRAN program (double precision, output format G14.7) produces the following result:

sinusoid	amplitude	frequency	phase
1	0.4580555E-14≅0 dB	356.0000	67.00000°
2	0.2676009E-13≅0 dB	631.0000	166.0000°
3	-60.27000 dB	533.0000	37.00000°

Comment: The weak signal is determined with optimal accuracy and by far better than by using the Fourier transform.

5.2. Second example

We consider again three superposed sinusoids but the two strong signals have almost identical frequencies; the weak signal has the same level as in the first example.

Input data:

sinusoid	amplitude	frequency	phase
1	0 dB	356 Hz	67°
2	0 dB	356.1 Hz	166°
3	-60.27 dB	533 Hz	37°

The analysis of 9 samples of these data with our method yields:

sinusoid	amplitude	frequency	phase
1	0.3672321E-07≅0 dB	356.0000 Hz	67.00000°
2	-0.3517563E-7≅0 dB	356.1000 Hz	166.0000°
3	-60.27000 dB	533.0000 Hz	37.00000°

Comment: The separation of the two almost identical frequencies is very good. The reduced accuracy of the amplitudes of the sinusoids number 1 and 2 is remarkable. The reason for this is the behaviour of relation (83) in the case of almost identical frequencies.

6. CONCLUSION

A new algorithm for determining the frequencies, the amplitudes and the phases of superposed discrete sinusoids has been presented. The mathematical foundations lie in the field of difference equations, and the numerical applications show excellent accuracy. We think that this method could be a very interesting complement to the usual procedures of signal processing, especially if a dedicated hardware is used.

The theory of this new method is not closed by far. Two main extensions are possible:

a) The spectral analysis of complex sinusoids.

b) The spectral estimation in the case of noisy signals.

This matter is under investigation and will be reported elsewhere.

7. REFERENCES

[1] CRUZ, J.R. (1986) 'Spectral Analysis Using the QD-Algorithm'. ICASSP 86, Tokyo, pp. 1345-1348.

[2] HORADAM, A.F. (1969) 'Tschebyscheff and Other Functions Associated with the Sequence $W_n(a,b;p,q)$'. The Fibonacci Quarterly 7, No. 1, pp. 14-22.

[3] KÜNG, R. (1986) 'FFT-Prozessoren für REAL-TIME-Signalverarbeitung'. SEV-Bulletin, 77(1986), 11, p. 646. (Bulletin of the Swiss Association of Electrical Engineers).

[4] LAHR, J.H.G. (1989) 'Recurrence Relations in Sinusoids and Their Applications to Spectral Analysis and to the Resolution of Algebraic Equations'. Proceedings of the Third International Research Conference on Fibonacci Numbers and Their Applications, University of Pisa, Italy, July 25-29, 1988.

[5] LAHR, J.H.G. (1985) 'Fibonacci and Lucas Numbers and the Morgan-Voyce Polynomials in Ladder Networks and in Electric Line Theory'. Proceedings of the First International Conference on Fibonacci Numbers and Their Applications, University of Patras, Greece, August 27-31, 1984.

[6] LAHR, J.H.G. (1981) 'Theorie Elektrischer Leitungen Unter Anwendung und Erweiterung der Fibonacci-Funktion'. Dissertation ETH Nr. 6958, Zürich.

[7] MORGAN-VOYCE, A.M. (1959) 'Ladder Network Analysis Using Fibonacci Numbers'. I.R.E. Trans. Circuit Theory 6, No. 3 (1959), pp. 321-322.

[8] d'OCAGNE, M. (1894) 'Mémoire sur les Suites Récurrentes'. Journal de l'Ecole Polytechnique 64(1894), Paris, pp. 151-224.

[9] SWAMY, M.N.S. (1966) 'Properties of the Polynomials defined by Morgan-Voyce'. The Fibonacci Quarterly 4, No. 1 (1966), pp. 73-81.

[10] SWAMY, M.N.S. (1966) 'More Fibonacci Identities'. The Fibonacci Quarterly 4, No. 4 (1966), pp. 369-372.

[11] SWAMY, M.N.S. (1968) 'Further Properties of Morgan-Voyce Polynomials'. The Fibonacci Quarterly 6, No. 2 (1968), pp. 167-175.

DIOPHANTINE INEQUALITIES AND SAMPLING RATES FOR MULTIBAND SIGNALS

M.M.DODSON
Department of Mathematics
University of York
York England YO1 5DD

ABSTRACT: Sampling rates for signals with equally spaced and equally wide frequency bands are obtained using Diophantine inequalities associated with a difference set condition on the spectrum.

1. INTRODUCTION

A physical (finite energy) signal can be regarded as a continuous square integrable function $f:\mathbb{R} \to \mathbb{R}:t \to f(t)$, where the real variable t can be thought of as time and f(t) as a voltage or a reading. A multiband signal is one which has frequencies consisting of a union of bands. The frequencies of a signal are essentially the support of the Fourier transform, which will here be defined for all real x by

$$\hat{f}(x) = \int_{\mathbb{R}} f(t)\, e^{2\pi ixt}\, dt.$$

As is well known, the Fourier transform of a physical (real valued) signal is symmetric. Reconstructing a signal from measurements or samples taken at regular intervals is a widely used technique in signal processing, particularly since the advent of digital computers.

J. S. Byrnes and J. F. Byrnes (eds.), Recent Advances in Fourier Analysis and Its Applications, 483–498.

In this paper, sampling rates of signals with spectra made up of a number of equally spaced bands of the same width are described. These bands can be thought of as being centred at a number of harmonics of a fundamental frequency or carrier wave. Signals of this kind, which will be called *harmonic*, can occur when the carrier wave is not perfectly sinusoidal or when an amplitude modulated carrier is rectified for detection.

The fundamental result in this area is Shannon's theorem (Shannon 1949), which states that any low band pass signal (i.e. a signal with one frequency band $(-\omega,\omega)$) can be reconstructed from samples taken at any rate at least 2ω. The reconstruction is given by the formula

$$f(t) = \sum_{k=-\infty}^{\infty} f\left(\frac{k}{2\omega}\right) \frac{\sin\pi(2\omega t-k)}{\pi(2\omega t-k)} .$$

It was first proved explicitly in 1914 by E.T.Whittaker from the point of view of interpolation theory and there is a connection with Paley-Wiener theory; further historical details are given in the interesting survey article (Higgins 1985). The sum on the right hand side is essentially Whittaker's cardinal series C(t) which assumes the values $f(k/2\omega)$, $k \in \mathbb{Z}$, and is free from oscillations above ω. The significance of the formula for communication theory was recognised by Shannon and the theorem is commonly known by his name, though Kotel'nikov (1933) in the USSR had obtained the result for radio communication independently in 1933. The result has been applied in many other fields, including optics, crystallography and spectroscopy and has an extensive literature (see for example the survey articles by Butzer (1983) and Jerri (1977).

The key to the Shannon theorem is a "disjoint translates" condition. The set $(-\omega,\omega)$ and its translates by integer multiples of 2ω repeat periodically along the frequency axis with period equal to the sampling rate, without overlapping and hence without aliasing. This allows a signal with frequencies less than ω to be reconstructed using (1) from samples taken every $1/2\omega$ seconds. There is a more general version of the Shannon theorem based on a more general disjoint translates condition. Let A be a subset of (frequency space) $\mathbb{R}$ satisfying for some number ρ and all non-zero integers k,

$$(A + k\rho)\cap A = \Phi, \tag{1}$$

where $A + k\rho = \{a + k\rho : a \in A\}$, i.e. the translate of the set A by any non-zero integer multiple of ρ is disjoint from A. Then any signal f with Fourier transform $\hat{f}$ vanishing outside A (i.e. if $\omega \notin A$, then $\hat{f}(\omega) = 0$) can be reconstructed from samples taken at a rate ρ, by means of the formula

$$f(t) = \frac{1}{\rho}\sum_{k=-\infty}^{\infty} f\left(\frac{k}{\rho}\right) \check{\chi}_A\left(t - \frac{k}{\rho}\right), \tag{2}$$

where $\check{\chi}_A$ is the inverse Fourier transform of the characteristic or indicator function of the set A, i.e.

$$\check{\chi}_A(t) = \int_{\mathbb{R}} \chi_A(x) e^{2\pi itx}\, dx = \int_{\mathbb{R}} e^{2\pi itx}\, dx$$

(Lloyd 1959, Powell 1983, Brown 1985, Dodson and Silva 1985, 1989, Higgins 1987). (In Shannon's theorem, $\rho = 2\omega$.) This "disjoint translates" condition is equivalent to the following "difference set" condition. For any set of real numbers X the *difference set* D(X) is defined by

$$D(X) = \{x - x' : x, x' \in X\}.$$

Note that D(X) always contains 0 and that D(X) is symmetric about the origin (i.e. if $y \in D(X)$, then $-y \in D(X)$). It is straightforward to verify that the disjoint translates condition (1) is equivalent to

$$D(A) \cap \rho\mathbb{Z} = \{0\}, \qquad (3)$$

i.e. for the positive number ρ, no nonzero integer multiple $k\rho$ of ρ lands in D(A). By symmetry, it suffices to consider only that part of D(A) on the positive real axis and positive integers k. Thus the number ρ is a sampling rate for the signal f with Fourier transform $\hat{f}$ vanishing outside A if, starting from 0, no "hops" of length ρ fall into the positive part of D(A). (Thus for a signal with all its frequencies below ω, ρ can be taken to be any number at least 2ω. For a high band pass signal, where the frequencies lie in the bands $(-\omega_1,-\omega_0)\cup(\omega_0,\omega_1)$, the sampling rate $\rho \geqslant 2\omega_1/[\omega_1/(\omega_1-\omega_0)]$ (Linden 1959, Dodson and Silva 1985).)

The representation (2) can be used with sampling rates lower than the Shannon rate when the spectrum of the signal has suitable gaps (Dodson and Silva 1985, 1989). The difference set condition provides a simple algorithm for obtaining efficient sampling rates (Dodson and Silva 1989), and is particularly suited for harmonic signals (with equally spaced, equally wide frequency bands).

A useful fact is that when ρ satisfies (1), ρ is at least $|A|$, the length or measure of the set A (or the bands in the Fourier transform of f), i.e.

$$|A| \leqslant \rho \qquad (4)$$

(see Dodson and Silva 1985, 1989). This is a special case of a principal familiar to electrical engineers and was proved by H J Landau (1967). A nice way to see this is to consider a wheel of radius $\rho/2\pi$ which is rolled along the real line $\mathbb{R}$ and to suppose that the points in A stick to the circumference. The difference set condition (3) (or the disjoint translates condition (1)) guarantees that distinct points in A will stick to different points on the wheel, so that the length $|A|$ of A is at most the circumference ρ.

2. SAMPLING RATES FOR HARMONIC SIGNALS

In practice sampling rates for a given signal with two or more frequency bands will not usually be obvious; there is no simple general formula for a sampling rate less than the Shannon rate. However the difference set of the frequencies in a harmonic signal has a relatively simple form. Because of this, a description of the sampling region and an explicit (if somewhat complicated) formula for the optimal sampling rate for harmonmic signals is possible.

Suppose that the Fourier transform of the signal consists of 2N+1 frequency bands

$$(-Nd-\tfrac{1}{2}w,-Nd+\tfrac{1}{2}w),\ldots,(-d-\tfrac{1}{2}w,-d+\tfrac{1}{2}w),(-\tfrac{1}{2}w,\tfrac{1}{2}w),(d-\tfrac{1}{2}w,d+\tfrac{1}{2}w),\ldots,(Nd-\tfrac{1}{2}w,Nd+\tfrac{1}{2}w),$$

each of width w. The positive part of the difference set is another set of 2N+1 bands, each centred as before but with width 2w:

$$[0,w),(d-w,d+w),\ldots,(Nd-w,Nd+w),\ldots,((2N+1)d-w,(2N+1)d+w).$$

For ρ to be a sampling rate, no hops of length ρ starting from 0 can land in any of the intervals (which can be considered to be "forbidden"). Thus ρ will be a sampling rate, i.e. will satisfy (3), if the following Diophantine inequalities

$$|md - n\rho| \geq w, \quad 0 \leq m \leq 2N, \quad n \geq 1 \tag{5}$$

hold. By (4), $\rho \geq (2N+1)w$ (the measure of the support of the Fourier transform of the signal). It follows that in (5), values of n satisfying

$$n \leq \frac{2Nd+w}{(2N+1)w} = \frac{2N(d/w) + 1}{2N+1} = \frac{d}{w} - \frac{(d/w)-1}{2N+1} < \frac{d}{w}$$

need only be considered. For convenience, divide by w and let $x = d/w$, $y = \rho/w$. Then instead of (5) we require x and y to satisfy the Diophantine inequality

$$|mx - ny| \geq 1, \tag{6}$$

for all integers m,n satisfying $0 \leq m \leq 2N$, $1 \leq n < x$. The solutions (x,y) of (6) correspond to a harmonic signal with band centres separated by a distance $d = xw$ and a sampling rate $\rho = yw$. For convenience we work with x and y, so that in particular we consider

$$A/w = \{a/w : a \in A\} =$$

$$(-Nx-½,-Nx+½)\cup\ldots\cup(-x-½,-x+½)\cup(-½,½)\cup(x-½,x+½)\cup\ldots\cup(Nx-½,Nx+½).$$

When $x < 2$ (corresponding to the distance $d = xw$ between the centres of adjacent bands satisfying $d < 2w$), the difference set $D(A/w) = D(A)/w$ is the interval $(-2Nx-1,2Nx+1)$. Thus

$$D(A/w) \cap y\mathbb{Z} = (-2Nx-1,2Nx+1)\cap y\mathbb{Z} = \{0\}$$

if and only if $y \geq 2Nx+1$ (corresponding to $\rho = yw = \geq 2Nd + w$), so that the interval $[2Nx+1,\infty)$ is the sampling set (this is essentially Shannon's theorem).

When $x \geq 2$, (corresponding to $d \geq 2w$), gaps appear between the bands in the difference set and it is possible to obtain sampling rates less than Shannon's. However in this case the sampling set is more complicated and for convenience the idea of a sampling region in the plane is introduced. This is a set of points (x,y) in the positive quadrant such that (6) holds for all integers m,n with $0 \leq m \leq 2N$, $1 \leq n < x$. One such region is the ***Shannon region***

$$\{(x,y) : y \geq 2Nx+1,\ x \geq 0\}.$$

For a low band pass signal, $N = 0$ and the only sampling region (with our normalisation) is the region $\{(x,y):y \geq 1,x \geq 0\}$.

The structure of the sampling regions becomes more complicated as the number of bands in the signal increases. The sampling rates depend on the ratio of the gap size to band width and as this ratio increases, new sampling rates appear. In figure 1, the sampling frequency y is plotted against band separation x for a signal with $N = 2$. It can be seen that the sampling region consists of ***sampling sectors*** (shown hatched in figure 1) with vertices at integer points, the sectors form a pattern radiating away from the origin, adjacent sectors have parallel edges and the sectors become more horizontal as x increases. A similar graph for a high band pass signal is given in Tauber and Schilling (1986), p 191, Figure 5.1-6. One aspect of these graphs is important in practical applications of multiband sampling techniques. As they lie in sectors, the sampling rates cannot be altered arbitrarily as they might then fall outside the sampling region. In particular when sampling multiband signals at less than the

Shannon rate, in contrast to common engineering practice, it will not always be possible to *increase* the sampling rate without causing aliasing.

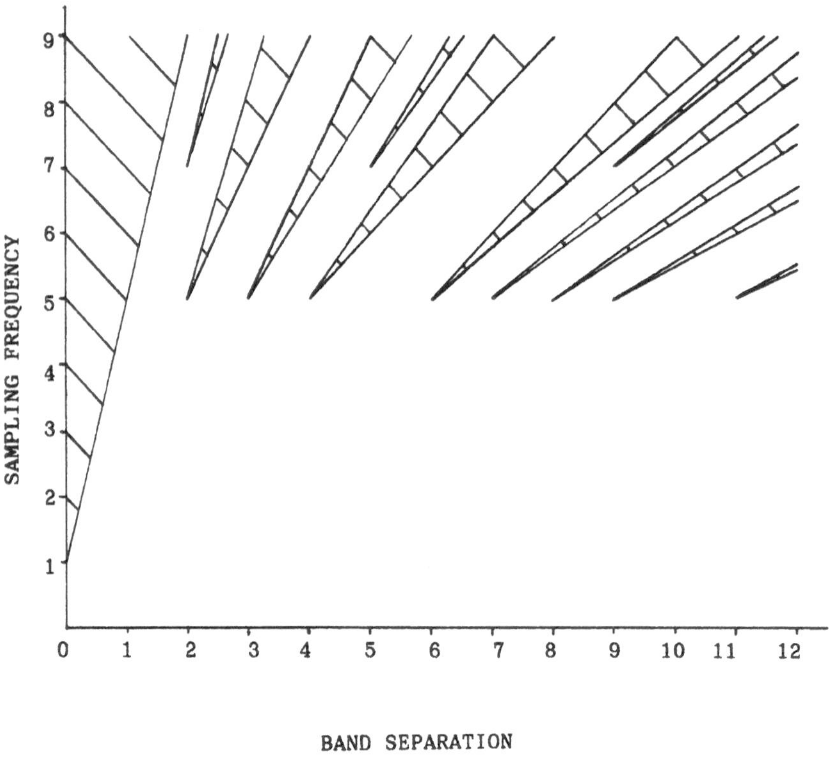

Fig. 1

In this paper we will discuss a couple of these features; more details will be given in (Beaty and Dodson 1989). To investigate the structure of the sampling sectors, i.e. to investigate the solutions of (6) (and hence of (5)), we first consider *critical* pairs (ξ,η) in the positive quadrant (i.e. $\xi,\eta \geqslant 0$) which, instead of (6), satisfy the equality

$$|m\xi - n\eta| = 1$$

for some pair of integers m,n satisfying $0 \leqslant m \leqslant 2N$, $1 \leqslant n < \xi$. This reduces to the pair of simultaneous equations

$$q\eta - p\xi = 1 \tag{7}$$

$$q'\eta - p'\xi = -1, \tag{8}$$

where $0 \leqslant p,p' \leqslant 2N$ and $1 \leqslant q,q' < \xi$. The solution (ξ,η) of (7) and (8) is given by

$$(\xi,\eta) = \left(\frac{q+q'}{\Delta}, \frac{p+p'}{\Delta} \right),$$

where $\Delta = p'q - pq'$ ($\in \mathbf{N}$). By (4), $\eta \geqslant 2N+1$. Suppose $\Delta \geqslant 2$. Then

$$\eta = \frac{p+p'}{\Delta} \leqslant \frac{p+p'}{2} \leqslant 2N,$$

a contradiction. Thus $\Delta = 1$ and ξ and η are positive integers and (7) and (8) are a pair of related Diophantine equations with integer solutions (ξ,η) satisfying

$$(\xi,\eta) = (p+p',q+q').$$

When ξ and η are coprime positive integers, the Diophantine equations (7) and (8) always have integer solutions. These can be found using continued fractions (Hardy and Wright 1954). If one particular solution (p',q') of (8) say is (a_0,b_0) then the general solution is $(a_0+k\eta, b_0+k\xi)$, $k \in \mathbf{N}$. Thus we can take $0 < a_0 < \eta$,

$0 < b_0 < \xi$ $(n < 2N)$ and similarly for (7). These least positive solutions to (7) and (8) will henceforth be written $(p,q) = (p(\xi,n),q(\xi,n))$ and $(p',q') = (p'(\xi,n),q'(\xi,n))$ respectively. Note that (7) corresponds to the qth "hop" of length n landing at the top of the pth band in the difference set, with centres of the bands a distance ξ apart. Equation (8) corresponds to the q'^{th} hop of length n landing at the bottom of the p'^{th} band. Clearly if ξ or n are altered arbitrarily, then (7) or (8) respectively will no longer hold and (6) will not be satisfied.

Now let ξ,n be two coprime positive integers with

$$qn - p\xi = 1, \quad q'n - p'\xi = -1.$$

The points $(\xi+u, n+pu/q)$ and $(\xi+u, n+p'u/q')$ also satisfy

$$q(n+pu/q) - p(\xi+u) = 1, \quad q'(n+p'u/q') - p'(\xi+u) = -1.$$

Numerical evidence suggests that the region $S(\xi,n)$ in the positive quadrant of the plane given by

$$S(\xi,n) = \{(x,y)\in\mathbb{R}^2 : \frac{px+1}{q} \leqslant y \leqslant \frac{p'x-1}{q'},\ x \geqslant \xi\} \tag{9}$$

is a sampling sector. It looks likely that this can be proved using properties of convergents, since we have to show that if the sampling rate y is increased from $(px+1)/q$ to $(p'x-1)/q'$, no allowable integer multiple of x will land in a forbidden interval.

If the numbers ξ,n are not coprime, then the set corresponding to (9) is not a sampling sector. Indeed if $(\xi,n) = c \geqslant 2$, write $\xi = ca$,

$\eta = cb$, where $(a,b) = 1$, so that $0 < b \leqslant \eta/2$, $0 < a \leqslant \xi/2$. Then

$$b\xi - a\eta = c(ba - ab) = 0.$$

Thus the a hop (of length η) lands in the bth forbidden interval $(b\xi-1, b\xi+1)$ (of width 2) and η cannot be a sampling rate.

3. STRUCTURE OF SAMPLING SECTORS

A sampling sector $S(\xi,\eta)$ has two edges (or boundary) consisting of the two rays

$$qx - py = 1, \; x \geqslant \xi, \; y \geqslant \eta$$

$$q'x - p'y = 1, \; x \geqslant \xi, \; y \geqslant \eta,$$

emanating from the vertex (ξ,η). The slope of the lower edge is p/q and that of the upper edge is p'/q'.

Since the approximation results are not affected, the transformation $T:(x,y) \to (x+ky,y)$, $k \in \mathbf{N}$, takes the sampling sector $S(\xi,\eta)$ to $S(\xi+k\eta,\eta)$. For example the vertex $(\xi+k\eta,\eta)$ satisfies the Diophantine equations

$$(q+kp)\eta - p(\xi+k\eta) = 1$$
$$(q'+kp')\eta - p'(\xi+k\eta) = -1.$$

Thus the family of sectors in $[2(2N+1),3(2N+1)]$ is the family in $[2N+1,2(2N+1)]$ but sheared (see figure).

In order to determine the sampling sets for varying sampling rates and distances between bands, we start with the Shannon region,

$\{(x,y): y \geqslant 2Nx+1, x \geqslant 0\}$. Then consider the point $(\xi, 2N+1)$, where ξ is the smallest positive integer coprime to $2N+1$ (in Figure $\xi = 2$) and construct the corrsponding sampling sector $S(\xi, 2N+1)$. Then holding ξ fixed construct sampling sectors at the points (ξ, n) where ξ, n are coprime until reaching the Shannon region. Continue in this way until $\xi = 2(2N+1)$ and then use the transformation T to obtain the rest.

Numerical evidence also suggests that sampling sectors form a net; i.e. either two such sectors are disjoint or one is contained in the other. Although the calculations are rather delicate, it should be possible to establish this also by using the properties of convergents.

4. OPTIMAL SAMPLING

The optimal sampling rate for a harmonic signal with 2N+1 frequency bands reconstructed by (2) can be determined, as follows. First assume that x is an integer ξ and that ξ and 2N+1 have no factors in common (this restriction will be dropped later). Then the best possible sampling rate is

$$\begin{cases} (2N+1) & \text{when } \xi \geqslant 2 \\ (2N+1)\xi + 1 & \text{when } \xi < 2 \end{cases}$$

For when $\xi \leqslant 2$, $D(A/w) = (-2N\xi-1, 2N\xi+1)$ and the sampling rate is $2N\xi+1$ by the Shannon theorem.

When $\xi \geqslant 2$, we require for all $k = 1,2,\ldots$ and $j = 1,2,\ldots,2N$,

$$|k(2N+1) - j\xi| \geqslant 1.$$

As before we only need to consider k up to $2N\ell/(2N+1) + 1$, i.e. up to ℓ. Since the highest common factor of 2N+1 and ℓ is 1, and since $1 \leqslant j \leqslant 2N$, $(2N+1)k \neq \ell j$ and so

$$|k(2N+1) - j\ell| \geqslant 1$$

for k = 1,2,... and j = 1,2,...2N. Hence 2N+1 is a sampling rate and is best possible by (4).

Next let x be any point in the interval $[\ell,\ell+1)$. Then the sampling rate has to be increased by an amount involving the continued fraction expansion of $\ell/(2N+1)$. The proof is based on "perturbing" the integer case and we write $\{x\}$ for the fractional part of x. For the moment we take 2N+1 and ℓ to have highest common factor 1, so that $(\ell,2N+1)$ is a critical point. Let p,q be the least positive integers satisfying $q(2N+1) - p\ell = 1$. Then for each $x \in [\ell,\ell+1)$, the optimal sampling rate y is given by

$$y = \begin{cases} 2Nx + 1 & \text{when } \ell = 1 \\ 2N+1 + \frac{p}{q}\{x\} & \text{when } \ell \geqslant 2. \end{cases}$$

When $\ell = 1$, D(A/w) reduces to the interval $(-2Nx-1,2Nx+1)$ and the result follows from Shannon's theorem.

When $\ell \geqslant 2$, write $x = \ell + \{x\}$ and put

$$y = 2N+1 + \frac{p}{q}\{x\}.$$

Then it suffices to prove that $|ky - jx| \geqslant 1$ for $1 \leqslant j \leqslant 2N$ and $1 \leqslant k < x$ and that for any $y' < y$,

$$|ky' - jx| < 1$$

for some j,k, $1 \leqslant j \leqslant 2N$ and $1 \leqslant k \leqslant (2Nx+1)/y$. Now when $j = p$ and $k = q$,

$$ky - jx = k(2N+1) + k\frac{p}{q}\{x\} - j(\xi + \{x\})$$
$$= 1,$$

since $q(2N+1) - p\xi = 1$. Next since p/q is the lower final convergent to $(2N+1)/\xi$, the inequalities

$$\frac{p}{q} < \frac{j}{k} < \frac{2N+1}{\xi}$$

cannot hold with $k < \xi$. Hence $k(2N+1) - j\xi$ and $kp - jq$ have the same sign. Thus for all integers j,k with $0 \leqslant j \leqslant 2N$, $1 \leqslant k \leqslant [x]$ and $(j,k) \neq (p,q)$,

$$|ky - jx| \geqslant 1.$$

It follows that y is a sampling rate and best possible since there are no integer points (ξ,n) below this sector and above the interval $[\xi,\xi+1)$.

When ξ and $2N+1$ have a common factor $\geqslant 2$, the sampling rate is obtained by taking the next integer M say greater than $2N+1$ which is coprime to ξ and then considering $x = \xi + \{x\}$ as above. This process is repeated until M exceeds the Shannon rate $2Nx+1$.

To illustrate this discussion, consider the graph of sampling frequency against band separation shown in figure 1. The sectors $S(\xi,5)$ with integer vertices $(\xi,5)$ are given by those points (x,y) satisfying simultaneously the two inequalities

$$qy - px \geqslant 1,\ q'y - p'x \leqslant -1,$$

where $\xi \geqslant 2$, ξ is not a multiple of 5 and satisfies $q5 - p\xi = 1$ and

$q'5 - p'\xi = -1$. For example, consider the sector S(2,5) with vertex at (2,5). The continued fraction expansion for 5/2 is

$$5/2 = 2 + \tfrac{1}{2} = [2,2] = [2,1,1]$$

(see Hardy and Wright 1954 for details). Hence $p/q = [2] = 2/1$, $p'/q' = [2,1] = 3/1$.

The sectors with vertices on the line $y = 7$ arise in a similar fashion. For instance the sector with vertex at (2,7) is bounded by the lines $\{(x,y):y = 3x + 1\}$ and $\{(x,y):y = 4x - 1\}$.

Points such as (7,7) or (3,9) have common factors so that it is impossible to prevent small multiples of the first coordinate equalling the second and so the pair cannot satisfy (6).

There are some similarities with results due to Powell (1983) who considers harmonic signals with varying band widths. However he assumes that all the useful information in the signal is carried in the base band at the origin and all other bands are unwanted. Consequently only sampling rates which protect the base band from aliasing are sought, in contrast with the present approach which recovers the complete signal. It is of interest that Powell uses Farey sequences, which are closely related to continued fractions (Hardy and Wright 1954).

Related ideas have been used to reduce sampling rates for speech signals which gives improved data compression. Of course speech signals are not harmonic but by ignoring frequencies with low amplitude, the signal can be regarded as multiband (Honari et al 1989).

5. REFERENCES

1. Beaty M.G. and Dodson M.M. (1989) 'Diophantine inequalities in frequency space and sampling rates for multiband signals', in preparation.
2. Brown J.L. (1985) 'Sampling expansions for multiband signals', IEEE Trans. Acoustics, Speech and Signal Processing, Vol ASSP-33, 312-5.
3. Butzer,P.L. (1983) 'A survey of the Whittaker-Shannon sampling theorem and some of its extensions', J. Math. Res. Exposition 3, 185-212.
4. Dodson M.M. and Silva A.M. (1985) 'Fourier analysis and the sampling theorem', Proc. Roy. Irish Acad. 85A, 81-108.
5. Dodson M.M. and Silva A.M. (1989) 'An algorithm for optimal sampling rates for multi-band signals', Signal Proc. 17, 169-174.
6. Hardy G.H. and Wright E.M. (1954) An Introduction to the Theory of Numbers (3rd edition), Oxford University Press.
7. Higgins J.R. (1985) 'Five short stories about the cardinal series', 12, 315-319.
8. Higgins J.R. (1987) 'Some gap sampling series for multiband signals', Signal Processing 12, 313-319.
9. Honari B, He W and M Darnell (1989), 'Adaptive rate sampling applied to the efficient encoding of speech waveforms', Proc. I.E.E. Conference, York (to appear).
10. Jerri A.J. (1977) 'The Shannon sampling theorem - its various extensions and applications: a tutorial review', Proc.IEEE. 65, 1565-1596.
11. Kotel'nikov V.A. (1933) 'On the carrying capacity of the "ether" and wire in telecommunications', Material for the First All-Union Conference on Questions of Communication, Izd. Red. Upr. Svyazi RKKA, Moscow (in Russian).
12. Landau H.J. (1967) 'Sampling, data transmission and the Nyquist rate', Proc. IEEE 55, 1701-1706.
13. Linden D.A. (1959) 'A discussion of sampling theorems', Proc. I.R.E., 47, 1219-1226.
14. Lloyd S.P. (1959) 'A sampling theorem for stationary (wide sense) stochastic processes', Trans. Amer. Math. Soc. 92, 1-12.
15. Powell F.D. (1983) 'Periodic sampling of Broad-band Sparse Spectra', IEEE Trans. Acoust.,Speech, Signal Processing, vol ASSP-31, 1317-1319.
16. Shannon C.E. (1949) 'Communication in the presence of noise', Proc.I.R.E. 37, 10-21.
17. Taub H. and Schilling D.L. (1986) Principles of Communication Systems (2nd edition), McGraw Hill.

SECOND-HARMONIC EFFECTS IN SCHOTTKY-BARRIER DIODES

A L CULLEN
Department of Electronic & Electrical Engineering
University College London
Torrington Place
London WC1E 7JE
United Kingdom

ABSTRACT. The paper describes the errors that can arise when a Schottky-barrier diode is used to measure the power level of a sinusoidal signal that contains an unwanted second-harmonic component.

1. INTRODUCTION

The first point to be made is that a small amount of unwanted second-harmonic power can lead to surprisingly large errors in the measured fundemental power. (Cullen and An, 1982), Cullen and An 1983). Later in this paper the way this error depends on the phase of the second-harmonic relative to the fundamental will be studied.

The circuit diagram of the Schottky-barrier diode power sensor to be analysed is shown in fig. 1 below. An r.f. voltage of

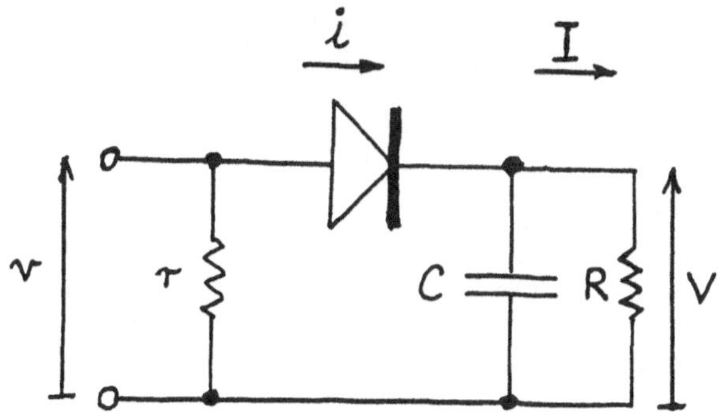

Figure 1. Circuit diagram of Schottky-barrier diode power sensor.

instantaneous value v is applied to the input terminals, and results in a rectified output current I producing an output voltage V across the d.c. load resistor R. The voltage v is

J. S. Byrnes and J. F. Byrnes (eds.), Recent Advances in Fourier Analysis and Its Applications, 499–506.

$$v=v_1+v_2 \qquad (1)$$

where $$v_1=E\cos\omega t \qquad (2)$$

and $$v_2=E\cos(2\omega t+\phi) \qquad (3)$$

Suppose that the current through the diode is represented by a power series thus

$$i=av+bv^2+cv^3 \qquad (4)$$

The critical term in this series is the cubic one. Expanding this term gives

$$v^3=v_1{}^3+3v_1^2v_2+\cdots$$

Substituting for v_1 and v_2 from (2) and (3) and taking the time average I of i gives

$$I=(1/2)bE^2+(1/2)b\beta^2E^2+(3/4)\beta cE^3\cos\phi \qquad (5)$$

Note that the third term in this expression depends on the relative phase ϕ of the second harmonic, but more importantly, note that it depends directly on the relative amplitude β of the second harmonic. Thus, if the second-harmonic power relative to the fundamental is 0.01, (1% second-harmonic power) the corresponding factor appearing in (5) will be 0.1. It is therefore understandable that 1% of second-harmonic power can lead to a ±10% error in the measured fundamental power. In the next section the simple theory outlined above will be replaced by a detailed theory of the Schottky-barrier diode power sensor with fundamental and second-harmonic power applied

2. SCHOTTKY-BARRIER DIODE POWER SENSOR

The circuit of fig. 1 still applies. The resistance r is a matching resistor, typically 50 ohms for a coaxial system. It will be assumed that the (non-linear) resistance of the diode is always very large in comparison with r. Let u be the instantaneous voltage across the diode itself, so that

$$u=v-V \qquad (6)$$

The voltage/current characteristic of the diode is given by

$$i=I_s[\exp(eu/nkT) - \exp(n-1)eu/nkT\ ] \qquad (7)$$

Substituting for v from (1), (2) and (3), and taking the time-average to find I, results in the following equations

$$I = I_s \exp\left(-\frac{eV}{nkT}\right)\Big[I_0(x)\,I_0(\beta x) + 2\,I_2(x)I_1(\beta x)\cos\phi + 2\,I_4(x)I_2(\beta x)\cos 2\phi + \cdots\Big]$$

$$- I_s \exp\left\{-\frac{(n-1)eV}{nkT}\right\}\Big[I_0(y)I_0(\beta y) - 2\,I_2(y)I_1(\beta y)\cos\phi + 2\,I_4(y)I_2(\beta y)\cos 2\phi + \cdots\Big] \qquad (8)$$

$$V = IR \qquad (9)$$

The symbols x and y in (8) are defined thus

$$x = eE/nkT \qquad (10)$$

$$y = (n-1)x \qquad (11)$$

In general, (8) and (9) must be solved as simultaneous equations for I (or V). However, there are two special cases in which a simple explicit solution is possible. These arise if R is very small, so that V may be put equal to zero in (8), or if R is very large, so that I may be put equal to zero in (8). These two cases will now be considered in turn.

3. SMALL D.C. LOAD RESISTANCE

In this special case, (8) simplifies, and leads to the following explicit expression for I.

$$I = I_s\Big[I_0(x)I_0(\beta x) + 2\,I_2(x)I_1(\beta x)\cos\phi + 2\,I_4(x)I_2(\beta x)\cos 2\phi + \cdots - I_0(y)I_0(\beta y) + 2\,I_2(y)I_1(\beta y)\cos\phi - 2\,I_4(y)I_2(\beta y)\cos 2\phi + \cdots\Big] \qquad (12)$$

A further simplification is possible for the case of an "ideal" diode, for which n=1. In this case (12) simplifies further, and can be written

$$I=I_s[F(\phi)-1] \tag{13}$$

where

$$F(\phi)=I_0(x)I_0(\beta x)+2I_2(x)I_1(\beta x)\cos\phi$$
$$+2I_4(x)I_2(\beta x)\cos 2\phi+ \ldots \tag{14}$$

For sufficiently small arguments (ie, small power levels), the function $F(\phi)$ can be written

$$F(\phi)=1+\frac{1}{4}x^2+\frac{1}{4}\beta^2x^2+\frac{1}{8}\beta x^3\cos\phi \tag{15}$$

Using (10), (13) and (15) the output current is found to be

$$I=I_s[(1/4)q^2E^2+(1/4)\beta^2q^2E^2+(1/8)\beta q^3E^3\cos\phi] \tag{16}$$

where q=(e/kT). This is formally in agreement with equation (5), which was derived from a simple power-series representation of the voltage/current characteristic of the diode.

For sufficiently large arguments (ie, large power levels), each of the Bessel functions appearing in (15) can be replaced by the first term of its asymptotic expansion. The result is

$$F(\phi)=\frac{\exp(x+\beta x)}{2\pi x\sqrt{\beta}}\left[1+2\cos\phi+2\cos 2\phi+\cdots\right] \tag{17}$$

The term in square brackets can be recognised as the Fourier series representation of a delta-function.

From the two extreme cases of small-signal and large-signal operation, it is clear that the error in power measurement due to unsuspected second-harmonic content may vary in a complicated way with the relative phase of the second-harmonic. At low power levels the variation is sinusoidal, butat high power levels the variation has the form of a sharp spike. It is interesting, therefore, to consider an intermediate case, using the exact formula (15) for the function $F(\phi)$. As an example, suppose that x=15. This corresponds to a power level of 1.41 mW; this is towards the high end of the range of powers usually used in microwave measurements. The second-harmonic level is taken to be 1% of the fundamental power. Fig. 2 shows how $F(\phi)$ varies with ϕ. The variation, though clearly not sinusoidal, is closer to a sinusoid than to a delta-function. There is an interesting interpretation of $F(\phi)$. It is in fact closely related to the antenna pattern of a linear array of infinite length of which the element excitations are

$$...I_4I_2 \,,\, I_2I_1,\, I_oI_o \,, I_2I_1,\, I_4I_2.....$$

in an obvious notation, in which the arguments of the Bessel functions are those of equation (14). The elements of the array are all in phase, and the aperture distribution is symmetrical. It is shown in Fig. 3 for the special case x=15 and 1% second-harmonic power used in Fig. 2. The quantity ϕ is now to be interpreted as the usual angle variable of antenna theory.

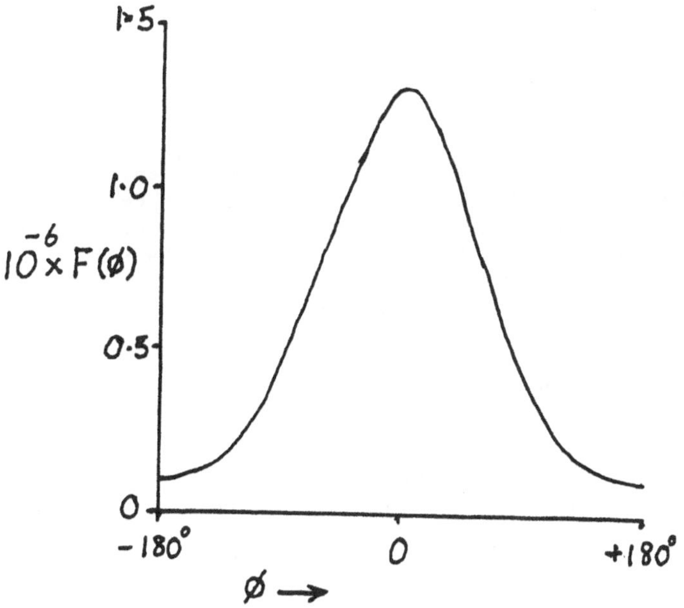

Figure 2. Variation of $F(\phi)$ with ϕ for x=15.

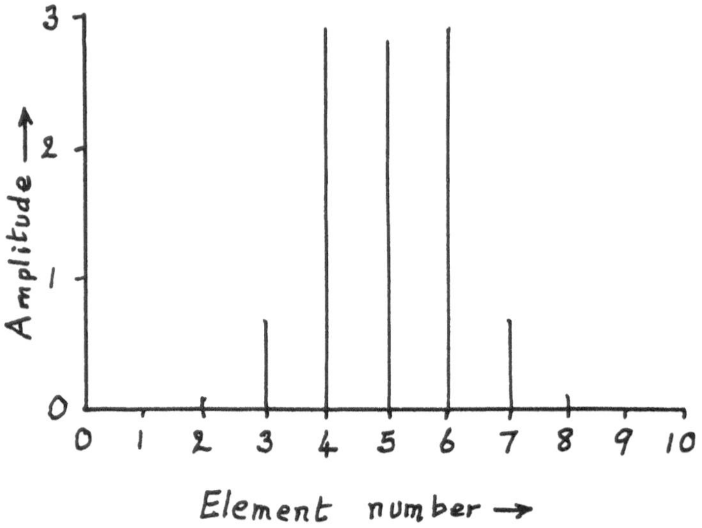

Figure 3. Aperture distribution of equivalent linear array.

It is useful to take numerical examples. For the small-signal case assume a power level of 10 microwatt into a 50 ohm load. At room temperature (290K) q=40. Using (16) it is possible to calculate the maximum and minimum value of I as the phase ϕ varies through 360°. The ratio is found to be 1.13. Next consider the higher power of 1.41 mW. The ratio now has the extraordinarily high value of 14 ! This is obviously quite unacceptable for a measuring instrument.

4. LARGE D.C. LOAD RESISTANCE

In the limiting case as R tends to infinity (8) can be simplified by putting I=0. If the further simplification n=1 is adopted, as before, (8) reduces to

$$V=(kT/e)[\ln F(\phi)] \qquad (18)$$

For small power levels, $F(\phi)$ is given by (15). Using the approximate formula $\ln(1+s) \doteqdot s$ for $s \ll 1$, the result is

$$V=(kT/e)\left[(1/4)q^2E^2+(1/4)\beta^2q^2E^2+(1/8)\beta q^3E^3\cos\phi)\right] \quad (19)$$

For larger power levels, $F(\phi)$ is given by (14).

Taking numerical examples again, for a power level of 10 microwatt the maximum to minimum voltage ratio given by (19) is 1.13. This is the same as the current ratio found in section 3 for this power level; comparing (19) and (16) makes this obvious. However, for a power of 1.41 mW, the ratio is found to be 1.226. In fact, at this power level, the diode circuit is beginning to act like a peak-voltage sensor, rather than a power (square-law voltage) sensor, as is the case at low power levels. This interpretation is confirmed by the following observation. A second-harmonic power level of 1% relative to the fundamental corresponds to a voltage ratio of 0.1. On a simple addition or subtraction basis (corresponding to a co-phase or anti-phase second-harmonic) the maximum/minimum ratio would be

$$(1+0.1)/(1-0.1)=1.222$$

This is in remarkably close agreement with the value 1.226 quoted above. From a practical measurement viewpoint, there is a vast improvement in comparison with the result for the same power level with a low value of load resistance.

5. CONCLUSIONS

The effect of a small amount of second-harmonic has been shown to have serious consequences in power measurement using Schottky-barrier diode sensors, and the seriousness has been shown to depend very strongly on the d.c. load resistance employed. The highest possible load resistance should be employed; not only is the second-harmonic problem brought under control, but also the sensitivity of the sensor is maximised. In both cases, the theory involves a certain function $F(\phi)$ of the phase of the second-harmonic relative to the fundemental. It is observed that this function can be regarded as the radiation pattern of a linear antenna array. This is mainly of academic interest, but for those familiar with the properties of such antennas, it may prove to be a helpful analogy. The theory given here is in principle general enough to cope with non-ideal diodes if the ideality factor is known, though here it has only been applied to ideal diodes. In this case, Fourier analysis of the variation of output voltage (or current) with the phase of the second-harmonic could give useful information about the ideality factor if this were independent of the barrier voltage; unfortunately this is never strictly true.

6. REFERENCES

Cullen, A.L. and An, T.Y., (1982), 'Microwave characteristics of the Schottky-barrier diode power sensor', IEE Proc. H., 129, pp 191-198.

An, T.Y. and Cullen, A.L., (1983), 'Double Schottky-barrier diode power sensor', IEE Proc. H., 130, pp 160-165.

HOLOGRAPHIC IMAGE CODING

AND

NEUROCOMPUTER ARCHITECTURES

Walter Schempp

Lehrstuhl fuer Mathematik I
University of Siegen
D-5900 Siegen
Germany

Contents

J. S. Byrnes and J. F. Byrnes (eds.), Recent Advances in Fourier Analysis and Its Applications, 507–559.

The digital computer is extremely effective at producing precise answers to well-defined questions. The nervous system accepts fuzzy, poorly conditioned input, performs a computation that is ill-defined, and produces approximate output. The systems are thus different in essential and fundamentally irreconcilable ways. Our struggles with digital computers have taught us much about how neural computation is not done; unfortunately, they have taught us relatively little about how it is done. Part of the reason for this failure is that a large proportion of neural computation is done in an analog rather than in a digital manner.

Carver A. Mead (1989)

1. ABSTRACT

The development of more powerful computers in recent years has been driven by a seemingly unending thirst for automation, control issues, information availability, and a yearning for new understanding of the self-organization principles of ourselves and our environment. The challenges of the future force to create and study new concepts of adaptive information processing and to implement novel computer architectures based on synergetic principles.

Up until now, the increased power has been driven largely by advancements in microelectronics, such as electronic switches (transistors) with higher switching speeds and integrated circuits (ICs) with increased levels of integration. Although the advancements in the IC wiring and packaging functions have been significant, their prospect for continuing at the same steady rates from very large scale integration (VLSI) to ultra large scale integration (ULSI) are being dimmed by physical limitations associated with further miniaturization.

As a result, computer architects are turning to the design of parallel processors to continue the drive toward more powerful computers. The system of interconnections by which the processing elements can share information among themselves is one of the most important characteristics of any parallel computer. The massively parallel organization principles which distinguish analog neural systems from the small scale interconnection architectures of special-purpose parallel electronic processors and even more from the von Neumann architecture of standard digital computer hardware are one of the main reasons for the largely emerging interest in neurocomputers. Just as photonics is becoming the technology of the future for telecommunications, it also will affect communications within computers, especially for parallel computer architectures and even more for neurocomputer architectures. Not only does coherent light emitted by a laser have a much higher information capacity than electrical wires, but optical beams can pass through one another without interfering, leading to a very high packing density of free-space holographic optical interconnects.

Holographic optical interconnect technology underlies the fundamental fact that in the quantized theory of the electromagnetic field the bosons (integral-spin particles) present in a beam of coherent light travelling in a well-defined direction are the photons. Based on a unified nilpotent harmonic analysis approach to artificial neural network models implemented with coherent optical, optoelectronic, or analog electronic neurocomputer architectures, the paper establishes a new identity for the matching polynomials of complete bichromatic graphs which connect neurons located in the neural plane. The key idea is to identify in a first step the hologram plane with the three-dimensional Heisenberg nilpotent Lie group quotiented by its one-dimensional center, then to restrict in a second step the holographic transform $\psi \otimes \varphi \mapsto H(\psi,\varphi;.,.)$ to the holographic lattices which form two-dimensional pixel arrays inside the hologram plane, and finally to recognize in a third step by canonical quantization the hologram plane as a neural plane. The quantum mechanical treatment of optical holography is imperative in micro-optics and nanotechnology since atoms coherently excited by short laser pulses may be as large as some transistors of microelectronic ICs and the pathways between them inside the hybrid VLSI neurochips of amacronics.

Not enough has been written about the philosophical problems involved in the application of mathematics, and particularly of group theory, to physics. On the one hand, mathematics is created to solve specific problems arising in physics, and, on the other hand, it provides the very language in which the laws of physics are formulated. One need only think of calculus or of Fourier analysis as examples of this dual relationship.

We are all familiar with the exploitation of symmetry in the solution of a mathematical problem. On the other hand, the very assertion of symmetry is often the most profound formulation of a physical law or the key step in the development of a new theory.

V. Guillemin and S. Sternberg (1984)

2. INTRODUCTORY COMMENTS

Real-time image analysis and processing, computer vision, automatic target recognition for intelligent robots, autonomous navigation, sound localization, speech processing and understanding, smart sensor processing, and various other application areas of artificial intelligence (AI) need to process an immense amount of data with very high velocity. The computational power required exceeds by many orders of magnitude the capabilities of sequential digital computers. The Space Station program's Earth Observing System (Eos) polar orbiting platforms, for instance, require to process data rates up to 1.5 gigabits per second. High resolution color images running at frame rates as low as 30 frames per second will require 10^8 bits per second. If one adds some form of autonomous feature identification to the system, the processing requirement will be between 10^{10} and 10^{13} operations per second. As a final example, a human-like speech recognizer must simultaneously perform phonetic, phonemic, syntactic, semantic, and pragmatic analyses of its inputs and match them to 50.000 words in real time. These processing throughputs exceed even the most optimistic projections for sequential supercomputers.

The problem of large-volume and high-speed computations can be solved by

o data compression techniques,
o parallel data processing.

Since their very beginning, artificial neural networks have been considered as massively parallel computing paradigms. Indeed, neural nets offer the potential of providing massive parallelism, adaption to dynamic data structures, and new algorithmic approaches to problems in image processing, speech recognition, robotics, knowledge processing, and other application fields of AI. Even the fastest sequential digital electronic computers typically require processing times ranging from many minutes to several hours for non-complex low-level image processing tasks on large image arrays.

The advantages of neural computation are now widely recognized and neural networks form one of the most rapidly expanding areas of contemporary research. In fact, research in neural networks, stimulated by major advancements in neurophysiological studies, neurosynergetic understanding, opto-electronic technology, molecular engineering, and bioelectronic material processing is currently in the midst of a gold rush period, an intense period of rapid discovery and exploitation. Everywhere new veins of gold are being uncovered and mined by thousands of prospectors, most of whom have crossed over into this exciting new research area from a diversity of other disciplines - neurobiology, neurosynergetics, quantum physics, electrical engineering, mathematics, and computer science.

The fundamental characteristics of all neurocomputer architectures are the synaptic interconnects of simple processing elements to form a concurrent distributed processing network of extensive connectivity. The processing units are arranged as two-dimensional arrays of neurons in the neural plane. Large scale (LS) collective systems like artificial neural networks exhibit many properties, including robustness, reliability, and fault tolerance, an ability to deal with ill-posed problems and noisy data, which conventional digital computer architectures do not. Adopting the synergetic point of view, neurobiology provides existence theorem on effectiveness of neural network parallel algorithms on appropriate problems.

For artificial neural networks to become ultimately useful, neuromorphic hardware must be developed. The effectiveness of neuromorphic hardware is in direct proportion to the attention it pays to the guiding neurobiological metaphor. Development efforts in the field of sixth generation computers have concentrated on one of two goals: to build

- efficient hardware that effectively executes software simulations,
- actual hardware emulators for specific neural network models.

Examples of the first are the Hecht-Nielsen Neurocomputer (HNC) accelerator board for conventional serial personal computers, and the Delta board by Science Applications International Corporation (SAIC). HNC is also pioneering a new computer language, *AXON*, that is designed for programming digital computers to simulate advanced neural networks. An important application of the SAIC neural network software simulation is the detection of explosives in checked airline baggage: the luggage is bathed in low energy (thermal) neutrons and the gamma rays resulting from neutron absorption by atomic elements in the luggage are analyzed. The artificial neural network software then searches for specific combinations of atomic elements that characterize explosives including dynamites and water gels.

Examples of the second can be viewed hierarchically. On the simplest level, the information is recorded and retrieved from an erasable magneto-optic disk by optical techniques. Higher-level building-blocks are two-dimensional arrays of coherent optical processors ([2], [3], [17], [31], [33], [35], [45], [49], [57]) for the analog implementation of neural network models by holographic optical interconnects, and neural network analog VLSI chips. For instance, the analog silicon models of the orientation-selective retina for pattern recognition ([1], [27], [28], [29]), and the analog electronic cochlea for auditory localization ([25], [27], [28]) belong to this category. The retinal and the cochleal VLSI neurochips are made with a standard complementary metal oxide semiconductor (CMOS) process ([55]).

Although the implementation of the various neural network models needs to overcome many difficult optical, optoelectronic, and analog electronic design problems, their performance is modest compared with the powerful organizing principles found in biological neural wetware. The visual system of a single human being does more image processing than do the entire world's supply of supercomputers, and the nervous system of even a very simple animal like the common house-fly (Musca domestica) contains computing paradigms that are orders of magnitude more effective than are those found in systems made by humans. Unlike conventional computer hardware, neural circuitry is not hard-wired or specified as an explicit set of point-to-point connections. Instead it develops under the influence of a genetic specification and epigenetic factors, such as electrical activity, both before and after birth. How this happens is in large part unknown.

Neurobiological development processes are far too complex to hope that a relatively complete understanding of how a perceptual system develops and functions will soon emerge. But we are familiar with complex synthetic systems whose principles of neural organization can be understood without one's knowing in detail how the components work. Furthermore, the same principles can be used to build neurocomputers in any of several different technologies. Presently the most advanced neural network analog CMOS VLSI chips model, to a first approximation, the time-frequency domain processing of two highly spectacular biological neural systems: the active echolocating system of the horseshoe bats (Rhinolophidae), and the passive auditory localization system of the barn owl (Tyto alba) which both produce complete maps of the auditory space from the time-frequency coding pathways. Continuing evolution, however, of hybrid submicron optoelectronic technology combined with neuromathematics, the highly promising and exciting new field of studying how computations can be carried out in extensive networks formed by two-dimensional arrays of heavily interconnected simple processing elements, will create advanced neuro-computers within the next decade which will be able to solve problems intractable for even the largest conventional digital computers.

This paper concentrates on a unified approach to massively parallel coherent optical, optoelectronic, and analog electronic neurocomputer architectures. A mathematical description of bosons is given by the Bargmann-Fock model of quantum mechanics ([6], [47]). Actually, the Bargmann-Fock model is based on harmonic analysis of the three-dimensional isotropic Heisenberg nilpotent Lie group. Because in the quantized theory of the electromagnetic field the bosons present in a beam of coherent light travelling in a well-defined direction are the photons, the Heisenberg nilpotent Lie group G allows to model, among other optical phenomena, optical holography and holographic optical interconnects. The key idea is to identify the hologram plane $\mathbb{R} \oplus \mathbb{R}$ with G quotiented by its one-dimensional center C and to recognize in this way the hologram plane as a neural plane. As a result, harmonic analysis on the central projection G-slice G/C provides the elementary holograms and the Gabor wavelets which form total families of approximating functions in $L^2(\mathbb{R} \oplus \mathbb{R})$ of decorrelating and correlating code primitives of artificial neural networks. Finally, a series of new identities for theta-null values which arises from artificial neural network identities shows that studies in computational mathematics combined with synthetic neurobiology may have an unexpected spin-off in pure mathematics.

Emphasis throughout the paper is placed on the application of harmonic analysis of the line bundle G over the hologram plane $\mathbb{R} \oplus \mathbb{R}$ to neural computer architectures. For the fairly deep details of the Mackey machinery and the Kirillov coadjoint orbit picture of G underlying the proofs, the reader is referred to the monograph [38]. Technological details of the implementations are described in the references indicated below and in the references listed in the monograph [15].

> *The various descriptions, Doppler filtering, aperture synthesis, holography, and cross-correlation, diverse as they are when described physically, become identically when formulated mathematically.*
>
> Emmett N. Leith (1978)

3. THE HOLOGRAPHIC IMAGE ENCODING PROCEDURE

Optical holograms are devices of central importance for the implementation of neurocomputer architectures by holographic optical interconnects. The starting point of the holographic principle is the fundamental fact that all detectors are phase-blind at the frequency range of the visible light. To overcome the phase-blindness, optical holography encodes in the writing step the phase information of the optical signals by a geometric encoding procedure. Then later on, it decodes in the readout step the phase information by the holographic reciprocity principle.

In order to get insight into the geometric encoding procedure of optical holography, let $\mathcal{S}(\mathbb{R})$ denote the Schwartz space of complex-valued $\mathcal{C}^{\infty}$ amplitude functions on the real line $\mathbb{R}$ rapidly decreasing at infinity. Consider $\mathcal{S}(\mathbb{R})$ as a dense vector subspace of the complex Hilbert space $L^2(\mathbb{R})$ under its natural isometric embedding with respect to the standard scalar product $\langle \, . \mid . \, \rangle$ and the associated total signal energy norm $\| \, . \, \|_2$. In optical holography, a square-law detector encodes in a massively parallel way the optical path length difference $x \in \mathbb{R}$ and the phase difference $y \in \mathbb{R}$ of two coherent signals having the same center frequency $\nu \neq 0$ and their amplitudes ψ and φ in the space $\mathcal{S}(\mathbb{R})$ by simultaneously recording the coordinates (x,y) of the interference pattern written by the coherent two-wave mixing $\psi \otimes \varphi$ into the hologram plane $\mathbb{R} \oplus \mathbb{R}$. In view of the phase-conjugate cross terms $\langle \psi | \varphi \rangle$ and $\langle \bar{\psi} | \bar{\varphi} \rangle$ of the total signal energy distribution identity or signal intensity relation

$$\|\psi + \varphi\|_2^2 = \langle \psi + \varphi \mid \psi + \varphi \rangle,$$

the sesquilinear extension to $\mathcal{S}(\mathbb{R}) \otimes \mathcal{S}(\mathbb{R})$ of the mapping

$$\psi \otimes \varphi \mapsto H(\psi, \varphi; x, y) = \int_{\mathbb{R}} \psi(t-x)\bar{\varphi}(t)e^{2\pi i y t}dt$$

describes by coherent superposition the angle image encoding procedure of optical holograms: each object to be globally stored by the coherent object signal beam in the hologram is encoded prior to its recording by mixing (or heterodyning) an unfocused linearly polarized coherent non-object-bearing reference signal beam having a particular angle between its wave vector and the normal vector of the hologram plane $\mathbb{R} \oplus \mathbb{R}$. Therefore the sesquilinear mapping $\psi \otimes \varphi \mapsto H(\psi, \varphi; .\,, .)$ is called the holographic transform of the writing amplitudes ([42], [43], [44]). It should be observed that unlike sequential data processing, the holographic transform $\psi \otimes \varphi \mapsto H(\psi,\varphi;x,y)$ does not treat time as a sequencer but as an expresser of information similarly to biological neural systems where time is used throughout as one of the fundamental representational coordinates. Moreover, it should be noticed that the coordinates (x,y) of the interference pattern are independent of the distance between the object to be globally recorded and the square-law detector located in the hologram plane.

The method of optical holography or coherent wavefront reconstruction applies to all waves: to electron waves, X rays, light waves, acoustic waves, and seismic waves, providing the waves are coherent enough to form the required interference patterns in the hologram plane ([44]). Therefore a laser is not really needed for optical holography; it is merely the use of solid, three-dimensional objects that calls for light waves whose coherence length exceeds the path differences due to the unevenness of such objects. In radar analysis, $\psi \otimes \varphi \mapsto H(\psi,\varphi;.,.)$ is called the cross-ambiguity function ([38], [4], [7], [13], [20], [32], [36]). In the following it will be convenient to define the auto-ambiguity function by $H(\psi;.,.) := H(\psi,\psi;.,.)$. The mapping $\psi \mapsto H(\psi;.,.)$ is called the holographic trace transform.

Remark 1. The only examples of strictly convex objects for which the scattering amplitude has been analysed fairly completely for high frequencies ($|\nu| \to \infty$) are the compact spheres in $\mathbb{R}^3$. According to the synergetic point of view, however, optical holography does not attempt to mathematically predict the scattering amplitudes but geometrically encodes and decodes the scattering amplitudes and their phases as an experimental result utilizing coherent reference beams.

Remark 2. A vital element of optical neurocomputer architectures is the medium for optical hologram recording because it plays the rôle of a holographic associative memory. Electro-optical photorefractive crystals (PRCs) are known to form reusable holographic storage materials that can be infinitely recycled and do not require additional processing. The crystals of the sillenite family, bismuth silicon oxide $Bi_{12}SiO_{20}$ (BSO), bismuth titanium oxide $Bi_{12}TiO_{20}$ (BTO), and bismuth germanium oxide $Bi_{12}GeO_{20}$ (BGO) exhibit the highest sensitivity to light among presently known PRCs ([52]). Optical holograms are recorded inside PRCs directly by illuminating the crystal with laser light. The light induces a charge redistribution inside the crystal ([14]) and in a certain characteristic time interval a dynamic equilibrium between distributions of the recording light intensity and internal electric charge is established. The electric charge induces an internal electrostatic field that changes the refractive index of the crystal by the electro-optical (Pockels or Kerr) effect and forms a volume holographic optical element (VHOE). As the interference pattern undergoes changes, a new charge distribution is formed, hence a new optical hologram is recorded. This charge distribution again comes to a dynamic equilibrium with the recording interference pattern. If the period during which the interference pattern changes is sufficiently long, the electro-optical crystal rerecords an optical hologram. Hence the electro-optical PRCs can adapt itself to varying external conditions, such as occasional temperature-induced changes of the phase difference between the writing object signal beam and reference signal beam, or mechanical instabilities. This is an extremely important feature because it allows more reliable storage of scattering objects by almost-real-time holography.

Research in the area of real-time holography in electro-optical PRCs needs to focus on materials in order to achieve a faster speed of photoresponse (< 1 msec), greater photorefractive sensitivity, control over image decay, and reduced fanning. Molecular engineering recently developed the highly interesting and promising organic crystals. As an alternative, bioelectronics or molecular electronics are offering photochemically sensitive materials like biopolymers of the chlorophyll-protein complex and the retinal-protein complex for real-time holographic recording. It has been discovered that specifically bacteriorhodopsin which belongs to the retinal-protein complex and which can be extracted from the purple membrane of Halobacterium halobium is a very attractive recording material for real-time optical signal processing. Depending on the preparation procedure, these materials have a very wide range of photoresponse time running from 100 sec down to 10 psec, and an extremely high spatial resolution limited by the dimensions of the molecules. However, research in this area is still in the early stage of development and for the present the studies are far from the practical implementation of potential biological neurocomputers. Nevertheless, investigations of the simplest optical processors and of associative memory elements based on biopolymers are being intensively developed in various laboratories all over the world so that it is expected that on the basis of purple membranes of halobacteria an optical memory with a capacity of 10^9 bits/cm^2 will be created in near future.

Remark 3. High-resolution radar imagery of the terrain and optical holographic imaging are closely related concepts. In fact, airborne and spaceborne synthetic aperture radar (SAR) imaging systems are active remote sensing systems which illuminate the terrain with electromagnetic energy at relatively long radar wavelengths as the platform moves with respect to the ground being mapped. SAR imaging systems coherently detect the signals returning from the terrain (called radar return) in order to store them both in amplitude and phase until all of the returns are collected. Simultaneous amplitude and phase recording is performed by heterodyning the radar returns with a reference signal in order to generate microwave holograms ([26]). The signals that are collected and coherently superposed in SAR systems by small antennas are not already focused as is the case in real aperture systems like radar altimeters. Because extensive processing is required to form the SAR image from the radar return, optical signal processing techniques have been applied to the collection and processing of SAR data. SAR imaging systems can be regarded as optical neurocomputers which implement a Doppler filter bank by a relatively static reflection pattern of the architecture mirror ([44]). The massive two-dimensional parallelism inherent to the optical data processing approach is in large part responsible for the success of SAR imaging.

Remark 4. Since the advent of optical holography there has been a strong interest in replacing lenses and other crucial parts of optical systems by high performance holographic optical elements (HOEs). In particular, optical SAR data processing systems may be realized by optical heads which include high performance hololenses. Many HOEs are fabricated by recording the interference pattern between two mixing laser beams. The use of digital computer-generated hologram (CGH) techniques, however, avoids the technological difficulties involved in the interferometric HOE fabrication. Moreover, one benefit that digital CGHs can offer that is not available with optical holography is the ability to deal with objects that exist only mathematically. Finally, high quality digital CGHs to implement holographic optical interconnects of high circuit density may be fabricated with the same technology used in the manufacture of CMOS VLSI circuit chips ([16], [17]). The geometrical CGH encoding computations for specific HOE pattern parameters are performed with a standard computer aided design (CAD) station. Upon completion of the HOE pattern database generation and conversion of the pattern by a subroutine to the required formatted data, a digital computer controlled output device such as a Perkin-Elmer electron-beam high-resolution microlithographic system then writes the desired geometric pattern on photoresist, which is subsequently processed to produce the finished transmissive or reflective holographic element. It is at this intermediate level of lower throughput requirements where sequential processors play a role in vision and image processing. Alternately, digital CGHs may be realized by writing the appropriate geometrical pattern on a spatial light modulator (SLM). In any case, digital CGHs are at the base of a technology transfer from microelectronics to microoptics or amacronics and form a bridge between digital computer and optical neurocomputer architectures. Since atoms coherently excited by short laser pulses (Rydberg atoms) may be as large as some transistors of microelectronic ICs and the pathways between them inside the CMOS VLSI chips ([58], [59], [5]), the quantum mechanical treatment of optical holography is of particular importance for amacronics (see Section 8 infra) and nanotechnology.

> *Perhaps the most rewarding aspect of analog computation is the extent to which elementary computational primitives are a direct consequence of fundamental laws of physics.*
>
> Carver A. Mead (1989)

4. CANONICAL QUANTIZATION

Let G denote the multiplicative group of all unipotent real matrices

$$\begin{pmatrix} 1 & x & z \\ 0 & 1 & y \\ 0 & 0 & 1 \end{pmatrix} \quad := \quad (x,y,z)$$

with unit element (0,0,0). Then G is a simply connected two-step nilpotent Lie group with one-dimensional center $C = \{(0,0,z) | z \in \mathbb{R}\}$. The polarized presentation

$$(x_1,y_1,z_1).(x_2,y_2,z_2) = (x_1+x_2, y_1+y_2, z_1+z_2+x_1y_2)$$

and the equivalent isotropic presentation

$$(x_1,y_1,z_1).(x_2,y_2,z_2) = (x_1+x_2, y_1+y_2, z_1+z_2+ \frac{1}{2}(x_1y_2-x_2y_1))$$

show that G is a realization of the three-dimensional Heisenberg group ([6], [38], [54]). The Haar measure of G is Lebesgue measure $dx \otimes dy \otimes dz$ of the underlying differential manifold $\mathbb{R}^3$. The Lie algebra $\mathfrak{g} = T_{(0,0,0)}(G)$ of G is formed by the upper triangular matrices $\{(x,y,z)-(0,0,0) | x,y,z \in \mathbb{R}\}$. In terms of the canonical basis $\{P, Q, Z\}$ of $\mathfrak{g}$ which is given by the matrices

$$P := \begin{pmatrix} 0 & 1 & 0 \\ 0 & 0 & 0 \\ 0 & 0 & 0 \end{pmatrix}, \quad Q := \begin{pmatrix} 0 & 0 & 0 \\ 0 & 0 & 1 \\ 0 & 0 & 0 \end{pmatrix}, \quad Z := \begin{pmatrix} 0 & 0 & 1 \\ 0 & 0 & 0 \\ 0 & 0 & 0 \end{pmatrix},$$

the Heisenberg commutation relations read as follows:

$$[P,Q] = PQ - QP = Z, \quad [P,Z] = 0, \quad [Q,Z] = 0.$$

Thus the center $\mathfrak{c} = \mathbb{R}.Z$ of the Heisenberg Lie algebra $\mathfrak{g}$ is one-dimensional and satisfies $\exp(\mathfrak{c}) = C$. The adjoint action $(x,y,z) \mapsto Ad_G(x,y,z)$ of G on $\mathfrak{g}$ is defined by conjugation:

$$\begin{pmatrix} 1 & x & z \\ 0 & 1 & y \\ 0 & 0 & 1 \end{pmatrix} \begin{pmatrix} 0 & a & c \\ 0 & 0 & b \\ 0 & 0 & 0 \end{pmatrix} \begin{pmatrix} 1 & x & z \\ 0 & 1 & y \\ 0 & 0 & 1 \end{pmatrix}^{-1} = \begin{pmatrix} 0 & a & c+bx-ay \\ 0 & 0 & b \\ 0 & 0 & 0 \end{pmatrix}$$

With respect to the basis $\{P, Q, Z\}$ of $\mathfrak{g}$ it follows

$$Ad_G(x,y,z) = \begin{pmatrix} 1 & 0 & 0 \\ 0 & 1 & 0 \\ -y & x & 1 \end{pmatrix} \qquad ((x,y,z) \in G).$$

If $\{P^*, Q^*, Z^*\}$ denotes the dual basis of $\{P, Q, Z\}$, the coadjoint action $(x,y,z) \mapsto CoAd_G(x,y,z)$ of G in the dual vector space $\mathfrak{g}^* = T^*_{(0,0,0)}(G)$ of $\mathfrak{g}$ is given by the formula

$$CoAd_G(x,y,z) = \begin{pmatrix} 1 & 0 & y \\ 0 & 1 & -x \\ 0 & 0 & 1 \end{pmatrix} \qquad ((x,y,z) \in G).$$

Hence

$$\mathrm{CoAd}_G(x,y,z)(\xi P^* + \eta Q^* + \nu Z^*) = (\xi + \nu y)P^* + (\eta - \nu x)Q^* + \nu Z^*,$$

where the triple (ξ,η,ν) denotes real coordinates. From this identity the Kirillov coadjoint orbit picture of G becomes apparent: For each center frequency $\nu \neq 0$ the orbit of the point $(0,0,\nu)$ under the CoAd_G-action of G is the affine plane $\mathcal{O}_\nu$ in $\mathfrak{g}^*$ through the point νZ^* parallel to the plane spanned by $\{P^*,Q^*\}$ through the origin of $\mathfrak{g}^*$. For $\nu = 0$ the points $(\xi,\eta,0)$ are zero-dimensional coadjoint orbits $\mathcal{O}_{(\xi,\eta)}$ of G located in the plane spanned by $\{P^*,Q^*\}$ through the origin of $\mathfrak{g}^*$.

Notice that the symplectic plane $\mathcal{O}_\nu$ $(\nu \neq 0)$ carries the canonical differential 2-form $\nu(dx \wedge dy)$ and that the point-orbit $\mathcal{O}_{(\xi,\eta)}$ $((\xi,\eta) \in \mathbb{R} \oplus \mathbb{R})$ can be identified with the Dirac measure $\varepsilon_{(\xi,\eta)}$ located at the point (ξ,η) of the "singular" plane $\nu = 0$.

In terms of the Heisenberg nilpotent Lie group G, canonical quantization means the choice of an irreducible unitary linear representation U of G acting in a complex Hilbert space $\mathcal{H}$. Recall that U is a continuous homomorphism of G into the unitary group $\mathbf{U}(\mathcal{H})$ of $\mathcal{H}$, i.e., $U: G \to \mathbf{U}(\mathcal{H})$ is a mapping such that

$$U((x_1,y_1,z_1).(x_2,y_2,z_2)) = U(x_1,y_1,z_1) \circ U(x_2,y_2,z_2),$$
$$U(0,0,0) = \mathrm{id}_{\mathcal{H}},$$

and such that the mapping $G \times \mathcal{H} \ni ((x,y,z),\psi) \to U(x,y,z)\psi \in \mathcal{H}$ is continuous. Irreducibility means that U is not obtained as a direct sum of two non-trivial linear subrepresentations of G. Equivalently, there exists no proper closed vector subspace $\neq \{0\}$ of $\mathcal{H}$ invariant under the action of G by U in $\mathcal{H}$.

According to the Stone-von Neumann-Mackey theorem ([6], [38]) the canonical quantization problem has a solution unique up to unitary isomorphy: For any given center frequency $\nu \neq 0$, the central character

$$\chi_\nu: C \ni (0,0,z) \mapsto e^{2\pi i \nu z}$$

determines up to a unitary isomorphism a unique infinite-dimensional irreducible unitary linear representation U_ν of G in the standard Hilbert space $\mathcal{H} = L^2(\mathbb{R})$ which acts on the vector subspace $\mathcal{S}(\mathbb{R})$ according to the rule

$$U_\nu(x,y,z)\psi(t) = e^{2\pi i\nu(z+yt)}\psi(t-x) \qquad (t \in \mathbb{R})$$

by time shifting and phasor multiplication of $\psi \in \mathcal{S}(\mathbb{R})$. The Kirillov correspondence assigns to the coadjoint orbit $\mathcal{O}_\nu \in \mathfrak{g}^*/\mathrm{CoAd}_G(G)$ $(\nu \neq 0)$ of the point $(0,0,\nu)$ in $\mathfrak{g}^*$ the isomorphy class of U_ν. Notice that this isomorphy class contains the Bargmann-Fock model of quantum mechanics describing bosons (cf. Section 1 supra), and also the linear lattice representation of G (see Section 8 infra).

Notice that the complex vector space of $\mathcal{C}^\infty$-vectors for the linear representation U_ν $(\nu \neq 0)$ of G acting on $\mathcal{H} = L^2(\mathbb{R})$ is given by the Schwartz space $\mathcal{S}(\mathbb{R})$ on $\mathbb{R}$, and that the differentiated form of U_ν reads

$$U_\nu(P) = \frac{d}{ds}U_\nu(\exp(sP))_{s=0} = -\frac{\partial}{\partial t},$$

$$U_\nu(Q) = \frac{d}{ds}U_\nu(\exp(sQ))_{s=0} = 2\pi i\nu t,$$

$$U_\nu(Z) = \frac{d}{ds}U_\nu(\exp(sZ))_{s=0} = 2\pi i\nu.$$

The linear operators $\{-\partial/\partial t,\ 2\pi i\nu t\}$ determine a representation of the Schrödinger operators by skew-symmetric operators. In particular, these operators satisfy the Heisenberg commutation relation ([6])

$$[P,Q] = PQ - QP = 2\pi i\nu.\mathrm{id} \qquad (\nu \neq 0).$$

Let $\bar{U}_\nu$ denote the contragredient representation of U_ν so that

$$\bar{U}_\nu(x,y,z) = {}^tU_\nu((x,y,z)^{-1})$$

holds for all elements $(x,y,z) \in G$. Obviously

$$U_\nu|C = \chi_\nu, \quad \bar{U}_\nu|C = \chi_{-\nu} \qquad (\nu \in \mathbb{R},\ \nu \neq 0).$$

In terms of neural network theory, $\bar{U}_\nu$ is the feedback or backprojection representation of G associated to U_ν $(\nu \neq 0)$. The flatness of the coadjoint orbits $\mathcal{O}_\nu \in \mathfrak{g}^*/\mathrm{CoAd}_G(G)$ and $\mathcal{O}_{-\nu} \in \mathfrak{g}^*/\mathrm{CoAd}_G(G)$ $(\nu \neq 0)$ in the dual vector space $\mathfrak{g}^*$ of the Heisenberg Lie algebra $\mathfrak{g}$ associated by the Kirillov correspondence with the isomorphy classes of U_ν and $\bar{U}_\nu$, respectively, is equivalent to the square integrability modulo C of U_ν and $\bar{U}_\nu$. From these facts the central projection and backprojection G-slice theorem follows:

Theorem 1. *The holographic transform is the coefficient function of the linear Schrödinger representation* U_1 *of the polarized Heisenberg group* G *projected along the center* C *onto* G/C, *i.e., the canonical quantization identities*

$$\begin{cases} H(\psi', \varphi'; x, y) = \langle U_1(x,y,0)\psi' \,|\, \varphi' \rangle \\ \bar{H}(\psi, \varphi; x, y) = \langle \bar{U}_1(x,y,0)\bar{\psi} \,|\, \bar{\varphi} \rangle \end{cases}$$

hold for all points $(x,y) \in \mathbb{R} \oplus \mathbb{R}$.

The irreducibility combined with the unitarity of the linear Schrödinger representation U_1 of G implies that the commutant of U_1 is isomorphic to the compact torus group $\mathbb{T}$. Hence from the central projection G-slice theorem

Corollary. *The holographic trace transform* $\psi \mapsto H(\psi;.,.)$ *extends to a mapping of* $L^2(\mathbb{R})$ *into* $L^2(\mathbb{R} \oplus \mathbb{R})$ *such that the identity* $H(\psi;.,.) = H(\varphi;.,.)$ *implies* $\psi = c\varphi$ *where* $c \in \mathbb{T}$ *denotes a constant phase factor.*

The cross-correlation viewpoint, however, better than any other, renders understandable the well known all-range-focusing capability of the synthetic aperture radar system, implied in our holographic viewpoint. Since the form of the recorded signal, as manifested in the quadratic phase factor, is a function of range, it is apparent that each range element must be processed differently, for example, by correlation with a reference function which is different for each range. Since the pulsing provides resolution in range, we can store the data from each range separately and process them differently, so that each range is cross-correlated with the reference function proper for that range. Thus, the synthetic antenna is in effect focused simultaneously at all ranges, a most remarkable feat when viewed in terms of the capabilities of conventional antennas.

Emmett N. Leith (1978)

5. METAPLECTIC COVARIANCE

The importance of the preceding result lies in the fact that the hidden symmetries of the holographic transform $\psi \otimes \varphi \mapsto H(\psi, \varphi; .,.)$ can be expressed by the group of automorphisms of the Heisenberg nilpotent Lie group G keeping the center C pointwise fixed. This group, the metaplectic group $\mathbf{Mp}(1,\mathbb{R})$, forms a twofold cover of the symplectic group $\mathbf{Sp}(1,\mathbb{R}) = \mathbf{SL}(2,\mathbb{R})$. Its natural action on the hologram plane $\mathbb{R} \oplus \mathbb{R}$ preserves the center frequencies $\nu \neq 0$ ([38], [47], [54]). Its action on the complex Hilbert space $L^2(\mathbb{R})$ is performed by the metaplectic representation σ. The representation σ of $\mathbf{Mp}(1,\mathbb{R})$ is a projective unitary linear representation of $\mathbf{Sp}(1,\mathbb{R})$ in $L^2(\mathbb{R})$ and satisfies the metaplectic covariance condition of the canonical quantization

$$\sigma(g)^{-1} \circ U_\nu(x,y,0) \circ \sigma(g) = U_\nu((g^{-1}(x,y),0)) \qquad ((x,y) \in \mathbb{R} \oplus \mathbb{R})$$

for all $g \in \mathbf{Sp}(1,\mathbb{R})$. It follows

Theorem 2. *The holographic transform satisfies the metaplectic covariance identity*

$$H(\sigma(g)\psi,\sigma(g)\varphi;x,y) = H(\psi,\varphi;g^{-1}(x,y)) \qquad ((x,y) \in \mathbb{R} \oplus \mathbb{R})$$

for all amplitudes ψ *and* φ *belonging to* $\mathcal{S}(\mathbb{R})$ *and all elements* $g \in \mathbf{Sp}(1,\mathbb{R})$.

Notice that the action of $\mathbf{Sp}(1,\mathbb{R})$ in $L^2(\mathbb{R})$ by the metaplectic representation σ includes the dilations by real scaling factors $a \neq 0$, and the one-dimensional Fourier transform $\mathcal{F}$, both being of importance for the Gabor wavelet transform. Indeed, for the element

$$g_a = \begin{pmatrix} a & 0 \\ 0 & a^{-1} \end{pmatrix}$$

of $\mathbf{Sp}(1,\mathbb{R})$, the identity

$$\sigma(g_a)\psi(t) = |a|^{-1/2}\psi(a^{-1}t) \qquad (t \in \mathbb{R})$$

follows, and similarly for the element

$$g_0 = \begin{pmatrix} 0 & 1 \\ -1 & 0 \end{pmatrix}$$

of $\mathbf{Sp}(1,\mathbb{R})$ the relation

$$\sigma(g_0)\psi(t) = \mathcal{F}\psi(t) \qquad (t \in \mathbb{R})$$

holds where $\psi \in \mathcal{S}(\mathbb{R})$. It follows as a special case of the metaplectic covariance identity of the holographic transform:

Corollary 1. *For the amplitudes* ψ *and* φ *in* $\mathcal{S}(\mathbb{R})$ *the identity*

$$H(\mathcal{F}\psi,\mathcal{F}\varphi;x,y) = H(\psi,\varphi;-y,x) \qquad ((x,y) \in \mathbb{R} \oplus \mathbb{R})$$

holds.

If $\psi^{-}: t \mapsto \psi(-t)$ denotes the amplitude of the time-reversed optical signal, the Fourier inversion theorem yields the identity

$$\sigma(g_0^2)\psi = \psi^{-}.$$

Thus the hologram plane rotated through 180° corresponds to the time-reversed writing signals.

It should be observed that the elements g_a $(a \neq 0)$, g_0, and

$$g^u = \begin{pmatrix} 1 & 0 \\ u & 1 \end{pmatrix}$$

for $u \in \mathbb{R}$ are generators of the group $\mathbf{Sp}(1,\mathbb{R})$. As a paraxial ray-transfer matrix, $g^u \in \mathbf{Sp}(1,\mathbb{R})$ defines a thin cylindrical lens of focal length $f = -\frac{1}{u}$ and $\sigma(g^u)$ defines the chirp modulation operator

$$\sigma(g^u)\psi(t) = e^{-i(u/2)t^2}\psi(t) \qquad (t \in \mathbb{R})$$

of chirp rate $u \neq 0$. For $u < 0$ the chirp modulation operator $\sigma(g^u) \in \mathbf{U}(L^2(\mathbb{R}))$ defines an up chirp amplitude modulation, and for $u > 0$ a down chirp amplitude modulation.

<u>**Corollary 2.**</u> *The chirp modulation* $\sigma(g^u)$ *of chirp rate* $u \neq 0$ *can be corrected by a thin cylindrical lens of focal length* $f = \frac{1}{u}$.

Finally, for

$$g = \begin{pmatrix} 1 & u \\ 0 & 1 \end{pmatrix}$$

the identity

$$\mathcal{F}(\sigma(g)\psi)(t) \;=\; e^{-i(\pi/4)\mathrm{sign}\, u}\sigma(g^{-u})\mathcal{F}\psi(t) \qquad (t \in \mathbb{R})$$

follows where $\mathrm{sign}\, u = u/|u|$. The phasor occurring in this formula arises by the Maslov index which is responsible for the fact that $\mathbf{Mp}(1,\mathbb{R}) = \tilde{\mathbf{Sp}}(1,\mathbb{R})$ forms a twofold cover of $\mathbf{Sp}(1,\mathbb{R})$.

Example 1. Let $T > 0$ be given and denote by

$$\psi_T(t) = \begin{cases} 1 & |t| \le \frac{1}{2}\,T, \\ 0 & |t| > \frac{1}{2}\,T, \end{cases}$$

the rectangular pulse of duration T. In terms of the triangular pulse

$$\Lambda(t) = \begin{cases} 1 - |t| & |t| \le 1, \\ 0 & |t| > 1, \end{cases}$$

and the cardinal sine mother wavelet

$$\operatorname{sinc} x = \begin{cases} \dfrac{\sin \pi x}{\pi x} & x \neq 0, \\ 1 & x = 0, \end{cases}$$

the holographic transform of ψ_T takes the form

$$H(\psi_T; x, y) = \Lambda(\tfrac{x}{T}) \operatorname{sinc} y(T - |x|) \quad ((x,y) \in \mathbb{R} \oplus \mathbb{R}).$$

An application of Theorem 2 supra shows that the chirp pulse $\sigma(g^u)\psi_T$ of duration T and chirp rate $u \neq 0$ admits the holographic transform

$$H(\sigma(g^u)\psi_T; x, y) = \Lambda(\tfrac{x}{T}) \operatorname{sinc} (y - ux)(T - |x|).$$

Satellite altimetry uses the ranging capability of radar sensors to measure the surface topographic profile. An example of an advanced-type system is the SEASAT altimeter which was put into orbit in June 1978. The satellite orbital altitude was 790 km and the platform velocity (ground track) v = 6.6 km/sec. SEASAT was in operation for a total of 105 days. During that time, the altimeter provided profiles of the ocean surface with an accuracy of a fraction of a meter. In the altimetry mode, SEASAT operated at a center frequency of 13.5 GHz. The stable local oscillator (STALO) generated a sequence of 12.5 nanosec pulses at a 250 MHz center frequency which has been applied to the chirp generator. The SEASAT chirp generator was a surface acoustic wave (SAW) device fabricated on a lithium tantalate substrate. The resulting chirp modulated pulse had a linearly decreasing frequency with an 80 MHz bandwidth and a pulse duration T = 3.2 μsec. The pulse repetition frequency (PRF) was 1020 Hz. During the transmit mode, the chirp pulse at 250 MHz has been upconverted to 3375 MHz, amplified to a 1 W level, and multiplied by 4 to 13.5 GHz. This also multiplied the bandwidth by 4 in order to achieve the desired 320 MHz bandwidth and height measurement accuracy of 0.47 m. In the receive mode the chirp pulse have been upconverted to 3250 MHz, amplified to 0.1 W, multiplied by 4 to 13.0 GHz, and used for mixing with the received echo signal.

Example 2. In SAR remote sensing systems (see Remark 3 supra), a target at distance r_0 with velocity v relative to the moving platform induces a chirp amplitude modulation $\sigma(g^u)$ of the received echo signal of chirp rate

$$u = \frac{4\pi v^2}{\lambda r_0}$$

where $\lambda = c/|\nu|$ denotes the wavelength of the coherent radar. The dependence of the chirp rate u of the range r_0 is called the range-azimuth coupling. It follows that the radar return focus a distance

$$f = -\frac{\lambda r_0}{4\pi v^2}$$

from the hologram plane. Thus the Doppler effect of the platform motion generates an axial astigmatism. To compensate the linear range variation of the focal length f, a wide-screen equalizer is introduced in the hologram plane. Such an equalizer takes the form of a conical lens or a tilted cylindrical lens which are parts of a correcting anamorphic optical system (cf. Corollary 2 of Theorem 2 supra). The recent developments of SLMs have supplied an attractive replacement for the holographic film as an input medium. Moreover, two-dimensional optical data processors using laser diode illumination, acousto-optic (AO) cell data input, and a charge-coupled device (CCD) detector array for the output have been designed. For each realization of the SAR data processor, however, it is important to notice that the spatial resolution of SAR imaging systems is independent of the range r_0 to the target and the velocity v of the radar platform.

In the imaging mode, SEASAT SAR operated at a center frequency of 1275 MHz (L-band, λ = 23.5 cm) with pulse duration T = 34 μsec and PRF selections of 1464, 1537, 1580, and 1647 Hz admitting a spatial resolution of 25 m. The depression angle ranged between 67° to 73° and produced an image-swath width of 100 km. The antenna was a 10.74 m by 2.16 m phased array system deployed after orbit insertion. The microwave holographic data for each 100 km wide image-swath have been optically processed to produce four film strips each of which covered a width of 25 km and a length of several thousand kilometers.

The first Shuttle imaging radar (SIR-A) experiment was launched on the second flight of Columbia in November 1981. The satellite orbital altitude was 250 km and the image-swath width 50 km in order to cover a surface area of about 10 million km^2. The SAR antenna of 9.44 m by 2.09 m radiating area was stowed inside the Shuttle cargo bay and operated when the Shuttle was in an inverted attitude. As in the SEASAT SAR, the transmitted pulse was a chirp pulse of 1275 MHz center frequency admitting a spatial resolution of 38 m. The image data were recorded as holographic film on board the Shuttle. The data film was developed and then processed at the laboratory by coherent laser light to generate the original image film at a scale of 1 : 500.000.

A second Shuttle imaging radar (SIR-B) experiment was conducted in October 1984. For SIR-B the SAR antenna was modified, however, to permit the depression angle to be changed during the mission within a range of 30° to 75°. The wavelength λ was the same as in the earlier missions. The satellite orbital altitude was 225 km and the spatial resolution improved to 25 m.

As an example, Figure 1 shows a SAR image of the Lakshmi region of Venus. It has been generated by the Soviet Union VENERA 15 and 16 orbiters through the cloud-shrouding atmosphere of Venus which is impenetrable for visible light.

Remark 5. The Heisenberg group G carries a sub-Riemannian metric and a sub-Laplacian ([50]). On the fibre $T^*_{(x,y,z)}(G)$ with base point $(x,y,z) \in \mathbb{G}$ of the cotangent bundle $T^*(G)$ of G, the associated bundled quadratic form Q is given by

$$Q_{(x,y,z)}(\xi,\eta,\nu) = (\xi + \nu y)^2 + (\eta - \nu x)^2.$$

Thus $Q_{(x,y,z)}$ is a parabolic quadratic form with one-dimensional null-space spanned by the vector $(y,-x,1)$. The sub-Laplacian $\square_G$ of G forms a sub-elliptic linear differential operator given by

$$\Box_G = (\frac{\partial}{\partial x} + y\frac{\partial}{\partial z})^2 + (\frac{\partial}{\partial y} - x\frac{\partial}{\partial z})^2.$$

The Heisenberg helix is the analog of a geodesic for the sub-Riemannian geometry of G defined by the sub-elliptic bundled quadratic form Q on $T^*(G)$. This fact corresponds to the "expansion theorem" discovered by Dennis Gabor in 1965 which says that information attached to an optical signal pattern is not carried by "rays", but by a certain "tube of rays" the cross-section of which is proportional to the square of the center wavelength λ of the optical signal ([18]).

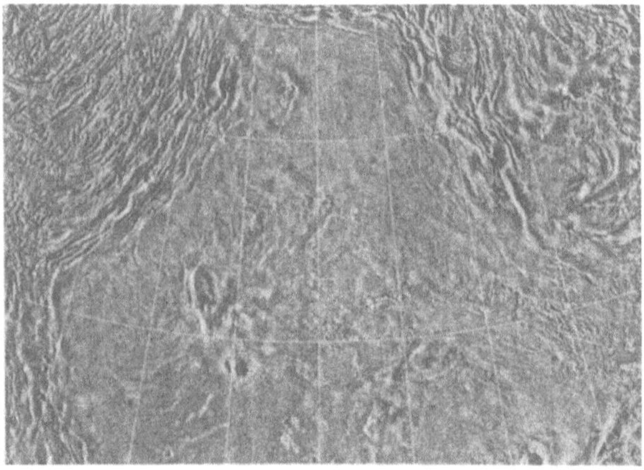

Figure 1

The coherence of laser light finds a spectacular application in the making of holograms. A typical hologram looks like a gray piece of plastic with no evident image on it. However, the hologram actually has a microscopically fine and highly complex pattern of lines and spaces. Now illuminate the developed hologram by the same laser system, except that the object has been removed. The pattern on the hologram converts the pure laser beam into a precise replica of the pattern of ordinary light that would be obtained if the object were still there. In this way, the hologram acts as a window. Each eye looking at the illuminated hologram sees a different point of view, thus creating a three-dimensional image by an illusion of depth and solidity. By changing one's vantage point, it is possible to see behind things and around corners, just as if one was looking at the real object through an ordinary window. The image has a realistic three-dimensional appearance. Holography resulted in a whole new concept in the development of imaging systems and technology.

Enders A. Robinson (1989)

6. THE HOLOGRAPHIC IMAGE DECODING PROCEDURE

In the following, the isomorphic G-manifolds

$$\mathcal{O}_1 \in \mathfrak{g}^*/\mathrm{CoAd}_G(G), \quad \mathcal{O}_{-1} \in \mathfrak{g}^*/\mathrm{CoAd}_G(G),$$

and the central projection G-slice G/C will be identified with the hologram plane $\mathbb{R} \oplus \mathbb{R}$. Then the Heisenberg nilpotent Lie group G is a line bundle over the symplectic hologram plane $\mathbb{R} \oplus \mathbb{R}$ which carries the canonical differential 2-form $dx \wedge dy$. In view of the square integrability of the irreducible unitary linear representations U_1 and $\bar{U}_1$ of G, an application of Schur's lemma provides the biorthogonality relations ([38], [37], [48])

$$\iint_{\mathbb{R}\oplus\mathbb{R}} H(\psi',\varphi';x,y)\bar{H}(\psi,\varphi;x,y)dxdy = \langle\psi' \otimes \varphi|\psi \otimes \varphi'\rangle$$

for $\psi',\varphi',\psi,\varphi$ in $\mathscr{S}(\mathbb{R})$. Therefore the dyads

$$\begin{cases} E(\psi',.;x,y): \varphi' \mapsto H(\psi',\varphi';x,y)\bar{U}_1(x,y,0)\bar{\psi}' \\ \bar{E}(\psi,.;x,y): \varphi \mapsto \bar{H}(\psi,\varphi;x,y)U_1(x,y,0)\psi \end{cases} \qquad ((x,y) \in \mathbb{R}\oplus\mathbb{R})$$

which embed $\psi' \in \mathscr{S}(\mathbb{R})$ and $\psi \in \mathscr{S}(\mathbb{R})$, respectively, into the Hilbert-Schmidt (HS) operators acting on $L^2(\mathbb{R})$, define a U_1-system $(E(.,.;x,y))_{(x,y)\in\mathbb{R}\oplus\mathbb{R}}$, and a $\bar{U}_1$-system $(\bar{E}(.,.;x,y))_{(x,y)\in\mathbb{R}\oplus\mathbb{R}}$ of coherent states based on the hologram plane $\mathbb{R}\oplus\mathbb{R}$ ([30], [34]). Observe that these coherent state systems provide a quantum mechanical description of nonspreading wave packets ([34], [58], [59]) and therefore of the Gabor tubes of rays (see Remark 5 supra).

Theorem 3. *For all writing amplitudes $\psi',\varphi',\psi,\varphi$ in $\mathscr{S}(\mathbb{R})$ the gain equations*

$$\iint_{\mathbb{R}\oplus\mathbb{R}} E(\psi',\varphi';x,y)dxdy = \|\psi'\|_2\,\bar{\varphi}'$$

$$\iint_{\mathbb{R}\oplus\mathbb{R}} \bar{E}(\psi,\varphi;x,y)dxdy = \|\psi\|_2\,\varphi$$

hold.

Remark 6. Similar inversion formulas can be established for the affine coherent states defined by the wavelet transform and the irreducible unitary linear representations of the non-unimodular affine Lie group

$$G_+ = \{(\alpha,\beta)|\alpha > 0,\ \beta \in \mathbb{R}\}$$

of the real line $\mathbb{R}$ ("$\alpha t+\beta$ group"). The affine wavelets are particularly useful code primitives for voice decomposition ([19]).

The non-abelian solvable Lie group G_+ has the presentation

$$(\alpha_1,\beta_1).(\alpha_2,\beta_2) = (\alpha_1\alpha_2, \alpha_1\beta_2 + \beta_1).$$

Of course, G_+ may also be represented as the group of real matrices

$$\begin{pmatrix} \alpha & \beta \\ 0 & 1 \end{pmatrix} \qquad (\alpha > 0,\ \beta \in \mathbb{R}).$$

under matrix multiplication. The left Haar measure of G_+ is given by $\frac{d\alpha \otimes d\beta}{\alpha^2}$ and the right Haar measure by $\frac{d\alpha \otimes d\beta}{\alpha}$. Apart from the trivial one-point coadjoint orbits located on the real line $\mathbb{R}$, the affine group G_+ of $\mathbb{R}$ admits exactly two non-trivial coadjoint orbits, the open upper half-plane $\mathcal{O}_+$ and the open lower half-plane $\mathcal{O}_-$. It follows from the Kirillov coadjoint orbit picture of G_+ that every irreducible unitary linear representation of G_+ of dimension > 1 is unitarily isomorphic to either U or its contragredient representation $\bar{U}$, where U can be realized on the complex Hilbert space $L^2(\mathbb{R})$ by the assignment

$$U(\alpha,\beta)\psi(t) = e^{i\beta e^t}\psi(t + \log \alpha) \qquad (t \in \mathbb{R}),$$

and $\bar{U}$ by the action

$$\bar{U}(\alpha,\beta)\psi(t) = e^{-i\beta e^t}\psi(t + \log \alpha) \qquad (t \in \mathbb{R})$$

on $\psi \in \mathcal{S}(\mathbb{R})$. More convenient are the realizations on $L^2(\mathbb{R}_+)$ given by

$$U(\alpha,\beta)\psi(t) = e^{i\beta t}\sqrt{\alpha}\ \psi(\alpha t) \qquad (t > 0),$$

and on $L^2(\mathbb{R}_-)$ given by

$$\bar{U}(\alpha,\beta)\psi(t) = e^{-i\beta t}\sqrt{\alpha}\ \psi(\alpha t) \qquad (t < 0).$$

Notice that the irreducible unitary linear representations U and $\bar{U}$ of G_+ are square integrable.

Remark 7. One of the most dramatic deployments of computer technology in medical diagnosis is the development of computer-aided tomography (CT). In this case and, more recently, in magnetic resonance imaging (MRI) systems, the computational capability made possible by the advent of high speed computers has been an absolutely essential ingredient in the process of image formation. Similarly to holography, the raw data provided by the physical imaging system in CT or MRI is in an encoded form which bears no discernible resemblance to the two-dimensional array of information comprising an image that can be visually perceived. CT generates an image of a cross-sectional slice of the body lying perpendicular to the long axis of the patient being examined. Unlike optical holography, in which the hologram plane is transversal to the direction of laser irradiation, CT is based upon the measurement of the attenuation of X ray beams lying entirely within the plane of the section being imaged. Turning from optical holography to CT ([12]), the preceding identities give rise by an application of the theory of the reductive dual pair $(\tilde{\mathbf{Sp}}(1,\mathbb{R}),\tilde{O}(n,\mathbb{R}))$, ([40], [22]), to the singular value decomposition of the Radon transform $\mathcal{R}: \mathcal{S}(\mathbb{R}^n) \to \mathcal{S}(\mathbb{R}\times \mathbf{S}_{n-1})$ acting on functions $f \in \mathcal{S}(\mathbb{R}^n)$ according to

$$\mathcal{R}f(r,\omega) = \int_{\mathbb{R}^n} f(x)\varepsilon_{(r-\langle\omega|x\rangle)}dx$$

(ε_p = Dirac measure located at the point $p \in \mathbb{R}$). It follows that the inversion problem for the Radon transform $\mathcal{R}$ which underlies CT, MRI, and tomographic reconstruction for geophysical applications is ill-posed. Neurocomputers, however, seem to be more appropriate to solve ill-posed problems than conventional digital computers.

As a special case we obtain from Theorem 3 supra the following result which describes the readout procedure of optical holograms, i.e., the retrieval of geometrically encoded information.

Corollary. *Let $\varphi \in \mathcal{S}(\mathbb{R})$ and assume that $\psi \in \mathcal{S}(\mathbb{R})$ satisfies the normalization condition $\|\psi\|_2 = 1$. If $\mathcal{F}$ denotes the Fourier transform acting on $\mathcal{S}(\mathbb{R})$ then the reproducing scattering integrals of degenerate coherent four-wave mixing*

$$\begin{cases} \displaystyle\iint_{\mathbb{R}\oplus\mathbb{R}} H(\psi,\varphi;x,y)e^{-2\pi iyt}\bar{\psi}(t-x)dxdy = \bar{\varphi}(t) \\ \displaystyle\iint_{\mathbb{R}\oplus\mathbb{R}} H(\psi,\varphi;y,x)e^{2\pi iyt}\mathcal{F}\bar{\psi}(t-x)dydx = \mathcal{F}\bar{\varphi}(t) \end{cases} \qquad (t \in \mathbb{R})$$

hold.

The preceding integral equations prove the holographic reciprocity principle which governs the angle image decoding procedure of optical holograms: The amplitude and the phase of the conjugate object signal can be read out simultaneously by illuminating the hologram with the unfocused conjugate reference signal beam. Thus the geometric encoding procedure of optical holography is able to overcome the phase-blindness of the detectors at the frequency range of the visible light.

The pair of reproducing scattering integrals describing the holographic filter bank is also at the basis of optical wavefront conjugation by means of real-time holography ([14]) in electro-optical PRCs. Two of the beams are referred to as pump beams and are arranged such that they are co-linear and counter-propagating and overlap both spatially and temporally in the hologram plane $\mathbb{R} \oplus \mathbb{R}$. The third beam, commonly called the probe beam, can interfere with each of the pump beams to generate transient phase gratings within the electro-optical PRC. These gratings arise because the refractive index of the PRC changes in response to the intensity of laser light: as the pump beams interfere with each other, the regions of constructive and destructive interference cause a corresponding modulation of the refractive index. The pump beams then entering the PRC can be deflected by the induced gratings to produce the fourth, wavefront conjugate beam. The wavefront conjugate beam propagates back along the path of the probe beam with a wave vector opposite to the wave vector of the probe beam.

Degenerate four-wave-mixing wavefront conjugate mirrors, as described above, provide retroreflection and optical tracking novelty filters. Therefore, Theorem 3 is at the basis of neural network models implemented by reconfigurable holographic optical interconnects in optical neurocomputer architectures ([2], [3], [17], [31], [33], [35], [45], [49], [57]). In the long term, real-time holography in PRCs appears to be the most appealing reconfigurable optical interconnection technique. If the holographic associative memory has net gain comparable with the losses in the resonator cavity, the output will converge to a real image of the globally stored object: the expanded conjugate reference signal beam acts as an optical scanner for readout of the associate information. In case of a linear resonator memory, gain is supplied by the wavefront conjugate mirror which provides regenerative feedback, whereas in case of a loop resonator memory, gain is supplied by an externally pumped electro-optical PRC.

The Soffer optical resonator forms an implementation of an optical neurocomputer architecture which includes two degenerate four-wave-mixing wavefront conjugate mirrors. For more details, the reader is referred to Section 10 infra.

Geometric quantization provides the structure for the geometric realizations of the irreducible unitary representations of the groups involved in physics.

Norman E. Hurt (1983)

7. RADIAL ISOTROPY

A writing amplitude $\psi \in \mathcal{S}(\mathbb{R})$ is called radially isotropic if the holographic trace transform $H(\psi;.,.)$ is a radial function on the hologram plane $\mathbb{R} \oplus \mathbb{R}$, i.e., if $H(\psi;.,.)$ is invariant under the natural action of the orthogonal group $O(2,\mathbb{R})$ in $\mathbb{R} \oplus \mathbb{R}$.

Theorem 4. *The amplitude* $\psi \in \mathcal{S}(\mathbb{R})$ *is radially isotropic if and only if it admits the form of Hermite-Gaussian eigenmodes*

$$\psi = c_n H_n$$

where $c_n \in \mathbb{C}$ *is a constant and* $H_n(t) = e^{-t^2/2} h_n(t)$ *is the Hermite function of degree* $n \geq 0$.

The proof follows by geometric quantization: There is a complete classification of the irreducible unitary linear representations of the diamond solvable Lie group $\mathbb{T} \times G$ having U_ν as their restrictions to G ([39]).

It follows from the preceding theorem that a quantum mechanical harmonic oscillator is equivalent to an assembly of bosons each having one polarization state. Notice that the Hermite-Gaussian eigenmodes $(H_n)_{n\geq 0}$ are crucial for the phenomenon of daydreaming in optical resonator neurocomputers ([3], [45]).

Corollary. *The elementary holograms* $(H(H_m, H_n; .,.))_{m\geq 0, n\geq 0}$ *form a Hilbert basis of the complex Hilbert space* $L^2(\mathbb{R} \oplus \mathbb{R})$.

The orthogonality of the elementary holograms $(H(H_m, H_n; .,.))_{m\geq 0, n\geq 0}$ in the complex Hilbert space $L^2(\mathbb{R} \oplus \mathbb{R})$ implies that in the Shannon sense the mutually information of the code coefficients is zero. Thus the code coefficients are non-redundant, they form a statistically independent ensemble, and image coding in terms of the decorrelating family of code primitives $(H(H_m, H_n; .,.))_{m\geq 0, n\geq 0}$ is optimally efficient.

Amacronics is a name coined for layered structures of processing electronics, binary micro-optics, and detector arrays, with applications in imaging systems with processing right at the focal plane. Amacronic structures are based on lessons that we learned from Mother Nature. Human beings live quite happily with a data trnsfer rate of a few kilocycles, massively parallel yes, but not very fast. All imaging systems suffer from the Von Neumann bottleneck in electro-optics (in computer systems all the processing functions go through a single CPU, the central processor unit; it slows down the overall system). Electo-optical systems are similar; all the optical information goes through a detector array at the focal plane which is the bottleneck. We are developing layered structures of optics and electronics in a parallel form (a processing unit per pixel), somewhat like what happens in front of the retina of your eye where you have similar amacrine clustered processing layers. The word "amacrine" comes from the Greek a macros *meaning short range. The idea is to couple dynamically clusters of detector arrays. With binary optics we may be able to build systems with peripheral vision much more motion-sensitive than on-axis fovial view, or systems tuned for edge detection or noise reduction.*

Wilfrid B. Veldkamp (1989)

8. AMACRONICS

The one-dimensional unitary linear representations of the polarized Heisenberg group G are given by the assignment

$$U_{(\xi,\eta)}(x,y,z)\psi(t) = e^{2\pi i(\xi x+\eta y)}\psi(t) \qquad (t \in \mathbb{R}).$$

Of course they are irreducible. Under the Kirillov correspondence they admit one-point coadjoint orbits $\{\mathcal{O}_{(\xi,\eta)} = \varepsilon_{(\xi,\eta)} | (\xi,\eta) \in \mathbb{R} \oplus \mathbb{R}\}$ in the "singular" plane $\nu = 0$ spanned by $\{P^*, Q^*\}$ in $\mathfrak{g}^* = T^*_{(0,0,0)}(G)$ which form a set of Plancherel measure zero. The Plancherel measure β_G of G is uniquely determined by the Haar measure $dx \otimes dy \otimes dz$ and concentrated on $\mathbb{R} - \{0\}$. It is given by

$$\beta_G = |\nu| \, d\nu.$$

The character formula of G ([38]) provides the radial distribution equation on the hologram plane $\mathbb{R} \oplus \mathbb{R}$:

$$\varepsilon_{(0,0)} = \int_{\mathbb{R}} \mathrm{Tr}_{G/C} \, U_\nu \, d\beta_G(\nu)$$

The tempered distributions $\mathrm{Tr}_{G/C} \, U_\nu$ $(\nu \neq 0)$ follow from the trace identity for the linear Schrödinger representation U_1 of the Heisenberg group G

$$\mathrm{Tr}_{G/C} \, U_1 = \sum_{n \geq 0} H(H_n; .,.)$$

by a simple scaling procedure. Figure 2 shows how to create microscopic multi-level surface relief patterns of high quality diffractive optical elements ([24], [51]). The design can be computed by the character formula of G in terms of the radial Gaussian $(x,y) \mapsto e^{-\pi(x^2+y^2)/2}$ on the hologram plane $\mathbb{R} \oplus \mathbb{R}$ and the matching polynomials $\Phi_{n,n}$ $(n \geq 0)$; see Section 9 infra. The focal length of the diffractive microlenses in the arrays is given by

$$f = |\nu|,$$

and therefore inversely proportional to the center wavelength λ. The plane $\nu = 0$ in $\mathfrak{g}^*$ forms therefore the focal plane layer of the amacrine structure. The quantization of the continuous phase profile into discrete phase levels is performed by using the VLSI ion-etching technology.

Both analog and symbolic optical computing requires sufficiently powerful and bright sources of radiation characterized by a small size and a highly efficient transformation of pumping energy into coherent radiation output. Diode or injection lasers provide the best choice in terms of power consumption and size. The coherence length of their radiation output is

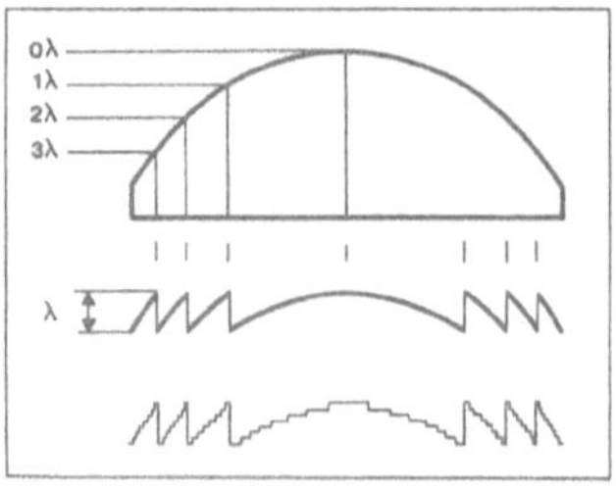

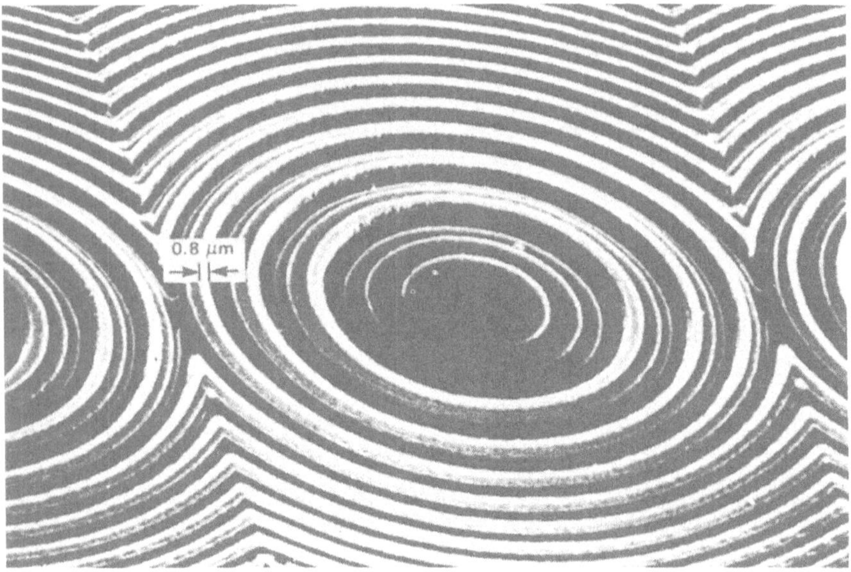

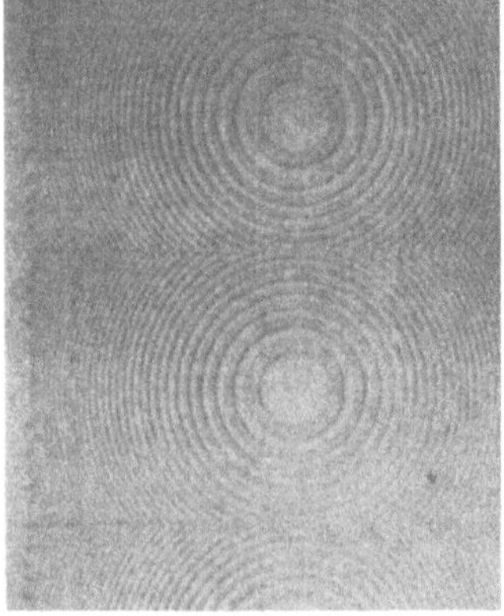

Figure 2

Section of a two-dimensional diffractive microlens array.

Figure 3

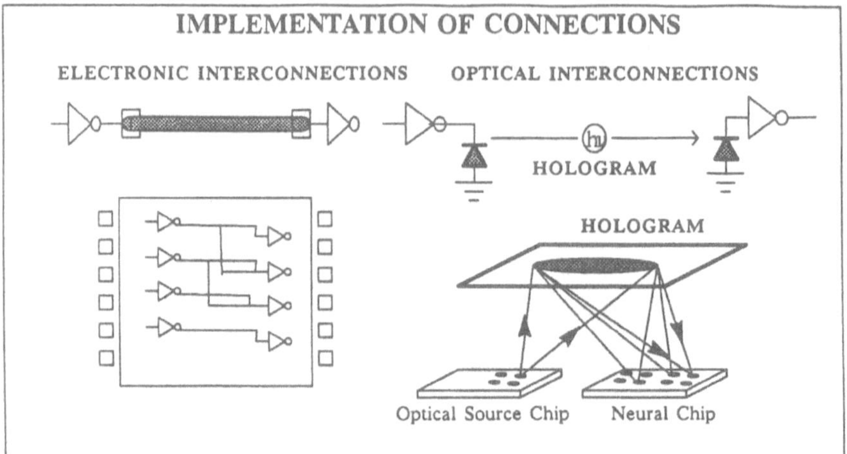

Implementations of Fixed Connections

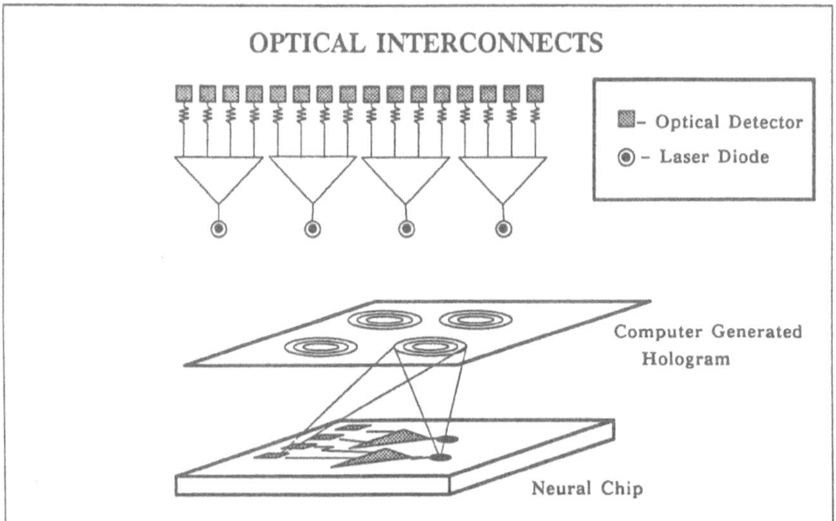

Optical Implementation of an Arbitrary Neural Network

Figure 4

sufficient for optical computing purposes. The most advanced implementation of two-dimensional arrays of laser diodes for free-space holographic optical interconnects is formed by a hybrid optoelectronic chip recently developed by the AT & T Bell laboratories and by Bellcore. It includes more than 2 millions of lasers arranged on a gallium arsenide GaAs substrate of size less than 1 cm^2. The active area emitting infrared laser light of about 8500 Å wavelength consists of thin indium gallium arsenide InGaAs layers sandwiched between more than 600 successive molecular beam epitaxial (MBE) or epi layers of GaAs and aluminum arsenide AlAs. Each of the cylindrical laser diodes has cross-section of about 5 μm and has been etched by a photolithographic process; see Figure 3. All the laser diodes are individually addressable, independently of the other ones by a current of about 1 mA and therefore are particularly suitable for performing the angle image encoding and decoding procedures of optical holograms. It is the aim of the present development in amacronics to integrate the optical source chip into a hybrid VLSI neurochip (cf. Figure 4).

The implementation of two-dimensional pixel arrays by holographic optical interconnects ([2], [3], [17], [31], [33], [35], [45], [49], [57]), and analog VLSI wavefront arrays ([27], [28], [29], [1], [25]) suggests to look at restrictions of the holographic transform $\psi \otimes \varphi \mapsto H(\psi,\varphi;.,.)$ to lattices inside the hologram plane ([8]). The quadratic lattice $\mathbb{Z} \oplus \mathbb{Z}$ embedded in the hologram plane $\mathbb{R} \oplus \mathbb{R}$ may be considered as the projection onto G/C of the 3-cubic lattice

$$L_o := \{(\mu,\mu',\zeta) \mid \mu \in \mathbb{Z},\ \mu' \in \mathbb{Z},\ \zeta \in \mathbb{Z}\}$$

and the normal subgroup $L := \mathbb{Z} \oplus \mathbb{Z} \oplus C$ inside the three-dimensional Heisenberg nilpotent Lie group G along its center C. Form the compact Heisenberg nilmanifold $L_o\backslash G$ associated to G which is a circle bundle over the two-dimensional compact torus $\mathbb{T}^2$. An application of the Weil-Zak isomorphism

$$W_1: \psi \mapsto ((x,y,z) \mapsto e^{2\pi i z}\sum_{m\in\mathbb{Z}} e^{2\pi i m y}\psi(m-x)) \qquad (\psi \in \mathcal{S}(\mathbb{R}))$$

allows to realize the linear Schrödinger representation U_1 of G as the linear lattice representation

$$\delta_1 = \mathrm{Ind}_L^G(\chi_1)$$

of G ([38], [6], [54]).

Remark 8. The Weil-Zak isomorphism w_1 combined with the one-dimensional Fourier transform $\mathcal{F} = \sigma(g_0)$ gives rise to the Poisson summation formula for the elements of $\mathcal{S}(\mathbb{R})$ and therefore to the Whittaker-Shannon-Nyquist-Kotel'nikov sampling theorem which allows the reconstruction of a band-limited signal from its uniformly distributed samples by means of translates of the cardinal sine mother wavelet (cf. Example 1 supra). For the affine Lie group G_+ of the real line $\mathbb{R}$ (see Remark 6 supra), however, there exists no analog of the linear lattice representation δ_1 of G.

It follows from the isomorphy performed by w_1 between the linear Schrödinger representation U_1 and the linear lattice representation δ_1 of G:

$$H(\psi,\varphi;x,y) = \langle\delta_1(x,y,0)w_1(\psi)|w_1(\varphi)\rangle$$

for all points (x,y) of the pixel $]-1/2,+1/2] \times]-1/2,+1/2]$ in the hologram plane $\mathbb{R} \oplus \mathbb{R}$. Therefore the Parseval-Plancherel type pixel identity

$$\sum_{(\mu,\mu')\in\mathbb{Z}\oplus\mathbb{Z}} H(\psi;\mu,\mu')\bar{H}(\varphi;\mu,\mu') = \sum_{(\mu,\mu')\in\mathbb{Z}\oplus\mathbb{Z}} |H(\psi,\varphi;\mu,\mu')|^2$$

holds for writing amplitudes ψ,φ in $\mathcal{S}(\mathbb{R})$. If the Hermite-Gaussian eigenfunctions H_m and H_n ($m\geq n\geq 0$) are inserted for ψ and φ, respectively, the radial symmetry of the terms of the left-hand side implies by a trace argument the following result ([42], [43], [57]).

Theorem 5. *The non-oriented lattices of two-dimensional pixel arrays in the hologram plane* $\mathbb{R} \oplus \mathbb{R}$ *have the crystallographic dihedral groups* D_k ($k \in \{1,2,3,4,6\}$) *of order* 2k *as their groups of symmetry.*

An application of the Weil-Zak isomorphism w_1 to the readout formulae of the Corollary of Theorem 3 supra shows that the scanout of the two-dimensional pixel arrays of the holographic lattices may be performed by a time-multiplexing procedure.

Remark 9. It is a highly remarkable observation of neurophysiology that the presynaptic vesicular grids of the mammalian brain are hexagonal holographic lattices. The thickness of the presynaptic membrane by which the synaptic vesicles emit their specific neurotransmitter substances is about 50 Å whereas the uncertainty of the position of a synaptic vesicle due to the Heisenberg uncertainty principle is about 50 Å per millisecond ([10], [11]). Of course, this observation and its consequences are also interesting from the philosophical point of view ([11], [56]).

Remark 10. The holographic lattices are at the basis of the detour phase method ([43], [46]) of writing digital CGHs of sampled images by use of the fast Fourier transform (FFT) algorithm. The height and the displacement of a single aperture centered at the sampling points of the holographic lattice are used to encode the amplitude and the phase of the complex wavefront. Thus the actual encoding of detour phase CGHs is performed without the explicit use of a reference beam. The holographic lattice corresponding to the crystallographic group D_6 of twelvefold symmetry offers substantial computational efficiency and a significant reduction of required data storage compared with rectangular sampling: the hexagonal FFT is 25% more efficient than the most efficient rectangular FFT algorithm. The scanout of the wavefront is achieved when the CGH is illuminated with a plane wave and focused with a Fourier-transforming lens.

Remark 11. The compact disks (CDs) may be regarded as one-dimensional digital CGHs that may be scanned out by the holographic optical read-head of a CD digital audio player. The scanning laser beam which is focused on the surface of the CD is focused on its return path on a quadrant detector located near the laser diode chip. The detector converts the arrays of minute optical holograms which are coherently encoded by mixing the scanning beam with the beam scattered by the pits into a sequence of electric pulses. Thus the massive amount of information arising by scanning the simple interference patterns of pits and lands has to serially pumped off the hologram plane and then to fed into the bit-stream chip or the multi stage noice shaping (MASH)

IC of the CD player's microelectronic circuitry. It is the focal plane of the collimating lens which forms the optoelectronic von Neumann bottleneck of the hybrid device. Erasable magneto-optic technology uses laser light both to record and to read data. A blank disk has all its magnetic domains oriented north-pole-down. To record information, a burst of a few nanoseconds of high-intensity light from an infrared laser heats a spot about 1μm across in one magnetic layer of the disk. The coercive force required to change the magnetic orientation of all the domains in the spot from north-pole-down to north-pole-up falls to almost zero as the temperature of the spot increases to 150°C, and the bias magnetic field created by a coil flips the magnetic field. The data are read by a lower-powered beam from the same laser, whose polarization depends by the Kerr magneto-optic effect on whether the magnetic orientation of the spot is north-pole-up or north-pole-down. Optoelectronic ICs in the magneto-optic write-read-head senses the polarization, and the magnetic orientation is interpreted as a digital 1 or 0. The magneto-optic technology suggests to consider the spin variables of an erasable CD as a one-dimensional artificial neural network.

> *Neural network models offer a data-driven unsupervised computational approach which is complementary to the algorithm-driven approaches of traditional information processing and artificial intelligence. The fine granularity, massive interconnectivity, and high degree of parallelism set neural network models apart from traditional electronic serial computing. These same features are the hallmarks of optical computing architectures which have led many workers to consider optical implementations of neural network models.*
>
> Bernard H. Soffer (1988)

9. THE SOFFER OPTICAL RESONATOR

In order to identify explicitly the terms of the Parseval-Plancherel type pixel identity indicated above, we denote by $K_{m,n}$ the complete bichromatic graph of m + n vertices. Define $c(K_{m,n},0) := 1$ and let $c(K_{m,n},l)$ denote the number of choices of $l \geq 1$ disjoint edges in $K_{m,n}$ each linking two vertices of different colours. Then

$$\Phi_{m,n}(X) := \sum_{0 \leq l \leq [(m+n)/2]} (-1)^l c(K_{m,n}, l) X^{m+n-2l}$$

denotes the matching polynomial ([21]) of variable X associated to the bipartite graph $K_{m,n}$.

Theorem 6. *The coefficients of the matching polynomial* $\Phi_{m,n}(X)$ *are the elementary synaptic weights* $(-1)^l c(K_{m,n}, l)$, $0 \leq l \leq [(m+n)/2]$, *where the matching coefficients* $c(K_{m,n}, l)$ *denote the number of disjoint synaptic interconnects of the neural network* $K_{m,n}$ $(m \geq n \geq 0)$ *activated by* l *simultaneously firing neurons.*

Example 3. In the case $m = n = 3$ the matching coefficients

$$c(K_{3,3}, 0) = 1$$
$$c(K_{3,3}, 1) = 9$$
$$c(K_{3,3}, 2) = 18$$
$$c(K_{3,3}, 3) = 6$$

arise. Thus the matching polynomial of the Thomsen graph $K_{3,3}$ is given by

$$\Phi_{3,3}(X) = X^6 - 9X^4 + 18X^2 - 6 .$$

Notice that the Thomsen graph $K_{3,3}$ is a non-planar graph.

The next theorem describes the relationship between the elementary holograms and the matching polynomials attenuated by the Gaussian $(H_0 \otimes H_0) \in L^2(\mathbb{R} \oplus \mathbb{R})$ with distance: the farther away an input is from a point in the neural network, the less weight it is given.

Theorem 7. *Let* $m \geq n \geq 0$. *Then the elementary holograms admit the form*

$$H(H_m, H_n; x, y) = \frac{(-1)^n}{\sqrt{m!n!}} e^{-\pi(x^2+y^2)/2} \Phi_{m,n}(\sqrt{\pi}(x+iy))$$

for all pairs $(x,y) \in \mathbb{R} \oplus \mathbb{R}$.

Example 4. According to the Corollary of Theorem 3 supra, scaled versions of the elementary hologram $H(H_0;.,.) = H_0 \otimes H_0$ can be implemented as diffraction HOEs (cf. Remark 4 supra) for the fundamental transverse mode of a coherent laser light beam. This implementation is performed with a CAD station by projecting layers of constant optical thickness of the rotationally symmetric Gaussian diffraction profile onto the hologram plane $\mathbb{R} \oplus \mathbb{R}$. In contrast to the Advanced Systems Analysis Package (ASAP) software procedure, however, the diffraction CGH is based on a quantum mechanical description of the diffraction profile and therefore adapted to amacronics purposes.

Corollary. *The hologram plane $\mathbb{R} \oplus \mathbb{R}$ can be realized as a neural plane.*

Presently one of the most successful implementations of the hologram plane $\mathbb{R} \oplus \mathbb{R}$ as a neural plane is the Soffer optical resonator built at the Hughes Research Laboratories ([33], [49]). The optical neurocomputer is formed by a coherent optical resonator cavity consisting of an optical hologram placed between two degenerate four-wave-mixing wavefront conjugate mirrors (PCMs) as shown schematically in Figure 5. One of the wavefront conjugate mirrors is linear, while the other one amplifies higher amplitude signals more than lower amplitude signals (see Theorem 3 supra). The optical hologram has multiple "example" images stored in it.

The neurocomputer is configured so that each example image is holographically encoded using a reference laser beam that impinges on the hologram plane at a slightly different angle than the reference beams utilized for the other example patterns. After the neural system has been prepared, one can enter any image into the cavity by impinging it onto the optical hologram. The net result is that the holographically encoded image causes partial reconstruction of the reference beams. The amplitude of each reconstructed reference beam is proportional to the L^2 distance between the entered image and the example image associated with the reference. As the reference beams reverberate through the cavity the strongest (highest amplitude in the L^2 sense) one is incrementally amplified and the others are incrementally attenuated so that before long only the reference beam corresponding to the best matching example is left. In other terms, the stored image with the smallest distance to the

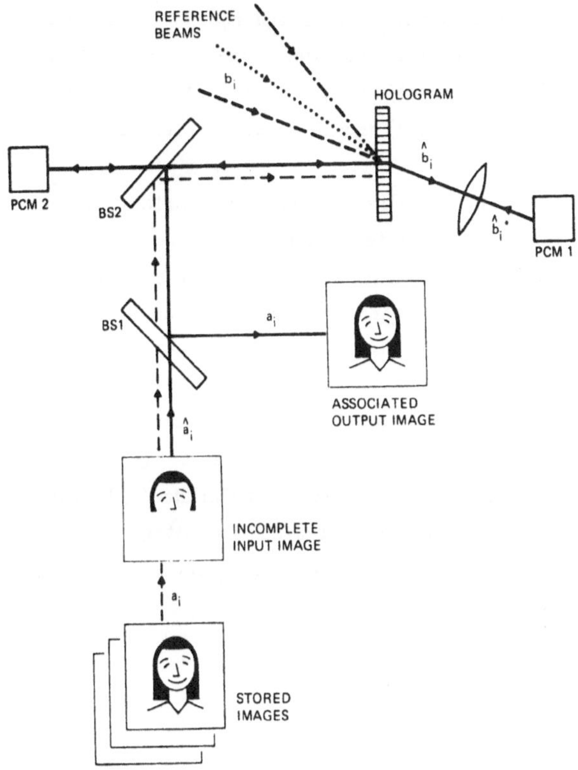
REFERENCE
BEAMS
b_i
HOLOGRAM
PCM 2
BS2
PCM 1
BS1
a_i
ASSOCIATED
OUTPUT IMAGE
INCOMPLETE
INPUT IMAGE
a_i
STORED
IMAGES

Figure 5

input pattern survives in the mode competition at the expense of the more distant images. At the output port, i.e., the reconstructed real image port of the optical hologram, the best L^2 fitting example pattern then appears. Thus the optical neurocomputer functions as a nearest neighbour classifier for holographic imagery by recalling through a competitive memory.

The Soffer optical resonator can be viewed as an infinite-dimensional version of the Hopfield network. Or alternatively, if one envisions the optical elements of the neural system as consisting of small discrete optical units, then the Soffer optical resonator can be thought of as simply a large Hopfield network.

Similar optical neurocomputers have also been recently built at the Department of Electrical Engineering of Caltech ([35]) and at the Joint Institute for Laboratory Astrophysics of the University of Colorado ([2], [3]). These neural systems have also successfully demonstrated recording multiple patterns and functioning as a nearest neighbor associative memory.

10. ARTIFICIAL NEURAL NETWORK IDENTITIES

Theorem 7 supra implies the shift register identity $(m \geq n \geq 0)$

$$H(H_m, H_n; \mu, \mu') = \langle \delta_1(\mu, \mu', 0) w_1(H_m) | w_1(H_n) \rangle = \frac{(-1)^n}{\sqrt{m!n!}} e^{-\pi(\mu^2+\mu'^2)/2} \Phi_{m,n}(\sqrt{\pi}(\mu + i\mu'))$$

for all points (μ, μ') of the quadratic holographic lattice $\mathbb{Z} \oplus \mathbb{Z}$. In particular, the following result obtains:

Theorem 8. *For* $m \geq n \geq 0$ *the identity*

$$\sum_{(\mu,\mu')\in\mathbb{Z}\oplus\mathbb{Z}} (-1)^{m+n} e^{-\pi(\mu^2+\mu'^2)} \Phi_{m,m}(\sqrt{\pi}(\mu+i\mu'))\Phi_{n,n}(\sqrt{\pi}(\mu+i\mu')) =$$

$$\sum_{(\mu,\mu')\in\mathbb{Z}\oplus\mathbb{Z}} e^{-\pi(\mu^2+\mu'^2)} |\Phi_{m,n}(\sqrt{\pi}(\mu+i\mu'))|^2$$

holds for the quadratic holographic lattice $\mathbb{Z} \oplus i\mathbb{Z}$ *of Gaussian integers inside the hologram plane* $\mathbb{R} \oplus \mathbb{R}$.

The preceding theorem gives rise to the following special identities for the odd powers of π in terms of theta-null values $\vartheta(0,1) = \sum e^{-\pi\mu^2}$ ([38], [39], [53]) where $\sum := \sum_{\mu\in\mathbb{Z}}$:

$m = 1,\ n = 0$

$$\pi = \frac{\sum e^{-\pi\mu^2}}{4 \sum \mu^2 e^{-\pi\mu^2}}$$

$m = 2,\ n = 1$

$$\pi^3 = \frac{15 \sum (8\pi^2\mu^4 - 1) e^{-\pi\mu^2}}{32 \sum \mu^6 e^{-\pi\mu^2}}$$

$m = 3,\ n = 2$

$$\pi^5 = \frac{45 \sum (16\pi^4\mu^8 - 140\pi^2\mu^4 + 21) e^{-\pi\mu^2}}{64 \sum \mu^{10} e^{-\pi\mu^2}}$$

$m = 4,\ n = 3$

$$\pi^7 = \frac{91 \sum (256\pi^6\mu^{12} - 15840\pi^4\mu^8 + 166320\pi^2\mu^4 - 25245) e^{-\pi\mu^2}}{1024 \sum \mu^{14} e^{-\pi\mu^2}}.$$

Theorem 7 supra shows that the preceding identities for the theta-null values $\vartheta(0,1)$ are of a combinatorial character.

Remark 12. The cardinal sine mother wavelet sinc mentioned in Example 1 supra, i.e., the univariate impulse response of the ideal lowpass filter, admits the Euler factorization

$$\operatorname{sinc} x = \prod_{n\geq 1} \left(1 - \frac{x^2}{n^2}\right) \qquad (x \in \mathbb{R}).$$

Its logarithmic derivative yields the identity

$$\pi x \operatorname{cotan} \pi x = 1 - 2x^2 \sum_{n\geq 1} \left(\frac{1}{n^2} + \frac{x^2}{n^4} + \frac{x^4}{n^6} + \ldots\right).$$

A comparison with the generating function of the Bernoulli polynomials $B_n(X)$ of degree $n \geq 0$

$$\frac{we^{wX}}{e^w - 1} = \sum_{n\geq 0} \frac{1}{n!} B_n(X) w^n \qquad (w \in \mathbb{C})$$

evaluated at $X = 0$ and $w = 2\pi i x$ yields the classical Euler formulae for the even powers of π:

$$\pi^{2n} = (-1)^{n+1} \frac{2(2n)!}{2^{2n} B_{2n}} \zeta(2n) \qquad (n \geq 1),$$

where ζ denotes the Riemann zeta-function and $B_{2n} = B_{2n}(0)$ are the Bernoulli numbers. In particular we get the special cases

$n = 1$	$\pi^2 = 6\,\zeta(2)$
$n = 2$	$\pi^4 = 90\,\zeta(4)$
$n = 3$	$\pi^6 = 945\,\zeta(6)$
$n = 4$	$\pi^8 = 9450\,\zeta(8)$

The first identity belongs to the nicest formulae established by Leonhard Euler. It has been explicitly reproduced in the Encyclopedia Britannica (1963).

Although a cell's response function is in general nonlinear, visual neurophysiologists have found that for many cells, a linear summation approximation is appropriate.

Ralph Linsker (1988)

11. GABOR WAVELETS ATTACHED TO A HOLOGRAPHIC LATTICE

In biological vision, the center-surround receptive field profiles of the retinal neurons ([9]) and the cells of the lateral geniculate nucleus are far from forming an orthogonal family in $L^2(\mathbb{R} \oplus \mathbb{R})$. Therefore the resulting neural representation remains highly correlated. Theorem 3 supra suggests to implement a matching filter bank by an adaptive artificial neural network model which is based for $(y,y') \in \mathbb{R} \oplus \mathbb{R}$ on the central projection and backprojection G-slice orbits

$$G^{(\mu,\mu')}_{(y,y')}: (x,x') \mapsto (U_1(x,y,0) \otimes \bar{U}_1(x',y',0)(H_0 \otimes H_0))(\mu,\mu') \quad ((\mu,\mu') \in \mathbb{Z} \oplus \mathbb{Z})$$

of the Gaussian mode $H_0 \otimes H_0$ in $L^2(\mathbb{R} \oplus \mathbb{R})$. The irreducibility of the linear Schrödinger representation U_1 of G combined with the Weil-Zak isomorphism W_1 implies:

Theorem 9. *The approximating family of Gabor wavelets*

$$\{G^{(\mu,\mu')}_{(y,y')} \mid (\mu,\mu') \in \mathbb{Z} \oplus \mathbb{Z},\ (y,y') \in \mathbb{R} \oplus \mathbb{R}\}$$

attached to the holographic lattice $\mathbb{Z} \oplus \mathbb{Z}$ inside the hologram plane $\mathbb{R} \oplus \mathbb{R}$ is total in the complex Hilbert space $L^2(\mathbb{R} \oplus \mathbb{R})$.

Notice that the Gabor wavelets form a non-orthogonal family in the Hilbert space $L^2(\mathbb{R} \oplus \mathbb{R})$. L^2 expansions in terms of Gabor wavelets offer high code compression rates appropriate for image processing purposes ([8]). Early stages of biological visual systems pay for keeping $m = n = 0$ by the non-orthogonality of the center-surround receptive field profiles. The family of Gabor wavelets give excellent fits in the chi-squared statistical sense to the correlating simple cell field profiles empirically studied in the cat striate cortex ([8], [23], [60]). The retina, an outpost of the central nervous system, and the lateral geniculate nucleus, however, act as decorrelators of the incoming signals. At the level of the mammalian visual cortex, the introduction of orientation selectivity through localized wave modulation combined with quadrature phase relations among paired cells results in a decorrelated neural representation with optimal image compression performance by the Hilbert basis of $L^2(\mathbb{R} \oplus \mathbb{R})$ of elementary holograms $(H(H_m, H_n; .,.))_{m\geq 0, n\geq 0}$. Signal preprocessing and processing in the auditory parts of the cortex follow similar basic lines.

Our ability to realize simple neural functions is strictly limited by our understanding of their organizing principles, and not by difficulties in implementation.

Carver A. Mead (1989)

12. SYNOPSIS

The computation of real world phenomena in real time requires computational power that exceeds by many orders of magnitude the capabilities of sequential digital computers presently available. Although the data transfer rate of biological neural networks is merely a few kilocycles, hence not very fast, biologial wetware is able to solve tasks such as real-time pattern recognition or sound localization because it operates in analog mode which allows simultaneous summing of many inputs from interconnected units and permits massively parallel data processing without the need for iterative procedures. Extrapolation from simulations of simple neural circuits indicate that a sequential digital computer would have to operate at speeds of more than 10^{18} floating point operations per second in order to match the performance limit of the human brain. The implementation of artificial neural network models based on coherent optical processors and analog electronic circuits of neurons and synapses is currently being pursued in a number of laboratories where several special purpose neurocomputer systems have been fabricated in holographic, optoelectronic, or CMOS VLSI electronic components. In the quantized theory of the electromagnetic field the bosons present in a coherent light beam travelling in a well-defined direction are the photons. The canonical quantization approach to the theory of the holographic transform $\psi \otimes \varphi \mapsto H(\psi, \varphi; ., .)$ as outlined in this paper implies a link between elementary holograms and artificial neural networks which allows to recognize the hologram plane $\mathbb{R} \oplus \mathbb{R}$ as a neural plane (Corollary of Theorem 7 supra). It is the quantum mechanical base of the holographic transform $\psi \otimes \varphi \mapsto H(\psi, \varphi; ., .)$ which establishes the universal validity of the holographic concept from amacronics to SAR image processing.

ACKNOWLEDGMENTS

The author has many people to thank for giving him insights from alternate points of view. He especially wants to thank Dr. Ephraim Feig (IBM Thomas J. Watson Research Center, Yorktown Heights, NY) for technical discussions on radar imagery, and to Dr. Arne Wunderlin (Institut für Theoretische Physik, Universität Stuttgart) for valuable conversations on synergetic computers.

REFERENCES

1. **T. Allen, C. Mead, F. Faggin, and G. Gribble,** "Orientation-selective VLSI retina," Visual Communications and Image Processing '88, T. Russell Hsing, Editor, Proc. SPIE 1001, 1040-1046 (1988).

2. **D.Z. Anderson,** "Coherent optical eigenstate memory," Optics Letters 11, 56-58 (1986).

3. **D.Z. Anderson, M.C. Erie,** "Resonator memories and optical novelty filters," Optical Engineering 26, 434-444 (1987).

4. **A. Berthon,** "Operator groups and ambiguity functions in signal processing," Wavelets, J.M. Combes, A. Grossmann, and Ph. Tchamitchian, Editors, 172-180, Springer-Verlag, Berlin, Heidelberg, New York, London, Paris, Tokyo, Hong Kong 1989.

5. **C.D. Cantrell, V.S. Letokhov, and A.A. Makarov,** "Coherent excitation of multilevel systems by laser light," Coherent Nonlinear Optics, M.S. Feld, V.S. Letokhov, Editors, 165-269, Springer-Verlag, Berlin, Heidelberg, New York 1980.

6. **P. Cartier,** "Quantum mechanical commutation relations and theta functions," Algebraic Groups and Discontinuous Subgroups, A. Borel, G.D. Mostow, Editors, 361-383, Amer. Math. Soc., Providence, Rhode Island 1966.

7. **C. Darmet, J.P. Gauthier, and F. Gourd,** "Elliptic and almost hyperbolic symmetries for the Woodward ambiguity function," Manuscript (to appear).

8. **J.G. Daugman,** "Relaxation neural network for complete discrete 2-D Gabor transforms," Visual Communications and Image Processing '88, T. Russell Hsing, Editor, Proc. SPIE 1001, 1048-1061 (1988).

9. **J.E. Dowling,** *The retina: an approachable part of the brain*, The Belknap Press of Harvard University Press, Cambridge, Massachusetts, and London 1987.

10. **J.C. Eccles,** "New light on the mind-brain problem: How mental events could influence neural events," Complex Systems - Operational Approaches, H. Haken, Editor, 81-106, Springer-Verlag, Berlin, Heidelberg, New York, Tokyo 1985.

11. **J.C. Eccles,** *Gehirn und Seele*, Piper, München, Zürich 1987.

12. **N.H. Farhat, C.L. Werner, and T.H. Chu,** "Prospects for three-dimensional projective and tomographic imaging radar networks," Radio Science 19, 1347-1355 (1984).

13. **E. Feig, C.A. Micchelli,** "L^2-synthesis by ambiguity functions," Multivariate Approximation Theory IV, C.K. Chui, W. Schempp, and K. Zeller, Editors, 143-156, Birkhäuser Verlag, Basel, Boston, Berlin 1989.

14. **J. Feinberg,** "Applications of real-time holography," Holography, Lloyd Huff, Editor, Proc. SPIE 532, 119-135 (1985).

15. **D.G. Feitelson,** *Optical Computing,* The MIT Press, Cambridge, Massachusetts, London 1988.

16. **M.R. Feldman, C.C. Guest,** "Holograms for optical interconnects for very large scale integrated circuits fabricated by electron-beam lithography," Optical Engineering 28, 915-921 (1989).

17. **J.W. Goodman,** "Optics as an interconnect technology," Optical Processing and Computing, H.H. Arsenault, T. Szoplik, and B. Macukow, Editors, 1-32, Academic Press, Boston, San Diego, New York, Berkeley, London, Sydney, Tokyo, Toronto 1989.

18. **P. Greguss,** "Manifestation of Gabor's holographic principle in various evolutionary stages of living material," International Conference on Holography Applications, Jingtang Ke, Ryszard J. Pryputniewicz, Editors, Proc. SPIE 673, 402-411 (1987).

19. **A. Grossmann, J. Morlet,** "Decomposition of functions into wavelets of constant shape, and related transforms," Mathematics and Physics, Lectures on Recent Results, Vol. 1, L. Streit, Editor, 135-165, World Scientific, Singapore, Philadelphia 1985.

20. **R.B. Holmes,** "Mathematical foundations of signal processing, II. The role of group theory," Technical Report 781, Massachusetts Institute of Technology, Lincoln Laboratory, Lexington, Massachusetts 1987.

21. **H. Hosoya,** "Matching and symmetry of graphs," Comp. and Maths. with Appls. 12B, 271-290 (1986).

22. **R. Howe,** "Dual pairs in physics: Harmonic oscillators, photons, electrons, and singletons," Applications of Group Theory in Physics and Mathematical Physics, M. Flato, P. Sally, and G. Zuckerman, Editors, 179-207, American Mathematical Society, Providence, Rhode Island 1985.

23. **J. Jones, L. Palmer,** "An evaluation of the two-dimensional Gabor filter model of simple receptive fields in cat striate cortex," J. Neurophysiol. 58, 1233-1258 (1987).

24. **V.P. Koronkevich,** "Computer synthesis of diffraction optical elements," Optical Processing and Computing, H.H. Arsenault, T. Szoplik, and B. Macukow, Editors, 277-313, Academic Press, Boston, San Diego, New York, Berkeley, London, Sydney, Tokyo, Toronto 1989.

25. **J. Lazzaro, C.A. Mead,** "A silicon model of auditory localization," Neural Computation 1, 47-57 (1989).

26. **E.N. Leith,** "Synthetic aperture radar," Optical Data Processing, D. Casasent, Editor, 89-117, Springer-Verlag, Berlin, Heidelberg, New York 1978.

27. **C. Mead,** *Analog* VLSI *and neural systems*, Addison-Wesley, Reading, Massachusetts 1989.

28. **C. Mead, M. Ismail,** *Analog* VLSI *implementation of neural systems*, Kluwer Academic Publishers, Boston, Dordrecht, London 1989.

29. **C.A. Mead, M.A. Mahowald,** "A silicon model of early visual processing," Neural Networks 1, 91-97 (1988).

30. **H. Moscovici,** "Coherent state representations of nilpotent Lie groups," Commun. math. Phys. 54, 63-68 (1977).

31. **J. Ohta, M. Takahashi, Y. Nitta, S. Tai, K. Mitsunaga, and K. Kjuma,** "A new approach to a GaAs/AlGaAs optical neurochip with three layered structure," Proc. IJCNN International Joint Conference on Neural Networks, Vol.II, 477-482 (1989).

32. **M. Ould-Beziou,** Thèse de doctorat, Université d'Orléans 1989.

33. **Y. Owechko, E. Marom, B.H. Soffer, and G. Dunning,** "Associative memory in a phase conjugate resonator cavity utilizing a hologram," IOCC-1986 International Optical Computing Conference, J. Shamir, A.A. Friesem, and E. Marom, Editors, Proc. SPIE 700, 296-7300 (1986).

34. **A. Peremolov,** *Generalized Coherent States and Their Applications,* Springer-Verlag, Berlin, Heidelberg, New York, London, Paris, Tokyo 1986.

35. **D. Psaltis, D. Brady, X. Gu, and K. Hsu,** "Optical implementation of neural computers," Optical Processing and Computing, H.H. Arsenault, T. Szoplik, and B. Macukow, Editors, 251-276, Academic Press, Boston, San Diego, New York, Berkeley, London, Sydney, Tokyo, Toronto 1989.

36. **G. Ries,** "Rotationssymmetrische Radar-Ambiguity-Funktion," Manuskript, Lehrstuhl für Elektrotechnik VIII - Hochfrequenztechnik, Universität Siegen 1989.

37. **W. Schempp,** "Radar ambiguity functions, the Heisenberg group, and holomorphic theta series," Proc. Amer. Math. Soc. 92, 103-110 (1984)

38. **W. Schempp,** *Harmonic analysis on the Heisenberg nilpotent Lie group, with applications to signal theory*, Pitman Research Notes in Math., Vol. 147, Longman Scientific and Technical, Harlow, Essex, and J. Wiley & Sons, New York 1986.

39. **W. Schempp,** "Group theoretical methods in approximation theory, elementary number theory, and computational signal geometry," Approximation Theory V, C.K. Chui, L.L. Schumaker, and J.D. Ward, Editors, 129-171, Academic Press, Boston, Orlando, San Diego, New York, Austin, London, Sydney, Tokyo, Toronto 1986.

40. **W. Schempp,** "The oscillator representation of the metaplectic group applied to quantum electronics and computerized tomography," Stochastic Processes in Physics and Engineering, S. Albeverio, Ph. Blanchard, M. Hazewinkel, and L. Streit, Editors, 305-344, D. Reidel, Dordrecht, Boston, Lancaster, Tokyo 1988.

41. **W. Schempp,** "Elementary holograms and 3-orbifolds," C.R. Math. Rep. Acad. Sci. Canada 10, 155-160 (1988).

42. **W. Schempp,** "Holographic grids," Visual Communications and Image Processing '88, T. Russell Hsing, Editor, Proc. SPIE 1001, 116-120 (1988).

43. **W. Schempp,** "The holographic transform," Numerical Methods and Approximation Theory III, G.V. Milovanovič, Editor, 67-91, University of Niš, Niš 1988.

44. **W. Schempp,** "Holographic image processing, coherent optical computing, and neural computer architecture for pattern recognition," Lie Methods in Optics II, K.B. Wolf, Editor, Lecture Notes in Physics 352, 19-45, Springer-Verlag, Berlin, Heidelberg, New York, London, Paris, Tokyo, Hong Kong 1989.

45. **W. Schempp,** "Optische Neurocomputer," Manuscript (to appear).

46. **D. Schreier,** *Synthetische Holografie*, Fachbuchverlag Leipzig 1984.

47. **I.E. Segal,** "Transforms for operators and symplectic automorphisms over locally compact abelian groups," Math. Scand. 13, 31-43 (1963).

48. **D.S. Shucker**, "Square integrable representations of unimodular groups," Proc. Amer. Math. Soc. 89, 169-172 (1983).

49. **B.H. Soffer, G.J. Dunning, Y. Owechko, and E. Marom,** "Associative holographic memory with feedback using phase-conjugate mirrors," Optics Letters 11, 118-120 (1986).

50. **R.S. Strichartz,** "Sub-Riemannian geometry," J. Diff. Geom. 24, 221-263 (1986).

51. **G.J. Swanson, W.B. Veldkamp,** "Infrared applications of diffractive optical elements," Holographic Optics: Design and Applications, Ivan Cindrich, Editor, Proc SPIE 883, 155-162 (1988).

52. **H.J. Tiziani,** "Real-time metrology with BSO crystals," Optica Acta 29, 463-470 (1982).

53. **J.C. Várilly, J.M. Gracia-Bondía, and W. Schempp,** "The Moyal representation of quantum mechanics and special function theory," Acta Applicandae Mathematicae

54. **A. Weil,** "Sur certains groupes d'opérateurs unitaires," Acta Math. 111, 143-211 (1964), also in Collected Papers, Vol. III, 1-69, Springer-Verlag, New York, Heidelberg, Berlin 1980.

55. **N. Weste, K. Eshraghian,** *Principles of* CMOS VLSI *design,* Addison-Wesley, Reading, Massachusetts 1985.

56. **K. Wilber,** *Das holographische Weltbild,* Scherz Verlag, Bern, München, Wien 1986.

57. **T. Yatagai,** "Cellular logic architectures for optical computers," Applied Optics 25, 1571-1577 (1986).

58. **J.A. Yeazell, C.R. Stroud,** "Rydberg-atom wave packets localized in the angular variables," Phys. Rev. A 35, 2806-2809 (1987).

59. **J.A. Yeazell, C.R. Stroud,** "Observation of spatially localized atomic electron wave packets," Phys. Rev. Lett. 60, 1494-1497 (1988).

60. **A.L. Yuille, D.M. Kammen, and D.S. Cohen,** "Quadrature and the development of orientation selective cortical cells by Hebb rules," Biol. Cybern. 61, 183-194 (1989).

Part IV: Complex Analysis and Fourier Analysis

A WEIGHTED L^1 ESTIMATE FOR THE LAPLACE OPERATOR

HAROLD S. SHAPIRO
Department of Mathematics
The Royal Institute of Technology
S-100 44 Stockholm, Sweden

ABSTRACT. Lavi Karp showed recently that the Poisson equation $\Delta u = f$, where f is a bounded measurable function on $\mathbb{R}^n$, has a solution that is $O(|x|^2 \log |x|)$ at infinity; the dual form of this result is a certain weighted L^1 estimate for smooth functions of compact support. Karp's proof was based on explicit calculations with the Newtonian kernel (fundamental solution). The goal of the present paper is a new proof of Karp's result based on Fourier analysis. Our method also works for the elliptic operator $P(\partial)$ instead of Δ, where P is a homogeneous polynomial vanishing only at 0.

1. Introduction

1.1. SOLUTIONS OF POISSON'S EQUATION OF SMALL GROWTH

Recently, Lavi Karp proved the following theorem (private communication; his work is part of a doctoral dissertation being done at the Technion, Haifa):

THEOREM A. Let $f \in L^\infty(\mathbb{R}^n)$. Then there is a solution u (in the sense of distributions) to

$$\Delta u = f \tag{1.1}$$

in $\mathbb{R}^n$ satisfying

$$|u(x)| \leq C\, W(x)\, \|f\|_\infty \tag{1.2}$$

where C depends only on n and

$$W(x) = (1+|x|)^2 \log (2+|x|) \tag{1.3}$$

J. S. Byrnes and J. F. Byrnes (eds.), Recent Advances in Fourier Analysis and Its Applications, 563–577.

<u>Remarks</u>. Notations are those commonly used in the literature of p.d.e. and harmonic analysis, as in Hörmander [5]. In particular, $\mathbb{R}^n$ is real Euclidean n-space, $L^\infty = L^\infty(\mathbb{R}^n)$ denotes the bounded measurable functions on that space normed by the sup norm $\|\cdot\|_\infty$, and Δ is the Laplace operator. For all other notations we refer to [5].

It is not hard to see that the result is sharp, in the sense that (1.1) may have no solution that is $o(W(x)$ at infinity. For example, this occurs when $n = 2$ and f is the characteristic function of $\{x \in \mathbb{R}^2 : x_1 > 0, x_2 > 0\}$. (We leave the verification to the reader).

Karp's method was to solve (1.1) by an integral formula

$$u(x) = \int_{\mathbb{R}^n} K(x,y)\, f(y)\, dy \tag{1.4}$$

Here, the obvious choice of $K(x,y)$, the fundamental solution of Δ, cannot be used because then (1.4) would not converge for general $f \in L^\infty(\mathbb{R}^n)$. Therefore, Karp modifies the fundamental solution so as to obtain a kernel $K(x,y)$ which, as a function of y, is $O(|y|^{-n-1})$ for large $|y|$. This technique was apparently first used by Hayman [4] to construct certain subharmonic functions, and independently by Nirenberg and Walker [7] who obtained results like that of Karp for weighted L^p spaces with $1 < p < \infty$. (See also McOwen [6].)

In this paper we give a new proof of Theorem A, based on Fourier methods, which does not require detailed knowledge about the fundamental solution. Our method is based on an idea ("generalized modulus of continuity") that we developed long ago to study the smoothness of functions in terms of degree of approximation (in [9,10,11]; see also Boman [1]); this technique seems not to have been used before to study the growth of solutions to p.d.e. Our method also yields the extension of Theorem A (which Karp also obtained) where Δ is replaced by $P(\partial)$, P being a homogeneous polynomial of degree m with real coefficients that is positive on the unit sphere, and W in (1.2) being replaced by a constant multiple of $(1+|x|)^m \log(|x|+2)$. Since this extension requires scarcely more than notational changes, we shall confine ourselves to proving Theorem A.

1.2. THE DUAL PROBLEM : A WEIGHTED L^1 ESTIMATE FOR Δ

Theorem A is equivalent to the following weighted L^1 estimate:

THEOREM B. With the same C and $W|x|$ as in Theorem A,

$$\int |\varphi| \, dx \leq C \int |\Delta\varphi| \cdot W(x) \, dx \tag{1.5}$$

for all $\varphi \in C_0^\infty(\mathbb{R}^n)$.

Proof. Assume Theorem A. Given $\varphi \in C_0^\infty(\mathbb{R}^n)$ choose $f \in L^\infty$ such that $\|f\|_\infty = 1$ and $f\varphi = |\varphi|$. If we now determine u as in Theorem A, we get

$$\int |\varphi| \, dx = \int \varphi f \, dx = \int \varphi \Delta u \, dx = \int u \Delta \varphi \, dx \leq \int |\Delta\varphi| \cdot CW(x) dx$$

proving (1.5).

Conversely, Theorem A can be deduced from (1.5) by a standard technique based on the Hahn-Banach theorem; we leave this deduction to the reader.

Now, (1.5) is one of a hierarchy of "a priori inequalities" that are used for various purposes in p.d.e., see for example Trèves [12] where estimates of the type

$$\int |\varphi|^2 \, dx \leq \int |P(\partial)\varphi|^2 \, w(x) \, dx \tag{1.6}$$

are proved for arbitrary polynomials P and suitable weight functions w(x). The best result I know for arbitrary P is that of Ortner [8], based on an elementary technique of Malgrange. Ortner shows (Satz 4, p. 22) that (1.6) holds with $w(x) = C(P,\epsilon)e^{\epsilon|x|}$ for every $\epsilon > 0$. The general problem of finding a significantly smaller w that works in (1.6) for a given operator $P(\partial)$ (hopefully, the "best possible" choice for the given $P(\partial)$) seems to be open. The results of Nirenberg-Walker and of Karp settle this question (also with L^p norms in (1.6)) for certain elliptic operators.

1.3. REDUCTION OF THE PROBLEM

In proving Theorem A, it is enough to do so under the assumption that f lies in any convenient subclass of $L^\infty(\mathbb{R}^n)$ that is weak* dense in the unit ball. Indeed, that is all that was needed to prove (1.5), which as we remarked implies Theorem A. In other words, it is sufficient to prove the estimate

$$|u(x)| \leq C \, W(x) \, \|\Delta u\|_\infty \tag{1.7}$$

for $u \in U$, where U is any subclass of $C_0^\infty(\mathbb{R}^n)$ such that $\{\Delta u: u \in U\}$ is weak* dense in the unit ball of L^∞. (Of course, it is essential here that the constant C depends only on n). A convenient choice for U is given by the following:

DEFINITION. U_0 is the set of $u \in C_0^\infty(\mathbb{R}^n)$ satisfying

$$\int_{B(0;1)} (\partial^\alpha u)\, dx = 0, \quad |\alpha| \leq 2 \tag{1.8}$$

Observe that we use $B(x^0;R)$ to denote $\{x:|x-x^0| < R\}$ and we use standard multi-index notation, as in [5]. It will also be convenient to denote

$$L_0^\infty = \{f \in L^\infty: \int_{B(0;1)} f\, dx = 0\} \tag{1.9}$$

LEMMA 1. The set $\{\Delta u: u \in U_0\} =: \Delta U_0$ is weak* dense in the unit ball of L_0^∞.

<u>Proof</u>. Note first that if $u \in U_0$, $\int_{B(0;1)} \Delta u\, dx = 0$ follows from (1.8) which is the reason we only can approximate to f in L_0^∞. Now, to prove the lemma, it is clearly enough to show that ΔU_0 is uniformly dense in

$$\{g \in C_0(\mathbb{R}^n): \int_{B(0;1)} g\, dx = 0\}$$

where $C_0(\mathbb{R}^n)$ denotes the continuous functions on $\mathbb{R}^n$ vanishing at infinity. Since the dual space of C_0 is $M = M(\mathbb{R}^n)$, the bounded measures on $\mathbb{R}^n$, we have then to show that $\mu \in M$ and

$$\int \Delta u\, d\mu = 0, \text{ all } u \in U_0 \tag{1.10}$$

imply

$$d\mu = c\chi_{B(0;1)}\, dx \tag{1.11}$$

for some constant c, where χ_E denotes the characteristic function of E.

We conduct the rest of the proof assuming $n \geq 3$, leaving to the reader the modifications needed when $n = 2$. From elementary potential theory we know that there is a solution v to

$$\Delta v = \chi_{B(0;1)} \tag{1.12}$$

in $\mathbb{R}^n$ which is asymptotic to a constant multiple of $|x|^{2-n}$ for large x (as a tour de force one could prove what will be needed about (1.12) by "pure" Fourier methods too, but it does not seem motivated to do this). We emphasize now that, in all that follows, all derivatives are intended in the distributional sense.

Now, U_0 is the set of $u \in C_0^\infty$ such that, for $|\alpha| \leq 2$

$$0= \int_{B(0;1)} (\partial^\alpha u)dx = \int (\partial^\alpha u)\Delta v \, dx = \int (-1)^{|\alpha|}(\Delta u)\partial^\alpha v dx$$

where integrals, if not otherwise specified, are over $\mathbb{R}^n$. Hence (1.10), and linear algebra imply that for suitable constants $c(\alpha)$

$$\int (\Delta u)\left[d\mu - \sum_{|\alpha|\leq 2} c(\alpha)\partial^\alpha v \, dx\right] = 0 \tag{1.13}$$

for all $u \in C_0^\infty$. Thus, by "Weyl's lemma" the bracketed object in (1.13) is identified with a harmonic function on $\mathbb{R}^n$. This means there is an integrable function $M = \frac{d\mu}{dx}$ such that

$$M(x) - \sum_{|\alpha|\leq 2} c(\alpha)(\partial^\alpha v)(x) =: H(x) \tag{1.14}$$

is harmonic on $\mathbb{R}^n$. Integrating (1.14) over $B(y;1)$ with respect to dx shows that $H(y) \to 0$ as $|y| \to \infty$, hence $H = 0$ and

$$M(x) = \sum_{|\alpha|\leq 2} c(\alpha)\partial^\alpha v \, . \tag{1.15}$$

Taking Fourier transforms and recalling (1.12) gives

$$|\xi|^2 \hat{M}(\xi) = \sum_{|\alpha|\leq 2} c(\alpha)(i\xi)^\alpha \hat{\chi}(\xi) \tag{1.16}$$

where we write X for $X_{B(0;1)}$. Now, since $\hat{X}$ is continuous and $\hat{X}(0) = 1$, (1.16) implies that

$$\sum_{|\alpha|\leq 2} c(\alpha)(i\xi)^{\alpha}/|\xi|^2$$

is continuous at $\xi = 0$. This clearly implies that $c(\alpha) = 0$ for $|\alpha| < 2$. but, if a homogeneous quadratic polynomial $Q(\xi)$ is such that $Q/|\xi|^2$ is continuous at $\xi = 0$ then Q is constant on the unit sphere, and hence Q is a constant multiple of $|\xi|^2$. Going back to (1.15) this means M is a constant multiple of $\Delta v = X_{B(0;1)}$, i.e. (1.11) holds and the lemma is proved.

We can summarize the situation thus far as: <u>to prove Theorem A it is sufficient to prove (1.7)</u> for all $u \in C_0^{\infty}(\mathbb{R}^n)$ <u>satisfying (1.8)</u>. Indeed, that will yield theorem A for f in $L_0^{\infty}(\mathbb{R}^n)$. But every $f \in L^{\infty}(\mathbb{R}^n)$ can be written as $f_0 + m$ where $f_0 \in L_0^{\infty}$ and m is the mean value of f over B(0;1). Thus, $|m| \leq \|f\|_{\infty}$, $\|f_0\|_{\infty} \leq 2\|f\|_{\infty}$ and the desired solution to $\Delta u = f$ satisfying (1.2) can be found as $u_0 + u_1$ where $\Delta u_0 = f_0$, with u_0 satisfying (1.2), and $\Delta u_1 = m$ where we may take e.g. the solution $u_1 = (2n)^{-1}m|x|^2$.

2. Second-difference estimates

Let us begin by showing "where the logarithm comes from" in Theorem A.

LEMMA 2. Let F be a continuous complex-valued function on $\{t\geq 0\}$ such that

$$|F(0) - 2F(a) + F(2a)| \leq 2Ba \tag{2.1}$$

holds for $a \geq 0$. Then, for $t \geq 0$

$$|F(t) \leq C_1B(1+t)\log(2+t) + C_2(1+t) \sup_{0\leq s\leq 1} |F(s)| \tag{2.2}$$

where C_1, C_2 are absolute constants.

<u>Proof</u>. Without loss of generality we may assume $F(0) = 0$. From (2.1) we then obtain, with $t = 2a$,

$$|F(t)| \leq Bt + 2|F(t/2)|$$

Iterating this yields for each $m \geq 1$

$$|F(t) \leq mBt + 2^m|F(2^{-m}t), \quad t \geq 0. \tag{2.3}$$

Now, in proving (2.2) we may clearly suppose $t > 1$. Choose m in (2.3) as the smallest positive integer for which $2^{-m} t < 1$, and (2.2) follows.

Let us now continue the proof of theorem A. From here on many "constants" variously denoted $C, C_1, \ldots$ will appear. We emphasize that the same symbol does not always denote the same constant, but all constants depend (at most) on n, the number of dimensions.

Fix $k \in \{1,2,\ldots,n\}$, and $y \in \mathbb{R}^n$, $|y| = 1$. We will prove the following estimates.

LEMMA 3. For $u \in C_0^\infty(\mathbb{R}^n)$,

$$|(\partial_k u)(x+2ay) - 2\partial_k u(x+ay) + \partial_k u(x)| \leq C\, a\|\Delta u\|_\infty \tag{2.4}$$

holds for $x \in \mathbb{R}^n$, $a \geq 0$. We emphasize that C does not depend on k nor y.

LEMMA 4. For $u \in C_0^\infty(\mathbb{R}^n)$ satisfying (1.8),

$$\sup_{x \in B(0;1)} |\partial^\beta u(x)| \leq C\, \|\Delta u\|_\infty \tag{2.5}$$

holds when $|\beta| \leq 1$.

From these the theorem follows. Indeed, (2.4) with $x = 0$, and Lemma 2, with $F(t) = (\partial_k u)(ty)$, together with the bound on $\partial_k u$ in $B(0;1)$ given by (2.5) shows, for $t \geq 0$

$$|(\partial_k u)(ty) \leq C(1+t)\ \log\ (2+t)\|\Delta u\|_\infty \tag{2.6}$$

and since moreover $|u(0)| \leq C\|\Delta u\|_\infty$ by Lemma 4, (2.6) can be integrated to yield $|u(ty)| \leq C(1+t)^2 \log(2+t)\|\Delta u\|_\infty$ which (since y is an arbitrary unit vector) completes the proof of Theorem A.

We now prove these two lemmas.

Proof of Lemma 4. Write $f = \Delta u$. For $|\alpha| = 2$ we have, taking Fourier transforms in the relation $\Delta(\partial^\alpha u) = \partial^\alpha f$, that $|\xi|^2 (\partial^\alpha u)^\wedge(\xi) = \xi^\alpha\, \hat{f}(\xi)$, or

$$|\xi|^2[(\partial^\alpha u)^\wedge(\xi) - (\xi^\alpha/|\xi|^2)\hat{f}(\xi)] = 0$$

Now, the bracketed expressions is in $L^2(\mathbb{R}^n)$ and has support $C\{0\}$, so it is 0, i.e. $\partial^\alpha u$ is obtained from f by a (generalized) Riesz transform, or, in the terminology of [3, Chapter 2] a singular integral operator which, as a Fourier multiplier, has symbol $\xi^\alpha/|\xi|^2$. Hence, by a theorem of Fefferman and Stein [2] (see also [3, p. 197, Theorem 5.7])

$$\|\partial^\alpha u\|_{BMO} \leq C \|f\|_\infty. \tag{2.7}$$

Also, since $\partial^\alpha u$ has mean value 0 over B(0;1) by (1.8), a well-known estimate [3, p. 166, Cor. 3.10] yields

$$\left(\int_{B(0;1)} |\partial^\alpha u|^p dx\right)^{1/p} \leq A_p \|\partial^\alpha u\|_{BMO} \tag{2.8}$$

for $0 < p < \infty$. We choose a fixed $p > n$ e.g. $p = 2n$ and then, from (2.7), (2.8)

$$\left(\frac{}{}\!\!\!\!\!-\!\!\!\!\!\!\int_{B(0;1)} |\partial^\alpha u|^{2n} dx)\right)^{1/2n} \leq C \|f\|_\infty, \quad |\alpha| = 2. \tag{2.9}$$

By a well-known embedding theorem of Morrey

$$|g(x) - g(x')| \leq C \sum_{j=1}^{n} \left(\int_B |\partial_j g|^{2n} dx\right)^{1/2n} \cdot |x-x'|^{1/2} \tag{2.10}$$

holds for $g \in C^\infty(B)$ where B is the unit ball and x,x′ are in B. Applying this with $g = \partial_k u$ and using (2.9) gives

$$|(\partial_k u)(x) - (\partial_k u)(x')| \leq C \|f\|_\infty \cdot |x-x'|^{1/2}$$

for x,x′ in B, so

$$\left|\int_B (\partial_k u)(x)dx' - \int_B (\partial_k u)(x')dx'\right| \leq C \|f\|_\infty$$

whence, using (1.8)

$$|(\partial_k u)(x)| \leq C \|f\|_\infty, \quad k \in \{1,2,\ldots,n\} \tag{2.11}$$

holds for $x \in B(0;1)$ which proves (2.5) when $|\beta| = 1$. Finally (2.11) and $\int_B u\, dx = 0$ imply (2.5) with $\beta = 0$. This completes the proof of Lemma 4.

<u>Proof of Lemma</u> 3. Write $v = \partial_k u$, and let $\nu = \nu_y$ denote the element of $M(\mathbb{R}^n)$ (bounded measures on $\mathbb{R}^n$) satisfying

$$\hat{\nu}(\xi) = (1-e^{-i(y\cdot\xi)})^2 . \tag{2.12}$$

Then (2.4) can be written

$$\|\hat{v}(\xi)\hat{\nu}(a\xi)\|_{FL^\infty} \lesssim C\, a\, \|f\|_\infty \tag{2.13}$$

where $f = \Delta u$ and $\|\hat{G}\|_{FX}$ denotes generically the X-norm of G.

Now, if $\hat{\varphi}$ is any function in $C_0^\infty(\mathbb{R}^n\backslash\{0\})$ and $a > 0$ we get from $\Delta v = \partial_k f$ by Fourier transformation

$$\hat{v}(\xi)\hat{\varphi}(a\xi) = -i\xi_k|\xi|^{-2}\hat{\varphi}(a\xi)\hat{f}(\xi)$$

whence

$$\|\hat{v}(\xi)\hat{\varphi}(a\xi)\|_{FL^\infty} \lesssim \|\xi_k|\xi|^{-2}\hat{\varphi}(a\xi)\|_{FL^1}\cdot\|\hat{f}\|_{FL^\infty}$$

and, since the norm in FL^1 is invariant under the "stretching" map $\xi \to a\xi$, the first term on the right equals $a\cdot\|\xi_k|\xi|^{-2}\ \hat{\varphi}(\xi)\|_{FL^1} = Ba$ where the constant $B = B_\varphi$ depends only on φ. Thus

$$\|\hat{v}(\xi)\hat{\varphi}(a\xi)\|_{FL^\infty} \lesssim B_\varphi a\|f\|_\infty \tag{2.14}$$

Thus, our task is to deduce (2.13) from (2.14). Deductions of this kind, which are important in approximation theory, were first treated within a general framework in [9,10,11] and later refined and extended in [1]. The present application is just a special case of Corollary 2.4 of [1], but since it is fairly simple we can give the detailed proof here; however, the reader should consult the cited works for motivation.

We make use of a well-known partition of unity (see e.g. [11, p. 220]. Namely, there is a function $\hat{\theta} \in C^{\infty}(\mathbb{R}^n)$ whose support is contained in $\{1/2 \leq |\xi| \leq 2\}$ and such that

$$\sum_{j=-\infty}^{\infty} \hat{\theta}(2^j\xi) = 1 \text{ for } \xi \neq 0 \tag{2.15}$$

Fix a function $\hat{\omega}$ in $C^{\infty}(\mathbb{R}^n)$ which equals 1 on B(0;1) and vanishes for $|\xi| \geq 2$. Then

$$\hat{\nu}(\xi) = \hat{\nu}(\xi)\hat{\omega}(\xi) + \hat{\mu}(\xi) \tag{2.16}$$

where $\hat{\mu}(\xi) := (1-\hat{\omega}(\xi))\hat{\nu}(\xi)$ vanishes for $|\xi| \leq 1$. Now, for all ξ

$$\hat{\mu}(\xi) = \sum_{j=-\infty}^{0} \hat{\theta}(2^j\xi)\hat{\mu}(\xi) \tag{2.17}$$

Indeed, for $\xi = 0$ both sides are 0, while for $\xi \neq 0$, (2.15) gives

$$\hat{\mu}(\xi) = \sum_{j=-\infty}^{\infty} \hat{\theta}(2^j\xi)\hat{\mu}(\xi)$$

and all summands with $j \geq 1$ vanish. Now, using (2.17)

$$\hat{v}(\xi)\hat{\mu}(a\xi) = \sum_{j=-\infty}^{0} \hat{v}(\xi)\hat{\theta}(2^j a\xi)\hat{\mu}(a\xi)$$

hence

$$\|\hat{v}(\xi)\hat{\mu}(a\xi)\|_{FL^{\infty}} \leq \sum_{j=-\infty}^{0} \|\hat{v}(\xi)\hat{\theta}(2^j a\xi)\|_{FL^{\infty}} \|\hat{\mu}\|_{FM}$$

and by (2.14) with $\varphi = \theta$,

$$\|\hat{v}(\xi)\hat{\theta}(2^j a\xi)\|_{FL^{\infty}} \leq C\, a\, 2^j \|f\|_{\infty} \tag{2.18}$$

so

$$\|\hat{v}(\xi)\hat{\mu}(a\xi)\|_{FL^{\infty}} \leq C\, a \|f\|_{\infty} .$$

Observing (2.16) we see that, to obtain the desired estimate (2.13), we only need show

$$\|\hat{v}(\xi)\hat{\nu}(a\xi)\hat{\omega}(a\xi)\|_{FL^\infty} \leq C\, a\|f\|_\infty \tag{2.19}$$

Now, let us fix a radially symmetric function $\hat{\varphi} \in C^\infty$ with $\hat{\varphi}(\xi)$ equal to 1 for $|\xi| \leq 2$ and 0 for $|\xi| \geq 3$. Then $\hat{\theta}(\xi) = \hat{\theta}(\xi)\hat{\varphi}(\xi)$ for all ξ. Now, $\hat{v}(\xi)\hat{\nu}(a\xi)\hat{\omega}(a\xi) = \sum_{j=-\infty}^{\infty} \hat{v}(\xi)\hat{\nu}(a\xi)\hat{\omega}(a\xi)\hat{\theta}(2^j a\xi)$ using (2.15) (note that $\hat{v}(\xi) = -i\xi_k\hat{u}(\xi)$, so $\hat{v}(0) = 0$). Also, $\hat{\omega}(\xi)\hat{\theta}(2^j\xi)$ vanishes identically unless $j \geq -1$, so in the last sum, terms with $j \leq -2$ can be neglected and we get

$$\hat{v}(\xi)\hat{\nu}(a\xi)\hat{\omega}(a\xi) = \sum_{j=-1}^{\infty} \hat{v}(\xi)\hat{\nu}(a\xi)\hat{\omega}(a\xi)\hat{\varphi}(2^j a\xi)\hat{\theta}(2^j a\xi)$$

whence

$$\|\hat{v}(\xi)\hat{\nu}(a\xi)\hat{\omega}(a\xi)\|_{FL^\infty} \leq \sum_{j=-1}^{\infty} \|\hat{v}(\xi)\hat{\theta}(2^j a\xi)\|_{FL^\infty} \cdot$$

$$\cdot \|\hat{\omega}\|_{FL^1} \cdot \|\hat{\nu}(a\xi)\hat{\varphi}(2^j a\xi)\|_{FL^1}$$

which, using (2.18) and stretching-invariance of the FL^1 norm, is

$$\leq C\, a\, \|f\|_\infty \cdot \sum_{j=-1}^{\infty} 2^j \|\hat{\nu}(\xi)\hat{\varphi}(2^j\xi)\|_{FL^1} .$$

Now, we claim, for $b > 0$

$$\|\hat{\nu}(\xi)\hat{\varphi}(b\xi)\|_{FL^1} \leq C\, b^{-2} \tag{2.20}$$

Assuming this the last estimate gives

$$\|\hat{v}(\xi)\hat{\nu}(a\xi)\hat{\omega}(a\xi)\|_{FL^\infty} \leq C\, a\, \|f\|_\infty \sum_{j=-1}^{\infty} 2^{-j} = C_1 a\|f\|_\infty$$

which is (2.19). Thus, Lemma 3 will be proved if we

demonstrate (2.20).

Now

$$\hat{\sigma}(\xi) := \frac{(1-e^{-i(y\cdot\xi)})^2}{(y\cdot\xi)^2} \tag{2.21}$$

is the Fourier transform of a measure $\sigma \in M(\mathbb{R}^n)$. Indeed, in view of the rotation-invariance of $M(\mathbb{R}^n)$ we need only verify this for $y = (1,0,\ldots,0)$, i.e. that

$$\hat{\sigma}_0(\xi_1,\ldots,\xi_n) = \xi_1^{-2}\left[1-e^{-i\xi_1}\right]^2$$

is in $FM(\mathbb{R}^n)$; the reader can check this. Observe that $\|\hat{\sigma}\|_{FM}$ does not depend on y ($|y| = 1$). Hence

$$\|\hat{\nu}(\xi)\hat{\varphi}(b\xi)\|_{FL^1} = \|\hat{\sigma}(\xi)(y\cdot\xi)^2\ \hat{\varphi}\ (b\xi)\|_{FL^1}$$
$$\leq \|\hat{\sigma}\|_{FM}\cdot\|(y\cdot\xi)^2\hat{\varphi}(\xi)\|_{FL^1}\cdot b^{-2} = Cb^{-2}$$

which proves (2.20). The proof of Lemma 3, and hence of Theorem A, is now complete.

3. Concluding remarks

3.1.

For F in certain subclasses of L^∞ the conclusion of Theorem A can be strengthened. For example, if $f \in FM$, (1.1) has a solution satisfying

$$|u(x)| \leq C\ |x|^2\cdot\|f\|_{FM} \tag{3.1}$$

where C is an absolute constant. Let us outline the proof. For $\lambda \in \mathbb{R}^n$,

$$u(x;\lambda) := [e^{-i(\lambda\cdot x)} - 1 + i(\lambda\cdot x)]\cdot|\lambda|^{-2} \tag{3.2}$$

satisfies $\Delta u(x;\lambda) = -e^{-i(\lambda\cdot x)}$ and

$$|u(x;\lambda)| \leq C(\lambda\cdot x)^2|\lambda|^{-2} \leq C|x|^2 \tag{3.3}$$

Hence, for $\mu \in M(\mathbb{R}^n)$, $u(x) := -\int u(x;\lambda)d\mu(\lambda)$ satisfies $\Delta u = \int e^{-i(\lambda\cdot x)}d\mu(\lambda) = \hat{\mu}(x)$ with $|u(x)| \leq C\,|x|^2\cdot\|\mu\|_M$ which, for $f = \hat{\mu}$, is (3.1).

One can also get the $O(|x|^2)$ estimate for other subclasses of L^∞; the subject appears to be of interest and we hope to return to it in another publication.

3.2.

The ellipticity of Δ entered the proof of theorem A in several essential ways. As already remarked, an analogous theorem to Theorem A is true with Δ replaced by $P(\partial)$ when P is a (real) homogeneous polynomial, $P(\xi) \neq 0$ for $|\xi| = 1$. If $P(\partial)$ is elliptic, but P not homogeneous, the above analysis is not applicable. Finding good weighted estimates analogous to (1.5), also for non-elliptic $P(\partial)$, seems still to be largely an unsolved problem, the importance of which has been emphasized e.g. by Trèves [12, p. 5 and p. 91]. Recently new weighted estimates (also called Carleman estimates) have been developed to deal with unique continuation and spectral problems (see e.g. the survey articles of C. Kenig in the proceedings of the 1986 International Congress (Berkeley) and in LNM vol. 1384).

3.3.

Implicit in the preceding analysis is the "local regularity theory" for $\Delta u = f$, $f \in L^\infty$. Note that since any two solutions differ by a harmonic function, all solutions u have essentially the same local regularity, namely each $\partial_k u$, restricted to any line, is locally of the Zygmund class Λ_* [13, p. 43]. (This is of course well known, and much easier than Theorem A.)

3.4.

From the preceding analysis follows a somewhat sharper result than Theorem A, namely

THEOREM C. Under the hypotheses of Theorem A, (1.1) has a solution satisfying (1.2) and, for $k \in \{1,2,\dots,n\}$

$$|(\partial_k u)(x)| \leq C(1+|x|)\log(2+|x|) \tag{3.4}$$

Let us sketch the proof. As shown in the proof of Theorem A, there is a sequence $\{u_j\} \subset C_0^\infty(\mathbb{R}^n)$ satisfying $\|\Delta u_j\|_\infty \leq \|f\|_\infty$, $\Delta u_j \to f$ (weak*) and

$$|\partial_k u_j(x)| \leq A(x) := C(1+|x|)\log(2+|x|) \tag{3.5}$$
$$|u_j(x)| \leq B(x) := C(1+|x|)^2\log(2+|x|) \tag{3.6}$$

By passing to a subsequence (still denoted by $\{u_j\}$) we may assume that the sequences $\{A^{-1}\partial_k u_j\}$ and $\{B^{-1}u_j\}$ converge weak* to functions denoted by v_k, v respectively with $\|v\|_\infty \leq 1$, $\|v_k\|_\infty \leq 1$. Then we have, for $\varphi \in C_0^\infty(\mathbb{R}^n)$,

$$\int Bv\Delta\varphi dx = \lim_j \int (B^{-1}u_j)B\Delta\varphi dx$$
$$= \lim_j \int \varphi(\Delta u_j)dx = \int \varphi f dx, \qquad \text{i.e.}$$

$u := Bv$ satisfies (1.1) and (1.2). Moreover, it is easy to check that, in the sense of distributions, $\lim_{j\to\infty} (\partial_k u_j) = Av_k$, whence $\partial_k u = Av_k$ and (3.4) holds.

3.5.

It seems worth pointing out that the essence of Theorem A is the à priori inequality implicit in our proof of it:

For $u \in C_0^\infty(\mathbb{R}^n)$ satisfying (1.8),

$$|\!|\!| u |\!|\!| \leq C \| \Delta u \|_\infty \qquad \text{where } C = C(n)$$

and

$$|\!|\!| u |\!|\!| := \sup_{x\in\mathbb{R}^n} [(1+|x|)^2\log(2+|x|)]^{-1}\cdot|u(x)|.$$

In particular this estimate holds for u in a vector subspace of C_0^∞ having finite codimension; of course it cannot hold in all of C_0^∞ since that would imply $|u(0)| \leq C\|\Delta u\|_\infty$ for all $u \in C_0^\infty$, which is manifestly false.

3.6.

In closing, I want to thank Lavi Karp for communicating to me his unpublished results, and Synnöve Kaxe for her rapid and skilful typing of the manuscript.

Bibliography

1. Boman, J. Equivalence of generalized moduli of continuity, Arkiv för mat. 18 (1980) 73-100.
2. Fefferman, R. and E.M. Stein, H^p spaces of several variables, Acta Math. 129 (1972) 137-193.
3. Garcia-Cuerva, J. and J.L. Rubio de Francia, Weighted Norm Inequalities and Related Topics, Math. Studies 116, North-Holland, 1985.
4. Hayman, W. and P. Kennedy, Subharmonic Functions, vol.1 Academic Press, 1976.
5. Hörmander, L. The Analysis of Linear Partial Differential Operators I, Springer-Verlag, 1983.
6. McOwen, R. The behaviour of the Laplacian on weighted Sobolev spaces, Comm. Pure Appl. Math. 32 (1979) 783-795.
7. Nirenberg, L. and H. Walker, The null spaces of elliptic partial differential operators in $\mathbb{R}^n$, J. Math. Anal. Appl. 42 (1973) 271-301.
8. Ortner, N. Fundamentallösungen und Existenz von schwachen Lösungen linearer Partieller Differentialgleichungen mit konstanten Koeffizienten, Ann. Acad. Sci. Fenn., Ser. A.I. Math. 4 (1978/1979) 3-30.
9. Shapiro, H.S. Smoothing and Approximation of Functions, Van Nostrand Reinhold, New York, 1969.
10. Shapiro, H.S. A Tauberian theorem related to approximation theory, Acta Math. 120 (1968) 279-292.
11. Shapiro, H.S. Topics in Approximation theory, Lecture Notes in Math. 187, Springer-Verlag, 1971.
12. Trèves, F. Linear Partial Differential Equations with Constant Coefficients, Gordon and Breach, 1966.
13. Zygmund, A. Trigonometric Series, Cambridge, 1968.

WEIGHTED EXTENSIONS OF RESTRICTION THEOREMS FOR THE FOURIER TRANSFORM

C. CARTON–LEBRUN
Department of Mathematics
University of Mons
7000 Mons
Belgium

and

H.P. HEINIG
Department of Mathematics
and Statistics
McMaster University
Hamilton, L8S 4K1
Ontario, Canada

ABSTRACT. Let $\hat{f}$ denote the Fourier transform of f, γ a smooth plane curve of curvature K and arclength measure ds. We establish estimates of the form.

$$\|\hat{f}|K|^{\delta}\|_{L^r(\gamma,ds)} \le A\|f\|_{L^p_V(\mathbb{R}^2)}$$

for certain classes of weights V and values of r, p, δ. These provide weighted extensions of a result of Sjölin and of a restriction theorem of Zygmund for circles. The particular case of power weights is also considered and further generalizations are discussed.

1. Introduction

Let I denote a compact interval of $\mathbb{R}$ and $\gamma(t) = (\gamma_1(t), \gamma_2(t))$ a smooth parametrized curve such that

$$|\gamma'(t)|^2 = \gamma_1'(t)^2 + \gamma_2'(t)^2 \neq 0$$

for $t \in I$. Denote by $(K\circ\gamma)(t)$ the curvature at the point $\gamma(t)$ and by ds the arclength measure of γ. In this paper we prove estimates of the form

$$\|\hat{f}|K|^{\delta}\|_{L^r(\gamma,ds)} \le A\|f\|_{L^p_V(\mathbb{R}^2)} \tag{1.1}$$

for certain weights V and values of p,r and δ.

When $V = 1$, $1 < p < 4/3$, $1 \le r \le p'/3$ and $\delta > 1/p'$, (1.1) reduces to a result of Sjölin [10].

Theorems 1 and 2 of Section 3 are our main results. They are proved in Section 4 by a procedure similar to that given in [10], i.e. by estimating the formal dual operator $S = T'$ of $T: f \to \hat{f}\circ\gamma$. Our contribution here is the introduction of

J. S. Byrnes and J. F. Byrnes (eds.), Recent Advances in Fourier Analysis and Its Applications, 579–596.

weights which are suitable to permit successive changes of variables, and the application of known results about weighted estimates of Fourier transforms and fractional integrals.

While Theorem 1 is concerned with general smooth curves, Theorem 2 deals with the case of circles and can thus be considered as a weighted extension of a certain restriction theorem of Zygmund [11]. We note here that major contributions were made in [3] as concerns spherical restriction theorems for functions in certain subspaces of $L^p_V(\mathbb{R}^n)$, $n \geq 2$. Our approach here, however, is quite different and leads to different classes of weights. The proof of Theorem 2 differs from that of Theorem 1 by the introduction of weights especially adapted to the geometry of the circle and leading to a fractional integral of lower singularity. Corollaries of Theorems 1 and 2 dealing with power weights are stated in Section 3 and proved in Section 4. Particular cases and examples are mentioned in Section 5. In Section 6, we consider further generalizations of the type

$$\|\hat{f}|K|^{\delta}U\|_{L^r(\gamma,ds)} \leq A\|f\|_{L^p_V(\mathbb{R}^2)} \tag{1.2}$$

for suitable weights U and V.

2. Notation

We use $\langle x,y\rangle = x_1y_1 + x_2y_2$, $|x|^2 = \langle x,x\rangle$ for $x = (x_1,x_2)$, $y = (y_1,y_2) \in \mathbb{R}^2$. If V is a non-negative function, L^p_V denotes the space

$$L^p_V = \{f: \|fV^{1/p}\|_p < \infty\},$$

with norm $\|f\|_{p,V} = \|fV^{1/p}\|_p$. We also define L^p_α, $\alpha > 0$ by

$$L^p_\alpha = \{f: |x|^\alpha f(x) \in L^p\} \text{ with norm } \|f\|_{p,\alpha} = \||x|^\alpha f\|_p.$$

If g is a Lebesgue measurable function defined on $\mathbb{R}^2$, then the equimeasurable decreasing rearrangement of g is defined by

$$g^*(t) = \inf\{y > 0: |\{x \in \mathbb{R}^2: |g(x)| > y\}| \leq t\}.$$

Here $|E|$ denotes the Lebesgue measure of $E \subset \mathbb{R}^2$. Rearrangements occur in our weight conditions.

Throughout the paper, I denotes a compact interval of $\mathbb{R}$ and $\gamma(t) = (\gamma_1(t), \gamma_2(t))$, $t \in I$ a parametrized curve which is at least in $C^2(I)$ and satisfies $|\gamma'(t)| \neq 0$ for every $t \in I$. The notation $\gamma \in C^k(I)$ means that γ is of class C^k in the interior of I and has one sided derivative of order k at the endpoints of I. The curvature of γ at the point $\gamma(t)$ is denoted by $(K\circ\gamma)(t)$ and arc length measure by ds.

If f is defined on $\mathbb{R}^2$, then the Fourier transform $\hat{f}$ of f is defined by

$$\hat{f}(x) = \int_{\mathbb{R}^2} e^{-i<x,y>} f(y)dy, \quad x \in \mathbb{R}^2$$

whenever the integral exists. In the estimates (1.1) and (1.2), the Fourier transform $\hat{f}$ of f is defined on γ by a suitable limit process. In fact, in each case where they occur, estimates of the form (1.1), (1.2) are first proved for functions in $L^p_V \cap L^1$. This allows one to define $\hat{f}$ on I by

$$\lim_m \|[\hat{f} - (\hat{f}_m\circ\gamma)]\,|K\circ\gamma|^{\delta}(U\circ\gamma)\|_{L^r(I,ds)} = 0,$$

where $\{f_m\}$ is any sequence of simple functions approximating f in L^p_V. Accordingly the left hand sides of (1.1) and (1.2) have the following meaning:

$$\|\hat{f}\,|K|^{\delta}U\|_{L^r(\gamma,ds)} = \lim_m \|(\hat{f}_m\circ\gamma)\,|K\circ\gamma|^{\delta}(U\circ\gamma)\|_{L^r(I,ds)}.$$

As usual, χ_E denotes the characteristic function of a set E and p' the conjugate index of p defined by the relation $1/p + 1/p' = 1$. Similarly for other indices. Finally, unless otherwise stated, constants are denoted by c. They may depend on the involved parameters and weights but are independent of the functions f or g under consideration. Moreover, their values are not necessarily the same at each occurrence.

3. Main Results

Theorem 1. a) Suppose $\gamma \in C^{\infty}(I)$ and the following assumptions are satisfied:

i) $4/3 < q < \infty$ and $1 < p \leq 6q/(3q+2)$;

ii) $V \in L^1_{loc}(\mathbb{R}^2)$ is such that

$$\sup_{s\geq c>0} \left[s^{-p'/(3q)} \int_0^s (V^{-p'/p})^*(t)dt\right] < \infty, \tag{3.1}$$

where $*$ denotes the decreasing rearrangement.

Then for $\delta > 1/(3q)$ and $1 \le r \le q$

$$\| \hat{f} |K|^{\delta} \|_{L^r(\gamma,ds)} \le c \|f\|_{p,V}. \tag{3.2}$$

b) If $K(t) \ge 0$, then the condition $\gamma \in C^2(I)$ is sufficient in order that (3.2) holds with p,q,r,δ and V as in a). Moreover, in this case (3.2) holds also for $\delta = 1/(3q)$.

<u>Corollary 1.</u> a) Suppose $\gamma \in C^{\infty}(I)$ and i) of Theorem 1 is satisfied. If

$$\max[0,2(1/p' - 1/(3q))] \le \alpha < 2/p' \tag{3.3}$$

holds, then for $\delta > 1/(3q)$ and $1 \le r \le q$,

$$\| \hat{f} |K|^{\delta} \|_{L^r(\gamma,ds)} \le c \|f\|_{p,\alpha} \tag{3.4}$$

for every $f \in L^p_{\alpha}$.

b) If $K(t) \ge 0$, $\gamma \in C^2(I)$ is sufficient for (3.4) under the same conditions on the indices as in a). Moreover, (3.4) holds also for $\delta = 1/(3q)$ in this case.

<u>Theorem 2</u>. Let C_{ρ} be the circle of radius $\rho > 0$ centered at the origin. Suppose conditions (B1) and (B2) hold:

$$\text{(B1)} \quad \begin{cases} 0 \le \lambda \le 1/[2(q-1)] & \text{if } 4/3 < q < \infty; \\ 0 < \lambda \le 3/2 & \text{if } q = 4/3; \\ (4-3q)/[4(q-1)] < \lambda \le 1/[2(q-1)] & \text{if } 1 < q < 4/3; \\ 1 < p \le (2R)' \text{ where } R = q(3+2\lambda)/(3q-2), \text{ i.e. } (2R)' = \\ \qquad 2q(3+2\lambda)/[q(3+4\lambda)+2] \end{cases}$$

and

$$\text{(B2)} \quad \begin{cases} V \in L^1_{loc}(\mathbb{R}^2) \text{ such that } \sup\limits_{s \ge c > 0} \left[s^{-p'/L} \int_0^s (V^{-p'/p})^*(t)\,dt \right] < \infty \\ \\ \text{with } L = 2q(3+2\lambda)/[2(1+\lambda)-\lambda q]. \end{cases}$$

Then for $1 \le r \le q$

$$\|\hat{f}\|_{L^r(C_\rho, ds)} \le c\rho^{1/r-2/p'} \|f\|_{p,V(\rho.)} \tag{3.5}$$

The last statement in this section is

Corollary 2. Let C_ρ be defined as in Theorem 2. If (B1) of Theorem 2 is satisfied and

$$\max[0, 2(1/p' - 1/L)] \le \alpha < 2/p',$$

where L is as in Theorem 2, then for $1 \le r \le q$

$$\|\hat{f}\|_{L^r(C_\rho, ds)} \le c\rho^{1/r+\alpha-2/p'} \|f\|_{p,\alpha}$$

for every $f \in L^p_\alpha$.

Note that if one chooses $\lambda = 0$, $p' = L$, $\alpha = 0$ and $r = q$, then by Corollary 2, $q = p'/3$, $4/3 < q < \infty$, $1 < p \le (2R)' = 6q/(3q+2)$. Consequently, $1 < p < 4/3$ and Corollary 2 reduces to Theorem 3 of Zygmund [11].

4. Proofs

In order to prove the theorems of Section 3 we require extension theorems for the dual operator. These results comprise weighted extensions of corresponding work of Sjölin [10] and may be of independent interest. They are given below as Proposition 1 and 2. In the sequel we also need to estimate the Fourier transform repeatedly. For this reason the following known result (c.f. [4], [5], [6], [7], [8]) is stated here:

Theorem A. Suppose $1 < R \le Q < \infty$ and μ and ν are nonnegative functions satisfying

$$\sup_{s>0} \left(\int_0^s \mu^*(t)dt\right)^{1/Q} \left(\int_0^{1/s} [(1/\nu)^*(t)]^{R'-1} dt\right)^{1/R'} < \infty . \tag{4.1}$$

Then for $F \in L^1(\mathbb{R}^n)$,

$$\Big(\int_{\mathbb{R}^n} |\hat{F}(x)|^Q \mu(x)dx\Big)^{1/Q} \le c\Big(\int_{\mathbb{R}^n} |F(x)|^R \nu(x)dx\Big)^{1/R}.$$

4.1 PROOF OF THEOREM 1.

We establish now our first extension result.

Proposition 1. (a) Suppose $\gamma(t) = (t, \psi(t))$ with $\psi \in C^2(I)$ and $\psi''(t) \ge 0$ for every $t \in I$. If i) and ii) of Theorem 1a) hold, then

$$\|V^{-1/p}Sg\|_{L^{p'}(\mathbb{R}^2)} \le c\|g(\psi'')^{-1/(3q)}\|_{L^{q'}(I)} \tag{4.2}$$

where

$$(Sg)(x) = \int_I e^{-i<x,\gamma(t)>}g(t)dt, \ x \in \mathbb{R}^2, \ g \in L^1(I).$$

(b) Suppose $\gamma(t) = (t, \psi(t))$ with $\psi \in C^\infty(I)$ and that the assumptions i) and ii) of Theorem 1a) are satisfied. Then

$$\|V^{-1/p}Sg\|_{L^{p'}(\mathbb{R}^2)} \le c\|g|\psi''|^{-\delta}\|_{L^{q'}(I)} \tag{4.3}$$

for every $\delta > 1/(3q)$.

Proof (a) We suppose that the right side of (4.2) is finite. Then as in [10] we obtain for $x \in \mathbb{R}^2$

$$(Sg)(x)^2 = 2\int_E e^{-i<x,\gamma(t)+\gamma(s)>}g(t)g(s)dtds$$

where $E = \{(t,s) \in I^2 : t < s\}$.

Setting $u = t+s$, $v = \psi(t) + \psi(s)$ then

$$(Sg)(x)^2 = 2\Big[\frac{(g\circ T)(g\circ S)\chi_D}{|(\psi'\circ T)-(\psi'\circ S)|}\Big]^{\wedge}(x), \quad x \in \mathbb{R}^2 \tag{4.4}$$

where D is the image of E under this mapping, and $(T,S)\colon D \to E\colon (u,v) \to (t,s)$ the inverse mapping. Since $(u,v) = \gamma(t) + \gamma(s)$ one has $|(u,v)| \le 2M$ for all $(u,v) \in D$, where

$$M = \max_{t \in I} |\gamma(t)|.$$

We now apply Theorem A with $\hat{F} = (Sg)^2/2$, $\mu = V^{-p'/p}$, $Q = p'/2$ and $\nu(x) = w(|x|)$ where

$$w(\tau) = 1 \text{ for } 0 < \tau \leq 2M \text{ and } +\infty \text{ for } \tau > 2M.$$

A straightforward calculation then shows that

$$(1/\nu)^*(t) = 1 \text{ if } 0 < t \leq 4\pi M^2 \text{ and equals 0 if } t > 4\pi M^2.$$

Condition (4.1) of Theorem A takes therefore the form

$$\sup_{s \geq c > 0} \left[s^{-p'/(2R')} \int_0^s (V^{-p'/p})^*(t)dt \right] < \infty \tag{4.5}$$

with $1 < R \leq p'/2 < \infty$. Under these conditions Theorem A yields

$$\|V^{-1/p}Sg\|_{p'}^2 \leq c\left(\int_E |g(t)|^R |g(s)|^R |\psi'(t) - \psi'(s)|^{1-R} dtds\right)^{1/R}.$$

Setting $\xi = \psi'(t)$, $\eta = \psi'(s)$ and denoting by α the inverse function of ψ', then the last inequality takes the form

$$\|V^{-1/p}Sg\|_{p'} \leq c\left(\iint_{J\times J} G(\xi)G(\eta)|\xi-\eta|^{1-R} d\xi d\eta\right)^{1/(2R)}.$$

where $J = \psi'(I)$ and $G = |g\circ\alpha|^R(\psi''\circ\alpha)^{-1}$. Now using the classical inequality for fractional integrals we see that the last integral is majorized by $\|G\|_{RP}$ if and only if

$$1 < 2/P = 3 - R, \text{ that is } P' = 2/(R-1). \tag{4.6}$$

If these conditions are satisfied, we deduce from the inverse change of variable $t = \alpha(\xi)$ that

$$\|V^{-1/p}Sg\|_{p'} \leq c\|g(\psi'')^{-1/(3q)}\|_{L^{q'}(I)}$$

where $q' = RP$. Since $1 < R \leq p'/2$ and (4.6) are equivalent to

$$2R' = 3q,\ 4/3 < q < \infty \text{ and } 1 < p \leq (2R)' = 6q/(3q+2),$$

(4.5) shows that Proposition 1a) is proved.

The proof of part b) of Proposition 1 is as in Sjölin [10]. If $\psi \in C^{\infty}(I)$, define Sg as above and

$$(S_n g)(x) = \int_{I_n} e^{-i<x,\gamma(t)>} g(t)dt, \quad n = 1,2,\ldots,$$

where I_n are the component intervals of $\{t \in I: \psi''(t) \neq 0\}$. Then by Proposition 1a) and Lemma 1 of [10]

$$\|V^{-1/p}(Sg)\|_{p'} \leq \sum_{n=1}^{\infty} \|V^{-1/p}(S_n g)\|_{p'} \leq c \sum_{n=1}^{\infty} \|g|\psi''|^{-1/(3q)}\|_{L^{q'}(I_n)}$$

$$\leq c \sum_{n=1}^{\infty} (\sup_{I_n}|\psi''|)^{\delta-1/(3q)} \|f|\psi''|^{-\delta}\|_{L^{q'}(I)} \leq c\|f|\psi''|^{-\delta}\|_{L^{q'}(I)}$$

if $\delta > 1/(3q)$.

This completes the proof of Proposition 1.

By a duality argument based on

$$\int_{\mathbb{R}^2} f(x)(Sg)(x)dx = \int_I g(t)(\hat{f}\circ\gamma)(t)dt$$

for $f \in L^1(\mathbb{R}^2)$ an $g \in L^1(I)$, we can now deduce Theorem 1 in the case $r = q$ for parametrized curves of the form $\gamma(t) = (t,\psi(t))$. It follows from this case and Hölder's inequality that (3.2) also holds for these parametrized curves when $1 \leq r < q$.

Consider now $\check{\gamma}(t) = (\psi(t),t)$, $t \in I$ and define $\bar{f}(x_1,x_2) = f(x_2,x_1)$ and $\bar{V}(x_1,x_2) = V(x_2,x_1)$. Since $(\bar{V})^* = V^*$, $\bar{V}$ satisfies ii) of Theorem 1 and (3.2) holds for $\check{\gamma}$, $\bar{f}$, $\bar{V}$ instead of γ, f, V. Since

$$\|\bar{f}\|_{p,\bar{V}} = \|f\|_{p,V} \quad \text{and} \quad (\bar{f})^{\wedge}\circ\gamma = \hat{f}\circ\check{\gamma},$$

it follows that (3.2) is also true for parameterized curves of the form $\check{\gamma}(t) = (\psi(t),t)$. Now, if $\gamma(t) = (\gamma_1(t), \gamma_2(t))$ with $|\gamma'(t)| \neq 0$, $t \in I$, then I can be covered by a finite union of compact intervals in each of which γ is equivalent to at least one of the two forms of parametrizations that have just been considered. Theorem 1 is thus completely established.

4.2 PROOF OF COROLLARY 1

In Theorem 1, choose $V(x) = |x|^{\alpha p}$ with $\alpha \geq 0$. Then a direct calculation shows that

$$(V^{-p'/p})^*(t) = \pi^{-1} t^{-p'\alpha/2}$$

and therefore condition (3.1) is equivalent to the conditions

$$0 \leq \alpha < 2/p' \quad \text{and} \quad \sup_{s \geq c > 0} \left[s^{-p'/(3q)+1-p'\alpha/2} \right] < \infty.$$

But this is clearly equivalent to (3.3) of Corollary 1. Hence the result.

4.3 PROOF OF THEOREM 2.

As in the case of Theorem 1 we require an extension result analogous to that of Proposition 1.

Proposition 2. Suppose $\gamma(t) = (t, \psi(t))$, $t \in I$ is an arc of the unit circle centered at the origin and assume that the conditions (B1) and (B2) of Theorem 2 are satisfied. Then

$$\|V^{-1/p}(Sg)\|_{L^{p'}(\mathbb{R}^2)} \leq c\|g\|_{L^{q'}(I)}. \tag{4.7}$$

Proof We proceed as in the proof of Proposition 1, but choose here a different auxiliary function $\nu:\mathbb{R}^2 \to [0,\infty]$ when applying Theorem A.

Since $|\gamma(t)| = 1$ for every $t \in I$ then again with $u = t + s$, $v = \psi(t) + \psi(s)$ we obtain $|(u,v)| = |\gamma(t) + \gamma(s)| \leq 2$ for every $(u,v) \in D$. Now define $\nu(x) = w(|x|)$ where

$$w(\tau) = (2-\tau)^\lambda, \ \lambda \geq 0, \ \text{for } 0 < \tau < 2 \text{ and } +\infty \text{ for } \tau > 2.$$

Since

$$|\{x \in \mathbb{R}^2 : 1/\nu(x) > y\}| = \begin{cases} \pi[4-(2-y^{-1/\lambda})^2] & \text{if } 2^{-\lambda} < y \\ 4\pi & \text{if } 0 < y \leq 2^{-\lambda} \end{cases}$$

for $\lambda > 0$, it follows that

$$(1/\nu)^*(t) = \begin{cases} 2^{-\lambda}[1-(1-t/(4\pi))^{1/2}]^{-\lambda} & \text{if } t < 4\pi \\ 0 & \text{if } t \geq 4\pi. \end{cases}$$

Hence

$$\left(\int_0^{1/s} (1/v)^*(t)^{R'-1}dt\right)^{1/R'} \le \begin{cases} cs^{-(1/R'-\lambda/R)} & \text{for } s > 1/(4\pi) \\ c & \text{for } s \le 1/(4\pi) \end{cases}$$

if $0 \le \lambda < R-1$.

We now apply Theorem A with $\hat{F} = (Sg)^2/2$, $\mu = V^{-p'/p}$, $Q = p'/2$ and ν defined above. Condition (4.1) of Theorem A takes then the form

$$\sup_{s\ge c>0} \left\{ s^{-p'/L} \int_0^s (V^{-p'/p})^*(t)dt \right\} < \infty$$

with

$$L = 2(1/R' - \lambda/R)^{-1},\ 1 < R \le p'/2,\ 0 \le \lambda < R-1,$$

so that under these conditions by Theorem A

$$\|V^{-1/p}Sg\|_{p'}^2$$

$$\le c\left(\int_E |g(t)g(s)|^R |\psi'(t) - \psi'(s)|^{1-R} w(|\gamma(t) + \gamma(s)|)dtds\right)^{1/R}$$

$$= c\left(\iint_{J\times J} |(g\circ\alpha)(\xi)(g\circ\alpha)(\eta)|^R |\xi-\eta|^{1-R} w(|\Gamma(\xi) + \Gamma(\eta)|)d\xi d\eta\right)^{1/R}. \tag{4.8}$$

Here the last equality follows from the change of variables $\xi = \psi'(t)$, $\eta = \psi'(s)$. Moreover, $\Gamma = \gamma\circ\alpha$ where α is the inverse function of ψ' and $|(\psi''\circ\alpha)(\xi)| \ge c > 0$ for all $\xi \in J$.

By definition

$$w(|\Gamma(\xi) + \Gamma(\eta)|) = (2 - |\Gamma(\xi) + \Gamma(\eta)|)^\lambda$$

for every $(\xi,\eta) \in J \times J$. Consequently the right side of (4.8) will reduce to a fractional integral if we show that

$$2 - |\Gamma(\xi) + \Gamma(\eta)| \le C|\xi-\eta|^2,\ (\xi,\eta) \in J \times J. \tag{4.9}$$

But if $\cos\theta = <\Gamma(\xi), \Gamma(\eta)>$, $\theta = \theta(\xi,\eta) \in [-\pi,\pi]$ then

$$2-|\Gamma(\xi)+\Gamma(\eta)| = 2-\sqrt{2+2\cos\theta} = 2-2\cos\frac{\theta}{2} = 4\sin^2\frac{\theta}{4}. \tag{4.10}$$

Since $<\Gamma(\xi), \Gamma'(\xi)> = 0$, we have the expansion

$$\cos\theta = 1 + (\eta-\xi)^2/2 <\Gamma(\xi), \Gamma''(\eta^*)>$$

for some η^* with $|\eta^*-\xi| \leq |\eta-\xi|$. This implies $\sin^2(\theta/2) \leq c|\xi-\eta|^2$ for all $(\xi,\eta) \in J \times J$. Together with (4.10) the latter inequality implies (4.9). From (4.8) we then have from the fractional integral theorem

$$\|V^{-1/p}(Sg)\|_{p'}$$

$$\leq c\left(\iint_{J\times J} |(g\circ\alpha)(\xi)(g\circ\alpha)(\eta)|^R|\xi-\eta|^{1-R+2\lambda}d\xi d\eta\right)^{1/(2R)}.$$

$$\leq c\left(\int_J |(g\circ\alpha)(\xi)|^{PR}d\xi\right)^{1/(PR)} = C\|g\|_{L^{q'}(I)}$$

where

$$0 \leq 2/P' = R-1-2\lambda < 1,\ q' = PR. \tag{4.11}$$

From (4.11) we deduce that $P = 2(3-R+2\lambda)^{-1}$, $R = q(3+2\lambda)/(3q-2)$, $(2R)' = 2q(3+2\lambda)/[q(3+4\lambda)+2]$ and remark that the conditions

$$0 \leq 2/P' < 1 \quad\text{and}\quad 0 \leq \lambda < R-1$$

are equivalent to the following:

- either $R < 2$ and $0 \leq \lambda \leq (R-1)/2$ – or, $R = 2$ and $0 < \lambda \leq \frac{1}{2}$
- or $R > 2$ and $(R-2)/2 < \lambda \leq (R-1)/2$.

Note also that then from (4.11) and the definition of R

$$0 \leq \lambda \leq (\lambda q+1)/(3q-2) < 1/2 + \lambda,$$

or equivalently $0 \leq \lambda \leq 1/[2(q-1)]$ and $0 < 3q-4+4\lambda(q-1)$. Since this is clearly satisfied if (B1) of Theorem 2 holds, Proposition 2 is proved.

By a duality argument, we next deduce that estimate (3.5) is true for $\rho = 1$, $C_1(t) = (t, \pm\sqrt{1-t^2})$ with $t \in [-A, A]$, $0 < A < 1$. When $A \to 1$, it results from the B. Levi theorem that (3.5) is still true for each closed half circle $C_1(t)$, $t \in [-1,1]$. Theorem 2 is thus proved for $\rho = 1$.

The case $\rho > 0$ can be deduced from the preceding one by a homogeneity argument. Let $f_\rho(x) = f(x/\rho)$, $\rho > 0$, then $\hat{f}_\rho = \rho^2\hat{f}(\rho.)$, $\hat{f}_\rho \circ C_1 = \rho^2(\hat{f}\circ C_\rho)$ and

$$\|\hat{f}\circ C_\rho\|_r = \rho^{-1/r}\|\hat{f}\|_{L^r(C_\rho,ds)}.$$

The result follows.

4.4 PROOF OF COROLLARY 2.

Take $V(x) = |x|^{\alpha p}$, $\alpha \geq 0$ in Theorem 2. Then

$$\|f\|_{p,V(\rho.)} = \rho^\alpha\|f|x|^\alpha\|_p$$

and Corollary 2 is thus proved.

5. Remarks and Examples

5.1 NOTE ON THE CIRCLE CASE.

In the proof of Theorem 2, i.e. Proposition 2, the inequality

$$w(|\Gamma(\xi) + \Gamma(\eta)|) \leq c|\xi - \eta|^{2\lambda},\ (\xi,\eta) \in J \times J,\ \lambda > 0 \tag{5.1}$$

is essential, in order to majorize the right side of (4.8) by a fractional integral. The proof of this inequality utilized particular geometric properties of the circle.

We now show that for certain types of weights w_1, the condition for Γ to be an arc of a circle is not only sufficient, but also necessary for (5.1) to hold.

Let $\gamma(t) = (t, \psi(t))$, $\psi \in C^2(I)$, $\psi''(t) \geq 0$ for $t \in I$ and $M = \max|\gamma(t)|$, $t \in I$. Suppose w_1: $[0,\infty) \to [0,\infty]$ is such that its restriction to $[0,2M]$ is strictly decreasing, satisfies

$$w_1(\tau) \leq C(2M - \tau)^{2\lambda},\ \tau \in [0,2M]$$

and in addition $w_1(\tau) = +\infty$, $\tau \in (2M, \infty)$. Let $\alpha: J \to I$ be a bijection, $\Gamma(\xi) = (\gamma \circ \alpha)(\xi)$, for every $\xi \in J$ and suppose (5.1) holds with $w = w_1$. Then γ is an arc of a circle of radius M, centered at the origin.

This follows at once from the fact that (5.1) with $w = w_1$ implies $w_1(2|\Gamma(\xi)|) = 0$ for every $\xi \in J$. But then the monotonicity condition on w_1 shows that $|\Gamma(\xi)| = M$ for every $\xi \in J$.

5.2 SJÖLIN'S RESULT AND COROLLARY 1.

Note that the condition i) of Theorem 1a), namely, $4/3 < q < \infty$ and $1 < p \leq 6q/(3q+2)$ is equivalent to either

$$4/3 < p \leq 6q/(3q+2) \ \text{ or } \ p \leq 4/3 < q. \tag{5.2}$$

If we consider $\gamma \in C^\infty(I)$, then we can paraphrase Corollary 1a) by saying that the estimate (3.4) holds for $\delta > 1/(3q)$ and $1 \leq r \leq q$ in anyone of the following cases:

$$4/3 < p \leq 6q/(3q+2) \text{ and } 0 < 2(1/p' - 1/(3q)) \leq \alpha < 2/p' \tag{5.3}$$

$$p = 4/3 < q \ \text{ and } 0 < 1/2 - 2/(3q) \leq \alpha < 1/2 \tag{5.4}$$

$$q > 4/3,\ p' = 3q, \ \text{ and } \ 0 \leq \alpha < 2/(3q) \tag{5.5}$$

$$1 < p < (3q)' < 4/3 < q \ \text{ and } \ 0 \leq \alpha < 2/p'. \tag{5.6}$$

The case (5.5) with $\alpha = 0$ reduces to Sjölin's Lemma 3 and hence Theorem 2 in [10]. The value $\alpha = 0$ is admissible in (5.6). In this case one has $3q < p'$, so we obtain the complete range of values in (3.4) included in Sjölin's results. In (5.5) and (5.6) the admissible values of p are the same as in the unweighted case, i.e. $1 < p < 4/3$, but $\alpha > 0$ is also admissible, which brings a first generalization of previous results.

The cases (5.3) and (5.4) have no analogues in the unweighted case and thus appear as further generalizations. Specific values satisfying (5.3) are

$$q = 8/3,\ p = 8/5,\ 1/2 \leq \alpha < 3/4 \ \text{ and } \ \delta > 1/8.$$

Similar remarks hold for $\gamma \in C^2(I)$, $K(t) \geq 0$.

In the case of the circle, Theorem 2 reduces to Theorem 1 if $\lambda = 0$. As already noted following Corollary 2, if $\lambda = \alpha = 0$, $q > 4/3$ and $r = q = p'/3$, Corollary 2 reduces to Theorem 3 of Zygmund [11]. Unlike Zygmund's method of proof, however, the core of the proof of Theorem 2 is obtained without an explicit parametrization of the circle.

6. Further Generalizations

6.1 In this section we obtain estimates of the type (1.2), that is, inequalities involving two weights. This can be done by interpreting the following double integral by

$$\iint_{J\times J} G(\xi)G(\eta)|\xi-\eta|^{1-R}d\xi d\eta = 2\int_J G(\xi)(I_\gamma G)(\xi)d\xi \tag{6.1}$$

where $\gamma = 2 - R$ and I_γ is the Riemann–Liouville fractional integral

$$(I_\gamma G)(\xi) = \int_a^\xi G(\eta)|\xi-\eta|^{\gamma-1}d\eta,\ \xi \in J = [a,b].$$

If w is a weight function and $k \geq 1$ then by Hölder's inequality the right side of (6.1) is dominated by

$$\|Gw\|_k\|w^{-1}I_\gamma G\|_{k'}.$$

In order to obtain the required estimate we must prove

$$\|w^{-1}I_\gamma G\|_{k'} \leq c\|wG\|_k. \tag{6.2}$$

However this result is a special case of [1, Theorem 2.3], namely the following:

Theorem B. Suppose $1 < k \leq 2$, $0 < \gamma < 1$. If there exists a $\beta \in [0,1]$, such that

$$\sup_{a<y<b}\left(\int_y^b (\xi-y)^{(\gamma-1)\beta k'} w(\xi)^{-k'} d\xi\right)\left(\int_a^y (y-\xi)^{(\gamma-1)(1-\beta)k'} w(\xi)^{-k'} d\xi\right) < \infty \tag{6.3}$$

then (6.2) is satisfied.

We also require the following:

Proposition 3. If (6.3) is satisfied for some $\beta \in [0,1]$, then

either $k' < 2/(1-\gamma)$
or $k' = 2/(1-\gamma)$ and $w^{-1} \in L^\infty[a,b]$.

Proof. For $h > 0$ sufficiently small, (6.3) implies

$$\left(h^{(\gamma-1)\beta k'}\int_y^{y+h} w(\xi)^{-k'}\,d\xi\right)\left(h^{(\gamma-1)(1-\beta)k'}\int_{y-h}^{y} w(\xi)^{-k'}\,d\xi\right) \le c$$

for every y satisfying $a < y < b$. But this is equivalent to

$$\left(h^{-1}\int_y^{y+h} w(\xi)^{-k'}\,d\xi\right)\left(h^{-1}\int_{y-h}^{y} w(\xi)^{-k'}\,d\xi\right) \le c\,h^{(1-\gamma)k'-2}. \tag{6.4}$$

Now if y is a Lebesgue point of $w^{-k'}$, then both factors of (6.4) tend to $w(y)^{-k'}$ as $h \to 0$. The right side of (6.4) implies therefore three cases:

If $(1-\gamma)k' - 2 > 0$, then $w(y)^{-k'} = 0$ for every Lebesgue point y and so $w(\xi)^{-k'} = 0$ a.e. on J. This case is excluded.

If $(1-\gamma)k' = 2$, then $w(y)^{-k'} \le c$ for every Lebesgue point y and so $\|w^{-1}\|_\infty \le c$.

Finally, the only remaining case is $(1-\gamma)k' < 2$, for which also (4.6) holds for sufficiently small $h > 0$. This proves the assertion.

We now can give the following generalization of Theorem 1.

Theorem 3. Suppose $\gamma(t) = (t, \psi(t)) \in C^2(I)$, $\psi''(t) \ge 0$, $t \in I$, and i), ii) of Theorem 1 are satisfied. Let $1 < k \le 2$, and $s' = k(3q/2)'$ and w a weight function on $J = \psi'(I) = [a,b]$ satisfying

$$\sup_{a<y<b}\left(\int_y^b (\xi-y)^{\beta k'(1-R)} w(\xi)^{-k'}\,d\xi\right)\left(\int_a^y (y-\xi)^{(1-\beta)k'(1-R)} w(\xi)^{-k'}\,d\xi\right) < \infty \tag{6.5}$$

for some $\beta \in [0,1]$ and $R = s'/k = (3q/2)'$. Then

$$\|\hat{f}\,|K|^{\frac{k-1}{s'}}U\|_{L^s(\gamma,ds)} \le c\|f\|_{L^p_V}$$

where $U\circ\gamma = (w\circ\psi')^{-k/s'}$.

Proof. Clearly (6.5) implies that Theorem B is applicable with $1 - \gamma = R - 1$. Therefore (6.2) holds and the argument of the proof of Proposition 1 shows that

$$\|V^{-1/p}Sg\|_{L^{P'}(\mathbb{R}^2)} \le c\|wG\|^{1/R}_{L^k(J)} = c\|(g\circ\alpha)^R(\psi''\circ\alpha)^{-1}w\|^{1/R}_{L^k(J)}$$

$$= c\Big(\int_I |g(y)|^{Rk}\psi''(y)^{1-k}(w\circ\psi')^k(y)dy\Big)^{1/(Rk)}$$

$$= c\|g(\psi'')^{1/s'-1/R}(w\circ\psi')^{1/R}\|_{s'}.$$

The result now follows from a duality argument.

Observe also that Proposition 3 implies that either

$$s < q, \text{ or } s = q \text{ and } w^{-1} \in L^\infty(J).$$

Indeed since $q' = RP$ and $P' = 2/(R-1)$ (c.f. proof of Prop. 1) we obtain $s' = k\cdot R = q'(k/P)$ and $P' = 2/(1-\gamma)$.

Similar generalizations hold for Theorem 2 and its corollary in the case of circles. One has, for instance, the following result relative to the parametrization $C_\rho(\theta) = (\rho\cos\theta, \rho\sin\theta)$, $\theta \in [-\pi/4, 7\pi/4]$:

<u>Theorem 4.</u> Suppose assumptions (B1), (B2) of Theorem 2 are satisfied. If $1 < k \le 2$, $s' = kR$, $(R = R(\lambda, q))$ and w a weight function on $[-1, 1]$ satisfying

$$\sup_{-1<y<1}\Big(\int_{-1}^{y}(y-\xi)^{\beta k'(1-R)}w(\xi)^{-k'}d\xi\Big)\Big(\int_{y}^{1}(\xi-y)^{(1-\beta)k'(1-R)}w(\xi)^{-k'}d\xi\Big) < \infty \qquad (6.6)$$

for some $\beta \in [0,1]$, then

$$\Big(\int_0^{2\pi}|\hat{f}(\rho\cos\theta,\rho\sin\theta)|^sU(\theta)^sd\theta\Big)^{1/s} \le c\rho^{1/s-2/p'}\|f\|_{p,V(\rho.)}, \qquad (6.7)$$

where

$$U(\theta) = \begin{cases} w(-\tan\theta)^{-k/s'} & \text{for } \theta \in [-\pi/4, \pi/4] \cup [3\pi/4, 5\pi/4], \\ w(-\cot\theta)^{-k/s'} & \text{for } \theta \in [\pi/4, 3\pi/4] \cup [5\pi/4, 7\pi/4]. \end{cases}$$

If in particular, assumption (B2) of Theorem 2 is replaced by

$$\max[0, 2(1/p' - 1/L)] \leq \alpha < 2/p'$$

where $L = 2q(3+2\lambda)/[2(1+\lambda) - \lambda q]$ (c.f. Corollary 2), then the left side of (6.7) is dominated by

$$\rho^{1/s+\alpha-2/p'} \|f|x|^{\alpha}\|_p.$$

Proof. Proceeding as in the proof of Theorem 3 one obtains on using the parametrization $(t, \pm\sqrt{1-t^2})$, $t \in [-A, A]$, $0 < A < 1$,

$$\left(\int_{-A}^{A} |\hat{f}(t, \pm\sqrt{1-t^2})|^s \, [w(\mp t/\sqrt{1-t^2}]^{-k(s-1)} \, (1-t^2)^{-1/2} dt\right)^{1/s} \leq c\|f\|_{p,V}.$$

Setting $t = \cos\theta$, $A = 1/\sqrt{2}$, this yields an estimate of the form (6.7) where the range of integration over $[0,2\pi]$ is replaced by integration over

$$E_1 = [\pi/4, 3\pi/4] \cup [5\pi/4, 7\pi/4].$$

But this estimate over E_1 is also true if we replace f and V by $\tilde{f}$ and $\tilde{V}$, respectively, where $\tilde{f}(x_1, x_2) = f(x_2, x_1)$ and $\tilde{V}(x_1, x_2) = V(x_2, x_1)$. As in the proof of Theorem 1 (c.f. §4.1) $\|f\|_{p,V} = \|\tilde{f}\|_{p,\tilde{V}}$ so that (6.7) holds also when restricted to the set of integration $[-\pi/4, \pi/4] \cup [3\pi/4, 5\pi/4]$. Here we used $\hat{\tilde{f}}(\cos\theta, \sin\theta) = \hat{f}(\sin\theta, \cos\theta)$ and the change of variable $\theta = \pi/2 - \theta^1$. Therefore (6.7) holds for the whole interval of integration $[0, 2\pi]$.

6.2 FURTHER REMARKS

Weighted estimates of the form (1.2) can also be obtained by using one–weight norm inequalities for estimating the double integral of (6.1) (c.f. [2] or [9]). However, such methods require additional applications of Hölder's inequality which reduces their effectiveness.

Another type of weighted extension theorem can be obtained by considering weights $v(x) = w(|x|)$, $x \in \mathbb{R}^2$, which satisfy,

$$w(|x+y|) \leq w(|x|)w(|y|)$$

when applying Theorem A in the proof of Proposition 1. This would lead to the required fractional integral and hence results of the type (1.2). However, such weight functions cannot vanish at a point, without vanishing identically.

ACKNOWLEDGEMENTS

This research was supported by NSERC grant A–4837. The first author also acknowledges the hospitality of McMaster University, where this research was initiated in the fall of 1988.

REFERENCES

[1] Andersen, K.F. and Heinig, H.P. (1983) 'Weighted norm inequalities for certain integral operators', SIAM J. Math. Anal. 14, 834–844.

[2] Andersen, K.F. and Sawyer, E.T. (1988) 'Weighted norm inequalities for the Riemann–Liouville and Weyl fractional integral operators', Trans. Amer. Math. Soc. 308, 547–558.

[3] Benedetto, J.J. and Heinig, H.P. (to appear) 'Fourier transform inequalities with measure weights', Advances in Math.

[4] Benedetto, J.J. Heinig, H.P. and Johnson, R. (1987) 'Weighted Hardy spaces and the Laplace transform II', Math. Nachr. 132, 29–55.

[5] Heinig, H.P. (1984) 'Weighted norm inequalities for classes of operators', Indiana Univ. Math. J. 33, 573–582.

[6] Jurkat, W.B. and Sampson, G. (1984) 'On rearrangement and weight inequalities for the Fourier transform', Indiana Univ. Math. J. 33, 257–270.

[7] Muckenhoupt, B. (1983) 'Weighted norm inequalities for the Fourier transform', Trans. Amer. Math. Soc. 276, 729–742.

[8] Muckenhoupt, B. (1983) 'A note on two weight function conditions for a Fourier transform norm inequality', Proc. Amer. Math. Soc. 88, 97–100.

[9] Muckenhoupt, B. and Wheeden, R.L. (1974) 'Weighted norm inequalities for fractional integrals', Trans. Amer. Math. Soc. 192, 261–274.

[10] Sjölin, P. (1974) 'Fourier multipliers and estimates of the Fourier transform of measures carried by smooth curves in $\mathbb{R}^2$', Studia Math. 51, 169–182.

[11] Zygmund, A. (1974) 'On Fourier coefficients and transforms of functions of two variables', Studia Math. 50, 189–201.

ATOMIC DECOMPOSITIONS

W. K. HAYMAN
Department of Mathematics,
University of York,
Heslington, York, YO1 5DD

0. INTRODUCTION

We consider a class B of functions $f(\xi)$ on an infinite compact metric space X. We shall in the main confine ourselves to the case when X is the unit circle $\{\xi = e^{i\theta}, -\pi \leqslant \theta \leqslant \pi\}$. Let B_0 be a subclass of B and consider representations of elements f of B in the form

$$(0.1) \quad f = \sum_1^\infty \lambda_k f_k$$

where $f_k \in B_0$ and the λ_k are real or complex numbers. If every f in X has a representation (0.1) in a suitable sense, B_0 will be called a basis.

Such a representation is called by Bonsall [1986] an atomic decomposition of f. We shall confine ourselves in the main to the case when X is the unit circle and B_0 consists of a class of Poisson kernels

$$P(z, \xi) = \frac{1 - |z|^2}{|\xi - z|^2}$$

We ask how big the subset E of the unit disk $D: |z|<1$, must be so that the class $f_z(\xi) = P(z, \xi)$ for $z \in E$ forms a basis.

In the first part of the talk I shall describe the results of Bonsall [1986 and 1987] where B is the class L^1, (0.1) holds in the sense of convergence in L^1 and the λ_k are real or complex numbers. In the second half I shall discuss the work of Hayman and Lyons [1990] where B is the class of positive continuous functions, and (0.1) is taken in the sense of pointwise, or equivalently uniform, convergence on X. Although our arguments in the two cases are rather different, the interplay between $f(\xi)$ on X and the harmonic extension $f^\dagger(z)$ of f into D plays a prominent role in both.

J. S. Byrnes and J. F. Byrnes (eds.), Recent Advances in Fourier Analysis and Its Applications, 597–611.

1. THE CASE L^1

The results I shall talk about are mainly due to Bonsall [1986 and 1987] though some go back to Brown, Shields and Zeller [1960]. The key condition here is the following:-

Definition 1. Let E be a subset of the unit disk D: $|z|<1$. We say that E is nontangentially dense (N.T.D.) if and only if, for almost all ξ in $|\xi|=1$, there exists a sequence z_k in E, tending to ξ nontangentially i.e.

$$z_k \to \xi, \text{ and } |\xi - z_k|/(1-|z_k|) = O(1), \text{ as } k \to \infty.$$

It turns out that this condition is necessary and sufficient for the $P(z,\xi)$, with $z \in E$, to form a basis in the L^1 sense, i.e. for every L^1 function on X to be expressible in the L^1 norm in the form (0.1) with $f_k(\xi) = P(z_k,\xi)$, where $z_k \in E$.

To introduce Bonsall's results we need the language of functional analysis. Let ℓ^1 be the space of absolutely convergent series of complex numbers $\lambda = \{\lambda_k\}$, with norm

$$\|\lambda\|_1 = \sum_{k=1}^{\infty} |\lambda_k|.$$

Again let $L^1(\partial D)$ be the space of Lebesgue integrable functions $f(e^{i\theta})$, $e^{i\theta} \in X$, with norm

$$\|f\|_1 = \frac{1}{2\pi} \int_{-\pi}^{\pi} f(e^{it})|dt.$$

We consider (0.1) with $f_k(\xi) = P(z_k,\xi)$, i.e.

$$(1.1) \quad f(\xi) = \sum_1^{\infty} \lambda_k P(z_k,\xi).$$

This corresponds to the case when E reduces to the sequence z_k. Then (1.1) represents a linear operator $f=T(\lambda)$ from ℓ^1 to L^1. The kernel ker T of this operator consists of all sequences λ such that $T(\lambda) = 0$. The range R(T) consists of all $f \in L^1$, such that $f=T(\lambda)$ for some $\lambda \in L^1$.

We need one further concept. For every $g \in L^\infty$ we write

$$(1.2) \quad g^\dagger(z) = \frac{1}{2\pi} \int_{-\pi}^{\pi} g(e^{it}) P(z,e^{it})dt.$$

The class of functions $g^\dagger(z)$ consists precisely of all complex valued bounded harmonic functions. For if $g^\dagger(z)$ is such a function the radial limits $g^\dagger(e^{it}) = g(e^{it})$ exist almost everywhere and the representation

(1.2) follows by dominated convergence. (See e.g. Hayman 1964, p.178 et seq.)

For $f \in L^1$ and $g \in L^\infty$, we write

$$[f,g] = \frac{1}{2\pi} \int_{-\pi}^{\pi} f(e^{it})\, g(e^{it})\, dt$$

and define the <u>dual</u> T^* of T from $L^\infty(\partial D)$ to ℓ^∞ by

$$[\lambda, T^* g] = [T\lambda, g] = \sum_1^\infty \lambda_k\, [P(z_k, \xi)] = \sum_1^\infty \lambda_k\, g^\dagger(z_k).$$

Thus

$$T^*(g) = \{g^\dagger(z_k)\}.$$

We can now state some of Bonsall's Theorems.

<u>Theorem 1. Let</u> z_k <u>be a sequence in</u> D: $|z|<1$.

<u>Then the following are equivalent</u>:

(i) z_k <u>is</u> N.T.D.
(ii) $T(\ell^1) = L^1(\partial D)$.
(iii) $T(\ell^1) = L^1(\partial D)$ <u>and</u> $\|f\|_1 = \inf\{\Sigma|\lambda_k|\}$ <u>in the representation</u> (1.1).
(iv) $\exists$ <u>a positive</u> κ <u>such that for all</u> $g \in L^\infty$ <u>or equivalently all bounded harmonic functions</u> $g^\dagger$ <u>in</u> D

$$\|g\|_\infty \leq \kappa \sup |g^\dagger(z_k)|.$$

(v) $\|g\|_\infty = \sup_k |g^\dagger(z_k)|$.

(vi) <u>We have</u> (iv) <u>or</u> (v) <u>for all bounded analytic functions</u> g(z) <u>in</u> D.

There is no time to prove all of these results but I should like to give Bonsall's [1986] deceptively simple argument for (i) $\Rightarrow$ (iii). He deduces it from the following general result itself derived from Banach's closed range theorem.

<u>Theorem 2. Let</u> X <u>be a Banach space, let</u> M_1, M_2 <u>be positive constants and let</u> u <u>be a mapping of a set</u> Δ <u>into</u> X <u>such that</u>

a) $\|u(\delta)\| \leq M_1 \quad (\delta \in \Delta)$

<u>and</u>

b) $\sup\{|\psi(u(\delta))| : \delta \in \Delta\} \geq M_2 \|\psi\|, \quad \psi \in X^*$.

<u>Here</u> M_1, M_2 <u>are positive constants.</u>

<u>Then every</u> f <u>in</u> X <u>is of the form</u>

$$(1.3) \quad f = \sum_{\delta\in\Delta} \lambda(\delta)u(\delta)$$

<u>with</u> $\lambda\in\ell^1(\Delta)$ <u>and also</u>

$$(1.4) \quad M_2 \inf\|\lambda\|_1 \leqslant \|f\| \leqslant M_1 \inf \|\lambda\|_1$$

This is Bonsall's proof. Let T be the bounded linear mapping of $\ell^1(\Delta)$ into X defined by

$$(1.5) \quad T(\lambda) = \sum_{\delta\in\Delta} \lambda(\delta)u(\delta).$$

Given $\epsilon\in\Delta$ we define e_ϵ on $\ell_1(\Delta)$ by $e_\epsilon(\epsilon)=1$, $e_\epsilon(\delta)=0$ $\delta\neq\epsilon$. Then

$$T(\ell_\epsilon) = u_\epsilon$$

and so for all $\psi\in X^*$ we have, using b),

$$(1.6) \quad M_2\|\psi\| \leqslant \sup\{\psi(u_\delta),\ \delta\in\Delta\} = \sup\{\psi(T(e_\epsilon)),\ \epsilon\in\Delta\} \leqslant \|T^*(\psi)\|.$$

Thus T^* has closed range and zero kernel and therefore, by Banach's closed range theorem $T\{\ell^1(\Delta)\}=X$ so that every f is of the form (1.3).

We write N=ker T, $Y=\ell^1(\Delta)/N$ and define S by

$$S(y) = T(\lambda), \text{ when } \lambda\in y\in Y.$$

Then S(y) is a bounded linear bijection of Y onto X and therefore has a bounded inverse. It follows that S^* is a bounded linear bijection of X^* onto Y^*. By (1.6) we have for all $\psi\in X^*$

$$M_2 \|\psi\|\leqslant\|T^*(\psi)\| = \sup\{|\psi(T(\lambda))| : \|\lambda\|_1 \leqslant 1\}$$

$$= \sup\{\psi\ (Sy)\ :\ \|y\|\leqslant 1\} = \|S^*\ \psi\|.$$

Therefore $\|S^{-1}\| = \|S^{*-1}\|\leqslant M_2^{-1}$. Thus, given $\eta > 0$, we can take λ in (1.4) so that

$$\|\lambda\|_1 \leqslant (M_2^{-1} + \eta)\|f\|$$

which implies the left hand inequality in (1.4). The right hand inequality is an immediate consequence of a).

Bonsall makes a number of deductions from his seminal Theorem 2, of which I only wish to give the first.

(i) ⇒ (iii) in Theorem 1.

We proceed to specialize the various concepts in Theorem 2. We take X to be the space $L^1(\partial D)$ of all integrable functions $f(e^{i\theta})$ on $|\xi| = 1$, with the norm

$$\|f\|_1 = \frac{1}{2\pi}\int_{-\pi}^{\pi}|f(e^{i\theta})|d\theta.$$

For Δ we take an N.T.D. sequence z_k in D. We define

$$u(z_k) = P(z_k,\xi).$$

Then (1.3) takes the form

$$f = \sum_k \lambda_k P(z_k, \xi). \tag{1.7}$$

Also

$$\|u(z_k)\| = \|P(z_k,\xi)\|_1 = 1,$$

so that a) is satisfied with $M_1 = 1$.

It remains to verify b).

We identify the dual X^* of X with $L^\infty(\partial D)$, i.e. with bounded complex valued integrable functions on ∂D. The elements ψ of X^* then take the form

$$\psi(f) = \frac{1}{2\pi}\int_{-\pi}^{\pi} f(e^{i\theta})\, g(e^{i\theta})d\theta.$$

Thus

$$\psi\{u(z_k)\} = \frac{1}{2\pi}\int_{-\pi}^{\pi} P(z_k,e^{i\theta})\, g(e^{i\theta})d\theta = g^\dagger(z_k),$$

while

$$\|\psi\| = \sup_{\|f\|=1} |\psi(f)| = \sup |g(e^{i\theta})|$$

where sup is the essential supremum. Thus b) takes the form

$$M_2 \sup |g(e^{i\theta})| \leqslant \sup |g^\dagger(z_k)|. \tag{1.8}$$

This is a consequence of Fatou's Theorem. (see e.g. Hayman 1964, Theorem 6.12, p.178). For $g^\dagger(z)$ is a bounded harmonic function in D and so has angular limits almost everywhere on ∂D. Since $\|g\|$ is the essential sup of $|g(e^{i\theta})|$ we can, given $\epsilon > 0$, find a set E_ϵ of positive measure on which $|g(e^{i\theta})|>\|g\|-\epsilon$. Hence there exists θ in E_ϵ such that $g^\dagger$ has an angular limit at $\xi = e^{i\theta}$, and such that z_k has a subsequence z_k converging nontangentially to ξ. Hence

$$|g^{\dagger}(z_k)| \to |g(\xi)| > \|g\| - \varepsilon,$$

so that

$$\sup |g^{\dagger}(z_k)| \geqslant \|g\| - \varepsilon$$

This proves (1.8) with $M_2 = 1$.

Hence we can apply Theorem 2 with $M_1 = M_2 = 1$ and this yields (1.7) with $\inf(\Sigma|\lambda_k|) = \|f\|$.

This proves that (i)$\Rightarrow$(iii) and clearly (iii)$\Rightarrow$(ii). The remaining implications in Theorem 1 are proved in Bonsall [1987].

2. POSITIVE BASES

When Walter Rudin heard Frank Bonsall talk about the above results at a conference in Lancaster in 1984, he asked whether the equation (1.1) holds with positive coefficients λ_k, if f is positive and lower semicontinuous. We now interpret the equation in the sense that everywhere the right hand side converges to $f(\xi)$ if ξ is finite, or diverges to $+\infty$ if $f(\xi) = +\infty$. The right hand side is the limit of an increasing sequence so that f must be lower semicontinuous. Rudin also asked what the conditions on E must be if the answer to his first question is positive.

The following result shows that in any space X the possibility of expressing an arbitrary positive continuous function f in the form (0.1) is precisely equivalent to uniform approximation of f by corresponding finite sums. (Hayman and Lyons [1990])

Theorem 3. Let X be an infinite compact metric space. Let B_0 be a class of continuous strictly positive functions on X and assume that $\lambda_k > 0$ for all k. The following are equivalent.

(i) Every positive continuous function $f(\xi)$ on X has a representation of the form (0.1), where $f_k \in B_0$ for all k.

(ii) Every lower semicontinuous $f(\xi)$ such that $0 < f(\xi) \leqslant \infty$ has a representation on (0.1) in the sense that everywhere both sides are equal and finite or both sides are $+\infty$.

(iii) Every positive continuous function can be uniformly approximated

on X by finite sums of the form $\sum_{1}^{N} \lambda_k f_k(x)$, $f_k \in B_0$.

(ii)⇒(i) is evident.

(i)⇒(iii). Suppose that f is positive and continuous on X and has a representation (0.1). Then partial sums $s_N(x) = \sum_1^N \lambda_k f_k(x)$ form an increasing sequence of continuous functions converging everywhere on X to the continuous limit f(x). It now follows from Dini's dominated convergence Theorem that the convergence is uniform on X and this implies (iii).

(iii)⇒(ii). This is also fairly straightforward. Suppose that f(x) is lower semicontinuous and positive on X. Then there exists an increasing sequence of positive continuous functions $g_n(x)$ such that

$$g_n(x) \to f(x) \text{ as } n \to +\infty, \; x \in X.$$

Let V be the class of finite sums

$$v(x) = \sum_1^N \lambda_k f_k(x)$$

with $f_k(x) \in B_0$ and $\lambda_k > 0$. We inductively define $v_n(x) \in V$ such that

$$(2.1) \qquad g_n - \epsilon\, 2^{-n} < \sum_{j=1}^n v_j < g_n, \; x \in X$$

where ϵ is chosen so that $g_1 > \epsilon > 0$, $x \in X$. For n=1, we choose v_1 so that

$$|g_1 - \frac{\epsilon}{4} - v_1| < \frac{\epsilon}{4}.$$

This is possible by (iii) since $g_1 - \frac{\epsilon}{4} > 0$.

Next if (2.1) holds for a value n, we choose $v_{n+1} \in V$ so that

$$-\epsilon\, 2^{-n-2} < g_{n+1} - \sum_{m=1}^n v_m - \epsilon\, 2^{-n-2} - v_{n+1} < \epsilon\, 2^{-n-2}.$$

Since $g_{n+1} - \sum_1^n v_m \geqslant g_n - \sum_1^n v_m > 0$, we can do this, if necessary decreasing ϵ so that

$$g_{n+1} - \sum_1^n v_m - \epsilon\, 2^{-n-2} > 0.$$

This yields (2.1) with n+1 instead of n, so that (2.1) holds for all n. Thus

$$\sum_1^\infty v_m(x) = \lim g_n(x) = f(x) \text{ for all } x \in X.$$

By a simple rearrangement of the left hand side of this equation we obtain (ii).

2.1 The case of Poisson kernels

We now specialise again to the case when X is the unit circle $|\xi|=1$, and B_0 consists of a class Poisson kernels

$$B_0 = \{P(z,\xi) \mid z \in E\}$$

where E is a subset of D. If B_0 satisfies the three equivalent conditions of Theorem 3, we shall say that E is basic. We note first that the unit disk D is basic, i.e. that the class of all Poisson kernels is sufficient for a representation (1.1). This answers positively Rudin's first question. It is sufficient to verify the conditions (iii) of Theorem 3. Suppose then that $f(\xi)$ is positive and continuous on ∂D. Then

$$f^\dagger(z) = \frac{1}{2\pi}\int_{-\pi}^{\pi} f(e^{i\xi})\, P(z,\, e^{i\xi})\, d\theta$$

gives a continuous extension from ∂D to D. Thus, given $\epsilon > 0$, we can find r, such that $0 < r < 1$ and

$$|f(r\xi) - f(\xi)| < \epsilon.$$

Next, if $\xi = e^{i\phi}$

$$2\pi f(r\xi) = \int_{-\pi}^{\pi} f(e^{i\theta})P(re^{i\phi},e^{i\theta})d\theta = \int_{-\pi}^{\pi} f(e^{i\theta})P(re^{i\theta},e^{i\phi})d\theta$$

$$= \int_{-\pi}^{\pi} f(e^{i\theta})P(re^{i\theta},\xi)\, d\theta$$

The integral on the right can be approximated, uniformly in ϕ, by a Riemann sum. Thus there exist θ_j, $-\pi=\theta_0<\theta_j<\ldots<\theta_N=\pi$, such that

$$|f(r\xi) - \sum_1^N(\theta_j-\theta_{j-1})\, f(e^{i\theta_j})P(re^{i\theta_j},\xi)| < \epsilon$$

or

$$|f(\xi) - \sum_1^N \lambda_j\, P(z_j,\xi)| < 2\,\epsilon$$

where $z_j = re^{i\theta_j}$ and $\lambda_j = (\theta_j - \theta_{j-1}) f(e^{i\theta_j}) > 0$. This yields (i).

The above method does not give very good sufficient conditions for E to be basic. It only shows that we can choose for E the union of a sequence of circles $|z|=r_\nu$, where $r_\nu \to 1$. However, since an arbitrary continuous function can be approximated by a sum of Poisson kernels, we may from now on confine ourselves to the question of whether the class of kernels $B_0 = \{P(z,\xi), z \in E\}$ is big enough so that postive linear combinations in B_0 can be used to approximate an arbitrary Poisson kernel.

2.2 Necessary and sufficient conditions for a set to be basic.

Theorem 4 If E is a subset of D, the following are equivalent

(i) E is basic.

(ii) Let H be the Hardy class of functions $h(z) = h_1(z) - h_2(z)$ where $h_1(z)$ and $h_2(z)$ are positive harmonic functions in D. Then wherever $h \in H$, we have $\sup_{z \in E} h(z) = \sup_{z \in D} h(z)$.

(iii) There is a sequence $z_j \in E$, $j=1,2,\ldots$ such that whenever $z_0 \in D$ there exist $c_j = c_j(z_0)$, such that $c_j \geq 0$ and for all $h \in H$

$$h(z_0) = \sum_1^\infty c_j\, h(z_j).$$

(iv) Suppose that $0 < \delta < 1$, and let $E(\delta)$ be a hyperbolic neighbourhood of E, i.e. $E(\delta)$ consisting of all points $z \in D$ such that

$$\left|\frac{z - z'}{1 - \bar{z}z'}\right| < \delta, \text{ for some point } z' \text{ in } E.$$

Then for some δ, or equivalently for all δ, $E(\delta)$ is not minimally thin at any point in ∂D.

(v) We write

$$z_{m,n} = (1-2^{-n}) \exp\{2\pi i m 2^{-n}\}.$$

Let Σ denote a sum over all pairs m,n, such that $1 \leq n < \infty$, $0 \leq m < 2^n$ and the box

(2.2) $$|z_{m,n}| \leq |z| \leq |z_{m,n+1}|, \quad \arg z_{m,n} \leq \arg z \leq \arg z_{m+1,n}$$

meets E. Then

$$(2.3) \qquad s(\xi) = \Sigma \left|\frac{1-|z_{m,n}|}{\xi - z_{m,n}}\right|^2 = \infty$$

<u>for every</u> ξ <u>in</u> ∂D.

For the proof of the above results we must refer to Hayman and Lyons [1990]. The condition (v) looks clumsy but is the easiest to work with. The equivalence of (iv) and (v) follows from a result in potential theory of Essen, Jackson and Rippon [1985]. The equivalence of (i) and (ii) is due to Bonsall and Walsh [1990], though a different proof is given in Hayman and Lyons [1990]. When we compare (ii) of Theorem 4 with (v) of Theorem 1, we see at once that the condition for E to be basic is more stringent than that for E to be N.T.D. This is not quite obvious from Theorem 4 (v). It is clear that if $z_k \to \xi$ nontangentially then

$$s(\xi) = \Sigma \left[\frac{1 - |z_k|}{|\xi - z_k|}\right]^2 = \infty$$

but the converse is false.

Walsh observed (in a letter to me) that a set E is N.T.D. if and only if the series (2.3) is infinite for almost all ξ in ∂D. This shows directly that "basic" implies N.T.D. The following argument yields Walsh's Theorem. It is almost obvious that if E has a subsequence converging nontangentially to ξ in ∂D, then the subsequence of the corresponding $z_{m,n}$, for which z_k lies in the box (2.2), also converges nontangentially to ξ, so that $s(\xi)$ is infinite in (2.3). Thus it is enough to show that if $s(\xi) = \infty$ for almost all ξ, then E is N.T.D. To see this let F be the set of all points $\xi = e^{i\theta}$, for which $s(\xi)<\infty$. By hypothesis F has measure zero. Let G_n be a sequence of open sets having measure less than 2^{-n} and such that $F \subset G_n$. Then each G_n consists of a sequence of disjoint intervals $(\alpha_{m,n}, \beta_{m,n})$, where

$$\sum_m (\beta_{m,n} - \alpha_{m,n}) < 2^{-n}.$$

We form the sequence

$$w_{m,n} = r_{m,n}\, e^{i\theta_{m,n}} \quad , \text{ where}$$

$$r_{m,n} = 1-(\beta_{m,n} - \alpha_{m,n}), \quad \theta_{m,n} = \tfrac{1}{2}(\alpha_{m,n} + \beta_{m,n}).$$

Then $w_{m,n}$ satisfies the Blaschke condition

$$\Sigma\Sigma\ (1 - |w_{m,n}|) = \sum_n \sum_m (\beta_{m,n} - \alpha_{m,n}) < \infty.$$

Thus the Blaschke product

$$\Pi(z) = \prod_{m,n=1}^{\infty} \left(\frac{w_{m,n} - z}{1-\bar{w}_{m,n} z}\right) \frac{\bar{w}_{m,n}}{|w_{m,n}|}$$

converges in D.

We define $E' = E \cup \{w_{m,n}\}$. Then if ξ_j is a point of ∂D, either $\xi \in F$, in which case $\xi = e^{i\theta}$, where $\alpha_{m,n} < \theta < \beta_{m,n}$ for $1 \leq n < \infty$, $m=m(n)$. In this case the corresponding $w_{m,n}$ tend to ξ nontangentially. Thus if we form the series $s(\xi)$ corresponding to the set E' then $s(\xi) = \infty$ for such points. By hypothesis $s(\xi) = \infty$ at all other points of ∂D so that E' satisfies the conditions for being basic and so E' is N.T.D.

Thus for almost all points ξ in ∂D there exists a subsequence z_k of either of E or of the sequence $w_{m,n}$ which tends nontangentially to ξ. Suppose e.g. the latter holds. Then $\Pi(z_k) = 0$ so that

$$\underline{\lim}\ |\Pi(z)| = 0$$

as $z \to \xi$ in some angle. By a Theorem of F. and M. Riesz (see e.g. Hayman [1964,p.178]) the set of such ξ has measure zero. Thus at almost all points ξ of ∂D the sequence z_k must belong to E, so that E is N.T.D. I know of no direct argument which shows that (2.3) holds almost everywhere or even everywhere, implies that E is N.T.D.

2.3 Outline of proof of Theorem 4

Rather than giving a full proof of Theorem 4 I would like to indicate just the methods on which our arguments depend, since these involve harmonic rather than functional analysis.

Suppose that E is basic. Then, for every fixed z_0 in D, there are z_j in E and positive constants $c_j = c_j(z_0)$, such that

$$(2.4) \qquad P(z_0,\xi) = \sum_{1}^{\infty} c_j\, P(z_j,\xi). \qquad \xi = e^{i\theta}$$

Also integrating both sides with respect to θ, we deduce that

$$(2.5) \qquad \sum_{1}^{\infty} c_j = 1.$$

Suppose now that h(z) is positive harmonic in D. Then it follows from the Riesz-Herglotz formula (see e.g. Hayman [1964,p.179]) that there exists a positive increasing $\mu(\theta)$ such that

$$h(z) = \int_{-\pi}^{\pi} P(z,e^{i\theta})\,d\mu(\theta).$$

Setting $z=z_0$, using (2.4) and changing the order of integration and summation, which is justified since $Pd\mu>0$, we obtain

$$h(z_0)= \sum_{1}^{\infty} \int_{-\pi}^{\pi} c_j P(z_j, e^{i\theta})d\mu(\theta) = \Sigma c_j h(z_j).$$

Thus (i)⇒(iii) when $h>0$ in D. If $h\in H$, then $h=h_1-h_2$ where h_1, h_2 are positive and we obtain the result by taking differences.

Next if (iii) holds and $h(z)\leqslant M$ on E, we deduce, using (2.5),

$$h(z_0) \leqslant \Sigma c_j M = M$$

so (iii) ⇒ (ii).

(ii)⇒(v). Here we use a construction of Beurling [1965]. We arrange the $z_{m,n}$ that feature in the sum Σ as a single sequence z_k and assume that there is $\xi = e^{i\Phi}$, such that

$$\Sigma \left| \frac{1-|z_k|}{\xi - z_k} \right|^2 < \infty.$$

Assuming that $z_k \neq 0$, we deduce

(2.6) $\qquad \Sigma\, g(0,z_k)\, P(z_k,\xi) < \infty,$

where g is Green's function in D. In this form the inequality is conformally invariant. It is convenient to map D onto the right half plane by means of the transformation

$$Z = \frac{\xi+z}{\xi-z} = X + i\,Y, \text{ where } X > 0,$$

so that z_k corresponds to $Z_k = X_k + iY_k$, ξ becomes $Z = \infty$, and $P(z_k, \xi)$ becomes X_k. Thus (2.6) yields

$$\Sigma\, X_k\, G(1,Z_k) < \infty,$$

where G is Green's function in the right half plane, or equivalently

$$\Sigma \frac{X_k^{\,2}}{1+X_k^{\,2}+Y_k^2} < \infty.$$

We now introduce the positive harmonic function

$$H(Z) = \Sigma \frac{X \, X_k^2}{(Y-Y_k)^2+X^2}, \text{ where } Z = X + iY,$$

and deduce, as $Z = X\to\infty$ through real positive values,

$$\frac{H(X)}{X} = \Sigma \frac{X_k^2}{X^2+Y_k^2} \to 0.$$

On the other hand

$$\frac{H(Z_k)}{X_k} = 1, \; k = 1,2,...$$

Transferring back to the unit disk we obtain a positive harmonic function h(z) in D, such that

$$\frac{h(z)}{P(z,\xi)} \to 0 \text{ as } z \to \xi \text{ radially}$$

while

$$\frac{h(z)}{P(z,\xi)} \geqslant 1, \quad z = z_k.$$

Since each z in E is hyperbolically close to one of the points z_k, and since h(z) and $P(z,\xi)$ are positive harmonic functions in D, we deduce from Harnack's inequality that there is a positive constant c, such that

$$\frac{h(z)}{P(z,\xi)} > c \text{ in } E,$$

while $h(z_0)/P(z_0,\xi) < c$, if $z_0 = r\xi$, and r is close to one.

Writing

$$u(z) = c\,P(z,\xi) - h(z)$$

we see that $u \in H$, and $u(z) < 0$ in E, while $u(z_0) > 0$. This contradicts (ii).

To finish the proof we need to show that (v)⇒(iv) and that (iv)⇒(i). The first implication follows from the result of Essen, Jackson and Rippon [1985]. The fact that $E(\xi)$ is not minimally thin at ξ is then used to obtain the equation

$$P(z_0,\xi) = \int_\gamma P(z,\xi)\, d\omega(z).$$

Here γ is a union of circles

$$(2.7) \qquad \left|\frac{z - z_j}{1-\bar{z}_j z}\right| = \delta$$

where z_j is a sequence of points of E, chosen so that the circles (2.6) are disjoint and ω is harmonic measure at the point z_0 of D outside the circles (2.7). If λ_j is the harmonic measure of the circle (2.7) this leads to

$$|P(z_0,\xi) - \Sigma\lambda_j \, P(z_j,\xi)| < \epsilon$$

provided that δ is small enough depending on ϵ and this proves that E is basic. For details we refer the reader to Hayman-Lyons [1990].

2.4 Some open questions

We have confined ourselves to the case when X is the unit circle and the functions $f_j(x)$ take the form of Poisson kernels. Using Beurling's invariant formulation (2.6) of the condition that (2.3) fails for some ξ it should not be difficult to extend the theory to the boundaries of smooth simply connected domains D in the plane and possibly in space and perhaps to more general domains. A more difficult question concerns the replacement of Poisson kernels by more general functions on X. After all atomic decomposition has on the face of it no connection with any space except X itself. What are corresponding Theorems when X is a more general set, such as a ball or a cube or a hypercube? Atomic decomposition of positive continuous functions is a topologically invariant property and so it should be possible to obtain purely topological conditions for it or, at least if X is a metric space, conditions depending only on the metric. When we refer to L^1 decomposition a suitable Lebesgue measure has of course to be introduced as well.

Is it true under more general conditions that the positive decomposition (0.1) implies an L^1 decomposition? Theorems 1,4 show that this is the case when X is a circle and the f_k (ξ) are Poisson-kernels.

What can we say about other decompositions? It follows from (0.1) with positive f, f_k and λ_k that every real continuous function f can be written in the form (0.1) with λ_k real where the series (0.1) is absolutely and uniformly convergent. For we may write $f^+ = \max(f,0)$, $f^- = \max(-f,0)$ and

$$f = 1 + f^+ - (1+f^-).$$

We then apply the above theory to $1+f^+$ and $1+f^-$ and subtract. A similar conclusion applies to uniform approximation by finite sums, with or without some restriction on $\Sigma|\lambda_k|$. Is there a converse result? In general positive decomposition appears to be a stronger property. Consider continuous functions f(x) on [0,1]. The positive powers x^k form a basis in the sense that f can be uniformly approximated by polynomials

$$f_N(x) = \sum_{k=1}^{N} \lambda_k x^k.$$

However if $\lambda_k > 0$, the right hand side is increasing and only increasing functions can be approximated by polynomials with positive coefficients. In particular the x^k do not form a positive basis on [0,1]. This example and Theorem 4 show that an analogue of the Stone-Weierstrass Theorem for positive decomposition is likely to be difficult.

References

1. Beurling, A. (1965) 'A minimum principle for positive harmonic functions', Ann. Acad. Sci. Fenn. Ser. A I No. 372, 7pp.

2. Bonsall, F.F. (1986) 'Decompositions of functions as sums of elementary functions', Quart. J. Math. Oxford (2),37, 129-136.

3. Bonsall, F.F. (1987) ' Domination of the supremum of a bounded harmonic function by its supremum over a countable subset', Proc. Edinburgh Math. Soc. 30, 471-477.

4. Bonsall, F.F. and Walsh, D. (1990) 'Vanishing ℓ^1-sums of the Poisson kernel, and sums with positive coefficients', Proc. Edinburgh Math. Soc. (to appear)

5. Brown, L. Shields, A. and Zeller, K. (1960) 'On absolutely convergent exponential sums', Trans. Amer. Math. Soc. 96, 162-183.

6. Essen, M, Jackson H.L. and Rippon, P.J. (1985) 'On minimally thin and rarefied sets in R^p, $p\geq 2$', Hiroshima. Math. J. 15(1985), 393-410.

7. Hayman, W.K. (1964) 'Meromorphic functions', Oxford University Press.

8. Hayman, W.K. and Lyons, T.J. (1990) 'Bases for positive continuous functions', J. London Math. Soc., to appear.

The Wiener-Plancherel Theorem on $\mathbb{R}^n$

Ward R. Evans
The MITRE Corporation
7525 Colshire Rd.
McLean, VA 22102
USA

ABSTRACT

This paper will discuss the recent proof of the n-dimensional Wiener-Plancherel formula [2] and some of the problems encountered in extending Wiener's Generalized Harmonic Analysis (GHA) to $\mathbb{R}^n$. A key feature of this theory is the definition of an appropriate s-function, which in turn depends on the convergence criteria selected. This paper considers convergence over rectangles.

1 Introduction

Norbert Wiener devised GHA to solve problems that could not be solved using classical techniques of harmonic analysis. In particular Wiener wanted a theory "adequate for the treatment of a ray of white light which is supposed to endure for an indefinite time" [5]. In modern terms, one would say that Wiener wanted a harmonic analysis of bounded functions. In [5], Wiener successfully solved this problem for functions bounded on $\mathbb{R}$.

A central theorem in GHA is the Wiener-Plancherel formula, an analogue of the Plancherel formula. It applies to functions having bounded quadratic means. Since the Wiener-Plancherel formula is a statement about the equality of means, one should not be surprised to find that the extension of the formula to $\mathbb{R}^n$ depends on convergence criteria determined by geometrical constraints. The generalization discussed in this paper depends on the local behavior of the function on n-dimensional rectangles.

While the work described in this paper is contained in more detail in [2], it seems useful to have a shorter exposition of the theory. I would like to thank my coauthors of [2], John Benedetto and George Benke, for their help, encouragement, and friendship.

2 The Wiener-Plancherel Formula on $\mathbb{R}$

In this section we shall examine the classical Wiener-Plancherel formula on $\mathbb{R}$, briefly discuss its place in Generalized Harmonic Analysis (GHA), and define terms and spaces needed for the extension to $\mathbb{R}^n$.

J. S. Byrnes and J. F. Byrnes (eds.), Recent Advances in Fourier Analysis and Its Applications, 613–622.

Definition 2.1

a. The space BQM (R^n) of functions having *bounded quadratic means* is the set of all functions $\varphi \in L^1_{\text{loc}}(\mathrm{R}^n) = \left\{ f : f \text{ measurable on } \mathrm{R}^n \text{ and } \int_K |f| < \infty \ \forall \text{ compact } K \subset \mathrm{R}^n \right\}$, where

$$\mathrm{B}(\varphi) \equiv \sup_{\mathrm{T}>0} \frac{1}{|\mathrm{R_T}|} \int_{\mathrm{R_T}} |\varphi(\mathrm{t})|^2 \, dt < \infty,$$

$\mathrm{T} = (\mathrm{T}_1, \ldots, \mathrm{T}_n) \in \mathrm{R}^n, \mathrm{T} > 0$ means $\mathrm{T}_i > 0$ for $i = 1, \ldots, n$, , and $|\mathrm{R_T}|$ denotes the Lebesgue measure of $\mathrm{R_T} = \{t \in \mathrm{R}^n : |t_j| \leq \mathrm{T}_j \ j = 1, \ldots, n\}$.

b. The *Wiener space*, $\mathrm{W}(\mathrm{R}^n)$, is the set of all functions $\varphi \in L^1_{\text{loc}}(\mathrm{R}^n)$ for which

$$W(\varphi) \equiv \int_{\mathrm{R}^n} \frac{|\varphi(t)|^2}{(1+t_1^2)\cdots(1+t_n^2)} dt < \infty.$$

Wiener [3] proved the first inclusion of the following theorem for $n = 1$; the theorem as stated is proved in [2].

Theorem 2.2 $\mathrm{BQM}(\mathrm{R}^n) \subseteq \mathrm{W}(\mathrm{R}^n) \subseteq \mathcal{S}'(\mathrm{R}^n)$, where $\mathcal{S}'(\mathrm{R}^n)$ is the class of tempered distributions on R^n.

A concept crucial to GHA is the concept of the s function. In this section we shall give the definition of the s function for $n = 1$; we shall extend the definition to n-dimensions in the next section.

Definition 2.3 Let $\varphi \in L^1_{\text{loc}}(\mathrm{R})$ and define

$$s(\gamma) = \int_{-\infty}^{\infty} e(t,\gamma)\varphi(t)\,dt, \quad \gamma \in \widehat{\mathrm{R}}$$

where

$$e(t,\gamma) = \begin{cases} \frac{\exp(-2\pi it\gamma)-1}{-2\pi it}, & |t| \leq 1 \\ \frac{\exp(-2\pi it\gamma)}{-2\pi it}, & |t| > 1. \end{cases}$$

Theorem 2.4 If $\varphi \in \mathrm{W}(\mathrm{R})$ then

a. $s \in L^2_{\text{loc}}(\mathrm{R})$

b. $\Delta_\lambda s \in L^2(\mathrm{R})$, where $\Delta_\lambda s(\gamma) = s(\gamma+\lambda) - s(\gamma-\lambda)$.

Proof:

a. Write $s(\gamma) = s_1(\gamma) + s_2(\gamma)$ where

$$s_1(\gamma) = \int_{-1}^{1} \varphi(t) \frac{e^{-2\pi it\gamma} - 1}{-2\pi it} dt,$$

$$s_2(\gamma) = \int_{|t|>1} \varphi(t) \frac{e^{-2\pi it\gamma}}{-2\pi it} dt.$$

Now

$$|s_1(\gamma)| \leq \int_{-1}^{1} \left|\gamma \frac{\sin(\pi t\gamma)}{\pi t\gamma}\right| |\varphi(t)|\, dt \leq |\gamma| \int_{-1}^{1} |\varphi(t)|\, dt < \infty.$$

For s_2 we calculate

$$\int_{|t|>1} \frac{|\varphi(t)|^2}{|t|^2} dt = \int_{|t|>1} (1+t^2) \frac{|\varphi(t)|^2}{t^2(1+t^2)} dt$$

$$= \int_{|t|>1} \left(1+\frac{1}{t^2}\right) \frac{|\varphi(t)|^2}{(1+t^2)} dt \leq 2 \int_{|t|>1} \frac{|\varphi(t)|^2}{(1+t^2)} dt \leq 2W(\varphi) < \infty$$

by theorem 2.2 . So $s_1 \in L^\infty\left(\widehat{\mathbb{R}}\right)$ and $s_2 \in L^2\left(\widehat{\mathbb{R}}\right)$ thus $s = s_1 + s_2 \in L^2_{\text{loc}}\left(\widehat{\mathbb{R}}\right)$.

b.

$$s(\gamma+\lambda) - s(\gamma-\lambda) = \int_{-\infty}^{\infty} \varphi(t)\left(e(t,\gamma+\lambda) - e(t,\gamma-\lambda)\right) dt$$

$$= \lambda \int_{-\infty}^{\infty} \varphi(t) \frac{\sin(2\pi\lambda t)}{\lambda t} e^{-2\pi i t\gamma} dt$$

and $\varphi(t)\frac{\sin(2\pi\lambda t)}{\pi\lambda t} \in L^2(\mathbb{R})$. qed

If $\varphi \in L^\infty(\mathbb{R})$ then $\varphi = \widehat{T}$ where $T \in A'\left(\widehat{\mathbb{R}}\right)$ is a pseudo-measure on $\widehat{\mathbb{R}}$. Thus, if $g \in C_c^\infty\left(\widehat{\mathbb{R}}\right)$ and s is defined as above, we have

$$\langle s', g\rangle = -\langle s, g'\rangle = \int_{\widehat{\mathbb{R}}} \int_{-1}^{1} \varphi(t) \frac{e^{-2\pi i t\gamma} - 1}{-2\pi i t} dt\, g'(\gamma)\, d\gamma +$$

$$\int_{\widehat{\mathbb{R}}} \int_{|t|>1} \varphi(t) \frac{e^{-2\pi i t\gamma}}{-2\pi i t} dt\, g'(\gamma)\, d\gamma = \int_{-1}^{1} \varphi(t)\widehat{g}(t)\, dt + \int_{|t|>1} \varphi(t)\widehat{g}(t)\, dt$$

$$= \langle T, g\rangle;$$

so $s' = T$ distributionally.

In [5], Wiener was specifically interested in the analysis of measurable functions, f, such that

$$\varphi(x) = \lim_{T\to\infty} \frac{1}{2T} \int_{-T}^{T} f(x+t)\overline{f(t)}\, dt$$

exists for every x. In this context, φ is positive definite and therefore, by Bochner's theorem, it is the Fourier-Stieltjes transform of a positive measure, μ_φ. Thus, as we saw above, $s' = \mu_\varphi$ distributionally. Wiener referred to s as the *integrated periodogram* of f.

Example 2.5

Let $f(x) = \sum_{k=1}^{N} a_k e^{i\lambda_k x}$ then

$$\varphi(x) = \lim_{T\to\infty} \frac{1}{2T} \int_{-T}^{T} \left(\sum_{k=1}^{N} a_k e^{i\lambda_k(x+t)} \right) \left(\sum_{k=1}^{N} \bar{a}_k e^{-i\lambda_k t} \right) dt$$

$$= \sum_{k=1}^{N} |a_k|^2 e^{i\lambda_k x},$$

$$\mu_\varphi = \sum_{k=1}^{N} |a_k|^2 \delta_{\lambda_k}.$$

Since $s' = \mu_\varphi$ distributionally, we have

$$s(\gamma) = \sum_{k=1}^{N} |a_k|^2 H(\gamma - \lambda_k),$$

where H is the Heaviside function.

The classical Plancherel formula, $\int |\varphi(t)|^2 dt = \int |\hat{\varphi}(\gamma)|^2 d\gamma$ not only defines $\hat{\varphi}$ for an arbitrary function φ having finite energy, but provides the means for obtaining quantitative estimates for $\hat{\varphi}$. Since the functions φ have Fourier transforms as distributions, the goal of the Wiener-Plancherel formula is to provide a means of obtaining quantitative estimates on $\hat{\varphi}$. In particular, if φ is an autocorrelation function, then the Wiener-Plancherel formula gives quantitative estimates of the power spectrum of a given signal, cf. [2], [1].

Theorem 2.6 (Wiener-Plancherel formula for $\mathbb{R}$) Given $\varphi \in L^\infty(\mathbb{R})$ then

$$\lim_{T\to\infty} \frac{1}{2T} \int_{-T}^{T} |\varphi(t)|^2 dt = \lim_{\lambda\to 0} \frac{2}{\lambda} \int |\Delta_\lambda s(\gamma)|^2 d\gamma.$$

Proof:

Since

$$\Delta_\lambda s(\gamma) = \frac{1}{2} \int \varphi(t) e^{-2\pi i t\gamma} \frac{\sin(2\pi\lambda t)}{\pi t} dt,$$

and since $\Delta_\lambda s \in L^2(\widehat{\mathbb{R}})$, we have by the Plancherel theorem

$$\frac{1}{2\lambda} \int \left| \varphi(t) \frac{\sin(2\pi\lambda t)}{\pi t} \right|^2 dt = \frac{2}{\lambda} \int |\Delta_\lambda s(\gamma)|^2 d\gamma$$

provided $\lambda \neq 0$. The theorem will be proved once we prove

$$\lim_{T\to\infty} \frac{1}{2T} \int_{-T}^{T} |\varphi(t)| dt = \lim_{\lambda\to 0} \frac{1}{2\lambda} \int_{-\infty}^{\infty} \left| \varphi(t) \frac{\sin(2\pi\lambda t)}{\pi t} \right|^2 dt.$$

Setting $\theta(t) = \frac{|\varphi(t)|^2 + |\varphi(-t)|^2}{2}$, we see that proving the above equality is equivalent to proving

$$\lim_{T\to\infty} \frac{1}{T} \int_0^T \theta(t)\, dt = \lim_{\lambda\to 0} \frac{1}{\lambda} \int_0^\infty \theta(t) \frac{\sin^2(2\pi\lambda t)}{\pi^2 t^2} dt. \tag{†}$$

Let $T = 1/\lambda = e^y$ and let $x = e^t$ so (†) becomes

$$\lim_{y\to\infty} \int_{-\infty}^{y} e^{-(y-x)} \theta(x)\, dx = \lim_{y\to\infty} \int_{-\infty}^{\infty} \theta(x)\, e^{y-x} \sin^2\left(e^{-(y-x)}\right) dx$$

It can be shown that if $f(x) = e^x \sin^2(e^{-x})$, then $\hat{f}(\gamma) \neq 0$, $\forall\gamma$ and the equality of limits follows from Wiener's Tauberian theorem.

3 The Wiener-Plancherel Formula in $\mathbb{R}^n$

The extension of the Wiener-Plancherel formula first depends on selecting an appropriate convergence criterion and an appropriate s-function. The selection of the convergence criterion and the definition of the s-function are intimately related.

Definition 3.1 Let $\varphi : \mathbb{R}^n \to \mathbb{C}$ then $\lim\limits_{x\to\infty} \varphi(x) = \varphi_0 \in \mathbb{C}$ means $\forall\epsilon > 0$ and $\forall c = (c_1, \ldots, c_n) \in \mathbb{R}^n$, where $|c| = 1$ and each $c_j > 0, \exists N(\epsilon, c) = N$ such that $x > N$ implies $|\varphi(xc_1, \ldots, xc_n) - \varphi_0| < \epsilon$.

$\lim_{x\to 0} \varphi(x) = \varphi_0 \in \mathbb{C}$ means $\forall\epsilon > 0$ and $\forall c = (c_1, \ldots, c_n) \in \mathbb{R}^n$, where $|c| = 1$ and each $c_j > 0$ $\exists\delta(\epsilon, c) = \delta$ such that $0 < x < \delta$ implies $|\varphi(xc_1, \ldots, xc_n) - \varphi_0| < \epsilon$.

Geometrically this notion of limit indicates convergence to $\varphi_0 \in \mathbb{C}$ along every straight line in $\mathbb{R}^{+n}$ through the origin except the coordinate axes

Definition 3.2

a. The Wiener s-function corresponding to $\varphi \in L^1_{\text{loc}}(\mathbb{R}^n)$ is

$$s(\gamma) = \int E(t,\gamma)\, \varphi(t)\, dt, \quad \gamma \in \widehat{\mathbb{R}}^n, \text{ where}$$

$$E(t,\gamma) = \prod_{j=1}^{n} e(t_j, \gamma_j), \; t \in \mathbb{R}^n, \gamma \in \widehat{\mathbb{R}}^n, \text{ and}$$

$$\forall t \in \mathbb{R}, \forall\gamma \in \widehat{\mathbb{R}},$$

$$e(t,\gamma) = \begin{cases} \frac{\exp(-2\pi i t\gamma) - 1}{-2\pi i t}, & |t| \le 1 \\ \frac{\exp(-2\pi i t\gamma)}{-2\pi i t}, & |t| > 1. \end{cases}$$

b.

$$\Delta_\lambda s(\gamma) = \frac{1}{2^n} \sum_{\omega\in\Omega} (-1)^{|\omega|} s(\gamma + \lambda\omega), \text{ where}$$

$$\Omega = \{-1, 1\}^n \subset \widehat{\mathbb{R}}^n, \text{ and}$$

$$|\omega| = \text{card}\,\{j : \omega_j = -1\}$$

Theorem 3.3 Given $\varphi \in L^1_{\text{loc}}(\mathbf{R}^n)$. If $W(\varphi) < \infty$ then $s \in L^2_{\text{loc}}(\mathbf{R}^n)$.

The proof of Theorem 3.3 requires the decomposition of $\mathbf{R}^n$ into rectilinear pieces defined in the following way:

a. Let $n = 2$ set

$$\begin{aligned} R_c &= \{t : |t_1| \le 1, |t_2| \le 1\}, \\ R_0 &= \{t : |t_1| \ge 1, |t_2| \ge 1\}, \\ R_1 &= \{t : |t_1| \le 1, |t_2| \ge 1\}, \\ R_2 &= \{t : |t_1| \ge 1, |t_2| \le 1\}. \end{aligned}$$

b. For arbitrary n set

$$\begin{aligned} R_c &= \{t : |t_j| \le 1, j = 1, \ldots, n\}, \\ R_0 &= \{t : |t_j| \ge 1, j = 1, \ldots, n\}, \end{aligned}$$

and define the regions:

$$\begin{aligned} &R_1, R_2, \ldots, R_n; \\ &R_{1,2}, R_{1,3}, \ldots, R_{1,n}; R_{2,1}, \ldots, R_{2,n}; \ldots, R_{n-1,n}; \\ &R_{1,2,3}, \ldots, R_{n-2,n-1,n}; \\ &\vdots \\ &R_{1,\ldots,n-1}; \end{aligned}$$

where

$$R_{j_1, j_2, \ldots, j_k}, \quad k < n$$

is the set of points

$$\begin{aligned} t = (t_1, \ldots, t_n) \in \mathbf{R}^n &\text{ such that } |t_{j_i}| \le 1, \\ i &= 1, \ldots, k \text{ and} \\ |t_j| \ge 1, \quad &j \notin \{j_i : i = 1, \ldots, k\}. \end{aligned}$$

Then $\mathbf{R}^n = \bigcup_\alpha R_\alpha$ where $\alpha = c, 0$, or $(j_1, \ldots, j_k) \in \mathbf{N}^k$, $k < n$.

The proof of Theorem 3.3 is technical and elementary [2] and involves making 3 different arguments depending on the value of α.

To prove Theorem 3.7 we need the notions of 1–lim and Segal algebras.

Definition 3.4 Given $\varphi : \mathbf{R}^n \to \mathbf{C}$, $1\text{–}\lim \varphi = \varphi_0$ means that $\forall \epsilon > 0$ and $\forall c = (c_1, \ldots, c_n) \in \mathbf{R}^n$, $\exists x_0 \in \mathbf{R}$ such that $x > x_0(\epsilon, c)$ implies $|\varphi(x + c_1, \ldots, x + c_n) - \varphi_0| < \epsilon$.

Definition 3.5

a. A *Segal algebra* $\mathcal{S}(\mathbb{R}^n)$ is a dense subalgebra of $L^1(\mathbb{R}^n)$ satisfying the following conditions:

i. $\mathcal{S}(\mathbb{R}^n)$ is a Banach algebra with norm $\|\cdot\|_S$ and the natural injection $\mathcal{S}(\mathbb{R}^n) \to L^1(\mathbb{R}^n)$ is continuous.

ii. $\mathcal{S}(\mathbb{R}^n)$ is translation invariant, and there exist constants $A, B > 0$ such that

$$\forall\varphi \in \mathcal{S}(\mathbb{R}^n) \text{ and } \forall t \in \mathbb{R}^n$$
$$A\|\varphi\|_S \leq \|\tau_t\varphi\|_S \leq B\|\varphi\|_S.$$

iii.

$$\forall\varphi \in \mathcal{S}(\mathbb{R}^n) \text{ and } \forall\epsilon > 0\ \exists\delta > 0$$
$$\text{such that } \forall t \in \mathrm{B}(0,\delta),$$
$$\|\tau_t\varphi - \varphi\|_S < \epsilon.$$

iv. If $\mathcal{S}(\mathbb{R}^n)$ is a *-algebra, i.e., $\varphi \in \mathcal{S}(\mathbb{R}^n)$ implies $\widetilde{\varphi} \in \mathcal{S}(\mathbb{R}^n)$ where $\widetilde{\varphi}(x) = \overline{\varphi(-x)}$, then $\mu\star\varphi(x) \equiv <\mu, \tau_x\widetilde{\varphi}> \in L^\infty(\mathbb{R}^n)$ for each $\mu \in \mathcal{S}'(\mathbb{R}^n)$ and $\varphi \in \mathcal{S}(\mathbb{R}^n)$.

The following theorem is the extension of Wiener's Tauberian theorem to Segal algebras needed.

Theorem 3.6 Let $\mathcal{S}(\mathbb{R}^n)$ be a *-Segal algebra. Suppose J is a subset of $\mathcal{S}(\mathbb{R}^n)$ with the property that for each $\gamma \in \widehat{\mathbb{R}}^n$ there is a $\kappa \in J$ such that $\widehat{\kappa}(\gamma) \neq 0$. If $\mu \in \mathcal{S}(\mathbb{R}^n)$ satisfies the condition,

$$\forall\kappa \in J,\ \ 1\text{–}\lim \mu\star\kappa = W_\mu \int \kappa(t)\,dt$$

for some constant W_μ, then

$$\forall\varphi \in \mathcal{S}(\mathbb{R}^n),\ \ 1\text{–}\lim \mu\star\varphi = W_\mu \int \varphi(t)\,dt.$$

Proof:

cf., [2].

We are now ready to discuss the proof of the Wiener-Plancherel theorem.

Theorem 3.7 Let $\varphi \in \mathrm{BQM}(\mathbb{R}^n)$ then

$$\lim_{T\to\infty} \frac{1}{|R_T|} \int_{R_T} |\varphi(t)|^2\,dt = \lim_{\lambda\to 0} \frac{2^n}{\lambda_1\cdots\lambda_n} \int |\Delta_\lambda s(\gamma)|^2\,d\gamma.$$

Proof:

Noting that $\varphi \in \mathbf{BQM}(\mathbf{R}^n)$, we have by the Plancherel theorem,

$$\frac{1}{2^n}\int\left|\varphi(t)\prod_{j=1}^{n}\frac{\sin(2\pi\lambda_j t_j)}{\pi t_j}\right|^2 dt = 2^n\int|\Delta_{\lambda^s}(\gamma)|^2\,d\gamma.$$

If each $\lambda_j > 0$ then

$$\frac{1}{2^n\lambda_1\cdots\lambda_n}\int\left|\varphi(t)\prod_{j=1}^{n}\frac{\sin(2\pi\lambda_j t_j)}{\pi t_j}\right|^2 dt = \frac{2^n}{\lambda_1\cdots\lambda_n}\int|\Delta_{\lambda^s}(\gamma)|^2\,d\gamma.$$

The result will be obtained once we prove that

$$\lim_{t\to\infty}\frac{1}{T_1\cdots T_n}\int_{R_T}|\varphi(t)|^2\,dt = \lim_{\lambda\to 0}\frac{1}{2^n\lambda_1\cdots\lambda_n}\int\left|\varphi(t)\prod_{j=1}^{n}\frac{\sin(2\pi\lambda_j t_j)}{\pi t_j}\right|^2 dt.$$

Now let $\theta(t) = \sum_{\omega\in\Omega}|\varphi(\omega t)|^2$ where Ω is defined in 3.2, and let

$$\Psi(t) = (2^n\lambda_1\cdots\lambda_n)^{-1}\left[\prod_{j=1}^{n}\frac{\sin(2\pi\lambda_j t_j)}{\pi t_j}\right]^2.$$

Now $\Psi(t) = \Psi(\omega t)\quad\forall\omega\in\Omega$ and the theorem will follow if we show that

$$\lim_{T\to\infty}\frac{1}{T_1\cdots T_n}\int_{R_T\cap\mathbf{R}^{+n}}\theta(t)dt = \lim_{\lambda\to 0}\frac{1}{\lambda_1\cdots\lambda_n}\int_{\mathbf{R}^{+n}}\theta(t)\left[\prod_{j=1}^{n}\frac{\sin(2\pi\lambda_j t_j)}{\pi t_j}\right]^2 dt$$

Now for $c \in \mathbf{R}^n$ with positive components, we write the ordinary limits,

$$A(c) = \lim_{T\to\infty}\frac{1}{T^n c_1\cdots c_n}\int_0^{c_nT}\cdots\int_0^{c_1T}\theta(t)\,dt,$$

$$R(c) = \lim_{\lambda\to 0}\frac{c_1\cdots c_n}{\lambda^n}\int_0^{\infty}\cdots\int_0^{\infty}\theta(t)\left[\prod_{j=1}^{n}\frac{\sin\left(2\pi\lambda c_j^{-1}t_j\right)}{\pi t_j}\right]^2 dt$$

We will be done if we prove that $A(c) = W$ (resp. $R(c) = W$) exists for all $c \in \mathbf{R}^{+n}$ then $R(c)$ (resp. $A(c)$) exists for these c and equals W.

We can assume, without loss of generality, $\theta = 0$ on $\mathbf{R}^n\backslash\mathbf{R}^{+n}$ and $\theta = 0$ on $I_0 = [0,1)^n$, cf. [2]. Now, for a given $c \in \mathbf{R}^n$ with positive components set $c_j = e^{d_j}$; also let $T = e^U$and

$\lambda = T^{-1}$. Define $\Psi_\varphi(u_1, \ldots, u_n) = \theta(t_1, \ldots, t_n)$ where $t_j = e^{u_j} > 0$ and $u \in \mathsf{R}^n$. We then obtain

$$A(c) = \lim_{U \to \infty} e^{-(nU + d_1 + \cdots + d_n)} \int_0^{e^{U+d_n}} \cdots \int_0^{e^{U+d_1}} \theta(t)\, dt$$

$$= \lim_{U \to \infty} \int_{-\infty}^{U+d_n} \cdots \int_{-\infty}^{U+d_1} \Psi_\varphi(u) \prod_{j=1}^{n} e^{-(U+d_j-u_j)} du$$

and

$$R(c) = \lim_{t \to \infty} T^n(c_1, \ldots, c_n) \int_0^\infty \cdots \int_0^\infty \theta(t) \left[\prod_{j=1}^{n} \frac{\sin\left(2\pi t_j (c_j T)^{-1}\right)}{\pi t_j} \right]^2 dt$$

$$= \lim_{U \to \infty} \frac{1}{\pi 2^n} \int \Psi_\varphi(u) \prod_{j=1}^{n} \frac{\sin^2\left(2\pi e^{-[(U+d_j)-u_j]}\right)}{e^{-[(U+d_j)-u_j]}} du.$$

Since $\varphi \in \mathbf{BQM}(\mathsf{R}^n)$, we can show

$$\sum_{m \in \mathbb{Z}^n} \int_{I_0+m} \Psi_\varphi(u)\, du < \infty.$$

Thus we can show $\Psi_\varphi \in \mathcal{S}'_W(\mathsf{R}^n)$ if we define $<\Psi_\varphi, \eta> = \int \overline{\eta(t)} \Psi_\varphi(t)\, dt \quad \eta \in \mathcal{S}_W(\mathsf{R}^n)$. Now let

$$\kappa_A(t) = \begin{cases} \prod_{j=1}^{n} e^{-t_j}, & \text{each } t_j \geq 0 \\ 0, & \text{some } t_j < 0 \end{cases}$$

$$\kappa_R(t) = \frac{1}{\pi^{2n}} \prod_{j=1}^{n} \left(e^{t_j} \sin^2\left(2\pi e^{-t_j}\right) \right).$$

It can be shown [2] that $\int \kappa_R(t)\, dt = 1$, $\kappa_R \in \mathcal{S}_W(\mathsf{R}^n)$, and $|\widehat{\kappa_R}(\gamma)| > 0$. It can also be shown that $\widehat{\kappa_A}(\gamma) \neq 0 \ \ \forall \gamma$ but $\kappa_A \notin \mathcal{S}_W(\mathsf{R}^n)$ since κ_A is not continuous on R. We have

$$A(c) = \lim_{t \to \infty} \Psi_\varphi \star \kappa_A(t + d)$$

and

$$R(c) = \lim_{t \to \infty} \Psi_\varphi \star \kappa_R(t + d)$$

where $c_j = e^{d_j}$and $t + d = (t + d_1, \ldots, t + d_n)$. If $A(c) = w$ (resp. $R(c) = w$) $\forall c \in \mathsf{R}^{+n}$ then the above limits are really 1–limits. We would be done at this point except for the fact that $\kappa_A \notin \mathcal{S}_W(\mathsf{R}^n)$. At this point Wiener's proof on R extends to R^n [4].

The first step is to verify that $R(c) = W \ \forall c$ with positive components if and only if

$$\forall \epsilon > 0, \ \lim_{t \to \infty} \Psi_\varphi \star (\kappa_A \star \delta_\epsilon)(t + d) = W \qquad (*)$$

where $\{\delta_\epsilon\}$ is the mean value approximate identity defined as $\delta_\epsilon(t) = \epsilon^{-n} \chi_{I_0}(t/\epsilon)$. This is a consequence of the Tauberian theorem since $\kappa_A \star \delta_\epsilon \in \mathcal{S}_W(\mathsf{R}^n)$ and $\forall \delta \in \widehat{\mathsf{R}}^n \exists \epsilon > 0$ for which $\widehat{\kappa_A \star \delta_\epsilon}(\gamma) \neq 0$.

Now it is relatively straightforward to show that if $A(c) = W \ \forall c$ with positive components, then (*) holds and hence $R(c) = W$. Conversely, if $R(c) = W$ then (*) holds and we can show $A(c) = W$. qed.

Bibliography

[1] John J. Benedetto. *Spectral Synthesis.* Academic Press, Inc., New York, 1975.

[2] John J. Benedetto, George Benke, and Ward R. Evans. An n-dimensional Wiener-Plancherel formula. *To appear in Advances in Applied Mathematics*, 1989.

[3] N. Wiener. Tauberian theorems. *Annals of Math.*, 33, 1932.

[4] N. Wiener. *The Fourier Integral and Certain of its Applications*. MIT Press, Cambridge, 1933.

[5] N. Wiener. Generalized harmonic analysis. *Acta Math.*, 50, 1950.

SUCCESSIVE DERIVATIVES OF ANALYTIC FUNCTIONS

J. G. CLUNIE
Department of Mathematics
University of York
York Y01 5DD, U.K.

ABSTRACT. This article attempts to convey an understanding of some of the basic ideas in the proofs of three recent results that deal with successive derivatives of complex analytic functions. These are:

(i) if ψ is of order 2, minimal type and lies in the Laguerre-Pólya class of entire functions, p is a real polynomial, i.e. real on the real axis, and $\varphi = p\psi$, then $\varphi^{(k)}$ is in the Laguerre-Pólya class for all large k.

(Craven, Csordas and Smith)

(ii) if f is a real entire function of finite order and f, f'' have only real zeros, then f is in the Laguerre-Pólya class.

(Sheil-Small)

(iii) if f is a real entire function of order exceeding 2 and its order on the real axis is smaller than its overall order, then each point of the real axis is a limit point of the set of the totality of zeros of $f, f', f'', \ldots$

(Clunie)

1. Introduction

I shall be concerned with three problems which greatly interested Pólya and which are referred to in his survey article of 1943 [10]. These problems have given rise to quite a lot of work over many years, but recently very significant results have been obtained and it is with these results that I shall deal. First of all I shall give some notation and definitions and then introduce the problems and state the recent results. In the later sections I shall give in brief outline some of the basic ideas in the proofs of these.

NOTATION AND DEFINITIONS

The Laguerre-Pólya class of entire functions, $\mathcal{L}-\mathcal{P}$ for short, is the set of functions φ of the form

$$\varphi(z) = z^M \prod_\nu \left\{\left(1-\frac{z}{\xi_\nu}\right) e^{\frac{z}{\xi_\nu}}\right\} \quad e^{-az^2+bz+c},$$

where $M \geq 0$ is an integer, the ξ_ν are real and $\sum_\nu \xi_\nu^{-2} < \infty$, and a, b, c are constants with $a, b \in \mathbf{R}$ and $a \geq 0$. A function is in $\mathcal{L}-\mathcal{P}$ if, and only if, it is the local uniform limit in $\mathbf{C}$ of a sequence of polynomials with only real zeros. Hence $\varphi \in \mathcal{L}-\mathcal{P} \Longrightarrow \varphi' \in \mathcal{L}-\mathcal{P}$.

$\mathcal{L}-\mathcal{P}^*$ is the class of functions φ of the form $p\psi$, where $\psi \in \mathcal{L}-\mathcal{P}$ and p is a real polynomial. If $Z_{\mathbf{C}}(\cdot)$ denotes the number of non-real zeros, then, by Rolle's theorem, for $\varphi \in \mathcal{L}-\mathcal{P}^*$, $Z_{\mathbf{C}}(\varphi') \leq Z_{\mathbf{C}}(\varphi)$.

Later we shall be particularly concerned with those $\varphi \in \mathcal{L}-\mathcal{P}^*$ of the form $\varphi = p\psi$, as above, having $a = 0$ in the representation of ψ. This subclass of $\mathcal{L}-\mathcal{P}^*$ is denoted by $\mathcal{L}-\mathcal{P}_0^*$.

J. S. Byrnes and J. F. Byrnes (eds.), Recent Advances in Fourier Analysis and Its Applications, 623–630.

THE RECENT RESULTS

Earlier work will be mentioned only insofar as it seems useful in providing some perspective. More information about such work can be found in the references given, particularly those that contain the recent results themselves.

Result 1. As pointed out above if $\varphi \in \mathcal{L}-\mathcal{P}^*$, then $Z_{\mathbf{C}}(\varphi') \leq Z_{\mathbf{C}}(\varphi)$. The Pólya-Wiman conjecture states:

If $\varphi \in \mathcal{L}-\mathcal{P}^*$, *then for some* $k \in \mathbf{N}$, $\varphi^{(k)} \in \mathcal{L}-\mathcal{P}$.

The first result we consider is due to Craven, Csordas and Smith, [CCS] for short, who proved:

If $\varphi \in \mathcal{L}-\mathcal{P}_0^*$, *then for some* $k \in \mathbf{N}$, $\varphi^{(k)} \in \mathcal{L}-\mathcal{P}$.

Details of proof with references to earlier work are in [2] and [3].

Remark. In a letter to me Wayne Smith mentions that a Korean mathematician, Young-One Kim, has proved the full Pólya-Wiman conjecture. I shall comment on this information later.

Result 2. From the properties of $\mathcal{L}-\mathcal{P}$ we have that if $\varphi \in \mathcal{L}-\mathcal{P}$ then $\varphi^{(k)} \in \mathcal{L}-\mathcal{P}$ for all $k \geq 0$. The question of possible converse results was raised by Pólya. In 1977, Hellerstein and Williamson, [8] and [9], characterized those real entire functions f, i.e. those that are real on the real axis, such that f, f' have only real zeros and $Z_{\mathbf{C}}(f'') = 2p$. In particular, they showed that if f is real entire and f, f', f'' have only real zeros, then $f \in \mathcal{L}-\mathcal{P}$. In 1983, Hellerstein, Shen and Williamson [7] gave some more results of a similar kind within a wider context.

In 1915, Wiman conjectured a result of the kind found by Hellerstein and Williamson for real entire functions of finite order with f having only real zeros, $Z_{\mathbf{C}}(f'') \leq 2p$ and no a priori restriction on the zeros of f'. This conjecture has been settled recently by Sheil-Small [12] and, in particular, if f is real entire of finite order and f, f'' have only real zeros, then $f \in \mathcal{L}-\mathcal{P}$. I shall deal with a very special case only which, nevertheless, contains a point of view essential to Sheil-Small's arguments.

Assuming the conjectured result of Wiman, if $q(z)$ is a real polynomial of degree at least 3 and $f(z) = e^{q(z)}$, then $f''(z) = (q''(z) + q'(z)^2)e^{q(z)}$ must possess non-real zeros. Putting $q'(z) = P(z)$ this result can be stated as a self-contained result for polynomials, viz., if $P(z)$ is a real polynomial of degree at least 2, then $P'(z) + P^2(z)$ possesses non-real zeros. Following the work of Hellerstein and Williamson the possibility of such a result became quite widely known and many of us spent a lot of time on it without success.

If the zeros of $P(z)$ are all real, then the result follows from the work of Hellerstein and Williamson. However, in this case it had been stated and proved much earlier in 'Pólya and Szegö' [11, vol.2, p.66], where it is shown that $P'(z) + P^2(z)$ has n or $n-1$ non-real zeros, where n is the degree of $P(z)$.

Sheil-Small settled the polynomial problem and showed that $P'(z) + P^2(z)$ has at least $n-1$ non-real zeros. Hence the case when $P(z)$ has all its zeros real is extremal. Subsequently Sheil-Small pointed out that if the degree of $P(z)$ is at least 2, then $P'(z) + P^2(z)$ has non-real zeros, whether $P(z)$ is real or not. His solution of the polynomial problem was the starting point for Sheil-Small's later work.

Result 3. The following is a hypothetical theorem of Pólya [10, p. 183].

If a real entire function of order greater than 1 remains bounded for real values of the variable, then its final set contains the whole real axis.

The final set is an idea of Pólya. Let f be analytic in $D \subseteq \hat{\mathbf{C}} = \mathbf{C} \cup \{\infty\}$. The final set of f, denoted here by $S(f)$, is the set of points $\zeta \in \bar{D}$ such that each neighborhood of ζ contains zeros of $f^{(n)}$ for infinitely many n.

A version of Pólya's theorem was proved by Edrei in 1956 [5] with additional hypotheses that are quite severe. I have shown [1] that if f is real entire of order exceeding 2, then $S(f)$ contains the real axis if its order on the real axis is less than its order (in the plane), i.e.

$$\limsup_{r\to\infty} \frac{\log^+\log^+(|f(r)|+|f(-r)|)}{\log r} < \limsup_{r\to\infty} \frac{\log^+\log^+ M(r,f)}{\log r},$$

where, as usual $M(r,f) = \max_{|z|=r} |f(z)|$. This result is a consequence of the following theorem.

Let F be a real entire function of order exceeding 2. Suppose that $F^{(n)}$ is free from zeros in the unit disc for all large n. Then F has the same order on the real axis as it does in the plane.

To see that this theorem implies the preceding result suppose that $S(f)$ does not contain the real axis and let $\xi \in \mathbf{R}\backslash S(f)$. For some $\delta > 0$, the disc $\{|z-\xi| < \delta\}$ contains no zeros of $f^{(n)}$ for all large n. One now considers the result of the theorem for $F(z) = f(\xi + \delta z)$.

2. Result 1: The Result of Craven, Csordas and Smith

2.1. The main result in the proof of [CCS] is the following:

If $\varphi \in \mathcal{L}-\mathcal{P}_0^*$ and for all γ in some non-empty open interval $I \subseteq \mathbf{R}$, $(D+\gamma)\varphi \in \mathcal{L}-\mathcal{P}$, then $\varphi^{(k)} \in \mathcal{L}-\mathcal{P}$ for some $k \in \mathbf{N}$.

Note that $D \equiv \frac{d}{dz}$ and that $(D+\gamma)\varphi = e^{-\gamma z} D(e^{\gamma z}\varphi)$.

The proof of this result is quite lengthy and requires some clever, delicate arguments. I shall give a rather crude indication of the proof which I hope conveys something of its flavour. The preliminary discussion deals with real polynomials, but extends to $\mathcal{L}-\mathcal{P}$ and $\mathcal{L}-\mathcal{P}^*$ when account is taken of their earlier descriptions. This discussion deals with real polynomials whose real zeros are all simple. The more general case can be dealt with by the coalescing of simple zeros.

Let p be a real polynomial with real zeros $\xi_1 < \cdots < \xi_m$. If $Z_{\mathbf{C}}(p) = 0$, then in each interval (ξ_j, ξ_{j+1}), $(-\infty, \xi_1)$, $(\xi_m, +\infty)$ the graph of $x \mapsto Y(x) = \frac{p'(x)}{p(x)}$ is strictly monotonic decreasing and $Y(-\infty) = Y(+\infty) = 0$ and $Y(\xi_j-) = -\infty$, $Y(\xi_j+) = +\infty$. When $Z_{\mathbf{C}}(p) > 0$, then $Y(x)$ will behave at $x = \pm\infty$ and at $x = \xi_j$ as before, but the graph of $x \mapsto Y(x)$ may no longer be monotonic in the intervals (ξ_j, ξ_{j+1}), $(-\infty, \xi_1)$, $(\xi_m, +\infty)$. If for some $\gamma \in \mathbf{R}$ the line $y = -\gamma$ cuts the graph of $x \mapsto Y(x)$ in more than one point in one, or more, of these intervals, then $Z_{\mathbf{C}}((D+\gamma)p) < Z_{\mathbf{C}}(p)$.

If $\alpha \pm i\beta (\beta > 0)$ is a pair of non-real zeros of p, the contribution they make to $Y(x)$ is

$$c(x) = \frac{2(x-\alpha)}{(x-\alpha)^2 + \beta^2}$$

and

$$c'(x) = \frac{2(\beta^2 - (x-\alpha)^2)}{((x-\alpha)^2+\beta^2)^2}.$$

Consequently $c(x)$ decreases in $(-\infty, \alpha-\beta)$, $(\alpha+\beta, +\infty)$ and increases in $(\alpha-\beta, \alpha+\beta)$ and $c(\alpha+\beta) - c(\alpha-\beta) = \frac{2}{\beta}$. Hence in the preceding discussion intervals of increase of $x \mapsto Y(x)$ lie in $\cup(\alpha-\beta, \alpha+\beta)$ and the total increase is bounded by $\sum \frac{2}{\beta}$. Note that in the situation considered in the preceding paragraph if $Z_{\mathbf{C}}((D+\gamma)p) = 0$, then $y = -\gamma$ intersects $x \mapsto Y(x)$ in $\frac{Z_{\mathbf{C}}(p)}{2}$ intervals of increase of $Y(x)$.

Consider the above discussion extended to $\mathcal{L}-\mathcal{P}^*$ and let $\varphi \in \mathcal{L}-\mathcal{P}^*$ and suppose that $Z_{\mathbf{C}}((D+\gamma)\varphi)=0$ for each $\gamma \in I$, a non-empty open interval of $\mathbf{R}$. If $\gamma_0 \in I$, then for all large m, $Z_{\mathbf{C}}\left(\left(\left(z+\frac{m}{\gamma_0}\right)^m \varphi\right)'\right)=0$ and hence $Z_{\mathbf{C}}\left(\left(\left(z+\frac{m}{\gamma_0}\right)^{m+k} \varphi^{(k)}\right)'\right)=0$ for $k=1,2,\ldots$ Suppose that the hypotheses of the result stated at the beginning of §2.1 are satisfied, but $Z_{\mathbf{C}}(\varphi^{(k)})=2d_0>0$ for all large k. One considers, as in the earlier observations, the intersections of the graphs $x \mapsto -\frac{m+k}{x+\frac{m}{\gamma_0}}$ and $x \mapsto \frac{\varphi^{(k+1)}(x)}{\varphi^{(k)}(x)}$. The arguments [CCS] used to cope with the case $d_0>1$ are clever and intricate and so I only deal with the case $d_0=1$.

Assume that $Z_{\mathbf{C}}(\varphi^{(k)})=2$ for all large k and let $\alpha_k \pm i\beta_k$ $(\beta_k>0)$ be the non-real zeros of $\varphi^{(k)}$. The graphs of $x \mapsto -\frac{m+k}{x+\frac{m}{\gamma_0}}$, $x \mapsto \frac{\varphi^{(k+1)}(x)}{\varphi^{(k)}(x)}$ intersect in the interval of increase of the latter and this interval lies in $(\alpha_k-\beta_k, \alpha_k+\beta_k)$.

A theorem of Jensen for real polynomials extends to $\mathcal{L}-\mathcal{P}^*$ and states: if $\varphi \in \mathcal{L}-\mathcal{P}^*$ and $\xi_\nu \pm i\eta_\nu$ are the non-real zeros of φ, then the non-real zeros of $\varphi^{(k)}$ lie in $\cup E_\nu$, where

$$E_\nu=\left\{z=x+iy\colon \frac{(x-\xi_\nu)^2}{k\eta_\nu^2}+\frac{y^2}{\eta_\nu^2}\le 1\right\}.$$

There is a choice of values for m, λ_0 and $|\alpha_k|\le \beta_1\sqrt{k}$, by Jensen's theorem, and hence, considering the positive increase of $x \mapsto \frac{\varphi^{(k+1)}(x)}{\varphi^{(k)}(x)}$, for some $a \in \mathbf{R}$ and $A_k=|\alpha_k+a|$, $\beta_k \le c\frac{A_k}{k}$ with c some constant. Using the theorem of Jensen again and this bound for β_k one finds that

$$A_{2^{k+1}} \le (1+c\,2^{-k/2})A_{2^k}.$$

From this inequality one shows that (α_n) converges by first considering the subsequence for $n=2^k$ and then using Jensen's theorem once more appropriately. Hence (β_n) also converges and one finds that for some $L \in \mathbf{R}$,

$$|L-(\alpha_n \pm i\beta_n)| \le \text{const.}\, n^{-\frac{1}{2}}.$$

A contradiction is obtained, and hence the main result of [CCS] established, by the following result of [CCS] which is a sharpened form of a result of Ålander.

Let $f(z)$ be an entire function of order 2 and minimal type, i.e. $\frac{\log M(r,f)}{r^2} \longrightarrow o(r\to\infty)$. *If $f^{(n)}(z_n)=0\,(n=0,2,\ldots)$ and $\omega \in \mathbf{C}$, then*

$$\limsup_{n\to\infty}\left(\sqrt{n}\,|z_n-\omega|\right)=\infty.$$

2.2 COMPLETION OF [CCS] PROOF

Assuming the result to be proved, i.e. $\varphi \in \mathcal{L}-\mathcal{P}_0^* \Longrightarrow \varphi^{(k)} \in \mathcal{L}-\mathcal{P}$ for some k, is false there is a smallest $d_0 \in \mathbf{N}$ and some $\varphi \in \mathcal{L}-\mathcal{P}_0^*$ such that $Z_{\mathbf{C}}(\varphi^{(k)})=2d_0$ for all $k \ge 0$. Let $\varphi=p\psi$, where $\psi \in \mathcal{L}-\mathcal{P}$ and $Z_{\mathbf{C}}(p)=2d_0$. Then ψ has infinitely many zeros and is of the form

$$\psi(z)=z^M\prod_{\nu=1}^{\infty}\left\{\left(1-\frac{z}{\xi_\nu}\right)e^{\frac{z}{\xi_\nu}}\right\}e^{bz+c}.$$

For $N \in \mathbf{N}$, set

$$\psi_N=\prod_{\nu=N}^{\infty}\left\{\left(1-\frac{z}{\xi_\nu}\right)e^{\frac{z}{\xi_\nu}}\right\}e^{\Lambda_N z+c},$$

where $\Lambda_N = b + \sum_{\nu=1}^{N-1} \frac{1}{\xi_\nu}$, and

$$\varphi_N = p \cdot \psi_N .$$

Consider

$$\Phi_N = p \cdot \Psi_N ,$$

where $\Phi_N = e^{-\Lambda_N z}\psi_N$. Given $K > 0$, for $-K \leq x \leq K$ the graph of $x \mapsto \frac{\Psi'_N(x)}{\Psi_N(x)}$ is 'flat' for large N. From the discussion in §2.1 the intervals of increase of $x \mapsto \frac{p'(x)}{p(x)}$ will put a wiggle in the graph of $x \mapsto \frac{\Phi'_N(x)}{\Phi_N(x)}$ provided first K and then N have been chosen sufficiently large. In this case there is an interval $I_0 \subseteq \mathbf{R}$ such that for $\gamma \in I_0$,

$$Z_{\mathbf{C}}((D+\gamma)\Phi_N) < 2d_0 .$$

If I_0 is translated by $-\Lambda_N$ to give I, then for $\gamma \in I$,

$$Z_{\mathbf{C}}((D+\gamma)\varphi_N) < 2d_0 .$$

By the choice of d_0, for some large fixed k_0,

$$D^{k_0}(D+\gamma)\varphi_N = (D+\gamma)\varphi_N^{(k_0)} \in \mathcal{L} - \mathcal{P}$$

and hence from the main result of §2.1, $\varphi_N^{(k)} \in \mathcal{L} - \mathcal{P}$ for large k. Now $\varphi = z^M \prod_{\nu=1}^{N-1} \left(1 - \frac{z}{\xi_\nu}\right) \varphi_N$ and one can reintroduce the factors of $z^M \prod_{\nu+1}^{N-1} \left(1 - \frac{z}{\xi_\nu}\right)$ one at a time into φ_N and so obtain the preceding conclusion for φ itself by using the following easily proved result.

If for some $m \in \mathbf{N}$, $f^{(m)} \in \mathcal{L} - \mathcal{P}$ and $\xi \in \mathbf{R}$, then $D^{m+1}((z+\xi)f) \in \mathcal{L} - \mathcal{P}$.

CONCLUDING REMARK

Apparently the proof of Young-One Kim follows the same lines as that of [CCS]. The sharpened form of Ålander's result is replaced by a result of Gontcharoff. I'm not clear how one copes with the factor $e^{-az^2} (a > 0)$ in the argument considered in §2.2 or what may replace this argument.

3. Result 2: The Result of Sheil-Small

3.1. The very special case of Sheil-Small's work that I shall deal with is the proof of the result that if P is a real non-constant polynomial and $Z_{\mathbf{C}}(P' + P^2) = 0$, then P is of degree 1. Let, then, $P(z)$ be a real non-constant polynomial and set

$$L(z) = P(z), \qquad Q(z) = z - \frac{1}{L(z)}$$

and note that

$$Q'(z) = \frac{P'(z) + P^2(z)}{P^2(z)} .$$

Define $H^+ = \{\operatorname{Im} z > 0\}$, $\Lambda = \{z \in H^+ : \operatorname{Im} L(z) > 0\}$, $K = \{z \in H^+ : \operatorname{Im} Q(z) > 0\}$.

Suppose that $\Lambda \neq \phi$ and let V be a component of Λ. Then V is simply connected, unbounded and $L(z) \to \infty$ as $z \to \infty$ in V. Each boundary component of ∂V is a curve that goes to ∞ in both directions and $\operatorname{Re} L(z)$ increases to $+\infty$ as $z \to \infty$ in one direction and

$\operatorname{Re} L(z)$ decreases to $-\infty$ as $z \to \infty$ in the other direction. There is a finite number of such boundary components and so for some $v \in \mathbf{N}$, each $c \in \mathbf{R}$ is assumed by $\operatorname{Re} L(z)$ precisely v times on ∂V.

Now $\Lambda \subseteq K$, and suppose $U \supseteq V$ is a component of K. There is at least one point $\zeta \in \partial V$ with $L(z) = 0$ and then $\zeta \in \partial U$. In U near ζ all large values in H^+ are taken by $Q(z)$. Since $P(z) \to \infty$ as $z \to \infty$ in U, Q is not univalent in U.

Assuming that $Q'(z)$ has no zeros in U and $Q(z_0) = w_0$ for some $z_0 \in U$, $Q^{-1}(w)$ can be analytically continued throughout H^+ from w_0. H^+ is simply connected and so, by the monodromy theorem, $Q(z)$ is univalent in U. However, this contradicts the behaviour of $Q(z)$ noted earlier and our initial assumption, i.e. $\Lambda \neq \phi$, is therefore false, and so $\operatorname{Im} L(z) < 0$ in H^+. It is easy to check that in this case $P(z) = -az + b$ with $a, b \in \mathbf{R}$ and $a > 0$.

3.2. If we return to the earlier discussion in §1, where we considered $f(z) = e^{q(z)}$, then $P(z) = q'(z)$ so that in this notation

$$L(z) = \frac{f'(z)}{f(z)}, \quad Q(z) = z - \frac{f(z)}{f'(z)}$$

and

$$Q'(z) = \frac{f(z), f(''(z)}{f'(z)^2}.$$

It is with such $L(z)$, $Q(z)$ that Sheil-Small deals in obtaining his results. It should not be thought, however that introducing $L(z)$ and $Q(z)$ easily leads to these more general results. Considerable skill and technical expertise is required to cope with the details of proof, but the use of $L(z)$ and $Q(z)$ and consideration of the sets Λ and K is the essential new idea in Sheil-Small's arguments.

4. Result 3: The Result of Clunie

The result considered here is the following theorem.

Let F be a real entire function of order exceeding 2. Suppose that $F^{(n)}$ is free from zeros in the unit disc for all large n. Then F has the same order on the real axis as it does in the plane.

4.1. In 1955, Edrei [4] proved the following result which is essential to our arguments and which I shall refer to a Edrei's lemma.

Suppose that $f(z)$ is entire and that

$$f(z) = 1 = +z + \alpha\frac{z^2}{2!} + \cdots \quad (\alpha \neq 1).$$

Then f. f'. f'' has a zero in

$$\left\{|z| \leq A\frac{|\log|1-\alpha|| + 1}{|1-\alpha|}\right\},$$

where $A > 0$ is absolute.

From Edrei's lemma we obtain the following important consequence.

Suppose that $f(z) = \sum a_n z^n$ is a real entire function and that $f^{(n)}$ has no zeros in the unit disc for all large n. For large n, put $b_n = n\frac{a_n}{a_{n-1}}$. If n is large and $a_{n-1}\ a_{n+1} < 0$, then $|b_n| \leq B$, where B is absolute, and

$$|b_{n-k}| \leq d(k+2)\log(k+2) \qquad (k = 0, 1, \ldots, n - n_0),$$

where d is absolute and $n_0 = n_0(f)$.

Here is a brief description of the proof of this result. Consider

$$f^{(n-1)}(z) = (n-1)!a_{n-1} + n!a_n z + \frac{(n+1)!}{2!} a_{n+1} z^2 + \cdots$$

and put $b_n z = \zeta$ to obtain

$$f^{(n-1)}(z) = (n-1)!a_{n-1} \left[1 + \zeta + \frac{b_{n+1}}{b_n} \frac{\zeta^2}{2!} + \cdots \right].$$

Now $\frac{b_{n+1}}{b_n} = \frac{n+1}{n} . \frac{a_{n-1,}a_{n+1}}{an^2} < 0$ and, assuming n is large so that $f^{(n-1)} . f^{(n)} . f^{(n+1)}$ has no zeros in $\{|\zeta| \le |b_n|\}$, it follows from Edrei's lemma that

$$|b_n| \le A \frac{\left[\log\left(1 + \frac{|b_{n+1}|}{|b_n|}\right) + 1\right]}{1 + \frac{|b_{n+1}|}{|b_n|}}$$

and hence $|b_n| \le B$.

Some fairly straightforward analysis now gives the bounds for $|b_{n-k}|$.

4.2. If $f(z) = \sum a_n z^n$ is entire, then to each r corresponds a largest integer N such that

$$|a_n| r^n \le |a_N| r^N (n = 0, 1, \ldots).$$

Then $N = N(r) = N(r, f)$ is called the central index of $f(z)$ for $|z| = r$ and $|a_N| r^N = \mu(r) = \mu(r, f)$ is called the maximum term for $|z| = r$. $N(r)$ and $\mu(r)$ are related by

$$\log \mu(r) - \log \mu(r_0) = \int_{r_0}^{r} \frac{N(t)}{t} dt \qquad (r > r_0 > 0).$$

From Wiman-Valiron [6] theory we obtain the following result:

Suppose that $f(z) = \sum a_n z^n$ *is entire of order* $\rho\,(0 < \rho \le \infty)$ *and that* $0 < \rho' < \rho$, $0 < \varepsilon < \frac{1}{2}$. *Then*

$$\left(\sum_{0}^{N - N^{\frac{1}{2}+\epsilon}} + \sum_{N + N^{\frac{1}{2}+\epsilon}}^{\infty} \right) |a_n| r^n = o(\mu(r))$$

as $r \to \infty$ *through some sequence for which* $N = N(r) > r^{\rho'}$, $\log \mu(r) > r^{\rho'}$.

OUTLINE OF PROOF OF THEOREM

Assume the hypotheses of the theorem and suppose that $2 < \rho' < \rho$ and $\varepsilon > 0$ is small. Consider $f(z) = F(z)$ and $r \to \infty$ through the sequence of the result in §4.2 and then

$$F(z) = \sum_{N - N^{\frac{1}{2}+\epsilon}}^{N + N^{\frac{1}{2}+\epsilon}} a_n z^n + o(\mu(r)) \qquad (|z| = r).$$

If for some subsequence of r, $a_{n-1} a_{n+1} > 0$ for all n in $[N - N^{\frac{1}{2}+\epsilon}, N + N^{\frac{1}{2}+\epsilon}]$, then

$$|F(r)| + |F(-r)| > \sum_{N - N^{\frac{1}{2}+\epsilon}}^{N + N^{\frac{1}{2}+\epsilon}} |a_n| r^n + o(\mu(r))$$
$$> (1 + o(1))\mu(r)$$

as $r \to \infty$ through this subsequence. In this case F is of order at least ρ' on the real axis.

Otherwise for all large r of the sequence there is a largest n in $[N - N^{\frac{1}{2}+\epsilon}, N + N^{\frac{1}{2}+\epsilon}]$ with $a_{n-1}a_{n+1} < 0$. Using the consequence of Edrei's lemma given in §4.1, if $\varepsilon > 0$ has been chosen small enough, depending on $\rho' > 2$, then

$$\sum_{N-N^{\frac{1}{2}+\epsilon}}^{n} |a_k| r^k = o(\mu(r)).$$

Therefore

$$\begin{aligned} |F(r)| + |F(-r)| &> \sum_{n+1}^{N+N^{\frac{1}{2}+\epsilon}} |a_k| r^k - 2 \sum_{N-N^{\frac{1}{2}+\epsilon}}^{n} |a_k| r^k + o(\mu(r)) \\ &> (1 + o(1))\mu(r) \end{aligned}$$

and in this case also F is of order at least ρ' on the real axis.

From the above F is of order at least ρ' on the real axis for all $\rho' < \rho$ and so F is of order ρ on the real axis. This establishes the theorem

REFERENCES

1. Clunie, J. G., On a hypothetical theorem of Pólya, Complex Variables, to appear.
2. Craven, T., Csordas, G., Smith, W., The zeros of derivatives of the entire functions and the Pólya-Wiman conjecture, Ann. of Math. **125** (1987), 405–431.
3. Craven, T., Csordas, G., Smith, W., Zeros of derivatives of entire functions, Proc. Amer. Math. Soc. **101** (1987), 323–326.
4. Edrei, A., On the zeros of successive derivatives, Proc. Amer. Math. Soc. (1955), 386–391.
5. Edrei, A., On a conjecture of Pólya concerning the zeros of successive derivatives, Scripta Mathematica, vol. XXII (1956), 1–30.
6. Hayman, W. K., The local growth of a power series: A survey of the Wiman-Valiron method, Can. Math. Bull. 17(3) 1974, 317–358.
7. Hellerstein, S., Shen, L.-C., Williamson, J., Reality of the zeros of an entire function and its derivatives, Trans. Amer. Math. Soc. **275** (1983), 319–331.
8. Hellerstein, S., Williamson, J., Derivatives of entire functions and a question of Pólya, Trans. Amer. Math. Soc. **227** (1977), 227–269.
9. Hellerstein, S., Williamson, J., Derivatives of entire functions and a question of Pólya II, Trans. Amer. Math. Soc. **234** (1977), 497–503.
10. Pólya, G., On the zeros of the derivatives of a function and its analytic character, Bull. Amer. Math. Soc. **49** (1943), 179–191.
11. Pólya, G., Szegö, G., *Problems and Theorems in Analysis*, Springer, 1976.
12. Sheil-Small, T., On the zeros of the derivatives of real entire functions and Wiman's conjecture, Ann. of Math. **129** (1989), 179–193.

Combinatorial dimension and measurements of interdependencies

Ron C. Blei
Department of Mathematics
University of Connecticut
Storrs, CT 06269 USA

1. Introduction

The concepts of fractional Cartesian products and combinatorial dimension were originally cast in a framework of multi–dimensional lattices [1], [2], [3], [6]. In this paper, I will describe the extension of these ideas to the J – dimensional Euclidean setting $\mathbf{R}^J$, taking into account its usual topological structure. As in the discrete framework of multi–dimensional lattices, the 'combinatorial dimension' of a set in the topological setting will be viewed as a measurement of interdependencies between coordinates of points in the given set.

The idea of interdependencies in a context of probability theory is of course not new. An assumption that a point $(X_1,...,X_J) \in \mathbf{R}^J$ is sampled according to some statistical law essentially means that interdependencies between the coordinates are given by the joint probability distribution

(1) $$\mu_{X_1,...,X_J}(A_1 \times \cdots \times A_J) = P(X_1 \in A_1,...X_J \in A_J), \quad A_1,..., A_J \subset \mathbf{R} \text{ Borel.}$$

We recognize two extremal cases: (i) complete statistical independence of $X_1,..., X_J$, in which case $\mu_{X_1,...,X_J}$ is the product of its J marginals; (ii) each of the r.v.'s $X_2,...,X_J$ depends deterministically on X_1, i.e., $X_i = f_i(X_1)$, where each f_i is a $\mathbf{R}$–valued function, $i = 1,..., J-1$.

J. S. Byrnes and J. F. Byrnes (eds.), Recent Advances in Fourier Analysis and Its Applications, 631–641.

To motivate our particular point of view, focusing on 'combinatorial dimension' as a gauge of interdependencies, we note that the support of $\mu_{X_1,\ldots,X_J}$ in (i) is a "J–dimensional" Cartesian product of the supports of the J marginals, while the support of $\mu_{X_1,\ldots,X_J}$ in (ii) is the "one–dimensional curve" $\{(x, f_1(x), \ldots, f_{J-1}(x)) : x \in \mathbb{R}\}$. In a context of probability theory, we are speculating that the 'combinatorial dimension' of the support of a multi–dimensional probability distribution effectively calibrates a scale of interdependencies between its respective marginals. As a start in this direction, one of our main results is an existence theorem:

Theorem 1

For every $\alpha \in (1,J)$, *there exist* J *random variables uniformly distributed on* $(0,1)$ *so that the support of their joint distribution has (combinatorial) dimension* α.

2. Definitions and some remarks

We recall first the definition of 'combinatorial dimension' of a subset of a multi–dimensional lattice. Let E be a set devoid of any structure, J a positive integer, and $F \subset E^J$. For all positive integers s, define

$$\Psi_F(s) = \max\{|F \cap (A_1 \times \cdots \times A_J)| : A_j \subset E, \ |A_j| = s, \ j = 1,\ldots,J\},$$

and for $a > 0$, define

$$d_F(a) = \sup_s (\Psi_F(s)/s^a). \tag{2}$$

The combinatorial dimension of F is given by

(3) $$\dim F = \inf\{a: d_F(a) < \infty\},$$

or equivalently,

$$\dim F = \lim_{s \to \infty} \inf(\ln \Psi_F(s) / \ln s).$$

Remark

The notion of combinatorial dimension surfaces naturally in a stochastic context. Suppose $\{X_j\}_{j \in N}$ is a system of statistically independent random variables each with zero mean, and assume that $|X_j| \leq 1$ for all $j \in N$. (We can think of N as a discrete time scale, and the X_j's as fluctuations over the j–th time slot.) By definition, the independence of the given system means that the joint probability distribution of $\{X_j\}_{j \in N}$ (a measure on R^∞) is a product of the probability distributions of the X_j's. Going beyond the definition, the actual effect of independence is the statistical abundance of cancellations. To be precise, we have (e.g., [10, p. 387])

(4) $$P\{ (1/N)^{1/2} | \sum_{j=1}^{N} X_j | > x \} \leq e^{-Kx^2}, \qquad N > 0 \text{ and } x > 0.$$

This trivially extends to

(5) $$P\{ (1/N)^{J/2} |(\sum_{j=1}^{N} X_j)^J| > x \} \leq e^{-Kx^{2/J}}, \qquad N > 0,\ x > 0,$$

which can be viewed as the effect of interdependencies within the system $\{X_{j_1} \cdots X_{j_J}\}_{(j_1,\dots,j_J) \in N^J}$.

Interpolating between 1 and J, given any $F \subset N^J$, we obtain that the 'combinatorial dimension' of F measures the amount of cancellations, in the sense just described, conveying the effect of interdependencies within the system $\{X_{j_1} \cdots X_{j_J}\}_{(j_1,\ldots,j_J)\in F}$.

Theorem 2

Let $F \subset N^J$ and $\dim F = \alpha$. Denote $F_N = F \cap \{1,\ldots,N\}^J$. Then, for all $\beta > \alpha$

$$(6) \qquad P\{\, (1/N)^{\beta/2} \Big| \sum_{j\in F_N} X_{\pi_1(j)} \cdots X_{\pi_J(j)} \Big| > x\} \leq e^{-Kx^{2/\beta}}, \qquad N > 0,\ x > 0$$

($K > 0$ is independent of $N > 0$ and $x > 0$, and may depend on β; $\pi_1,\ldots,\pi_J$ denote the canonical projections from N^J onto N).

The proof of the theorem is based on [2, Theorem 7.1] (it will soon be written up in the form of a note; for a discussion of related issues, see [4]).

We now define the combinatorial dimension of $F \subset R^J$. Let $\tau = \tau_1 \times \cdots \times \tau_J$ be a grid of R^J, where $\tau_1, \ldots, \tau_J$ denote partitions of R consisting of contiguous intervals, and let the mesh τ be given by

$$\|\tau\| = \max\{\, \|\tau_j\| : j = 1,\ldots,J \,\},$$

where $\|\tau_j\|$ on the right hand side is the usual mesh of a partition. We denote

$$(7) \qquad F_\tau = \{c : c \in \tau,\ c \cap F \neq \phi\}.$$

Define

(8) $$D_F(a) = \lim_{\epsilon\to 0} \inf\{ d_{F_\tau}(a) : \|\tau\| < \epsilon \}, \quad a > 0,$$

and

(9) $$\mathrm{Dim}F = \inf\{ a : D_F(a) < \infty \}.$$

Remarks

(a) Answering an obvious question, we observe that Dim in (9) and the usual Hausdorff dimension $\mathrm{Dim}_{\mathscr{H}}$ are different measurements. Combinatorial dimension is based on a measurement of "interdependencies" between coordinates, while Hausdorff dimension is based on a measurement of "content;" $\mathrm{Dim}F \neq \mathrm{Dim}_{\mathscr{H}}F$ is indeed possible. To illustrate $\mathrm{Dim}_{\mathscr{H}}F > \mathrm{Dim}F$, take $f : \mathbf{R} \to \mathbf{R}$ continuous, observe that $\mathrm{Dim}(\mathrm{graph}(f)) = 1$, and note that for every $\alpha \in (1,2)$ there are f for which $\mathrm{Dim}_{\mathscr{H}}(\mathrm{graph}(f)) = \alpha$ [7, Theorem 8.2]. To illustrate $\mathrm{Dim}F > \mathrm{Dim}_{\mathscr{H}}F$, take F countable, note that $\mathrm{Dim}F \geq 1$ (which is true for any infinite F), while $\mathrm{Dim}_{\mathscr{H}}F = 0$.

(b) We have defined in (9) a dimension relative to the underlying topological product structure of $\mathbf{R}^J$. Moving to a general topological setting, let $\mathscr{X}$ be a locally compact Hausdorff space, and suppose $F \subset \mathscr{X}^J$ is closed. Let $\tau_1, \ldots, \tau_J$ be open covers of $\mathscr{X}$, $\tau = \tau_1 \times \cdots \times \tau_J$, and define F_τ as in (7). Given a natural partial ordering of open covers ($\tau < \sigma$ if for every $O \in \tau$ there is $U \in \sigma$ so that $O \subset U$), define

(10) $$D_F(a) = \sup_{\sigma} \inf_{\tau<\sigma} d_{F_\tau}(a),$$

and obtain DimF, as in (9). When $\mathscr{X} = \mathbf{R}$, it is easy to see that $D_F(a)$ given by (10) is finite if and only if $D_F(a)$ in (8) is finite, i.e., the two definitions of DimF coincide.

The step leading from (2) to (10) is archtypical. For example, take $\mathscr{X}$ to be a measurable space, and let F be a measurable subset of $\mathscr{X}^J$. In (10) above, replace open covers by measurable partitions (partial ordering is given by refinement) and obtain the dimension of F relative to the *measurable* product structure. We thus have in the setting $\mathbf{R}^J$ three natural measurements: (i) dimF (assuming no structure in R), (ii) DimF relative to the usual topological structure, and (iii) DimF relative to the usual Borel structure. Note that (i) is invariant under *all* rearrangements of R, (ii) is invariant under all homeomorphisms of R, and (iii) is invariant under all Borel measurable rearrangements of R. We know only some of the relationships between (i), (ii), and (iii): for all $F \subset \mathbf{R}^J$, dimF $\leq$ DimF (topological or Borel), and there exist closed sets $F \subset [0,1]^2$ for which dimF = 1 and topological DimF = 2 [Tom Körner, Private communication]. I do not know of closed sets F for which topological DimF $\neq$ Borel DimF. In this paper, DimF will refer to the dimension defined by (9) in the setting $\mathbf{R}^J$.

(c) Let μ be a bounded J–linear form on $C_0(\mathbf{R})$. By a classical theorem of Frechet [8], μ is canoncially identified with a function on measurable rectangles in $\mathbf{R}^J$ which is a finite Borel measure in each of its J coordinates. In this particular context, the notion of combinatorial dimension makes precise a classical observation of Littlewood [9], that μ need not be extendible to a finite Borel measure on $[0,1]^J$. Let F be a closed subset of $\mathbf{R}^J$, let $p > 0$, and define the pth–variation of μ over F by

$$|\mu|_p(F) = \lim_{\epsilon\to 0} \inf\{(\sum_{c\in F_\tau} |\mu(c)|^p)^{1/p} : \tau \text{ a grid of } \mathbf{R}^J, \|\tau\| < \epsilon\}.$$

It is clear that μ is extendible to a measure on $\mathbf{R}^J$ if and only if $|\mu|_1(\mathbf{R}^J) < \infty$. The connection between the combinatorial dimension of F and the pth–variation of bounded J–linear forms on $C_0(\mathbf{R})$ is given by

***Theorem 3** [3, Theorems 5.6, 5.7]*

Let $F \subset R^J$, and suppose $DimF = \alpha$. Then, (i) for all bounded J–linear forms μ on $C_0(R)$

$$|\mu|_p(F) < \infty \quad \text{for all } p > 2\alpha/(\alpha+1),$$

and (ii) if $DimF = dimF = \alpha$, then there are bounded J–linear forms μ on $C_0(R)$ so that $|\mu|_p(F) = \infty$ for all $p < 2\alpha/(\alpha+1)$.

It is an open question whether (ii) remains valid under the weaker and more natural assumption $\mathrm{DimF} = \alpha$.

3. α – dimensional supports of probability distributions

We now come to a basic question: do α – dimensional sets exist for every $\alpha \in (1,J)$? In the discrete framework N^J, explicit designs and random constructions of fractionally dimensioned sets appeared, respectively, in [1] and [6]. Adapting the designs of [1] to the topological framework R^J, we obtain J/K – dimensional Cartesian products of R for all integers $0 < K \leq J$. However, as in [6], a complete solution to the existence question in R^J is provided by random constructions:

Theorem 4

Let $\alpha \in (1,J)$ be arbitrary. There exist random perfect sets $F \subset [0,1]^J$ so that for every $x \in F$ and every neighbourhood V of x,

$$Dim(F \cap V) = dim(F \cap V) = \alpha.$$

We give here only a general over–view of the proof; the details are worked out in [5]. The archtypical case $J = 2$ is based on a recursive random selection of a decreasing tower of closed subsets in $[0,1]^2$. For each positive integer n, let σ_n denote the square grid of $[0,1]^2$ given by

$$\sigma_n = \{[y/n, (y+1)/n]\times[z/n, (z+1)/n]: \ y, z = 0, \ldots, n-1\}.$$

By induction, we produce a sequence of integers $(n_j)_{j=1}^{\infty}$ tending very fast to infinity and a sequence of randomly chosen collections of cells $F_j \subset \sigma_{n_j}$, $j = 1, \ldots,$ so that

(i) the number of cells in F_j is (approximately) $(n_j)^{\alpha}$,

(ii) $\Psi_{F_j}(s) \leq Ks^{\alpha}$ for all $s > 0$, where $K > 0$ depends only on α,

and

(iii) each cell in F_j is contained in a cell belonging to F_{j-1}.

With the F_j's at hand, we then verify that

$$F = \bigcap_{j=1}^{\infty} \{c: c \in F_j\}$$

is the desired α – dimensional set.

To obtain for every $\alpha \in (1,J)$ J random variables uniformly distributed in $(0,1)$ the support of whose joint probability distribution has dimension α (Theorem 1), we retrace the inductive procedure that led to Theorem 4 above. At the j–th step of the induction, we require that each column and each row of F_j contain approximately $(n_j)^{\alpha-1}$ cells. We then take a probability measure μ_j which is 'uniform' on F_j, and verify that a weak* – limit of the μ_j's is practically the desired probability distribution with uniform marginals and α–dimensional support.

The non–random α–dimensional designs, which are produced in [5], are essentially fractional Cartesian products of locally compact abelian groups on which the Haar measures are the probability measures of Theorem 1. Given integers $J \geq K > 0$, we construct in I^J ($I = [0,1]$) a perfect set $I^{J/K}$ and a probability measure μ on I^J so that

(i) $\text{Support}(\mu) = I^{J/K}$,

(ii) $\text{Dim}(I^{J/K}) = J/K$

and

(iii) each of the marginals of μ is Lebesgue measure on I.

I will illustrate here the case $J = 3$, $K = 2$. For each $n = 1, \ldots$, let $Y_n = \{0, 1\}^n$, and denote (for typographical convenience) $H_n = (Y_n)^2$. Imbed H_n in I via the standard 'binary expansion':

$$(m_1, m_2) \in H_n \mapsto \sum_{i=1}^{n} m_1(i)/2^{2i-1} + \sum_{i=1}^{n} m_2(i)/2^{2i}$$

Take H_n to be an indexing set for the partition

(11) $$\tau_n = \{[j/4^n, (j+1)/4^n] : 0 \leq j < 4^n\},$$

and define

$$(H_n)^{3/2} = \{((m_1, m_2), (m_2, m_3), (m_3, m_1)) : (m_1, m_2, m_3) \in (Y_n)^3\},$$

which we view as a collection of cells in the grid $(\tau_n)^3$. We have

$$\bigcup\{c: c \in (H_{n+1})^{3/2}\} \subset \bigcup\{c: c \in (H_n)^{3/2}\}, \tag{12}$$

and

$$|\{c \in C: c \in (H_{n+1})^{3/2}\}| = 2^3, \quad C \in (H_n)^{3/2}. \tag{13}$$

Let

$$I^{3/2} = \bigcap_{n=1}^{\infty} \{c: c \in (H_n)^{3/2}\}.$$

Theorem 5

(i) $Dim(I^{3/2}) = dim(I^{3/2}) = 3/2$.

(ii) There is a probability measure μ *on* I^3 *whose support is* $I^{3/2}$ *and each of whose marginals is Lebesgue measure on* I.

The proof of (i) is based on [3, Theorem 2.5]. To establish (ii), for each $n > 0$ define probability measures

$$\mu_n = (1/2^{3n}) \sum_{h \in (H_n)^{3/2}} \delta_h,$$

and

$$\nu_n = (1/2^{2n}) \sum_{h \in H_n} \delta_h .$$

By the construction of $I^{J/K}$, (12) and (13), μ_n converges weak* to a ("uniform") probability measure μ whose support is $I^{3/2}$. Fixing $h \in H_n$, observe that for each $j = 1, 2, 3$

$$\mu_n(\pi_j^{-1}(h) \cap (H_n)^{3/2}) = 1/2^{2n} = \nu_n(h),$$

deduce that ν_n is the marginal of μ_n on each of the 3 axes, and thus obtain (ii), since ν_n converges weak* to Lebesgue measure.

References

1. R. C. Blei, Fractional Cartesian products of sets, Ann. Inst. Fourier, Grenoble 29, 2 (1979), 79–105.
2. _______________, Combinatorial dimension and certain norms in harmonic analysis, Amer. J. of Math., Vol. 106 (1984), 847–887.
3. _______________, Fractional dimensions and bounded fractional forms, Memoirs of the American Math. Soc., 331, September 1985.
4. _______________, α – Chaos, J. of Functional Analysis, Vol 81 (1988), 279–297.
5. _______________, Combinatorial dimensions and fractional Cartesian products in $[0,1]^J$.
6. _______________, and T. W. Körner, Combinatorial dimension and random sets, Israel J. of Math., Vol. 47 (1984), 65–74.
7. K. J. Falconer, **The Geometry of Fractal Sets**, Cambridge University Press, 1985.
8. M. Fréchet, Sur les fonctionnelles bilineaires, Trans. Amer. Math. Soc., Vol. 16 (1915), 215–234.
9. J. E. Littlewood, On bounded bilinear forms in an infinite number of variables, Quart. J. Math. Oxford 1 (1930), 164–174.
10. A. Renyi, **Probability Theory**, North–Holland, 1970.

COMPARISON OF SPECTRAS IN SOME FUNCTIONAL SPACES

A. TURAN GÜRKANLI
Ondokuz Mayıs University
Faculty of Art and Sciences
Department of Mathematics
Samsun, Turkey

ABSTRACT. In this work we define two different spectras denoted by $sp_{w^*}f$, spf in the space $(S(\mathbb{R}^n))'$ and investigate some properties of these spectras, where $S(\mathbb{R}^n)$ is the Segal algebra containing the vector space of rapidly decreasing functions $\mathcal{S}(\mathbb{R}^n)$ as a dense subspace and $(S(\mathbb{R}^n))'$ is the topological dual of $S(\mathbb{R}^n)$. Finally we prove that $spf = sp_{w^*}f$. In addition to this, some applications of this work are also given.

1. INTRODUCTION

Throughout G will be a locally compact Abelian group with Haar measure dx. $L^p(G)$ and $L^p_{loc}(G)$ denotes the usual Lebesgue spaces and the space of (equivalence classes of) functions f on G such that f restricted to any compact subset of G belongs to $L^p(G)$ respectively. A Banach function space (BF space) on G is a Banach space $(B,\|.\|_B)$ which is continuously embedded into $L^1_{loc}(G)$. A BF-space is called solid if any measurable function g, for which there exists $f\in B$ such that $|g(x)|\le|f(x)|$ l.a.e. belongs to B, with $\|g\|_B\le\|f\|_B$. A Segal algebra S (G) on G is a translation invariant L^1-dense ideal in $L^1(G)$ that is a Banach algebra with respect to a norm $\|.\|_S$ and such that for every g in S the mapping $v\to L_v g$ from G to S is continuous and $\|L_v g\|_S=\|g\|_S$. The

J. S. Byrnes and J. F. Byrnes (eds.), Recent Advances in Fourier Analysis and Its Applications, 643–651.

symbol $L_y g$ denotes the translate of g by y, that is $L_y g(t) = g(t-y)$. For $t\varepsilon\hat{G}$ the multiplication operator M_t is given by

$$M_t f(x) = \langle x,t\rangle . f(x),\ x\varepsilon G,$$

where $\hat{G}$ is the character group of G. A Segal algebra S(G) is called character invariant if $||M_t f||_S = ||f||_S$ for all $t\varepsilon\hat{G}$, $f\varepsilon S(G)$. Let $\mathcal{A}(G)$ be a topological standard algebra [20]. If the functions with compact supports are dense in $\mathcal{A}(G)$ then $\mathcal{A}(G)$ will be called a Wiener algebra.

In this paper we will assume that $S(IR^n)$ is a Segal algebra with the norm $||.||_S$ containing $\mathcal{S}(IR^n)$ as a dense subspace, where $\mathcal{S}(IR^n)$ is the vector space of all functions f which are defined and infinitely differentiable in IR^n, and all their derivatives tends to zero at infinity faster than any power of $\frac{1}{|x|}$. If we denote the topological duals of $S(IR^n)$ and $\mathcal{S}(IR^n)$ by $(S(IR^n))'$ and $\mathcal{S}'(IR^n)$ respectively, we find $(S(IR^n))' \subset \mathcal{S}'(IR^n)$. That means every element $T\varepsilon(S(IR^n))'$ a temperate distribution. Let $T\varepsilon\mathcal{S}'(IR^n)$. The closed set supp $\hat{T}$ is called the spectrum of T and denoted by spT, where $\hat{T}$ is the Fourier transform and supp $\hat{T}$ is the support of $\hat{T}$. Evidently $spT = \emptyset$ if and only if $T = 0$ [3].

Given an element $f\varepsilon(S(IR^n))'$, we denote by [f] the smallest translation invariant subspace of $(S(IR^n))'$ that contains f. We denote by $\overline{[f]}_{w^*}$ the weak-star closure of [f] in $(S(IR^n))'$. We will define the weak-star spectrum by

$$sp_{w^*} f = \{\xi\varepsilon IR^n \mid e^{i\langle\xi,t\rangle}\varepsilon\overline{[f]}_{w^*}\}.$$

The purpose of this paper is to introduce the concept spf, $sp_{w^*}f$ of given any element $f\varepsilon(S(IR^n))'$ and compare these spectra.

2. THE ALGEBRA $A_s(\mathbb{R}^n)$ AND SPECTRAL ANALYSIS IN THE SPACE $(S(\mathbb{R}^n))'$

We denote by $A_s(\mathbb{R}^n)$ (or shortly A_s) the vector space of all functions on $\mathbb{R}^n$, which are the Fourier transforms of functions in $S(\mathbb{R}^n)$. Since $S(\mathbb{R}^n)\subset L^1(\mathbb{R}^n)$, it is clear that every element of $A_s(\mathbb{R}^n)$ is a continuous function vanishing at infinity. Now we introduce a norm to $A_s(\mathbb{R}^n)$ by transferring to it the norm of $S(\mathbb{R}^n)$, that is we write $\|\hat{f}\| = \|f\|_S$.

<u>Theorem 2.1:</u> $A_s(\mathbb{R}^n)$ is a semisimple regular Banach algebra with pointwise multiplication and the norm $\|.\|$.

<u>proof</u>. To the proof of this theorem is easy because for any Segal algebra and $A_s(\mathbb{R}^n)\approx S(\mathbb{R}^n)$ is a Wiener algebra ([20], p. 129).

<u>Lemma 2.1.</u> Let f and $f_1 \varepsilon A_s(\mathbb{R}^n)$ be such that the support of f_1 is compact and f is bounded away from zero on it. Then $f_1 = g.f$ with $g\varepsilon A_s(\mathbb{R}^n)$.

<u>Proof</u>. Denote by I the principal ideal generated by f; then the set

$$zI = \{t \mid f(t) = 0\}$$

is disjoint from the support of f_1, where zI is the zero set of I. Therefore by the ([15], corollary 5.7), write $f_1 \varepsilon I$, which means $f_1 = g.f$ for some $g\varepsilon A_s(\mathbb{R}^n)$.

<u>Theorem 2.2</u>: Let $f\varepsilon(S(\mathbb{R}^n))'$. Then $\xi\varepsilon \, sp_{w^*} f$ if and only if $\hat{h}(\xi) = 0$ for all $h\varepsilon[f]^{\perp}$, where $[f]^{\perp}$ is the annihilator of the set $[f]$.

<u>Proof</u>. Let $\xi \notin sp_{w^*} f$. By the Hahn-Banach theorem there exists a function $h\varepsilon S(\mathbb{R}^n)$ such that

$$<h, e^{i<\xi,t>}> = \int h(t)\overline{e^{i<\xi,t>}}dt = \hat{h}(\xi) = 1$$

and h=0 on $\overline{[f]}_{w^*}$. That means $h\varepsilon(\overline{[f]}_{w^*})^{\perp}$ and $h\neq 0$. Therefore $h\varepsilon[f]^{\perp}$.

For the converse consider $\xi\varepsilon\mathbb{R}^n$ and assume that there exists a function $h\varepsilon[f]^{\perp}$ such that $\hat{h}(\xi)\neq 0$. It is easy to see

$e^{i<\xi,t>} \notin [f]$. We will show that $e^{i<\xi,t>} \notin \overline{[f]}_{w^*}$. If $e^{i<\xi,t>} \in \overline{[f]}_{w^*}$, then there exists a net $(g_n)_{n\in I} \subseteq [f]$ which converges to $e^{i<\xi,t>}$ in the weakstar topology. Therefore $<h,g_n>$ converges to $\hat{h}(\xi)$. But this is a contradiction because $<h,g_n>= 0$ for all $n\in I$. Then $e^{i<\xi,t>} \notin \overline{[f]}_{w^*}$.

<u>Theorem 2.3</u>: If $f\in (S(IR^n))'$ then $spf = sp_{w^*}f$.

<u>Proof</u>. Assume $\xi_o \notin spf$. Then there exists an open neighborhood U of ξ_o such that $U \cap spf = \emptyset$ and hence if $h\in S(IR^n)$ and supp $\hat{h} \subset U$, we have $<\hat{f},\hat{h}>=<f,h>= 0$. It is easily seen that $h\in [f]^{\perp}$. But since $\mathcal{G}(IR^n) \subset A_S(IR^n)$, there exists a function $h \in \mathcal{G}(IR^n) \subset S(IR^n)$ such that $\hat{h}(\xi_o) \neq 0$ and supp $\hat{h} \subset U$. Hence by the Theorem 2.2, $\xi_o \notin sp_{w^*}f$. This proves $sp_{w^*}f \subseteq spf$.

Conversely assume that $\xi_o \notin sp_{w^*}f$. Since $S(IR^n) \simeq A_S(IR^n)$ is a Wiener algebra [20], there exists a function $f_1 \in S(IR^n)$ such that $\hat{f}_1(\xi_o) \neq 0$, $f_1 \in [f]^{\perp}$ and $\hat{f}_1$ is bounded away from zero on some neighborhood U of ξ_o. Let $h\in \mathcal{G}(IR^n)$ and supp$\hat{h} \subset U$. By the Lemma 2.1 and Corollary 2.1.1.,there exists $g\in S(IR^n)$ such that $\hat{h} = \hat{g}.\hat{f}_1$ (or $h = g * f_1$). Then we have

$$<f,h>=<f,g*f_1>=<f_x \otimes \overline{g(t)},f_1(x-t)>=<\overline{g(t)},<f_x,f_1(x-t)>>=0,$$

which means $\hat{f}$ vanishes in U. Then $\xi_o \notin spf$. That is $spf \subseteq Sp_{w^*}f$. This completes the proof.

<u>Theorem 2.4</u>: Let $f\in (S(IR^n))' - \{0\}$. Then the following statements are equivalent:

(a) $spf = IR^n$,

(b) $[f]^{\perp} = \{0\}$

(c) $\overline{[f]}_{w^*} = (S(IR^n))'$.

<u>Proof</u>. Statement (b) implies statement(a). Assume that $spf \neq IR^n$. Since $spf = sp_{w^*}f$, there exists a $\xi_o \in IR^n$ such that $e^{i<\xi_o,x>} \notin \overline{[f]}_{w^*}$. Also by the H-B theorem, there exists a

function $h \in S(\mathbb{R}^n)$ such that $\hat{h}(\xi_o) = 1$ and $h=0$ on $\overline{[f]}_{w^*}$. It is easy to see $h \in [f]^{\perp}$ and $h \neq 0$. Therefore $[f]^{\perp} \neq \{0\}$.

Statement (a) implies statement (b). If $[f]^{\perp} \neq \{0\}$, therefore there is a function $h \neq 0$, $h \in [f]^{\perp}$. Then $\hat{h}(\xi_o) \neq 0$ for some $\xi_o \in \mathbb{R}^n$. That means $\xi_o \notin spf = sp_{w^*} f$. Consequently find $spf = sp_{w^*} f = \mathbb{R}^n$.

Statement (c) implies statement (b). Let $\overline{[f]}_{w^*} = (S(\mathbb{R}^n))'$. It is clear that $e^{i<\xi,t>} \in \overline{[f]}_{w^*}$ for each $\xi \in \mathbb{R}^n$. Assume that there exists a function $h \in [f]^{\perp}$ such that $h \neq 0$. Given any $\xi \in \mathbb{R}^n$, there exists a net $(g_n) \subset [f]$ which converges to $e^{i<\xi,t>}$ in the weak-star topology. That means the net $<g_n, h>$ converges to $\hat{h}(\xi)$. But that is impossible for some $\xi \in \mathbb{R}^n$ because $<g_n, h> = 0$ for all $n \in \mathbb{N}$. This is a contradiction. Therefore $[f]^{\perp} = \{0\}$.

To prove the statement (b) implies (c) is easy using the Hahn Banach Theorem. Because (a)$\Longleftrightarrow$(b) and (b)$\Longleftrightarrow$(c), we also have (a)$\Longleftrightarrow$(c).

Remark. One can show that these results are also true for some Banach spaces including $\mathcal{S}(\mathbb{R}^n)$ as a dense subspace.

3. APPLICATIONS

The following theorem will be used in the applications considered in this paper.

Theorem 3.1. If $S(\mathbb{R}^n)$ is a Segal algebra and $\mathcal{S}(\mathbb{R}^n) \subset S(\mathbb{R}^n)$ then $\mathcal{S}(\mathbb{R}^n)$ is every where dense in $S(\mathbb{R}^n)$.

Proof. If I is closed ideal in $S(\mathbb{R}^n)$ then

$$I = \bar{I}^{L^1} \cap S(\mathbb{R}^n),$$

where $\bar{I}^{L^1}$ is the L^1-closure of I. Let $I = \overline{\mathcal{S}(\mathbb{R}^n)}^{S}$. Then I is S- closed and translation invariant. Indeed, we know that $\mathcal{S}(\mathbb{R}^n)$ is translation invariant. For every $f \in I$ there exists a sequence $(f_m) \subset \mathcal{S}(\mathbb{R}^n)$ such that $\|f_m - f\|_S \to 0$. Let $x \in \mathbb{R}^n$. Then

$$\| L_x f_m - L_x f \|_S = \| L_x (f_m - f) \|_S = \| f_n - f \|_S \to 0.$$

Since $L_x f_m \in \varphi(\mathbb{R}^n)$ for all $m \in \mathbb{N}$, we have $L_x f \in I$. We know that every closed translation invariant subset of $S(\mathbb{R}^n)$ is a closed ideal [20] in $S(\mathbb{R}^n)$. Therefore I is closed ideal in $S(\mathbb{R}^n)$. Using the equality

$$\overline{I}^{L^1} = \left(\overline{\overline{\varphi(\mathbb{R}^n)}^{S}}\right)^{L^1} = \overline{\varphi(\mathbb{R}^n)}^{L^1} = L^1(\mathbb{R}^n),$$

One obtains

$$I = L^1(\mathbb{R}^n) \cap S(\mathbb{R}^n) = S(\mathbb{R}^n).$$

Consequently we find

$$\overline{\varphi(\mathbb{R}^n)}^{S} = S(\mathbb{R}^n)$$

<u>Corollary 3.1.</u> If $S(\mathbb{R}^n)$ is strongly character invariant Segal algebra then $\varphi(\mathbb{R}^n)$ is everywhere dense in $S(\mathbb{R}^n)$.

<u>Proof</u>. Feichtinger proved that ([7]) $S_o(\mathbb{R}^n) \subset S(\mathbb{R}^n)$. Poguntke ([18]) showed that $\varphi(\mathbb{R}^n) \subset S_o(\mathbb{R}^n)$. Using Theorem 3.1 we find that $\varphi(\mathbb{R}^n)$ is everywhere dense in $S(\mathbb{R}^n)$.

1. Applications to Banach Convolution Algebras of Wiener Type:

Among the various applications of this work the spaces of Wiener's type seems to be most interesting one which is due to Feichtinger [5]. He gives the definition of these space in the following way: Let G be a locally compact group and B be a BF-space on G. Assume that there exists a homogeneous Banach space $(A, \|.\|_A)$, continuously embedded into the Banach algebra $(C_b(G), \|.\|_\infty)$ which is a regular Banach algebra with respect to pointwise multiplication and which is closed under complex conjugation, such that $(B, \|.\|_B)$ is continuously embedded into $A_o' = (A \cap C_c(G))'$, as well as a Banach module over A with respect to point wise multiplication, i.e $\|h.f\|_B \leq \|h\|_A . \|f\|_B$ for $h \in A$, $f \in B$. B_{loc} denote the space of all elements in A_o' such that $h.f \in B$ for all

$h \in A_o$. Let Q be any open subset of G with compact closure. Then the space of Wiener's type $W(B,C)$ consist of all elements $f \in B_{loc}$ such that the function $F=F_f: Z \to \|f\|_{B(ZQ)}$ belongs to the solid, translation invariant BF-space C; $\|f\|_{W(B,C)} = \|F\|_C$. B is called the local, and C the global component of $W(B,C)$. The space $W(A,L^1)(G)=S_o(G)$ has introduced in [6] and investigated some different properties in [7], [21], where $A(\hat{G})$ is the algebra of Fourier transforms of the functions in $L^1(G)$. Since $S_o(IR^n)$ is a strongly character invariant Segal algebra [7] on IR^n, by the Corollary 3.1., $\mathcal{S}(IR^n)$ is everywhere dense in $S_o(IR^n)$. Therefore $S_o(IR^n)$ satisfies our initial conditions. Then the theorems (2.2), (2.3), (2.4) applicable to the dual space, $S_o'(IR^n)$.

2. Let G be a locally compact Abelian group, $\hat{G}$ be the dual group of G and denote by $A_p(G)$, $(1 \le p < \infty)$, those f in $L^1(G)$ whose Fourier transform $\hat{f}$ belongs to $L^p(G)$, [17]. For any fixed $(1 \le p < \infty)$, $A_p(G)$ is a Segal algebra [18] with the norm

$$\|f\|_p = \|f\|_1 + \|\hat{f}\|_p = \int |f(x)|\,dx + \left(\int |\hat{f}(x)|^p dx\right)^{\frac{1}{p}} .$$

Since $A_p(IR^n)$ is a strongly character invariant Segal algebra, our initial conditions are satisfying. Therefore our work applicable to the topological dual space $A_p(IR^n)'$.

3. Let G be locally compact group. For $1<p<\infty$ the intersection $L^1(G) \cap L^p(G)$ is a strongly character invariant Segal algebra with the norm $\|f\|_s = \|f\|_1 + \|f\|_p$, [19]. Therefore the dual space $(L^1(IR^n) \cap L^p(IR^n))'$ is also an example for our work.

4. Let G be a locally compact Abelian group. The intersection $L^1(G) \cap C_o(G)$ is a strongly character invariant Segal algebra with the norm $\|f\|_s = \|f\|_1 + \|f\|_\infty$, [19],

where $C_o(G)$ denotes the vector space of all continuous functions on G which vanish at infinity. The space $(L^1(IR^n) \cap C_o(IR^n))'$ is another example to our work.

ACKNOWLEDGEMENT
The Author wishes to thank to Doz.Dr. Hans G.Feichtinger for his valuable discussions. Financial support from NATO during the meeting in Il Ciocco greatfully acknowledged.

REFERENCES:

1. A.BENEDEK, R.PANZONE, The space L^p, with mixed norm. Duke Math. J.28, 301-324(1961).

2. J.BENEDETTO, Spectral synthesis. NewYork 1975.

3. W.F. DONOGHUE, Distributions and Fourier transforms. NewYork, 1969.

4. H.G. FEICHTINGER, A characterization of Wiener's algebra on locally compact groups. Arch. d.Math.29, 136-140 (1977).

5. H.G. FEICHTINGER, Banach convolution algebras of Wiener type. Proc. Conf. Functions, series, operators, Budapest(1980) 509-524.

6. H.G. FEICHTINGER, Un espace de Banach de distributions tempérées sur les groupes localement compacts abéliens. C.R. Acad. Sci. Paris, Sér. A290, 791-794(1980).

7. H.G. FEICHTINGER, On a new Segal algebras, Mh.Math. 92, 269-289(1981).

8. J.J.F. FOURNIER, Lacunarity for amalgams. Rocky Mountain J.Math. 17/2, 277-294(1985).

9. J.J.F. FOURNIER, J.STEWART, Amalgams of L^p and ℓ^q. Bull. Amer. Math. Soc. 13, 1-21(1985).

10. R.R. GOLDBERG, On a space of functions of wiener. Duke Math. J. 34, 683-691(1967).

11. A.T. GÜRKANLI, On narrow spectral analysis, Glas. Mat. Ser III, 19(39), 271-274(1984).

12. A.T. GÜRKANLI, Spectral analysis of bounded functions. Karadeniz Üniv. Math. J.5., No. 1., 285-292(1982).

13. E.HEWITT and K.A. ROSS, Abstract harmonic analysis I, II, Springer-Verlag 1970-1979.

14. F. HOLLAND, Harmonic analysis on amalgams of L^p and ℓ^q. J.London Math. Soc.(2). 10, 295-305(1975).

15. Y.KATZNELSON, An Introduction to harmonic analysis. NewYork, (1968).

16. J. STEWART, S.WATSON, which amalgams are convolution algebras? Proc. Amer.Math. Soc.(93), Number 4, 621-627(1985).

17. R. LARSEN, T.S.LIU and J.K.WANG, On functions with Fourier transforms in L^p. Michigan Math. J.11, 369-378 (1964).

18. D. POGUNTKE, Gewisse Segal'sche Algebren auf lokalkompakten gruppen. Arch. Math. 33(1980), 454-460.

19. H.REITER, L^1-Algebras and Segal algebras, Lecture Notes in Mathematics 231, Springer-Verlag(1971).

20. H.REITER, Classical harmonic analysis and locally compact groups. Oxford University Press(1968).

21. M.L.TORRES DE SQUIRE, Multipliers for amalgams and the algebra $S_o(G)$. Can. J.Math. Vol. 39, No.1, 123-148(1987).

ON A CLASS OF RIESZ PRODUCTS ON METRIZABLE GROUPS

C. KARANIKAS
Department of Mathematics
University of Thessaloniki
54006 Thessaloniki ,Greece

ABSTRACT: We establish a complete system of Walsh functions $(w_n)_{n=0}^{\infty}$ on a subset E of the group having positive measure. We deal with certain Riesz products $\Pi_{n=1}^{\infty}(1+\alpha_n w_n(x))dx$ e.g. continuous singular measures with absolutely continuous convolution squares. We also deal with a Wiener's type characterization of continuous measures in terms of Walsh functions.

§1. INTRODUCTION

In this talk, we deal with Riesz products based on a Walsh type system which can be constructed on any metrizable non discrete group G.
We shall denote by M(G) the convolution measure algebra on G. Let $M_c(G)$ and $M_a(G)$ be the spaces of continuous and absolutely continuous measures (with respect to the left Haar measure), respectively. It is well known that on the circle group $\mathbb{T}$, we can obtain Riesz product measures having various properties. We recall some of these classical examples.

At 1917, F. Riesz produced the first continuous singular measure $d\mu = \Pi_{n=1}^{\infty}(1+\cos 4^n x)dx$, with Fourier Stieltjes coefficients $\hat{\mu}(n) = \langle e^{-int}, \mu \rangle$, $n \in \mathbf{Z}$, not tending to zero ($n \to \infty$). In other words, $\mu \in M_c(\mathbb{T}) \setminus M_0(\mathbb{T})$, where $M_0(G)$ is the space of all measures on the abelian group G, whose Fourier Stieltjes transforms vanish at infinity. This was the first "Riesz Product" measure.

At 1938 Wiener and Wintner [W,W], proved that there exists,singular measure $\mu \in M(\mathbb{T})$, such that the convolution square $\mu * \mu \in M_a(\mathbb{T})$; this is a finer result of the famous example of Menshov (cf, [B]), which proved at 1914 that there are measures $\mu \in M_0(\mathbb{T}) \setminus M_a(\mathbb{T})$.

The Wiener and Wintner result was extended for any non discrete abelian group by Hewitt and Zuckerman (1966)[H,Z]. The proof, for the case where $G = \mathbb{T}$, is a simple application of the Riesz products;allow me to sketch it.

Let $f_n(x) = \Pi_{k=1}^{n}(1+\alpha_k \cos n_k x)dx$, where $\alpha_k \in [-1,1]$. We denote by $d\mu = \Pi_{k=1}^{\infty}(1+\alpha_k \cos n_k x)dx$ the weak* limit in $M(\mathbb{T})$ of the polynomials $f_n(x)$. A widely known Theorem, (cf. in [Z])states that $\mu \in M_a(\mathbb{T})$ if and only if $(\alpha_n) \in \mathscr{L}_2$.

J. S. Byrnes and J. F. Byrnes (eds.), Recent Advances in Fourier Analysis and Its Applications, 653–659.

Now, one can observe that the measure $\mu*\mu$ is the Riesz product $\Pi_{k=1}^{\infty}(1+\alpha_k^2\cos n_k x)dx$, and so if $(\alpha_n) \in \mathscr{L}_4\setminus\mathscr{L}_2$, μ is as we claimed. On this class of singular measures with absolutely continuous convolution squares see also, Saeki's works [S1] and [S2].

Afterword of this example we have a plenitude of works on Riesz products. So actually it is difficult to make a reasonable list of authors, with contribution on the develloping of this field.

In our work here we give, with some details, the construction of a Wiener-Wintner type measure on metrizable groups (see §3). With less details we shall discuss (in §4) examples of continuous Riesz products, and a characterization of the space $M_c(G)$, using a system of Walsh functions.

This system is deficient of any group structure, in contrast with a system of characters on G, appearing in the classical Riesz product, as well as in the well known Wiener's characterization of $M_c(G)$. Notice that our Walsh system is L^2-complete (see §2). Thus it seems that the important property of these examples and of the spaces $M_c(G)$ and $M_o(G)$ are the completeness of the dealing system and not its algebraic structure.

This encourages an investigation of Riesz products in a space of measures defined on a set without a group structure. One can easily realize that this process will be apparently more difficult.

We hope to deal with this problem in another occasion.

§2. THE WALSH SYSTEM

Let E be a measurable subset of G with Haar measure one. A system of Rademacher functions associated with E is any sequence of functions (r_j) which are zero off E, are independent as radom variables, and take values 1 and -1 on subsets of E of equal measure.

In fact we divide E into two subsets of equal measure to divide similarly each of the two subsets to define the second Rademacher function e.t.c. The n-partition of E will be indexed: $E_{n,1}, E_{n,2},\dots,E_{n,2^n}$. Furthermore we require that the partition of E satisfies:

$$\max_{1\le\kappa\le 2^n}\operatorname{diam}(E_{n,\kappa}) \to 0, \quad n \to \infty$$

The Walsh system (w_n) associated with this Rademacher system will be indexed: $w_0 = \chi_E$, and each w_n, $n \ge 1$, is a finite product of Rademacher such that as is the classical Walsh system on $\mathbb{T}$, $r_{n+1}(x) = w_{2^n}$, $n=1,2,\dots$

A Walsh system having these properties can be easily constructed on a matrices group, but in general E may be a compact totally disconected perfect set of positive measure having an homeomorphism with the Cantor group D, as follows.

We start with a compact set V of positive measure and diam V = L. The refinement process on V will be on stages and substages. Starting the stage n we divide the sets of the previous stage such that:

Each set we refine has a partition in 2^{κ} (some κ depending of the stage) Borel sets of maximum diameter less than L/n. Their distances (mutual distances) are positive and the total mass of the refined sets is $(1-2^{-n})$ times the mass of the sets we refined. Therefore the total mass of the sets in each stage is $\Pi_{\kappa=1}^{n}(1-2^{-\kappa})|V|$ and so $|E|>0$.

The 1-1 map between E and D can be established, if on each substage with 2^{κ} sets, we correspond $2^{\kappa-1}$ sets with +1 and the rests with -1. Now, by the homeomorphism between M(E) and M(D) we consider the continuous measure μ on D, corresponding to the left Haar measure on E. A partition on D: $D_{n,1}, D_{n.2} \ldots D_{n,2^n}$ such that $\mu(D_{n,1}) = 2^{-n}$, $(1 \leq 2^n)$ yields a partition $(E_{n,1})$ on E. Because of the homeomorphism and since μ is continuous we have: $\max\operatorname{diam} E_{n,j} \to 0$ $(n \to \infty)$. This property seems to be important because one can shows that:

(I) The system of the Walsh functions on E is L^2-complete.
(II) If $f_n(x) = \Pi_{\kappa=1}^{n}(1+\alpha_{\kappa}r_{\kappa}(x))$, $|\alpha_{\kappa}| \leq 1$ then $f_n(x)$ converges weak* in M(G) to a measure μ.
(III) μ is in $M_{\alpha}(G)$ if and only if $(\alpha_n) \in \mathscr{L}_2$.

§3. EXAMPLE OF A SINGULAR MEASURE WITH ABSOLUTELY CONTINUOUS CONVOLUTION SQUARE

We shall describe that given any sequence $(\alpha_n) \notin \mathscr{L}_2$ and $\Sigma_{n=2}^{\infty}\alpha_n^4 \Pi_{\kappa=1}^{n-1}(1+\alpha_{\kappa}^2) < \infty$, (e.g. $\alpha_n = (n\ln n)^{-1/2}$), then we can find a subsequence (r_n') of Rademacher functions such that if $d\mu=\Pi_{n=1}^{\infty}(1+\alpha_n r_n'(x))dx$, we have $\mu*\mu \in M_{\alpha}(G)$. Because $(\alpha_n) \notin \mathscr{L}_2$ it is clear that $\mu \notin M_{\alpha}(G)$ i.e. It is a singular measure with absolutely continuous convolution square. For more details we refer [K,K]. Our proof based on the following two observations:

(1) If $f \in L_1(G)$ and $\mu_n \to 0$ weak* in M(G) then for any $p \geq 1$, we have $\|f*\mu_n(x)\|_p \to 0$ $n \to \infty$. Thus since $w_n \to 0$ weak*, we have $\|f*w_n(x)\|_p \to 0$ $n \to \infty$.

(2) If $\mu_n \to \mu$ weak* in M(G), where μ_n, μ are probability measures with compact support, then $\mu_n*\mu_n \to \mu*\mu$ weak* in M(G).

In fact (1) can be seen easily because for any $x \in G$, $f*w_n(x)= =\langle f_x,w_n\rangle$, where f_x is the translation of f by x, and since $w_n \to 0$, $n \to \infty$, weak* in $L^{\infty}(G)$ $\langle f_x,w_n\rangle \to 0$. The proof can be completed using the fact that the map $x \to f_x$: $G \to L_1(G)$ is uniformly continuous.

For (2) we consider any continuous function f with compact support. Since $\mu_n \to \mu$ weak* by Observation (1) we have $\|\mu_n*f - \mu*f\|_{\infty} \to \infty$ $n \to \infty$. By the fact that $\langle\mu_n*\mu_n,f\rangle = \langle\mu_n,\mu_n*f\rangle$ one can shows (2).

We sketch the proof. Let $\mu_n(x) = \Pi_{\kappa=1}^{n}(1+\alpha_{\kappa}r_{\kappa}'(x))$ by the second observation it suffices to show that $\|\mu_N*\mu_N-\mu_M*\mu_M\| \to 0$, $N,M \to \infty$. It is easy to see that

$$\|\mu_M*\mu_M - \mu_N*\mu_N\| \leq \Sigma_{\kappa=M+1}^{N}|\alpha_{\kappa}|\,\|\mu_{\kappa-1}*\mu_{\kappa-1}r_{\kappa}'\| +$$

$$+ \Sigma_{\kappa=M+1}^{N}|\alpha_{\kappa}|\,\|\mu_{\kappa-1}r_{\kappa}'*\mu_{\kappa-1}\| +$$

$$+\left\| \sum_{\kappa=M+1}^{N} \alpha_\kappa^2(\mu_{\kappa-1}r'_\kappa * \mu_{\kappa-1}r'_\kappa)\right\|$$

by the observation (1) for any κ, we can find $r'_\kappa(x)$ such that:

$$\|\mu_{\kappa-1}*\mu_{\kappa-1}r'_\kappa\| \leq \kappa^{-2}$$

and so by the first and the second term of the sum in the R.H.S.of (3) are $\leq$ const. $(M^{-1}-N^{-1}) \to 0$, $N,M \to \infty$. For the third term we apply some elementary inequalities to obtain:

$$\left\| \sum \alpha_\kappa^2(\mu_{\kappa-1}r'_\kappa * \mu_{\kappa-1}r'_\kappa)\right\| \leq \tag{4}$$

$$\leq \int \sum_{\kappa=1}^{M} \alpha_\kappa^4(\mu_{\kappa-1}r'_\kappa * \mu_{\kappa-1}r'_\kappa)^2 dx +$$

$$+ 2\int \sum_{M+1\leq\kappa\leq\lambda\leq N} \alpha_\kappa^2 \cdot \alpha_\lambda^2(\mu_{\kappa-1}r'_\kappa * \mu_{\kappa-1}r'_\kappa)(\mu_{\lambda-1}r'_\lambda * \mu_{\lambda-1}r'_\lambda)dx$$

Now we use a refined version of observation (1) to find r'_n such that for each $\kappa < \lambda$

$$\int (\mu_{\kappa-1}r'_\kappa * \mu_{\kappa-1}r'_\kappa)(\mu_{\lambda-1}r'_\lambda * \mu_{\lambda-1}r'_\lambda)dx < \frac{1}{\lambda^3}$$

The last term in (4) is $\leq$const $\sum_{\kappa=M+1}^{N} \kappa^{-3} \leq$ const$(M^{-1}-N^{-1}) \to 0$, $N,M \to \infty$.

Now one can shows that the first term of the sum in (4) gets smaler than $\sum_{\kappa=M+1}^{N} \alpha_\kappa^4 \prod_{j=1}^{\kappa-1}(1+\alpha_j^2)$ and so by our hypothesis $(\mu_N*\mu_N)$is Cauchy in $M_\alpha(G)$.

§4. CONTINUOUS MEASURES

A well known therorem of Wiener ([Z] III9.6) states that if μ is in $M(\mathbb{T})$, then $\mu \in M_c(\mathbb{T})$ if and only if

$$\frac{1}{2N+1} \sum_{-N}^{N} \left|\int e^{-int}d\mu(t)\right|^2 \to 0, \quad N \to \infty$$

i.e. the middle square of the Fourier-Stieltjes transforms of μ tends to zero at infinity. This is a weak Riemann-Lebesque-type property, namely $\langle e^{-int},\mu\rangle \to 0$, $|n| \to \infty$, for all $\mu \in M_\alpha(\mathbb{T})$. Since for our Walsh system, as in §2, and for any $\mu \in M_\alpha(G)$, we have $\langle w_n,\mu\rangle \to 0$, $n \to \infty$, one can expect that $\mu \in M_c(E)$ if and only if

$$\frac{1}{N+1} \sum_{n=0}^{N} \left|\int w_n(t)d\mu(t)\right|^2 \to 0, \quad N \to \infty$$

We shall describe this proof; for more details we refer [B,K].

The main idea of this proof is in the following combinatorial argument.

Lemma 1. Let $E_{n_0,\kappa}$ $(1 \leq \kappa \leq 2^{n_0})$ be a given subset of the n_0-th partition of E. For any $N = 0,1,2,\ldots$ the function: $A_N(x,y) = \sum_{j=0}^{N} w_j(x)w_j(y)$ where $(x,y) \in \cup\{E_{n_0,i}: i \neq \kappa\} \times E_{n_0,\kappa}$ is

$$|A_N(x,y)| \leq 2^{n_0-1}$$

Proof Let $x_0 \in \bigcup_{i\neq\kappa} E_{n_0,i}$ and $y_0 \in E_{n_0,\kappa}$, then there is a $n \in \mathbb{N}$, such that

$$r_j(x_0)r_j(y_0) = 1 \quad j = 0,1,2,\ldots,n \text{ and}$$

$$r_{n+1}(x_0)r_{n+1}(y_0) = -1 \tag{1}$$

It is clear that this $n = n(x_0,y_0) \leq n_0-1$ and of course $n < n_0$. Let

$$B_\lambda(x) = \sum_{j=(\lambda-1)2^{n+1}}^{\lambda 2^{n+1}-1} w_j(x)w_j(y_0), \quad \lambda = 1,2,\ldots$$

We observe that

$$B_\lambda(x) = \Pi_{j=1}^{n+1}(1+r_j(x)r_j(y_0))w(x)_{(\lambda-1)2^{n+1}}w(y_0)_{(\lambda-1)2^{n+1}} \tag{2}$$

By (1) we have $B_1(x_0) = 0$ and so by (2) $B_\lambda(x_0) = 0$, $\lambda = 1,2,\ldots$ Now, given any N, there is a λ such that $\lambda 2^{n+1} \leq N < (\lambda+1)2^{n+1}$. Thus

$$A_n(x_0,y_0) = \left|\sum_{\rho=1}^{A} B_\rho(x_0) + \sum_{j=\lambda 2^{n+1}}^{N} w_j(x_0)w_j(y_0)\right| \leq 2^{n_0-1}$$

One can use this Lemma and the fact that the $\max\operatorname{diam}(E_{n,\kappa}) \to 0$, $n \to \infty$, to prove the following:

Lemma 2. If $\mu \in M(G)$ then for each $y \in E$.

$$\mu(\{y\}) = \lim_{N\to\infty} \frac{1}{N+I} \sum_{n=0}^{N} \mu(w_n)w_n(y)$$

Notice: In order to have this result whenever $y \in \bigcup_{n=1}^{\infty} \bigcup_{j=1}^{2^n} \operatorname{bdr}E_{n,j}$, we provide in the construction of the Walsh system, that the boundary points belong on the sets $(E_{n,j})$. Thus if y is a boundary point, then for some κ, $y \in E_{n,\kappa}$ and $r_n(y) = r_n(t)$ for any $t \in E_{n,\kappa}$. Following the ideas of the original Wieners Theorem and by Lemma 2 we obtain. If $\mu \in M(E)$ then:

$$\sum_{y\in E}|\mu(\{y\})|^2 = \lim_{N\to\infty} \frac{1}{N+I} \sum_{n=0}^{N} |\mu(w_n)|^2$$

and this implies our Wiener Theorem: $\mu \in M_c(E)$ if and only if

$$\frac{1}{N+1}\sum_{n=0}^{N}|\mu(w_n)|^2 \to 0, \quad N \to \infty$$

Example Let $d\mu = \Pi(1+r_{n_j}(x))dx$ where $n_{j+1}/n_j \geq 2$ clearly

$$\frac{1}{2^{2M}+1}\sum_{n=0}^{2^{2M}} \mu(w_n)^2 \leq \frac{2^M}{2^{2M}+1} \to 0, \ M \to \infty$$

and so μ is a continuous measure.

We complete this talk with remarks on certain continuous Riesz products. We recall, that the space $M_0(G)$ is the completion of $M_\alpha(G)$ under the norm of the left reqular repesentation. Because the Walsh system is complete, one can shows that $\mu \in M_0(G)$ if and only if given $\varepsilon > 0$, $\sup\|\mu*(w_n)_y\|_2 < \varepsilon$, $n \geq n_0(\varepsilon)$, where the supremum is taken over all translations $(w_n)_y$, $y \in G$, of w_n. This allows as to establish the following.

Example Let $\mu \in M_0(G)$, we can find a singular Riesz product ν, such that $\mu*\nu \in M_\alpha(G)$.

Proof Let $d\nu = \Pi_{n=1}(1+\alpha_n r_n'(x))dx$, where $(\alpha_n) \notin \mathscr{L}_2$ and (r_n') be a sequence of Rademacher functions satisfying:

$$\|\mu*(q_j(x)r_j'(x))\|_2 < 3^{-n} , \quad j = 1,2,\dots,2^n ,$$

where $q_j(x)$ is a Walsh functions in terms of $r_1',\dots,r_n'$. Now, if $\nu_N(x) = \Pi_{n=1}^{N}(1+\alpha_n r_n')$ we have

$$\|\mu*\nu_N - \mu*\nu_M\| \leq \sum_{j=N}^{M-1}|q_{j+1}|\,\|\mu*(\nu_j r_{j+1}')\|$$

$$\leq \text{const} \sum_{j=N}^{M-1} \|\mu*\nu_j r_{j+1}'\|_2$$

$$\leq \text{const} \sum_{N}^{M-1}(\tfrac{2}{3})^n \to 0 \quad N, M \to \infty$$

Thus $\{\mu*\nu_N\}_N$ is a Cauchy sequence in $M_\alpha(G)$.

Remarks 1. Following the spirit of the previous example we have: If K is a norm σ-compact subset of $M_0(G)$ then we can find a singular Riesz product ν such that $K*\nu \subseteq M_\alpha(G)$.

2. This set K in Remark 1 can be a σ-compact subset of $M_c(G)$. We only need the fact that: if $\mu \in M_c(G)$, then $\frac{1}{N+1}\sum_{n=1}^{N}\|\mu*w_n(x)\| \to 0$, $N \to \infty$ (cf. [B,K]).

3. It seems that it is not difficult to obtain singular Riesz products inside or outside of $M_0(G)$. One can use Lemma (5.5) and methods of Theorem (5.6) in [B,K,W].

4. An interesting problem in this direction seems to be the following: If m is a continuous measure, find a Riesz product ν such that

v is singular with respect to m and $v*v \ll m$.

§5. REFERENCES

[B] N. Bary. A Treatise on Trigonometric Series. New York: Macmillan 1964.

[B,K] A.Bisbas and C. Karanikas. On the Wiener Characterization of Continuous measures (submited).

[B,K,W] G.Brown, C. Karanikas and J.H. Williamson. The asymmetry of $M_0(G)$. Proc. Camb. Philos. Soc. 91, 407-433, 1982.

[H,Z] E.Hewitt and H. Zuckermann. Singular measures with absolutely continuous convolution squares. Proc. Camb. Philos. Soc. 62, 399-420, 1966.

[K,K] C.Karanikas and S. Koumandos. Continuous singular measures with absolutely continuous convolution square on locally compact groups.(submited).

[S1] S.Saeki. Singular measures having absolutely continuous convolution powers. Illinois J. Math. vol. 21, 395-412, 1977.

[S2] On convolution squares of singular measures. Illinois J. Math. vol. 24, num. 2, 225-232, 1980.

[W,W] W.Wiener and A. Wintner. Fourier-Stieltjes transforms and singular infinite convolution. Amer. J. Math. 60, 513-522 (1938).

[Z] A. Zygmund. Trigonometric series. (2nd edition, two vols; Cambridge Univ. Press, 1959.

GROUPOIDS AND FOURIER APPROXIMATIONS

K. SEITZ
Department of Mathematics
Technical University of Budapest
1111 Budapest XI. Műegyetem rakpart 9,
H.ép. V.em. 5.
HUNGARY.

ABSTRACT. In this lecture we show that groupoid theoretical methods could be useful tools in the theory of Fourier approximation. On a special set of functions a groupoid operation is defined and two structure theorems are proved.

1. INTRODUCTION

Let $H(n,\varepsilon)$, $(\varepsilon>0,\ n=0,1,2,\ldots)$ be the set of all $f(x)$ functions which satisfy the following conditions:

1° $f(x)$ continuous on $[0,2\pi)$ and

$$f(x+2\pi)=f(x),\quad(-\infty<x<\infty),$$

2° $$|f(x)-\hat{f}_n(x)|\leqq\varepsilon,$$ where

$$(1)\qquad \hat{f}_n(x)=\frac{1}{2\pi}\int_0^{2\pi}\frac{\sin\frac{2n+1}{2}t}{\sin\frac{t}{2}}f(x+t)\,dt.$$

Let $\alpha(x)\geqq 0$, $\beta(x)\geqq 0$ be fixed continuous functions on $[0,2\pi)$ and we suppose that

$$(2)\qquad \int_0^{2\pi}[\alpha(s)+\beta(s)]\,ds=1.$$

Now on $H(n,\varepsilon)$ we define a "$\circ$" operation as follows:

$$(3)\qquad f(x)\circ g(x)=\int_0^{2\pi}[f(x+s)\alpha(s)+g(x+s)\beta(s)]\,ds.$$

J. S. Byrnes and J. F. Byrnes (eds.), Recent Advances in Fourier Analysis and Its Applications, 661–668.

If $f(x), g(x) \in H(n, \varepsilon)$ then

$$(4)\quad \left(\widehat{f(x)\circ g(x)}\right)_n = \frac{1}{2\pi}\int_0^{2\pi} \frac{\sin\frac{2n+1}{2}t}{\sin\frac{t}{2}}\left[f(x+t)\circ g(x+t)\right]dt =$$

$$= \frac{1}{2\pi}\int_0^{2\pi} \frac{\sin\frac{2n+1}{2}t}{\sin\frac{t}{2}}\left\{\int_0^{2\pi}\left[f(x+t+s)\alpha(s)+g(x+t+s)\beta(s)\right]ds\right\}dt =$$

$$= \int_0^{2\pi}\left[\hat{f}_n(x+s)\alpha(s)+\hat{g}_n(x+s)\beta(s)\right]ds.$$

From (1), (2), (3), (4) we have that

$$(5)\quad \left|f(x)\circ g(x) - \left(\widehat{f(x)\circ g(x)}\right)_n\right| \leq \left|\int_0^{2\pi}\left[f(x+s)-\hat{f}_n(x+s)\right]\alpha(s)ds + \right.$$

$$\left. + \int_0^{2\pi}\left[g(x+s)-\hat{g}_n(x+s)\right]\beta(s)ds\right| \leq \int_0^{2\pi}\left|f(x+s)-\hat{f}_n(x+s)\right|\alpha(s)ds +$$

$$+ \int_0^{2\pi}\left|g(x+s)-\hat{g}_n(x+s)\right|\beta(s)ds \leq \varepsilon\int_0^{2\pi}\left[\alpha(s)+\beta(s)\right]ds = \varepsilon.$$

Therefore $(H(n,\varepsilon), \circ)$ is a groupoid.

In this lecture structure theorems are proved for $(H(n,\varepsilon), \circ)$.

The material of this paper is strongly related to the works [1], [2], [3], [4].

2. BASIC PROPERTIES OF OPERATION

We can see that $(H(n,\varepsilon), \circ)$ generally is not a semigroup but the following proposition is true.

Proposition 1. If $f, g, h \in H(n,\varepsilon)$, then

$$(6)\quad (f\circ g)\circ h = f\circ(g\circ h)$$

if and only if

$$(7)\ (-f)\circ h = [(-f)\circ 0]\circ[0\circ h].$$

Proof. If $f, g, h \in H(m,\varepsilon)$ and (6) holds then

$$(8)\ \left(f(x)\circ g(x)\right)\circ h(x) = \int_0^{2\pi}\Big\{\int_0^{2\pi}[f(x+u+s)\alpha(s)+g(x+u+s)\beta(s)]\,ds\cdot \alpha(u) + h(x+u)\beta(u)\Big\}du = \int_0^{2\pi}\Big\{f(x+u)\alpha(u) + \int_0^{2\pi}[g(x+u+s)\cdot \alpha(s) + h(x+u+s)\beta(s)]\,ds\,\beta(u)\Big\}du,$$

from which

$$(9)\ \int_0^{2\pi} h(x+u)\beta(u)\,du + \int_0^{2\pi}\int_0^{2\pi}[f(x+u+s)\alpha(s)+g(x+u+s)\beta(s)]\,ds\cdot \alpha(u)\,du = \int_0^{2\pi} f(x+u)\alpha(u)\,du + \int_0^{2\pi}\Big\{\int_0^{2\pi}[g(x+u+s)\alpha(s) + h(x+u+s)\cdot \beta(s)]\,ds\Big\}\beta(u)\,du$$

follows.

From (9) we have that

$$(10)\ \int_0^{2\pi}[h(x+u)\beta(u) - f(x+u)\alpha(u)]\,du = \int_0^{2\pi}\int_0^{2\pi}[h(x+u+s)\beta(s)\beta(u) - f(x+u+s)\alpha(s)\alpha(u)]\,ds\,du.$$

We can see that (10) has the form

$$(11)\ (-f(x))\circ h(x) = [(-f(x))\circ 0]\circ[0\circ h(x)].$$

If $f, g, h \in H(n, \varepsilon)$ and (11) holds then (10) is true from which (7) follows.

Next two propositions will be proved.
Proposition 2. If $f, g, h \in H(n, \varepsilon)$ and $f \circ g = f \circ h$ then $p \circ g = p \circ h$ where p is an arbitrary element of $H(n, \varepsilon)$.

Proof. If $f, g, h \in H(n, \varepsilon)$ and $f \circ g = f \circ h$ then

$$(12) \quad \int_0^{2\pi} g(x+s)\beta(s)\,ds = \int_0^{2\pi} h(x+s)\beta(s)\,ds$$

from which

$$(13) \quad p \circ g = \int_0^{2\pi} [p(x+s)\alpha(s) + g(x+s)\beta(s)]\,ds =$$

$$= \int_0^{2\pi} [p(x+s)\alpha(s) + h(x+s)\beta(s)]\,ds = p \circ h$$

follows, where $p(x)$ is an arbitrary element of $H(n, \varepsilon)$.

On a similar way we can prove that if $g \circ f = h \circ f$, then $g \circ p = h \circ p$.
Proposition 3. If $f, g, h \in H(n, \varepsilon)$ and $f \circ g = g \circ f$ further $f \circ h = h \circ f$ then $g \circ h = h \circ g$.
Proof. If $f, g, h \in H(n, \varepsilon)$ and

$$(14) \quad f \circ g = \int_0^{2\pi} f(x+s)\alpha(s)\,ds + \int_0^{2\pi} g(x+s)\beta(s)\,ds = \int_0^{2\pi} g(x+s)\alpha(s)\,ds +$$

$$+ \int_0^{2\pi} f(x+s)\beta(s)\,ds = g \circ f$$

further

$$(15) \quad f \circ h = \int_0^{2\pi} f(x+s)\alpha(s)\,ds + \int_0^{2\pi} h(x+s)\beta(s)\,ds = \int_0^{2\pi} h(x+s)\alpha(s)\,ds +$$

$$+ \int_0^{2\pi} f(x+s)\beta(s)\,ds = h \circ f$$

then

$$(16)\quad \int_0^{2\pi}[g(x+s)\beta(s)+h(x+s)\alpha(s)]\,ds=\int_0^{2\pi}[g(x+s)\alpha(s)+h(x+s)\beta(s)]\,ds,$$

from which

$$(17)\qquad g\circ h=h\circ g$$

follows.

3. DECOMPOSITIONS AND SPECIAL SUBSTRUCTURES OF $(H(n,\varepsilon),\circ)$.

Let A be the set of all $f(x)\in H(n,\varepsilon)$ functions for which

$$(18)\quad \int_0^{2\pi} f(x)\,dx\leqq k$$

where k is a fixed constant.

If $f(x), g(x)\in A$, then

$$(19)\quad \int_0^{2\pi}[f(x)\circ g(x)]\,dx=\int_0^{2\pi}\left\{\left[\int_0^{2\pi} f(x+s)\,dx\right]\alpha(s)+\left[\int_0^{2\pi} g(x+s)\,dx\right]\beta(s)\right\}ds\leqq$$

$$\leqq k\int_0^{2\pi}[\alpha(s)+\beta(s)]\,ds=k,$$

from which $f(x)\circ g(x)\in A$ follows.

We can see that if $f(x), g(x)\in H(n,\varepsilon)$, $p, q\geqq 0$, $p+q=1$ then $p\,f(x)+q\,g(x)\in H(n,\varepsilon)$.
If $f(x), g(x)\in A$ then $p\,f(x)+q\,g(x)\in A$, because

$$(20)\quad \int_0^{2\pi}[p\,f(x)+q\,g(x)]\,dx\leqq k(p+q)=k.$$

Let C be the set of all $f(x)\in H(n,\varepsilon)$ functions for which

$$(21)\quad \int_0^{2\pi} f(x)\,dx \geq k,$$

where k is a fixed constant.

If $f(x), g(x) \in \mathcal{C}$ then

$$(22)\quad \int_0^{2\pi}[f(x)\circ g(x)]\,dx = \int_0^{2\pi}\Big\{\Big[\int_0^{2\pi} f(x+s)\,dx\Big]\alpha(s) + \Big[\int_0^{2\pi} g(x+s)\,dx\Big]\cdot$$

$$\cdot\beta(s)\Big\}\,ds \geq k\int_0^{2\pi}[\alpha(s)+\beta(s)]\,ds = k,$$

from which $f(x)\circ g(x) \in \mathcal{C}$ follows.

If $p, q \geq 0$, $p+q = 1$ then $p\,f(x) + q\,g(x) \in \mathcal{C}$.

Therefore we have proved the following result.

Theorem 1. The convex groupoid $(H(n,\varepsilon),\circ)$ has the following decomposition

$$(23)\quad (H(n,\varepsilon),\circ) = (A,\circ) \cup (\mathcal{C},\circ),$$

where $(A,\circ)$, $(\mathcal{C},\circ)$ are convex subgroupoids of $(H(n,\varepsilon),\circ)$.

Now on $H(n,\varepsilon)$ we define a σ relation as follows: $f(x)\,\sigma\,g(x)$, $(f(x), g(x) \in H(n,\varepsilon))$ if and only if

$$(24)\quad \int_0^{2\pi} f(x)\,dx = \int_0^{2\pi} g(x)\,dx.$$

We can see that σ is an equivalence relation.

Let $\mathcal{T}$ be the set of all σ classes.

Theorem 2. $(H(n,\varepsilon),\circ)$ has the following decomposition

$$(25)\quad (H(n,\varepsilon),\circ) = \bigcup_{K\in\mathcal{T}} (K,\circ),$$

where $(\mathcal{K}, \circ)$ is a subgroupoid of $(H(m, \varepsilon), \circ)$ further if $\tilde{\mathcal{K}} \in \mathcal{C}$ then $\mathcal{K} \circ \tilde{\mathcal{K}}$ is a groupoid which is contained by a σ class.

Proof. If $f(x), g(x) \in \mathcal{K} \in \mathcal{C}$ then

$$(26) \int_0^{2\pi} (f(x) \circ g(x))\, dx = \int_0^{2\pi}\left\{ \int_0^{2\pi} [f(x+s)\alpha(s) + g(x+s)\beta(s)]\, ds \right\} dx =$$

$$= \int_0^{2\pi}\left\{ \left[\int_0^{2\pi} f(x+s)\, dx \right] \alpha(s) + \left[\int_0^{2\pi} g(x+s)\, dx \right] \beta(s) \right\} ds =$$

$$= \int_0^{2\pi} f(x)\, dx \cdot \int_0^{2\pi} [\alpha(s) + \beta(s)]\, ds = \int_0^{2\pi} f(x)\, dx.$$

If $\mathcal{K}, \tilde{\mathcal{K}} \in \mathcal{C}$ and $f, g \in \mathcal{K}$; $\tilde{f}, \tilde{g} \in \tilde{\mathcal{K}}$, then

$$(27) (f \circ \tilde{f}) \circ (g \circ \tilde{g}) = \int_0^{2\pi}\left\{ \int_0^{2\pi} [f(x+t+s)\alpha(s) + \tilde{f}(x+t+s)\beta(s)]\, ds\, \alpha(t) + \right.$$

$$\left. + \int_0^{2\pi} [g(x+t+s)\alpha(s) + \tilde{g}(x+t+s)\beta(s)]\, ds \cdot \beta(t) \right\} dt =$$

$$= \int_0^{2\pi}\left\{ \int_0^{2\pi} [f(x+t+s)\alpha(t) + g(x+t+s)\beta(t)]\, dt\, \alpha(s) + \right.$$

$$\left. + \int_0^{2\pi} [\tilde{f}(x+t+s)\alpha(t) + \tilde{g}(x+t+s)\beta(t)]\, dt\, \beta(s) \right\} ds =$$

$$= (f \circ g) \circ (\tilde{f} \circ \tilde{g}) \in \mathcal{K} \circ \tilde{\mathcal{K}},$$

and

$$(28)\quad \int_0^{2\pi} \left(f(x) \circ \tilde{f}(x)\right) dx = \int_0^{2\pi} \left\{ \int_0^{2\pi} \left[f(x+s)\,\alpha(s) + \tilde{f}(x+s)\,\beta(s) \right] ds \right\} dx =$$

$$= \int_0^{2\pi} \left\{ \left[\int_0^{2\pi} f(x+s)\,dx \right] \alpha(s) + \left[\int_0^{2\pi} \tilde{f}(x+s)\,dx \right] \beta(s) \right\} ds =$$

$$= \int_0^{2\pi} \left\{ \left[\int_0^{2\pi} g(x+s)\,dx \right] \alpha(s) + \left[\int_0^{2\pi} \tilde{g}(x+s)\,dx \right] \beta(s) \right\} ds =$$

$$= \int_0^{2\pi} \left(g(x) \circ \tilde{g}(x)\right) dx.$$

Therefore $(f \circ \tilde{f})\, \sigma\, (g \circ \tilde{g})$.

Finally we set up two problems:

Problems 1. Determine all subgroupoids of $(H(n, \varepsilon), \circ)$.

Problems 2. Determine all generator systems of $(H(n, \varepsilon), \circ)$.

References

1. Cohn, P.M. (1981) Universal Algebra, D. Reidel Publishing Company, Boston.

2. Balakrishnan, A.V. (1976) Applied Functional Analysis, Springer-Verlag, Berlin and New York.

3. Yosida, K. (1966) Functional Analysis, Springer-Verlag, Berlin and New York.

4. Grätzer, G. (1979) Universal Algebra, Springer-Verlag, New York.

Part V: Problems

Problems

1. (due to Tom Körner) Let $\Lambda \subseteq \{2\pi k/N: 0 \leq k \leq N\}$ have $|\Lambda|$ elements. If $\mu = |\Lambda|^{-1}\sum_{x\in\Lambda}\delta_x$, then $\hat{\mu}(r) = |\Lambda|^{-1}\sum_{x\in\Lambda}\exp(irx)$ has period N and $\hat{\mu}(0) = 1$. I am interested in

$$\phi(N,M) = \inf\{\sup_{1\leq r\leq N-1} |\hat{\mu}(r)|: |\Lambda| = M\}.$$

Automatically $\phi(N,M) \geq M^{-1/2}$ by Parseval's equality. If $M \geq N^{\alpha}$ for some fixed $\alpha > 0$, then a simple probabilistic argument gives $\phi(N,M) \leq K(\alpha)(\log N)^{1/2}M^{-1/2}$. If M is very close to N, I can improve this, otherwise I can do nothing. See "A Rudin-Shapiro Type Polynomial" in *Illinois Journal of Mathematics*.

2. (due to Donald Newman) Let $S \subset C[0,1]$, S be a Chebyshev system (i.e., any linear combination of k of them has at most $k-1$ zeros), $1 \in S$. Must $\{(f/g): f, g \in S\}$ span $C[0,1]$?

3. (due to H.S. Shapiro) Can one give an algorithm to determine the number of roots in an interval $[a,b]$ of the real axis of a function of the form

$$\sum_{j=1}^{n} c_j e^{\lambda_j x} \quad (c_j, \lambda_j \in \mathbf{R})?$$

In case $\lambda_j = j-1$, the function is a polynomial in e^x and the problem reduces after change of variables to the corresponding one for polynomials, which is solved by the well-known algorithm of Sturm. The algorithm of Budan-Fourier gives an upper bound for the number of roots (indeed, this works for all smooth functions), but good lower bounds seem much harder to get.

Similar questions arise for linear combinations of other simple functions, e.g.,

$$\sum_{j=1}^{n} c_j \log(x - x_j) \quad \text{with given } c_j, x_j \in \mathbf{R}.$$

4. Let

$$f(x) = \sum_{j=1}^{n} P_j(x)e^{\lambda_j x}$$

where the P_j are polynomials, and $\{\lambda_j\}$ distinct complex numbers. Must there exist $\epsilon > 0$ and $m > 0$ such that

$$\int_{-\infty}^{\infty} |f(x)|^{-\epsilon}(1 + |x|)^{-m} < \infty?$$

(I am also interested in the corresponding multi-variable problem.)

J. S. Byrnes and J. F. Byrnes (eds.), Recent Advances in Fourier Analysis and Its Applications, 671–672.

5. If f, F are functions in $L^1(\pi)$ with respective Fourier coefficients $\{\hat{f}(n)\}$, $\{\hat{F}(n)\}$, one says F *majorizes* f (written $f \ll F$) if $|\hat{f}(n)| \leq \hat{F}(n)$, for all $n \in \mathbf{Z}$.

The following are known to hold when $f \ll F$:

a) If $F \in L^p = L^p(\Pi)$ and $p \geq 2$ is an even integer, then $f \in L^p$.

b) If $F \in L^p$, $p \geq 1$, and p is not an even integer, then f may fail to be in L^p.

I propose the following strengthening of case b: If $p < r$ and the closed interval $[p, r]$ contains no even integers there exists a pair f, F with $f \ll F$, $F \in L^r$ and $f \notin L^p$ (note that, in view of case a above and Hölder's inequality, the restriction imposed on the interval $[p, r]$ is necessary).

In view of a well-known technique (so-called "encapsulation"), it is easy to reduce the problem to this: with p, r as above, find $f \ll F$ with $\|f\|_p > \|F\|_r$. (Here L^p always denotes $L^p(\Pi)$ where $\Pi = \mathbf{R}/2\pi\mathbf{Z}$ and $\|f\|_p$ denotes $\left(1/2\pi \int_0^{2\pi} |f(x)|^p \, dx\right)^{1/p}$.)

6. Let $\Lambda \in \mathbf{R}$ and let f be an infinitely differentiable real-valued function on $[a, b]$. It is easy to show that for there to exist a sequence $\{f_j\}$ of finite linear combinations of $\{e^{\lambda x}\}_{\lambda \in \Lambda}$ with *positive* coefficients such that f_j converges to f uniformly on $[a, b]$ (or even in the sense of distributions on (a, b)), the following condition is necessary:

$$P(d/dx)f(x) \geq 0 \quad \text{on } (a, b)$$

for every polynomial P with real coefficients which is nonnegative on Λ. *Question:* is this condition also sufficient?

Remark: Although the above condition may appear rather abstract, it is fairly easy to apply in some situations. For example, it implies that when $\Lambda = \mathbf{R} \backslash (\lambda_0 - \epsilon, \lambda_0 + \epsilon)$ with $\epsilon > 0$, $e^{\lambda_0 x}$ cannot be approximated on any interval by linear combinations of $\{e^{\lambda x} : \lambda \in \Lambda\}$ with positive coefficients. (Note that this is radically different from the case of real coefficients!) Indeed, the quadratic polynomial P vanishing at $\lambda_0 - \epsilon$ and $\lambda_0 + \epsilon$ and equal to -1 at λ_0 shows this, in view of the above necessary condition.

A PROBLEM FROM THE THEORY OF WAVE PROPAGATION IN A CURVED, LAYERED MEDIUM

ANTON G. TIJHUIS
Laboratory of Electromagnetic Research
Department of Electrical Engineering
Delft University of Technology
P.O. Box 5031, 2600 GA Delft, The Netherlands
Electronic Mail TIJHUIS@ET.TUDELFT.NL

This problem originates from the theory of wave propagation in a layered medium. The propagating wave may be of an electromagnetic, acoustic or elastodynamic nature, and the layering may be cylindrical, spherical or elliptical. The wave propagates in a symmetry direction along which the configuration is curved. Applications include the propagation along bends in optical waveguides, and the investigation of boreholes in geophysics.

We restrict ourselves to formulating the most simple form, which shows up in the introduction of the Watson transformation in Subsection 5.1 of this author's contribution elsewhere in the present volume. Starting point is the Fourier- and Laplace-transformed wave equation for $E_m(\rho, s)$:

$$\left[\partial_\rho^2 + \frac{1}{\rho}\partial_\rho - \frac{m^2}{\rho^2} - s^2\,\epsilon_r(\rho) - s\,\sigma(\rho)\right] E_m(\rho, s) = 0. \tag{1}$$

In this equation, we have $m \in \mathbb{Z}$, $s \in \mathbb{C}$, $\mathrm{Im}(s) \geq 0$. The material parameters $\epsilon_r(\rho)$ and $\sigma(\rho)$ are real-valued, piecewise continuously differentiable functions of ρ, with

$$\begin{aligned} &1 \leq \epsilon_r(\rho) < \infty \quad \text{for } 0 \leq \rho < 1, \\ &\epsilon_r(\rho) = \epsilon_{1r} \geq 1 \quad \text{for } 1 < \rho < \infty, \end{aligned} \tag{2}$$

and

$$\begin{aligned} &0 \leq \sigma(\rho) < \infty \quad \text{for } 0 \leq \rho < 1, \\ &\sigma(\rho) = 0 \qquad\quad \text{for } 1 < \rho < \infty. \end{aligned} \tag{3}$$

We are interested in the solution on the interval $0 \leq \rho < \infty$ that

- satisfies the radiation condition as $\rho \to \infty$:

$$E_m(\rho, s) = I_m(s\epsilon_{1r}^{\frac{1}{2}}\rho) + b_m(s)\,K_m(s\epsilon_{1r}^{\frac{1}{2}}\rho) \quad \text{for } 1 \leq \rho < \infty; \tag{4}$$

J. S. Byrnes and J. F. Byrnes (eds.), Recent Advances in Fourier Analysis and Its Applications, 673–675.

- remains bounded as $\rho \downarrow 0$:

$$\lim_{\rho \downarrow 0} \rho^{-|m|} E_m(\rho, s) < \infty. \tag{5}$$

In (4), $b_m(s)$ is a complex constant, and $I_m(z)$ and $K_m(z)$ are modified Bessel functions of integer order and complex argument. For the physical background of the various conditions and quantities, the reader is referred to the discussion elsewhere in this volume.

In the *Watson transformation*, the function $E_m(\rho, s)$ is continued analytically into the complex-order plane, i.e. $m \in \mathbb{Z}$ is replaced by $\nu \in \mathbb{C}$. Obviously, there is some ambiguity in the choice of $E_\nu(\rho, s)$. The way in which this ambiguity is removed raises a the following questions.

Question 1

Introduction It can be shown [1] that imposing a boundedness condition for $\rho \downarrow 0$ on the spectral constituents of the incident and total fields, i.e. choosing

$$\begin{aligned} I_m(s\epsilon_{1r}^{\frac{1}{2}}\rho) &\rightarrow I_{\sqrt{\nu^2}}(s\epsilon_{1r}^{\frac{1}{2}}\rho), \\ \lim_{\rho \downarrow 0} \rho^{-\sqrt{\nu^2}} E_\nu(\rho, s) &< \infty, \end{aligned} \tag{6}$$

with $\mathrm{Re}\sqrt{\nu^2} \geq 0$, yields a solution $E_\nu(\rho, s)$ that, for fixed ρ and s, decays faster than exponentially as $|\nu| \rightarrow \infty$, unless $\arg(\nu) = \pm\pi/2$. Establishing this decay is essential in deriving field representations involving guided-wave modes.

As an illustration, consider the special case where the medium in $0 \leq \rho < 1$ is homogeneous, i.e.

$$\epsilon_r(\rho) = \epsilon_{2r}, \quad \sigma(\rho) = \sigma_2 \quad \text{for } 0 \leq \rho < 1. \tag{7}$$

In that case, the solution to (1)–(5) for $0 \leq \rho \leq 1$ assumes the form

$$E_m(\rho, s) = c_m(s)\, I_m(sn_2(s)\rho), \tag{8}$$

where $c_m(s)$ is a complex constant, and where

$$n_2(s) \stackrel{\text{def}}{=} \sqrt{\epsilon_{2r} + \sigma_2/s}, \quad \text{with } \mathrm{Re}(n_2) \geq 0. \tag{9}$$

The constants $b_m(s)$ and $c_m(s)$ in (4) and (8) follow from the continuity of $E_m(\rho, s)$ and $\partial_\rho E_m(\rho, s)$ at $\rho = 1$. Replacing, in the expressions for $E_m(\rho, s)$ thus obtained,

$$\begin{aligned} I_m(s\epsilon_{1r}^{\frac{1}{2}}\rho) &\rightarrow I_\nu(s\epsilon_{1r}^{\frac{1}{2}}\rho) \\ I_m(sn_2(s)\rho) &\rightarrow I_\nu(sn_2\rho) \end{aligned} \tag{10}$$

leads to a solution $E_\nu(\rho, s)$ that increases faster than exponentially as $|\nu| \rightarrow \infty$ for $|\arg(\nu)| > \pi/2$, while replacing

$$\begin{aligned} I_m(s\epsilon_{1r}^{\frac{1}{2}}\rho) &\rightarrow I_{\sqrt{\nu^2}}(s\epsilon_{1r}^{\frac{1}{2}}\rho) \\ I_m(sn_2(s)\rho) &\rightarrow I_{\sqrt{\nu^2}}(sn_2(s)\rho) \end{aligned} \tag{11}$$

does lead to the correct decay for almost all values of $\arg(\nu)$.

Question From this example, it would appear that the boundedness condition is not only a *sufficient* but also a *necessary* condition for attaining the desired decay. This leads to the obvious question whether this conjecture can be proven rigorously.

Question 2

From the the principle of causality, it follows that all poles of $E_\nu(\rho, s)$ in the complex ν-plane are, for $\text{Re}(s) \geq 0$, located in the second and fourth quadrant of that plane. In the first and third quadrant, $E_\nu(\rho, s)$ is analytic in ν (see [1]).

This poses the problem of deriving this "physical" result directly from the generalized second-order differential equation and the generalized boundary conditions for $\rho \downarrow 0$ and $\rho \to \infty$.

Question 3

As mentioned in the introduction, a similar analysis can be carried out for spherically or elliptically layered geometries. For such geometries, the wave equation is separable as well.

It would be helpful, therefore, to know whether more general results exist pertaining to second-order differential equations of the type (1).

References

[1] Tijhuis, A.G. (1986) "Angularly propagating waves in a radially inhomogeneous, lossy, dielectric circular cylinder and their connection with the natural modes," *IEEE Trans. Antennas Propagat.* *34*, 813–824 (discussion of the cylindrical problem).

[2] Felsen, L.B. and Marcuvitz, N. (1973) *Radiation and Scattering of Waves*, Prentice-Hall, Englewood Cliffs, New Jersey (general introduction to the theory of wave propagation in stratified media).

Problems on Polynomials with Restricted Coefficients Arising from Questions in Antenna Array Theory

J.S. Byrnes

We present mathematical formulations of some classical problems in antenna design. For references concerning these problems, see the paper of Newman and Giroux in this volume.

To begin, we give precise mathematical interpretations to two standard electrical engineering definitions, that of *maximum sidelobe level* and *beamwidth*, which occur in the analysis and synthesis of a line array of equally-spaced, identical, omnidirectional antenna elements. As is well known, the model for such an array is a polynomial, where its degree is one less than the number of elements, and the coefficients are the *weights*, or *shading coefficients*, of the array.

Throughout this discussion, let $z = e^{i\theta}$ lie on the unit circle Γ, let $P(z)$ denote a polynomial with coefficients a_j, at least two of which are nonzero, and let θ_0 be the smallest nonnegative value of θ where the maximum of $|P(z)|$ occurs. Now let

$$\delta_1 = \min\{\, \delta : \delta < \theta_0 \text{ and } |P(e^{i\theta})| \text{ is increasing on } (\delta, \theta_0)\,\},$$

$$\delta_2 = \max\{\, \delta : \delta > \theta_0 \text{ and } |P(e^{i\theta})| \text{ is decreasing on } (\theta_0, \delta)\,\},$$

$$\sigma_P = \max_{\theta \in [\theta_0 - \pi, \theta_0 + \pi]/(\delta_1, \delta_2)} |P(e^{i\theta})|, \qquad \text{and}$$

$$\beta_P = \delta_2 - \delta_1.$$

The graph of (one full period of) $|P(e^{i\theta})|$ is the (modified) *beam pattern*, σ_P is the (modified) *maximum sidelobe level*, and β_P is the (modified) *beamwidth* of the array. We use the adjective "modified" because we have omitted the change of variable, normalization, and expression in terms of decibels that one finds in the engineering literature, but these differences are of no import here.

All problems presented will be of the following form:

Determine the existence and construction of a P which lies in a certain class, whose beam pattern satisfies certain properties, and which is optimum (or nearly so) with respect to one of its parameters. These classes are:

$$\mathcal{U} = \{\, P : |a_j| = 1 \,\} \quad \text{(the *unimodular* polynomials)},$$

$$\mathcal{U}_0 = \{\, P : |a_j| = 1 \text{ or } 0 \,\},$$

$$\mathcal{E} = \{\, P : a_j = \pm 1 \,\},$$

$$\mathcal{E}_0 = \{\, P : a_j = \pm 1, 0 \,\},$$

$$\mathcal{D}_K = \{\, P : \max \left|\frac{a_j}{a_m}\right| \le K \,\} \quad \text{(this maximum is the *dynamic range* of the array)},$$

$$\mathcal{D}_{K,0} = \{\, P : \max_{a_j \times a_m \neq 0} \left|\frac{a_j}{a_m}\right| \le K \,\},$$

$$\mathcal{C} = \text{any one of the above classes.}$$

Although a reasonable bound on the dynamic range is ordinarily important (an old rule of thumb is that "$K = 2$ and everyone is happy, $K = 10$ and some are happy, $K = 100$ and nobody is happy"), which would seem to render zero coefficients fatal, this is not usually the case. Since a zero coefficient simply means that that element is turned off (either by

J. S. Byrnes and J. F. Byrnes (eds.), Recent Advances in Fourier Analysis and Its Applications, 677–678.

choice or because it is not functioning properly), this does not affect the dynamic range in a meaningful way. Furthermore, it appears that in many cases allowing zero coefficients enables a significant reduction in the degree of P (see problem 4 below).

The time has come to state specific problems:

1. Given a finite (possibly empty) subset S of Γ, $q = e^{i\theta_0} \notin S$, $\sigma > 0$, $\beta > 0$, and a class $\mathcal{C}$, find $P \in \mathcal{C}$ of minimum degree (if it exists at all) such that $P(z) = 0$ for all $z \in S$, $\sigma_P \leq \sigma$, and $\beta_P \leq \beta$.

2. Given $\epsilon > 0$ and a class $\mathcal{C}$, find $P \in \mathcal{C}$ of minimum degree such that

$$\big||P(z_1)| - |P(z_2)|\big| \leq \epsilon \quad \text{for all } z_1, z_2 \in \Gamma$$

(a so-called *ultra-flat* polynomial).

3. Given $S = \{e^{i\theta_1}, e^{i\theta_2}, \ldots, e^{i\theta_m}\}$ nonempty, $\epsilon > 0$, $\delta > 0$, and a class $\mathcal{C}$, find $P \in \mathcal{C}$ of minimum degree such that $P(z) = 0$ for all $z \in S$ and

$$\big||P(z_1)| - |P(z_2)|\big| \leq \epsilon \quad \text{for all } z_1 = e^{i\tau_1}, z_2 = e^{i\tau_2}$$

with the property that $(\tau_1, \tau_2) \cap (\theta_j - \delta, \theta_j + \delta)$ is empty for $1 \leq j \leq m$ (a *notch filter*).

It is appropriate to observe that, although optimization of the degree is called for in problems 1–3, near-optimal solutions are definitely of interest. For example, the problem of most practical interest is 1, in the case where the cardinality of S is very small (even 0, 1, or 2), and this appears mathematically interesting as well. Also along these lines, we have:

4. Prove or disprove the conjecture: The $P \in \mathcal{U}$ of minimum degree with an n-fold 0 at $z = 1$ is

$$\prod_{m=0}^{n-1} (1 - z^{2^m}).$$

In fact, even a proof that the degree of such a P must be exponential in n would be of interest. Note that if the class $\mathcal{U}$ is replaced by $\mathcal{U}_0$ (actually even $\mathcal{E}_0$), then there is a P of degree less than n^3 satisfying the required property.

Analogous problems, when one of the other parameters aside from the degree is optimized, are of interest as well. In this case there is an additional class of polynomials to be considered:

$$\mathcal{M}_n = \{\, P : \text{degree } P = n \text{ and } |a_j|, 0 \leq j \leq n, \text{ are preassigned} \,\}.$$

Thus, another typical problem is:

5. Given $q = e^{i\theta_0}$, $\beta > 0$, the degree n (often between 10 and 40, although values can be as much as a few thousand), and the class $\mathcal{C}$ of P, find P such that $\beta_P \leq \beta$ and σ_P is minimized.

The simple job of stating other varieties of these problems is left to the reader. Their solution, a much more daunting task, also awaits the reader's efforts.

Index